Precision Machining Technology

Precision Machining Technology

Third Edition

Peter J. Hoffman

Berks Career & Technology Center, West Campus, Leesport, PA

Eric S. Hopewell

Berks Career & Technology Center, West Campus, Leesport, PA

CENGAGE

Australia • Brazil • Mexico • Singapore • United Kingdom • United States

Precision Machining Technology,
Third Edition
Peter Hoffman/Eric Hopewell

SVP, GM Skills & Global Product Management: Jonathan Lau

Product Director: Matthew Seeley

Senior Product Manager: Katie McGuire

Product Assistant: Kimberly Klotz

Executive Director of Development: Marah Bellegarde

Learning Design Director: Juliet Steiner

Senior Learning Designer: Mary Clyne

Vice President Strategic Marketing Services: Jennifer Ann Baker

Marketing Director: Sean Chamberland

Marketing Manager: Andrew Ouimet

Senior Director Content Delivery: Wendy Troeger

Manager, Content Delivery: Alexis Ferraro

Senior Content Manager: Sharon Chambliss

Senior Digital Delivery Lead: Amanda Ryan

Design Director: Jack Pendleton

Senior Designer: Angela Sheehan

Cover image(s): Dmitry Kalinovsky/ShutterStock.com

For product information and technology assistance, contact us at Cengage Customer & Sales Support, 1-800-354-9706 or **support.cengage.com**.

For permission to use material from this text or product, submit all requests online at **www.cengage.com/permissions**.

Library of Congress Control Number: 2018960474

Book Only ISBN: 978-1-3377-9530-2

Cengage
20 Channel Center Street
Boston, MA 02210
USA

Cengage is a leading provider of customized learning solutions with employees residing in nearly 40 different countries and sales in more than 125 countries around the world. Find your local representative at **www.cengage.com**.

Cengage products are represented in Canada by Nelson Education, Ltd.

To learn more about Cengage platforms and services, register or access your online learning solution, or purchase materials for your course, visit **www.cengage.com**.

NOTICE TO THE READER
Publisher does not warrant or guarantee any of the products described herein or perform any independent analysis in connection with any of the product information contained herein. Publisher does not assume, and expressly disclaims, any obligation to obtain and include information other than that provided to it by the manufacturer. The reader is expressly warned to consider and adopt all safety precautions that might be indicated by the activities described herein and to avoid all potential hazards. By following the instructions contained herein, the reader willingly assumes all risks in connection with such instructions. The publisher makes no representations or warranties of any kind, including but not limited to, the warranties of fitness for particular purpose or merchantability, nor are any such representations implied with respect to the material set forth herein, and the publisher takes no responsibility with respect to such material. The publisher shall not be liable for any special, consequential, or exemplary damages resulting, in whole or part, from the readers' use of, or reliance upon, this material.

Printed in the United States of America
Print Number: 03   Print Year: 2019

CONTENTS

SECTION 7
Grinding 559

SECTION 8

Computer Numerical Control 592

Unit 1   CNC Basics 594

Unit 2   Introduction to CNC Turning ... 608

Unit 3   CNC Turning: Programming 623

PREFACE

Precision Machining Technology introduces students, both at the secondary and postsecondary levels, to the exciting world of precision machining technology as it is practiced in the 21st century. In writing this text, the authors' main goal is to provide a deep understanding of the fundamental and intermediate machining skills needed for career success in a rapidly changing manufacturing environment. In line with this objective, the author team has taken special care to ensure that the text:

- Has a down-to-earth, practical orientation that covers what students need to know about the field of precision machining as it is practiced today
- Develops modern interpersonal skills that are demanded by the job market
- Covers current career information and trends
- Includes modern shop practices
- Contains specific instructions and examples, with images showing many step-by-step applications
- Provides in-depth knowledge as a base for strong foundational skills without becoming difficult to read or comprehend
- Includes current computer numerical control (CNC) content with various programming examples.

This text is written for students of precision machining at the secondary and postsecondary levels who have the opportunity and desire to learn skills required by the precision machining industry and to obtain NIMS certifications. The book is written in such a way that the student needs no prior knowledge of machining to benefit.

Precision Machining Technology has been sponsored and endorsed by NIMS. The text and its supporting supplements fill the need of comprehensively covering all of the material encountered by a student during the NIMS certification process, and were written with the Machining Level I Standards in mind. The text's close adherence to NIMS's nationally recognized skills standards will be especially useful for schools and school districts that wish to comply with the funding requirements of the Carl D. Perkins Career and Technical Education Act of 2006 (Perkins IV).

How the Text Was Developed

In order to create a truly new set of teaching and learning tools, *Precision Machining Technology* was launched with no preconceived notion of how the text should be designed. A large number of instructors at NIMS-accredited programs participated in the initial development of the table of contents, which then led to the recruitment of the author team, also from NIMS-accredited programs. During the development of the project, over a dozen instructors reviewed drafts of the manuscript and provided useful feedback to the authors. Their input has played a major role in improving the final product. Last, the publisher and NIMS committed to an extra developmental step, class-testing the manuscript at multiple institutions, in order to assure the highest level of accuracy and teaching effectiveness. Reviewers and class-testers are listed in the Acknowledgments section.

To enhance the teaching and learning experience, the authors developed the text with the following objectives in mind:

- Achieve an easy-to-read writing style that assumes the student has no prior knowledge of machining and takes the student all the way through to the intermediate stage
- Include many images to clarify explanations and procedures so students can make visual connections
- Identify key and secondary terms throughout the text to guide students to important points

- Assume that students are taking or have already taken basic geometry, basic algebra, and have good proficiency in computation of fractions, decimals, and order of operations
- Allow for the companion *Workbook/Projects Manual* to provide a beneficial measure of practice to prepare the student for NIMS product creation and the knowledge examination

Organization of the Text

In designing *Precision Machining Technology*, the authors followed the typical progression through the NIMS certifications. For many of the sections, a student should have sufficient knowledge to obtain a NIMS certification at the completion of the sections.

The text is divided into eight major sections, as follows:

Section 1—Introduction to Machining

Section 2—Measurement, Materials, and Safety

Section 3—Job Planning, Benchwork, and Layout

Section 4—Drill Press

Section 5—Turning

Section 6—Milling

Section 7—Grinding

Section 8—Computer Numerical Control (CNC)

Each section of the text contains multiple "bite-sized" units, which provide the following teaching and learning aids: learning objectives, key terms, caution safety checks, chapter summary, and review questions.

Special care was taken to make each unit progress in a logical presentation of content for someone with no prior knowledge. The authors took steps to ensure that no new terminology was presented prior to a complete explanation of each term. Each unit builds on another, and many sections build on previous sections. As the text progresses, topics are explored more deeply. Previous knowledge is reinforced through new application of previous information.

What's New in This Edition

- Increased CNC programming content including additional canned cycles
- CNC programming examples with graphics presented in both Fanuc and Haas formats, two of the most widely used industry formats
- Multi-step procedures modified to a bulleted format to make following steps easier
- Expanded details in many areas including optical comparator use, hazardous materials, key and keyseat machining, and more
- Updated images to complement textual content
- This edition of Precision Machining Technology is also correlated to Precision Exams' Machining I and Machining II exams, part of the Manufacturing Career Cluster's Production Pathway.

A Note for Students: How to Use This Text

Do not become overwhelmed with all of the information. The text is arranged so that you may take each piece step by step. Pause and think about key and secondary terms while reading.

Supplements

Workbook and Projects Manual

The student Workbook and Projects Manual contains helpful review material to ensure that students have mastered key concepts in the text, and guided practice operations and projects on a wide range of machine tools that will enhance their NIMS credentialing success. All projects are keyed to NIMS Duties and Standards.

Instructor Companion Website

The Instructor Companion Website, found on cengage.com, includes the following components to help minimize instructor preparation time and engage students:

PowerPoint® lecture slides, which present the highlights of each chapter.

An **Image Gallery,** which offers a database of hundreds of images in the text. These can easily be imported into the PowerPoint® presentations.

An **Answer Key** file for the Core text and Workbook, which provides the answers to all end-of-chapter questions and the questions in the Workbook.

Cengage Learning Testing Powered by Cognero—a flexible online system that allows you to:

- Author, edit, and manage test bank content from multiple Cengage Learning solutions.
- Create multiple test versions in an instant.
- Deliver tests from your LMS, your classroom, or wherever you want.

MindTap for Precision Machining Technology

MindTap is a personalized teaching experience with relevant assignments that guide students to analyze, apply, and improve thinking, allowing you to measure skills and outcomes with ease.

- *Personalized Teaching*: Becomes YOURS with a Learning Path that is built with key student objectives.
- Control what students see and when they see it—match your syllabus exactly by hiding, rearranging, or adding your own content.
- *Guide Students*: Goes beyond the traditional "lift and shift" model by creating a unique learning path of relevant readings, multimedia, and activities that move students up the learning taxonomy from basic knowledge and comprehension to analysis and application.
- *Measure Skills and Outcomes*: Analytics and reports provide a snapshot of class progress, time on task, engagement, and completion rates.

ACKNOWLEDGMENTS

The authors and publisher thank the following individuals for their assistance in the reviewing process.

Reviewers:

Kurt J. Billsten, Harper College, Palatine, IL

Robert G. Cantin, Chicopee Comprehensive High School, Chicopee, MA

Stephen Hadwin, Southern Oklahoma Technology Center, Ardmore, OK

David R. Hall, Bladen Community College, Dublin, NC

David R. McGough, Knox County Career Center, Mount Vernon, OH

Elliot Shuler, Gordon Cooper Technology Center, Shawnee, OK

Robert T. Tanchak, Onondaga Community College, Syracuse, NY

Thank you to the following individuals and companies who provided support, photography, and illustrations for the text:

3M Company

American Coatings Association

American Metal Treating Co.

Armstrong Tool

AVK Industrial Products

Baldor Electric Company

Behringer Saws, Inc.

Berks Career and Technology Center

Bilz Tool Co.

Brady Corporation

Brown & Sharpe division of Hexagon Metrology, Inc.

Cincinnati Millicron

Clausing Industrial Inc.

Dake Corporation

DE-STA-CO

Edward Drapatin, Kaynor Technical High School, Waterbury, CT

Dykem

Everett Industries

Flexbar Machine Corp.

Flow International Corporation

Fred V. Fowler Company, Inc.

Gage Crib Worldwide Inc.

Getty Images

Grizzly Industrial, Inc.

Haas Automation, Inc.

Hardinge, Inc.

Industrial Press, Inc.

Ingersoll Cutting Tools

ITW Vortec

Jet-Wilton

Justrite Mfg. Co.

Kasto, Inc.

Kennametal, Inc.

Kool Mist Corporation

Kurt Mfg.

Lenox

LPS Laboratories

Lucifer Furnaces, Inc.

Lyndex-Nikken, Inc.

Mag Americas

Manluk Industries

Master Chemical Corporation

Mazak Corp.

Mitee-Bite Products

Mitutoyo Corporation

NFPA

Niagara Cutter, Inc.

Oil-Dri Corporation of America

Palmgren, Inc.

Penske Racing Shocks

Pernell Parts

PTC Instruments

Royal Products

Sandvik AB

Sentry Air Systems, Inc.

Servo Products Co.

SFS Intec

South Bend Lathe Co.

Suburban Tool, Inc.

Tapmatic Corporation

The Dunham Tool Company

The Grieve Corporation

The L. S. Starrett Company

The Red Wing Shoe Company, Inc.

Tray-Pak Corp.

Trimetric Mold & Design

TruTech Systems/Harig Products

Vermont Gage

Walker Magnetics Group

Walton Tools

Weldon Solutions

Weldon Tool Co., a division of Talbot Holdings Co.

Wellsaw, Inc.

Mack Williams, Radford High School and Dalton Intermediate School, Radford, VA

Wilson Instruments

Yuasa International

FROM THE AUTHORS

The authors express their gratitude to the entire production team for their patience, support, and guidance. Special thanks to Sharon Chambliss for coordinating the countless aspects of the project and maintaining patience with all of us.

Pete Hoffman

I thank my loving wife, Reba, who graciously supported me during this project. Thanks to my children Lucas, Niklas, Autumn, and Summer for allowing me to be away from them while I was working on the project. And thanks to my parents, Carl and Laureen, who have taught me many great things. I have been richly blessed to have all of you in my life.

Eric Hopewell

I thank my wife, Peg, and my daughters Selena, Lindsey, and Haley for their understanding and patience throughout this project. I appreciate them more than they know, and am truly blessed to have such a family that unconditionally loved and supported me during the entire process.

ABOUT THE AUTHORS

Peter J. Hoffman, Instructor, Berks Career & Technology Center, West Campus, Leesport, PA

Peter Hoffman holds an associate degree in Toolmaking Technology from Pennsylvania College of Technology and a permanent Vocational Education II teaching certification in Pennsylvania. He has several level I and II NIMS certifications, was a 2001 postsecondary Skills USA National Gold Medalist in Precision Machining Technology, and a National Silver Medalist in 2000. He owns and operates a small machining business.

Eric S. Hopewell, Instructor, Berks Career & Technology Center, West Campus, Leesport, PA

Eric Hopewell has over 30 years of combined experience in the machine tool and education fields. He holds an AAS degree in Machine Tool Technology from Pennsylvania College of Technology, a BS in Business Administration from Albright College, an MS in Education from Temple University, a permanent Pennsylvania Vocational Education II certification, and is a Pennsylvania Department of Labor certified tool & die maker/moldmaker. He also holds several NIMS machining certifications.

INTRODUCTION TO MACHINING   SECTION 1

UNIT 1 Introduction to Machining

Learning Objectives

After completing this unit, the student should be able to:

- Define the term *machining*
- Define a machine tool
- Discuss the evolution of machining and machine tools
- Identify the role of machining in society
- Discuss the principles of the basic types of machining processes

Key Terms

Abrasive machining	End product	Milling machine
Computer Numerical Control (CNC)	Laser machining	Numerical Control (NC)
	Lathe	
Drill press	Machine tool	Sawing machine
Electrical Discharge Machining (EDM)	Machining	Water jet machining
	Manufacturing	

INTRODUCTION

The word *machining* probably has very little meaning to the typical person today. However, nearly all people depend on that word more than they could ever imagine. How can that be? What *is* machining and how does it influence everyday life?

The answers to these questions, and many others that will come up along the journey to discover the world of machining, involve exploring several different related topics.

First, the terms *machining* and *machine tool* need to be defined and many details of their definitions explained.

Next, a realization of how machining is connected to people's daily lives is needed. Connections will be made to a wide variety of consumable and durable goods and even services used by millions of people worldwide.

Discussion of the equipment, tools, processes, and technology used in the world of machining is necessary to begin to understand the role of machining in society. A brief history of machining and how it has progressed over time also helps to portray the importance of the machining field in the past, present, and future.

Once an overview of these topics is complete, the journey into the complex world of machining will have begun.

MACHINING DEFINED

What is *machining*?

Merriam-Webster's Dictionary defines **machining** in this way:

"to process by or as if by machine; especially: to reduce or finish by or as if by turning, shaping, planing, or milling by machine-operated tools."[1]

This definition may not give a very clear picture of machining. It is from the year 1853, and its basic meaning is still correct, but that definition does not tell the whole story of machining.

Beginning with Merriam-Webster's definition is a fine start. First, "to process by machine" means to use a machine to perform a task.

The second part of this definition, "to reduce or finish," means to change size and/or shape by cutting a piece of material. Turning, shaping, planing, and milling are cutting methods. Materials that are machined are usually metals, but other materials, including plastics and graphite, can also be machined.

Finally, the "machine-operated tools" used to perform the cutting are called **machine tools**.

---

[1]By permission. From Merriam-Webster's Collegiate® Dictionary, 11th Edition © 2013 by Merriam-Webster, Inc. (http://www.merriam-webster.com/).

All of these factors add to a definition of machining that is well suited for the topics discussed throughout this text:

Machining: Using machine tools to cut materials to desired sizes and shapes.

HISTORY OF MACHINING

Humans have used machine tools for centuries, beginning with very primitive forms and advancing to the high levels of technology, precision, and efficiency that exist today. The earliest machine tools were hand powered, and progressed to being powered by animals or water, then steam, and finally electricity.

Simple Machine Tools

The bow drill is the simplest and most likely the first machine tool. The cord of a bow was wrapped around a round cutting tool and, when the bow was moved back and forth, the cutting tool rotated and produced a hole. Similar to the bow drill is another hand-powered machine tool called the pump drill. It was developed around the time of the Roman Empire and was common until the 18th century. In the pump drill, a cord still rotates the round cutting tool, but motion is up and down and more easily creates rotary cutting action to produce holes. **Figure 1.1.1** shows these simple hand-powered tools.

The spring pole lathe was developed in the 13th century to produce cylindrical wooden parts. One end of a rope was connected to the part being cut and the other end to a spring pole, and power was produced by use of a foot

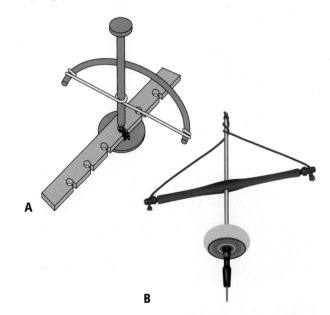

FIGURE 1.1.1 Examples of the earliest hand-powered machine tools. (A) The bow drill; and (B) the pump drill.

FIGURE 1.1.2 The Great Wheel lathe.

pedal. Cutting tools were then held against the rotating part to create cylindrical surfaces. Early settlers in North America used live saplings to build this type of machine tool at their home sites. Later a metal cutting version was developed.

The spring pole lathe had one drawback: its motion was not continuous. In the mid-18th century, John Smeaton developed the Great Wheel lathe that was powered by a drive cord or belt attached to a large wheel. One person spun the wheel to create power, and another performed the machining. **(See Figure 1.1.2.)**

Industrial Revolution

Machine tools began to drastically improve with the beginning of the Industrial Revolution in the late 18th century. More products were being produced from metals, and better machine tools were needed.

In England in 1775, John Wilkinson developed a water wheel–powered boring machine to machine the inside of cannons. **(See Figure 1.1.3.)** Soon the machine began to bore cylinders for Boulton and Watts's steam engines. That began the era of steam-powered machine tools.

In 1797, Henry Maudslay developed a machine that was able to accurately cut screw threads. This revolutionized manufacturing because interchangeable threaded parts could be produced.

In 1818, Eli Whitney produced the first milling machine. This machine tool was able to produce flat surfaces

FIGURE 1.1.3 John Wilkinson's boring machine. It was first used to machine cannon bores, then cylinders for steam engines.

Horner, Joseph Gregory, Modern Milling Machines, C. Lockwood and Son,© 1906

FIGURE 1.1.4 An early milling machine from around 1860. Horner, Joseph Gregory. *Modern milling machines.*

more easily than by hand with filing and scraping tools. Over the next several years, several individuals made improvements on Whitney's machine and different models became available. **Figure 1.1.4** shows an early milling machine from around 1860.

The post drill produced holes by turning a crank by hand. The crank turned gears that rotated the cutting tool and advanced it into the part being drilled. It was commonly used into the early 20th century before electricity became widely available.

Throughout the 18th and 19th centuries, steam-powered machine tools were driven by a series of belts that were connected to a large centralized wheel powered by a steam engine. During the Industrial Revolution, many companies began producing machine tools as metal cutting operations became more common.

20th-Century Machining

In the early part of the 20th century, electric power began to replace steam power, and machine tools continued to become more complex, more precise, and more efficient. Better machine tools were able to produce more accurate parts, which in turn produced even better machine tools, in a cycle of constant improvement.

In the early 1900s, Henry Ford's creation of the assembly line for mass production of automobiles relied heavily on machining. Parts needed to be machined efficiently to keep up with automobile assembly.

World War I and World War II both created huge growth in the machining industry in the United States as the country produced war-related materials.

Up until the 1940s, machine tool movements were controlled by levers, hand wheels, and geared transmissions. After World War II, great economic growth took place in the United States. Consumerism began, and the machining industry needed to become more efficient to support manufacturing. The invention of **numerical control (NC)** greatly improved machine tool performance. A language of machine code was developed and loaded on a punch card or tape and then fed into the machine tool to automatically guide the motions of the machine and change tools without the need of an operator.

In the 1970s, the NC punch card or tape began to be replaced with **computer numerical control (CNC)**. Instead of machine code being punched on the tape or card, code was entered through an integrated computer on the machine tool. Continued advancement in computer technology and machine tool construction has resulted in machine tools that can produce intricate, complex shapes with extreme accuracy and efficiency. When properly configured, they can also perform many operations with many different types of cutting tools while running without the need of an operator. **Figure 1.1.5** shows an ultra-modern CNC machine tool in operation.

Courtesy of Haas Automation, Inc

FIGURE 1.1.5 Today's state-of-the-art CNC machine tools can be programmed to run unattended and machine extremely complex shapes.

THE ROLE OF MACHINING IN SOCIETY

Nearly every person depends either directly or indirectly on machining in some way. Without machining, very few goods and services used every day would exist. How is that possible? Some exploration is needed to find the answer.

People, Manufacturing, and Machining

Many think of manufacturing in terms of big-ticket items like cars and televisions, but everyone uses manufactured items every day. **Manufacturing** simply means to produce something. Paper is a manufactured item. Plastic bags are manufactured items. So are tissues, clothing, and many foods. **End products** are final manufactured items used by consumers. The machining industry produces end products and components that are assembled as end products, and supports manufacturing for the products used by people throughout the world every day.

Machining also normally involves producing sizes and shapes to high levels of precision. Some machining operations can produce sizes with variations of 0.0001 inch or less of the desired size. This one ten-thousandth of an inch (0.0001) is approximately 1/50 of the *thickness* of an ordinary piece of paper. Why do parts need to be produced with such precision? The answer is performance and interchangeability of parts. When mating parts are assembled, high accuracy ensures proper fit and long life. Further, mating parts can be mass produced and interchangeable because they are manufactured to standard sizes, instead of needing to be custom fit to each other.

Some common connections to machining can be made fairly easily, while others require more careful investigation. It is more obvious that machining is connected to manufacturing of durable goods in a wide variety of industries, such as automotive, aerospace, and motor-sports, than to the paper, computer, or food industries.

Manufacturing in the United States

Recent history has convinced the vast majority of people that manufacturing is a dead industry in the United States. While it is certainly true that manufacturing has experienced some decline since the last several years of the 20th century, the United States is still the leading manufacturing nation of the world. U.S. manufacturing was valued at $2.1 trillion in 2016. There were only eight other countries in the world whose entire economies were larger than the U.S. manufacturing sector. Further, more than three-fourths of the research and development activities conducted in the

United States are performed by manufacturing companies, leading the way for technological advancements in many different fields.

Manufacturing also plays a major role in supporting the American workforce. U.S. Bureau of Labor and Statistics 2017 data show that there are more than 12.5 million people directly employed in manufacturing jobs in the United States. With a total workforce of approximately 154 million, manufacturing provides employment to over 8 % of all U.S. workers. At the end of the first quarter of 2018, the average American worker in the manufacturing sector earned an hourly wage of $26.86. That equates to more than $55,600 annually based on the average 40.8-hour workweek. When benefits such as health insurance and retirement plans are included, the average manufacturing worker's hourly wage rises to just under $39. That equates to slightly more than $81,000 annually. These figures show that manufacturing is alive and well in the United States, the country is still a global leader in manufacturing, and manufacturing careers provide excellent salaries and benefits.

Aerospace, Automotive, and Motorsports

Automotive and aerospace industries rely heavily on machining and machine tools. Consider cars and planes as examples. These highly complex and technologically advanced vehicles contain parts that were produced by machining operations. Engine, drive-train, and suspension components, as well as wheels, gears, and instrumentation, are just a few examples, not to mention the countless variations of nuts, bolts, and washers used for assembly. Machining operations produce all of these parts precisely. The motor-sports industry also uses many of the same types of parts as those used in the automotive and aerospace industries. **(See Figure 1.1.6.)**

Courtesy of Penske Racing Shocks

FIGURE 1.1.6 Machined shock absorber components and an assembled shock absorber used in motorsports racing.

Medical Fields

Other high-tech fields that are not easily seen as related to machining still depend on machining and machine tools for their existence. The medical field is one major example. Machine tools produce many medical devices that are used in today's high-tech surgical procedures. Surgical and dental tools, heart catheters; intravenous and hypodermic needles; joint replacement parts for knees, hips, and elbows; replacement discs for the spinal cord; and even artificial hearts are produced by high-tech machining operations. By manufacturing these types of components, machining operations and machine tools play key roles in medical and surgical advancements. **Figure 1.1.7** shows some machined parts used in the medical industry.

Plastics

In today's society, plastic plays a role in nearly every aspect of life. Electronic items such as CD and DVD players, televisions, portable digital music players, mobile phones, and computers all contain plastic parts. Plastic bottles, cups, and other containers are everywhere. Most children's toys are plastic. All of these objects depend on the machining

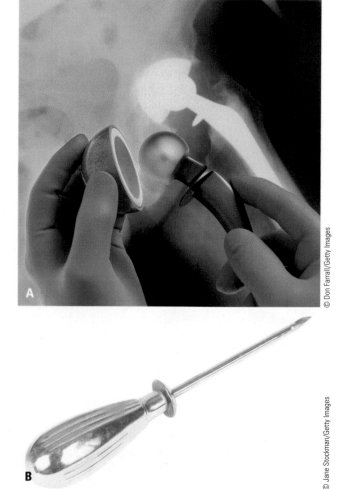

© Don Farrall/Getty Images

© Jane Stockman/Getty Images

FIGURE 1.1.7 (A) Machining produces medical products such as this hip implant and (B) surgical tool.

Courtesy of Trimetric Mold and Design

FIGURE 1.1.8 One-half of a machined mold and the plastic molded part produced by the mold.

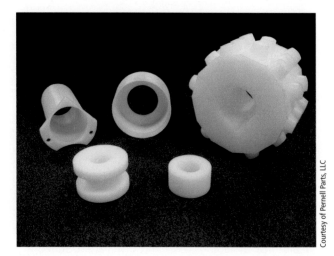

Courtesy of Pernell Parts, LLC

FIGURE 1.1.9 Some machined plastic components.

industry. Many finished plastic parts are produced from molds when high numbers are required and sizes do not need to be extremely precise. Molds contain cavities that are made to the shape of the desired part and then filled with molten plastic. When the plastic cools, the mold is opened and the part is removed. Machine tools machine the molds for these types of plastic parts that are used by millions every day. **Figure 1.1.8** shows a mold and a molded plastic part. Plastic parts that require very precise sizes or those produced in low numbers are often machined from solid pieces of plastic instead of being molded. **Figure 1.1.9** shows some machined plastic components.

Other Durable Goods

Products in daily use, such as hand tools, metal cans, metal pans, and drinking glasses, as well as countless metal components used throughout the world, also need the machining industry to exist. Molds similar to those for plastics are used to create glass products. Tools called dies are used to cut or form sheet metal into desired shapes. Cans and pans are examples of parts made with one type of die called a forming die. A flat piece of metal is squeezed between two parts of the die to stretch it to the desired shape. Forging dies use great force to form

hot pieces of metal into countless products, including wrenches, hammers, gears, crankshafts, hitches, and door hardware. Blanking and piercing dies punch out sheet metal products using the same principle as a paper hole punch. Sometimes several dies that pierce and form metal are mounted in one assembly called a progressive die. At each stage, operations are performed on the part, resulting in a finished end product. The components of these types of dies are made with machine tools. **Figure 1.1.10** shows a progressive die assembly and the part produced by the die.

Consumable Goods

It is becoming clearer how machining is connected to durable goods that are manufactured, but how does machining relate to consumable items such as food, paper, and clothing? The plastic packaging that holds many food items is made in molds **(see Figure 1.1.11)**, and molds also actually form the shape of some foods, such as ravioli and snack foods.

Other consumables such as clothing and paper rely on machining indirectly, just like durable goods do. The equipment and machinery that produce clothing, paper, or any manufactured consumable product are built from parts that were created through machining processes. Hydraulic cylinders, shafts, conveyor rollers, and bearings are a few examples of machined components used to produce equipment that in turn produces every manufactured product used today.

MAJOR MACHINE TOOLS

There are some basic types of machine tools or machining operations. Each is designed and suited for specific applications. Many are greatly improved versions of the earliest handheld and hand-powered tools using the same basic principles of operation.

FIGURE 1.1.10 (A) This progressive die has eight stages. Each stage shapes the part further from a strip of material. At the final stage, the die cuts the finished part off the strip. (B) The final product.

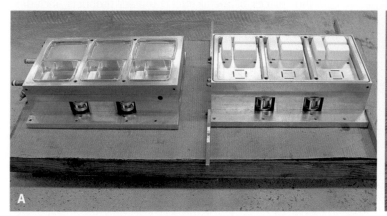

FIGURE 1.1.11 (A) Machining produced this mold that forms plastic food packaging. (B) The package produced by the mold.

Others are completely different technologies that have been developed recently. Machine tools are referred to as either conventional (or manual) or CNC types. Conventional machine tools require an operator to use hand-operated wheels or levers or engage geared transmissions to perform machining operations. They can usually perform only straight-line movements in one plane, or direction, at a time. Movements of CNC machine tools are directed by computerized controls. They can produce intricate, complex shapes with extreme accuracy and efficiency. When properly configured, they can also perform many operations with many different types of cutting tools while running without the need of an operator. Some types of machine tools are available in either conventional or CNC versions, while others are strictly CNC versions.

Sawing Machines

Sawing machines, often just called saws, use multi-tooth saw blades to perform cutting. These machines are usually used to cut material to rough lengths or to remove large sections of material quickly in preparation for other machining operations. Two types of blades are generally used. Band saws use a band-type blade and are available in horizontal or vertical machine types, as shown in **Figure 1.1.12**. **Figure 1.1.13** shows a saw that uses a circular-type blade. Some saws can be equipped with CNC controls to automatically cut multiple pieces of material very efficiently, as shown in **Figure 1.1.14**.

The Drill Press

The **drill press** performs holemaking operations by feeding various types of rotating cutting tools into the work. It is normally used when precise hole locations are not necessary. **Figure 1.1.15** shows a typical conventional drill press.

The Lathe

The **lathe** is used to produce cylindrical parts and operates on the principle of moving a cutting tool across the surface of a rotating piece of material. Operations that can be performed on the lathe include machining of external and internal diameters, lengths, threads, grooves, and

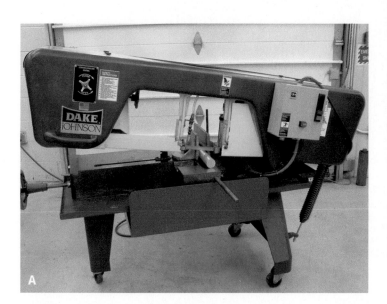

FIGURE 1.1.12 (A) A horizontal band-sawing machine; and (B) a vertical band-sawing machine.

Courtesy of Behringer Saws, Inc

FIGURE 1.1.13 A circular-blade-type sawing machine.

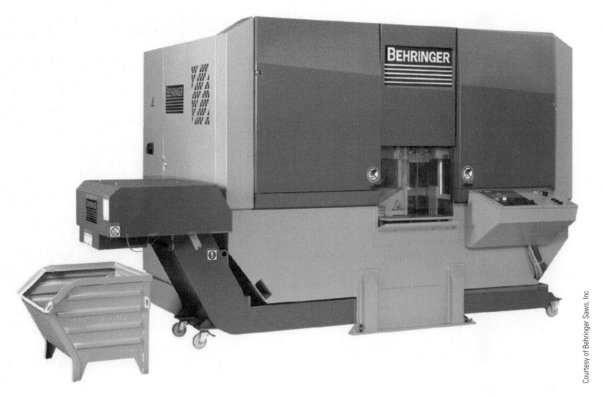

Courtesy of Behringer Saws, Inc

FIGURE 1.1.14 A CNC-controlled sawing machine.

tapers. **Figure 1.1.16** shows conventional and CNC lathes and a few sample parts produced by these machine tools.

The Milling Machine

Milling machines use rotating cutters moved across a part to remove material. These machine tools can use either a vertical spindle or a horizontal spindle and are available in conventional or CNC versions. Conventional types perform accurate holemaking operations and produce primarily flat surfaces. CNC versions can cut curves and contours. **Figure 1.1.17** shows conventional and CNC milling machines and some examples of the types of parts produced by these machines.

Abrasive Machining

Abrasive machining refers to using grinding wheels in either a nonprecision or precision manner. Noncritical operations are usually performed by hand on pedestal- type grinders like the one shown in **Figure 1.1.18**.

Precision grinders produce very accurate dimensions and very smooth surfaces. The most common types are surface grinders and cylindrical grinders. Surface grinders produce flat surfaces like milling machines, but with higher precision. Cylindrical grinders produce cylindrical parts like lathes, but with higher precision. Both surface and cylindrical grinders are available in conventional and CNC versions. **Figure 1.1.19** shows two types of precision grinders.

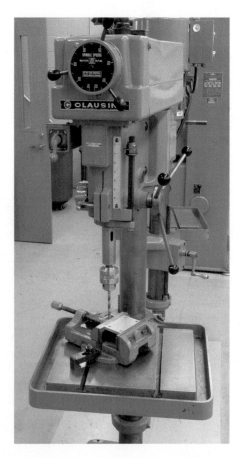

FIGURE 1.1.15 The drill press is often used for holemaking operations.

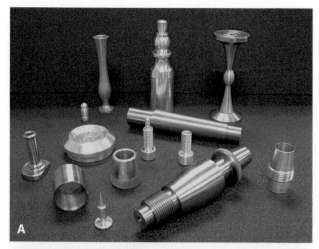

FIGURE 1.1.16 (A) Cylindrical parts are normally machined on either a (B) conventional lathe, or (C) a CNC lathe.

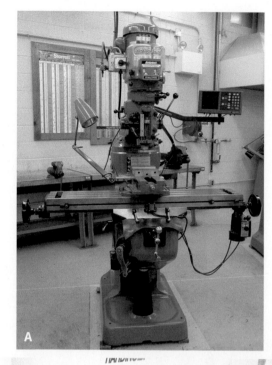

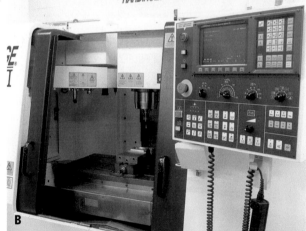

FIGURE 1.1.17 (A) A conventional vertical spindle milling machine and (B) a CNC vertical spindle milling machine. (C) Some parts produced on milling machines.

FIGURE 1.1.18 A pedestal grinder can be used to perform shaping operations when high accuracy is not needed.

Electrical Discharge Machining

Electrical discharge machining, or **EDM**, uses electrical current to machine any material that will conduct electricity. No contact is made between tool and the part, but material is eroded away by the spark created when the tool comes in close proximity to the part. The ram- or sinker-type EDM uses an electrode that is the opposite of the form desired. It is commonly employed in mold making because a mold often requires a small internal radius, or a narrow feature, that cannot be machined by other methods. However, the opposite form can often be easily machined on the electrode in a milling machine or a lathe. Then the electrode can be used to produce the final desired shape in the mold. **(See Figure 1.1.20.)** The wire-type EDM uses a small-diameter wire to produce very intricate shapes and sharp inside corners that could not be machined by other methods. Its ability to maintain those shapes through relatively thick material is another added benefit. **(See Figure 1.1.21.)** EDM machines are classified as CNC machine tools, but some sinker/ram types perform only simple movements. A sinker/ram EDM can create pockets that do not pass entirely through a part. Features produced by wire EDM must pass completely through the part because the wire must be able to pass from a spool, through the part, and then be collected in a container. One disadvantage of the EDM is that the process is generally much slower than using lathes and milling machines.

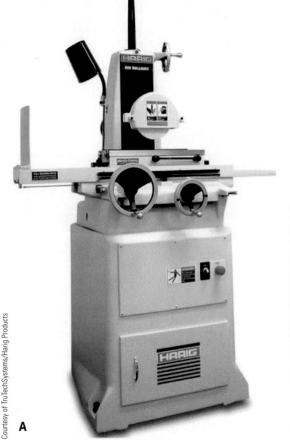

A

B

FIGURE 1.1.19 (A) A conventional surface grinder. (B) A CNC cylindrical grinder.

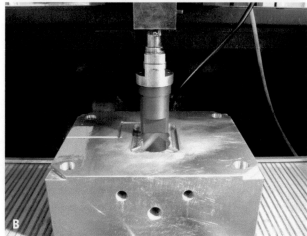

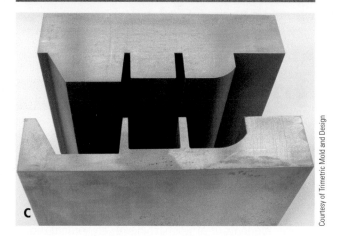

FIGURE 1.1.20 (A) A machined graphite electrode that acts as the cutting tool in a sinker EDM. (B) The electrode mounted in the EDM to produce a mold cavity. (C) The work being machined by the EDM must be submerged in a liquid called dielectric fluid during operation.

FIGURE 1.1.21 (A) The wire EDM uses a very small-diameter electrically charged wire to cut through material. A flow of water helps to wash away the eroded material. (B & C) Examples of intricate parts than can be produced by the wire EDM.

Laser Machining

Laser machining uses a highly concentrated beam of light with temperatures up to 75,000°F to cut, groove, or engrave metals. Lasers are classified as CNC machine tools and can cut very hard materials easily. **Figure 1.1.22** shows a laser cutting operation.

Water Jet Machining

In **water jet machining**, abrasive grit is introduced into a very high-pressure, focused jet of water to perform cutting. Such jets can cut through 6 inches of steel. This type of machine tool was developed in the 1960s but is becoming more widely used in the machining industry. **Figure 1.1.23** shows a water jet cutting operation.

FIGURE 1.1.23 A water jet machine uses a high-pressure water stream containing an abrasive to cut materials.

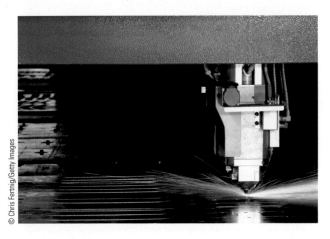

FIGURE 1.1.22 A laser machine tool in operation.

SUMMARY

- The machining industry is a widespread and complex, but little-known, field. It uses machine tools to machine, or cut, material to desired shapes and sizes.

- Manufacturing is simply the production of durable and consumable goods used by everyone, and machining has an impact on nearly every manufacturing process.

- Machining can directly produce end products or components used in end products. Machining can also produce components of other machines or equipment required for the manufacture of end products.

- Simple, hand-powered machine tools have existed for centuries and have been improved over time into highly efficient pieces of equipment capable of achieving extreme levels of precision.

- Several major machine tools are widely used today, and many are controlled by integrated computerized controls that allow them to perform complex operations with little human input once they are set up.

- Even though machining is not well understood, it has a great impact on many aspects of our lives.

REVIEW QUESTIONS

1. Define the term *machining*.
2. What is a machine tool?
3. What is manufacturing?
4. List four industries that depend on machining.
5. What does the abbreviation *CNC* stand for?
6. What is the purpose of the drill press?
7. What machine tool produces cylindrical parts?
8. Briefly describe the primary purpose of sawing machines.
9. What machine tool is available with either horizontal or vertical spindle orientation and uses rotating cutting tools to primarily produce flat surfaces?
10. Abrasive machining makes use of _____ to remove material.
11. What does the abbreviation *EDM* stand for?
12. What are the two types of EDM machines?
13. _____ uses a high-pressure stream of water containing abrasive particles to cut material.
14. Briefly describe the principle of laser machining.

UNIT 2

Careers in Machining

Learning Objectives

After completing this unit, the student should be able to:

- Identify and discuss careers in the machining industry
- Identify and discuss careers in fields related to machining
- Discuss the job outlook in the machining field
- Understand and explain effective job-seeking skills

Key Terms

CNC machinist
Computer-Aided
   Design (CAD)
Computer-Aided
   Manufacturing
   (CAM)
Conventional
   machinist
Die maker
Engineering drawing
Industrial salesperson
Inspector

Machine tool service
   technician
Machinist
Manufacturing
   engineer
Manufacturing
   engineering
   technician
Mechanical designer
Mechanical engineer
Mechanical engineering
   technician

Metrology
Mold maker
Operator
Print
Programmer
Quality control
   technician
Set-up technician
Supervisor/manager
Toolmaker

INTRODUCTION

Throughout the Industrial Revolution and into the early part of the 20th century, people who worked with machine tools were usually referred to as mechanics. Their jobs were varied, as they performed manual labor in addition to operating machinery. With the great requirement for hands-on skills, these mechanics were highly skilled craftspeople.

As the machining industry grew through the 20th century, the term **machinist** replaced *mechanic*. The new term connected the person with the machine tool. The machinist was one who made a living at and was skilled in the use of machine tools.

Work for the machinist into the early 20th century was a combination of physical labor and mental effort, and working conditions were often dangerous and dirty.

As time progressed, machine tools, machining, and related industries became more complex. Specialty jobs in different areas of machining started to develop. Today, the title of machinist is often misunderstood and brings to mind antiquated visions of hard physical labor in dark, dangerous, sweatshop environments.

MODERN MACHINING CAREERS

Careers in the modern machining industry still require a combination of hands-on and mental skills but are far safer and far less physically demanding. Jobs are frequently in very clean, climate-controlled, high-tech environments. Workers in today's machining industry are highly skilled professionals with a combination of hands-on and theoretical talent.

Due to an aging workforce in the machining field, there are shortages of qualified candidates in many fields of durable and consumable goods manufacturing. Those shortages equal opportunities for good-paying jobs with benefits and excellent working conditions. According to the Occupational Outlook Handbook website prepared by the U.S. Bureau of Labor Statistics (BLS), there should be good job opportunities for CNC programmers and operators and tool and die makers through the year 2026, largely due to few people pursuing those career paths. This means that there should be more openings than qualified candidates. These conditions create an environment ripe with career opportunities.

Today there are many different jobs in the machining field and many jobs in related fields that require knowledge of machine tool operations and capabilities. All of these jobs require at least some knowledge of reading **engineering drawings** or **prints**. These are the drawings of parts to be machined and show shapes, sizes, and specifications that must be met. **Figure 1.2.1** shows an example of some engineering drawings.

The next sections give brief descriptions of primary responsibilities and tasks of these jobs. It must be stressed that there is frequently an overlap of skills, depending on the particular position and company. All of these jobs may be available in three basic roles: large-scale production manufacturing, small-volume custom manufacturing, and support positions for manufacturing.

The following websites are resources that can provide additional information and details about careers connected to machining, as well as national statistics about career outlook and wage data.

http://online.onetcenter.org/ is the home page of **O*NET**, the Occupational Information Network (O*NET) developed under the sponsorship of the U.S. Department of Labor/Employment and Training Administration (USDOL/ETA).

http://www.careeronestop.org/ is the home page of **Career One Stop**, another website that is also sponsored by the U.S. Department of Labor.

http://www.bls.gov/audience/jobseekers.htm is a page of the BLS, which is also a part of the U.S. Department of Labor.

http://www.dol.gov/dol/location.htm lists links to individual state agencies that provide career and employment data and resources specific to each U.S. state and territory.

Operator

Operator positions are available in both conventional and CNC machining environments but are more common in CNC facilities. They are often filled by entry-level employees who are beginning careers in machining, and they require little prior knowledge of machining. **Operators** place parts in machines and continually run a set operation or group of operations. They are often also responsible for measuring sizes to ensure parts meet specifications shown on engineering drawings.

Depending on the environment, operators may be responsible for keeping more than one machine tool running and checking measurements on parts from each machining process.

Set-up Technician

Set-up technicians prepare or set up machine tools so that operators can run them. These positions are more common in CNC environments than in conventional machining facilities. Set-up technicians may specialize in preparing only one type of machine tool, such as the lathe or milling machine. Others may be skilled in setting up multiple types of machine tools. In some jobs, set-up technicians may be responsible for a certain number or group of machines. Set-up positions require previous

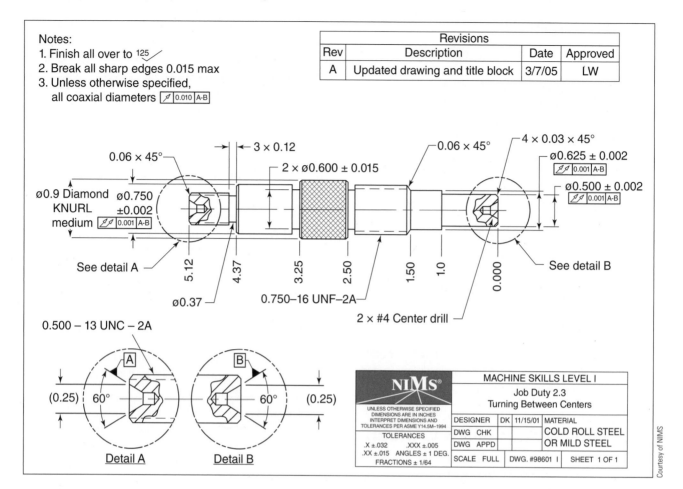

Notes:
1. Finish all over to ^{125}
2. Break all sharp edges 0.015 max
3. Unless otherwise specified,
   all coaxial diameters ⟋ 0.010 A-B

Revisions			
Rev	Description	Date	Approved
A	Updated drawing and title block	3/7/05	LW

0.06 × 45°
3 × 0.12
2 × ø0.600 ± 0.015
0.06 × 45°
4 × 0.03 × 45°
ø0.625 ± 0.002
⟋⟋ 0.001 A-B
ø0.750 ±0.002
ø0.500 ± 0.002
⟋⟋ 0.001 A-B
ø0.9 Diamond KNURL medium ⟋⟋ 0.001 A-B

See detail A
5.12
4.37
3.25
2.50
1.50
1.0
0.000
See detail B

ø0.37
0.750–16 UNF–2A
2 × #4 Center drill

0.500 – 13 UNC – 2A

(0.25) 60° A
B 60° (0.25)

Detail A   Detail B

NiMs®
UNLESS OTHERWISE SPECIFIED DIMENSIONS ARE IN INCHES INTERPRET DIMENSIONS AND TOLERANCES PER ASME Y14.5M–1994
TOLERANCES
.X ±.032 .XXX ±.005
.XX ±.015 ANGLES ± 1 DEG.
FRACTIONS ± 1/64

MACHINE SKILLS LEVEL I			
Job Duty 2.3 Turning Between Centers			
DESIGNER	DK	11/15/01	MATERIAL
DWG CHK			COLD ROLL STEEL
DWG APPD			OR MILD STEEL
SCALE FULL	DWG. #98601 I		SHEET 1 OF 1

FIGURE 1.2.1 An example of some engineering drawings, or prints, showing the shape and specifications of components to be produced by machining operations.

machining experience and/or education provided by secondary career and technical education (CTE) programs or certificate or associate degree programs.

Set-up technicians select proper cutting tools and devices to hold parts and mount them in the machine tool. They will often load computer programs into CNC machine tools and modify machining operations or replace tools as needed to maintain part specifications. Set-up technicians work closely with operators who run CNC machines as well as those who program CNC machines. **Figure 1.2.2** shows a set-up technician at work.

Conventional Machinist

Conventional machinists, or manual machinists, are highly skilled workers who usually have experience running almost every type of conventional machine tool. They normally do not specialize in one type of machine tool and work in environments where they must use many different machines to complete specific projects. **Figure 1.2.3** shows a conventional machinist.

Conventional machinists will examine prints, select materials, establish process plans, and then perform all of the machining operations needed to complete a project to meet print specifications. Training through high school vocational education programs, certificate or associate degree programs, or apprenticeships is often needed for these jobs.

CNC Machinist

CNC machinists are very similar to conventional machinists except that they are normally skilled in the set-up and operation of CNC machine tools. They possess the skills of both the set-up technician and the operator. CNC machinists, like set-up technicians, may specialize in the use of one type or multiple types of CNC machine tools.

Many CNC machinists will, like conventional machinists, study prints, select material, and establish plans for machining operations. They will then use CNC machine tools to perform those machining operations and measure machined parts to ensure engineering drawing specifications are met.

Usually, but not always, CNC machinists will have experience in conventional machining, set-up, and operation as well and may use both conventional and CNC machines to perform their duties. Training for CNC machinists begins like that for conventional machinists, but also requires additional training in CNC applications.

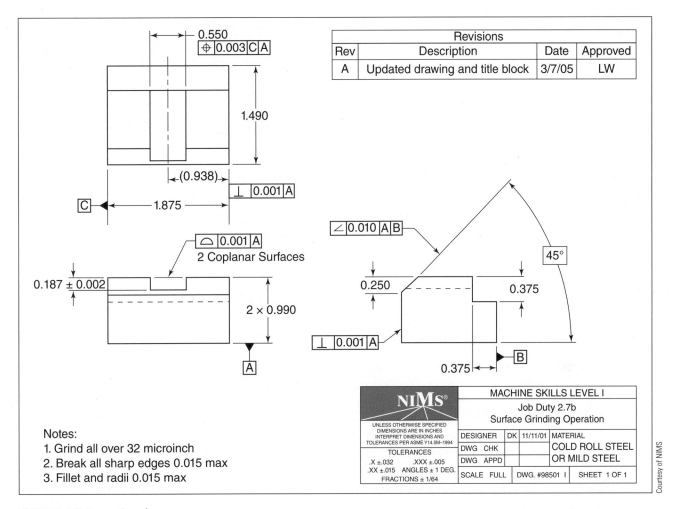

Notes:
1. Grind all over 32 microinch
2. Break all sharp edges 0.015 max
3. Fillet and radii 0.015 max

	Revisions		
Rev	Description	Date	Approved
A	Updated drawing and title block	3/7/05	LW

NIMS®	MACHINE SKILLS LEVEL I			
UNLESS OTHERWISE SPECIFIED DIMENSIONS ARE IN INCHES INTERPRET DIMENSIONS AND TOLERANCES PER ASME Y14.5M-1994	Job Duty 2.7b Surface Grinding Operation			
	DESIGNER	DK	11/11/01	MATERIAL
TOLERANCES	DWG CHK			COLD ROLL STEEL
.X ±.032 .XXX ±.005	DWG APPD			OR MILD STEEL
.XX ±.015 ANGLES ± 1 DEG.	SCALE FULL	DWG. #98501 I		SHEET 1 OF 1
FRACTIONS ± 1/64				

Courtesy of NIMS

FIGURE 1.2.1 *continued.*

Courtesy of Penske Racing Shocks

FIGURE 1.2.2 This set-up technician is setting up the cutting tools on a CNC lathe.

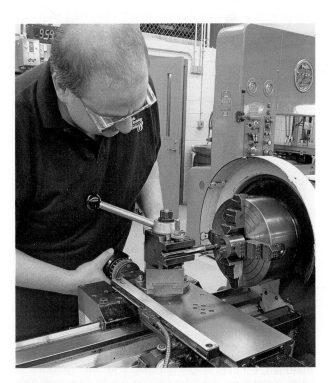

FIGURE 1.2.3 This machinist is cutting an internal thread in a custom part on a conventional lathe.

Programmer

Programmers, sometimes called CNC or NC programmers, are the people who write programs consisting of machine code for CNC machine tools. They typically have previous experience as operators, set-up technicians, or CNC machinists and have advanced into programming positions. Programmer training is also similar to that needed by conventional and CNC machinists, but with additional training required in understanding machine programming languages.

Programmers may enter code directly into the machine tool using the machine's computerized control or write programs with a typical word processing program that can then be loaded into the machine's control system.

More complicated CNC programs require the use of **computer-aided manufacturing (CAM)** software. The programmer uses the computer software to select tools and cutting operations using a computer-generated virtual model of a part. **(See Figure 1.2.4.)** The CAM software then automatically generates the complex machine code that can be downloaded into the machine tool's computerized control.

Die Maker, Mold Maker, Toolmaker

Die makers, mold makers, and toolmakers are very highly skilled specialists in the machining field. They are nearly always expert machinists, either in the use of conventional machine tools or both conventional and CNC machine tools. Those with CNC experience will likely be able to perform CNC programming online or offline using CAM software. They may also use **computer-aided design (CAD)** software to design some of their projects. Assembly and troubleshooting of dies, molds, and tools are also responsibilities in these positions. Formal training through certificate

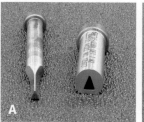

FIGURE 1.2.5 A die maker machined these punch and die components that pierce out small pieces of sheet metal. They are part of a larger die assembly.

degree programs, associate degree programs, or apprenticeships prepares die makers, mold makers, and toolmakers.

Die Makers

Die makers specialize in making cutting tools consisting of punches and dies that are used to either bend, form, or pierce metal parts. **Figure 1.2.5** shows some typical punch and die components machined by a die maker.

Mold Makers

Mold makers specialize in machining molds and mold components for either plastic or die-cast metal parts. Molds contain cavities that are made in sections, which when assembled will have either molten plastic or metal forced into the molds. When the material cools, the mold is opened and the parts are removed. **Figure 1.2.6** shows machining of a mold on a CNC milling machine.

Toolmakers

Toolmakers specialize in machining complex tools, jigs, fixtures, or machinery used to manufacture other parts. Jigs are devices that hold parts and guide tools for manufacturing processes. Fixtures hold parts for manufacturing

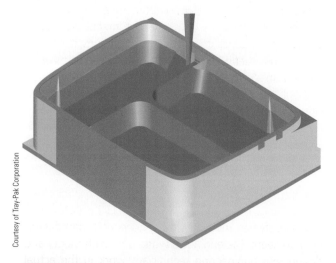

FIGURE 1.2.4 CNC programmers often use CAM software to create complex programs for CNC machining operations. This software is simulating the program for the machining of a mold.

FIGURE 1.2.6 CNC machining of this prototype mold is typical of the type of work performed by mold makers.

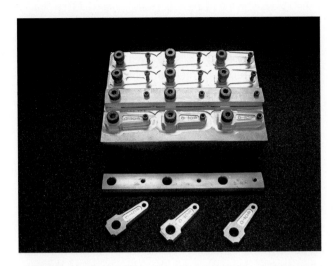

FIGURE 1.2.7 A fixture used for holding parts that was designed and machined by a toolmaker. A strip of material is mounted on the fixture and a CNC milling machine cuts the profile of the finished parts.

processes. **Figure 1.2.7** shows an example of a fixture designed and machined by a toolmaker.

Die makers, mold makers, and toolmakers typically work from highly detailed engineering drawings, plan and perform required machining operations, and then perform the final assembly and fitting of all the machined components to ensure that the final die, mold, or tool performs as required. Die makers and toolmakers are often called tool and die makers, as their work is very closely related, and many toolmakers and die makers are experienced in both specialty areas.

Supervisory Positions

After sufficient practical experience is gained in the machining field, there can be opportunities for advancement into a position of a **supervisor** or **manager**. People in these positions normally do not perform machining operations, but are responsible for planning, scheduling, purchasing, budgeting, and personnel issues.

RELATED CAREERS

There are some careers that are closely related to or that support machining. They include design, engineering, service, sales, and inspection positions.

Mechanical Designer

Mechanical designers use CAD software to draw models or engineering drawings that are used as reference for machining operations. **(See Figure 1.2.8.)** Designers are normally given overall parameters or specifications from engineers, and then they design components or subassemblies for use in larger assemblies.

A general knowledge of machining capabilities is important to the designer so designs can be machined or

Courtesy of Tray-Pak Corporation

FIGURE 1.2.8 A mechanical designer used CAD software to design this mold used to produce plastic food storage containers.

manufactured efficiently and economically. Mechanical designer positions usually require at least an associate's degree and often require a 4-year degree.

Engineering Positions

Mechanical Engineer

Mechanical engineers understand the theory of many topics, including selection and properties of materials. They usually design overall projects, assemblies, or systems using high-end computer software. An understanding of machining is helpful for the same reason as it is important to the mechanical designer.

Mechanical engineers can also work in areas of machine tool, cutting tool, or machine tool accessory design. Mechanical engineers usually need to have completed a bachelor's degree.

Manufacturing Engineer

Manufacturing engineers establish manufacturing processes and continually study and improve on those processes. Often these are high-tech automated production lines. If a process includes machining operations, it is crucial that manufacturing engineers understand machining operations. If manufacturing engineers understand machining, they will be more efficient in determining needed components to build production lines. These engineers will often work with machinists and/or toolmakers and die makers who produce components for manufacturing equipment.

Engineering Technician

Mechanical engineering technicians and **manufacturing engineering technicians** could be considered a cross between mechanical engineers and manufacturing engineers. The engineers work more with theory and design, but engineering technicians work in the actual construction or testing of mechanisms, machinery, and equipment. Engineering technician jobs usually require an associate's degree, and some require bachelor's degrees.

These technicians will, like manufacturing engineers, frequently work with machinists and toolmakers and die makers when building equipment. Basic machining knowledge is important so they can interact with those types of people and complete needed tasks accurately and efficiently.

Sometimes engineering technicians will actually have machining backgrounds and may perform some machining operations as they construct projects. In these cases, working knowledge of machine tools is even more important.

Machine Tool Service Technician

Given the variety and number of machine tools in service in the machining industry, there is certainly the need for repairs to the pieces of equipment. **Machine tool service technicians** travel into the field to perform repairs as needed. They are like the appliance repairpersons of the machining industry. Service technicians frequently complete apprenticeships and may receive specialized training provided by machine tool manufacturers.

Machine tool service technicians can specialize in one type or many types of machines. They may work for a particular machine tool manufacturer or for a company that services many different manufacturer brands. Depending on the employer, machine tool service

technicians may travel around the country or the world. These service technicians need to have troubleshooting and problem-solving skills as well as understand the function and repair of mechanical, electrical, electronic, hydraulic, and pneumatic systems.

Quality Control Technician/ Inspector

Whenever and wherever machining operations are being performed, there is always a need for inspection of parts to be sure they meet specifications. Some companies employ **quality control technicians** or **inspectors** skilled in **metrology**, the science or practice of measurement, to perform this function. Training for these positions can vary greatly depending on the complexity of the job. Some may require only company training while others require 4-year degrees in metrology.

Inspectors may move from machine to machine to inspect parts or parts may be brought to a station or an inspection lab. The inspector's job is to confirm all specifications and report to the person responsible for maintaining those specifications. Inspectors will often work closely with operators or set-up technicians to quickly identify errors so corrections can be made. They often use highly specialized measuring equipment. **(See Figure 1.2.9.)**

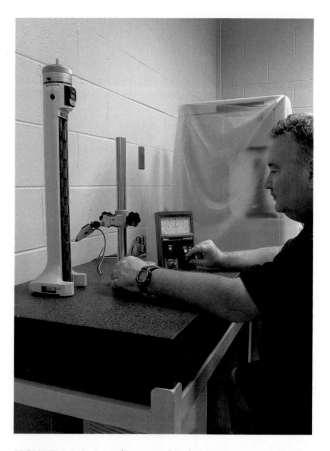
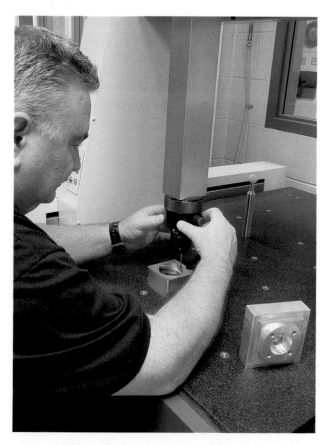

FIGURE 1.2.9 A quality control technician uses specialized measuring instruments to inspect machined components.

Industrial Salesperson

Sales jobs exist in a few different areas. Machine tool manufacturers, cutting tool or machine accessory manufacturers, and industrial suppliers can employ salespeople. In each case the **industrial salesperson** visits customers and prospective customers to discuss their needs in order to gain their business. They may travel locally or nationally and often deal with supervisors, managers, machinists, and toolmakers and die makers.

SUMMARY

- Many different careers are available in machining and related fields. These occupations have evolved from mostly hands-on jobs to those that also require the use of new technology.

- The machining and related career fields offer many different jobs requiring different levels of skill and education. Jobs in these fields often offer opportunities for career advancement.

- Today many careers in the machining industry pay well and offer excellent working environments and there are currently opportunities for successful careers in the machining field in the United States.

- Because of the wide variety of industries and jobs that rely on machining, the field can provide a variety of rewarding careers.

REVIEW QUESTIONS

1. What is an engineering drawing and what is its purpose?
2. What is the primary duty of a machine tool operator?
3. What occupation involves preparing tools and machines for machining operations?
4. Briefly describe a CNC programmer's responsibilities.
5. What is CAM software?
6. What is CAD software?
7. What career advancement opportunities exist for employees in the machine tool industry?
8. What career area related to machining deals with designing, establishing, and improving products and/or manufacturing processes?
9. What occupation requires knowledge of several different types of systems in order to troubleshoot and repair machine tools?
10. What is metrology?

UNIT 3 Workplace Skills

Learning Objectives

After completing this unit, the student should be able to:

- Identify and understand personal skills needed for success in the machining field
- Identify and understand technical skills needed for success in the machining field
- Show understanding of training opportunities and methods available to gain skills required for the machining field
- Create a career plan
- Create a resume
- Create a cover letter
- Compile a list of references
- Create a thank-you letter
- Describe a portfolio and its importance
- Use different methods to find job opportunities
- Conduct a practice interview

Key Terms

Apprentice
Apprenticeship
Associate degree
Baccalaureate degree
Career and Technical Education (CTE)

Certificate
Journeyperson
National Institute for Metalworking Skills (NIMS)

On-the-Job Training (OJT)
Personal skills
Technical skills
Vocational education

INTRODUCTION

Every career requires certain skills for success, and careers in the machining field are no different. Since there is often a lack of understanding of machining careers, there is also often a lack of understanding of the skills and knowledge needed to be a successful professional in the industry. All machining careers require a combination of mental and hands-on skills, although the blend may differ in different positions. Many of these abilities can be labeled as personal skills, or soft skills, that are largely part of someone's personality or nature but that can be honed and improved over time with practice. Others are technical skills, or practical skills, that are largely learned through various methods of formal and informal training and practice. There is, however, some overlap between personal and technical skills.

Once someone has the skills that are needed for industry employment, the search for a job must begin. There are several steps to take along this path, and there are some key job-seeking activities to pursue.

This unit will discuss the personal and technical skills commonly needed for success in machining careers and provide some job-seeking suggestions that can help someone find the job that is a good fit.

Once again, the following resources can provide additional information about skills required for specific jobs as well as data on wages and the outlook for machining jobs.

- http://online.onetcenter.org/ is the home page of **O*NET**, the Occupational Information Network (O*NET) developed under the sponsorship of the U.S. Department of Labor/Employment and Training Administration (USDOL/ETA).
- http://www.careeronestop.org/ is the home page of **Career One Stop**, another website that is also sponsored by the U.S. Department of Labor.
- http://www.bls.gov/audience/jobseekers.htm is a page of the Bureau of Labor and Statistics, which is also a part of the U.S. Department of Labor.
- http://www.dol.gov/dol/location.htm lists links to individual state agencies that provide career and employment data and resources specific to each U.S. state and territory.

PERSONAL SKILLS

Personal skills are largely part of an individual's personality or natural ability, but they can be honed or improved with practice. Some are purely physical, some are purely psychological, and some are a combination of both. All play a key role in achieving success in machining careers.

Mechanical Aptitude

Mechanical aptitude is a combined mental and physical skill. It refers to the ability to visualize and understand basic laws of how things work and move. That includes the relationship between moving parts and the concept of cause and effect.

This skill is critical in the machining field, as there are many relationships occurring at the same time between machine tools, cutting tools, and the materials they cut. Those kinds of connections also exist in complex tool or machinery assembly. The talent to assess many factors and predict results is a daily occurrence for many machinists, mold makers, toolmakers, and die makers, and it is important for success in those machining jobs.

Manual Dexterity and Eye-Hand Coordination

Manual dexterity and eye-hand coordination describe the physical ability to precisely control hand motions. Performing intricate operations involving small movements to make fine adjustments is common in the machining field. This occurs during hand tool and machine tool operations as well as during assembly procedures.

Problem-Solving, Troubleshooting, and Decision-Making Skills

Problem solving, or troubleshooting, means being able to recognize when something is incorrect and then making corrections to fix errors. Because of the complexity of many processes in machining, it is not always easy to identify causes of problems. It is also an asset to be able to look at a situation and predict areas where problems might arise before they occur. This skill can be improved by training in machining principles, but the base capability of making judgments using many pieces of information is largely an instinctive skill.

Once problems are identified, decisions need to be made to correct them. To make good decisions, analyze as much information as is available. Then identify possible solutions. By projecting and comparing the outcomes of each possible solution, you can make decisions that provide the best expected result.

Focus and Concentration with Attention to Detail

To become successful in the machining industry, the machine tool professional must have a high level of concentration as well as an eye for detail. Because of the high-precision nature of machining, even small lapses in attention can lead to large errors and huge losses of time and money. The complexity of normal daily tasks also calls for attention to minor details to ensure that specifications and goals are being met.

Because of the highly automated and powerful equipment used today, loss of concentration and attention can create unsafe situations leading to personal injury or even death.

Persistence and Patience

Machine tool professionals must have the mind-set to stay on task until projects are complete. They also must take the time and precautions to make sure that work is done correctly. There are many tasks, especially in intricate part machining, programming, and mold, die, and toolmaking, that take long periods of time to complete, and often little visible progress is seen on a daily basis. Instant gratification and completion is not usually the norm for those in the machining field, so it is necessary to possess or develop long-term vision and goals and be able to persist at work that can be time consuming and tedious.

Personal Responsibility and Reliability

People in positions such as set-up technician; programmer; machinist; die, mold, and toolmaker; and especially manager are frequently given responsibility for progress and project completion with very little supervision along the way. For this reason, these jobs require personal accountability to meet both short-term and long-term goals. These people must strictly meet specifications and ensure that final products are correct, so it is necessary for a person to take ownership of his or her own work.

Ability to Perform Multi-Step Processes

Due to the lengthy procedures and multiple steps needed to perform even common everyday operations, the machining industry worker must be capable of performing those steps accurately. Following written or verbal instructions is also crucial to complete daily duties.

Ability to Use Technical Reference Materials

Machine tool technical manuals and complex reference books are very common in the field, so the skill of finding information in these different forms is fundamental to accomplishing many machining objectives. Reference materials can be textbooks or technical manuals. The Internet is also a valuable resource for obtaining technical information through educational, manufacturer, and machining industry associations and forums. This skill can be partially learned by becoming familiar with terminology and the format of types of sources, but it also requires a solution- and detail-oriented mind-set to know what to search for.

Interpersonal Skills

In the machining field, there is often the need to work with others. Many times, you must work with people who are skilled in areas that you are not. There is a need to communicate effectively to share information. When working in a team setting, all members should value each other's input and cooperate to meet required goals and objectives. Respect for others and their opinions creates a positive environment that will promote continuous improvement and success.

Significant Memory Use

There is an incredibly large amount of information required to perform machining operations and become highly skilled and successful in the field. No one can remember every small piece of information, but there are many of these small, but very important, pieces of information that are used every day in the industry and are needed for even small tasks. There are certainly times when reference materials need to be used, but many, many mathematical formulas, machining principles, and concepts used on a daily basis need to become second nature in order for you to perform efficiently and effectively.

TECHNICAL SKILLS

Technical skills are those that can be learned and improved with practice. Many are the "hands-on" abilities that need to be combined with personal skills in order to build a successful career in machining.

Ability to Interpret Engineering Drawings

Engineering drawings or prints are the plans or maps to creating parts through machining operations. They are a two-dimensional representation of three-dimensional parts and contain many important facts about types of materials to be used, part dimensions, required degrees of precision, surface and finish requirements, and other engineering specifications.

A significant amount of time needs to be invested to become proficient in understanding the language of engineering drawings. This usually involves studying sample prints and performing mathematical calculations using decimal and fractional numbers.

Knowledge of English and Metric Systems of Measurement

In today's global economy, there is a great deal of manufacturing that must meet specifications in both the United States and other countries around the world. For that reason, workers in the machine tool industry need to be able to recognize, compare, and convert measurements between the English, or inch, system and the metric system. Fortunately there are many tables and conversion charts available, but the skilled machining professional should be able to learn to use memory to reasonably visualize sizes in both inches and millimeters.

Proficient Math Skills

Whether planning or performing machining operations or conducting measurements, math plays a major role in the daily duties of machining professionals. Fractional and decimal operations as well as conversions between the two are needed every day. Basic skills in algebra, geometry, and right-angle trigonometry are also vital to performing common tasks.

Use of Hand Tools, Measuring Tools, and Machine and Cutting Tools

Different projects have different requirements according to specifications given on engineering drawings; so different tools will be used depending on those specifications. There are a very large number of different specialty hand, measurement, and cutting tools and machines in the industry, so those in machining careers must be able to select the proper tool for any given situation. Most of the tools used in the machining field are also very expensive, and many are very delicate, so proper use and care is necessary to avoid damage that leads to loss of time and money. Learning about the many tools in machining also requires a major investment of time.

Understanding of Metals and Other Materials and Their Properties

Machined parts can be manufactured from many different types of metals or other materials such as plastics, graphite, carbon fiber, or fiberglass. Metallurgy is the science of metals, and the basics must be learned to understand the characteristics of many different types of metals and how they will react during machining operations. That knowledge can also be applied in selecting the proper metal for a given application. The same can be said for nonmetal materials. There are many different compositions and grades of plastic, graphite, carbon fiber, fiberglass, and other materials.

Knowledge and Skill in the Use of Computer Technology

The computer is becoming a larger part of more and more occupations today. It is understandable that CNC programmers, designers, engineers, and managers need computer skills. But many other machining jobs require computer use also. Software is used in the field to perform tasks such as communication via email between co-workers, companies, departments and divisions of larger companies, and customers. Computer programs can be used to track orders, hours, and projects as they progress through different stages. The Internet can be a valuable resource to find and order tools and materials. For these reasons, today's machining professionals would be wise to learn basic computer skills.

TRAINING OPPORTUNITIES/ METHODS

The technical skills required for successful machining careers can be learned through several different methods, ranging from programs provided by public schools, community colleges, and universities to employer provided education. Some of the methods also aid in developing and improving personal skills as a complement to technical training.

Secondary School (High School) Programs

Many public school districts in the United States offer opportunities for basic to intermediate training related to the machining industry. The major benefit of training provided by the public school system is that there is no tuition cost to the student. Some high schools offer very basic hand and machine tool exposure through technology educational programs. These courses usually take place during one daily class period and last for one quarter or one semester. They can provide a brief introduction of machining to students and act as a gateway to other options.

Career and technical education (CTE) or **vocational education** historically has provided hands-on training in the trades to high school students to prepare them for career paths in various industries. These types of programs can provide training that covers topics in the machine tool field more broadly and deeply. These elective programs offer far more hours of education than standard high school technical courses and are usually offered at a common location for students in a particular geographic region. These specialty schools have many different names, including career centers, career and technology centers, and vocational high schools, and operate in a few different formats. Some of these schools are part time, where students split their attendance between their regular high schools and these specialty schools on a split-day schedule or an alternating weekly schedule over a period of 2 to 3 years. Others utilize senior-year-only systems where students attend full time and spend most of the day in a machining lab and only one to two class periods daily in academic courses.

The major benefit of this type of training is that there is no tuition cost to students, and some CTE programs provide instruction that is comparable to the first year of some post-secondary technical education programs. Further, this type of education can provide a solid background for further education or immediate entry into the workforce with above-average wage earning potential.

Post-Secondary Training

Post-secondary education in machining skills is offered through technical schools and colleges, community colleges, and universities. Many different schools offer training programs ranging from general machining to specialty areas of CNC programming; die making, mold making, and toolmaking; metrology; engineering; and engineering technologies. These programs vary in length from approximately 18 months to 4 years or longer.

Technical Schools and Community Colleges

Technical schools and community colleges usually offer **certificate** and/or **associate degree** programs. Certificate programs focus primarily on practical lab application courses and applied or practical academics. Associate degree programs normally require the same lab courses, but also call for more theoretical academic courses than certificate programs. Both of these programs can generally be completed within 2 years.

Universities

Universities normally offer associate (2-year) and **baccalaureate** (4-year) **degrees**. The 4-year programs normally offer more theoretical education and training in the specialty areas such as engineering disciplines.

Employer-Provided Training

Some employers provide training to employees while they are receiving wages. Companies will sometimes hire inexperienced employees at low wage levels as operators and move them into different positions as their skills increase. Further, companies may also need to provide training for specific specialized areas and may send employees off-site or bring trainers on-site to meet that need. This type of education is called **OJT** or **on-the-job training**. OJT can either be unstructured or structured.

Unstructured employer training is when a company teaches an employee only the skills that are needed to perform his or her current job or perhaps a future job in the company. The instruction occurs during the course of the normal operations during the work day and is usually given to meet only immediate needs.

Structured training exists in companies that have more formal, established training programs, but again these programs are frequently specific to the individual company's needs. Employees may receive instruction over a specified time period, and when they learn and demonstrate new, higher-level skills, they will likely receive wage increases and different job titles.

Apprenticeships

Some companies' training programs are called **apprenticeships**. Company trainees, called **apprentices**, receive a certain number of hours of practical training in machining operations during normal working hours. Apprenticeships can be internal, or relating only to a company. Others are sponsored by either a state labor department or the U.S. Department of Labor.

Internal or Company-Provided Apprenticeships

These are similar to the structured training programs discussed earlier, but upon completion of these apprenticeships, the employee receives the title of a **journeyperson** and is expected to be able to perform any machining operation required by that company. Company apprenticeships may or may not require classroom training outside regular work hours. They can be as short as 1 year or as long as 5 years in duration. Skills learned through completion of a company-sponsored apprenticeship are usually recognized by other companies in the same local area, but completion may not carry as much weight as state or nationally sponsored programs.

State or Nationally Recognized Apprenticeships

These apprenticeships are more formal programs that combine theoretical and practical education for a person while the person is employed in the machining industry. Such apprenticeships are sponsored by and accountable to either a state's labor department or the U.S. Department of Labor. Companies agree to provide a certain number of hours of practical training in machining operations during normal working hours. Apprentices must also attend classes outside their working day to learn theoretical aspects of the machining field. These classes are usually offered at a local technical or community college. As apprentices progress through these programs, they receive wage increases. Upon successful completion of an apprenticeship, the apprentice earns the title of a journeyperson and is expected to have a strong set of skills and knowledge related to the profession as a whole, not just a particular company. They also receive certification from the state or the United States. Credentials earned through these apprenticeships are usually more widely accepted than those earned through a company's internal apprenticeship. Apprenticeships can vary in length, but they average between 4 and 6 years. They are available in general machining, as well as specialties such as CNC programming, mold making, and toolmaking and die making.

NIMS

The **National Institute for Metalworking Skills (NIMS)** is an organization that plays a vital role in training for the machining industry. NIMS has established national benchmarks, or standards, for performance and knowledge related to several different areas of the machining industry. Many educational institutions and machining companies across the United States offer opportunities for individuals to earn certifications in those areas. The certifications are competency based, which means a person will gain certification only if rigid standards of performance are met.

NIMS has also developed a competency-based apprenticeship program that requires apprentices to meet nationally recognized standards before receiving journeyperson status. This is different from many apprenticeships, where completion is based on the number of hours spent in the program instead of meeting any specific levels of achievement.

NIMS certifications can be an asset to any person desiring to build a successful career in the world of modern machining. The topics covered in this text address the knowledge and skill areas required to achieve competency in the field and earn NIMS machining certifications. To learn more about NIMS, visit http://www .nims-skills.org.

JOB SEEKING

Specific skills needed for success in machining careers and methods to gain those skills have been identified. Next, there are some key steps to finding the job that is right for you. The first step is to have a career plan. Creating a resume and cover letter, along with a list of references, summarizes your skills and goals. Then you must find opportunities, show your interest in those opportunities, and be prepared to pursue them.

Career Plan

The purpose of a career plan is to help you pursue a career path that will give you satisfaction, not frustration. The plan is an evaluation of yourself and a list of your career goals. A goal is like a target; if you do not have one, you will never hit it.

Keep the plan simple so you can actually use it. Being honest with yourself is important. No one else needs to see your career plan, but you could ask someone to review it if you like. Listing strengths and weaknesses is a good idea. Then you know where you can excel and where you need to improve. Everyone can improve in some areas.

Update your plan every year, because you and your goals will probably change over time. Referring back to your career plan will help guide you in the direction you want to go. Here are a few questions that should be answered in your career plan.

- What duties or tasks required in the industry do I like to do, and which ones do I not like?
- What duties or tasks do I perform best, and which ones do I need more practice at?
- Where do I want to work and live? What state or country? In a rural or urban setting?
- What kind of company do I want to work for? Large or small?
- Do I want to own my own business?
- Do I want to continue my education?
- Do I want to advance or move into other areas of the industry?
- Where do I want to be a year from now? Three years from now? Ten years from now?

By answering these types of questions, you will look for opportunities that fit your wants and needs. **Figure 1.3.1** shows an example of a simple career plan.

Resume

Everyone should have a resume to show skills and education that are relevant to their chosen career fields. It is best to limit your resume to one or two pages. Prospective employers are distracted by having to flip through multiple pages when reviewing resumes. Resumes can be written in many different styles, but they should all contain the same key parts. **Figure 1.3.2** shows a sample resume layout of someone with no work experience seeking a first job. Refer to it while reviewing the following resume elements.

Personal Information

Across the top of your resume, write your name, address, phone number, and possibly email address. Make your name the largest text on the resume for easy identification.

Career Objective

A career objective is a short statement about your career goal. It can also list one or two of your important skills. Wording your objective in a way that shows how

---

John Doe's Career Plan

Things I like to do:
- Precision Measuring/Inspection
- Lathe Operations
- Vertical Milling Operations
- CAM Programming
- CNC Lathe Set-up/Operation
- CNC Mill Set-up/Operation

Things I don't like to do:
- Hacksawing
- Filing
- Deburring
- Pedestal Grinding
- Assembly
- Cutting Raw Material
- Surface Grinding

What am I good at?
- Programming
- CNC Mill Set-up
- Short Projects
- Inspection

Where I do need more practice?
- CNC Lathe set-up
- Long projects-patience
- CAM programming

I want to live in a very rural area like Wyoming or Montana.
I want to work within a 45-minute drive of my home.
I only want to work a day-shift job.
I want to work for a small company with about 10–50 employees.
I want to get a 4-year degree in engineering or design.
Within 3 years, I want to start working on my 4-year degree.
Within 5 years, I want to become a foreman or lead person.
Within 10 years, I want to finish my 4-year degree and move into design.

FIGURE 1.3.1 Simple career plan.

---

you would benefit the potential employer is a good idea. Avoid overused phrases such as "hard-working" and "dedicated." An employer expects these qualities, and has probably seen those words on countless other resumes. Make yours different, so it stands out.

Skills/Work Experience

List your abilities here. Start with those that are directly related to the job. Then list additional skills that might set you apart from other candidates for the job. For example, even if you are applying for an entry-level machine operator position, you could list computer or communication skills. These might show an employer that you have skills that could lead to advancement.

If you have employment experience, list your jobs in reverse chronological order, starting with your current or most recent one. List the dates worked, company name and location (city and state), and job title. Then list your job responsibilities using action words.

Education

If you have formal education related to the job you are applying for, list that information next. Start with the current (or most recent). List the name of the institution (school, college, university, etc.), its location (city and state), what type of credential you earned—such as a diploma, certificate, or degree—and the date you earned that credential.

You can also list any special achievements you accomplished during your education, such as honor roll, dean's list, or extracurricular activities.

Other

If you have earned any special awards or certifications that are not related to the previous topics, list them here. Also include membership in organizations or clubs. Examples include those related to community, social, or religious organizations. This section can have a title such as "Achievements" or "Community Activities."

John Doe

123 Main Street • Anytown, ZZ 54321 • Phone: 123-456-7890 • E-Mail: johndoe@anyserver.com

Objective

Goal-oriented student seeking an entry-level CNC machining position with a company that will allow me to apply my training, grow with the company, and help the company succeed.

Skills

- Solid knowledge of g-code programming and program formats for CNC milling machines and lathes
- Ability to select proper tooling for CNC machining operations
- Capable of set-up and operation of CNC machine tools equipped with ABC and XYZ controls
- Skilled in the use of semi-precision and precision measuring tools
- Capable of following verbal and written instructions
- Working knowledge of word processing, spreadsheet, email, and Internet software

Education

Well-Known Technical College   Bigtown, BB   2011–2014

- Received Associate Degree in CNC Machining Technology in June 2009
- Member of Student Machine Tool Safety Committee
- Earned NIMS Level 1 Machining Certifications in CNC Milling and CNC Turning

Anytown Career Center   Anytown, ZZ   2009–2011

- Received Certificate in Precision Machining Technology in June 2007
- Member of National Student Organization
- Earned the following NIMS Level 1 Machining Certifications
  - Measurement, Materials, and Safety
  - Job Planning, Benchwork, and Layout
  - Drill Press Operations
  - Turning Between Centers Operations
  - Turning/Chucking Operations
  - Vertical Milling Operations

Community Service

- Volunteer at Anytown Retirement Community
- Member of Anytown Volunteer Fire Department

References

Available on request

FIGURE 1.3.2 Sample resume.

References

References are people you know who can provide positive information about your attitude, work ethic, and skills. References should be people who are not family members. Teachers, coaches, guidance counselors, pastors, and neighbors can all serve as references. Always ask permission before using someone as a reference and record their names, addresses, and phone numbers. Compile a list of about six references, because you may want to use different people as references for different jobs. Let your references know that you are actively seeking a job so if employers contact them, they will not be surprised and unprepared to answer questions about you. **Figure 1.3.3** shows a sample list of references.

Keep your list of references separate from your resume and write "References available on request" at the bottom of your resume.

Cover Letter

A cover letter introduces you to a prospective employer. It needs to create enough interest for the person reading it to look at your resume. A poor cover letter might prevent a resume from ever being read. Like resumes, cover letters can be written in many styles, but they should contain the same basic elements. A cover letter should always be typed or prepared using word processing software and printed. Never submit a handwritten cover letter. Refer to the sample cover letter in **Figure 1.3.4** while reviewing the following elements.

Greeting

Try to find out the name of the person receiving the letter. If it is not known, try searching for a company website and directory. You can also call the company and ask for the name of the person in charge of human resources or personnel. Personally addressing your cover letter is much better than using a generic greeting. Address the person as either Mr. or Ms., not by first name, and be sure to spell the name correctly. If the person's name cannot be determined, use a reference line and greeting like this:

RE: CNC Operator Position

Greetings:

Then begin the body of the letter. This may not be ideal, but it is better than the impersonal, outdated "Dear Sir or Madam" or "To whom it may concern."

Body

If you are applying for a specific job, refer to it in the first paragraph. Also state how you learned about the job. Write something specific about the company to show that you are interested in its business. A little research

---

References for John Doe

Dave Johnson
CNC Machining Instructor
Well-Known Technical College
1 Education Drive
Bigtown, BB
321-123-5555, Extension 105

Mike Davis
Machining Instructor
Anytown Career Center
100 Career Lane
Anytown, ZZ
123-321-4321, Extension10

John Smith
Guidance Counselor
Anytown Career Center
100 Career Lane
Anytown, ZZ
123-321-4321, Extension 12

Steve Michael
Science Instructor
Anytown High School
12 Main Street
Anytown, ZZ
123-321-1234, Extension 302

Tim Edwards
Director
Anytown Retirement Community
50 Country Road
Anytown, ZZ
123-321-1111, Extension 101

Bob Jones
Pastor
Anytown Community Church
4321 Pine Street
Anytown, ZZ
123-321-4444

Ron Thomas
Chief
Anytown Volunteer Fire Department
Anytown, ZZ
123-321-7777

FIGURE 1.3.3 Sample list of references.

Machining Supervisor
Precision Fixture Design, Co.
50 Tool Drive
Ourtown, ZZ 12345

Re: CNC Set-Up Technician Position

Greetings:

This letter is in response to your company's advertisement in the *Ourtown Times* on June 21, 2018, for a CNC Set-Up Technician. Your main business of producing fixtures and components for the aerospace industry is a very interesting and challenging field.

This position fits well with my skills and goals because I have just graduated from Well-Known Technical School with an Associate Degree in Machine Tool Technology and am interested in CNC machining related to the aerospace industry. I have earned all NIMS Level I Machining certifications, including CNC milling and CNC turning. Thorough knowledge of g-code programming for CNC milling machines and lathes and basic training in CAM software are skills that I believe can allow me to play a role in the success of Precision Fixture Design and allow me to grow with the company.

After reviewing my enclosed resume, feel free to contact me phone by between 1:00 and 5:00 PM at 123-456-7890 or through email at johndoe@anyserver.com with any questions or to schedule an interview. Thank you for your time and consideration. I look forward to meeting with a representative of Precision Fixture Design in the near future to discuss this position in detail.

Sincerely,

John Doe

John Doe

FIGURE 1.3.4 Sample cover letter.

might be needed to accomplish that. If you are not applying for a specific job, note that you are interested in learning about the company and opportunities in your area of interest.

In the second paragraph, briefly mention a few key points about your experience, training, skills, and attitude. State why you think you are a good candidate for the position and what benefit you can provide to the company. Though most companies are willing to compensate qualified employees fairly, they are not interested in hearing what you expect them to do for you.

In the third and final paragraph, state that your resume and references are included and give a method for the employer to contact you with questions or to schedule an interview. Thank the recipient for his or her time and say that you look forward to meeting him or her and learning more about the position and the company.

Closing

Finish the cover letter with a complimentary closing, and then sign your name in addition to it being shown in print.

Career Portfolio

A career portfolio is a more visual summary of your experience and achievements. Anything that showcases your achievements and sets you apart from other job candidates can be placed in your portfolio. It can include photographs of projects you have created or were involved with. Industry certifications or awards, such as those awarded by NIMS, should be included as well. Other items to add could be academic or attendance awards or letters of recommendation. Keep your portfolio items in a neat binder or in electronic form and show your portfolio to prospective employers to give them a better picture of you and your abilities.

Finding Opportunities

Job opportunities can be found in many different ways. The traditional method is to check newspaper classified ads and respond by mail or telephone. When responding by mail, always include your cover letter, resume, and references.

Applying directly to a company you know of is also acceptable, even if it is not advertising open positions. This can be done through the mail by sending your cover letter, resume, and references. You can also show up in person to submit your cover letter, resume, and references and complete a job application. Always bring your resume when applying in person, because many prospective employers will want you to transfer your resume information onto an application form.

Today there are many job-searching websites that allow job seekers to browse jobs from all around the country and even the world. Searches can be refined by specific types of industries, geographic locations, key words, specific companies, and other criteria. Many job-searching websites allow job seekers to post resumes on the site so they can be viewed by hiring companies. The companies can then contact people whom they feel are good candidates for their positions.

Often, some of the best job opportunities are not advertised. They are made known through networking. Networking is simply the distribution of information through people. One example of networking is employees spreading the word about openings at their company by speaking to people they know. Another might be when employees in charge of hiring at a company communicate opportunities through their list of contacts. These contacts can include other companies, schools or other training providers, or even guidance counselors and teachers.

Because of the many informal methods of finding job openings, be sure to tell people that you are actively looking, because they may learn about opportunities through their personal networks. Also take advantage of chances to meet people who work in the machining industry, because they may know of opportunities as well.

Interviewing

The interview is when you meet with a representative of the prospective employer to discuss the job opportunity. There are a few basic steps to follow before, during, and after an interview.

Pre-Interview

Dress in a neat and professional manner. Do not wear any revealing or offensive clothing.

Allow plenty of extra time when traveling to the interview and arrive 15 minutes early. Being late for an interview shows an employer you are unreliable. The employer will think that if you are late for the interview, you will be late for work.

Bring a few extra copies of your resume and references to your interview. The employer may ask for them again. Also bring your career portfolio. If you are carrying it with you, the interviewer might even ask what it is and that gives you the opportunity to show it.

During the Interview

Introduce yourself to the interviewer and greet him or her with a firm but not overpowering handshake. Speak clearly at all times and avoid using any offensive language or profanity. Do not bring a mobile phone or pager to an interview. You want to give the interviewer your full attention.

Take paper and a pen or pencil to the interview so you can write down notes or questions about the company or the position. The interviewer will probably write notes about you, so expect that.

Maintain good eye contact with the interviewer and speak clearly. Learn about the company and ask questions. Pause and think for a moment before answering questions and always be honest. Do not say you can do something if you cannot. That type of lie might get you a job, but it will be quickly revealed that you were not truthful. If you are expected to perform and you cannot, you will not be in the position for long. Here are a few questions that might be asked by the interviewer:

• Tell me a little about yourself.

• Why do you want to work for this company?

• What is your greatest strength?

• What is your greatest weakness?

• Why should we hire you instead of another applicant?

At the conclusion of the interview, thank the interviewer for his or her time and know what the next step will be. Are you to contact the company or will someone contact you?

Post-Interview

Write a thank-you letter to the interviewer within a few days after the interview. Thank her or him for taking the time to meet with you. Refer back to what you discussed during the interview and mention why you feel you would be a good fit for the position. **Figure 1.3.5** shows a sample thank-you letter.

If you are to contact the company within a certain time frame, be sure to do so. Missing a follow-up call or calling late will most likely eliminate you from the running.

123 Main Street
Anytown, ZZ 54321

June 25, 2019

Mr. Jim Smith
Machining Supervisor
Precision Fixture Design, Co.
50 Tool Drive
Ourtown, ZZ 12345

Dear Mr. Smith

Thank you for taking time to meet with me on June 24 to discuss the CNC Set-Up Technician position at Precision Fixture Design. It was a pleasure to meet you and to tour your facility. The shop is very impressive and I was especially interested in your new 5-axis milling center and CAM software package. Those additions will certainly allow the company to efficiently machine the pump bodies for the new contract you mentioned.

Precision Fixture Design is a company that I would be very interested in working for. I believe I could be a productive employee, learn a great deal, and provide even more benefit as the company expands. You can contact me at 123-456-7890 to further discuss my possibility of working for your company.

Sincerely,

John Doe

John Doe

FIGURE 1.3.5 Sample thank-you letter.

SUMMARY

- Machining careers require a combination of both technical and personal skills. Personal skills are related to a person's personality or nature and include abilities such as focus, responsibility, memory use, and patience. Technical skills can be learned through instruction and practice and include skills in print reading, measurement, math, and specialty tool use.

- Secondary education in public schools sometimes offers basic training in machining through elective courses and/or career centers or vocational high schools.

- Post-secondary education through technical colleges, community colleges, and universities can offer higher, specialized, in-depth training for machining and related careers through certificate programs, associate degrees, and baccalaureate degrees.

- Apprenticeships can provide practical training through employers during regular work hours and often require classroom instruction in theory outside of regular work hours.

(Continued)

- The nonprofit organization called NIMS has established national standards for skills required by the machining industry. NIMS provides testing of machining theory and practical skills and provides certification to people who demonstrate competency in those skill areas.
- A career plan will help you keep track of career goals and find the job that best fits your goals, skills, and personality.
- Keep your resume and references up to date and always notify references when you are job hunting so they will be prepared if a prospective employer contacts them.
- A cover letter should refer to a specific job opening and be personally addressed whenever possible.
- A career portfolio, including reference letters, awards, certificates, and photographs, can make you stand out from other job candidates.
- Always act professionally during the interviewing process and send a thank-you letter to the interviewer and refer to a specific topic you discussed during the interview.
- Career opportunities can be found in many ways including through newspaper and online classified advertisements, job-seeking websites, and networks of people. By following some key job-seeking steps, opportunities can be discovered and successful employment secured.

REVIEW QUESTIONS

1. Briefly compare and contrast technical and personal skills.
2. List four personal skills beneficial to achieving success in a machining career.
3. Explain why two of the four personal skills from question 2 are important in the machining field.
4. List four technical skills beneficial to achieving success in a machining career.
5. Explain why two of the four technical skills from question 4 are important in the machining field.
6. What is the purpose of CTE or vocational education?
7. What are the similarities and differences between certificate and associate degree programs?
8. What is an apprenticeship?
9. What are two different ways that apprenticeships are structured?
10. What is a journeyperson?
11. What does NIMS stand for?
12. Discuss how NIMS certification is different from traditional methods of school- or employer-based training for the machining industry.
13. Briefly describe a portfolio and give an example of what it might contain.

MEASUREMENT, MATERIALS, AND SAFETY
SECTION 2

Introduction to Safety

Learning Objectives

After completing this unit, the student should be able to:

- Define *OSHA* and describe its purpose
- Define *NIOSH* and describe its purpose
- Describe appropriate clothing for a machining environment
- Identify appropriate PPE used in a machining environment
- Describe proper housekeeping for a machining environment
- Describe the purpose of lockout/tagout (LOTO) procedures
- Define the terms *GHS, NFPA* and *HMIS*
- Identify and interpret *GHS,* NFPA and HMIS labeling systems
- Define the term *SDS*
- Identify and interpret SDS terms
- Interpret SDS information
- Select the proper fire extinguisher application

Key Terms

Acute	**Globally Harmonized**	**Hazardous**
Chronic	**System (GHS)**	**Communication**
Environmental	**of Classification**	**Standard (HCS)**
Protection Agency	**and Labeling of**	**Hazardous Material**
(EPA)	**Chemicals**	**Identification**
Flash point		**System (HMIS)**

Lockout/Tagout (LOTO)
Lockout device
Lower Explosive Limit (LEL) or
Lower Flammability Limit (LFL)
National Fire Protection
Association (NFPA)
National Institute for
Occupational Safety and
Health (NIOSH)

Occupational Safety and Health
Administration (OSHA)
Parts per million (PPM)
Parts per billion (PPB)
Parts per trillion (PPT)
Permissible Exposure Limit (PEL)
Personal Protective Equipment
(PPE)
Pictogram

Safety Data Sheet (SDS)
Short-Term Exposure Limit
(STEL)
Tagout device
Threshold Limit Value (TLV)
Time-Weighted Average (TWA)
Upper Explosive Limit (UEL)
or Upper Flammability Limit
(UFL)

INTRODUCTION

Safety is the primary concern when working in any machining environment. Sharp tools, heavy objects, and powerful machinery can create situations that can cause serious injuries or even death. That does not mean that people should be afraid of working in that environment. Compare a machining environment to the beach. The beach can be a very dangerous place also. Strong currents, cold and deep water, jellyfish, and even sharks can possibly cause injury or death. But millions of people enjoy the beach without any problems because they use caution, good judgment, and common sense; have respect for potential dangers; and avoid dangerous situations. Working in a machining setting is the same.

Different machining operations require different safety precautions, and those specifics will be discussed during those operations throughout this text. But there are some standard safety rules that will apply in any situation. Learning to be aware of surroundings and following those rules from the beginning will create good, safe work habits and a safe, enjoyable working environment. Always stay alert and focused on the current task. Many accidents are caused by not paying attention and becoming distracted.

 CAUTION

If you notice any unsafe situations or actions, report them immediately to the proper person so they can be resolved. If any type of injury occurs, immediately notify your supervisor, instructor, or other superior so that proper medical or emergency procedures can begin as soon as possible.

GENERAL SAFETY GUIDELINES

When working in a machining environment around sharp tools, heavy items, and powerful machinery, there is no room for any type of horseplay. Pushing, shoving, tripping, or startling others, running, and similar actions should never be tolerated.

Machining work requires focus and concentration, so many times someone may not even notice others

walking or standing nearby. For that reason, when approaching another person in the shop, move toward him or her from the front and avoid quick movements or sudden noises. Wait for a "break in the action" before starting to talk. Remember that it only takes a split second for "accidents" to happen, but once they happen it is impossible to erase them.

Some materials create hazardous dust particles, and improperly used tools and machines that may seem harmless can cause serious or fatal injuries if certain steps are not performed correctly. For those reasons:

- Never work with any material, tool, or piece of equipment you are not familiar with.
- Always stop machines before cleaning or making adjustments. Never perform those tasks while machinery or equipment is running.
- Do not operate machinery if you are under the influence of alcohol or medication. If you are ill or taking medication, consult with your doctor to see if there are any side effects that would prevent you from working around machinery.
- Know the location of first aid kits and supplies.
- For cuts and other bleeding injuries, wear gloves from a first aid kit and apply pressure to the wound to slow bleeding.
- In cases of severe injuries or emergencies, someone should remain with the injured person and another should call 911 immediately.
- Know evacuation routes and locations of power switches, emergency stop switches, and fire extinguishers.
- Learn any specific guidelines for your company or school, or specific work area or machining lab, including procedures for reporting emergency situations.

OSHA AND NIOSH

The **Occupational Safety and Health Administration**, or **OSHA**, is an agency of the federal government that sets and enforces regulations for workplace safety. Companies are required to follow OSHA regulations and provide safe working environments for their employees. OSHA performs periodic on-site inspections to check that regulations are being followed. OSHA also works with business and other organizations to provide training and improve working conditions. To learn more about OSHA, visit http://www.osha.gov.

NIOSH, the **National Institute for Occupational Safety and Health**, is the federal agency responsible for conducting research and making recommendations for the prevention of work-related injury and illness. Unlike OSHA, NIOSH does not enforce any regulations. It only assesses hazards and works on prevention methods. To learn more about NIOSH, visit http://www.cdc.gov/niosh/.

GENERAL CLOTHING FOR A MACHINING ENVIRONMENT

Choosing appropriate clothing to wear when working in a machining environment is important.

Wear long pants to protect legs from sharp or hot metal shavings. Shorts, skirts, or dresses are not recommended. Short-sleeved, close-fitting shirts are best. If wearing long sleeves, roll sleeves up past the elbows. Shirts or sweat-shirts with hoods or drawstrings should be removed. If a shop coat or apron is worn, it should fit snugly and any ties or strings secured so they do not hang loosely.

Hard, flat-soled work shoes should always be worn because metal shavings, or chips, often fall onto the shop floor and can possibly cut through the soles of other shoes and cause serious cuts. Slip-resistant soles should also be worn because oils and other fluids may spill onto the floor and create slippery conditions. Leather work shoes are preferred, and sometimes safety-toe shoes or even those with metatarsal shields are required. See **Figure 2.1.1** for some examples of appropriate footwear for use in a machining environment.

FIGURE 2.1.2 Long hair should be secured safely in a machining environment.

See **Figure 2.1.2** for securing long hair.

Most machining tasks should be performed without wearing gloves because gloves can also get caught in moving parts. Gloves can be worn to perform some specific tasks, as discussed next under the topic heading of Personal Protective Equipment (PPE)

© 2010 The Red Wing Shoe Company, Inc.

FIGURE 2.1.1 Some examples of appropriate footwear for a machining environment. Different heights are available with either (A) soft toes, (B) safety toes, or (C) metatarsal shields, which help to protect the entire foot from heavy objects.

! CAUTION

- Avoid any loose-fitting clothing that can get caught in moving machinery.
- Casual shoes, open-toed shoes, sandals, high-heeled shoes, and flip-flops should never be worn.
- Specific types of jewelry should not be worn while machining. Necklaces, dangling earrings, bracelets, watches, and rings can get caught in moving parts of equipment and pull a person into a machine, causing serious injury or even death.
- Long hair should always be secured with a hat, cap, or hairnet so that it does not get caught in moving machinery.
- Never wear gloves when working around running machinery or any moving parts.

PERSONAL PROTECTIVE EQUIPMENT (PPE)

PPE, or **personal protective equipment**, refers to safety equipment that is to be worn to protect a person from potential dangers. Become familiar with PPE requirements for the situations that you will be working in.

 CAUTION

Always use the appropriate PPE for every situation.

Eye Protection

Eye protection is the most common type of PPE used in the machining industry. Safety glasses with side shields must always be worn when entering a shop environment, even if just visiting or observing. Always wear safety glasses that meet the standard for protection set by the American National Standards Institute (ANSI), as required by OSHA regulations. The label of *ANSI Z87.1* identifies eyewear that meets this standard. Some situations may also require use of a face shield in addition to standard safety glasses. Ordinary prescription eyeglasses are not adequate protection. Their lenses may shatter when impacted and send lens particles into the eyes along with the foreign object that broke the lens. For those who wear prescription eyeglasses, there are safety glasses that can be worn over the prescription glasses. Prescription safety glasses are also available. Some operations may also require the additional protection of a face shield. A face shield does not replace standard safety glasses, but provides added protection. **Figure 2.1.3** shows a few of the many different styles of eye and face protection.

If any foreign objects enter the eyes, do not rub the eyes because the particles can cause more damage if moved around. If small particles enter the eye, gently pull the eyelid away from the eye, flush with water as recommended by NIOSH, and see a doctor if irritation persists. If a larger foreign object is stuck in the eye, do not try to remove or flush with water, but seek immediate emergency medical attention.

 CAUTION

Always wear safety glasses when in any machining environment.

Hearing Protection

Some machining environments may involve high noise levels. OSHA requires some type of noise control or hearing protection if sound levels are above 115 dB (decibels) for ¼ hour or over 90 dB for 8 hours. Protection is also required if people are exposed for any length of time to a noise level above 140 db. In such situations, hearing protection in the form of earplugs or earmuffs can be used. **Figure 2.1.4** shows some different types of hearing protection.

 CAUTION

Personal music players do not provide hearing protection and prevent a worker from hearing machine crashes and people who may be injured. These devices should never be used in a machining environment.

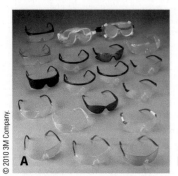

FIGURE 2.1.3 (A) Safety glasses and goggles come in many different styles. Note that some can be worn over ordinary eyeglasses. Face shields (B) do not take the place of safety glasses, but are worn in addition to them when performing certain operations.

© 2010 3M Company.

FIGURE 2.1.4 Some examples of hearing protection.

© 2010 3M Company.

Courtesy of Sentry Air Systems, Inc. Houston, TX USA www.sentryair.comç

FIGURE 2.1.5 This hood provides ventilation by drawing harmful fumes out of the work area.

Respirators

Occasionally the air in a machining environment may contain hazardous particles or gases. Fumes are airborne solid particles, while vapors are gases. In some instances, a central ventilation system using hoods or exhaust vents provides protection by removing contaminants from the air. (**See Figure 2.1.5**.)

When hoods or exhaust vents are not strong enough to clean the air, a personal respirator may be required for protection from those contaminants. There are two basic types of respirators: air purifying and atmosphere supplying. Air-purifying respirators use fine mesh or canister-type filters to remove particles or gases from the air so it is safe to breathe. (**See Figure 2.1.6**.)

© 2010 3M Company.

FIGURE 2.1.7 This worker is wearing an atmosphere-supplying respirator. Note the hose that would be connected to an external oxygen source.

An atmosphere-supplying respirator supplies clean air from a separate source, as shown in **Figure 2.1.7**. Be sure to receive proper training before using any type of respirator.

Gloves

There are a few situations in machining where the use of gloves is acceptable. Canvas or leather gloves can be worn to protect hands from cuts when handling raw materials, saw blades, or large cutting tools.

Many chemicals require wearing a specific type of glove while handling. "Rubber gloves" is an everyday term that some people use, but gloves are also made from many other materials such as nitrile, neoprene, latex, or polyvinyl chloride (PVC). They come in various lengths as well. **Figure 2.1.8** shows a few of the many types of gloves

© 2010 3M Company.

FIGURE 2.1.6 These air-purifying respirators use replaceable filters or cartridges that filter contaminants from the air.

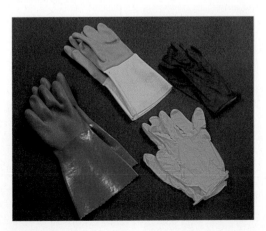

FIGURE 2.1.8 Different gloves are available for different uses. Note the different lengths. When handling chemicals, always follow the chemical manufacturer's recommendation for the proper type of glove.

available. Always check the chemical manufacturer's recommendation so that the correct type of glove is used. Using the wrong type of glove may provide no protection at all and can lead to exposure and chemical burns.

 CAUTION

Never wear gloves when working around moving parts of running machinery.

Hard Hats

Some machining environments deal with very large materials and parts. When working in an area where items are stored overhead or moved overhead by hoists or cranes, hard hat usage should be implemented. OSHA regulations require hard hats to be worn whenever there is a danger of head injury from falling items.

HOUSEKEEPING

Housekeeping refers to keeping the working environment clean to prevent dangerous situations. A few simple guidelines can keep a shop clean and safe.

Organizing materials and equipment while working and after job completion is the first step in creating a clean, safe setting. Clutter on benches, on machinery, and in storage areas creates situations where falling items can cause injury, and can damage tools as well. Clean tools and always store them appropriately after use to avoid injury and to keep tools in good condition. Bars of material leaning against walls and machinery can also fall and cause injury. Do not stack round bars of material against walls, because the stack can collapse and roll uncontrollably. Store material in racks like those shown in **Figure 2.1.9**. Items piled on floors can create tripping hazards.

FIGURE 2.1.10 Dispose of oil- and solvent-soaked rags in safety cans like these to avoid spontaneous combustion.

Clear workbenches and machinery of chips with a brush, and use rags to wipe fluids from machinery. Never remove chips or clean machinery while it is running. Place rags soaked with oil or solvents in a safety container like the one shown in **Figure 2.1.10** to avoid spontaneous combustion.

Keeping floors clean of other debris such as metal chips, scraps, oils, and other fluids is also important to create a safe environment. Aisles and walkways also need to be kept clear of obstructions. Do not store materials or equipment in places that can block access to these areas. Daily floor sweeping (or more often if needed) is a

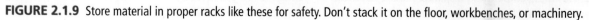

FIGURE 2.1.9 Store material in proper racks like these for safety. Don't stack it on the floor, workbenches, or machinery.

FIGURE 2.1.11 (A) Oil and solvent spills can be absorbed with granular absorbents, then the absorbent can be swept up. (B) Pads or "socks" can also be used to soak up spills.

good habit to develop. Clean up liquid spills as soon as they occur to minimize the potential of falling hazards from slippery floors. Small liquid spills can often simply be mopped. Large spills can be absorbed with granular absorbents, mats, or rolls like those shown in **Figure 2.1.11**. If using a granular absorbent, sweep up the particles after the liquid has been absorbed. Be sure to properly dispose of the absorbents and mop the area with a quality floor cleaner afterward. A well-organized and clutter-free space helps to create a safe work environment.

FIGURE 2.1.12 Never operate machinery without all guards and covers properly in place.

Machines frequently have physical guards surrounding cutting tools or other danger areas to prevent hands or fingers from reaching a cutting tool or other moving parts. (**See Figure 2.1.12.**) A switch may keep a machine from operating if a guard is not in its proper place. A light beam may take the place of a physical barrier or guard. If someone crosses into a danger zone and breaks the light beam, the machine is disabled.

 CAUTION

- Always store material, tools, and other equipment in safe locations to prevent tripping and falling hazards.
- Never use your hands to remove chips from equipment.

GUARDS AND BARRIERS

Some environments will contain warning signs or physical barriers to prevent people from entering a danger area. These provide protection during machine operation.

 CAUTION

Always respect warnings and barriers, and never operate machinery without all safety devices in place. Never attempt to bypass or override any safety device.

HANDLING AND LIFTING

Moving raw materials, finished components, and machine tool accessories is very common in the machining field. Small handheld items present no concerns, but some simple precautions need to be taken when moving larger, heavier items.

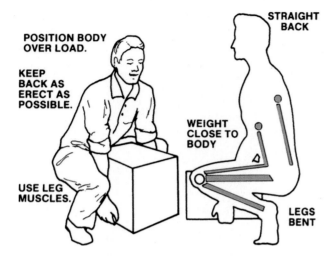

FIGURE 2.1.13 Use the legs and keep the back straight when lifting.

When lifting, always bend at the knees and keep the back straight, as shown in **Figure 2.1.13**. If there is any concern that an item may be too heavy for one person to move, ask for help. It is far better to take a few extra moments to find help than to live with a long-term painful back injury.

Raw material may contain sharp edges and be covered with oil. Wear gloves for protection from cuts and to secure a better grip. Long bars may not be very heavy but are often awkward for one person to move. Ask for help to keep bar ends from swinging out of control and possibly hitting other people or equipment.

Some very heavy items need to be moved by forklifts, or other types of lifts, hoists, or cranes. Always receive proper training before operating any lifting device, and never walk or stand under any item suspended in the air. As discussed under the topic heading of Personal Protective Equipment (PPE), always use hard hat protection as required by OSHA when in an area using any overhead lifting equipment.

COMPRESSED AIR SAFETY

Do not use compressed air to remove metal chips, oils, solvents, or other chemicals from machinery or workbenches. Always use a chip brush to clean these areas. Air can blow chemicals or chips into the eyes. Debris in air lines can also be discharged during use and cause injury.

Misdirected compressed air at only 12 pounds per square inch (psi) can pop an eyeball from its socket. Compressed air can enter the body through the navel and burst intestines. A 40-psi blast within 4 feet of the eardrum can rupture the eardrum. If an air bubble enters the bloodstream through an open cut, an embolism can form, causing coma or even death. If compressed air must be used for a specific cleaning purpose under controlled conditions, use a nozzle that limits pressure to 30 psi or less and ensure that other guards and PPE are used according to OSHA regulations.

CAUTION

An air nozzle should never be used to clean a person's body.

LOCKOUT/TAGOUT

In the machining and related industries, there are often times that equipment must be shut down if it is faulty or for maintenance or repair. Extra precautions need to be taken so that workers are protected from hazards that could occur if that equipment was powered during those times. OSHA has a standard that calls this the *Control of Hazardous Energy*. Hazardous energy sources include electrical, mechanical, hydraulic, pneumatic, chemical, thermal, or other sources in machines and equipment. Examples include the dangerous movement of machinery parts, release of hazardous materials, or release of extremely hot high-pressure steam. When machinery and equipment are faulty or undergoing maintenance or repair, this hazardous energy must be controlled because its unexpected release can cause serious injury or even death to workers. This can include the disabling of an electrical switch, preventing access to a compartment of a machine, or preventing opening of a container or chamber containing a hazardous material. OSHA has developed a procedure called **lockout/tagout (LOTO)** to prevent release of this hazardous energy. It requires the use of locks and/or tags to disable faulty equipment or equipment undergoing maintenance or repair.

Main power switches on machines usually have a small loop where a lock or a tag (or both) can be placed to prevent the switches from being moved. This prevents machinery from being powered during maintenance. (**See Figure 2.1.14**.) For example, if a large machine tool

FIGURE 2.1.14 A lock or tag (or both) can be placed through a machine's main power switch to prevent the machine from being powered.

is being serviced, a maintenance worker may be performing a repair near the spindle motor and be unseen by others. If the machine is not locked or tagged out, someone could start it and unknowingly cause serious injury or death to that maintenance worker. If an electrical component is being repaired or replaced in a machine, a maintenance worker will lock out or tag out the electrical panel. This will prevent anyone from engaging the main power, which could lead to injury or death by electrocution.

 **CAUTION**

Never remove or override another person's tagout or lockout device or attempt to operate any piece of equipment that is locked out or tagged out.

Tagout

In tagout, maintenance personnel use **tagout devices** to prevent powering up equipment. The tagout device consists of a highly visible tag that is secured to a machine with a metal or plastic wire. The tag states that no one can remove it except the person who placed it. The person who attaches the tag writes his or her name, the date the tag was attached, and possibly details about the reason for the tagout. Several people can attach tags to a piece of equipment. Perhaps an electrician, a machinist, and an engineering technician all have tasks to perform on a machine and it cannot be operated until each one performs his or her specific task. Individual tags prevent operation until each person has given approval. **Figure 2.1.15** shows some tagout devices. Tagout stations are often used as a central location for workers to obtain tagout devices.

FIGURE 2.1.15 A tagout device provides space for the person who placed the tag to write specific details. The tag is then secured through a machine's power switch with a wire or plastic tie.

FIGURE 2.1.16 A lockout hasp allows several different people to lock out or tag out a piece of equipment. This machine has been locked out by two different people.

Lockout

In lockout, the procedure is the same as tagout, except that tags are replaced with padlocks or other **lockout devices**. Lockout is more secure, because locks are more tamper proof than tags. Lockout hasps allow several locks to be placed by different people as needed. (**See Figure 2.1.16**.) Lockout stations are also often used to provide workers access to lockout devices.

HAZARDOUS MATERIALS

As stated under the topic heading of Personal Protective Equipment (PPE), some machining environments may have exposure to hazardous materials. Hazardous materials can take the form of liquids, solids, or gases. OSHA's **Hazard Communication Standard (HCS)** requires manufacturers of hazardous materials to provide information about these types of materials to users. Employers must also label hazardous materials and provide training to employees who work around them. In the past, HCS stated that workers had a right to know what hazards existed in the workplace and that employers must provide knowledge of those hazards. These activities were often also enforced by individual states as *Right to Know* laws. The HCS now states that workers have a right to understand those hazards, not just know that they exist, so Right to Know is being replaced with *Right to Understand*. Companies often conduct special training sessions and safety meetings to provide this information to employees.

Hazardous Material Labeling

OSHA's HCS is aligned with the **Globally Harmonized System of Classification and Labeling of Chemicals (GHS)**. GHS is an internationally recognized system for labeling hazardous materials. One requirement of HCS is for manufacturers to place a label on hazardous material that shows specific details. In addition to OSHA's HCS required label, there are two other systems of labeling hazardous materials that are widely used in the machining industry: **NFPA (National Fire Protection Association)** and **HMIS (Hazardous Material Identification System)**.

OSHA GHS Compliant Label

OSHA's HCS requires a GHS compliant label on all hazardous materials. That label must contain the following six elements.

- Product identifier: This includes the specific product name as well as a common or trade name. There can also be a number used by either the UN (United Nations) or the CAS (Chemical Abstracts Service) to identify the hazardous material or its components.

- *Signal word:* There are two signal words that can be used: *Danger* or *Warning*. Danger is used for more severe hazards and Warning is for less severe hazards.

- Hazard statements: These statements briefly summarize the type of hazard the material may pose.

- Precautionary statements: These are more detailed statements about the handling and use of the material. They can include information on storage, PPE required, first aid, and firefighting measures.

- Supplier identification: Name, address, and emergency phone number of the manufacturer, importer, or other party responsible for the material.

- **Pictogram:** These are highly visual images that represent specific types of hazards. There are nine GHS approved pictograms, but OSHA only enforces eight because one is related to the environment and enforced by the EPA. The border for the pictogram can be either red or black.

Figure 2.1.17 shows a sample GHS compliant label. **Figure 2.1.18** shows and describes all nine GHS pictograms.

NFPA

NFPA standard 704 can be used to quickly identify specific types of hazards and their levels of danger. The

FIGURE 2.1.17 A sample GHS compliant label required by OSHA's Hazardous Communication Standard (HCS).

OSHA® QUICK CARD™

Hazard Communication Standard Pictogram

The Hazard Communication Standard (HCS) requires pictograms on labels to alert users of the chemical hazards to which they may be exposed. Each pictogram consists of a symbol on a white background framed within a red border and represents a distinct hazard(s). The pictogram on the label is determined by the chemical hazard classification.

HCS Pictograms and Hazards

Health Hazard	Flame	Exclamation Mark
• Carcinogen • Mutagenicity • Reproductive Toxicity • Respiratory Sensitizer • Target Organ Toxicity • Aspiration Toxicity	• Flammables • Pyrophorics • Self-Heating • Emits Flammable Gas • Self-Reactives • Organic Peroxides	• Irritant (skin and eye) • Skin Sensitizer • Acute Toxicity (harmful) • Narcotic Effects • Respiratory Tract Irritant • Hazardous to Ozone Layer (Non-Mandatory)
Gas Cylinder	Corrosion	Exploding Bomb
• Gases Under Pressure	• Skin Corrosion/Burns • Eye Damage • Corrosive to Metals	• Explosives • Self-Reactives • Organic Peroxides
Flame Over Circle	Environment (Non-Mandatory)	Skull and Crossbones
• Oxidizers	• Aquatic Toxicity	• Acute Toxicity (fatal or toxic)

For more information:

OSHA® Occupational Safety and Health Administration

www.osha.gov   (800) 321-OSHA (6742)

U.S. Department of Labor

OSHA 3491-01R 2016

FIGURE 2.1.18 The GHS approved pictograms used in OSHA's Hazard Communication Standard (HCS).

standard uses the multi-colored diamond-shaped symbol shown in **Figure 2.1.19**.

The smaller, different-colored diamonds in an NFPA label show different types of hazards. Numbers ranging from 0 to 4 in each colored diamond rate the hazard level from least to most severe.

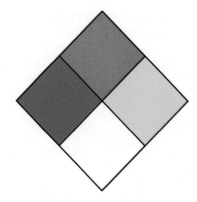

FIGURE 2.1.19 The NFPA (National Fire Protection Association) hazardous material label.
Reprinted with permission from NFPA 704-2012. *System for the Identification of the Hazards of Materials for Emergency Response*, Copyright ©2012, National Fire Protection Association.[1]

The red diamond rates flammability of a material. The blue diamond rates health hazards. The yellow diamond rates the hazard of instability, or how easily the substance can explode. The white diamond does not use the numbers 0 to 4, but uses symbols to identify special hazards. **Figure 2.1.20** explains the ratings for each type of hazard as well as symbols used to denote special hazards. **Figure 2.1.21** shows the details for two sample NFPA hazard labels.

Additional information can be obtained by visiting the NFPA's website at http://www.nfpa.org

HMIS® III®

HMIS III is the most recent version of the HMIS® method. This system is similar to that of the NFPA but uses colored *bars* and numbers from 0 (minimal hazard) to 4 (severe hazard). It also can be used to quickly identify hazards. An example of an HMIS® III label is shown in **Figure 2.1.22**.

The blue bar at the top in an HMIS® label rates the health hazard of a substance. An asterisk (*) shows that the material has been shown to cause **chronic** (long-term) health hazards. Flammability hazards are rated in the red second bar. The orange third bar rates physical hazards. A white bar at the bottom of the label uses letters and symbols to identify PPE recommended when using the product. Details of the rating system are shown in **Figure 2.1.23**. **Figure 2.1.24** gives details for two sample HMIS® labels.

Additional information about HMIS® can be obtained by visiting the American Coatings Association website at http://www.paint.org/programs/hmis.html.

[1]This reprinted material is not the complete official position of the NFPA on the referenced subject, which is represented solely by the standard of its entirety. The classification of any particular material within this system is the sole responsibility of the user and not the NFPA. NFPA bears no responsibility for any determinations of any values for any particular material classified or represented using this system.

HAZARDOUS MATERIALS CLASSIFICATION

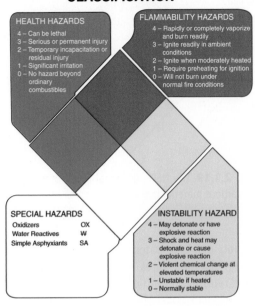

HEALTH HAZARDS
4 – Can be lethal
3 – Serious or permanent injury
2 – Temporary incapacitation or residual injury
1 – Significant irritation
0 – No hazard beyond ordinary combustibles

FLAMMABILITY HAZARDS
4 – Rapidly or completely vaporize and burn readily
3 – Ignite readily in ambient conditions
2 – Ignite when moderately heated
1 – Require preheating for ignition
0 – Will not burn under normal fire conditions

SPECIAL HAZARDS
Oxidizers   OX
Water Reactives   W
Simple Asphyxiants   SA

INSTABILITY HAZARD
4 – May detonate or have explosive reaction
3 – Shock and heat may detonate or cause explosive reaction
2 – Violent chemical change at elevated temperatures
1 – Unstable if heated
0 – Normally stable

FIGURE 2.1.20 Explanation of the NFPA hazardous material rating system.
Reprinted with permission from NFPA 704-2012. *System for the Identification of the Hazards of Materials for Emergency Response,* Copyright ©2012, National Fire Protection Association.[2]

 **CAUTION**

- When using any product, it is a good idea to check for a hazardous material label.
- If any product is stored in a container other than its original container, a GHS compliant, NFPA, or HMIS label should always be placed on the new container.

SDS

OSHA HCS, NFPA, or HMIS labels are meant to quickly and easily identify potential hazards, but OSHA's Hazard Communication Standard also requires detailed documentation of hazardous materials used in the workplace. That detailed hazardous material information is contained in a document known as an **SDS**, or **Safety Data Sheet**. The SDS is another component of *Right to Understand*. Companies rely on SDS information to guide training and protect workers. They are required to train employees on hazardous materials and may have regular safety

meetings to reinforce the use of hazardous materials and SDS information. It is important to become familiar with interpreting basic information on an SDS to protect yourself and others, and to know what to do in case of an emergency involving a hazardous material. Many manufacturers provide online access to SDSs for their products.

OSHA's HCS requires every SDS to follow a specific 16-section format that aligns with the internationally recognized format used by GHS. An overview of those 16 sections is explained below. *Appendix A shows a sample SDS.*

OSHA will directly enforce the content of Sections 1 to 11 and 16. SDSs are required to contain Sections 12 to 15 to align with GHS, but OSHA will not enforce that content because it is related to regulations enforced by other federal government agencies such as the **Environmental Protection Agency (EPA)**, whose purpose is to protect human health and the environment, and the *Department of Transportation (DOT)*, because DOT regulates the transport of hazardous materials.

 **CAUTION**

If there are any questions or concerns about the use of any potentially hazardous material, request to view the product's SDS. Remember, you have a "right to understand" potential hazards.

For additional, more detailed information on the HCS, SDS, and GHS search the OSHA website at http://www.osha.gov/.

Section 1: Identification

Section 1 must have the product name and any other common names used to identify it. The company name, address, phone number, and an emergency phone number must also be listed. This section must also give a brief overview of the product's intended use and restrictions on its use.

Section 2: Hazard(s) Identification

Section 2 lists all hazards related to the product. The same pictograms that are used for product labeling can be used to show specific types of hazards on the SDS.

Section 3: Composition/Information on Ingredients

Section 3 lists all of the ingredients that are contained in the material. That includes chemical and common names, as well as any other information that may be used to identify the ingredients. Percentages of each ingredient must also be specified.

Section 4: First Aid Measures

Section 4 contains symptoms that result from exposure to the material and first aid instructions to be followed by

[2]This reprinted material is not the complete official position of the NFPA on the referenced subject, which is represented solely by the standard of its entirety. The classification of any particular material within this system is the sole responsibility of the user and not the NFPA. NFPA bears no responsibility for any determinations of any values for any particular material classified or represented using this system.

[3]We recommend caution when using the HMIS® III and including the following: HMIS® ratings are based on a 0 to 4 rating scale, with 0 representing minimal hazards or risk, and 4 representing significant hazards or risks. Although HMIS® ratings are not required on SDSs under 29 CFR 1910.1200, the preparer may choose to provide them. HMIS® ratings are to be used with a fully implemented HMIS® program. HMIS® materials may be purchased exclusively from J. J. Keller at 800-327-6868.

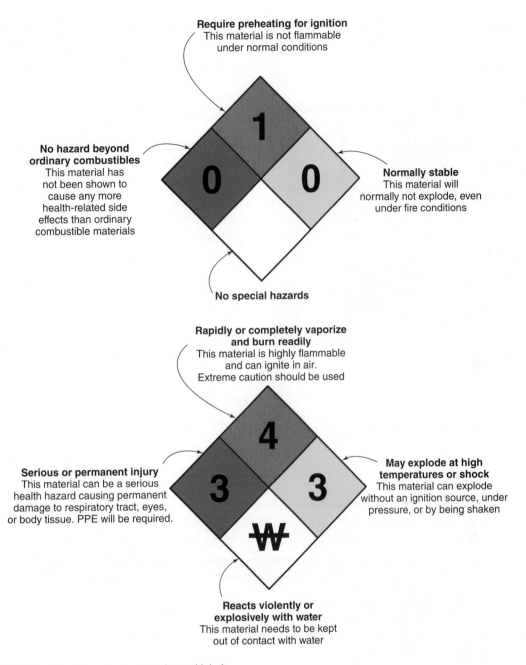

Require preheating for ignition
This material is not flammable under normal conditions

No hazard beyond ordinary combustibles
This material has not been shown to cause any more health-related side effects than ordinary combustible materials

Normally stable
This material will normally not explode, even under fire conditions

No special hazards

Rapidly or completely vaporize and burn readily
This material is highly flammable and can ignite in air. Extreme caution should be used

Serious or permanent injury
This material can be a serious health hazard causing permanent damage to respiratory tract, eyes, or body tissue. PPE will be required.

May explode at high temperatures or shock
This material can explode without an ignition source, under pressure, or by being shaken

Reacts violently or explosively with water
This material needs to be kept out of contact with water

FIGURE 2.1.21 Details of two sample NFPA hazard labels.
Reprinted with permission from NFPA 704-2012. *System for the Identification of the Hazards of Materials for Emergency Response,* Copyright ©2012, National Fire Protection Association.[4]

people without formal first aid or medical training. These are the first steps to take if exposure occurs. Recommendations for medical attention, if required, are also listed in Section 4.

Section 5: Firefighting Measures

Section 5 lists the type of equipment and fire extinguisher that should be used if the material is burning. Protective equipment for use by firefighters may also be listed, as well as additional chemical hazards that may be caused by the burning material.

Section 6: Accidental Release Measures

Section 6 provides steps to take in case of a material spill. They can include protective equipment, emergency procedures, and methods to contain a spill.

[4]This reprinted material is not the complete official position of the NFPA on the referenced subject, which is represented solely by the standard of its entirety. The classification of any particular material within this system is the sole responsibility of the user and not the NFPA. NFPA bears no responsibility for any determinations of any values for any particular material classified or represented using this system.

FIGURE 2.1.22 The HMIS III hazard label.

Section 7: Handling and Storage

Section 7 lists precautions to take when handling or storing the material.

This section may also give instructions about steps to take after product use, such as washing hands or clothing before eating or drinking.

Section 8: Exposure Controls/Personal Protection

Section 8 lists exposure limits as either a **PEL (permissible exposure limit)** or a **TLV (threshold limit value)**. PEL and TLV are measures of the maximum amount of a material that someone can be exposed to before steps to limit exposure must be taken. These exposure limits can be listed in two different ways. A ceiling is a level that should never be exceeded and is shown with the letter "C." A **TWA** value, or **time-weighted average**, states an amount that should not be exceeded within an 8-hour time frame. Sometimes an **STEL (short-term exposure limit)** is also shown. An STEL is a concentration of exposure that is permissible for 15 minutes at a time. OSHA allows four 15-minute STEL periods per 8-hour shift so long as there is a minimum of 60 minutes of nonexposure in between. This may sound a bit confusing, so in summary this is what OHSA permits: A ceiling, "C," exposure can never be exceeded. Four STEL periods spaced by at least 60-minute breaks are allowed. Finally,

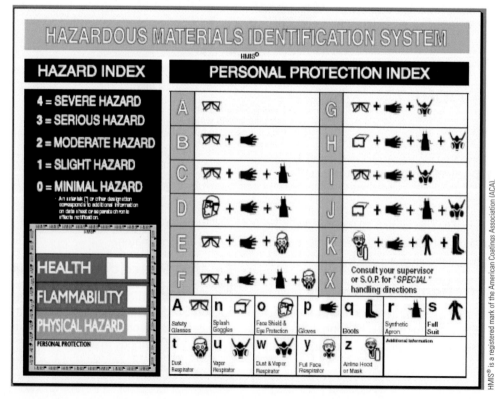

FIGURE 2.1.23 Explanation of the HMIS[3] III rating system.

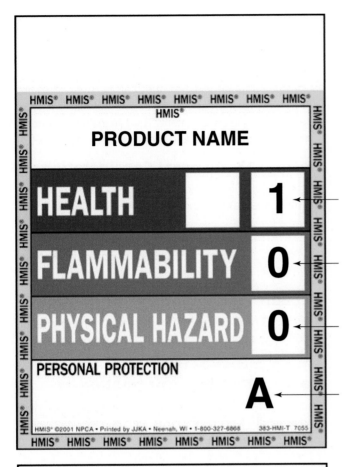

Slight health hazard from exposure to this material. No long-term hazard.

Material will not burn.

Material is very stable.

Safety glasses required during use.

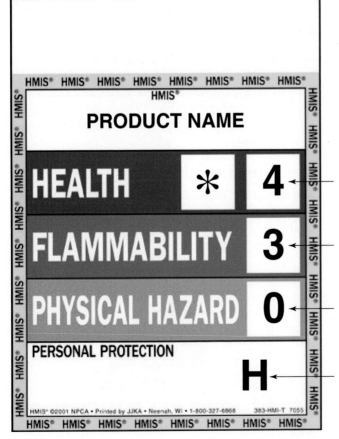

Severe, possibly deadly hazard. Long-term health hazard shown by ∗.

Highly flammable material.

Material is very stable.

Requires use of splash goggles, gloves, protective apron, and vapor respirator during use.

FIGURE 2.1.24 Explanation of two sample HMIS hazard labels.

the total of all of the STEL periods cannot exceed the TWA in any 8-hour period.

These exposure limits can be expressed as mg/m³, which means milligrams of substance per cubic meter of air. They can also be listed as **PPM (parts per million), PPB (parts per billion)**, or **PPT (parts per trillion)**. Those are the number of parts of the material per million, billion, or trillion parts of air.

If the user of the material is at risk of exceeding any of the exposure limits, proper PPE and/or engineering controls must be used to reduce exposure levels below allowable limits. Engineering controls are processes that are designed to reduce or eliminate exposure of the material during use. Examples of engineering controls would be providing ventilation or confining the material use to a specific, enclosed area. Recommended PPE and/or engineering controls are listed in Section 8. Keep in mind that it is good practice to make use of recommended PPE and engineering controls even before exposure limits are reached.

Section 9: Physical and Chemical Properties

Section 9 tells what the substance looks like and/or smells like. It also contains information such as melting points, freezing points, boiling points, evaporation rates, and the **flash point** or lowest temperature where the substance can be ignited.

The product's **LEL (lower explosive limit)** and **UEL (upper explosive limit)** are shown here as well. The LEL is the lowest concentration of the substance in air that will burn or explode if ignited, while the UEL is the highest concentration of the substance in air that will burn or explode if ignited. LEL is sometimes called **LFL (lower flammable limit)**, and UEL is sometimes called **UFL (upper flammable limit)**.

LEL (or LFL) and UEL (or UFL) are usually given as g/m³ or g/ft³, meaning grams of substance per cubic meter or cubic foot of air. For example, a product with an LEL of 2 g/m³ and a UEL of 8 g/m³ means that a mixture between 2 and 8 grams per cubic meter of air will burn or explode if ignited.

Section 10: Stability and Reactivity

Section 10 talks about the product's stability and conditions to avoid. Other substances and conditions that can cause hazardous reactions with the material are listed. Examples could be avoidance of confined spaces, high temperatures, exposure to flame, other chemicals, or even water.

Section 11: Toxicological Information

Section 11 lists ways that the material can enter the body. Can it be inhaled, absorbed through the skin, or ingested? Acute (short-term) and chronic (long-term) health effects are identified. If the substance is a known

or suspected carcinogen, that information is also provided. Symptoms of exposure will also be listed here.

Section 12: Ecological Information

Section 12 lists information that can be used to evaluate the environmental impact of the material if it were released into the environment. This can include its effects on plants, animals, soil, and groundwater.

Section 13: Disposal Considerations

Section 13 lists instructions on disposal, recycling, or reclaiming the material and/or its containers.

Section 14: Transport Information

Section 14 gives information about shipping the material by road, air, railroad, or ship.

Section 15: Regulatory Information

Section 15 provides any safety, health, or environmental regulations about the material that are not covered in any other section of the SDS.

Section 16: Other Information

Section 16 tells when the SDS was written or last updated, and what changes might have been made since the last update. Any other useful information not included in other sections may also be contained in Section 16.

FIRE SAFETY

Fire safety is important in any setting, and a machining environment is no different. The first goal in fire safety is prevention, and conditions where fire hazards exist will be reviewed throughout this text. Knowledge of the factors needed for a fire to burn will help in understanding how to extinguish fires.

The Fire Triangle

The fire triangle shown in **Figure 2.1.25** is a simple symbol that shows the factors needed for a fire. Its sides represent fuel, heat, and oxygen. All three of those elements are needed for a fire to burn.

FIGURE 2.1.25 The fire triangle. All three sides must exist for a fire to burn. The job of a fire extinguisher is to remove one of those three elements.

Fire Extinguishers

Fire extinguishers work by cooling the fuel, removing the oxygen, or stopping the reaction. A fire extinguisher sprays a pressurized substance, or media, out of a nozzle. The spray should be aimed at the fuel, which is at the base of a fire, not at the flames.

OSHA identifies five basic types of fires and extinguishers for those particular types of fires. Four may be found in or near a machining setting. Refer to **Figure 2.1.26** for the symbols for these different types while reading their descriptions below. Using the wrong type of extinguisher can fuel a fire and make conditions worse, so be sure to understand the different types and when to use them. Always check fire extinguisher labels to be sure the proper type is available for the given situation.

If a fire is beyond the control of a fire extinguisher, evacuate the area according to an established plan. Notify the proper emergency authorities after safe evacuation.

Class A

Class A fires are what OSHA defines as those involving ordinary combustibles, such as paper, cloth, wood, rubber, and many plastics. Extinguishers labeled for Class A fires use pressurized water as an extinguishing media. Only use a water-based Class A extinguisher on a Class A fire, never on any other type of fire.

Class B

OSHA defines fires in oils, gasoline, some paints, lacquers, grease, solvents, and most other flammable liquids as Class B fires. Class B extinguishers use carbon dioxide (CO_2) as an extinguishing media.

Class C

OSHA defines Class C fires as electrical fires, such as in wiring, fuse boxes, energized electrical equipment, and computers. Class C extinguishers use a dry chemical as an extinguishing media. Class C extinguishers can also use a dry powder or foam media.

Class D

Metal fires are defined as Class D fires by OSHA. They are fires of small particles, such as chips or filings of combustible metals such as magnesium, titanium, potassium, and sodium.

Multipurpose Extinguishers

CO_2 extinguishers may be labeled as usable for both Class B and C fires. Some dry chemical extinguishers may be labeled for use on Class A, B, and C fires. These extinguishers are probably most common in machining because of the possibility of fires of each class or fires fueled by a combination of each class of combustibles.

CAUTION

- Be familiar with the location and types of fire extinguishers in your machining area.
- Never use a water-based Class A extinguisher on a Class B fire, as it will splatter the burning liquid and expand the fire, leading to serious injury or death.
- Never use a water-based Class A extinguisher on a Class C fire because the water can increase the risk of electrocution, leading to serious injury or death.
- Never use a water-based Class A extinguisher on a Class D fire because the water can intensify some metal fires.

SAFETY DOCUMENTATION

Some shops maintain safety checklists and use them to perform daily, weekly, monthly, and yearly safety inspections. Different items are inspected at different intervals:

- *Daily inspection* items might include checking that machine guards and barriers are in working order, cleanliness of floors, or proper storage of hazardous materials.

- *Weekly inspections* might include checking first aid supplies and ensuring that all hazardous materials are labeled appropriately.

- *Monthly inspections* could cover checking functionality of PPE such as respirators and fire extinguishers, or taking inventory of lockout/tagout supplies.

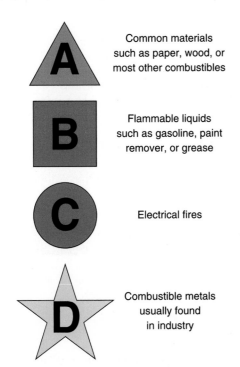

Common materials such as paper, wood, or most other combustibles

Flammable liquids such as gasoline, paint remover, or grease

Electrical fires

Combustible metals usually found in industry

FIGURE 2.1.26 Identification of the different classes of fire extinguishers.

Daily Safety Inspection   Week Ending _____ Inspector _____

Check box as items are completed

	Monday	Tuesday	Wednesday	Thursday	Friday
Floors swept and oil free					
All machine guards in place					
Main power shut off					
All chemicals in proper storage location					
Material storage area properly secured					
Cutting tools properly stored					
Measuring tools properly stored					
Workbenches clear					

Comments:

FIGURE 2.1.27 A sample of a simple safety inspection checklist.

- *Annual inspections* and activities could be the retesting of employees about SDS or company safety policies, or third-party inspection and certification of lifts or cranes.

Every shop's safety inspection reports will probably look different, but be sure to become familiar with those used in your work area so you can be more aware of potential hazards and watch for missed inspections. **Figure 2.1.27** shows an example of a safety inspection checklist.

SUMMARY

- Safety is the number-one priority in a machining environment, and showing respect for potential danger and staying alert and focused lay the foundation for a proper attitude about safety.
- Know the location of first aid supplies and in the event of an accident, notify the proper emergency personnel and superiors.
- OSHA is a federal agency that regulates workplace safety, and NIOSH is a federal agency that researches hazards and recommends prevention methods. Both help to create safe environments in the workplace.

- Choice of clothing and footwear is extremely important in remaining safe in a machining setting. There are many forms of PPE, and it is critical to know proper PPE and its use for every situation in areas where you work.

- Safety glasses should always be worn when in a machining setting, even if just observing.

- Housekeeping, or keeping the working environment clean and clutter free, is also important in maintaining a safe working area.

- Safety barriers or guards are often used to help prevent injuries by restricting access to danger areas. Never cross barriers during machine operation or operate machinery without proper guarding installed.

- Always use proper equipment and methods when lifting to avoid painful injuries.

- Compressed air should not be used to clear debris from machinery or to clean a person's body because debris can become dangerous projectiles and compressed air entering the body can cause severe injuries or even death.

- Use of lockout and tagout procedures prevents machinery from being operated during maintenance or repair to prevent serious injury or death and no one should ever attempt to remove or override someone else's lockout or tagout device.

- Hazardous materials that may be used in a machining setting can be quickly identified by GHS, NFPA and HMIS labeling systems. An SDS shows important detailed information about composition, handling, and dangers of hazardous materials.

- A fire can occur in a machining environment, so be familiar with the location and types of fire extinguishers available in your area and know the fire evacuation plan for any area where you must work.

- Following basic safety guidelines will go a long way in keeping the machining area safe for everyone.

REVIEW QUESTIONS

1. Briefly describe what you believe are the two most important actions everyone *should observe* to remain safe in a machining environment.
2. Briefly describe what you believe are the two most important actions that *should never be tolerated* in a machining environment.
3. What should be done in the case of a personal injury emergency?
4. What does OSHA stand for and what is OSHA's purpose?
5. List three specific clothing items that should not be worn in a machining setting.
6. If someone working around machinery has long hair, how should it be worn?
7. Never operate machinery without proper _____ in place.
8. What is PPE?
9. What is the most common and important piece of PPE that should always be used in a machining environment?
10. What does HCS stand for?
11. What does GHS stand for?
12. What is a pictogram?
13. What does NFPA stand for?
14. What does HMIS stand for?
15. What is the purpose of GHS, NFPA and HMIS labeling?
16. What does SDS stand for, and what is the purpose of SDSs?
17. Class A fire extinguishers use _____ as a media to put out fires.
18. What class of fire extinguisher should be used on an electrical fire?
19. What class of fire extinguisher should be used on flammable liquids?
20. What is the purpose of lockout and tagout procedures?

HMIS® is a registered mark of the American Coatings Association (ACA).

UNIT 2

Measurement Systems and Machine Tool Math Overview

Learning Objectives

After completing this unit, the student should be able to:

- Understand English and metric (SI) measurement systems and perform conversions between the two
- Demonstrate understanding of fractional and decimal math and conversions between fractions and decimals
- Demonstrate ability to solve formulas and equations using basic algebra
- Identify and use properties of basic geometry
- Demonstrate understanding of angular relationships
- Perform conversions between angular measurements in decimal degrees and degrees, minutes, and seconds
- Perform addition and subtraction of angular measurements
- Demonstrate ability to locate and identify points in a Cartesian coordinate system
- Demonstrate ability to locate and identify points in a polar coordinate system
- Demonstrate ability to use the Pythagorean theorem
- Demonstrate the ability to solve right triangles using sine, cosine, and tangent trigonometric functions

Key Terms

Adjacent side	**International System of**	**Pythagorean theorem**
Arc	**Units (SI)**	**Radius**
Cartesian coordinate system	**Metric system**	**Ratio**
Complementary angle or	**Millimeter**	**Sine**
complement	**Opposite side**	**Square**
Cosine	**Order of operations**	**Supplementary angle or**
Diameter	**Parallel**	**supplement**
English system	**Perpendicular**	**Tangent**
Geometry	**Pi (π)**	**Trigonometry**
Hypotenuse	**Polar Coordinate System**	
Inch	**Proportion**	

INTRODUCTION

Every occupation in the machine tool industry requires use of measurement and math skills every day. Successful machining professionals must understand measurement systems using both English and metric units. They must also become proficient in several areas of math, including fractional/decimal operations, English/metric conversions, and basic algebra, geometry, and trigonometry. This unit is meant to be an overview and describes English and metric values and gives some examples of the basic math principles and types of operations typically used in the machining field. Skill in math operations involving whole and decimal numbers is required for understanding these concepts. Later in the text, specific applications of these principles and operations will be explored.

> **NOTE:**
>
> An excellent resource for an in-depth and more comprehensive study of these and other math topics in specific machining contexts is *Mathematics for Machine Technology* by Robert D. Smith and John C. Peterson.

MEASUREMENT SYSTEMS OF THE MACHINING WORLD

The English System (Inches)

The standard system of measurement in the United States is the **English system**. That system of measurement is based on the **inch**. When dealing with inches in the machining industry, fractional and decimal measurements and math are both very common, so understanding these concepts is crucial for success. Inch sizes are commonly followed by the symbol ". For example, 1-1/4" means 1 and 1/4 inches, and 2.250" means 2.250 inches.

The Metric System or SI

The meter is the basic unit used in the metric system of measurement, but the **millimeter** is most commonly used in the machining industry. The millimeter is 1/1000 of a meter and is denoted by the abbreviation mm. In global manufacturing and in the machining industry, the **metric system** is sometimes referred to as the **International System of Units**, or **SI**. SI is the abbreviation for the French le Systeme international d'unités. SI usually does not use fractional units, but instead uses decimal millimeters such as 1.5 mm, 3.05 mm, and 12.022 mm. The metric system's use of decimals is another reason for proficiency in understanding decimal math operations.

MACHINING MATHEMATIC CONCEPTS AND OPERATIONS

Fractional Operations

What is a fraction? It is a part of a whole. In machining, inches are commonly divided into 2, 4, 8, 16, 32, or 64 fractional parts, resulting in halves, quarters, eighths, sixteenths, thirty-seconds, and sixty-fourths.

Examples of fractional operations will be shown using these divisions. Modern calculators can perform these operations, but an understanding of the concepts of fractional math is still a valuable skill. Regardless of the operation performed, answers involving improper fractions should always be converted to mixed numbers, and all answers should be reduced to lowest terms.

Here is a summary of some fractional terms and beginning rules.

- $\frac{5}{8}$ means 5 of 8 parts; 5 is the numerator and 8 is the denominator. $\dfrac{\text{Numerator} \rightarrow}{\text{Denominator} \rightarrow} \dfrac{5}{8}$

- $3\frac{5}{8}$ is called a mixed number because the whole number 3 is mixed with the fraction $\frac{5}{8}$.

- $\frac{28}{8}$ is called an improper fraction because the numerator is larger than the denominator. All improper fractions should be converted to mixed numbers.

- To convert an improper fraction to a mixed number, divide the numerator by the denominator, and write the remainder as a fraction. Consider $\frac{28}{8}$.

  $\frac{28}{8} = 28 \div 8 = 3$ with a remainder of 4, which is $3\frac{4}{8}$.

- All fractions must be reduced to lowest terms. In machining, if the numerator is even, the fraction can be reduced. Consider again $3\frac{4}{8}$. The whole number will remain. To reduce the fraction, find the number that can be divided into both numerator and denominator evenly. To reduce $\frac{4}{8}$, choose to divide by 4. $\frac{4 \div 4}{8 \div 4} = \frac{1}{2}$ so the final answer is $3\frac{1}{2}$.

- Sometimes, it is necessary to convert a mixed number back to an improper fraction. To do that, multiply the denominator by the whole number and add the numerator to that answer. Then write that answer over the original denominator. Consider $2\frac{3}{4}$.

- Denominator $4 \cdot$ Whole Number $2 +$ Numerator $3 = 11$. Then write that answer over the original denominator of 4. So, $2\frac{3}{4} = \frac{11}{4}$.

Let's move on to some other fractional operations.

Fractional Comparison

Comparing fractions and being able to determine smallest or largest of a group of fractions will be very important to those in machining careers. In order to practice doing this, you can convert fractions to equivalent fractions having the same denominator. Then you can compare numerators to determine ascending or descending order. (After some practice, memorization of many commonly used fractions will strengthen the ability to recall or mentally compare fractional sizes.)

EXAMPLE: Order these fractions from smallest to largest: $\frac{7}{32}, \frac{1}{4}, \frac{5}{8}, \frac{37}{64}, \frac{13}{16}$.

Begin by converting all the fractions to equivalents using the largest denominator in the group. Choose 64. Next divide 64 by each denominator. Then multiply the numerator by that answer.

$\frac{7}{32} = \frac{?}{64}$ $64 \div 32 = 2.$ $2 \cdot 7 = 14,$ so $\frac{7}{32} = \frac{14}{64}$

$\frac{1}{4} = \frac{?}{64}$ $64 \div 4 = 16.$ $16 \cdot 1 = 16,$ so $\frac{1}{4} = \frac{16}{64}$

$\frac{5}{8} = \frac{?}{64}$ $64 \div 8 = 8.$ $8 \cdot 5 = 40,$ so $\frac{5}{8} = \frac{40}{64}$

$\frac{37}{64} = \frac{37}{64}$ No change is needed here.

$\frac{13}{16} = \frac{?}{64}$ $64 \div 16 = 4.$ $4 \cdot 13 = 52,$ so $\frac{13}{16} = \frac{52}{64}$

Then order the fractions according to their numerators and change back to their original form.

$$\frac{14}{64} = \frac{7}{32}$$

$$\frac{16}{64} = \frac{1}{4}$$

$$\frac{37}{64} = \frac{37}{64}$$

$$\frac{40}{64} = \frac{5}{8}$$

$$\frac{52}{64} = \frac{13}{16}$$

Another method of comparing fractions is to first convert them to decimal form. Dividing the numerator by the denominator performs this conversion. Then decimals can be compared. This example uses the same fractions as before.

$$\frac{7}{32} = 7 \div 32 = 0.21875$$

$$\frac{1}{4} = 1 \div 4 = 0.25$$

$$\frac{5}{8} = 5 \div 8 = 0.625$$

$$\frac{37}{64} = 37 \div 64 = 0.578125$$

$$\frac{13}{16} = 13 \div 16 = 0.8125$$

(Continued)

EXAMPLE: (continued)

Then order by decimal form.

$$0.21875 \quad = \frac{7}{32}$$

$$0.25 \qquad = \frac{1}{4}$$

$$0.578125 = \frac{37}{64}$$

$$0.625 \qquad = \frac{5}{8}$$

$$0.8125 \qquad = \frac{13}{16}$$

Fractional Addition

Addition of fractions also requires all of the fractions to be converted to equivalents having the same denominator. Then numerators are added. If the answer is an improper fraction, it must be converted to a mixed number.

EXAMPLE: $\frac{3}{8} + \frac{5}{32} + \frac{7}{16} + \frac{1}{4}$.

First convert all fractions to their equivalents using the largest common denominator of the group.

$$\frac{3}{8} = \frac{12}{32}$$

$$+\frac{5}{32} = \frac{5}{32}$$

$$+\frac{7}{16} = \frac{14}{32}$$

$$+\frac{1}{4} = \frac{8}{32}$$

Then add the numerators.

$$\frac{3}{8} = \frac{12}{32}$$

$$+\frac{5}{32} = \frac{5}{32}$$

$$+\frac{7}{16} = \frac{14}{32}$$

$$+\frac{1}{4} = \frac{8}{32}$$

$$\overline{\qquad \frac{39}{32}}$$

Then convert to a mixed number.

$$\frac{39}{32} = 1\frac{7}{32}$$

Fractional Subtraction

Subtraction of fractions also requires all of the fractions to be converted to equivalents having the same denominator. Then numerators are subtracted. Sometimes borrowing is required.

EXAMPLE ONE: Evaluate $\frac{3}{4} - \frac{5}{32}$.

Convert one fraction to its equivalent using the larger denominator of the other fraction, and then subtract only the numerators.

$$\frac{3}{4} = \frac{24}{32}$$

$$-\frac{5}{32} = \frac{5}{32}$$

$$\overline{\qquad \frac{19}{32}}$$

EXAMPLE TWO: Evaluate $2\frac{1}{8} - 1\frac{11}{16}$.

First convert, just like the last example.

$$2\frac{1}{8} = 2\frac{2}{16}$$

$$-1\frac{11}{16} = 1\frac{11}{16}$$

But 11 cannot be subtracted from 2, so borrow 1 from the 2 in the form of $\frac{16}{16}$. Then subtract only the numerators and the whole numbers.

$$2\frac{1}{8} = 2\frac{2}{16} = 1\frac{16}{16} + \frac{2}{16} = 1\frac{18}{16}$$

$$-1\frac{11}{16} = 1\frac{11}{16} \qquad \rightarrow \qquad 1\frac{11}{16}$$

$$\overline{\qquad \qquad \qquad 0\frac{7}{16} = \frac{7}{16}}$$

Fractional Multiplication

Multiplication of fractions does not require any conversion. Numerators are multiplied and denominators are multiplied.

EXAMPLE ONE:

$$\frac{1}{4} \cdot \frac{3}{16} = \frac{3}{64}$$

$$\frac{1}{2} \cdot \frac{3}{4} = \frac{3}{8}$$

$$\frac{1}{8} \cdot \frac{7}{8} = \frac{7}{64}$$

If mixed numbers are involved, there are two methods that can be used. One way is to convert them to improper fractions first. After multiplication, be sure to convert back to a mixed number.

EXAMPLE TWO:

$$\frac{1}{2} \cdot 2\frac{9}{16}$$

$$\frac{1}{2} \cdot \frac{41}{16} = \frac{41}{32} = 1\frac{9}{32}$$

Another method is to separate the whole number from the fraction and multiply both by the other fraction. Then add the two together. Consider the same problem.

EXAMPLE THREE:

$$\frac{1}{2} \cdot 2\frac{9}{16} \text{ is the same as}$$

$$\left(\frac{1}{2} \cdot 2\right) + \left(\frac{1}{2} \cdot \frac{9}{16}\right)$$

$$\frac{1}{2} \cdot 2 = 1 \text{ and } \frac{1}{2} \cdot \frac{9}{16} = \frac{9}{32}, \text{ so}$$

$$1 + \frac{9}{32} = 1\frac{9}{32}$$

Fractional Division

Division does not require conversion either. To divide, multiply the first fraction by the reciprocal (switching numerator and denominator) of the second fraction. This is often called cross multiplying. Then follow all multiplication rules.

EXAMPLE:

$$\frac{1}{2} \div \frac{1}{4}$$

$$\frac{1}{2} \cdot \frac{4}{1} \text{ (Change to multiplication by the reciprocal of } \frac{1}{4})$$

$$\frac{1}{2} \cdot \frac{4}{1} = \frac{4}{2} \text{ (Multiply)}$$

$$\frac{4}{2} = 2 \text{ (Reduce)}$$

Fractional/Decimal Conversion

Another common practice is converting fractions to their decimal equivalents and determining the closest fraction to a given decimal. In theory, to convert a fraction to its decimal equivalent, divide the numerator by the denominator, and to determine the fraction closest to a decimal, you can use a proportion. A calculator may be used to perform these conversions, but it is common practice to use a conversion chart like the one shown in **Figure 2.2.1** before becoming familiar with and memorizing the more common fractional/decimal equivalents.

Basic Algebra

Numerous formulas are used in the machining field that require an understanding of basic algebra principles and concepts. Most important is the **order of operations**. Order of operations is the principle that certain mathematical operations must be done in a certain order to correctly perform calculations. Another is solving for a variable in an equation or formula.

Order of Operations

The first letters of each word in the phrase "**P**lease **E**xcuse **M**y **D**ear **A**unt **S**ally" can help in remembering the order of operations.

P—Parentheses: Perform all calculations inside parentheses from innermost to outermost. This also applies to groupings created with a fraction bar or root sign.

E—Exponents: Perform all exponent operations next.

M—Multiplication } Perform all multiplication and
D—Division } division from left to right.

A—Addition } Perform all addition and
S—Subtraction } subtraction from left to right.

Many modern handheld calculators perform order of operations correctly when formulas are entered, but some require the use of parentheses at particular times. Further, pressing the "=" button at the incorrect time can lead to errors. Here are a few examples of some formulas using basic algebra and showing the proper order of operations.

EXAMPLE ONE: Calculate the value for *A* when *B* = 150 and *C* = 0.5, using the formula:

$$A = \frac{3.82 \cdot B}{C}$$

$$A = \frac{3.82 \cdot 150}{0.5}$$

$$A = \frac{573}{0.5} = 1146$$

Fractional/Decimal Equivalents			
Fraction	**Decimal**	**Fraction**	**Decimal**
1/64	0.015625	33/64	0.515625
1/32	0.03125	17/32	0.53125
3/64	0.046875	35/64	0.546875
1/16	0.0625	9/16	0.5625
5/64	0.078125	37/64	0.578125
3/32	0.09375	19/32	0.59375
7/64	0.109375	39/64	0.609375
1/8	0.125	5/8	0.625
9/64	0.140625	41/64	0.640625
5/32	0.15625	21/32	0.65625
11/64	0.171875	43/64	0.671875
3/16	0.1875	11/16	0.6875
13/64	0.203125	45/64	0.703125
7/32	0.21875	23/32	0.71875
15/64	0.234375	47/64	0.734375
1/4	0.250	3/4	0.750
17/64	0.265625	49/64	0.765625
9/32	0.28125	25/32	0.78125
19/64	0.296875	51/64	0.796875
5/16	0.3125	13/16	0.8125
21/64	0.328125	53/64	0.828125
11/32	0.34375	27/32	0.84375
23/64	0.359375	55/64	0.859375
3/8	0.375	7/8	0.875
25/64	0.390625	57/64	0.890625
13/32	0.40625	29/32	0.90625
27/64	0.421875	59/64	0.921875
7/16	0.4375	15/16	0.9375
29/64	0.453125	61/64	0.953125
15/32	0.46875	31/32	0.96875
31/64	0.484375	63/64	0.984375
1/2	0.500	1	1

FIGURE 2.2.1 To find the decimal equivalent of a fraction, find the fraction in the left column. Its decimal equivalent is in the column just to the right of the fraction. To find a fraction closest to a decimal, work in the opposite direction.

EXAMPLE TWO: Calculate *A* when *B* = 2.375, *C* = 1.6, and *D* = 3.25, using the formula:

$$A = \frac{B - C}{D}$$

$$A = \frac{2.375 - 1.6}{3.25}$$

$$A = \frac{0.775}{3.25}$$

$$A = 0.2385$$

Note that 2.375 − 1.6 must be calculated before you divide by 3.25. Depending on the type of calculator used, 1.6 might first be divided by 3.25, and then that answer subtracted from 2.375, resulting in the incorrect answer shown below. Use of parentheses and grouping is crucial to correctly perform calculations. There is quite a difference in these two answers!

$$A = \frac{B - C}{D}$$

$$A = \frac{2.375 - 1.6}{3.25}$$

$$A \neq 2.375 - 1.6 \div 3.25$$

$$A \neq 2.375 - 0.4923$$

$$A \neq 1.8827 \text{ INCORRECT}$$

EXAMPLE THREE: Solve for *a* when *b* = 0.322 and *c* = 0.5, using the formula:

$$a = \sqrt{c^2 - b^2}$$

$$a = \sqrt{0.5^2 - 0.322^2}$$

$$a = \sqrt{0.25 - 0.103684}$$

$$a = \sqrt{0.146316}$$

$$a = 0.3825$$

Note that the values under the square root sign must first be squared and then subtracted before evaluating the square root.

Solving for a Variable

Sometimes formulas are shown like the prior examples, where the variable is already isolated, or by itself, on one side of the formula or equation. In these cases, following the order of operations is all that is required to find the answer. At other times, the variable is somewhere else in the equation and needs to be isolated before solving for its value.

Any operation can be performed to both sides of the equation (the parts on each side of the =) to get that variable by itself. Whatever operation is being performed on the variable can be *undone* by performing the opposite operation. Subtraction undoes addition; division undoes multiplication; and so on. Here are just a few examples of how to isolate a variable.

EXAMPLE ONE: Isolate B in the following equation.

$$A = \frac{3.82 \cdot B}{C}$$

First, since the side with the B is being divided by C, multiply both sides by C.

$$AC = 3.82 \cdot B$$

Next, since the B is being multiplied by 3.82, divide both sides by 3.82.

$$\frac{AC}{3.82} = B$$

Now values can be substituted for A and C, and the equation solved for B.

EXAMPLE TWO: Isolate B in the following equation.

$$A = \frac{B - C}{D}$$

First, since the side with the B is being divided by D, multiply both sides by D.

$$AD = B - C$$

Next, since C is being subtracted from B, add C to both sides.

$$AD + C = B$$

Now values can be substituted for A, C, and D, and the equation solved for B. Remember that AD must be multiplied together before you add C.

EXAMPLE THREE: Isolate b in the following equation.

$$c^2 = a^2 + b^2$$

First, since a^2 is being added to the side with the b, subtract a^2 from both sides.

$$c^2 - a^2 = b^2$$

Next, since b is being squared, take the square root of both sides.

$$\sqrt{c^2 - a^2} = \sqrt{b^2}$$

Then simplify $\sqrt{b^2}$ as b.

$$b = \sqrt{c^2 - a^2}$$

Now values can be substituted for c and a, and the equation solved for b.

Ratios and Proportions

A **ratio** is a comparison between two numbers. It can be written in the form of 1:2 or as a fraction $\frac{1}{2}$. This can be read as 1 to 2 or 1 per 2. A speed limit of 60 mph is an everyday ratio and can be rewritten as 60 miles:1 hour or $\frac{60 \text{ miles}}{1 \text{ hour}}$. A **proportion** is two ratios that are equal to each other. An example would be $\frac{60 \text{ miles}}{1 \text{ hour}} = \frac{1 \text{ mile}}{1 \text{ minute}}$.

Uses of ratios and proportions in machining include converting units or mixing concentrates with water for cutting fluids.

English/Metric and Metric/English Conversions

Because of today's global market, there is often the need to move back and forth between English and metric (SI) measurements. The following examples show where standard conversion factors come from while explaining how ratios and proportions can be used.

- 1 inch = 25.4 mm and can be stated by the ratio

$$\frac{1 \text{ inch}}{25.4 \text{ mm}}$$

Think of this ratio as "1 inch is to 25.4 mm."

- 1 mm = 0.03937 inches and can be stated by the ratio

$$\frac{1 \text{ mm}}{0.03937 \text{ inches}}$$

Think of this ratio as "1 mm is to 0.03937 inches."

Converting English (Inches) to Metric (mm)

There are two methods using the two ratios above to find the answer. Both use proportions.

EXAMPLE ONE: Convert 2.63 inches to mm.

Method 1:
First, set up a proportion with x representing the unknown value. Think of the following proportion as "2.63 inches is to x mm as 1 inch is to 2.54 mm."

$$\frac{2.63 \text{ inches}}{x} = \frac{1 \text{ inch}}{25.4 \text{ mm}}$$

Cross multiply (extremes · means) and solve for x

$$x \text{ (inches)} = 2.63(\text{inches}) \cdot 25.4(\text{mm})$$

The inch units cancel out and the answer is in mm

$$x = 66.802 \text{ mm}$$

(Continued)

EXAMPLE: (continued)

Method 2:

Again, set up a proportion with x representing the unknown value. Think of the following proportion as "2.63 inches is to x mm as 0.03937 inches is to 1 mm."

$$\frac{2.63 \text{ inches}}{x} = \frac{0.03937 \text{ inches}}{1 \text{ mm}}$$

Cross multiply (extremes · means) and solve for x.

$$0.0397x(\text{inches}) = 2.63(\text{inches}) \cdot 1(\text{mm})$$

$$x = \frac{2.63(\text{inches}) \cdot 1(\text{mm})}{0.03937(\text{inches})}$$

The inch units cancel out and the answer is in mm.

$$x = \frac{2.63 \text{ mm}}{0.03937}$$

$$x = 66.802 \text{ mm}$$

This example's methods show that there are two ways to convert inches to millimeters.

Multiply the inch value by 25.4

Or

Divide the inch value by 0.03937

EXAMPLE TWO: What is the inch equivalent of 37.3 mm? Again, two different proportions can be used to find the answer.

Method 1:

$$\frac{37.3 \text{ mm}}{x} = \frac{1 \text{ mm}}{0.03937 \text{ inches}}$$

Cross multiply (extremes · means) and solve for x.

$$x(\text{mm}) = 373(\text{mm}) \cdot 0.03937(\text{inches})$$

The mm units cancel out and the answer is in inches.

$$x = 1.4685 \text{ inches}$$

Method 2:

$$\frac{x}{37.3 \text{ mm}} = \frac{1 \text{ inch}}{25.4 \text{ mm}}$$

Cross multiply (extremes · means) and solve for x.

$$25.4x(\text{mm}) = 37.3(\text{mm}) \cdot 1(\text{inch})$$

solve for x.

$$x = \frac{37.3(\text{mm}) \cdot 1(\text{inch})}{25.4(\text{mm})}$$

The mm units cancel out and the answer is in inches.

$$x = 1.4685 \text{ inches}$$

This example's methods show that there are two ways to convert millimeters to inches.

Multiply the mm value by 0.03937

Or

Divide the mm value by 25.4

When seeking a standard metric value for a given inch value, or a standard inch value for a given metric value, a simpler method is to use a conversion chart when available. The method is similar to using a fractional/decimal equivalent chart. Find the inch value on the chart, and its metric equivalent will be next to the inch value, or find the metric value on the chart, and its inch equivalent will be next to the metric value. **(See Figure 2.2.2.)**

Basic Geometry

Certain geometric concepts are also valuable to people pursuing machining careers. **Geometry** is math involving shapes and their relationships.

Parallelism and Perpendicularity

The term **parallel** refers to lines or surfaces that, in theory, will never intersect, or touch, if extended. The distance between them is always the same. **Figure 2.2.3** shows examples of parallel lines and surfaces.

Perpendicular lines or surfaces are those that intersect at a 90-degree angle. In machining, this is commonly called **square**. The statement that two surfaces are square means that they are at 90-degree angles to each other. **(See Figure 2.2.4.) Figure 2.2.5** shows an illustration of perpendicular or "square" surfaces.

Circles

Properties of circles are important geometric concepts in the machining industry too. The **diameter** of a circle is the distance from one side to another through the center, while the **radius** is the distance from the center point to any point on the circle. **Figure 2.2.6** shows the relationship between diameter and radius.

The circumference of a circle is the distance around the periphery of the circle. This is also an important feature to understand. A constant number known as **Pi** (shown by the symbol π) is used in the following formulas to calculate the circumference of a circle. The approximate value of Pi is 3.14159. If you use a calculator with a Pi key during calculations, an even more accurate value will be used.

A

Starrett® METRIC EQUIVALENTS

MILLIMETERS TO DECIMALS

MM	DECIMAL	MM	DECIMAL	MM	DECIMAL	MM	DECIMAL	MM	DECIMAL
0.01	.0004	0.41	.0161	0.81	.0319	21	.8268	61	2.4016
0.02	.0008	0.42	.0165	0.82	.0323	22	.8661	62	2.4409
0.03	.0012	0.43	.0169	0.83	.0327	23	.9055	63	2.4803
0.04	.0016	0.44	.0173	0.84	.0331	24	.9449	64	2.5197
0.05	.0020	0.45	.0177	0.85	.0335	25	.9843	65	2.5591
0.06	.0024	0.46	.0181	0.86	.0339	26	1.0236	66	2.5984
0.07	.0028	0.47	.0185	0.87	.0343	27	1.0630	67	2.6378
0.08	.0032	0.48	.0189	0.88	.0347	28	1.1024	68	2.6772
0.09	.0035	0.49	.0193	0.89	.0350	29	1.1417	69	2.7165
0.10	.0039	0.50	.0197	0.90	.0354	30	1.1811	70	2.7559
0.11	.0043	0.51	.0201	0.91	.0358	31	1.2205	71	2.7953
0.12	.0047	0.52	.0205	0.92	.0362	32	1.2598	72	2.8346
0.13	.0051	0.53	.0209	0.93	.0366	33	1.2992	73	2.8740
0.14	.0055	0.54	.0213	0.94	.0370	34	1.3386	74	2.9134
0.15	.0059	0.55	.0217	0.95	.0374	35	1.3780	75	2.9528
0.16	.0063	0.56	.0221	0.96	.0378	36	1.4173	76	2.9921
0.17	.0067	0.57	.0224	0.97	.0382	37	1.4567	77	3.0315
0.18	.0071	0.58	.0228	0.98	.0386	38	1.4961	78	3.0709
0.19	.0075	0.59	.0232	0.99	.0390	39	1.5354	79	3.1102
0.20	.0079	0.60	.0236	1.00	.0394	40	1.5748	80	3.1496
0.21	.0083	0.61	.0240	1	.0394	41	1.6142	81	3.1890
0.22	.0087	0.62	.0244	2	.0787	42	1.6535	82	3.2283
0.23	.0091	0.63	.0248	3	.1181	43	1.6929	83	3.2677
0.24	.0095	0.64	.0252	4	.1575	44	1.7323	84	3.3071
0.25	.0098	0.65	.0256	5	.1969	45	1.7717	85	3.3465
0.26	.0102	0.66	.0260	6	.2362	46	1.8110	86	3.3858
0.27	.0106	0.67	.0264	7	.2756	47	1.8504	87	3.4252
0.28	.0110	0.68	.0268	8	.3150	48	1.8898	88	3.4646
0.29	.0114	0.69	.0272	9	.3543	49	1.9291	89	3.5039
0.30	.0118	0.70	.0276	10	.3937	50	1.9685	90	3.5433
0.31	.0122	0.71	.0280	11	.4331	51	2.0079	91	3.5827
0.32	.0126	0.72	.0284	12	.4724	52	2.0472	92	3.6220
0.33	.0130	0.73	.0287	13	.5118	53	2.0866	93	3.6614
0.34	.0134	0.74	.0291	14	.5512	54	2.1260	94	3.7008
0.35	.0138	0.75	.0295	15	.5906	55	2.1654	95	3.7402
0.36	.0142	0.76	.0299	16	.6299	56	2.2047	96	3.7795
0.37	.0146	0.77	.0303	17	.6693	57	2.2441	97	3.8189
0.38	.0150	0.78	.0307	18	.7087	58	2.2835	98	3.8583
0.39	.0154	0.79	.0311	19	.7480	59	2.3228	99	3.8976
0.40	.0158	0.80	.0315	20	.7874	60	2.3622	100	3.9370

The L.S. Starrett Company — World's Greatest Toolmakers

Source: Starrett

B

Starrett® METRIC EQUIVALENTS

DECIMALS TO MILLIMETERS				FRACTIONS TO DECIMALS TO MILLIMETERS					
DECIMAL	MM	DECIMAL	MM	FRACTION	DECIMAL	MM	FRACTION	DECIMAL	MM
.001	0.03	.470	11.94	1/64	.0156	0.40	33/64	.5156	13.10
.002	0.05	.480	12.19	1/32	.0313	0.79	17/32	.5313	13.49
.003	0.08	.490	12.45	3/64	.0469	1.19	35/64	.5469	13.89
.004	0.10	.500	12.70						
.005	0.13	.510	12.95	1/16	.0625	1.59	9/16	.5625	14.29
.006	0.15	.520	13.21						
.007	0.18	.530	13.46						
.008	0.20	.540	13.72	5/64	.0781	1.98	37/64	.5781	14.68
.009	0.23	.550	13.97	3/32	.0938	2.38	19/32	.5938	15.08
.010	0.25	.560	14.22	7/64	.1094	2.78	39/64	.6094	15.48
.020	0.51	.570	14.48						
.030	0.76	.580	14.73	1/8	.1250	3.18	5/8	.6250	15.88
.040	1.02	.590	14.99						
.050	1.27	.600	15.24	9/64	.1406	3.57	41/64	.6406	16.27
.060	1.52	.610	15.49	5/32	.1563	3.97	21/32	.6563	16.67
.070	1.78	.620	15.75	11/64	.1719	4.37	43/64	.6719	17.07
.080	2.03	.630	16.00						
.090	2.29	.640	16.26	3/16	.1875	4.76	11/16	.6875	17.46
.100	2.54	.650	16.51						
.110	2.79	.660	16.76	13/64	.2031	5.16	45/64	.7031	17.86
.120	3.05	.670	17.02	7/32	.2188	5.56	23/32	.7188	18.26
.130	3.30	.680	17.27	15/64	.2344	5.95	47/64	.7344	18.65
.140	3.56	.690	17.53						
.150	3.81	.700	17.78	1/4	.2500	6.35	3/4	.7500	19.05
.160	4.06	.710	18.03						
.170	4.32	.720	18.29	17/64	.2656	6.75	49/64	.7656	19.45
.180	4.57	.730	18.54	9/32	.2813	7.14	25/32	.7813	19.84
.190	4.83	.740	18.80	19/64	.2969	7.54	51/64	.7969	20.24
.200	5.08	.750	19.05						
.210	5.33	.760	19.30	5/16	.3125	7.94	13/16	.8125	20.64
.220	5.59	.770	19.56						
.230	5.84	.780	19.81	21/64	.3281	8.33	53/64	.8281	21.03
.240	6.10	.790	20.07	11/32	.3438	8.78	27/32	.8438	21.43
.250	6.35	.800	20.32	23/64	.3594	9.13	55/64	.8594	21.83
.260	6.60	.810	20.57						
.270	6.86	.820	20.83	3/8	.3750	9.53	7/8	.8750	22.23
.280	7.11	.830	21.08						
.290	7.37	.840	21.34	25/64	.3906	9.92	57/64	.8906	22.62
.300	7.62	.850	21.59	13/32	.4062	10.32	29/32	.9063	23.02
.310	7.87	.860	21.84	27/64	.4219	10.72	59/64	.9219	23.42
.320	8.13	.870	22.10						
.330	8.38	.880	22.35	7/16	.4375	11.11	15/16	.9375	23.81
.340	8.64	.890	22.61						
.350	8.89	.900	22.86	29/64	.4531	11.51	61/64	.9531	24.21
.360	9.14	.910	23.11	15/32	.4688	11.91	31/32	.9688	24.61
.370	9.40	.920	23.37	31/64	.4844	12.30	63/64	.9844	25.00
.380	9.65	.930	23.62						
.390	9.91	.940	23.88	1/2	.5000	12.70	1	1.0000	25.40
.400	10.16	.950	24.13						
.410	10.41	.960	24.38						
.420	10.67	.970	24.64						
.430	10.92	.980	24.89						
.440	11.18	.990	25.15						
.450	11.43	1.00	25.40						
.460	11.68								

Bulletin 1318   50M/S 06/09   starrett.com

FIGURE 2.2.2 Chart A can be used to find the decimal inch equivalent of a mm value. The left side of chart B can be used to find the mm equivalent of a decimal inch value. The right side of chart B can be used to find the mm equivalent of a fractional inch value. Notice the decimal inch equivalent of the fraction is also shown.

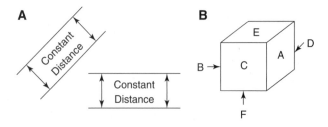

FIGURE 2.2.3 The distance between parallel lines is constant. In both Illustrations in (A), both pairs of lines are parallel. Measurements across parallel surfaces are equal. In this sketch of a block (B), surfaces A and B are parallel, and so are C and D. What other pair of surfaces is parallel?

$$Circumference = 2\pi r, \text{ where } \pi \text{ is } pi$$
(approximately 3.14159) and *r* is the circle's radius

or

$$Circumference = \pi D, \text{ where } \pi \text{ is } pi$$
(approximately 3.14159) and *D* is the circle's diameter

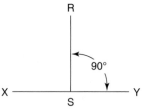

FIGURE 2.2.4 Perpendicular lines create 90-degree corners. Lines RS and XY are perpendicular, or square, to each other.

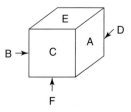

FIGURE 2.2.5 In this block, surfaces A and C are perpendicular, or square, to each other. So are A and F, as well as B and F. What other pairs of surfaces are perpendicular, or square, to each other?

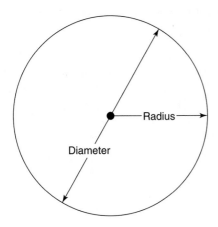

FIGURE 2.2.6 Diameter = 2 × Radius. Radius = Diameter ÷ 2.

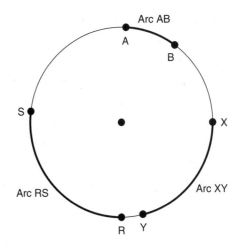

FIGURE 2.2.8 An arc is a portion of a circle. It is usually labeled by its endpoints.

EXAMPLE: Find the circumference of a 24.5-inch-diameter circle.

$$Circumference = \pi \cdot 24.5$$
$$Circumference = 3.14159 \cdot 24.5$$
$$Circumference = 76.969 \text{ inches}$$

Figure 2.2.7 illustrates the circumference and the formula for circumference.

An **arc** is a portion of a circle and is a frequently used term in the machining field. **Figure 2.2.8** gives a visual explanation of arcs.

The term **tangent** refers to a line, circle, or arc that touches a circle or an arc at only one point. The location of that point is called the *point of tangency*. **Figure 2.2.9** gives a visual explanation of some tangent situations.

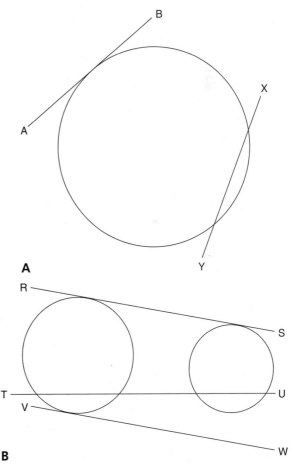

FIGURE 2.2.9 In (A), line AB is tangent to the circle. Line XY is not. In (B), line RS is tangent to both circles. Line TU is not tangent to either circle. Line VW is tangent to only the circle on the left.

Angles

Angles are common in the machining industry also, so understanding some relationships of angles is important. A circle contains 360 degrees, and angles can be created by dividing circles. The fractional part of a circle is multiplied by 360 degrees to obtain the angle in degrees.

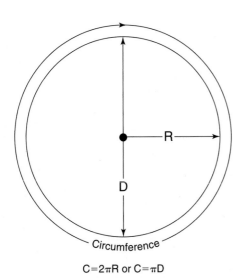

$$C = 2\pi R \text{ or } C = \pi D$$

FIGURE 2.2.7 The circumference of a circle is the distance around its periphery and is calculated by either C = 2πr or C = πD.

EXAMPLE ONE: 1/2 of a circle is 180 degrees, because $\frac{1}{2} \cdot 360 = 180$.

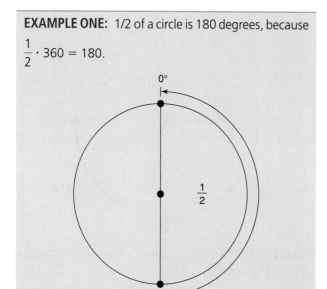

EXAMPLE TWO: 1/4 of a circle is 90 degrees, because $\frac{1}{4} \cdot 360 = 90$.

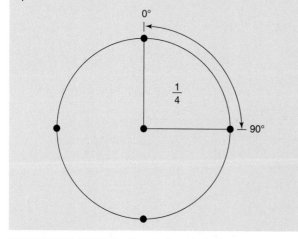

EXAMPLE THREE: How many degrees are in 1/6 of a circle?

$\frac{1}{6} \cdot 360 = 60$, so 1/6 of a circle is 60 degrees

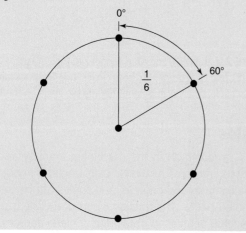

Angular Measurement and Conversions

It is also important to expect that angles will not always be whole numbers. They can be shown as decimals or divided into smaller units known as minutes and seconds. It is important to keep in mind that in angular measurement, minutes and seconds are not related to time. They just happen to share the same name as time units.

One degree can be divided into 60 smaller units called minutes, so one minute is 1/60 of a degree. Minute values are identified by the symbol '. For example, 21' means 21 minutes. That 21' is actually a fraction of a degree with a denominator of 60 or $\frac{21}{60}$ degrees. This can be converted to decimal degrees like any other fraction: so $\frac{21}{60} = 21 \div 60 = 0.35°$. So 21' = 0.35 degrees.

One minute can then be divided into 60 smaller units called seconds, so one second is 1/60 of a minute. Second values are identified by the symbol ". For example, 36", means 36 seconds. That 36" is actually a fraction of a minute with a denominator of 60 or $\frac{36}{60}$ minutes. This can be converted to decimal minutes like any other fraction, so $\frac{36}{60} = 36 \div 60 = 0.6$ minutes. This 0.6 minutes is actually a fraction of a degree and can be viewed as $\frac{0.6}{60}$ degrees. This can also be converted like the previous fractions: $\frac{0.6}{60} = 0.6 \div 60 = 0.01$ degree. So 36" = 0.01 degree.

Considering those smaller angular units, an angular measurement in the form 30°21'36" would be read as 30 degrees, 21 minutes, 36 seconds. This can also be expressed in decimal form, since those conversions have already been performed. Since 21 minutes = 0.35 degrees and 36 seconds = 0.01 degrees, adding 0.35 and 0.01 results in 0.36 degrees. This must then be added to the 30 degrees. The decimal equivalent of 30°21'36" is 30.36 degrees.

Sometimes, angles are given in decimal degrees and may need to be converted to degrees, minutes, and seconds using proportions. Let's convert the 30.36 degrees as an example and to verify the process.

STEP 1: The 30 degrees is subtracted and recorded leaving the 0.36. Next, choose a ratio for conversion of the 0.36. Since one degree contains 60 minutes, use the ratio $\frac{1 \text{ degree}}{60 \text{ minutes}}$. Then set up a proportion and solve.

$$\frac{1 \text{ degree}}{60 \text{ minutes}} = \frac{0.36 \text{ degrees}}{x}$$

$$x \text{ degrees} = 0.36 \text{ degrees} \cdot 60 \text{ minutes}$$

Degree units cancel out and the answer is in minutes.

$$x = 21.6 \text{ minutes}$$

Adding the 21.6' to the original 30° gives the value of 30°21.6'. But seconds are still needed so the conversion needs to continue.

STEP 2: Subtract 21 from the 21.6 and record the 30°21'. Choose a ratio for conversion of the 0.6 just like before. Since one minute contains 60 seconds, use the ratio $\dfrac{1 \text{ minute}}{60 \text{ seconds}}$. Then set up a proportion and solve as before.

$$\frac{1 \text{ minute}}{60 \text{ seconds}} = \frac{0.6 \text{ minutes}}{x}$$
$$x \text{ minutes} = 0.6 \text{ minutes} \cdot 60 \text{ seconds}$$

Minute units cancel out and the answer is in seconds.
$$x = 36 \text{ seconds}$$

This shows that 30.36 degrees and 30°21'36" are equivalent measurements of the same angle.

These conversions can be summarized by the following:

Subtract the whole number of degrees and multiply the answer by 60. This is the minutes.

Subtract the whole number of minutes from this answer and multiply the answer by 60. This is the seconds.

EXAMPLE: Convert 32.34 degrees to degrees, minutes, and seconds.

$$32.34 - 32 = 0.34 \text{ degrees}$$

Record the 32 degrees, then convert remaining decimal degrees to minutes:

$$0.34 \times 60 = 20.4 \text{ minutes}$$
$$20.4 - 20 = 0.4 \text{ minutes}$$

Record the 20 minutes, then convert remaining decimal minutes to seconds:

$$0.4 \times 60 = 24 \text{ seconds}$$

32°20'24" is the equivalent of the decimal angle 32.34 degrees.

Complementary and Supplementary Angles

When two angles add up to 90 degrees, they are called **complementary angles** (or simply **complements**). When two angles add up to 180 degrees, they are called **supplementary angles** (or simply **supplements**). To find the complement of a known angle, subtract that known angle from 90. To find the supplement of a known angle, subtract that known angle from 180. This concept is important because depending on how angles are designated on prints, their complements or supplements may need to be considered during measurement or machine tool setup and operation. **Figure 2.2.10**

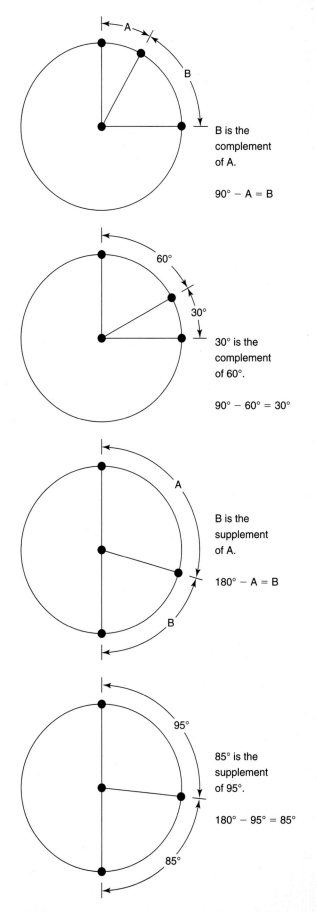

B is the complement of A.

$$90° - A = B$$

30° is the complement of 60°.

$$90° - 60° = 30°$$

B is the supplement of A.

$$180° - A = B$$

85° is the supplement of 95°.

$$180° - 95° = 85°$$

FIGURE 2.2.10 Complementary and supplementary angles.

illustrates this concept of complementary and supplementary angles and how they can be found.

Addition and Subtraction of Angles

Adding and subtracting angles must frequently be done when using complementary and supplementary angles. When angles are in decimal form, these operations are just like operations with any decimal numbers. Knowing how to add and subtract angles when they are displayed in degrees, minutes, and possibly seconds is also an important math concept used in the machining field. Addition sometimes requires reducing units, similar to fractional addition when numerators in answers become larger than denominators. Subtraction sometimes requires borrowing units, also similar to fractional subtraction when a larger numerator is subtracted from a smaller numerator. An example of each follows.

EXAMPLE ONE: Addition results requiring reducing. Add 34° 38' 40" and 20° 28' 35".

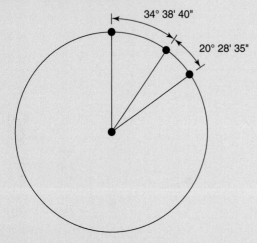

Arrange the problem vertically with some space between units for clarity.

$$\begin{array}{rrr} 34° & 38' & 40" \\ +20° & 28' & 35" \\ \hline 54° & 66' & 75" \end{array}$$ Add the columns.

Since we know that 60' = 1' and 60' = 1°, any sum in the minutes or seconds column over 60 is like an improper fraction. Starting with the seconds column and moving left, units need to be converted, and the whole number added to the next column on the left.

The current answer is 54° 66' 75".

To convert the seconds, use factor $\dfrac{1\ minute}{60\ seconds}$.

$$75\ seconds \cdot \dfrac{1\ minute}{60\ seconds}$$

$$\dfrac{75\ seconds}{1} \cdot \dfrac{1\ minute}{60\ seconds}$$ Seconds units cancel out.

$$\dfrac{75}{60}\ minutes = 1\dfrac{15}{60}\ minutes,\ which = 1\ minutes\ 15\ seconds$$

Replace the 75" with 15", then add the 1' to the 66' for 67'. The answer is then:

$$54°\ 67'\ 15"$$

Next, convert the 67' using the factor $\dfrac{1\ degree}{60\ minutes}$.

$$67\ minutes \cdot \dfrac{1\ degree}{60\ minutes}$$

$$\dfrac{67\ minutes}{1} \cdot \dfrac{1\ degree}{60\ minutes}$$ Minutes units cancel out.

$$\dfrac{67}{60}\ degrees = 1\dfrac{7}{60}',\ which = 1\ degree\ 7\ minutes$$

Replace the 67' with 7', then add the 1° to the 54° for 55° for a final answer of:

$$55°\ 7'\ 15"$$

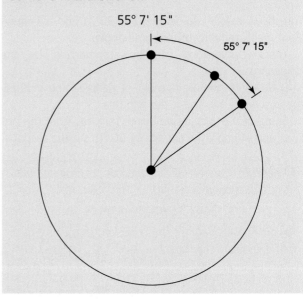

EXAMPLE TWO: Subtraction requiring borrowing.

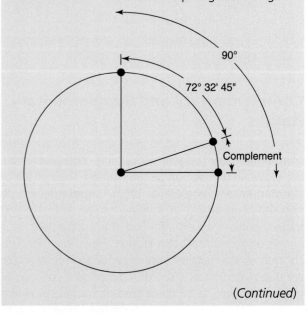

(Continued)

EXAMPLE TWO: (continued)

Find the complement of 72° 32' 45".

Recall that the complement is the given angle subtracted from 90.

Ninety degrees can be rewritten as 90° 0' 0", then set up the subtraction like the previous addition example.

$$\begin{array}{rrr} 90° & 0' & 0" \\ -72° & 32' & 45" \\ \hline \end{array}$$

Begin on the right like any subtraction problem. The 45 cannot be subtracted from 0, so borrowing from the next column of units to the left is needed. That 0 cannot lend either, so moving to the next column of units to the left is required. The 90° can lend, so, if 1° is borrowed from 90°, that column becomes 89°. Since the 1° needs to be deposited in the minutes column, that 1° needs to be written as 60' because 1° = 60'. The problem now looks like this:

$$\begin{array}{rrr} 89° & 60' & 0" \\ -72° & 32' & 45" \\ \hline \end{array}$$

Now 1' can be borrowed from the 60' for deposit in the seconds column. The 60' becomes 59' and the borrowed 1' is deposited in the seconds column as 60" because 1' = 60". The problem now looks like this:

$$\begin{array}{rrrl} 89° & 59' & 60" & \text{Now each column} \\ -72° & 32' & 45" & \text{can be subtracted.} \\ \hline 17° & 27' & 15" & \text{is the final answer.} \end{array}$$

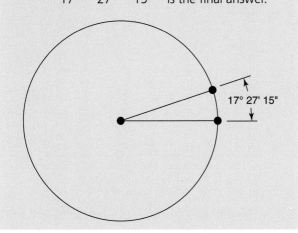

Cartesian Coordinates

In machining, parts and their features are often located using two- or three-dimensional **Cartesian coordinate systems**. Cartesian coordinate systems use X and Y or X, Y, and Z values for location. This is like graphing using X and Y or X, Y, and Z values. An origin is established at X0, Y0, Z0, and machining operations are located by moving in either positive or negative directions from that origin. In the two-dimensional (2D) system, the X and Y axes

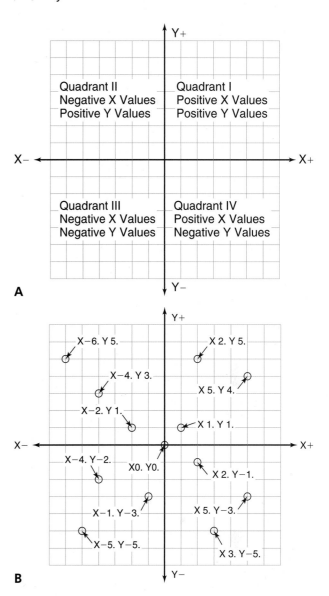

FIGURE 2.2.11 Examples of points in the 2D Cartesian coordinate system. Always record the points in X, Y format.

divide the system into four quadrants. **Figure 2.2.11** shows the sign (+ or −) of X and Y values in each quadrant and labels some coordinate locations.

Polar Coordinates

Another method of showing location in machining that is sometimes used is called the **Polar Coordinate System**. Instead of using X and Y distances, the polar coordinate system identifies location by an angle and a straight-line distance from the origin. The starting point for an angle is on the right side of the origin. **Figure 2.2.12** shows how a location is defined by polar coordinates.

Basic Trigonometry

Trigonometry can be defined as math using triangles. In machining, application is normally limited to right-angle

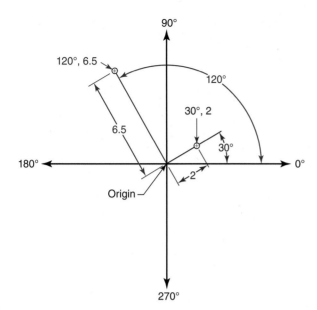

FIGURE 2.2.12 Polar coordinate positions are defined by an angle and a straight-line distance from the origin.

trigonometry, where one of the angles of the triangle is 90 degrees. Right-angle trigonometry has a wide variety of uses in the machining field, ranging from material layout to positioning on machine tools to high-precision inspection tasks.

Sum of Angles

The sum of the angles of every triangle is always 180 degrees. In right-angle trigonometry, since one angle is 90 degrees, the other two must add up to 90 degrees also. In some cases one of those other two angles is known, so the final angle can be found by subtracting the one non-90-degree angle from 90 degrees. This is just like finding a complementary angle.

EXAMPLE: What is the value of the angle shown at letter "B"?

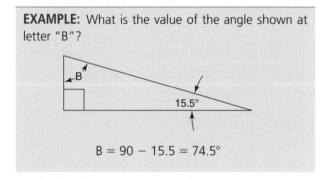

$$B = 90 - 15.5 = 74.5°$$

Pythagorean Theorem

The **Pythagorean theorem** is a formula that relates the lengths of the sides, or legs, of right triangles. It is written as $c^2 = a^2 + b^2$, where a, b, and c are the lengths of the triangle sides. With this formula, when any two sides are known, the third side can be found. **Figure 2.2.13** shows sides a, b, and c and angles named A, B, and C of a right triangle.

There are three different forms of the Pythagorean theorem frequently used depending on which side needs to be calculated.

EXAMPLE ONE: Find the length of side c when a and b are known.

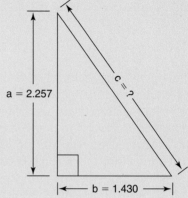

Begin with the original form of $c^2 = a^2 + b^2$.

Solving for c gives $c = \sqrt{a^2 + b^2}$. Then substitute and evaluate.

$$c = \sqrt{a^2 + b^2}$$
$$c = \sqrt{2.257^2 + 1.430^2}$$
$$c = \sqrt{5.094049 + 2.0449}$$
$$c = \sqrt{7.138949}$$
$$c \approx 2.6719$$

EXAMPLE TWO: Find the length of side a when b and c are known.

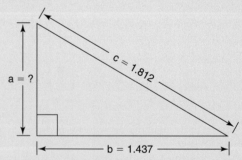

Begin with the original form of $c^2 = a^2 + b^2$.

Solving for a gives $a = \sqrt{c^2 - b^2}$. Then substitute and evaluate.

$$a = \sqrt{c^2 - b^2}$$
$$a = \sqrt{1.812^2 - 1.437^2}$$
$$a = \sqrt{3.283344 - 2.064969}$$
$$a = \sqrt{1.218375}$$
$$a = 1.1038$$

Solving for b is very similar to solving for side a. The variables a and b are switched, resulting in the new form of $b = \sqrt{c^2 - a^2}$.

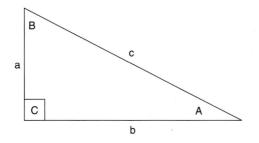

FIGURE 2.2.13 Sides *a, b,* and *c* and angles A, B, and C of a right triangle.

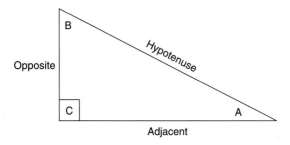

FIGURE 2.2.14 The names for the sides of a right triangle related to angle A.

Sides and Angles of a Right Triangle

Recall that the angles of the right triangle shown in **Figure 2.2.14** are also labeled. In right-angle trigonometry, angle C is always 90 degrees and side *c* is called the **hypotenuse**. The hypotenuse can be labeled as *Hyp* and is the longest side of a right triangle. The side across from any given angle is called the **opposite side**. It can be labeled as *Opp.* The side next to a given angle that is not the hypotenuse is called the **adjacent side**. It can be labeled as *Adj.* **Figure 2.2.15** shows the labeling of the hypotenuse, opposite, and adjacent sides of a right triangle related to angle A.

Trigonometric Functions Definitions: Sine, Cosine, and Tangent

When performing calculations of right triangles, three trigonometric functions are also often used: sine, cosine, and tangent. Each of these values is a ratio between the lengths of two sides of the right triangle. These ratios can be remembered using the phrase SOHCAHTOA:

S is for **sine** (usually abbreviated as *sin*) and is a ratio between the lengths of the opposite side (O) and the hypotenuse (H):

$$Sine = \frac{Opposite}{Hypotenuse}$$

C is for **cosine** (usually abbreviated as *cos*) and is a ratio between the lengths of the adjacent side (A) and the hypotenuse (H):

$$Cosine = \frac{Adjacent}{Hypotenuse}$$

T is for tangent (usually abbreviated as *tan*) and is a ratio between the lengths of the opposite side (O) and the adjacent side (A):

$$Tangent = \frac{Opposite}{Adjacent}$$

Many reference resources contain tables that show the sine, cosine, and tangent ratios using the letter labels of a, b, and c for the side lengths. It also shows different forms of the sine, cosine, and tangent formulas as well as different forms of the Pythagorean theorem. **(See Figure 2.2.16.)** Refer to this table while reading the following examples of the uses of sine, cosine, and tangent.

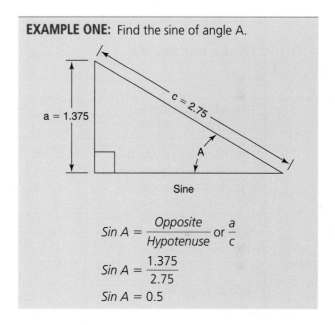

EXAMPLE ONE: Find the sine of angle A.

$$Sin\ A = \frac{Opposite}{Hypotenuse}\ or\ \frac{a}{c}$$

$$Sin\ A = \frac{1.375}{2.75}$$

$$Sin\ A = 0.5$$

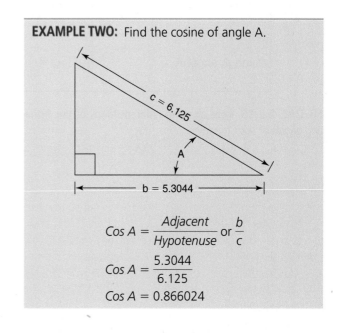

EXAMPLE TWO: Find the cosine of angle A.

$$Cos\ A = \frac{Adjacent}{Hypotenuse}\ or\ \frac{b}{c}$$

$$Cos\ A = \frac{5.3044}{6.125}$$

$$Cos\ A = 0.866024$$

Solution of Right-Angled Triangles

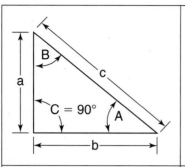

As shown in the illustration, the sides of the right-angled triangle are designated *a* and *b* and the hypotenuse, *c*. The angles opposite each of these sides are designated A and B, respectively.

Angle C, opposite the hypotenuse *c* is the right angle, and is therefore always one of the known quantities.

Sides and Angles Known	Formulas for Sides and Angles to be Found		
Sides *a*; side *b*	$c = \sqrt{a^2 + b^2}$	$\tan A = \dfrac{a}{b}$	$B = 90° - A$
Sides *a*; hypotenuse *c*	$b = \sqrt{c^2 - a^2}$	$\sin A = \dfrac{a}{c}$	$B = 90° - A$
Sides *b*; hypotenuse *c*	$a = \sqrt{c^2 - b^2}$	$\sin B = \dfrac{b}{c}$	$A = 90° - B$
Hypotenuse *c*; angle B	$b = c \times \sin B$	$a = c \times \cos B$	$A = 90° - B$
Hypotenuse *c*; angle A	$b = c \times \cos A$	$a = c \times \sin A$	$B = 90° - A$
Side *b*; angle B	$c = \dfrac{b}{\sin B}$	$a = b \times \cot B$	$A = 90° - B$
Side *b*; angle A	$c = \dfrac{b}{\cos A}$	$a = b \times \tan A$	$B = 90° - A$
Side *a*; angle B	$c = \dfrac{a}{\cos B}$	$b = a \times \tan B$	$A = 90° - B$
Side *a*; angle A	$c = \dfrac{a}{\sin A}$	$b = a \times \cot A$	$B = 90° - A$

FIGURE 2.2.15 Table showing different forms of the Pythagorean theorem and several formulas used to calculate side lengths and angles of right triangles.

EXAMPLE THREE: Find the tangent of angle *A*.

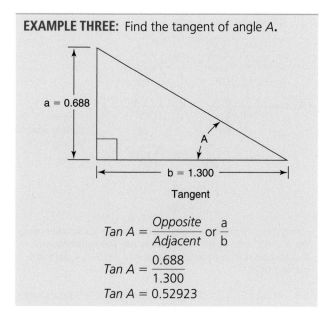

Tangent

$$Tan\ A = \frac{Opposite}{Adjacent}\ or\ \frac{a}{b}$$

$$Tan\ A = \frac{0.688}{1.300}$$

$$Tan\ A = 0.52923$$

Using Sine, Cosine, and Tangent Values to Determine an Angle

When a sine, cosine, or tangent value is calculated as shown in the previous examples, it can then be used to determine the measurement of a specific angle in degrees. A calculator can find the angle by using the 2nd or INV key with either the *sin, cos,* or *tan* key. Check calculator manuals to see which keystroke order is required for any specific model. The result will be a decimal angle. Some calculators can convert that decimal angle to degrees, minutes, and seconds and display it with one keystroke. If that option is not available, perform conversions as shown under the topic heading of Angles earlier in this unit.

The sine of angle *A* in *Example One* above was 0.5. Using the 2nd and *sin* keys on a calculator shows that the decimal angle is exactly 30 degrees. No conversion is needed.

The cosine of angle *A* in *Example Two* was 0.866024. Using the 2nd and *cos* keys on a calculator shows that the decimal angle is 30.0001 degrees. The small variation from 30 degrees results from a slight rounding during calculations, so it is safe to say that 30 degrees is the actual angle.

The tangent of angle *A* in *Example Three* was 0.52923. Using the 2nd and *tan* keys on a calculator shows that the decimal angle is 27.88914 degrees. Conversion defines this angle as 27°53'20.9". Slight variation may result from rounding during calculations. The required degree of accuracy needs to be considered when performing calculations, but variations smaller than a few seconds will rarely have any effect on applications.

More Right-Triangle Uses of Sine, Cosine, and Tangent Functions

The sine, cosine, and tangent formulas can also be rearranged to solve for different unknown values. If any one side length and angle of a right triangle are known, the other side lengths and angles can be calculated. A calculator can quickly evaluate the sine, cosine, or tangent value for the given angle. Again, check each calculator's manual to determine the correct order of keystrokes.

EXAMPLE ONE: Find the length of side *a* of the following triangle.

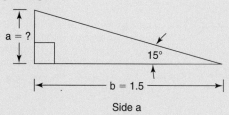

Side a

First select the function formula that uses the known side, side *b* (the adjacent side) and the unknown side, side *a* (the opposite side). Tangent must be used.

$$Tan\ A = \frac{a}{b}$$

Solve for the unknown variable, side *a*.

$$a = b \cdot Tan\ A$$

Substitute and evaluate. Side *b* = 1.5, and a calculator shows the tangent of 15 degrees to be 0.267949.

$$a = 1.5 \cdot Tan\ 15°$$

$$a = 1.5 \cdot 0.267949$$

$$a = 0.40192$$

EXAMPLE TWO: Find the length of side *b* of the following triangle.

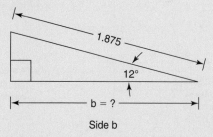

Side b

Select the formula that uses the known side, side *c* (hypotenuse) and the unknown side, side *b* (adjacent side). Cosine must be used.

$$Cos\ A = \frac{b}{c}$$

Solve for the unknown variable side, *b*.

$$c \cdot Cos\ A = b$$

$$b = c \cdot Cos\ A$$

Substitute and evaluate. Side *c* = 1.875, and a calculator shows the cosine of 12 degrees to be 0.97815.

$$b = 1.875 \cdot Cos\ 12°$$

$$b = 1.875 \cdot 0.97815$$

$$b = 1.834$$

Sample Application

Here is a real-world scenario showing application of several of the concepts from this unit.

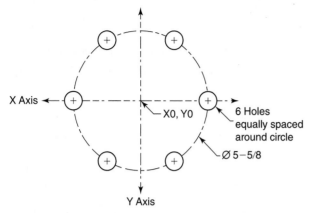

The 6 holes are equally spaced around a 5−5/8 diameter circle that is centered on the origin (X0, Y0) of a Cartesian coordinate plane.

Since the center of the hole on the far left is located directly on the X-axis, its Y value is 0.

The distance from the origin to the center of the hole is equal to the radius of the 5−5/8 diameter circle.

Converting 5−5/8 to a decimal using a chart gives 5.625.

Dividing that by 2 gives a radius value of 2.8125. The X coordinate of that hole is X −2.8125.

So that hole has coordinates of X −2.8125, Y0

The hole on the right side is the same distance from the origin, but has a positive X value, so its coordinates are X 2.8125, Y0

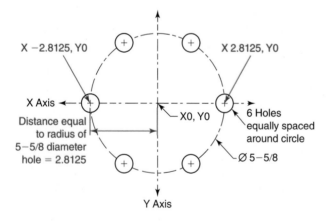

Since there are 360 degrees in a circle and the holes are equally spaced, divide 360 by 6. This shows that there are 60 degrees between each of holes.

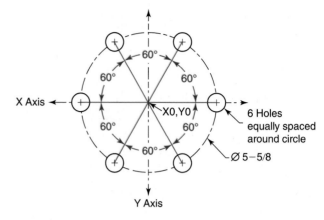

Using that information, a right triangle can be constructed using the X and Y axes as the 90 degree angle. The hypotenuse of that triangle is also equal to the radius of the 5−5/8 diameter circle, 2.8125.

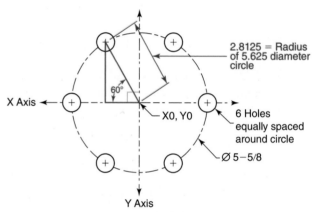

With that triangle, the X and Y coordinates can be found for solving for the opposite and adjacent sides using the cosine and sine formulas.

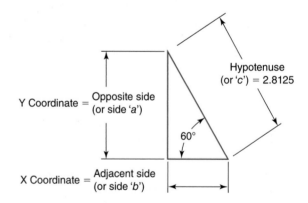

Use Cosine to find the adjacent side of the triangle (side *b*). This will be the X coordinate.

$$\text{Cosine} = \frac{Adjacent}{Hypotenuse} \text{ or Cosine} = \frac{b}{c}$$

$$\text{Cos } 60 = \frac{b}{2.8125}$$

The cosine of 60 is .5, so

$$.5 = \frac{b}{2.8125}$$

$$a = (.5 \cdot 2.8125)$$

$$a = 1.40625$$

So the X coordinate for the hole is X −1.40625

Use Sine to find the opposite side of the triangle (side *a*). This will be the Y coordinate.

$$Sine = \frac{Opposite}{Hypotenuse} \quad \text{or} \quad Sine = \frac{a}{c}$$

$$Sine\ 60 = \frac{a}{2.8125}$$

The sine of 60 is .86603, so

$$.86603 = \frac{a}{2.8125}$$

$$a = (.86603 \cdot 2.8125)$$

$$a = 2.4357$$

So the Y coordinate for the hole is Y 2.4357

Since the holes are equally spaced, the other coordinates values are the same, but because they are in different quadrants of the Cartesian plane, their signs (+ and −) are different. Coordinate locations have now been determined for all six hole locations.

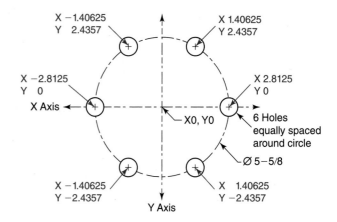

SUMMARY

- The machining field requires the use of both the English-and metric-based measurement systems and performing conversions between both depending on print specifications and types of measuring tools used.

- Machining careers also require application of many math concepts. The most commonly used math concepts include fractional and decimal operations and conversions between the two and the use of ratios and proportions.

- Machining professionals must also be able to perform basic algebra such as using the order of operations when using machining formulas to solve for one variable.

- An understanding of geometry, including parallelism, perpendicularity, angles, and properties of circles, is an important skill in the machining field as well.

- The machining industry frequently requires the use of Cartesian and polar coordinate systems.

- Application of basic right-angle trigonometry, including use of the Pythagorean theorem and sine, cosine, and tangent functions, is often required in machining careers.

- Many of these concepts will become more familiar with practice and repeated use.

REVIEW QUESTIONS

1. What is the inch equivalent of 1 millimeter?
2. What is the inch equivalent of 32.5 mm to the nearest 0.0001"?
3. Use the decimal equivalent chart on page 63 to find the fraction nearest 0.267".
4. Use the decimal equivalent chart on page 63 to find the fraction nearest 0.69".
5. Use the decimal equivalent chart on page 63 to find the fraction nearest 0.345".
6. What does parallel mean?
7. What does perpendicular mean?
8. What is the radius of a 3.65-inch-diameter circle?
9. What is the circumference of the circle in the previous question?

10. What are the Cartesian coordinates of the four points shown at A, B, C, and D?

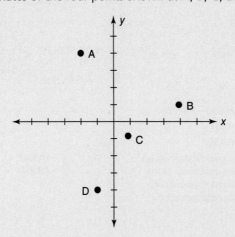

11. Label the hypotenuse, adjacent side, and opposite side of the given triangle as related to angle A.

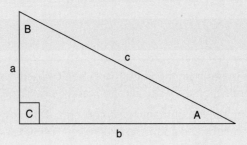

Semi-Precision Measurement   UNIT 3

Learning Objectives

After completing this unit, the student should be able to:

- Define comparative measurement
- Demonstrate understanding of care of common semi-precision measuring instruments
- Read an English rule to within 1/64 of an inch
- Read an English (decimal) rule to within 1/100 of an inch
- Read a metric rule within 0.5 mm
- Identify and explain the uses of semi-precision calipers
- Identify and explain the uses of squares
- Identify and explain the uses of the combination set
- Identify and explain the uses of protractors
- Read protractors within 1 degree
- Identify and explain the uses of common semi-precision fixed gages

Key Terms

Adjustable square	Die maker's square	Screw pitch gage
Angle gage	Dimension	Semi-precision
Caliper	Fixed gage	measurement
Combination set	Graduation	Square
Comparison or	Protractor	Tolerance
comparative	Radius and fillet gage	Transfer or helper-
measurement	Rule	type measuring tool

INTRODUCTION

Performing machining operations is only one part of a career in the machining field. Measurement is another core duty that is performed many times every day in the industry. During machining, **dimensions**, or specified print sizes, must be measured to be sure they are within required specifications. Every dimension shown on a print is given a certain amount of allowable variation. That allowable variation is called a **tolerance**. A tolerance gives a size range that is acceptable. Why do tolerances exist? They are an engineer's or designer's way of saying "so long as this print dimension is within this range, the machined part will function correctly." Tolerances will be discussed in detail in the Job Planning, Benchwork, and Layout section of this text. For now, it is important to know that they exist and that the size of a tolerance will determine what type of measuring tool will be needed to measure a dimension.

Keep in mind that tolerances are key in determining whether a machined component is acceptable and usable. A part may have many dimensions, and even one dimension that is not within its specified tolerance renders the part unacceptable and useless. What might seem like one meaningless small error will cause the part to be rejected, or scrapped. In the machining field, parts that do not meet print specifications are called rejects or scrap. Scrap can have far-reaching consequences because it causes lost time, efficiency, and money. It affects the efficiency and profitability of a company, and your paycheck too!

WHAT IS SEMI-PRECISION MEASUREMENT?

Semi-precision measurement usually refers to measurement when results do not need to be more precise than 1/64" or 1/100", .5 mm, or 1 degree. This unit will explore several types of semi-precision measuring instruments or tools. Always remember to treat these tools with care so as not to cause damage that will affect their accuracy and performance. Do not pile measuring tools on top of each other or drop them, and keep them away from moving parts of machinery and metal chips. Use measuring tools only for their intended purposes. Unintended use and abuse can cause damage and make tools unreliable or worthless. When not in use, measuring tools should be cleaned and gently stored in safe locations away from moisture to prevent rust. A light application of appropriate lightweight lubricating oil can help prevent rust.

RULES

The **rule** is by far the most common semi-precision measuring tool utilized in the machining field. It is a flat piece of steel with graduations that divide inches or millimeters into fractional parts. Rules are often incorrectly called scales. Everyone working in the industry needs to be able to read and use different types of rules. They are available in many different styles, widths, and lengths to suit individual needs. **Figure 2.3.1** shows some different types of rules. A hook rule is helpful when

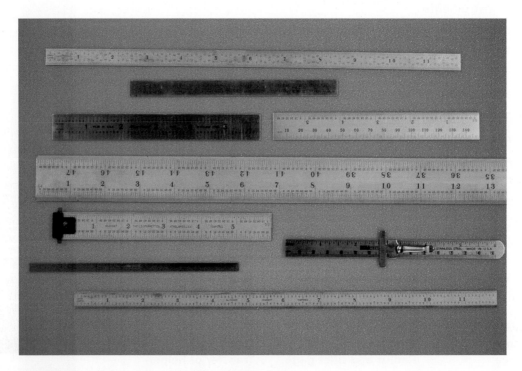

FIGURE 2.3.1 A few of the many different styles of rules.

FIGURE 2.3.2 A hook rule makes it easier to measure from an edge.

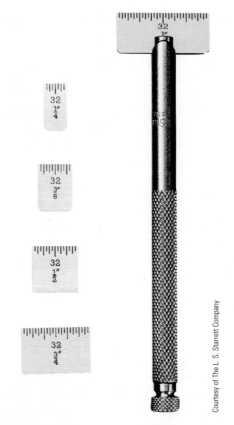

Courtesy of The L. S. Starrett Company

FIGURE 2.3.3 Small rules can be useful in tight spaces.

measuring from the edge of a part (**Figure 2.3.2**). Small rules (**Figure 2.3.3**) can be used to measure in confined spaces. **Graduations** are the divisions or spaces on a rule. Rules are available with English and metric graduations. Some include both inch and millimeter graduations.

Rules can be used to measure linear (straight-line) distances, as shown in **Figure 2.3.4**. When you use rules, measurements are more easily taken when the rule is placed with its edge on the part being measured instead of laying it flat. When the rule is flat, it is more difficult to align and read graduations. **(See Figure 2.3.5.)** Aligning a major graduation on a part's edge also results in more accuracy than attempting to place the end of the rule flush with the end of the part. When using this method, be sure to subtract that major graduation from the overall reading. For example, if aligning the 1mk, graduation, be sure to subtract 1" from the reading.

(See Figure 2.3.6.) It is also important to keep the rule parallel with the dimension being measured, or errors will result. **(See Figure 2.3.7.)**

Reading English Rules

There are many styles of graduations for English rules. The most common graduation style divides inches into 1/8 (one-eighth), 1/16 (one-sixteenth), 1/32 (one-thirty-second), and

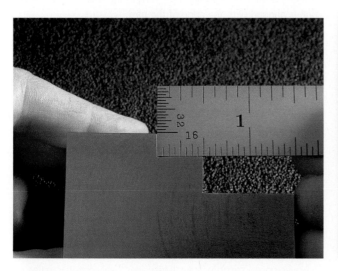

FIGURE 2.3.4 Rules being used to measure linear distances.

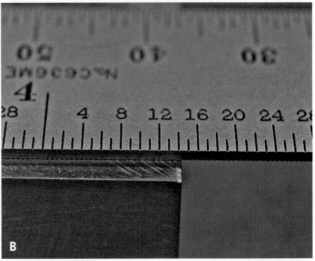

FIGURE 2.3.5 (A) By holding the edge of the rule against the surface being measured, (B) graduations are more easily read than when the rule is lying flat.

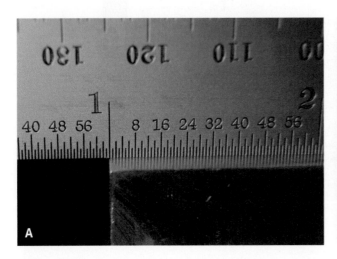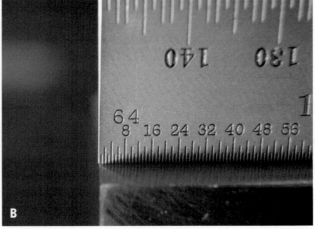

FIGURE 2.3.6 (A) Aligning a major graduation with an edge is easier than (B) aligning the end of the rule. In this example, the actual measurement would be 1 " less than the reading since the 1 " graduation is aligned with the edge of the part.

1/64 (one-sixty-fourth) divisions. When you read a rule with these graduations, measurements should always be reduced to lowest terms. Most of these rules have what are called quick-reading graduations. Quick-reading graduations label some of the 1/32 and 1/64 graduations so the user does not need to count every graduation. **Figure 2.3.8** shows some sample measurements using rules with these fractional graduations.

Another frequently used graduation type is one that divides inches into 1/10 (one-tenth), 1/50 (one-fiftieth), and 1/100 (one-one-hundredth) divisions. This type of rule is usually used when a decimal value is desired and

readings are normally expressed in decimal form. For that reason, they are sometimes called decimal rules. Decimal rules also normally have quick-reading graduations to make reading easier. **Figure 2.3.9** shows some sample measurements using rules with these decimal graduations.

Reading Metric Rules

Metric rules are graduated in millimeter and one-half (0.5) millimeter divisions. Quick-reading numbers every 5 or 10 millimeters simplify measurement. **Figure 2.3.10** shows some sample measurements using metric rules.

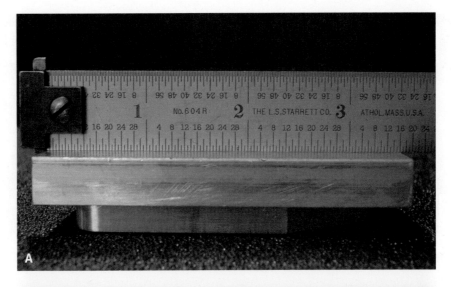

FIGURE 2.3.7 (A) Hold a rule parallel to the dimension being measured. (B) Because of the angle of the rule, a different, incorrect, measurement will result.

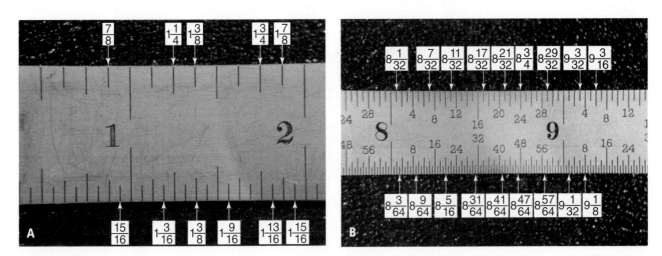

FIGURE 2.3.8 (A) Some sample measurements using 1/8 and 1/16 graduations and (B) 1/32 and 1/64 gradations. Notice the quick-reading graduations. Remember to always reduce fractional readings to lowest terms.

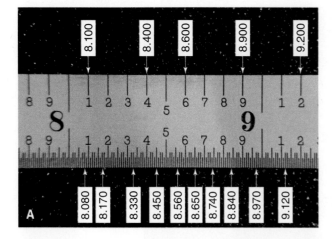 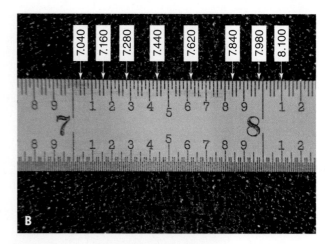

FIGURE 2.3.9 (A) Some sample measurements using 10th and 100th graduations and (B) 50th graduations.

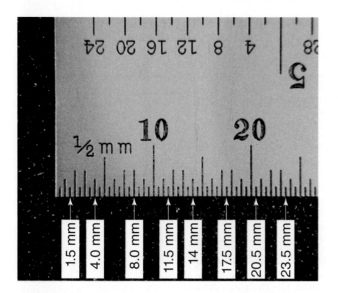

FIGURE 2.3.10 Some sample metric rule measurements.

CALIPERS

Calipers used in semi-precision measurement have two legs that make contact with part surfaces to obtain measurements. One type of caliper that is very similar to the rule is the slide caliper. It can be used to measure external or internal dimensions. Readings are made where a reference line on the moveable jaw aligns with a graduation on the scale. **Figure 2.3.11** shows an example of a slide caliper.

External dimensions can also be measured with another type of caliper called an outside caliper. An inside caliper can be used to measure internal dimensions. **Figure 2.3.12** shows both an outside caliper and an inside caliper. These types of calipers are especially useful when measuring diameters. **Figure 2.3.13** shows the use of an outside caliper, and **Figure 2.3.14** shows the use of an internal caliper. Both external and internal calipers are

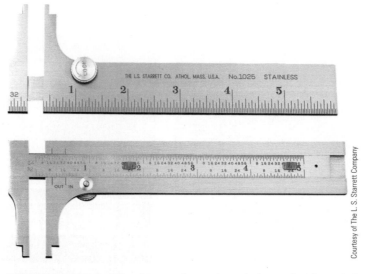

Courtesy of The L. S. Starrett Company

FIGURE 2.3.11 A slide caliper can be used to take internal and external measurements. Notice that a different reference line is used for external and internal measurements.

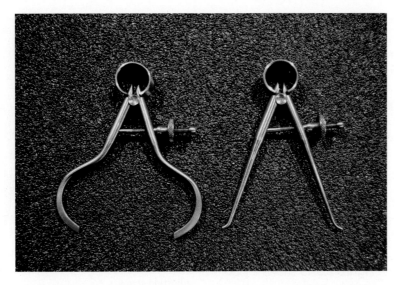

FIGURE 2.3.12 An outside caliper and an inside caliper. Notice the difference in the shape of the legs.

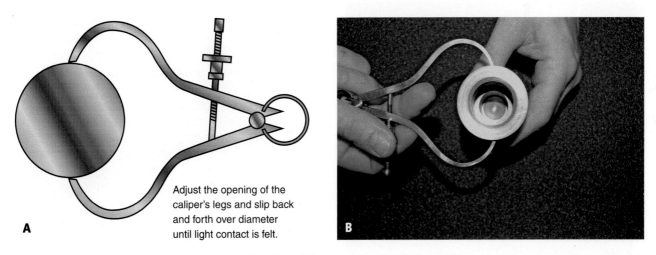

Adjust the opening of the caliper's legs and slip back and forth over diameter until light contact is felt.

A

B

FIGURE 2.3.13 (A) Method for setting an outside caliper. (B) An outside caliper being used to measure an external diameter.

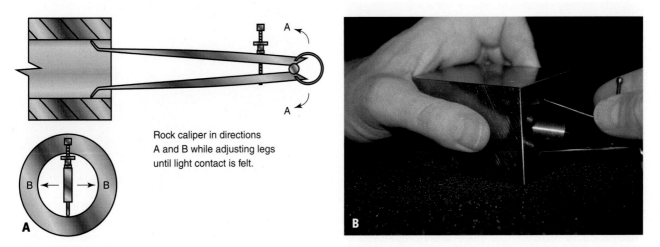

Rock caliper in directions A and B while adjusting legs until light contact is felt.

B ⟷ B

A

B

FIGURE 2.3.14 (A) Method for setting an internal caliper. (B) An internal caliper being used to measure a hole diameter.

adjusted so their legs lightly contact part surfaces. Some practice is needed to obtain the proper contact pressure. Then the measurement must be transferred to and measured with a rule, as shown in **Figure 2.3.15** and **Figure 2.3.16**. That is why these are called **transfer** or **helper-type measuring tools**.

Outside and inside calipers are available in spring-type, firm-joint, and lock-joint styles. The spring-type caliper's legs are contained by a spring and adjustments are made with a screw. **(See Figure 2.3.17.)** The legs of the firm-joint caliper are together and pivot at the joint, as shown in **Figure 2.3.18**. The legs of the lock-joint type move freely and are locked after adjusting by tightening the nut at the joint. A lock-joint caliper may also have a fine adjustment to slightly move the legs after the caliper is set near the desired size. **Figure 2.3.19** shows a lock-joint outside caliper with a fine adjustment.

FIGURE 2.3.17 The jaws of the spring-type caliper are held under tension by a spring and opened or closed by turning the adjusting screw.

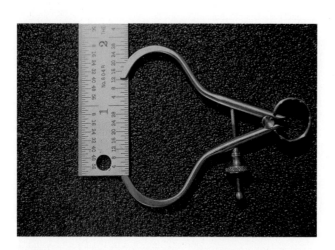

FIGURE 2.3.15 Transferring an outside caliper measurement to a rule.

FIGURE 2.3.18 The firm-joint caliper has no fine adjustment.

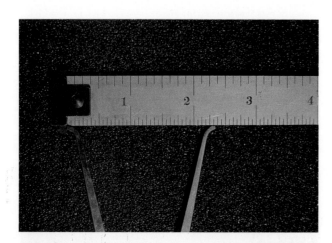

FIGURE 2.3.16 Transferring an inside caliper measurement to a hook rule.

FIGURE 2.3.19 The lock-joint caliper can be set to an approximate size and locked in place by turning the large finger screw lock clockwise. Then the smaller fine-adjustment finger screw can be used to obtain the final setting.

ADJUSTABLE SQUARES

Adjustable squares are made in two pieces and can easily be adjusted and disassembled. The beam has a clamping mechanism that holds the blade in place when tightened. Blades are available with graduations like rules and may have commonly used angles on their ends. **Figure 2.3.20** shows an example of an adjustable square.

Squares can be used to check for perpendicularity. This is often called "checking for squareness" or "checking for square." Making checks such as these are **comparison** or **comparative measurements** because no actual numerical measurement is taken. Part surfaces are *compared* to the 90-degree corner of the square. **(See Figure 2.3.21.)** With a graduated blade, an adjustable

FIGURE 2.3.20 An adjustable square with different blades.

FIGURE 2.3.21 Using an adjustable square to "check for square."

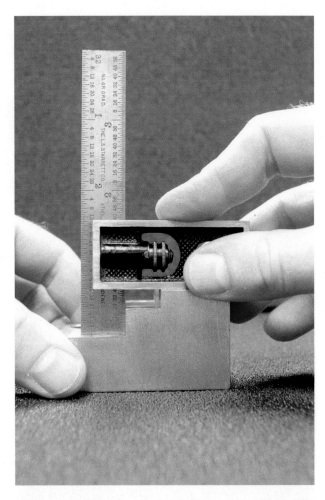

FIGURE 2.3.22 Measuring depth with an adjustable square. The blade can be locked in place after positioning by tightening the clamping screw. The reading is then taken where the blade intersects the beam.

square can also be used to measure depths or heights, as shown in **Figure 2.3.22**.

The Combination Set

The **combination set** consists of a blade, square head, center head, and protractor head. **(See Figure 2.3.23.)** The different heads are mounted to the blade by tightening a clamping screw.

When the square head is mounted to the blade, the tool is called a combination square. This tool can be used to make comparative measurements of 90-degree and 45-degree angles, as shown in **Figure 2.3.24**. It can also be used to check heights or depths like other adjustable squares with graduated blades.

The center of round material can be found with the use of the center head and blade, as shown in **Figure 2.3.25**.

The protractor head is explained under the topic heading of Angular Measurement.

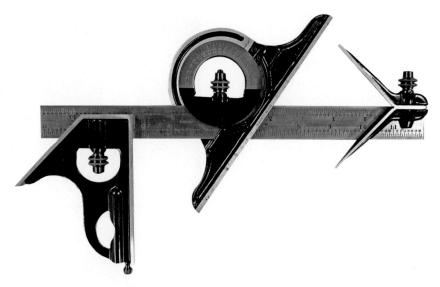

FIGURE 2.3.23 The combination set.

FIGURE 2.3.24 Making a comparative measurement of a 45-degree angle with a combination square. Hold one surface tightly against the square head and check for gaps between the blade and the second surface.

FIGURE 2.3.25 Finding the center of a round piece of material using the center head and blade of the combination set.

ANGULAR MEASUREMENT

Protractors

A plain **protractor** is used to measure angles and has two sets of one-degree graduations to allow measurement from either side of the protractor. The arm rotates and can be locked in place by tightening the screw. **Figure 2.3.26** shows a plain protractor in use. A protractor depth gage adds a graduated rule and can be used for measuring depths as well as angles. Depths are measured as with an adjustable square when the rule is set at 90 degrees. **Figure 2.3.27** shows a protractor depth gage.

Angular measurements can also be made when the protractor head of the combination set is mounted to the blade. **(See Figure 2.3.28.)** It also has two sets of 1-degree graduations.

Figure 2.3.29 shows some sample protractor readings using a plain protractor and the protractor from the combination set.

Bevels

Universal and combination bevels (**Figure 2.3.30**) can only be used to compare angles or bevels, because they have no graduations. The locking screws are loosened, the blades are positioned, and then they are locked in place by retightening the screws.

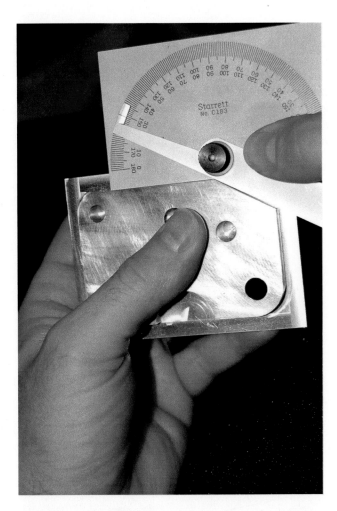

FIGURE 2.3.26 Measuring an angle with a plain protractor. The arm can be locked in place by the finger screw after positioning.

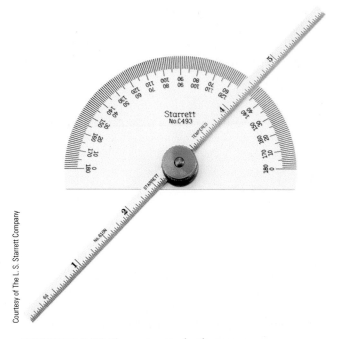

Courtesy of The L. S. Starrett Company

FIGURE 2.3.27 The protractor depth gage.

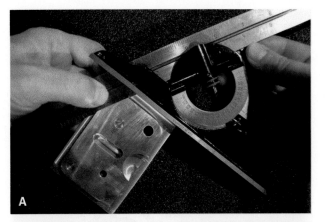

FIGURE 2.3.28 (A) Making an angular measurement with the protractor head and blade of the combination set. (B) After positioning, two locking screws on the back of the protractor head will prevent it from rotating.

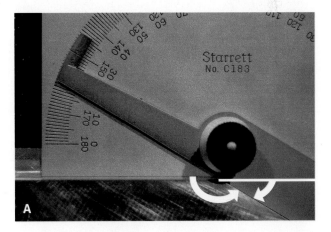

FIGURE 2.3.29 (A) Some sample angular measurements using protractors. In all cases, there are two possible solutions. Print specifications will dictate which angle is to be measured. (A) The head of the plain protractor is placed on one surface and serves as the "0" or starting point for taking angular measurements. The arm was rotated clockwise to contact the second surface. Then the reading is taken where the mark on the end of the arm aligns with the angular scale. The left arrow shows that the angle between the two surfaces is 150°. The right arrow shows 30°, the supplement of 150°. That is how far the arm was rotated.

FIGURE 2.3.29 *Continued.* (B) The blade of the combination set is placed on one surface and serves as the "0" or starting point for taking angular measurements. The protractor head was then rotated to make contact with the second surface. The reading is taken where the "0" mark on the protractor head aligns with the angular scale. The right arrow shows that the angle between the two surfaces is 60°. That is how far the head was rotated. The left arrow shows 120°, the supplement of 60°.

FIGURE 2.3.30 (A) A universal bevel, and (B) a combination bevel.

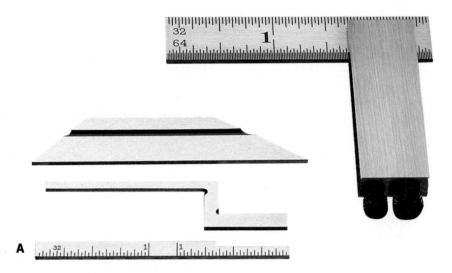

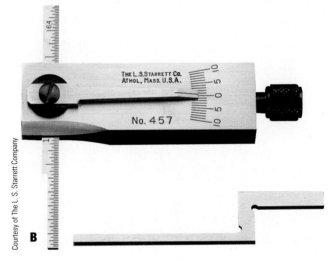

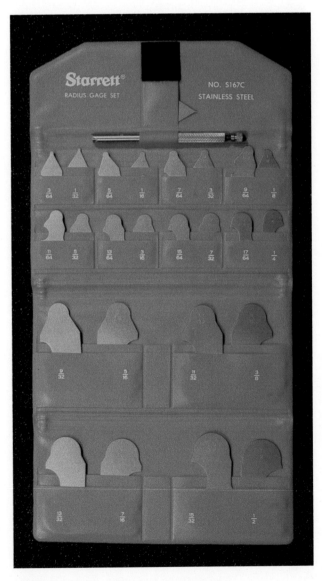

Courtesy of The L. S. Starrett Company

FIGURE 2.3.31 Two types of die maker's squares. They can be used to check perpendicularity and angles up to 10°.

Die Maker's Square

A **die maker's square** is a hybrid between a square and a protractor. It can be used like a standard adjustable square, but its blade can also be tilted by turning an adjusting screw to make comparative measurements of angles up to 10 degrees. **(See Figure 2.3.31.)**

FIXED GAGES

Fixed gages are tools without adjustment that are used for comparative measurement of parts to a particular size. The gage is a certain size, and the size of the part is compared to the size of the gage.

Radius and Fillet Gages

Radius and fillet gages are used to check outside corner and insider corner (fillet) radii. A set is shown in **Figure 2.3.32**. **Figure 2.3.33** shows how radius and fillet gages can be used.

FIGURE 2.3.32 A radius and fillet gage set.

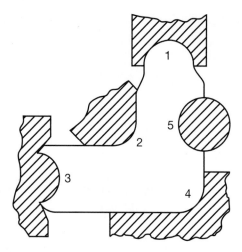

FIGURE 2.3.33 Uses of radius and fillet gages.

Angle Gages

Angle gages (**Figure 2.3.34**) can compare part angles to standard angles similar to the way radius and fillet gages are used.

Screw Pitch Gage

A **screw pitch gage** determines the distance between threads. Each leaf is used for a different size. They are available in inch and metric versions. **Figure 2.3.35** shows a typical screw pitch gage and how it is used.

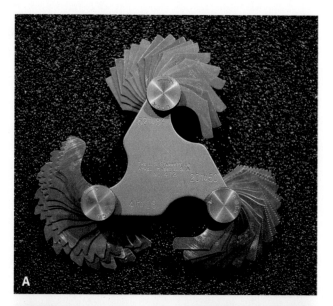

FIGURE 2.3.35 (A) An example of a screw pitch gage. (B) Checking threads with a screw pitch gage.

FIGURE 2.3.34 A set of angle gages.

SUMMARY

- Semi-precision measurement is a daily activity in machining, so understanding the use and care of tools needed to perform those measurements is important for success in the field.

- Rules are one of the most commonly used measuring tools and come in a variety of styles and lengths. They are used to make semi-precision linear measurements within 1/64 or 1/100 of an inch or 0.5 millimeter.

- Different styles of protractors are frequently used for obtaining semi-precision angular measurement with an accuracy of one degree.

- Adjustable squares are constructed of two parts, the beam and the blade, and can be used to check for perpendicularity between two sides of a workpiece.

- The combination set is a versatile tool that can be used to measure lengths, heights, depths, and angles and to find the center of round parts.

- Transfer tools are set to contact surfaces of a part feature, and then measured with a rule to determine size.

- Fixed gages such as radius, fillet, and angle gages are often used to perform comparison measurements to known sizes.

- Becoming familiar with the use and care of these measuring tools is an important step in creating a strong foundation of machining skills.

REVIEW QUESTIONS

1. Define *semi-precision measurement*.
2. What is comparison measurement?
3. List three rules to follow when using or storing semi-precision measuring tools.
4. List the parts of the combination set.
5. Describe four uses of the combination set.
6. What semi-precision tool is used to take angular measurements?
7. What type of square has a blade that can be tilted to angular settings?
8. Identify the following tools.

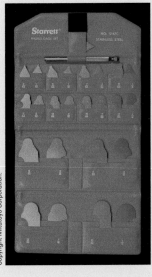

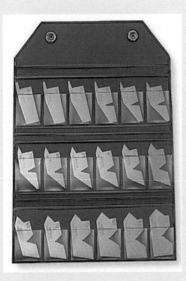

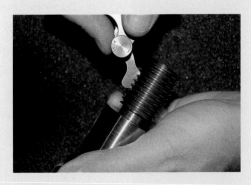

Copyright Mitutoyo Corporation.

UNIT 4   Precision Measurement

Learning Objectives

After completing this unit, the student should be able to:

- Explain the care of precision measuring tools
- Define calibration and explain its purpose
- Identify and explain the use of common precision fixed gages
- Explain the principle of the micrometer
- Identify the parts of an outside micrometer
- Describe the process of micrometer calibration
- Identify and describe uses of micrometer-type measuring tools
- Read an English micrometer
- Read a metric micrometer
- Identify and describe uses of vernier measuring tools
- Read English vernier scales
- Read metric vernier scales
- Read a vernier bevel protractor
- Identify and explain uses of precision transfer-type measuring instruments
- Identify features of dial indicators and explain their uses
- Explain the purpose of a surface plate
- Identify gage blocks and their uses, and calculate gage block builds
- Identify and explain the uses of simple and compound sine tools
- Discuss methods for measuring surface finishes

- Identify and discuss the use of a toolmaker's microscope
- Identify and discuss the use of an optical comparator

Key Terms

Adjustable parallels	Optical comparator	Surface plate
Calibration	Pin or plug gage	Telescoping gage
Coordinate Measuring Machine (CMM)	Precision fixed gage	Thickness or feeler gage
Depth micrometer	Precision measurement	Toolmaker's microscope
Dial indicator	Profilometer	Transfer or helper-type measuring tool
Digital indicator	Ring gage	Vernier
Gage block	Sine tool	Vernier bevel protractor
Go/no-go plug gage	Small hole gage	Vernier caliper
Height gage	Snap gage	
Micrometer	Solid square	
	Straight edge	

INTRODUCTION

Precision measurement refers to measurement when desired results must be more precise than 1/64" or 1/100", .5 mm, or 1 degree. When tolerances are in these ranges, semi precision tools are not adequate to measure dimensions, so a new family of measuring tools is required. Remember the importance of meeting tolerances and the consequences of not meeting tolerances. Precision measurement helps those in machining careers meet very small (called close or tight) tolerances. That directly aids in reducing scrap or rejected parts. It also increases efficiency and profitability for companies and their employees.

WHAT IS PRECISION MEASUREMENT?

When performing precision measuring using English measurements (inch-based), expected accuracy is generally between 0.001 and 0.0001 inches. In SI (metric-based), expected accuracy is generally between 0.01 and 0.002 mm. Expected angular accuracy is generally within 5 minutes (1/12) of a degree. The idea of such small increments can be difficult to understand. An ordinary piece of paper or an average human hair is about 0.003 inch or 0.1 mm in *thickness*. Precision measurement is literally splitting hairs, as 0.001" or 0.025 mm is about 1/3 the diameter of a human hair and 0.0001" or 0.002 mm is about 1/30 the diameter of a human hair.

Since precision is common in the machining field, the language of precision dimensions is a bit different from everyday language. Three decimal places are the norm, so sizes are called thousandths of an inch or just "thousandths." Because thousandths are the standard, accepted, unit in inch-based precision measurement, frequently the word *thousandths* is not even stated. When numbers go to the fourth decimal place (ten-thousandths), they are simply stated as "tenths" because thousandths are understood. Even when dimensions are only in one or two decimal places, they are normally referred to as thousandths by zeroes being added mentally to reach the third decimal place. **Figure 2.4.1** gives a

Examples of Sizes Using the Machining Language	
Size/Dimension	**Spoken as:**
1.245	One inch, two hundred forty-five thousandths or *just* one inch, two-forty-five
0.137	One hundred thirty-seven thousandths or *just* one-thirty-seven
0.7588	Seven hundred ffty-eight and eight tenths
3.25	Three inches, two hundred fifty thousandths or *just* three inches, two-fifty
0.1	One hundred thousandths or *just* one hundred
0.32	Three hundred twenty thousandths or *just* three-twenty
0.2204	Two hundred twenty and four tenths

FIGURE 2.4.1 Sizes in the language of machining.

few examples of how sizes are spoken using machining language.

GENERAL CARE AND USE OF PRECISION TOOLS

Precision measuring tools are very delicate and expensive. Great care must be taken during use and storage. Dropping precision tools will greatly affect their accuracy and will frequently damage them beyond repair. Even small, seemingly harmless bumps can cause them to not function properly. Keep precision tools away from dust, grit, and moving machine parts during use. It is good practice to store precision tools away from dust, dirt, and moisture and to store them so that they do not touch other tools. When provided, use cases in which to store precision tools and keep them clean and properly lubricated to prevent rust and to keep them in smooth working order. Precision tools treated with respect and care will provide a lifetime of reliable service, while one moment of carelessness or misuse can quickly end their lives.

When using precision measuring tools, also keep in mind some factors that can affect accuracy. Both the measuring tool and the part surfaces to be measured must be clean of debris and free of any rough edges. Since measurement is commonly performed within .001", items such as a hair or small metal or dirt particles can affect the accuracy of the measurement. Always wipe the part surfaces and the tool measuring surfaces with fingers to remove any type of debris that can affect measurement results.

Another factor that can affect measurement is temperature. When metals are machined, heat is often created. Heat causes metals to expand, so a measurement taken when part temperature is elevated can be different from a measurement taken at room temperature. For example, suppose room temperature is 70°F. If a 5" diameter piece of aluminum is 100°F immediately after machining, that is 30°F higher than room temperature. That temperature can increase the part size by almost 0.002". The measurement taken at that elevated temperature may be within tolerance, but after cooling to room temperature, its size change might cause it to be outside of the tolerance. Always allow machined parts to cool to room temperature before measuring them with precision measuring tools.

Making sure that precision measuring tools are working correctly before use is also important. **Calibration** is the process of verifying the accuracy of a measuring tool with another tool having a higher degree of precision that is known to be properly functioning and accurate, and making adjustments if needed. Many companies and schools have calibration plans that require periodic calibration of measuring tools. That plan can list the tools, the methods of calibration, and time intervals between calibrations. Some companies and schools will perform this calibration on site, while others might have a third party perform this service. Smaller tools are usually sent to that third party for calibration, but very large ones might require an on-site visit. If a third party is used, they will provide a of the result of the calibration check, including whether it meets the required level of accuracy. A documented calibration plan may be a requirement if a company is machining/manufacturing parts for certain industries such as automotive, aerospace, and medical. Even with a calibration plan, it is a good idea to take a few short steps before each use of a tool to check calibration or verify that a measuring tool is giving correct measurements. Some of these will be discussed as this unit continues. If a tool has been dropped or subject to use that may have caused damage, it should be recalibrated.

STRAIGHT EDGES

When checking flatness, a **straight edge** can be used. They are bars with one edge that is extremely flat and are available in steel, cast iron, or granite. Steel straight edges may have beveled edges or graduations like a rule. A straight edge can be placed across a surface to see if there is space between the straight edge and the surface. Any space shows that the surface is not flat. **Figure 2.4.2** shows some straight edges. Straight edges may be as long as 72" or 1800 mm.

CAUTION

Large straight edges can be heavy and awkward to move and use. Get help when using large straight edges to avoid damage to the tool and personal injury.

PRECISION FIXED GAGES

Like semi-precision fixed gages, **precision fixed gages** are used to make comparative measurements against known sizes, but with higher accuracy. They are not usually used when accuracy better than 0.0005" is required.

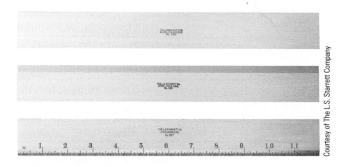

FIGURE 2.4.2 Some examples of straight edges.

Courtesy of The L. S. Starrett Company

FIGURE 2.4.3 Examples of thickness gages.

Thickness Gages

Thickness gages, or **feeler gages**, are strips of metal available in different thicknesses that can be used to check small clearances and spaces. They are available in individual pieces or grouped together as sets. **(See Figure 2.4.3.)** They can be used with straight edges to check for flatness by determining the amount of space between the tool's flat edge and the surface of a part. Thickness gages can also be used as shims to fill spaces during machine tool setup or assembly operations.

Pin or Plug Gages

Pin or **plug gages** are cylindrical rods with very accurate diameters that can be used to check hole dimensions. They frequently come in sets in 0.001-inch increments but are also available in 0.0001-inch increments. **Figure 2.4.4** shows a set of pin gages.

These smooth plug or pin gages are called *plain plug or pin gages*, because they are used to check plain, or unthreaded, holes. There are different classes of pin gages depending on the accuracy of their diameters.

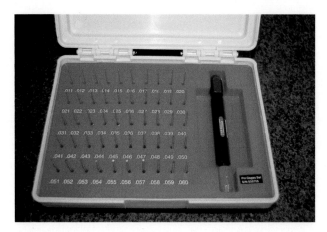

FIGURE 2.4.4 A set of pin gages in 0.001" increments from 0.011" to 0.060".

Those classes are listed in **Figure 2.4.5** according to their degrees of accuracy. The chart titled "Inch and Metric Plain Gage Diameter Tolerances" in Appendix B shows the actual tolerances for plain gages as dictated by ANSI/ASME (American National Standards Institute/American Society of Mechanical Engineers). Most pin and plug gages will come with a certificate of accuracy when purchased. This shows that they have been inspected to prove that their sizes are within these standards. With use, pin and plug gages can wear, so they are often included in a calibration plan. When they fall outside of acceptable standards, they cannot be adjusted, and they need to be replaced.

Each pin gage or gage set should be identified by its class and tolerance amount. Pin gages are also identified as either "plus" or "minus" sizes. A plus size is guaranteed to be between the ideal size and the ideal size plus its tolerance. For example, a 0.200 Class ZZ plus size pin would be between 0.2000 and 0.2002. A minus size is between its ideal sizes and its ideal size minus its tolerance. For example, a 0.200 Class ZZ minus size pin would be between 0.1998 and 0.2000. Keep this in mind when selecting and using them.

Gage Classes	
Gage Class	**Accuracy**
Class ZZ	(Least accurate)
Class Z	
Class Y	
Class X	
Class XX	
Class XXX	(Most accurate)

FIGURE 2.4.5 Gage classes.

FIGURE 2.4.6 A double-end go/no-go plug gage made from pins from a set. The green end indicates the go member and the red end indicates the no-go member.

Go/No-Go Plug Gages

Two pin gages can be used in a holder to create what is called a double-end **go/no-go plug gage**. **(See Figure 2.4.6.)** A go/no-go plug gage is used to check whether a hole diameter is within tolerance. Each end is called a member. The go member verifies that a hole size is not below its minimum allowable size and should always enter the machined hole. The no-go member verifies that a hole is not over its maximum allowable size and should never enter the hole. If either of these conditions is not met, the hole is not within tolerance and the part should be considered a reject.

Go/no-go plug gages can also be purchased for a particular hole size and tolerance instead of being made from pins from a set. With this type of gage, the go member is always longer than the no-go member for easy identification, as shown in **Figure 2.4.7**.

A go/no-go plug gage can also be a single-end progressive type. This style contains two different diameters separated by a groove. The first diameter is the go member and the second diameter is the no-go member. **(See Figure 2.4.8.)** A progressive plug gage allows the go and no/go inspection to be performed without the need to turn the gage end for end.

Because of gage tolerances, a "plus" size pin should be used for the go member. That ensures that an undersized

FIGURE 2.4.8 A single-end progressive-type go/no-go plug gage can more quickly check hole sizes.

hole will not be accepted because it is checked with an undersized gage. In contrast, a "minus" size pin should be used for the no-go member. That ensures that an oversize hole will not be accepted because it is checked with an oversize gage.

The following example shows how to select pin sizes for go and no-go members and explains why a plus or minus pin should be used.

EXAMPLE: What pins should be selected to create a go/no-go gage for a 1/4"-diameter hole with a tolerance that allows the hole to be either 0.003" larger or smaller than 1/4"?

First, convert the fraction to a decimal: 1/4 = 0.2500.

Next, subtract the tolerance from the decimal to find the smallest acceptable hole size:

$$0.2500 - 0.003 = 0.2470.$$

This is the *go* member size. Select a 0.247"-diameter plus size pin. If a minus pin is used, the gage itself would be under 0.247. It could be argued that it could allow the pin to fit in a hole smaller than 0.247. Such a hole would be outside the given tolerance.

Then add the tolerance to the decimal to find the largest acceptable hole size:

$$0.2500 + 0.003 = 0.2530.$$

This is the *no-go* member size. Select a 0.247"-diameter minus size pin. If a plus pin is used, the gage itself would be over 0.253. It could be argued that it could allow the pin to fit in a hole larger than 0.253. Such a hole would be outside the given tolerance.

Thread Go/No-Go Plug Gages

Thread plug gages are similar to ordinary plug gages except that the members are threaded to check whether threads meet required tolerances. Again, the go member should always enter the threaded hole, and the no-go member should never enter the threaded hole. If either condition is not met, the threaded hole is not within tolerance. **Figure 2.4.9** shows thread go/no-go plug gages.

FIGURE 2.4.7 A go/no-go plug gage purchased to check a particular hole size. Note that the go member is longer than the no-go member.

FIGURE 2.4.9 Two typical go/no-go thread plug gages. The go member is either marked as green or is longer than the no-go member, just like a plain plug gage.

Taper Plug Gages

Taper plug gages are single-ended plug gages that match a particular rate of diameter change. When using a taper plug gage, first apply a non-drying thin paste known as Prussian blue to a tapered hole. The plug gage is inserted, twisted, and then removed. If the Prussian blue transfers evenly onto the gage, the taper is correct. **(See Figure 2.4.10.)** Some taper plug gages have a step on the front to show the minimum and maximum limits of insertion into a hole. **(See Figure 2.4.11.)**

Ring Gages

Ring gages are used to inspect external diameters similar to the way pin or plug gages are used to check internal diameters. Plain ring gages are used to check plain, or unthreaded, diameters. They are available in the same classes as plug gages, depending on the required level of accuracy.

Like pin gages, each ring gage should be identified by its class and tolerance amount. Ring gages will also usually come with a certificate of accuracy when purchased. Again, this shows that they have been inspected to prove that their sizes are within tolerance standards. They can wear like plug and pin gages, so they are often included in a calibration plan as well. When plain ring gages fall outside of acceptable standards, they cannot be adjusted, and they need to be replaced.

Ring gages are available in either "plus" or "minus" sizes just like pin gages. The plus sizes are only over the ideal size by their tolerance, while the minus sizes are only under the ideal size by their tolerance. Again, just as with pin gages, keep this in mind when selecting and using them.

Go/No-Go Ring Gages

Ring gages can be used as go and no-go sets also. The go gage should always fit over the diameter while the no-go gage should never fit. If either condition is not met, the diameter is not within tolerance and the part

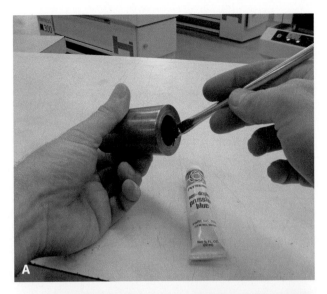

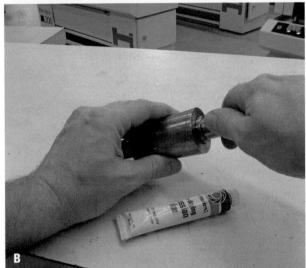

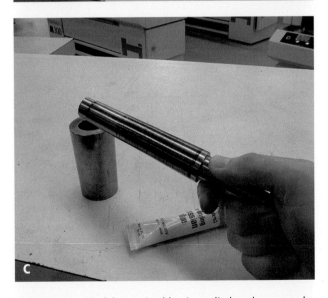

FIGURE 2.4.10 (A) Prussian blue is applied to the part to be checked. (B) The taper plug gage is then inserted in the hole and twisted. (C) The even transfer of Prussian blue to the gage indicates a proper fit and accurate taper.

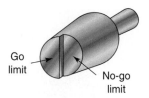

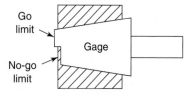

FIGURE 2.4.11 A step on the front of a taper plug gage can indicate the go and no-go limits. If the end of the part to be checked is between the two end surfaces, the part is within tolerance.

should be considered scrap. The no-go ring gage is easily identifiable by a groove around its circumference. **Figure 2.4.12** shows a go/no-go ring gage set.

The process of selecting "plus" or "minus" gages for ring gages is the opposite of that for plug gages. A "minus" size ring should be used for the go member. That ensures that an undersized diameter will not be accepted because it fits through an undersized ring gage. In contrast, a "plus" size ring should be used for the no-go member. That ensures that an oversize diameter will not be accepted because it fits through an oversize ring gage.

The following example shows how to select ring gage sizes for go and no-go members and explains why a plus or minus ring gage should be used.

FIGURE 2.4.12 Go and no-go plain ring gages. Note the groove around the circumference to identify the no-go gage.

EXAMPLE: Select ring gage sizes for go and no-go gages for a 7/8" diameter with a tolerance that allows the diameter to be either 0.003" larger or smaller than 7/8". First, convert the fraction to a decimal:

$$7/8 = 0.875$$

Next, add the tolerance to the decimal to find the largest acceptable diameter:

$$0.875 + 0.003 = 0.878.$$

This is the *go* gage size. A 0.878"-diameter minus gage should be used. If a plus size is used, the hole in the gage itself will be over 0.878. It could be argued that the gage would fit over a diameter larger than 0.878. Such a diameter would be outside the given tolerance.

Then subtract the tolerance from the decimal to find the smallest acceptable diameter:

$$0.875 - 0.003 = 0.872.$$

This is the *no-go* gage size. A 0.872"-diameter plus gage should be used. If a minus size is used, the hole in the gage itself will be under 0.872. It could be argued that the gage would fit over a diameter smaller than 0.872. Such a diameter would be outside the given tolerance.

Thread Go/No-Go Ring Gages

Thread ring gages have internal threaded diameters. They are used to check the tolerances of external threaded diameters. They are purchased according to detailed thread specifications shown on engineering drawings. Again, the go gage or member should always spin on the thread, while the no-go gage or member should not spin on the thread. If either condition is not met, the thread is not within tolerance and the part is considered scrap. The no-go thread ring gage has a groove around its circumference, like the plain ring gage, for easy identification. **Figure 2.4.13** shows typical go and no-go

FIGURE 2.4.13 Go and no-go thread ring gages. The no-go gage has a ring around its circumference for identification, just like a no-go plain ring gage.

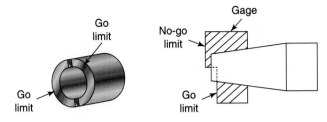

FIGURE 2.4.14 The go and no-go limit can be shown by a step on a taper ring gage. A tapered part passes inspection when the end is between the stepped surfaces.

thread ring gages. If thread ring gages fall outside of acceptable standards when calibration is checked, they can be reset by using the adjusting screws and fitting them to a master thread plug gage. This is usually the responsibility of a specially trained person.

Taper Ring Gages

Similar to a plain ring gage is the taper ring gage. It can be used with Prussian blue to check an external taper in the same manner as a taper plug gage is used to check a tapered hole. Some taper ring gages have a step like some taper plug gages to show the go and no-go limits. **(See Figure 2.4.14.)**

Snap Gages

Snap gages are c-shaped gages that can be used to check external dimensions. They have one fixed face and two adjustable faces, called anvils, which can be set to desired sizes using plug gages. **Figure 2.4.15** shows two typical snap gages and explains how the anvils are adjusted so it can be used as a go/no-go gage.

Two variations of the snap gage are the dial and digital snap gage. Instead of two adjustable solid anvils, a dial or digital gage is added to show variation from a required dimension. **(See Figure 2.4.16.)**

Snap gages and dial snap gages are also available to check external threads. They are sometimes used in production settings because threads can be checked faster than with ring gages. These thread snap gages may have the thread shape on solid anvils or on rollers, as shown in **Figure 2.4.17**. Snap gages can wear or fall out of adjustment, so they might be included in a calibration plan. They can normally be adjusted to be brought back within required levels of accuracy.

SURFACE PLATES

A **surface plate** is a flat plate that is used as an accurate reference surface with other precision tools to aid with some measurement tasks. Some older surface plates are cast iron, but today most are made of granite because granite is stable, and the surface will not expand or

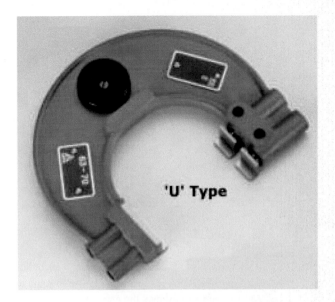

Courtesy of Flexbar Machine Corp.

FIGURE 2.4.15 Two styles of snap gages. The outer anvils of the snap gages are set to the go limit and should always pass over the part being checked. If they do not pass, the part is too large. The inner anvils are set to the no-go limit and should not pass over the part. If they do, the part is too small.

contract with changes in temperature. Further, unlike cast iron plates, granite will not rust or develop surface irregularities from scratches or damage if something is accidentally dropped on it. Surface plates are available in a wide range of sizes. Some can even have threaded holes and slots for securing parts during measuring. **Figure 2.4.18** shows two granite surface plates.

Before using a surface plate, clean it with a quality surface-plate cleaner. Then use a clean hand to wipe away dust or lint so it does not interfere with contact between the plate and other precision tools. Carefully

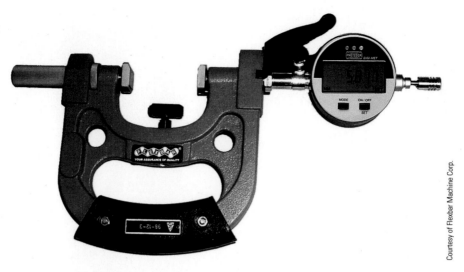

Courtesy of Flexbar Machine Corp.

FIGURE 2.4.16 A digital snap gage. A "0" point can be set at any size and the digital readout shows the amount of variation from that size.

Courtesy of Gage Crib Worldwide Inc., www.gagecrib.com

FIGURE 2.4.17 Some thread snap gages used for checking external thread dimensions. Note that some have solid anvils and some have rollers. The front gage is only a go gage.

place other precision tools on surface plates to avoid damage to both. Rough tools such as hammers, files, and wrenches should not be placed on surface plates. Never hammer on surface plates or drop heavy items on surface plates, especially near the corners, because this can cause cracks or breaks. Do not use surface plates for

storage shelves, and protect them with a cover when not in use.

Surface plates are manufactured in three different grades of flatness. From highest to lowest accuracy, they are AA—Laboratory Grade, A—Inspection Grade, and B—Shop Grade. Even though Grade B has the lowest level of precision, a standard 18" square Grade B surface plate is still flat within 0.0003" (1/10 the thickness of a standard piece of paper) over its entire surface. Since surface plates are sometimes large and very heavy, if calibration is required it will likely be performed on site by a specialty company.

SOLID SQUARES

Solid squares are similar to semi-precision squares except that because of their solid construction, the 90-degree surfaces between the blade and the beam are much more precise. Some squares with 6" blades are perpendicular within 0.0001" of the beam. Squares are available in a wide variety of sizes, and some have beveled edges on the blade. **Figure 2.4.19** shows some examples of solid squares. The calibration of solid squares can be checked like other precision measuring tools. Once they are outside of acceptable range, having them refurbished may approach the cost of a replacement.

To use a square, hold the beam against one surface of a part and then slide the square down until the blade makes contact with the adjacent side. If there is no space between the blade and either corner of the part, the two sides are square within 0.0005–0.001". If a gap is seen at either corner, the two sides are not square, or are "out of square." (A light source or white piece of paper behind the part and square can make it easier to see any

FIGURE 2.4.18 Examples of typical surface plates.

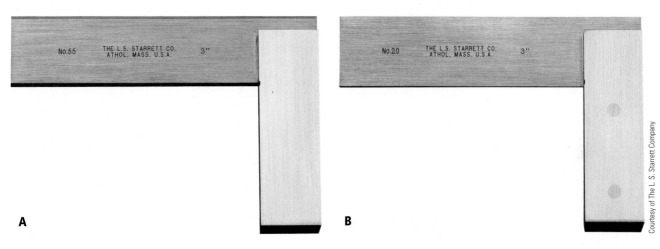

FIGURE 2.4.19 Two types of solid squares. The blade of the model on the left has beveled edges.

gaps.) **Figure 2.4.20** shows a part being checked for square with a solid square.

Figure 2.4.21 shows an illustration of the two possible "out-of-square" conditions. This shows that surfaces are not perpendicular, it but does not indicate how far "out of square" they are. To quantify the amount of error, refer to **Figure 2.4.22** while following these steps:

• First place the part with one of the surfaces down on a surface plate.

• Then stand up the square with the beam on the surface plate and slide it until the blade nearly touches the adjacent part surface.

• Place two feeler gages of 0.0005" thickness between the blade and the part as shown and slide either the part or the square until contact is made. If both feeler gages are tight, the perpendicularity error of the two surfaces is less than 0.0005".

• If only one feeler gage is tight, remove it and check again with only the one that was loose in its original location. If it is now tight, the two surfaces are square within 0.0005" because the feeler gage is taking up the gap between the part and the blade of the square.

• If it is still loose, repeat with a 0.001" feeler gage.

FIGURE 2.4.20 Checking two surfaces for square (perpendicularity) with a solid square. The light showing between the part and the blade at the inner corner indicates the two surfaces are "out of square."

• Replace the feeler gage with ones that are .0005" or 0.001" thicker until the feeler gage is snug between the part and the blade of the square.

• When a feeler gage is found that is tight, its thickness indicates how "square" the two part surfaces are because that feeler gage is taking up the gap between the part and the blade of the square.

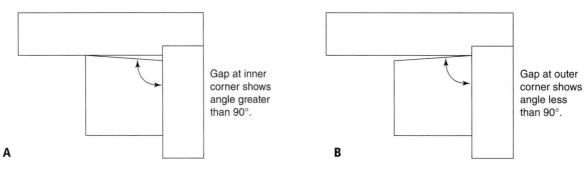

A   **B**

Gap at inner corner shows angle greater than 90°.

Gap at outer corner shows angle less than 90°.

FIGURE 2.4.21 The two possible "out of square" conditions.

A   **B**

FIGURE 2.4.22 Checking for the amount of perpendicularity error.

GAGE BLOCKS

Gage blocks are extremely accurately sized blocks with very smooth surfaces that can be used for part inspection or to check the accuracy of other precision measuring tools. They are normally purchased in sets with a certain number of blocks of various sizes and are available in rectangular and square versions. **Figure 2.4.23** shows a typical gage block set.

Gage blocks are available in different grades of accuracy, similar to the classes of plug and ring gages. The allowable deviations are measured in millionths of an inch and larger deviations are allowed for larger sizes. The symbol +/− means that the block's stated size can be either above or below by the indicated amount. The allowable deviations are given in millionths of an inch (0.000001 = one-millionth).

Grades AS-1 and AS-2 are normally for general shop and inspection use. Grade 0 blocks may be used by quality control technicians or inspectors. Grade 00 blocks are often used as "masters" that are used in highly controlled environments to inspect lower grades during

Courtesy of The L.S. Starrett Company

FIGURE 2.4.23 A typical rectangular gage block set.

calibration. These grades and their degrees of accuracy are shown in **Figure 2.4.24**. Gage blocks are available in inch and metric sizes.

Gage blocks can be made of steel, carbide, or ceramic. Steel blocks are the least expensive, but they can rust and are most subject to wear. Carbide blocks are more costly than steel, but they are more resistant to rust and are more durable. Ceramic blocks are even more expensive than carbide, but they will not rust and are extremely wear resistant. Due to their high hardness, they are somewhat brittle and may shatter if dropped.

All gage blocks should be handled with great care. They should not be dropped, bumped, or hit against other surfaces. After use, gage blocks should be oiled and stored in their cases.

The most common gage block sets are the 81- and 88-piece sets. **Figure 2.4.25** lists the block sizes in the 88-piece set. Different size blocks can be *assembled* to create nearly any size needed. *Wringing* is the process of causing two gage blocks to stick to each other by displacing the air from between the gaging surfaces of the blocks. These wrung assemblies of gage blocks are called *gage block builds*. **Figure 2.4.26** shows the proper method for wringing gage blocks.

Single blocks or builds can be used to measure the widths of slots or other spaces, as shown in **Figure 2.4.27**.

Selecting Gage Blocks for Builds

When creating gage block builds, the fewest number of blocks should be used to create the desired size. This eliminates accumulating error from individual block tolerances and is easier because using fewer blocks requires less wringing and reduces chances of them falling apart. Follow these steps to eliminate decimal values from right to left using blocks from the 88-piece set.

Individual Gage Block Sizes in Inches	Tolerance in Millionths of an Inch (0.000001) for Each Gage Block Grade			
	Grade 00	Grade 0	Grade AS1	Grade AS2
Thru 0.0500	+4 / −4	+6 / −6	+12 / −12	+24 / −24
Over 0.0500 through 0.4000	+3 / −3	+5 / −5	+8 / −8	+18 / −18
Over 0.4000 through 1.0000	+3 / −3	+6 / −6	+12 / −12	+24 / −24
Over 1.0000 through 2.0000	+4 / −4	+8 / −8	+16 / −16	+32 / −32
Over 2.0000 through 3.0000	+5 / −5	+10 / −10	+20 / −20	+40 / −40
Over 3.0000 through 4.0000	+6 / −6	+12 / −12	+24 / −24	+48 / −48

FIGURE 2.4.24 Gage block grades and their tolerances.

0.050	0.102	0.114	0.126	0.138	0.150	0.750
0.100	0.103	0.115	0.127	0.139	0.200	0.800
0.1001	0.104	0.116	0.128	0.140	0.250	0.850
0.1002	0.105	0.117	0.129	0.141	0.300	0.900
0.1003	0.106	0.118	0.130	0.142	0.350	0.950
0.1004	0.107	0.119	0.131	0.143	0.400	1.000
0.1005	0.108	0.120	0.132	0.144	0.450	2.000
0.1006	0.109	0.121	0.133	0.145	0.500	3.000
0.1007	0.110	0.122	0.134	0.146	0.550	4.000
0.1008	0.111	0.123	0.135	0.147	0.600	
0.1009	0.112	0.124	0.136	0.148	0.650	
0.101	0.113	0.125	0.137	0.149	0.700	

0.0625 (1/16)	0.078125 (5/64)	0.093750 (3/32)	0.109375 (7/64)	0.100025	0.100050	0.100075

FIGURE 2.4.25 Block sizes of an 88-piece inch-based gage block set.

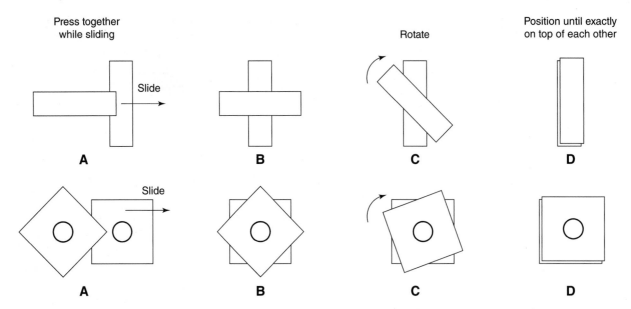

FIGURE 2.4.26 Steps for wringing gage blocks together. Using this process, virtually any size build can be created.

EXAMPLE ONE: Create a gage block build for 1.6453".

Build size	1.6453
Select and subtract first block	0.1003 block
Remaining size	1.5450
Select and subtract next block	0.1450 block
Remaining size	1.4000
Select and subtract next block	0.4000 block
Remaining size	1.0000
Select and subtract next block	1.0000 block
When zero is reached, build is complete	0.0000

FIGURE 2.4.27 Measuring a slot with gage blocks.

EXAMPLE TWO: Create a gage block build for 2.0775".

Keep in mind that the stack with fewer blocks is best. Compare these two options.

5-Block Build		4-Block Build	
Build size	2.0775	Build size	2.0775
Select and subtract first block	–0.1005	**Select and subtract first block**	–0.1005
Remaining size	1.9770	Remaining size	1.9770
Select and subtract next block	–0.1070	**Select and subtract next block**	–0.1270
Remaining size	1.8700	Remaining size	1.8500
Select and subtract next block	–0.1200	**Select and subtract next block**	–0.8500
Remaining size	1.7500	Remaining size	1.0000
Select and subtract next block	–0.7500	**Select and subtract next block**	–1.0000
Remaining size	1.0000	When zero is reached, build is complete	0.0000
Select and subtract next block	–1.0000		
When zero is reached, build is complete	0.0000		

FIGURE 2.4.28 An external dimension being measured with a vernier caliper.

VERNIER MEASURING TOOLS

A **vernier** measuring tool contains a main scale and a secondary sliding scale called the *vernier scale*. The vernier scale divides the smallest increment on the main scale into smaller increments. The number of divisions, or parts, in the vernier scale determines the number of divisions of the main scale increments. For example, a vernier scale with 10 parts will divide the smallest main scale increment into 10 pieces. A vernier scale with 25 parts will divide the smallest main scale increment into 25 pieces. This principle will become clearer when sample vernier readings are explained, but let us first look at the types of tools that use vernier scales. Before using a vernier measuring tool, it is a good practice to verify that the tool itself is accurate. A common practice is to measure some gage blocks of different sizes to ensure that the vernier tool provides the correct readings. If it does not, notify someone who can provide direction on how to have the tool calibrated.

Vernier Calipers

A **vernier caliper** is similar to a semi-precision slide caliper, but its vernier scale allows it to be used for measurements as small as 0.001" or 0.02 mm. A vernier caliper has a solid jaw and a moveable jaw that are brought in contact with part surfaces to measure external dimensions, as shown in **Figure 2.4.28**. Internal dimensions are measured by placing the "nibs" between two surfaces, as shown in **Figure 2.4.29**. Some vernier calipers also have a rod for depth measurement. **(See Figure 2.4.30.)** Vernier calipers are available in many sizes and measuring ranges. Smaller calipers may have a range of 0–6" (0–150 mm), while large models may have a range from 0" to 72". On some vernier calipers, there

FIGURE 2.4.29 Measuring an internal dimension with a vernier caliper.

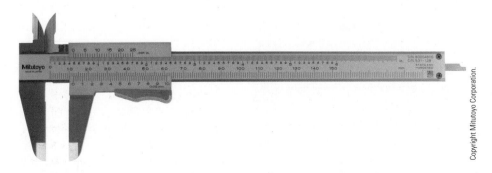

FIGURE 2.4.30 A vernier caliper with a rod for depth measurement.

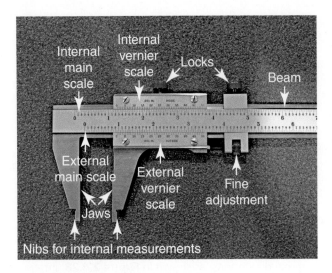

FIGURE 2.4.31 Parts of a vernier caliper.

are separate scales for internal and external measurements. Others can have both English and metric scales. **Figure 2.4.31** shows the parts of a typical vernier caliper.

Vernier Height Gage

A vernier **height gage** is like a caliper mounted on a solid base for use on a surface plate. It measures vertical dimensions from the reference zero created by the surface plate's horizontal plane and has 0.001" or 0.02-mm graduations like the vernier caliper. Small vernier height gages might have a measuring range of 0–12" (0–300 mm), while large inch-based models can reach measurements up to 72" and metric models up to 900 mm. The base takes the place of the solid jaw and a removable point takes the place of the moveable jaw. The point acts just like the moveable jaw of a caliper. When the point is removed, other measuring attachments can be mounted to the tool. The pointer can also be used as a scribe for marking purposes. **Figure 2.4.32** shows an example of a vernier height gage.

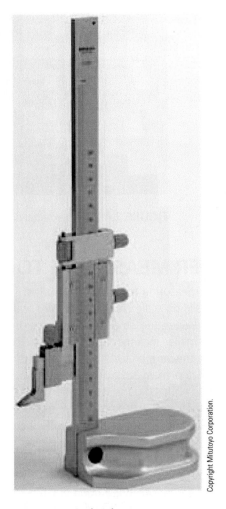

FIGURE 2.4.32 A vernier height gage.

Vernier Depth Gage

A vernier depth gage uses a sliding rod similar to a vernier caliper with a depth rod. Also using 0.001" or 0.02-mm graduations, it usually has a measuring range of 0–6" (0–150 mm) or 0–12" (0–300 mm). A vernier depth gage is shown in **Figure 2.4.33**.

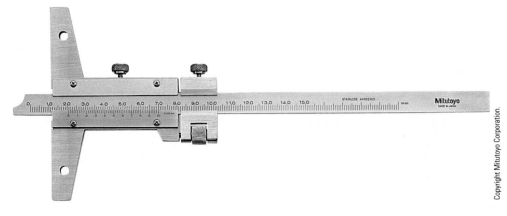

FIGURE 2.4.33 A vernier depth gage.

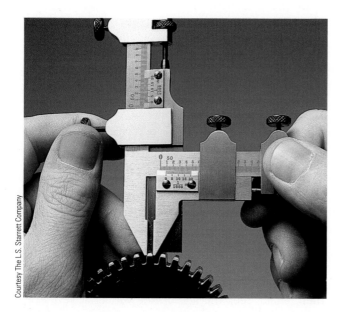

FIGURE 2.4.34 A vernier gear tooth caliper measures gear tooth width at a preset depth. Note the vertical and horizontal scales.

FIGURE 2.4.35 A vernier bevel protractor.

Vernier Gear Tooth Caliper

The vernier gear tooth caliper combines two vernier scales with 0.001" or 0.02-mm graduations. It is used to measure the thickness of gear teeth at a certain depth. One movement is used to set the depth, similar to the vernier depth gage. Then the second movement is used to measure gear tooth width, similar to the vernier caliper. **(See Figure 2.4.34.)**

Vernier Protractor

When angles need to be measured more accurately than within 1 degree, the **vernier bevel protractor** can be used. Usually referred to as just a vernier protractor, it has 5 minute (1/12 degree) graduations. With the acute angle attachment, the protractor can more easily measure angles less than 90 degrees. **Figure 2.4.35** shows a vernier protractor.

Reading Vernier Scales

All linear vernier tools (caliper, height gage, and depth gage) are available with either a 25- or 50-part vernier scale. Both require recording a reading from the tool's main scale and then adding the reading from the vernier scale to that main scale measurement. That is done by looking for a line on the vernier scale that directly aligns, or coincides, with a line on the main scale. Since these scales can be small, sometimes a magnifying glass is helpful to determine the correct line. It is also important to look straight down on a line when checking for alignment. Looking at an angle will lead to reading errors.

> **NOTE:**
>
> Because of the nature of reading vernier scales, there are possibilities of errors in readings. Even though they are graduated to 0.001" or 0.02 mm, the user should only expect reliability within a range of about two graduations. An English vernier scale can be expected to be reliable within about 0.002". A metric vernier scale can be expected to be reliable within about 0.04 mm.

Reading a 25-Part English Vernier Scale

On tools with a 25-part vernier scale, each inch is divided into 40 parts measuring 0.025" each. Every fourth graduation represents 0.100" and is numbered. Finally, every graduation on the vernier scale is 0.001". (This is because the 25-part vernier scale divides the 0.025" main scale graduations into 25 pieces: 0.025 ÷ 25 = 0.001.) Every fifth line on the vernier scale is numbered to be easier to recognize—0.005, 0.010, 0.015, and so on. **Figure 2.4.36** labels these graduations on a typical vernier scale.

To take a reading, first record the main scale reading by observing the inches, 0.100" graduations, and 0.025" graduations to the left of the vernier scale zero line. Then seek the line on the vernier scale that is directly aligned with any line on the main scale and add that amount to the main scale reading. **Figure 2.4.37** shows a few sample readings from a 25-part English vernier caliper scale.

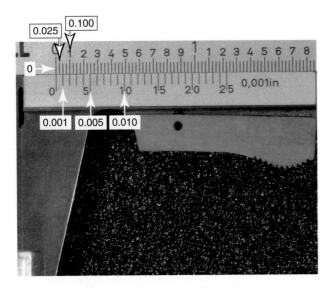

FIGURE 2.4.36 The 25-part inch vernier scale.

Reading a 50-Part English Vernier Scale

On tools with a 50-part vernier scale, each inch is divided into 20 parts measuring 0.050" each. Every other graduation represents 0.100" and is numbered. Finally, every graduation on the vernier scale is 0.001". (This is because the 50-part vernier scale divides the 0.050" main scale graduations into 50 pieces: 0.050 ÷ 50 = 0.001.) Every fifth line on the vernier scale is numbered to be easier to recognize—0.005, 0.010, 0.015, and so on, just like on the 25-part scale. **Figure 2.4.38** labels these graduations on a typical vernier scale.

The steps for reading are the same as with the 25-part scale. First the main scale reading is recorded by observing the inches, 0.100" graduations, and 0.050" graduations to the left of the vernier scale zero line. Then, again, seek the line on the vernier scale that is directly aligned with any line on the main scale and add

FIGURE 2.4.37 Examples of 25-part inch vernier scale readings. (A) 0.021"; (B) 0.075 + 0.011 = 0.086"; (C) 0.550 + 0.005 + 0.555"; (D) 1.025 + 0.015 + 1.040".

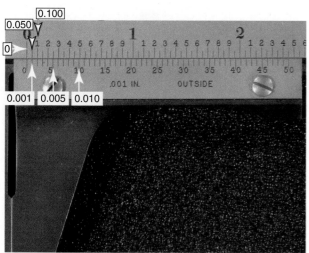

FIGURE 2.4.38 The 50-part inch vernier scale.

that amount to the main scale reading. **Figure 2.4.39** shows some sample readings using a 50-part English vernier caliper scale.

Reading a 25-Part Metric Vernier Scale

Refer to **Figure 2.4.40** while reading about the 25-part metric vernier scale. The smallest graduations on the main scale are 0.5 mm. The slightly longer graduations are 1 mm. They are numbered every 10 mm at 10, 20, 30, and so on. Each graduation on the vernier scale is 0.02 mm. (This is because the 25-part vernier scale divides the main scale's 0.5 mm graduations into 25 parts: $0.5 \div 25 = 0.02$.) Every fifth line on the vernier scale is numbered to be easier to recognize—0.10, 0.20, 0.30, and so on.

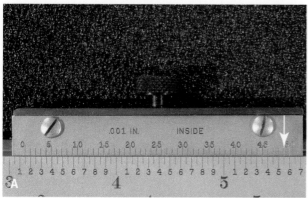

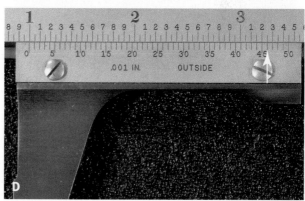

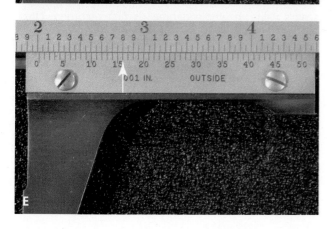

FIGURE 2.4.39 Examples of 50-part inch vernier scale readings. (A) 3.100 + 0.049 = 3.149"; (B) 3.600 + 0.013 = 3.613"; (C) 0.600 + 0.035 = 0.635"; (D) 0.950 + 0.046 = 0.996"; (E) 2.000 + 0.016 = 2.016".

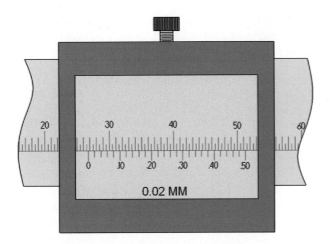

FIGURE 2.4.40 The 25-part metric vernier scale.

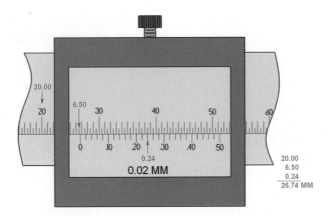

FIGURE 2.4.41 A sample reading using a 25-part metric vernier scale.

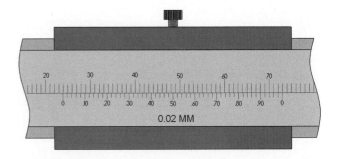

FIGURE 2.4.42 The 50-part metric vernier scale.

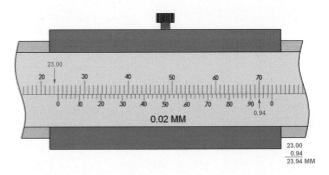

FIGURE 2.4.43 A sample reading using a 50-part metric vernier scale.

First, the main scale reading is recorded by observing the main scale graduations to the left of the vernier scale zero line. Then find the line on the vernier scale that is directly aligned with any line on the main scale and add that to the main scale reading. **Figure 2.4.41** shows a sample reading using a 25-part metric vernier scale.

Reading a 50-Part Metric Vernier Scale

Refer to **Figure 2.4.42** while reading about the 50-part metric vernier scale. Each main scale graduation represents 1 mm and they are numbered every 10 mm for easier reading. Every graduation on the vernier scale is again 0.02 mm. (This is because the 50-part vernier divides the 1 mm main scale graduations into 50 parts: 1 ÷ 50 = 0.02.) Every fifth line on the vernier scale is numbered to be easier to recognize—0.10, 0.20, 0.30, and so on.

First the main scale reading is recorded by observing the 1 mm graduations to the left of the vernier scale zero line. Then find the line on the vernier scale that is directly aligned with any line on the main scale and add that to the main scale reading. **Figure 2.4.43** shows a sample reading from a 50-part metric vernier scale.

Reading the Vernier Protractor

The vernier protractor is read using the same principle as other vernier tools and can measure angles within 1/12 degree, or 5 minutes. **Figure 2.4.44** shows some uses of a vernier protractor. The main scale is graduated in whole degrees from both sides of "0" up to 90. This allows measurement in either direction from that zero. The vernier scale on the protractor has 12 parts on each side of its "0." That 12-part vernier divides the 1-degree graduations on the main scale into 12 parts: 1 ÷ 12 = 1/12 degree or 5 minutes. Each third graduation is labeled for easier reading.

To take a reading with the bevel protractor, the main scale's value in whole degrees is first recorded. Next, either the left or the right side of the vernier scale must be selected and read. If the main scale's numbers are increasing moving counterclockwise, or to the left, the left side of the vernier scale is used. If the main scale's numbers are increasing moving clockwise, or to the right, the right side of the vernier scale is used. Look for the line on the 5-minute vernier scale that aligns with any line on the main scale. That value is added to the whole degree value of the main scale for the total reading. **Figure 2.4.45** shows a sample vernier protractor reading.

FIGURE 2.4.44 Some uses of the vernier protractor.

FIGURE 2.4.45 A sample vernier protractor reading. The whole degree reading is 50. Since the degree graduations are increasing moving to the left, read the left side of the vernier scale. The 20-minute graduation aligns, so the full reading is 50 degrees, 20 minutes (50° 20').

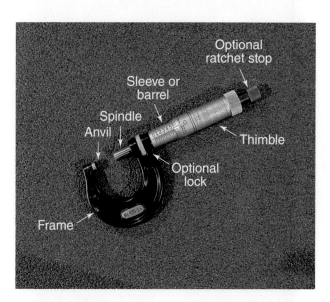

FIGURE 2.4.46 Parts of the outside micrometer, or "mic."

MICROMETERS

The **micrometer** is a precision measuring tool that uses a very accurate screw thread to perform measurement. It has one stationary point of contact and another that moves as the screw is rotated. A graduated scale on the tool then shows the distance between the two points of contact equal to the measurement. Micrometers come in many different sizes and styles for different applications.

Outside Micrometer

The outside micrometer is the most common type of micrometer and is usually called a "mic" (pronounced like the name Mike). It is used to measure external dimensions and is available in both English and metric versions. Refer to **Figure 2.4.46** to become familiar with the parts of a typical mic.

To make measurements with a micrometer, place the part between the anvil and spindle. **Figure 2.4.47** shows the proper method of holding and using micrometers when measuring handheld parts. Then the spindle is rotated until it lightly contacts the part. Some practice will be needed to become comfortable with holding a micrometer and developing proper feel when taking measurements. The option of a ratchet stop or friction thimble can help to achieve proper contact pressure, as they will slip when the spindle contacts the part being measured. **(See Figure 2.4.48.)**

Larger Micrometers

Most micrometers have only 1" or 25 mm of measuring range, so different-size tools are needed for different size part measurements. Inch micrometers are generally available in 1-inch increment sizes. For example, a 0–1" mic measures between 0 and 1 inch, and a 4–5" measures between 4 and 5 inches. Metric micrometers are generally available in 25-mm increment sizes. A 0–25 mm mic measures between 0 and 25 mm, and a 100–125 mm size measures between 100 and 125 mm.

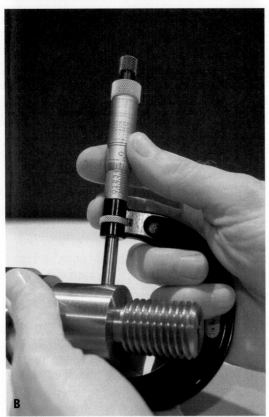

FIGURE 2.4.47 (A) Wrap the little finger around the frame when using a 0–1" micrometer. Then rotate the thimble with the thumb and forefinger. (B) Two fingers can be wrapped around the frames of larger micrometers.

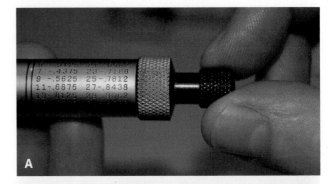

FIGURE 2.4.48 (A) A ratchet stop or (B) friction thimble will slip so that contact pressure is the same for every measurement.

Outside micrometers can be purchased in sizes up to 30" or 750 mm, with larger sizes available on a custom-order basis. When you use larger micrometers, the frame is usually held in the left hand and the thimble rotated with the right hand. Actual hand placement depends on the specific micrometer size. See **Figure 2.4.49** for one example of how to hold and use a larger micrometer.

 CAUTION

Very large micrometers can be bulky and difficult to handle. Get help when using these micrometers to prevent tool damage and personal injury.

Larger micrometers can become very expensive, so some styles provide the cost-effective option of different-length interchangeable anvils to cover a range of sizes with just one frame. **(See Figure 2.4.50.)**

Multiple-Anvil Micrometers

The multiple-anvil micrometer features a clamp opposite the spindle that allows different types of anvils to be used. **(See Figure 2.4.51.)** Measurements can then be made from flat surfaces to the edges of holes or in places with limited clearance for the micrometer's anvil end. With the clamp removed, the tool can be used to measure heights. **Figure 2.4.52** shows some uses of a multiple-anvil micrometer.

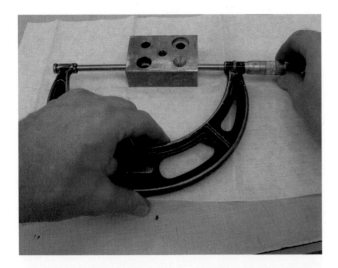

FIGURE 2.4.49 Measuring with a micrometer with a range of 3–4". Gently pivot the micrometer while turning the thimble so that the anvil and spindle become parallel with part surfaces.

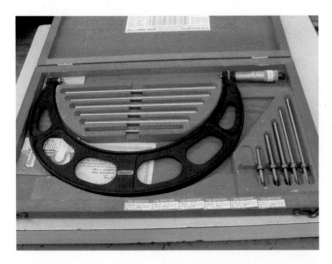

FIGURE 2.4.50 A micrometer with interchangeable anvils that can measure from 6" to 12" with one frame size.

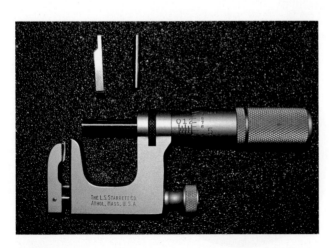

FIGURE 2.4.51 A multiple-anvil micrometer with the flat and rod anvils.

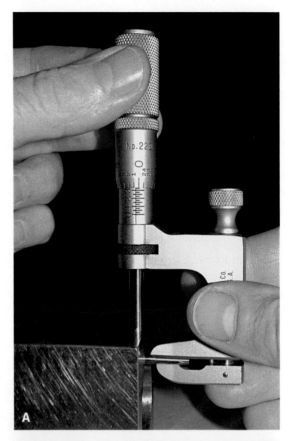

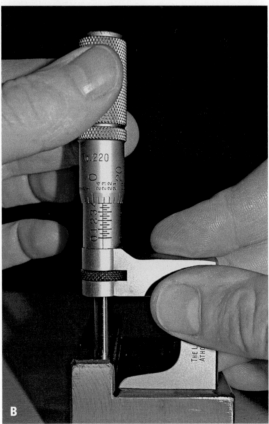

FIGURE 2.4.52 (A) A multiple-anvil micrometer can be used to measure from the edge of a hole to a flat surface, or (B) to measure height.

Other Specialty Micrometers

A screw thread micrometer has a v-shaped anvil and a pointed spindle to match the shape of a thread. It is used to inspect the dimension of a thread called the pitch diameter. **(See Figure 2.4.53.)**

A tube micrometer's anvil is replaced with a vertical pin that is 90 degrees to the spindle. It can be used to measure the wall thickness of tubing. **(See Figure 2.4.54.)**

An outside micrometer can also have anvil and spindle faces configured as discs, blades, and points for special applications. **Figure 2.4.55** shows examples of these micrometers.

A v-anvil micrometer is useful for checking roundness or measuring cylindrical parts with an odd number of grooves running along their lengths. **(See Figure 2.4.56.)**

A ball attachment can be snapped onto the anvil or spindle of a micrometer (or both) to provide two spherical contacts for measuring to concave surfaces. Note: the diameter of the ball(s) must be subtracted to get the final measurement. **(See Figure 2.4.57.)**

FIGURE 2.4.53 The anvil and spindle of a screw thread micrometer fit the form of the thread.

FIGURE 2.4.54 A tube micrometer.

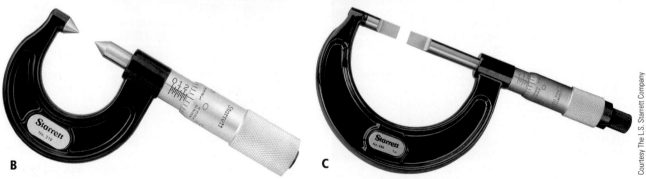

FIGURE 2.4.55 (A) A disc micrometer. (B) A conical micrometer. (C) A blade micrometer.

Courtesy The L.S. Starrett Company

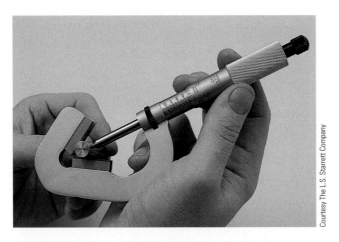

FIGURE 2.4.56 A v-anvil micrometer.

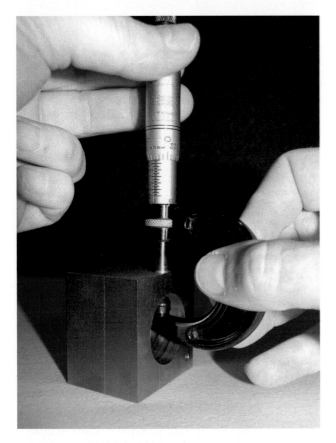

FIGURE 2.4.57 A ball attachment can be added to the anvil, spindle, or both to allow measuring concave surfaces or from a flat surface to an edge of a hole.

Reading Outside Micrometers

All of the previously described micrometers are variations of the outside micrometer caliper. For that reason, all of the inch versions are read in the same way and all of the metric versions are read in the same way. All micrometers can record measurements to 0.001" or 0.01 mm. When equipped with an optional vernier scale, an inch-based

micrometer can measure to 0.0001" and a metric micrometer can measure to 0.002 mm. As with other precision measuring tools, it is a good practice to verify that a micrometer is providing correct measurements before use by measuring some gage blocks of different sizes. If it does not, notify someone who can provide direction on how to have the tool calibrated.

Reading an English (Inch) Micrometer

Once the micrometer anvil and spindle properly contact the part to be measured, the micrometer scale must be read to determine size. An inch-based mic uses 40 threads per inch, so each rotation of the thimble moves the spindle 1/40" or 0.025" (25 thousandths).

Each vertical line on the sleeve represents one of these rotations and movement of 0.025". Each fourth vertical line is numbered for ease of reading and represents an increment of 0.100" (100 thousandths).

The horizontal lines around the circumference of the thimble divide each 0.025"-turn into 25 increments of 0.001" each.

To read the micrometer, the sleeve divisions (0.025" each) are first recorded. Then the thimble graduations (0.001" each) are added to the sleeve reading. **Figure 2.4.58** shows some examples of how to read a 0.001" graduated (plain) micrometer.

With the addition of a vernier scale, readings can be made to 0.0001" (1 ten-thousandth). This scale on the sleeve divides the 0.001" thimble graduations into 10 parts. The reading is taken as before, and then the vernier reading is added to that measurement. **Figure 2.4.59** shows how to read a vernier micrometer to 0.0001".

Reading a Metric Micrometer

The principle of reading a metric micrometer is the same as when reading an inch micrometer. The only difference is the value of the divisions. A metric micrometer spindle advances 0.5 mm for every rotation.

Each vertical line on the sleeve represents one of these rotations and movement of 5 mm. Notice that unlike the inch micrometer, these lines are staggered above and below the index line. Those below the index line are the 0.5-mm divisions and those above the index line are the 1-mm divisions. Alternating above and below the index line, the graduations can be read 0.5 mm, 1 mm, 1.5 mm, and so on. Every fifth line is numbered for ease of reading and represents an increment of 5 mm.

The horizontal lines around the circumference of the thimble divide each 0.5-mm turn into 50 increments of 0.01 mm each.

To read the micrometer, the sleeve divisions (0.5 mm) are first recorded. Then the thimble graduations (0.01 mm) are added to the sleeve reading. **Figure 2.4.60** shows some sample readings to 0.01 mm.

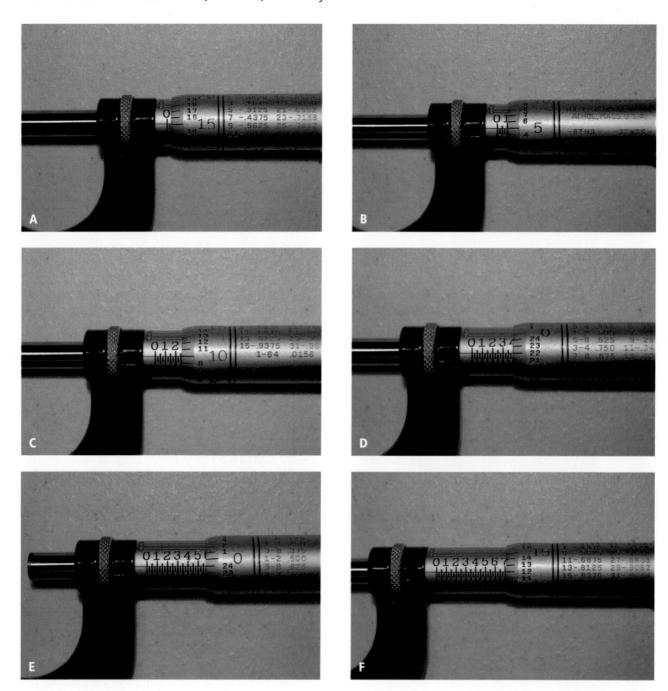

FIGURE 2.4.58 Some sample inch micrometer readings to 0.001". (A) 0.025 + 0.015 = 0.040"; (B) 0.100 + 0.005 = 0.105"; (C) 0.250 + 0.010 = 0.260"; (D) 0.375 + 0.022 = 0.397"; (E) 0.550 + 0.024 = 0.574"; (F) 0.700 + 0.012 = 0.712".

With the addition of a vernier scale, readings can be made to 0.002 mm. This scale on the sleeve divides the 0.01-mm thimble graduations into five parts. The reading is taken as before, and then the vernier reading is added to that measurement. **Figure 2.4.61** shows how to read a metric micrometer to 0.002 mm using the vernier scale.

Calibration of the Outside Micrometer

Remember that calibration is checking a tool with a more accurate tool, and it should be performed periodically to

check for proper function. Follow these steps to calibrate an outside micrometer.

- Clean the anvil and spindle. For a 0–1" micrometer, this can be done by bringing the anvil and spindle together with a piece of paper between them, then removing the paper. **(See Figure 2.4.62.)** For larger sizes, simply open the micrometer and wipe any debris from the anvil and spindle with your fingers.

- For a 0–1" micrometer, bring the anvil and spindle together with light contact as if performing a

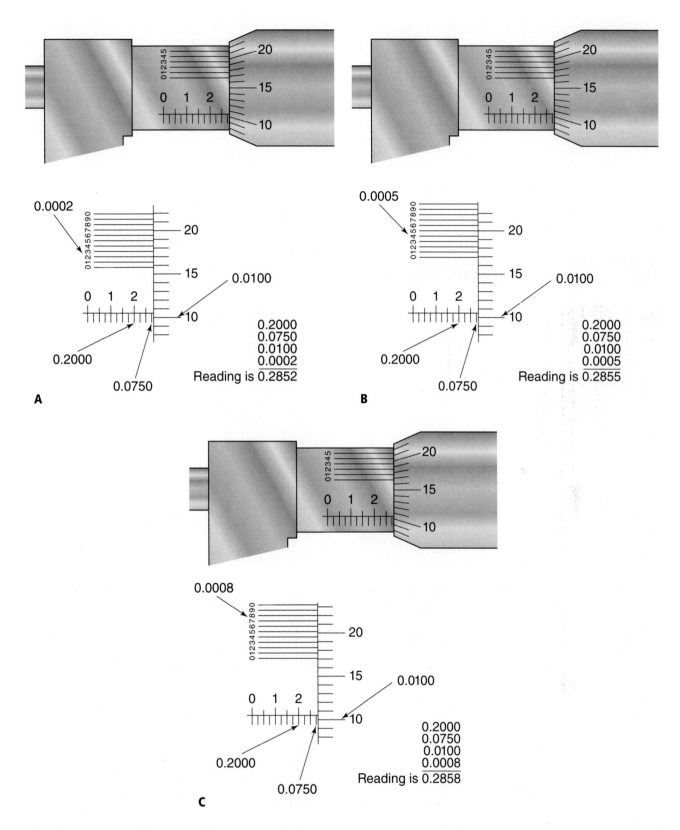

FIGURE 2.4.59 A few sample readings to 0.0001 " from an inch micrometer with a vernier scale. The reading is first taken just like with a plain micrometer. Then the vernier reading is added.

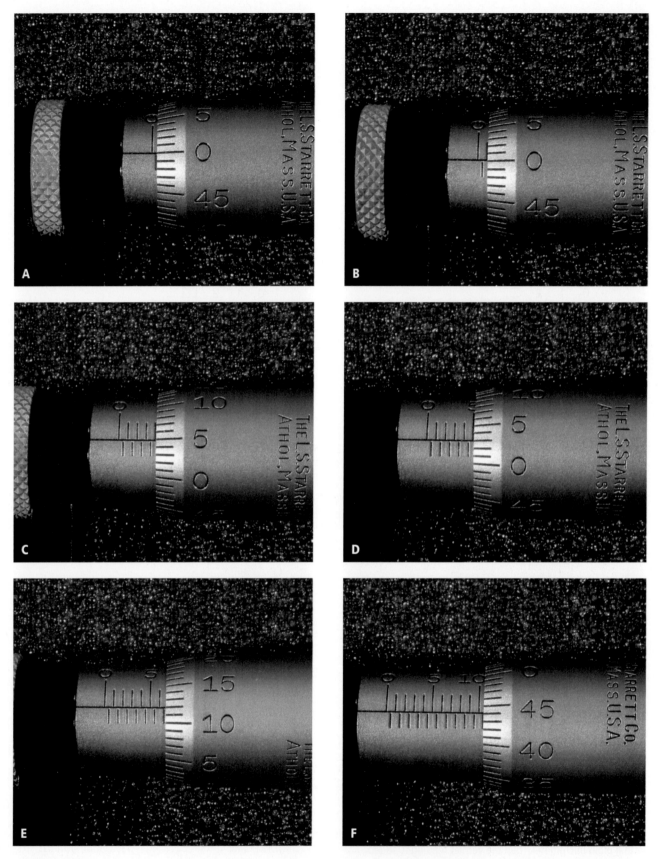

FIGURE 2.4.60 Some sample metric micrometer readings to 0.01 mm. (A) 0.50 mm; (B) 1.00 mm; (C) 4.05 mm; (D) 5.03 mm; (E) 6.62 mm; (F) 10.44 mm.

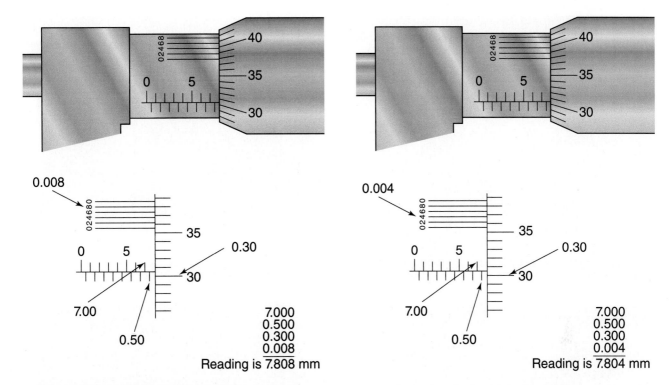

0.008

0.30

7.00

0.50

30

35

Reading is 7.808 mm

7.000
0.500
0.300
0.008

0.004

0.30

7.00

0.50

30

35

Reading is 7.804 mm

7.000
0.500
0.300
0.004

FIGURE 2.4.61 How to read a metric micrometer with a vernier scale to 0.002 mm. The reading is first taken just with a plain micrometer. Then the vernier reading is added.

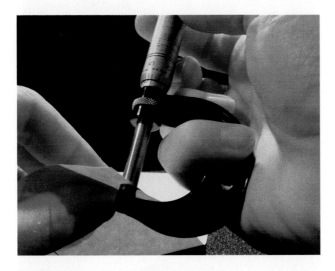

FIGURE 2.4.62 Using a piece of paper to clean the anvil and spindle of a 0–1" micrometer.

measurement. For larger micrometers use a gage block equal to the low end of the measuring range. For example, use a 2" gage block for a 2–3" micrometer.

- The zero on the thimble should align with the index line on the sleeve indicating a measurement of "0" or the smallest size of the micrometer's range.

- If the zero on the thimble does not align with the index line, a special wrench can slightly rotate the sleeve to correct alignment, as shown in **Figure 2.4.63**.

- Over time, a micrometer thread may wear and become a bit loose. To compensate for this wear, open the micrometer to the extreme end of its travel to expose the adjusting nut.

- Using the micrometer wrench, slightly tighten the adjusting nut to eliminate play in the thread. **(See Figure 2.4.64.)** Do no over-tighten, or the thimble will not rotate freely.

- Check the micrometer at three to four places throughout its range and at the far end of the range with gage blocks.

- Measurements should be accurate throughout the range. If not, the micrometer is worn beyond adjustment.

Inside Micrometers

Inside micrometers are used to measure internal dimensions and are available in three major types. The tubular or rod styles shown in **Figure 2.4.65** have micrometer heads with interchangeable ends to cover a large range of sizes. Adjust the head while pivoting the tool inside a hole or slot to make contact with two surfaces, as shown in **Figure 2.4.66**. The smallest dimension that can be measured by these inside micrometers is approximately 1-1/2". These types of inside micrometers are read just like outside micrometers.

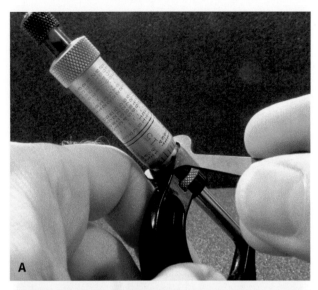

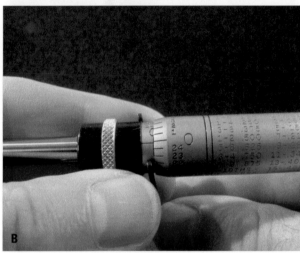

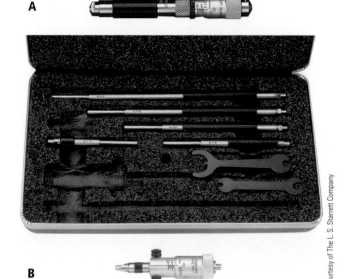

FIGURE 2.4.63 (A) Place the pin of the micrometer wrench in the small hole on the sleeve. (B) Then rotate the sleeve to align the "0" on the thimble with the index line.

FIGURE 2.4.65 (A) A tubular-style inside micrometer. (B) A rod-style inside micrometer.

Courtesy of The L. S. Starrett Company

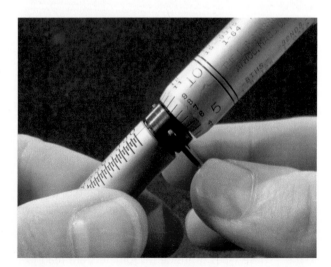

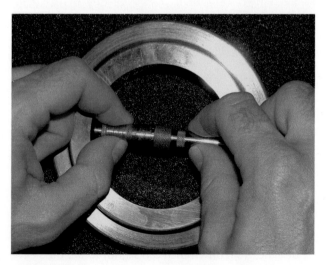

FIGURE 2.4.64 Tightening the adjusting nut to compensate for thread wear.

FIGURE 2.4.66 Taking a measurement using an inside micrometer.

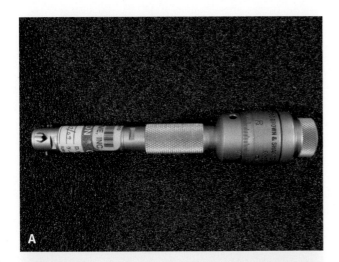

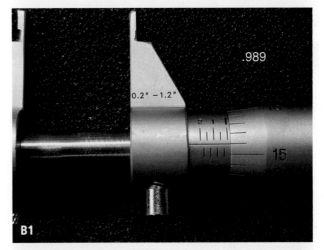

FIGURE 2.4.67 (A) Note the three contact points of a bore micrometer. (B) The measurement shown is 0.5314.

Another type of inside micrometer for measuring hole diameters uses three telescoping legs to make contact inside a hole. It is sometimes called a bore micrometer. A ratchet slips when there is proper contact pressure. Reading it is similar to reading other micrometers, except that the divisions on the thimble are further divided into smaller units of 0.0002" or 0.005 mm. This type of inside micrometer provides very reliable, accurate measurements. **Figure 2.4.67** shows this type of inside micrometer and a sample reading from an inch-based model.

The inside micrometer caliper has jaws and is similar to the outside micrometer, except that the graduations on the sleeve are numbered and read from right to left, and the graduations on the thimble are in reverse order. They can normally measure down to approximately 0.200" or 5 mm. **Figure 2.4.68** shows an inside micrometer caliper and some sample readings.

Before using, inside micrometer function and accuracy can be verified much like outside micrometers,

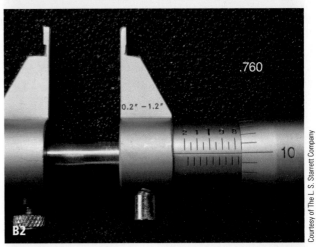

FIGURE 2.4.68 An inside micrometer caliper (A). Two sample readings from an inside micrometer caliper (B).

but by using ring gages of known sizes instead of gage blocks. The tubular type, rod type, and inside micrometer caliper are calibrated much like outside micrometers as well. Follow specific manufacturer instructions for calibrating a bore micrometer.

Depth Micrometers

The micrometer depth gage, or **depth micrometer**, features a base and interchangeable rods for different size ranges. The rods are produced in 1" or 25-mm increments. A depth micrometer is shown in **Figure 2.4.69**. The graduations are numbered and read from left to right as with an inside micrometer caliper. **Figure 2.4.70** shows some sample readings from an inch-based depth micrometer. A depth micrometer should be periodically calibrated like other micrometers. Follow these steps for calibration.

- Install the 0–1" rod.
- Clean the surface of the base and the end of the rod.
- Place the base on a surface plate.
- Rotate the thimble until the rod makes light contact against the surface plate as if performing a measurement.

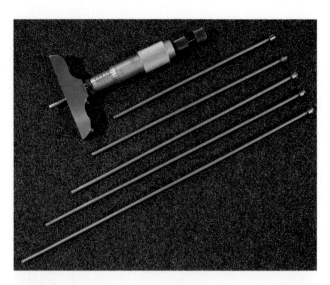

FIGURE 2.4.69 A depth micrometer with rods for different size ranges.

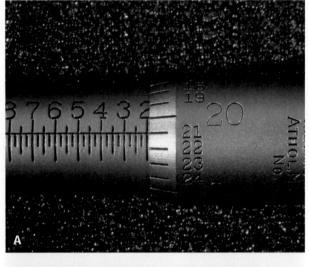

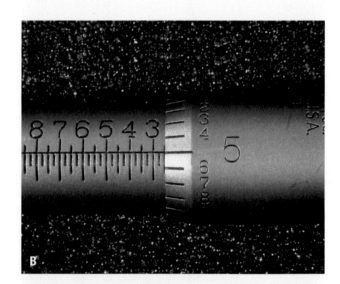

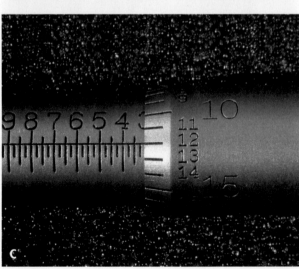

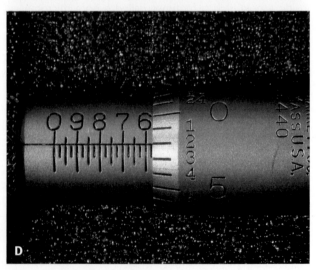

FIGURE 2.4.70 Some sample readings from an inch-based depth micrometer. (A) 0.175 + 0.021 = 0.196"; (B) 0.250 + 0.005 = 0.255"; (C) 0.300 + 0.012 = 0.312"; (D) 0.575 + 0.002 = 0.577".

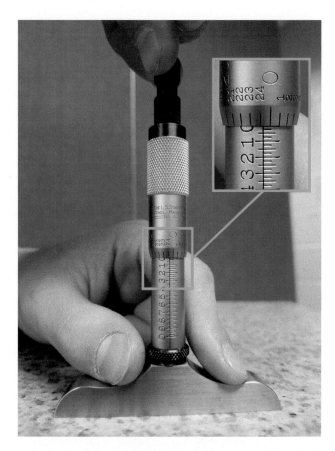

FIGURE 2.4.71 Checking the "0" measurement of a depth micrometer during calibration with the tool. Notice the reading 0.001" less than the "0" and needs adjustment.

- The zero on the thimble should align with the index line on the sleeve indicating a measurement of "0". If it does not, adjustment is needed. **(See Figure 2.4.71.)** Some depth micrometers can be adjusted by rotating the sleeve just like an outside micrometer (**Figure 2.4.63**). If this is not possible, the rod itself needs to be adjusted. Continue with these steps to adjust the rod.

- Remove the rod.

- Secure the rod in a way that will not damage its surface and use the small end of the special micrometer wrench to turn the adjusting nut on the end of the rod. Clamping between two pieces of wood in a vise or in a precision vise with smooth jaws will work (**Figure 2.4.72**). Some models have two nuts that are locked together, so both will have to be turned, then locked together again after moving them.

- Turn clockwise to "shorten" the rod if the measurement is too small.

- Turn counterclockwise to "lengthen" the rod if the measurement is too large.

- Insert the rod and recheck the "0".

- Repeat as needed until the "0" line on the thimble aligns with the index line on the sleeve.

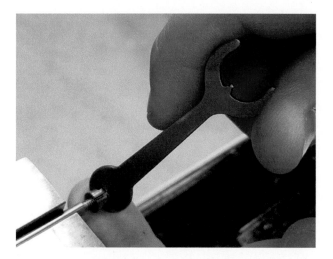

FIGURE 2.4.72 Adjusting the nut on the end of a depth micrometer rod.

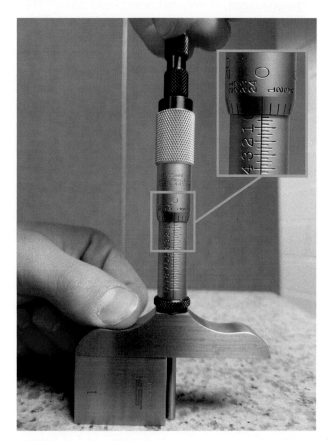

FIGURE 2.4.73 Checking the "0" of a depth micrometer 1–2" rod with a 1" gage block.

- Check the micrometer at three to four places throughout the 0–1" range with gage blocks. Measurements should be accurate throughout the range.

- Repeat the process and adjust each rod as needed. Use a gage block equal to the low end measurement of the range for each rod. For example, use a 1" gage block for the 1–2" rod, a 2" gage block for a 2–3" rod, and so on (**Figure 2.4.73**).

- There is no need to check at different places throughout the range for each rod, because the actual micrometer thread accuracy was already verified when calibrating the 0–1" rod. Only adjust the individual rods if needed.

- If the threads of the micrometer wear and become loose over time, slightly tighten the adjusting nut just like with an outside micrometer. **(See Figure 2.4.64.)** Then perform a calibration check for each rod again.

DIAL AND DIGITAL MEASURING TOOLS

Several measuring tools are available in versions with dials or digital readouts and are becoming more commonly used in the machining field today. Their popularity is largely because they are easier to read than the conventional-type tools. Dial-based calipers, height gages, and depth gages are similar to vernier models with a main scale reading, but then the smaller graduations are read from the dial instead of using a vernier scale. Digital micrometers, calipers, height gages, and depth gages may or may not have a main scale, but always directly show the total measurement in a digital format on a small LCD display. Digital tools can usually be switched between inch and metric units by pressing a button. Some digital tools have a data output port that can be used to transmit measurements to a computer for record keeping. One disadvantage of digital measuring tools is the fact that they are battery powered. When the battery power is gone, measurements cannot be taken. For that reason, some manufacturers have begun to produce solar-powered digital calipers. Dial and digital tools should be checked before using by the same methods as their conventional versions, and if measurements are not accurate, tool calibration should be performed.

Dial and Digital Calipers

Dial and digital calipers are beginning to replace vernier calipers in the industry. They are very versatile because they have the ability to measure external, internal, depth, and step measurements quickly and easily. **(See Figure 2.4.74.)** Some newer models are made of nonmetal materials and are water and chemical proof for use in harsh machining environments. These calipers can have measuring ranges from 0" to 3" (0–75 mm) all the way up to 0–40" (0–1000 mm). Dial calipers usually have 0.001" or 0.02-mm graduations. Digital calipers usually read to 0.0005" or 0.01 mm. **Figure 2.4.75** shows a dial caliper.

Reading an Inch-Based Dial Caliper

The first step in reading a dial caliper is the same as when reading a vernier caliper. First, whole inches are read by

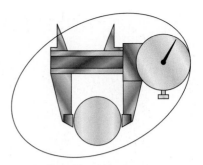

1. Outside measurement

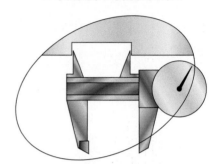

2. Inside measurement

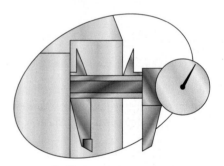

3. Step measurement

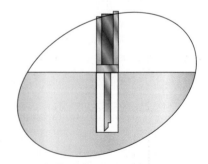

4. Depth measurement

FIGURE 2.4.74 Uses of dial and digital calipers.

the largest numbers on the main scale. Then the smaller 0.100" graduations on the beam are recorded. Finally, the dial reading is added to the main scale reading. **Figure 2.4.76** shows a few sample readings from an inch-based dial caliper.

Reading a Metric Dial Caliper

Similar steps are followed when reading a metric dial caliper. The main scale is read from the beam, and

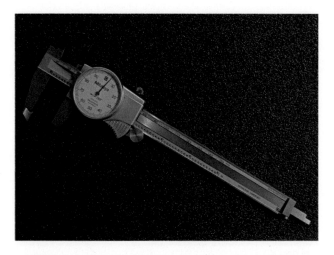

FIGURE 2.4.75 A 0–6" dial caliper. This model's dial shows 0.100" per revolution.

then the dial reading is added to the main scale value. **Figure 2.4.77** shows some sample readings from one type of metric dial caliper.

Reading a Digital Caliper

Most digital calipers measure in both English and metric units. To use, bring the jaws together and power the caliper on. This sets "0." If the caliper is already powered on, press a button marked "zero" or "0" to set the zero point. A "zero" location can be set at any position in the caliper's travel. When a measurement is taken, the entire measurement is displayed on the digital readout. To switch between inch and millimeter display, simply press an "inch/mm" or similar conversion button. **Figure 2.4.78** shows a digital caliper.

Dial and Digital Height Gages

Dial and digital height gages are used in the same way as vernier height gages. They are available in many sizes and ranges in English, metric, and combination English/metric styles. They are read just like dial and digital calipers. **Figure 2.4.79** shows examples of dial and digital height gages.

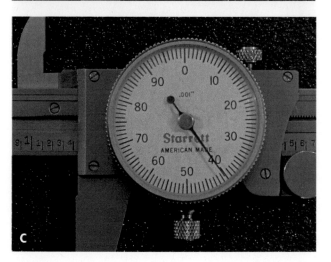

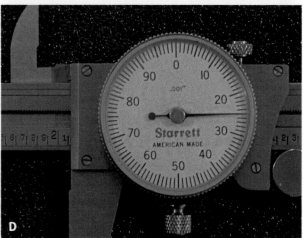

FIGURE 2.4.76 Sample readings from an inch-based dial caliper. (A) 0.980"; (B) 1.005"; (C) 1.440"; (D) 2.125".

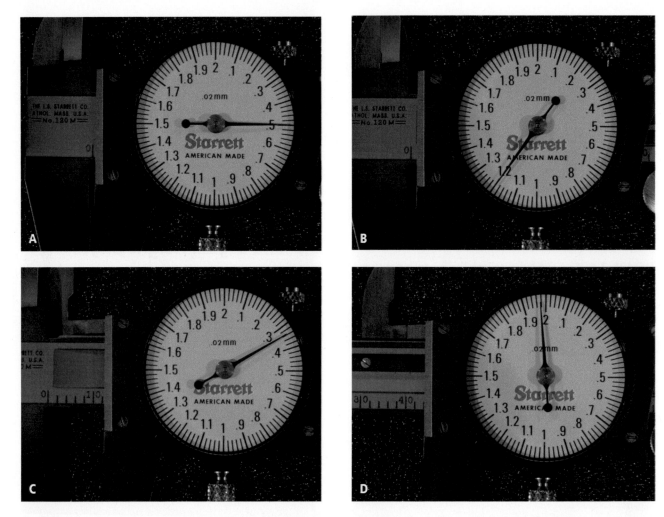

FIGURE 2.4.77 Sample readings from a metric dial caliper. (A) 0.50 mm; (B) 1.20 mm; (C) 12.34 mm; (D) 43.98 mm.

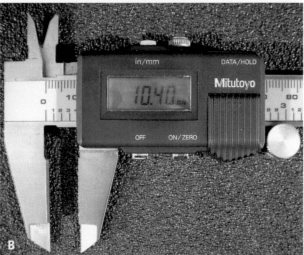

FIGURE 2.4.78 A 0–6" (0–150 mm) digital caliper. Pressing the zero button will set a reference zero at any position. The in/mm button switches the measurement display between inches and millimeters.

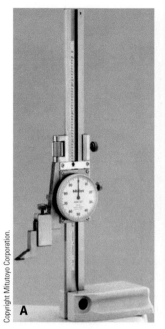

FIGURE 2.4.79 (A) A dial height gage, and (B) a digital height gage.

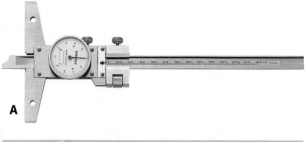
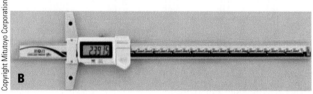

FIGURE 2.4.80 (A) A dial depth gage, and (B) a digital depth gage.

Dial and Digital Depth Gages

Dial and digital depth gages are used in the same way as vernier depth gages and have similar measuring ranges. They are available in English, metric, and combination English/metric styles. They are read just like dial and digital calipers and height gages. **Figure 2.4.80** shows a few examples of the many styles of dial and a digital depth gages.

Dial and Digital Bore Gages

When measuring internal diameters to close tolerances, a dial or digital bore gage can be used. **Figure 2.4.81** shows some examples of bore gages. A "zero" is normally

FIGURE 2.4.81 (A) A digital bore gage, and (B) a dial bore gage.

set with a ring gage equal to the desired hole size. Then the gage is placed in the hole to be checked and variation from that size is read on the dial or digital readout.

On digital bore gages, the "plus" or "minus" variation from the ideal size is automatically displayed on a digital readout. When set with a ring gage, a digital bore gage can also be set to display the desired hole size. Then during measurement, the actual hole size would be displayed instead of the amount of variation.

On dial bore gages, if the needle moves to the "plus" side of the dial, the hole is larger than the ideal size. If the needle moves to the "minus" side of the dial, the hole is smaller than the ideal size. **Figure 2.4.82** shows some sample readings from a dial bore gage.

PRECISION TRANSFER OR HELPER-TYPE MEASURING TOOLS

Transfer or helper-type measuring tools work similarly to those used for semi-precision measurement. They must be used with another tool because they have no graduated scales of their own. The difference here is that these tools are more suited to maintaining their settings and to be measured with other precision tools.

Small Hole Gages

Small hole gages feature a split ball end that expands when an adjusting screw is tightened. They are useful for

FIGURE 2.4.82 The dial bore gage is set to "0" for the desired hole size. (A) A hole that is 0.0008" smaller than the set size. (B) A hole that is between 0.0007" and 0.0008" larger than the set size.

measuring holes or slots and are available in full-ball and half-ball styles. **Figure 2.4.83** shows some small hole gages. The gage is adjusted while being gently rocked to make contact with part surfaces and then measured with a micrometer. **Figure 2.4.84** shows the use of a small hole gage.

Telescoping Gages

A **telescoping gage** is shaped like a "T" and has two arms that expand when the locking screw is loosened. The arms lock in place when the screw is tightened. Telescoping gages are shown in **Figure 2.4.85**. **Figure 2.4.86** shows the method for using a telescoping gage to measure internal diameters. Use one motion and do not place the gage back in the hole after tightening because it will collapse the arms and give a false reading. After being set, telescoping gages are usually measured with a micrometer to obtain measurements, as shown in **Figure 2.4.87**. When measuring the telescoping gage,

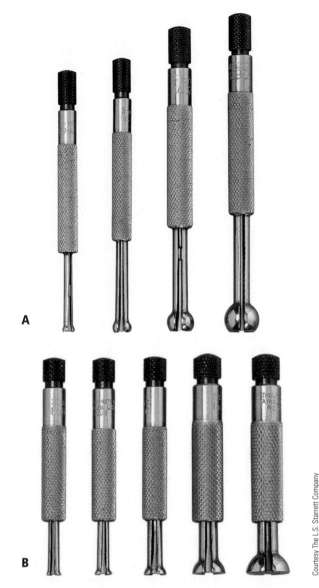

Courtesy The L.S. Starrett Company

FIGURE 2.4.83 Small hole gages are available in (A) full-ball styles and (B) half-ball styles.

be careful not to use too much pressure and collapse the gage, as this will change the reading.

Adjustable Parallels

Adjustable parallels are two-piece blocks with parallel surfaces that can be locked in position by a clamping screw. They can be used to measure slot widths by placing the parallel between two surfaces, adjusting to obtain a fit, and then tightening the clamping screw. The parallel can then be measured with a micrometer. **(See Figure 2.4.88.)** Adjustable parallels can also be preset to certain sizes and used much like go or no-go gages.

DIAL AND DIGITAL INDICATORS

A **dial indicator**, sometimes called a dial indicator gage, is a tool that shows small movements by displaying them

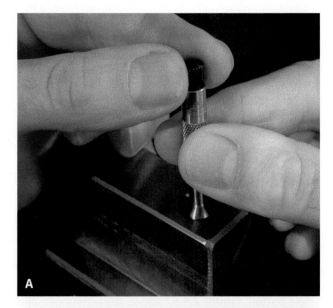

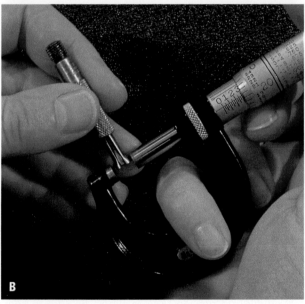

FIGURE 2.4.84 (A) Measuring with a small hole gage. (B) Transferring a small hole gage measurement to a micrometer.

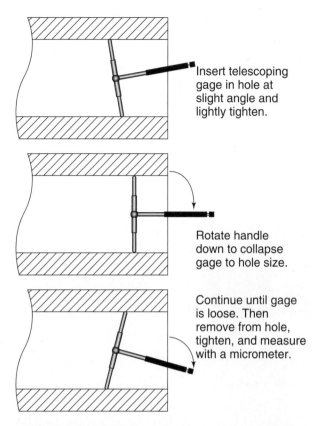

Insert telescoping gage in hole at slight angle and lightly tighten.

Rotate handle down to collapse gage to hole size.

Continue until gage is loose. Then remove from hole, tighten, and measure with a micrometer.

FIGURE 2.4.86 Method for measuring with a telescoping gage.

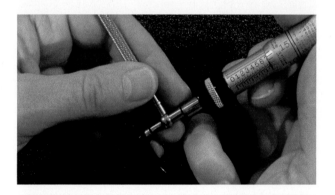

FIGURE 2.4.87 Transferring a telescoping gage measurement to a micrometer.

Courtesy The L.S. Starrett Company

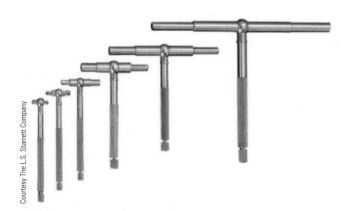

FIGURE 2.4.85 Telescoping gages.

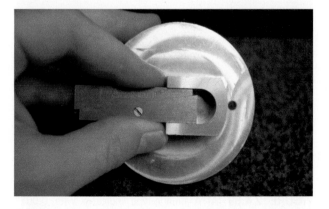

FIGURE 2.4.88 Measuring a slot width with an adjustable parallel.

with a needle on a graduated dial face. Dial indicator graduations usually range from 0.001" to 0.00005". The smallest graduation will be listed on the face of the indicator. A dial indicator looks and works much like a car's speedometer. The part of the tool that touches a part to register movement is called the contact. There are two general types of dial indicators. One type uses a probe- or plunger-type movement and is often called a plunge- or plunger-type indicator. The other is called a *test* indicator and uses a lever-type movement. These two basic types are shown in **Figure 2.4.89**.

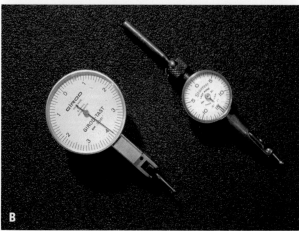

FIGURE 2.4.89 (A) A plunge-type indicator with 0.001" graduations. (B) Test-type indicators with 0.0001" and 0.0005" graduations.

The faces of dial indicators come in two designs: balanced and continuous. The balanced dial's graduations are equally numbered on both sides of a zero mark, while the continuous dial is numbered in one direction like a speedometer. A continuous dial may often have a second set of numbers in the opposite direction. When using an indicator with a large measuring range, the needle can make more than one revolution around the dial face. To keep track of those revolutions, these indicators will have a smaller counter-dial in addition to the main dial. **Figure 2.4.90** shows balanced and continuous dial faces and counter-dials.

Digital indicators are used just like dial indicators, except that instead of the dial face and needle being read, movements are displayed on a digital readout. Like other digital tools, a single button can set a zero reference position. Metric and inch units can also be switched with a single button. **Figure 2.4.91** shows digital plunger- and test-type indicators.

Dial and digital indicators are often used during machine tool set-up for machine and workpiece alignment, and those uses will be discussed later in this text. The following applications are examples of some ways indicators are used when performing precision measurement.

Applications of Plunge-Type Indicators

Plunge-type indicators can travel longer distances than test indicators and are sometimes called travel indicators. On dial versions, the needle can normally make at least 2.5 complete revolutions around the dial face. Some models can measure travel up to 6" or more. These features make these indicators well suited for measuring linear distances. Sideways movement of plunge-type indicators should be avoided, as it can cause unwanted movement and provide false readings. Be sure to keep contact of the tool perpendicular to the surface being measured to ensure accurate readings. The illustration in **Figure 2.4.92** will help explain these steps.

These indicators can be fitted with different-shaped contacts depending on where the tip needs to reach. **Figure 2.4.93** shows some different types of contact points.

A dial or digital plunge-type indicator can be mounted on a base and used to make comparative measurements. The indicator can be set to a zero reference using gage blocks. When using a dial indicator, it is good practice to "pre-load" by applying enough pressure for the needle to move about 1/4 turn. Then the face can be rotated so the needle is aligned with the "0" mark. When using digital models, just be sure the indicator registers enough movement so that any part that is measured will make contact. Then press the "0" button. This process of setting a reference zero is called *mastering*. When a

A

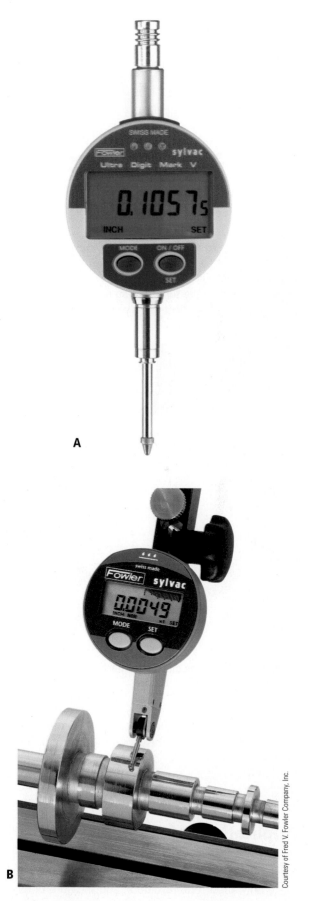

A

B

FIGURE 2.4.90 (A) A balanced-type dial indicator face with 0.0005" graduations. (B) A continuous-type dial indicator face with 0.001" graduations. (C) A continuous face with 0.001" graduations and a second set of counterclockwise numbers. Note the counter-dial on the continuous face dials to keep track of the revolutions of the main needle.

FIGURE 2.4.91 (A) A digital plunge-type indicator. (B) A digital test-type indicator.

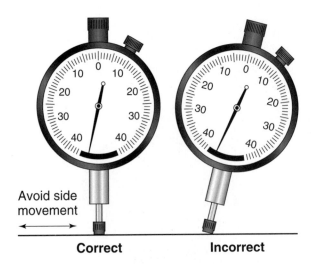

Avoid side movement

Correct   Incorrect

FIGURE 2.4.92 The contact of a plunge-type indicator must be perpendicular to the surface being measured or errors will result. Avoid side movement with a plunger indicator.

Courtesy The L.S. Starrett Company

FIGURE 2.4.93 Some different contact points for use on dial indicators.

dimension is inspected, the indicator shows the variation from the gage block. In this type of application, the indicator is sometimes called a drop gage. **Figure 2.4.94** shows mastering and measuring with a dial indicator gage.

Applications of Test Indicators

Test indicators have smaller measuring ranges than plunge-type indicators. The needles of dial models usually make only about one revolution and measure between 0.008" and 0.030" of movement. Test indicators are ideal for measuring variation while in constant contact

with a part or when making adjustments to machine attachments.

When using a test indicator, keep the contact as parallel as possible to the surface being indicated. The lever contact of a test indicator can be adjusted to create the best possible position. Angles between the surface and the contact will create false readings because motion is actually at an angle to the indicator's movement ability. As the angle increases, the amount of error is multiplied. **Figure 2.4.95** shows some examples of correct and incorrect lever contact positions.

When using test indicators, "pre-loading" the indicator with about 1/4 of a turn of contact pressure before taking measurements is recommended, as with plunge-type indicators.

A test indicator can be mounted on a height gage to be used for comparison measurement the same way a plunge-type indicator can. Master (zero) the indicator with gage blocks and then measure variation between the gage blocks and machined part sizes. **(See Figure 2.4.96.)**

This setup can also be used to check for parallelism. Place a part to be checked on a surface plate. Then bring the indicator in contact with the surface to be checked and zero the dial face. No mastering is required because no actual size is being measured, only variation of a single surface. Slide or sweep the indicator across the surface of the part by sliding the height gage on the surface plate. To check smaller parts, it is usually better to keep the height gage stationary and move parts under the indicator. **Figure 2.4.97** shows this procedure. Movement on the indicator will show errors in parallelism.

A height-gage-mounted test indicator can be used in another capacity. It can take the place of the measuring point of the height gage. This works particularly well with a digital height gage because a zero position can be set at any location. The benefit is that instead of relying on contact pressure, the test indicator is returned to the same reading for each measurement taken. The result is a more precise measurement. To use this method, refer to **Figure 2.4.98**.

Indicators should also be periodically calibrated. The process is rather specialized and involved, but it is a good idea to verify an indicator's accuracy before use. Follow these steps for a test indicator.

- Mount the indicator on a surface gage or height gage that is placed on a surface plate.

- Place a gage block on the surface plate.

- Bring the indicator contact point against the gage block and apply pressure to move the indicator needle 2–3 graduations.

- Turn the dial face so the needle is at the "0" on the dial face.

FIGURE 2.4.94 (A) Mastering a dial indicator with a gage block build. (B) Checking a measurement with the drop gage. Here a 0.0001 " graduated indicator is being used and the part being measured is 0.0007 " over the size set with the gage block build. Note the lever that can be used to raise and lower the indicator contact.

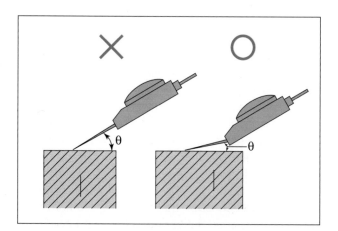 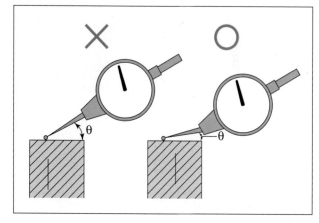

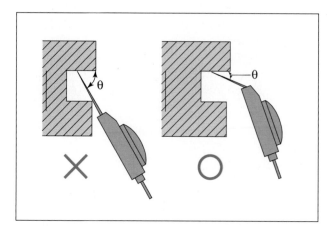 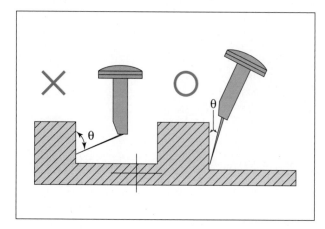

FIGURE 2.4.95 Correct (O) and incorrect (X) settings of test indicator contact points. A small angle between the contact and the work will minimize errors and create more accurate readings.

FIGURE 2.4.96 A height-gage-mounted test indicator can be used for comparison measurements after mastering with a gage block build.

FIGURE 2.4.97 Using a height-gage-mounted test indicator to check for parallelism errors by sliding a part beneath the indicator.

- Slide the gage block out from under the indicator.
- Slide the gage block under the indicator again.
- The needle should return to the same "0" position.
- Change to a gage block that will move the indicator needle approximately ¼ of a rotation on the dial face.
- Slide that gage block under the indicator.
- Verify that the indicator movement amount is equal to the change in the gage block build. **(See Figure 2.4.99.)**
- Repeat this with increasing gage block build sizes at every ¼ turn of a rotation throughout the range of the indicator.
- If the return to "0" of the first check or the readings of the gage block builds is more than ½ of a graduation, it is a good idea to have the indicator calibrated by a specialist.

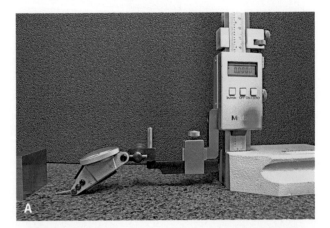

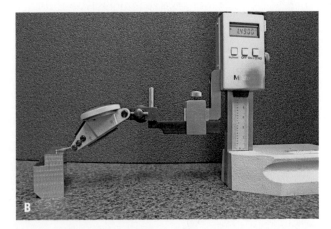

FIGURE 2.4.98 To avoid measurement errors from varying contact pressure, use a test indicator in place of the standard height gage pointer. (A) Adjust the height gage to bring the indicator in contact with the surface plate to pre-load the indicator. Then zero the indicator and the height gage. (B) To take a measurement, adjust the height gage until the indicator reaches the same zero indicator reading when contacting the surface of the part and record the measurement from the height gage.

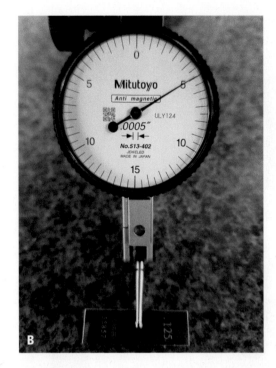

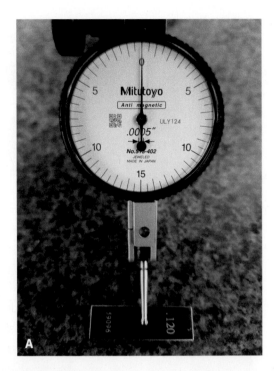

FIGURE 2.4.99 Verifying the accuracy of a dial test indicator using gage blocks. A shows the initial "0" setting using a 0.120" gage block. B shows 0.005" of indicator movement when the 0.120" gage block is replaced with a 0.125" gage block.

SINE TOOLS

Sometimes angles must be measured more accurately than with a vernier protractor. In those cases, **sine tools** are frequently used. They can be used to measure angled surfaces relative to a reference surface (normally a surface plate). Sine tools have two equal-sized cylinders, called rolls, mounted near each end. The diameters of the rolls and the distance between their centers are held to very tight tolerances (usually within 0.0001–0.0002"). With one roll on a surface plate and the other roll on a gage block build, the bar can be inclined to very accurate angular settings. Sine tools get their name from the fact that the trigonometric sine function is used to determine the size of gage block builds used to incline the tools. **Figure 2.4.100** gives an illustration of this principle.

The center-to-center distance of sine tool rolls is usually a standard dimension of 3, 5, 10, 15, or 20 inches. This center-to-center distance is representative of the hypotenuse, or side c, of a right triangle. To calculate a gage block build for a sine tool setting, determine the sine of the desired angle and multiply it by the length of the tool's roll center-to-center distance. The gage block build is then placed under one of the tool's rolls to raise the bar to the angular position.

Sine values change more rapidly at smaller angles and more slowly at larger angles. To avoid errors because of this factor, it is good practice to use an angle's complement if it is greater than 45 degrees.

Once the sine tool has been set up on the surface plate using the correct gage block build, an angled part can be placed on the sine tool and a test indicator moved across the part's surface to measure variation from the desired angle. **(See Figure 2.4.101.)**

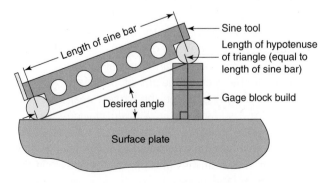

FIGURE 2.4.100 The basic principle of sine tools.

FIGURE 2.4.101 Indicating an angular surface using a sine tool. Note the gage blocks used to raise the one end of the tool.

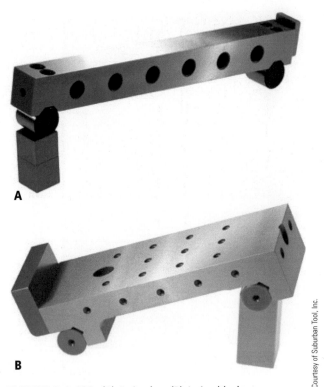

FIGURE 2.4.102 (A) A sine bar. (B) A sine block.

Sine Bars and Sine Blocks

A sine bar is a narrow bar with rolls near each end and is used to measure angles on relatively small parts. A sine block is a wider version of a sine bar. **Figure 2.4.102** shows a sine bar and a sine block.

Sine Plates

A sine plate is similar to a sine bar except with a larger area to hold larger parts. The plate usually has threaded holes for the use of clamps to secure parts. A compound sine plate is like one sine plate on top of another with pivot points located at 90 degrees to each other. It is used for measuring compound angles. **Figure 2.4.103** shows a sine plate and a compound sine plate.

Sine Vises

A sine vise is a sine bar with an integral vise for securing parts. **Figure 2.4.104** shows a sine vise.

SURFACE FINISH MEASUREMENT

Surface finish refers to the texture of the surface of a machined part. Roughness and waviness are the two factors generally considered when discussing surface finish. Roughness is the peaks and valleys created by the cutting action of a machining process. Waviness is the variation of those peaks and valleys over a larger distance. **Figure 2.4.105** shows the difference between roughness and waviness.

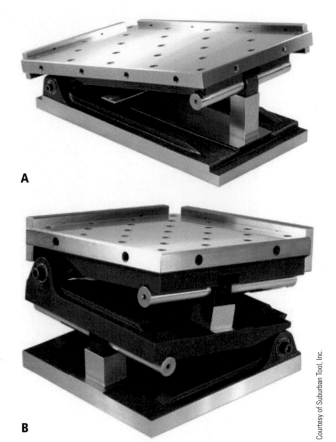

FIGURE 2.4.103 (A) A sine plate. (B) A compound sine plate.

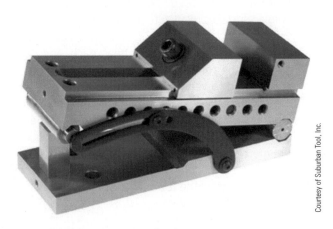

FIGURE 2.4.104 An example of a sine vise.

There are several systems for measuring surface finish, but the most widely used system measures roughness and is called Ra, or arithmetical average. It is a measure of the average height of a surface above a line that is midway between the highest peak and lowest valley of a surface within a given waviness length. **Figure 2.4.106** illustrates Ra roughness measurement. The Ra system measures this average height in the English system in microinches (millionths of an inch). In the metric system, the height is specified in micrometers (millionths of a meter). Microinch finishes range from

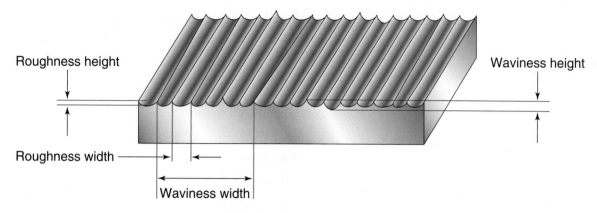

Surface characteristics

FIGURE 2.4.105 Surface finish roughness and waviness.

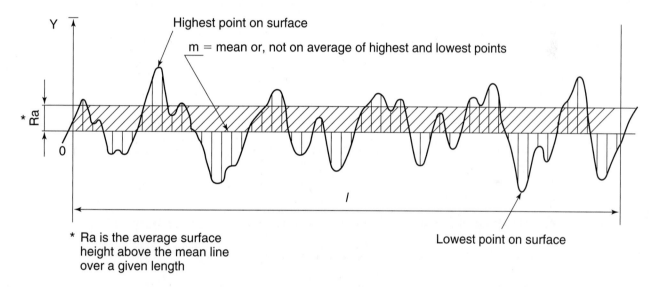

FIGURE 2.4.106 The basic concept of Ra surface finish measurement.

0.5 to 1000 microinches, and micrometer finishes range from 0.012 to 50 micrometers. Smaller values indicate smoother surfaces, and smoother surfaces are usually more difficult and expensive to produce. **Figure 2.4.107** shows surface finish ranges that can be produced by some common machining operations.

Different machining operations produce different texture patterns. The pattern and direction of the texture is called the lay. This is directly related to the machining process that produces the surface. **Figure 2.4.108** illustrates some different lay patterns and shows the symbols or letters that can be used to specify those specific processes on a print.

Ra surface finish requirements can be simply shown as a general note on a print like "125 microinch finish." If a particular machining process is required the note may read "125 milled microinch finish." Surface finish specifications may also be shown using a symbol and

values. They can be very simple and show only the maximum allowable Ra, or much more detailed and include maximum and minimum Ra, maximum waviness height and spacing, a sampling length (to be used to measure Ra), a machining allowance (amount of material to be removed by machining), and a lay symbol. **(See Figure 2.4.109.)**

Surface Roughness Comparator

Ra surface roughness is normally inspected by one of two different methods. A surface roughness comparator gage shows examples of different levels of Ra. Machined surfaces are visually compared to these samples to determine if they are within acceptable limits. **(See Figure 2.4.110.)** This method does not give an actual measurement in microinches or micrometers, but it is simple, highly portable, cost effective, and adequate for many applications.

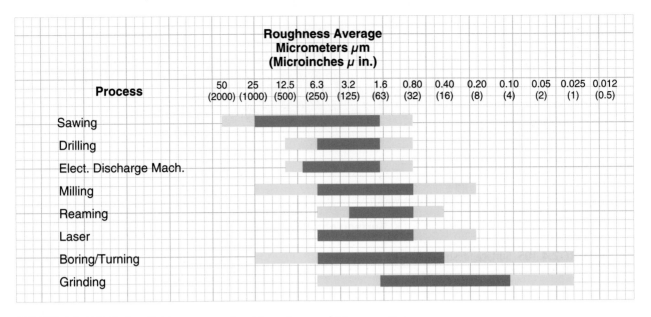

FIGURE 2.4.107 Surface finish ranges produced by various machining operations.

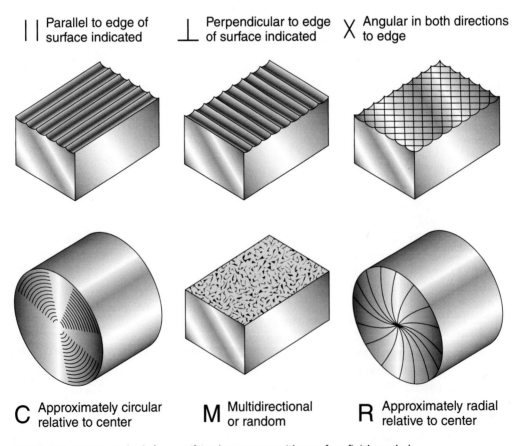

| | Parallel to edge of surface indicated

⊥ Perpendicular to edge of surface indicated

X Angular in both directions to edge

C Approximately circular relative to center

M Multidirectional or random

R Approximately radial relative to center

FIGURE 2.4.108 Methods for specifying lay patterns with a surface finish symbol.

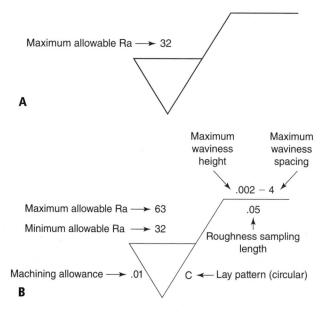

A

Maximum allowable Ra ⟶ 32

Maximum waviness height

Maximum waviness spacing

.002 – 4

Maximum allowable Ra ⟶ 63

Minimum allowable Ra ⟶ 32

.05

Roughness sampling length

Machining allowance ⟶ .01

C ⟵ Lay pattern (circular)

B

FIGURE 2.4.109 Surface finish symbols in the machining industry can range from simply showing maximum Ra (A) to showing very detailed specifications (B).

FIGURE 2.4.111 A portable profilometer measuring the surface finish of the hole in a machined component.

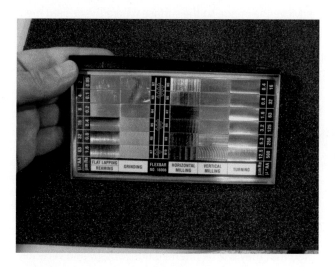

FIGURE 2.4.110 A surface finish comparator gives examples of machined surfaces at different levels of roughness.

Profilometer

A more accurate measurement of Ra surface roughness can be obtained with a **profilometer**. This electronic tool moves a stylus, or contact point, across a surface and actually measures the height of the peaks and valleys. An actual value in microinches or micrometers is then automatically calculated and shown on a display. **Figure 2.4.111** shows a portable model.

OPTICAL COMPARATORS

An **optical comparator** projects a magnified image of a part on a screen for measurement. They are very useful for measuring small parts and features that are difficult to see with the naked eye. The image is projected as a shadow of the part's profile on a screen. Different levels of magnification are available, and most models offer interchangeable lenses to change magnification levels. Some optical comparators also have an option called surface illumination. This feature uses fiber optics to show an actual surface of a part for visual inspection or to show edges where light cannot pass through to create a shadow cast image. Because the optical comparator uses a magnifying lens and a mirror to project the part image on a screen, the user must keep in mind that the part image is inverted on the screen. Some models have a method to correct this, but they are much more expensive and not very common. This must be considered when measuring using an optical comparator.

The work area of the comparator has three movements for part positioning. One movement is to adjust focus of the part image on the screen. The other two move the part so the image on the screen can be measured. One is an elevating knob to move the part up and down. The other is a table knob to move the part left and right. They can contain micrometer scales or be connected to a digital display with a readout that shows the distance moved on a display. Some displays can also perform mathematical calculations to measure various shapes and sizes. **Figure 2.4.112** shows an optical comparator.

The screen of an optical comparator usually has two lines etched on it that represent an X- and Y-axis. These lines can be used to align part surfaces for measurement. The screen can also be rotated to align those axis lines with angular part surfaces for measuring angles. The angular scale commonly contains a vernier scale graduated

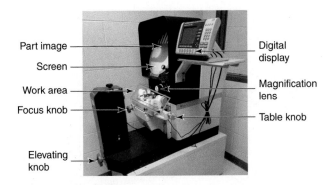

FIGURE 2.4.112 An optical comparator magnifies and projects images of small parts on a screen for measurement.

in 2-minute increments that is read like a vernier protractor. Linear measurements can be made by moving the part using a hand-operated knob. A digital readout or a micrometer scale shows the distance moved. Follow these general steps to measure a linear distance using an optical comparator.

- Place the part on the comparator table and align it so the surface to be measured is parallel to the table and screen.
- Position the part so the image is shown on the screen and focus the display.
- Set the X- and Y-axis lines to "0" angular setting so that they will be oriented horizontally and vertically.
- Align the part surface where the desired measurement will start with the X- axis or Y-axis line.
- Make note of the micrometer reading or set the display readout to "0" for the axis that will be used to measure. See **Figure 2.4.113**.

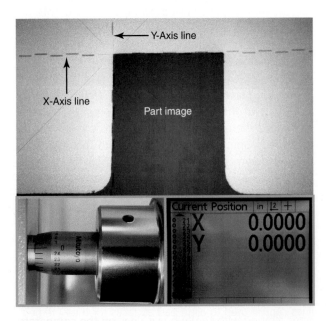

FIGURE 2.4.113 Setting the position of a part to begin measurement on an optical comparator. The micrometer and display both show "0".

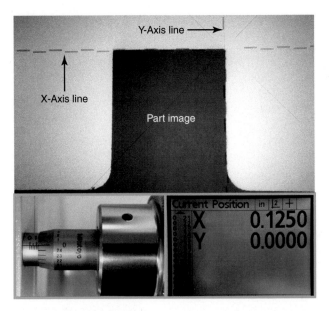

FIGURE 2.4.114 The ending measurement position of a part on an optical comparator. The micrometer and display both show a 0.125" measurement.

- Move the image so the surface where the desired measured will end is aligned with the appropriate axis line.
- Subtract the two micrometer readings for the measurement, or simply read the value shown on the display readout screen. See **Figure 2.4.114**.

To perform an angular measurement, follow these general steps.

- Place the part on the comparator table and align it so the surface to be measured is parallel to the table and screen.
- Position the part so the image is shown on the screen and focus the display.
- Set the X- and Y-axis lines to the "0" angular setting so that they will be oriented horizontally and vertically.
- Align the part surface where the desired angular measurement will start with the X-axis or Y-axis line.
- Rotate the screen until the appropriate axis line is aligned with the second surface of the angle to be measured and read the angular scale. The X or Y positions (or both) may need to be adjusted to obtain this alignment. Read the measurement from the angular scale (**Figure 2.4.115**).

Since the optical comparator normally requires light to pass through an area to create images, some part areas cannot be measured by the standard shadow image method because light cannot pass through the area. The surface illumination feature can be used in cases like this because it displays the actual part surface on the screen. Measurements are then performed the same as when using normal projection. **Figure 2.4.116** shows an example of surface illumination.

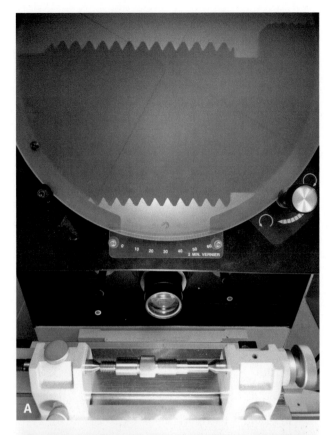

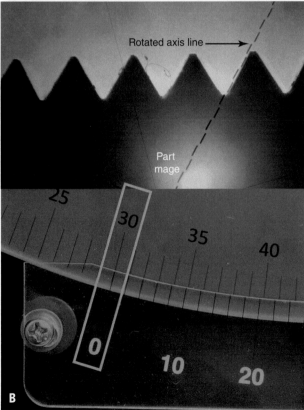

FIGURE 2.4.115 (A) Performing an angular measurement of the shape of a thread with an optical comparator. (B) The magnified thread image and its 30-degree angular measurement. Note the 2-minute graduations on the vernier scale.

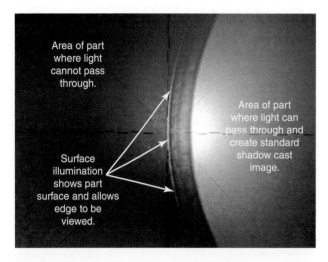

FIGURE 2.4.116 The surface illumination feature of an optical comparator shows the actual part surface on the screen and can be used when measuring areas where light cannot pass through.

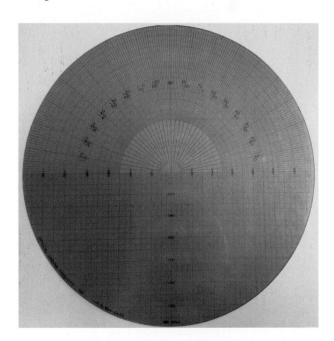

FIGURE 2.4.117 A comparator overlay chart.

Another method for making measurements on an optical comparator is to use a clear overlay chart that is placed on the comparator screen. The chart material is transparent but has gridlines at 0.001" increments for linear measurements, curved lines at 0.001" increments for radius measurements, and angular lines at 1 degree increments for angular measurements. **Figure 2.4.117** shows an overlay chart.

To use an overlay chart, it must first be mounted on the comparator screen and the chart's center axis lines aligned with the two axis lines on the comparator screen. When part images are projected on the screen, the lines on the chart can be used for measurement by aligning them with part edges. The comparator screen can still be rotated if needed to position the lines on the chart with part edges. **Figure 2.4.118** shows an overlay chart being used to measure the width of a slot and a radius at the end of the slot. **Figure 2.4.119** shows an overlay chart being used to measure an angle.

As mentioned previously, some digital displays can perform mathematical calculations. Point locations can be saved, then the display can calculate measurements using those points such as radius/diameter size, angles, or converting X and Y coordinates to polar coordinates. **Figure 2.4.120** shows a display that has converted X and Y coordinates to polar coordinates.

If a display to perform conversion is not available, polar coordinates can also be found using basic trigonometry. The straight-line distance can be found using the Pythagorean theorem with the X and Y values as sides *a* and *b* of a triangle. Sine or cosine formulas can be used to determine the angle.

TOOLMAKER'S MICROSCOPE

Another tool that is sometimes used to measure or inspect very small parts is the **toolmaker's microscope**. These microscopes feature movement in two or three directions through the use of micrometer dials

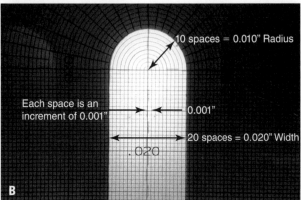

FIGURE 2.4.118 (A) Using an overlay chart on an optical comparator to measure the width of a slot and a radius. Note that the image is inverted on the screen. (B) The lines on the chart show the slot width is 0.020" and the radius is 0.010".

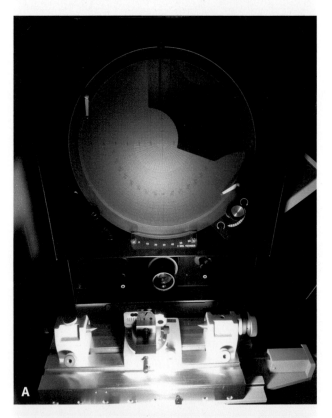

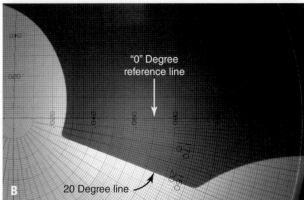

FIGURE 2.4.119 (A) Using an overlay chart on an optical comparator to measure an angle. Note that the image is inverted on the screen. (B) The lines on the chart show that the angle is 20 degrees.

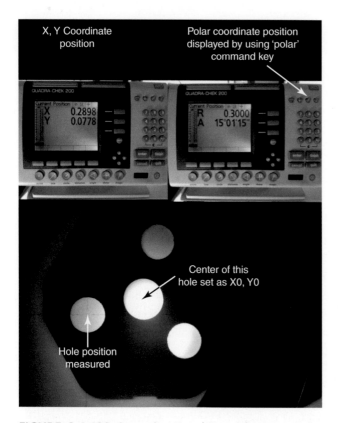

FIGURE 2.4.120 Converting X and Y coordinates to polar coordinates on the display of an optical comparator.

to obtain measurements. Some models can be connected to cameras to take still photos. Others can be connected directly to a computer monitor to display real-time video of parts during inspection. The part to be measured is placed in the work area. After focusing the image while looking through the eyepiece, the micrometer dials are used to position the part. Measurements can be taken directly from the micrometer dials or by using software on a connected computer. **(See Figure 2.4.121.)**

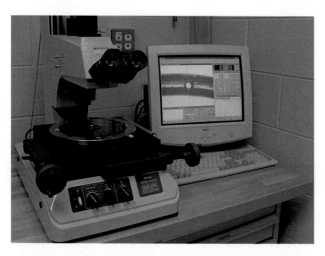

FIGURE 2.4.121 A toolmaker's microscope with video output for display on a computer.

COORDINATE MEASURING MACHINE

A **coordinate measuring machine (CMM)** identifies locations in an X, Y, Z coordinate system. The foundation of a CMM is usually a granite surface plate where parts can be secured for inspection. A bridge holds a moveable electronic probe that is used to touch surfaces of a part. When the probe touches a part, locations are transmitted to a computer that calculates dimensions and displays them on the computer monitor. The user establishes an origin (X0, Y0, Z0) on the workpiece first. Then a desired feature type is selected in the CMM software, such as a hole location. The probe is then touched against the inside surface of the hole at a few points, and the software can calculate the hole size, location, and if the hole is out of round. Some CMMs are controlled by CNC systems and can be programmed so that the probe automatically moves across a part's surfaces and compares the part to a computer model. A CMM is shown **Figure 2.4.122**.

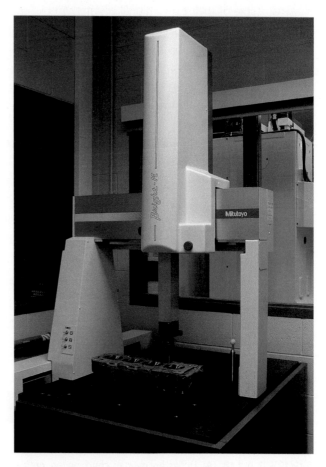

FIGURE 2.4.122 A coordinate measuring machine can inspect dimensions of complex parts like this engine cylinder head by moving the probe to touch part surfaces.

SUMMARY

- Use and care of precision measuring tools is a vital step in training for the machining industry.
- Periodic calibration of measuring tools is important to ensure they are properly functioning and will provide accurate measurements.
- Fixed gages such as pin gages and go/no-go plug and ring gages are often used to check that internal and external straight and threaded diameters are within required specifications.
- Straight edges are very flat bars used to check the accuracy of flat surfaces and are often used together with feeler gages.
- When performing precision measurements, a surface plate provides an accurate reference surface.
- A solid square is a very useful tool for checking perpendicularity of workpiece surfaces and is also often used together with feeler gages to determine perpendicularity errors.
- Vernier scales are used in several different types of precision measuring tools including calipers, depth gages, height gages, and protractors. A linear English vernier scale is graduated in 0.001" increments and a linear metric vernier scale is graduated in 0.02-mm increments. A vernier protractor is graduated in 5-minute (1/12-degree) increments.
- Micrometers use a very accurate screw thread to perform measurement and several different types are available to measure different types of dimensions, including outside, inside, and depth dimensions. English (inch) micrometers can measure within 0.001", or 0.0001" when outfitted with a vernier scale and metric micrometers can measure within 0.01 mm, or 0.002 mm when outfitted with a vernier scale.
- Most types of precision measuring tools are also available with dial or digital scales, making them easier to read than some of their conventional counterparts.
- Precision transfer tools, such as telescoping and small hole gages, are used to gage sizes and then must be measured using other tools to determine actual dimensions.
- Dial indicators perform measurement by displaying movements of a contact point with a needle on a dial face. The two basic styles of dial indicators are the plunge indicator and the test indicator. Each style can have either a balanced or continuous dial face and graduations ranging from 0.001" to 0.00005" Digital indicators display movements on a small screen on the indicator face.
- Gage blocks are extremely accurate blocks that can be stacked together to create desired sizes and are often used with indicators to inspect variation from a given size.
- A sine tool represents the hypotenuse of a right triangle and is used with gage blocks to measure angles very accurately.
- Surface finish is a measure of the roughness or smoothness of a part's surface and is measured in microinches or micrometers using a surface finish comparator or a profilometer.
- An optical comparator magnifies a part and projects its image on a screen for measurement.
- A toolmaker's microscope can be used to measure very small parts and dimensions that are nearly impossible to inspect using the naked eye.
- A CMM utilizes a probe to calculate dimensions using X, Y, Z coordinates.

REVIEW QUESTIONS

1. List three key points in caring for precision measuring tools.
2. What type of fixed gage can be used to check hole diameters?
3. What type of fixed gage can be used to check a threaded hole?

4. How is the "go" member of a go/no-go ring gage set identified?

5. What type of fixed gage can be used to check external threads?

6. A _____ can provide a reference plane for taking precision measurements.

7. What two other tools could be used with the answer to the previous question to check for perpendicularity?

8. _____ is the process of attaching gage blocks to each other.

9. What is the smallest graduation on an English vernier caliper?

10. What is the smallest graduation on a metric vernier caliper?

11. A micrometer uses an accurate _____ to perform measurement.

12. What is the smallest graduation on an inch micrometer with a vernier scale?

13. What is the smallest graduation on a metric micrometer with a vernier scale?

14. What is calibration and why is it important?

15. What is a transfer-type measuring tool?

16. What are the two basic types of indicator movements?

17. Briefly describe the main difference between the two indicator types from the previous question.

18. A sine tool uses the trigonometric function of sine. The length of a sine tool represents which side of a right triangle?

19. List the two most common methods for measuring surface finish.

20. What are one advantage and one disadvantage of each surface finish measurement method from the previous question?

21. A(n) _____ displays a magnified image of a part on a screen for conducting measurement.

22. What other tool can be useful for measuring very small parts?

23. What does CMM stand for?

Quality Assurance, Process Planning, and Quality Control

Learning Objectives

After completing this unit, the student should be able to:

- Define quality assurance
- Discuss the purpose of a process plan and describe its major parts
- Define and discuss the purpose of quality control
- Discuss the purpose of an inspection plan and describe its key points
- Define SPC and its purpose
- Identify and discuss the features of X-bar and R-charts

Key Terms

Control charts	Quality Assurance	Sampling plan
Inspection plan	(QA)	Statistical Process
Mean	Quality Control (QC)	Control (SPC)
Process plan	R-chart or range chart	X-bar chart
	Range	

INTRODUCTION

In the machining industry, many different secondary, or supportive, actions take place. Some companies develop formal methods and processes to be as effective and efficient as possible. Sometimes supervisors, managers, or engineers create these plans of action, but sometimes they are developed by people who perform machining, including machine operators, machinists, programmers, or others. These methods can be driven by a company's commitment to certain standards and take place in the stages of planning, operation, and evaluation.

QUALITY ASSURANCE

During the planning stage, a machining company will decide how to perform to meet customers' needs. **Quality assurance**, or **QA**, can be defined as the creation of a system of activities used to make sure that happens. It is an overall view or commitment to meeting demands, or a master plan. It is not the performance of specific tasks or actions, but the plan that guides tasks and actions. Supervisors and managers in machining companies may develop methods for ensuring that products will be acceptable to customers. Those are quality assurance activities.

THE PROCESS PLAN

During the planning stage, methods for producing machined components can be developed. A **process plan** can be defined simply as a strategy of steps needed to perform a machining operation or operations. It begins with selecting the proper material and ends when a part is ready for final inspection. Probably, no two process plans from any two companies will be identical. Some may be very simple or only verbal, and others may be very detailed written documents, but all process plans contain the same basic information. It is important to develop or use an existing plan to make sure products are produced to specifications and no required steps are missed.

The process plan is implemented during operation phases of machining. Keep in mind that nothing is ever perfect, and that even an established process plan might be improved. Continuous improvement should always be on the minds of machining professionals, and if you see a way that might be better, talk to your supervisors or instructors. They are usually willing to look at ideas that could improve a process.

Work orders, job cards, routers, lot travelers, and standard operating procedures (SOPs) are a few examples of process plans. Often these documents travel with parts through machining steps and require information to be recorded and signatures of people performing particular steps. This keeps track of the process, so if errors occur they can be identified and corrected.

The process plan's major steps involve material selection, machine selection, tooling selection, and speed and feed calculation. Process plans establish the steps needed to answer these questions: What is the part made from? What machine or machines will make the part? What tools will be used in those machines to make the part? How fast will those tools and machines operate to make the part? There may also be details about cleaning, packaging, or other required steps that are not directly related to machining. Refer to the sample process plan in **Figure 2.5.1** during discussion of these factors to locate information that answers these questions.

Material Selection

This part of the process plan lists the material or materials needed to complete a job. In a small company, material description might be as simple as "mild steel." Different options may be given for a material, such as *AISI/SAE 303 or 304 stainless steel*. One material type may be listed, such as *6061-T6 aluminum* or *AISI/ SAE P20 mold steel*. Companies with very controlled operations may be so specific as to list a company part number for a material. Sometimes a specific lot number of a material needs to be recorded so that if a part fails, it can be traced all the way back to its beginning to see if there was a problem with the raw material.

Machine Selection and Workholding

This section lists the pieces of equipment that will be used to perform specific operations to machined parts. Some of today's complex components require several different types of machining operations. A part may start at a band saw, and then progress through operations on a lathe, a milling machine, a heat-treating furnace, a precision grinder, and a wire EDM. If there are hand-tool operations, they will usually be listed here also. If these operations are not listed or are followed in the wrong order, parts can be ruined.

Workholding devices can be described here too. These are the methods used to secure parts during machining.

Tooling Selection

The tooling selection section gives instructions about particular cutting tools that will be used on machines to perform each machining step. It may list general sizes or types or a company part number for a specific tool. Hand tools might also be listed if their use is required.

PMT Inc. Manufacturing Process Router

Part name	Spacer block
Part #	001-22
Customer	XYZ Enterprises
Order #	1032
Quantity	2700
Material	AISI/SAE 4140, $1'' \times 2''$
Material lot #	2123-02

Operation 1	*Description*	Saw blanks
	Machine	Horizontal band saw
	Workholding	Saw vise
	Tooling	6–8 pitch bi-metal blade
	Speed/feed data	250 SFPM
	Date/time completed	
	Initial	

Operation 2	*Description*	CNC milling
	Machine	VMC 1, 2, or 3
	Workholding	1st position vise, 1″ parallels
	Program	001-22-1
	Tooling	2.5″ carbide face mill
		1″ carbide endmill
		1/2″ Spot drill
		13/32″ HSS drill
	Speed/feed data	Controlled in program
	Date/time completed	
	Initial	

Operation 3	*Description*	Deburr 13/32 holes
	Machine	VMC 1, 2, or 3
	Workholding	2nd position vise, 1″ parallels
	Program	001-22-2
	Tooling	1/2″ 90-deg. carbide C' sink
	Speed/feed data	Controlled in program
	Date/time completed	
	Initial	

Operation 4	*Description*	Cleaning
	Machine	Parts washer
	Workholding	N/A
	Tooling	N/A
	Speed/feed data	N/A
	Date/time completed	
	Initial	

FIGURE 2.5.1 A fairly simple example of a process plan for producing a machined component.

Speed and Feed Calculation

Different machine tools, operations, cutting tool types, and materials determine how fast a machine will operate. Speed refers to rotational speed, and feed refers to how fast a tool advances through or across a workpiece. Speeds and feeds must be calculated and applied properly or costly damage to parts, tooling, and machinery can occur.

Other Information

Other details and directions might also be included in a process plan. Specific methods of cleaning components at different stages may be noted. Use of common fasteners and assembly instructions may be required. Packaging of parts might even be included.

QUALITY CONTROL

Wherever and whenever machined parts are produced, there is a need for some type of inspection of dimensions. While quality assurance sets methods for doing that to evaluate the results of the process plan, **quality control (QC)** is the name given to the actions of inspecting dimensions to make sure tolerances are met. Think of quality control as another action part of a quality assurance plan. Inspection may be performed using basic conventional measuring tools such as micrometers or indicators, or highly sophisticated specialty instruments such as CMMs.

Sampling Plan

Quality control is used to inspect dimensions, but how many parts should be inspected and how often? The **sampling plan** states how many parts should be inspected from a given batch or during a given time period. The number of parts inspected is called the sampling group or subgroup. These two terms will both be used as we continue. This sample of parts from a machining operation provides a very good representation of all of the parts being produced and can reasonably confirm that all parts are within tolerances without the need for 100 percent inspection.

Inspection Plan

The sampling plan tells how many and/or how often to inspect dimensions. An **inspection plan** tells what dimensions to inspect and what measuring tools and processes to use during inspection. Just like process plans and sampling plans, inspection plans can be very different from company to company, but they all accomplish the same task. Formal inspection plans are generally used in CNC or other production environments where large numbers of the same parts are machined or when complex shapes are produced.

Selection of Critical Dimensions

The first task when creating an inspection plan is selection of critical dimensions. The definition of a critical dimension is open to interpretation, but often they are dimensions with the smallest or tightest tolerances. Another point to consider is whether multiple operations or multiple machines are used to machine a part. If so, relationships between dimensions produced by different operations may be called critical because of variations in how the part is secured during those operations. A critical dimension might also be one that aligns mating parts. If different cutting tools perform machining operations, the first and/or last dimension produced by each tool may also be considered critical to be sure that each tool is performing well enough to meet print tolerances.

Selection of Measuring Tools and Procedures

Once decisions are made on critical dimensions, measuring tools and procedures must be selected. Tool choice and inspection methods depend on the tolerance of the dimension being inspected. Consider a linear dimension with a tolerance or allowable variation of 0.0001". Certainly a rule could never measure accurately enough to inspect this tolerance. A vernier or dial caliper would not be suitable either since their scales only read to 0.001". Even a micrometer with a vernier scale would be questionable. Even though its scale reads to 0.0001", the concern would be that the dimension might not be measured consistently each time. In the case of a 0.0001" tolerance, a test indicator, gage blocks, and surface plate might be the best choice. On the other hand, if a tolerance were 1/16", a rule would be sufficient and use of a dial test indicator, gage blocks, and surface plate would require much more time and effort than needed. That would be inefficient, unnecessary, and too costly.

Figure 2.5.2 shows an example of a combination sampling and inspection plan. The document tells how many parts to inspect and when, as well as what dimensions to inspect and the tools and methods to perform the inspection. It also provides space for the inspector's initials and to record measurements, dates, and times. The subgroup summary sections showing average and range will be discussed shortly.

Statistical Process Control (SPC)

Statistical process control (SPC) is a sophisticated method for tracking variation in sizes of machined parts. SPC makes use of the dimensions specified in an inspection plan from the sample parts outlined in a sampling plan. SPC plots data on two types of charts to analyze trends in part variation. They are called **control charts**. This information helps to predict the consistency of an operation and guides adjustments to the operation

PMT Inc. Sampling/Inspection Plan

Part #	00143
Part name	Spacer
Order #	101

Sampling interval	2 hours
Sampling/subgroup size	3

Critical dimensions	Inspection method
0.882 height +/– 0.002	Micrometer
1.750 width +/– 0.005	Micrometer

Subgroup #	Dimension		Dimension			Inspector	Date	Time
Part #	Print	Actual	Print		Actual			
1-1	0.882 +/– 0.002	0.881	1.750 +/– 0.005		*1.749*			
1-2	0.882 +/– 0.002	0.882	1.750 +/– 0.005		*1.750*			
1-3	0.882 +/– 0.002	0.882	1.750 +/– 0.005		*1.750*	*John*	*11-3-09*	*7:35 AM*
Subgroup #1 summary	**Average**	**0.8817**	**Subgroup #1 summary**	**Average**	**1.7497**			
	Range	**0.001**		**Range**	**0.001**			
2-1	0.882 +/– 0.002	0.882	1.750 +/– 0.005		*1.750*			
2-2	0.882 +/– 0.002	0.881	1.750 +/– 0.005		*1.750*			
2-3	0.882 +/– 0.002	0.882	1.750 +/– 0.005		*1.751*	*Jim*	*11-3-09*	*9:20 AM*
Subgroup #2 summary	**Average**	**0.8817**	**Subgroup #2 summary**	**Average**	**1.7503**			
	Range	**0.001**		**Range**	**0.001**			
3-1	0.882 +/– 0.002	0.881	1.750 +/– 0.005		*1.752*			
3-2	0.882 +/– 0.002	0.882	1.750 +/– 0.005		*1.753*			
3-3	0.882 +/– 0.002	0.882	1.750 +/– 0.005		*1.753*	*Jim*	*11-3-09*	*11:25 AM*
Subgroup #3 summary	**Average**	**0.8817**	**Subgroup #3 summary**	**Average**	**1.7527**			
	Range	**0.001**		**Range**	**0.001**			
4-1	0.882 +/– 0.002	0.882	1.750 +/– 0.005		*1.754*			
4-2	0.882 +/– 0.002	0.8815	1.750 +/– 0.005		*1.754*			
4-3	0.882 +/– 0.002	0.882	1.750 +/– 0.005		*1.754*	*Jane*	*11-3-09*	*1:25 PM*
Subgroup #4 summary	**Average**	**0.8818**	**Subgroup #4 summary**	**Average**	**1.754**			
	Range	**0.0005**		**Range**	**0**			

FIGURE 2.5.2 An example of a sampling/inspection plan. It shows that two dimensions of three parts should be inspected every 2 hours and what inspection methods should be used.

before it starts to produce parts outside of tolerances. SPC software can automatically create these charts by inputting measurement data. The data port on some measuring tools (usually digital models) can be connected with a special cable to a computer running SPC software. Then measurements can be sent to the software and be automatically processed for creation of these charts.

X-Bar Charts

The **X-bar chart** graphs the average size, or **mean** size, of each sampling. It contains an upper control limit (UCL) and a lower control limit (LCL) line. According to a sampling plan, the average of the sampling's part sizes is recorded. This is done at each sampling labeled along the X-axis, and size values are plotted on the Y-axis. A mean line shows the average of the subgroup's average sizes. The X-bar chart shows the average sizes

produced by the operation over time. **Figure 2.5.3** shows an X-bar chart and labels its major parts.

When the X-bar chart stays between the UCL and the LCL, the process is said to be "in control." **Figure 2.5.4** shows an X-bar chart using the subgroup summary data for averages of the 0.882" dimensions from the sampling/inspection plan in **Figure 2.5.2**.

When the line of the X-bar chart shows a trend approaching the UCL, steps should be considered to adjust the machining operation to keep parts closer to the desired size and flatten out the line. When movements like this happen, the mean line will also shift upward, because sample part sizes are increasing. (The trend is moving toward the UCL.) The subgroups using the 1.750" print dimension from the sampling/inspection plan from Figure 2.5.2 would result in this type of X-bar chart, as shown in **Figure 2.5.5**.

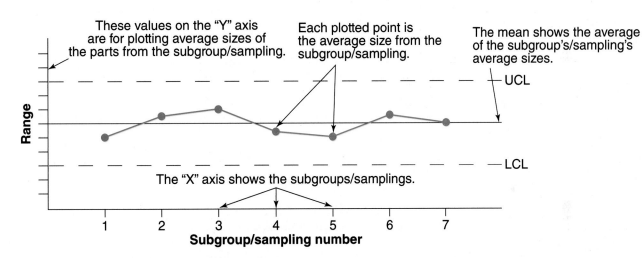

FIGURE 2.5.3 Explanation of the parts of an X-bar chart.

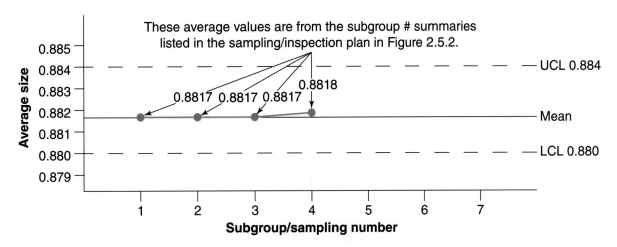

FIGURE 2.5.4 This X-bar chart shows that the machining process is "in control" because the average of the subgroups is well within the UCL and the LCL. Note that the mean line is close to the midway point between the UCL and the LCL.

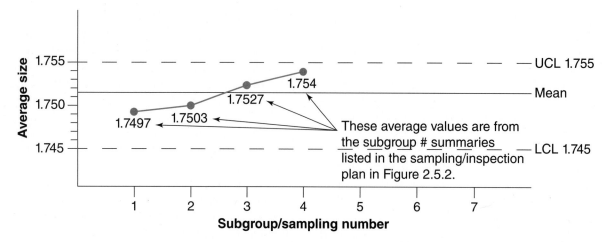

FIGURE 2.5.5 If an X-bar chart shows a trend of moving toward the UCL, steps should be taken to adjust the machining process to bring the process closer to the desired size. Note that the mean line has also moved toward the UCL. That is because the mean (or average) sizes of the sample parts are increasing.

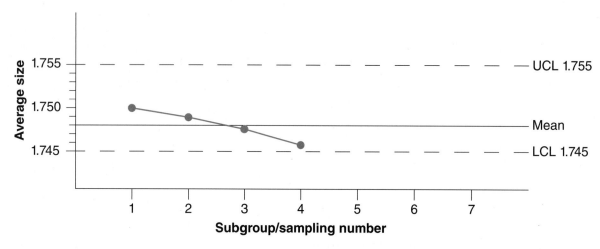

FIGURE 2.5.6 If an X-bar chart shows a trend of moving toward the LCL, steps should be taken to adjust the machining process to bring the process closer to the desired size. Note that the mean line has also moved toward the LCL. That is because the mean (or average) sizes of the sample parts are decreasing.

The opposite situation is when the line of the X-bar chart shows a trend approaching the LCL. In cases like these, steps should also be considered to adjust the machining process. The mean line will instead shift downward, because sample part sizes are decreasing. (The trend is moving toward the LCL.) **Figure 2.5.6** shows an X-bar chart depicting this type of situation.

Large spikes that alternately approach the UCL and LCL also point to the need for investigation. Even though the mean appears to be in a good location, these large variations show that the process is not very stable. The X-bar chart in **Figure 2.5.7** shows this situation.

When the line crosses the UCL or LCL, the process is said to be "out of control." Machining processes must be stopped because parts are being produced that are out of tolerance. Then adjustments need to be made

to bring the process back "in control." Recall that the X-bar chart in Figure 2.5.5 was approaching the UCL. If no adjustments were made at the time of sampling 4, it is likely that part sizes would have continued to increase and the graph would cross the UCL, as shown in **Figure 2.5.8**.

R-Charts

The **R-chart** or **range chart** shows the amount of variation of each sampling. The **range** is calculated by subtracting the smallest dimension from the largest dimension in the sampling. Those range values are plotted on the Y-axis at the sampling intervals shown by the X-axis. The R-chart shows the size variation (taken from the sampling) at any given time during an operation. **Figure 2.5.9** shows an R-chart and labels its major parts.

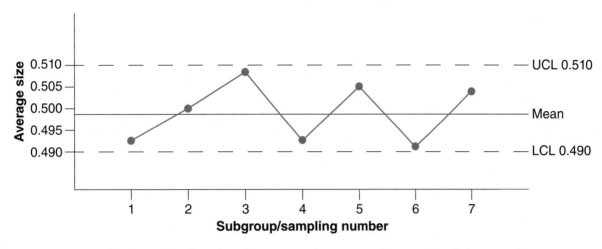

FIGURE 2.5.7 Large spikes in an X-bar chart show that a process is not very stable, even though the mean size seems acceptable. Steps should be considered to stabilize this operation because of the large variation and because sizes are getting close to both the UCL and LCL.

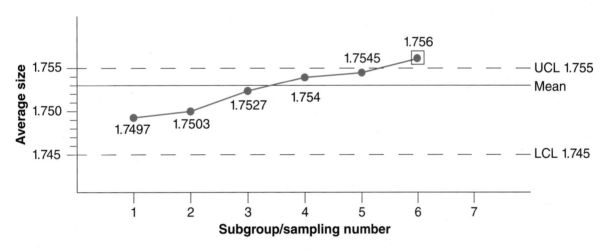

FIGURE 2.5.8 This X-bar chart is most likely what would have happened with the process from **Figure 2.5.5** if no corrective actions were taken. Sizes have continued to increase and the graph has risen above the UCL. The machining process is "out of control." Note that the mean has also moved closer to the UCL.

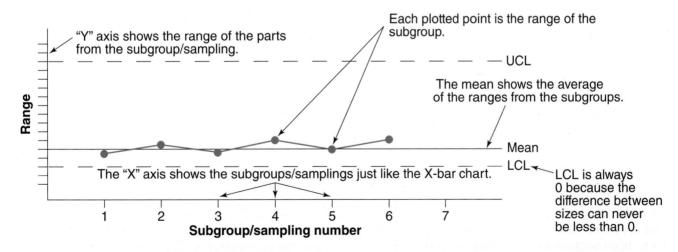

FIGURE 2.5.9 Explanation of the parts of an R-chart.

Consider the ranges of the 0.882" dimensions from the sampling/inspection plan in Figure 2.5.2. From subgroups 1–4, the ranges are 0.001, 0.001, 0.001, and 0.0005 inch. **Figure 2.5.10** shows these values plotted on an R-chart. The graph of low values close to the LCL is good and shows that there is little variation in part sizes within the sampling.

An R-chart of the ranges from the 1.750" dimension would produce a similar graph. While the low amount of variation is good within each sampling, it only tells part of the story of the process. Remember that the X-bar chart showed this dimension was moving toward the UCL, indicating that part sizes were increasing. For that reason, R-charts need to be used together with X-bar charts to track machining processes. Both the X-bar and the R-chart graphs must be within their UCL and LCL for the process to be "in control."

When an R-chart shows continual increase, as shown in **Figure 2.5.11**, the operation is becoming unreliable at producing consistent sizes. There is too much variation in part sizes, and the operation should be examined for possible changes to produce parts that are more uniform in size. When the graph crosses the UCL, the process must be stopped and adjusted.

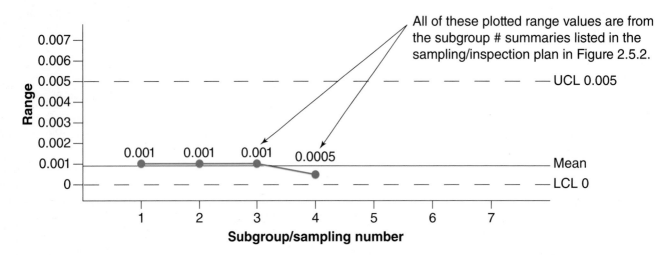

FIGURE 2.5.10 This R-chart shows low variation because the graph and the mean line are close to the LCL (0). Remember that a decreasing or low-value R-chart graph does not always mean the machining process is "in control." Analysis of the X-bar chart must also be used.

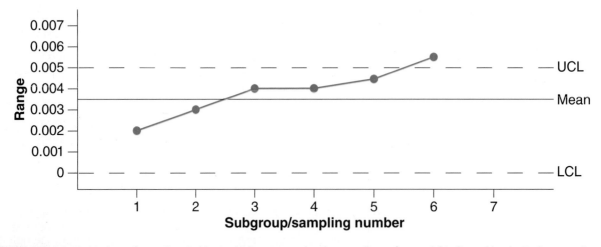

FIGURE 2.5.11 At the time of sampling 6, this machining process has become "out of control," indicated by the R-chart graph crossing above the UCL. Note that the mean is also near the UCL. This increasing R-chart graph should have caused an alert by subgroup 5. Then adjustments might have been made to prevent the range of subgroup 6 from rising above the UCL.

SUMMARY

- A basic understanding of how to plan for and meet engineering and customer requirements is important in the machining industry.

- Quality assurance is an overall plan and commitment to manufacturing products that will be functional and meet customer requirements.

- Quality control is the process of, or activities involved in, inspecting part dimensions specified on engineering drawings and is part of a quality assurance program.

- A process plan shows the method for performing a machining operation or series of operations. It can list items such as machines and tools required, material selection, and machine settings. Work orders, job routers, and SOPs are all forms of process plans.

- A sampling plan tells how many parts and how often to inspect parts produced by a machining process and is used to ensure that machined parts are within specifications without the need for 100% inspection.

- An inspection plan shows the appropriate tools required to measure critical part dimensions to ensure they are within required tolerances.

- SPC utilizes X-bar charts and R-charts to track trends in machining operations by graphing data recorded according to the sampling and inspection plans.

- QA, process planning, QC, and SPC are tools used in machining to ensure that required levels of accuracy and customer satisfaction are met.

REVIEW QUESTIONS

1. What are the four basic parts of a process plan?
2. Briefly define quality control and explain its purpose.
3. Briefly explain the purpose of a sampling plan.
4. What is the purpose of an inspection plan?
5. What is the most important factor to consider when choosing the proper measuring tool to inspect a dimension?
6. What does SPC stand for?
7. What does an X-bar chart track?
8. What does an R-chart track?
9. If an X-bar chart graph is between the LCL and the UCL, a machining process is said to be _____.
10. If an X-bar chart graph falls below the LCL or rises above the UCL, a machining process is said to be _____.

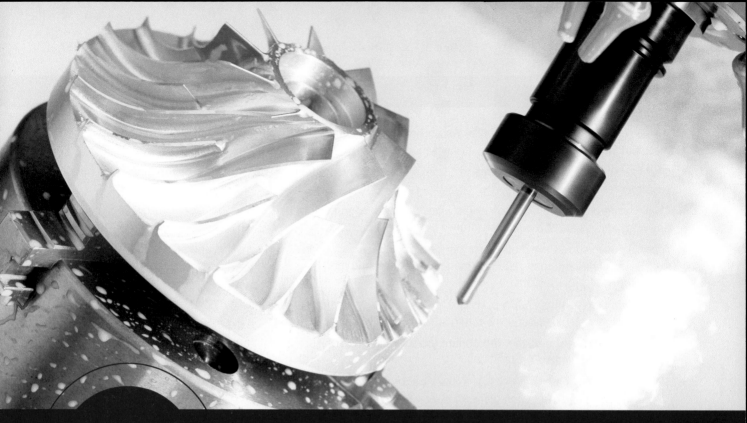

UNIT 6

Metal Composition and Classification

Learning Objectives

After completing this unit, the student should be able to:

- Describe the difference between ferrous and nonferrous metals
- Compare and contrast low-, medium-, and high-carbon steels
- Define an alloy and an alloying element
- Describe the differences/similarities between steel and cast iron
- Demonstrate understanding of the AISI/SAE system of classification for steels
- Demonstrate understanding of UNS classification of carbon and alloy steels
- Demonstrate understanding of AA/IADS classification of aluminum alloys
- Identify UNS designations for stainless steels
- Identify UNS designations for cast iron
- Identify UNS designations for nonferrous alloys

Key Terms

Alloy
Alloy steel
Aluminum alloy
Aluminum
  Association of the
  United States (AA)

American Iron and
  Steel Institute (AISI)
American Society
  for Testing and
  Materials (ASTM)
Brass

Bronze
Cast iron
Copper alloy
Ferrous

International Alloy Designation System (IADS)
Iron
Magnesium
Nonferrous metal
Plain carbon steel

Society of Automotive Engineers (SAE)
Stainless steel
Superalloy
Titanium alloy
Tool steel

Unified Numbering System (UNS)
Wrought iron

INTRODUCTION

There are many different types of metals used in the machining industry, depending on the product being produced. Because of this large variety of metal types, workers in the field may be exposed to a wide range of those metals with very different characteristics. A good starting point is to first divide metals into two major categories: ferrous and nonferrous.

Ferrous metals are metals that contain **iron**. Most ferrous metals are magnetic. One way to remember ferrous is to think about the symbol for the element iron, Fe, since fe is the beginning of the word ferrous. Iron ore is rock found naturally in the earth that contains the element iron. It is mined and then goes through several refining processes to create steel or cast iron.

Nonferrous metals are metals that contain no iron. Aluminum, copper, magnesium, and titanium are examples of nonferrous metals.

Several organizations have developed systems for classifying or designating metals over several decades. One is the **American Iron and Steel Institute (AISI)**. The **Society of Automotive Engineers (SAE)** is another. Some systems are a combination of both AISI and SAE efforts. The **American Society for Testing and Materials (ASTM)** plays a part in metal classification as well. Another system is the **International Alloy Designation System (IADS)**. All of these systems use different methods for classification.

The differences between AISI, SAE, IADS, and other systems for identifying metals have become difficult to interchange and cross-reference. The **Unified Numbering System (UNS)** was developed to overcome that confusion. It is becoming more widely used because it uses one standard numbering method instead of several different methods. A letter prefix designates a particular type of metal, such as steel, aluminum, or copper. The letter is followed by a five-digit number that specifies a certain alloy within the group. An **alloy** is a metal that is a combination of two or more metals or a metal and a nonmetal element. **Figure 2.6.1** shows the general UNS method of classification for all metals.

> **NOTE:**
>
> Throughout this unit, many tables will list the elements contained in different metals. Those elements are listed by their chemical symbols in those tables. **Figure 2.6.2** shows the names of elements listed by their chemical symbols.

FERROUS METALS

Wrought Iron

The first refinement of iron ore produces **wrought iron**. This very soft iron with only a small amount of carbon is normally used for decoration because it is easily shaped and formed. It is not commonly used in the machining industry.

Plain Carbon Steels

When molten iron is heated and the appropriate amount of carbon and impurities are removed, the resulting metal is called **plain carbon steel**. Steel is made by either hot-rolled or cold-rolled (or cold-drawn) methods. Hot-rolled steel, called HRS, is formed to size when it is red hot. When it cools, it develops a nearly black outer scale. When machining HRS, it is good practice to penetrate the scale with the first cutting operations because the scale is abrasive and can cause higher-than-normal tool wear. Cold-rolled steel, called CRS, is formed below red-hot temperature by being pulled through rollers and has a smoother, gray surface. CRS contains more internal stress than HRS and will warp and twist more during machining. Careful consideration needs to be taken when deciding how to machine CRS to minimize distortions. **Figure 2.6.3** shows the difference in appearance between HRS and CRS.

Steel is generally defined as low-, medium-, or high-carbon steel. Low-carbon or mild steels contain about 0.05–0.3 percent carbon and can be bent and formed easily. Their uses include body panels, frame components, and suspension components for the automotive industry.

Table 1: Unified Numbering System (UNS) for Metals and Alloys

UNS Series	Metal
A00001 to A99999	Aluminum and aluminum alloys
C00001 to C99999	Copper and copper alloys
D00001 to D99999	Specified mechanical property steels
E00001 to E99999	Rare earth and rare earthlike metals and alloys
F00001 to F99999	Cast irons
G00001 to G99999	AISI and SAE carbon and alloy steels (except tool steels)
H00001 to H99999	AISI and SAE H-steels
J00001 to J99999	Cast steels (except tool steels)
K00001 to K99999	Miscellaneous steels and ferrous alloys
L00001 to L99999	Low-melting metals and alloys
M00001 to M99999	Miscellaneous nonferrous metals and alloys
N00001 to N99999	Nickel and nickel alloys
P00001 to P99999	Precious metals and alloys
R00001 to R99999	Reactive and refractory metals and alloys
S00001 to S99999	Heat-and corrosion-resistant (stainless) steels
T00001 to T99999	Tool steels, wrought and cast
W00001 to W99999	Welding filler metals
Z00001 to Z99999	Zinc and zinc alloys

From Machinery's Handbook 28, Industrial Press, New York, 2008

FIGURE 2.6.1 Classification of metals according to the Unified Numbering System (UNS).

CHEMICAL SYMBOL	ELEMENT	CHEMICAL SYMBOL	ELEMENT
Al	Aluminum	O	Oxygen
C	Carbon	P	Phosphorous
Co	Cobalt	Pb	Lead
Cr	Chromium	Pd	Palladium
Cu	Copper	S	Sulfur (Sulphur)
Fe	Iron	Si	Silicon
H	Hydrogen	Sn	Tin
Mg	Magnesium	Ti	Titanium
Mn	Manganese	V	Vanadium
Mo	Molybdenum	W	Tungsten
N	Nickel	Zn	Zinc
Nb	Niobium	Zr	Zirconium

FIGURE 2.6.2 Chemical symbols for names of elements frequently found in metals.

Medium-carbon steels contain about 0.3–0.5 percent carbon and are not as easily bent or formed. They are used in applications that require more strength but still need some degree of flexibility. Automobile crankshafts, camshafts, and axles are frequently machined from medium-carbon steels.

The carbon content of high-carbon steels is about 0.5–1.5 percent. They are very hard and will more easily break instead of bending. Springs, knives, and cutting tools are made from high-carbon steels.

Plain carbon steels are usually machined fairly easily, but higher-carbon steels are more difficult to cut than lower-carbon steels.

Alloy Steels

Different elements are added to steels to create different characteristics, just the way changing ingredients in a recipe changes the result. Steels that have other elements added to them are called **alloy steels**. The element or elements added to the steel are called the *alloying elements*. Some common alloying elements are chromium, manganese, molybdenum, nickel, tungsten, and vanadium.

Alloy steels have countless applications, depending on their compositions. Gears, axles, bicycle frames, and race car chasses are often machined from chromium-molybdenum (chrome-moly) steel. Drive shafts are often machined from nickel-chromium-molybdenum steels. Alloy steels are available in CRS and HRS varieties. These alloys are usually a little more difficult to machine than carbon steels because the alloying elements make them harder.

In some cases, alloying elements make steels softer instead of harder. For example, lead can be added to steel to make it easier to machine.

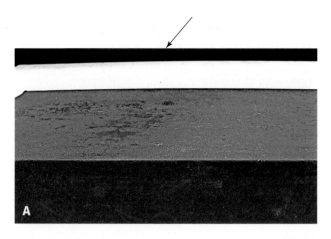

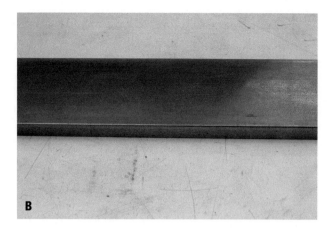

FIGURE 2.6.3 (A) Hot-rolled steel (HRS) has a rough, scaly, oxidized surface. (B) Cold-rolled steel (CRS) has a dull gray, smoother surface.

AISI/SAE Designation of Plain Carbon and Alloy Steels

The AISI/SAE numbering system is a widely used method for identifying plain carbon and alloy steels. Most steels are identified by four-digit numbers. The first two numbers show that the steel contains certain alloying elements. The last two digits stand for the amount of carbon in hundredths of a percent. Some alloy steels are identified by five digits because the carbon content is over 1 percent. An "L" in the middle means the steel contains lead and a "B" in the middle means the steel contains boron. Refer to **Figure 2.6.4**, AISI/SAE Numbering System, during the explanation of the following examples.

EXAMPLE ONE: AISI/SAE 4135

The chart shows that 41xx steels are chromium-molybedenum steels, sometimes called chrome- moly steels. The 35 means that the steel contains 0.35 percent carbon. This would be considered a medium-carbon steel.

EXAMPLE TWO: AISI/SAE 12L14

The chart shows that 12xx steels have been resulfurized and rephosphorized. Small amounts of sulfur and phosphor have been added to make the steel easier to machine. The "L" in the middle of 12L14 means that lead has also been added. Lead makes the steel even easier to machine. The steel contains 0.14 percent carbon and would be considered a low-carbon or mild steel.

EXAMPLE THREE: AISI/SAE 50105

The chart shows that 50xxx steels are alloyed with chromium. The 105 shows a carbon content of 1.05 percent. This would be a high-carbon steel.

UNS Designation of Plain Carbon and Alloy Steels

UNS numbers for plain carbon and alloy steel are closely based on the AISI/SAE numbers. For many steels, the only change is that a "G" is placed before the AISI/SAE number. Since all UNS numbers are five digits, if the AISI/SAE number is less than five digits, zeros are frequently placed at the end. Sometime those zeros are replaced with numbers to indicate special metal treatments. For example, AISI/SAE 4135 would be UNS G41350 and AISI/SAE 50105 would be UNS G50105. The top two sections of the chart in **Figure 2.6.5** show the UNS designations for carbon and alloy steels cross-referenced to AISI and SAE numbers.

Tool Steels

Tool steel refers to steels used to make tools that will bend, cut, form, or somehow "work" other metals. They contain alloying elements that make them well suited for particular applications. Molds, punches, dies, and cutting tools such as drills are made from tool steels. Tool steels are generally tougher to machine than both plain carbon and alloy steels.

AISI Designation of Tool Steels

AISI numbers for tool steels are different from those used for carbon and alloy steels. **Figure 2.6.6** shows the major categories of tool steels in this system. A one- or two-digit number would follow the prefix letters. Those numbers classify the tool steel according to the amounts of specific alloying elements, but do not stand for any specific amounts of elements like the carbon and alloy steel numbers. For example, M1 would be a high-speed tool steel with molybdenum as the major alloying element. D2 would be a high-carbon cold-work tool steel with chromium as the major alloying element. A steel used for making molds to produce plastic parts might have a number like P20. **Figure 2.6.7** shows a sample table of the M series of high-speed tool steels and their compositions.

TABLE 3: AISI-SAE SYSTEM OF DESIGNATING CARBON AND ALLOY STEELS

AISI-SAE Designation*	Type of Steel and Nominal Alloy Content (%)
	Carbon Steels
10xx	Plain Carbon (Mn 1.00% max.)
11xx	Resulpherized
12xx	Resulpherized and Rephosphorized
15xx	Plain Carbon (Max. Mn range 1.00 to 1.65%)
	Manganese Steels
13xx	Mn 1.75
	Nickel Steels
23xx	Ni 3.50
25xx	Ni 5.00
	Nickel-Chromium Steels
31xx	Ni 1.25; Cr 0.65 and 0.80
32xx	Ni 1.75; Cr 1.07
33xx	Ni 3.50; Cr 1.50 and 1.57
34xx	Ni 3.00; Cr 0.77
	Molybdenum Steels
40xx	Mo 0.20 and 0.25
44xx	Mo 0.40 and 0.52
	Chromium-Molybdenum Steels
41xx	Cr 0.50, 0.80, and 0.95; Mo 0.12, 0.20, 0.25, and 0.30
	Nickel-Chromium-Molybdenum Steels
43xx	Ni 1.82; Cr 0.50 and 0.80; Mo 0.25
43BVxx	Ni 1.82; Cr 0.50; Mo 0.12 and 0.35; V 0.03 min.
47xx	Ni 1.05; Cr 0.45; Mo 0.20 and 0.35
81xx	Ni 0.30; Cr 0.40; Mo 0.12
86xx	Ni 0.55; Cr 0.50; Mo 0.20
87xx	Ni 0.55; Cr 0.50; Mo 0.25
88xx	Ni 0.55; Cr 0.50; Mo 0.35
93xx	Ni 3.25; Cr 1.20; Mo 0.12
94xx	Ni 0.45; Cr 0.40; Mo 0.12
97xx	Ni 0.55; Cr 0.20; Mo 0.20
98xx	Ni 1.00; Cr 0.80; Mo 0.25
	Nickel-Molybdenum Steels
46xx	Ni 0.85 and 1.82; Mo 0.20 and 0.25
48xx	Ni 3.50; Mo 0.25
	Chromium Steels
50xx	Cr 0.27, 0.40, 0.50, and 0.65
51xx	Cr 0.80, 0.87, 0.92, 0.95, 1.00, and 1.05
50xxx	Cr 0.50; C 1.00 min.
51xxx	Cr 1.02; C 1.00 min.
52xxx	Cr 1.45; C 1.00 min.
	Chromium-Vanadium Steels
61xx	Cr 0.60, 0.80, and 0.95; V 0.10 and 0.15 min.
	Tungsten-Chromium Steels
72xx	W 1.75; Cr 0.75
	Silicon-Manganese Steels
92xx	Si 1.40 and 2.00; Mn 0.65, 0.82, and 0.85; Cr 0.00 and 0.65
	High-Strength Low-Alloy Steels
9xx	Various SAE grades
xxBxx	B denotes boron steels
xxLxx	L denotes leaded steels

*xx in the last two digits of the carbon and low-alloy designations indicates that the carbon content (in hundredths of a percent) is to be inserted.

FIGURE 2.6.4 The AISI/SAE system for classifying plain carbon and alloy steels. The first two digits of a steel's number designate the major alloying elements and the last two or three represent the carbon content in hundredths of a percent.

Table 2: AISI AND SAE NUMBERS AND THEIR CORRESPONDING UNS NUMBERS FOR PLAIN CARBON, ALLOY, AND TOOL STEELS

AISI-SAE Numbers	UNS Numbers	AISI-SAE Numbers	UNS Numbers	AISI-SAE Numbers	UNS Numbers	AISI-SAE Numbers	UNS Numbers
			Plain Carbon Steels				
1005	G10050	1030	G10300	1070	G10700	1566	G15660
1006	G10060	1035	G10350	1078	G10780	1110	G11100
1008	G10080	1037	G10370	1080	G10800	1117	G11170
1010	G10100	1038	G10380	1084	G10840	1118	G11180
1012	G10120	1039	G10390	1086	G10860	1137	G11370
1015	G10150	1040	G10400	1090	G10900	1139	G11390
1016	G10160	1042	G10420	1095	G10950	1140	G11400
1017	G10170	1043	G10430	1513	G15130	1141	G11410
1018	G10180	1044	G10440	1522	G15220	1144	G11440
1019	G10190	1045	G10450	1524	G15240	1146	G11460
1020	G10200	1046	G10460	1536	G15260	1151	G11510
1021	G10210	1049	G10490	1527	G15270	1211	G12110
1022	G10220	1050	G10500	1541	G15410	1212	G12120
1023	G10230	1053	G10530	1548	G15480	1213	G12130
1025	G10250	1055	G10550	1551	G15510	1215	G12150
1026	G10260	1059	G10590	1552	G15520	12L14	G12144
1029	G10290	1060	G10600	1561	G15610	...	...
			Alloy Steels				
1330	G13300	4150	G41500	5140	G51400	8642	G86420
1335	G13350	4161	G41610	5150	G51500	8645	G86450
1340	G13400	4320	G43200	5155	G51550	8655	G86550
1345	G13450	4340	G43200	5160	G51600	8720	G87200
4023	G40230	E4340	G43406	E51100	G51986	8740	G87400
4024	G40240	4615	G46150	E52100	G52986	8822	G88220
4027	G40270	4620	G46200	6118	G61180	9260	G92600
4028	G40280	4626	G46260	6150	G61500	50B44	G50441
4037	G40370	4720	G47200	8615	G86150	50B46	G50461
4047	G40470	4815	G48150	8617	G86170	50B50	G50501
4118	G41180	4817	G48170	8620	G86200	50B60	G50601
4130	G41300	4820	G48200	8622	G86220	51B60	G51601
4137	G41370	5117	G51170	8625	G86250	81B45	G81451
4140	G41400	5120	G51200	8627	G86270	94B17	G94171
4142	G41420	5130	G51200	8630	G86300	94B30	G94301
4145	G41450	5132	G51320	8637	G86370	...	...
4147	G41470	5135	G51350	8640	G86400	...	...
			Tool Steels (AISI and UNS Only)				
M1	T11301	T6	T12006	A6	T30106	P4	T51604
M2	T11302	T8	T12008	A7	T30107	P5	T51605
M4	T11304	T15	T12015	A8	T30108	P6	T51606
M6	T11306	H10	T20810	A9	T30109	P20	T51620
M7	T11307	H11	T20811	A10	T30110	P21	T51621
M10	T11310	H12	T20812	D2	T30402	F1	T60601
M3-1	T11313	H13	T20813	D3	T30403	F2	T60602
M3-2	T11323	H14	T20814	D4	T30404	L2	T61202
M30	T11330	H19	T20819	D5	T30405	L3	T61203
M33	T11333	H21	T20821	D7	T30407	L6	T61206
M34	T11334	H22	T20822	O1	T31501	W1	T72301
M36	T11336	H23	T20823	O2	T31502	W2	T72302
M41	T11341	H24	T20824	O6	T31506	W5	T72305
M42	T11342	H25	T20825	O7	T31507	CA2	T90102
M43	T11343	H26	T20826	S1	T41901	CD2	T90402

FIGURE 2.6.5 Cross-reference of AISI/SAE and UNS numbers for steel classification.

From Machinery's Handbook 28, Industrial Press, New York, 2008

Table 2: AISI AND SAE NUMBERS AND THEIR CORRESPONDING UNS NUMBERS FOR PLAIN CARBON, ALLOY, AND TOOL STEELS

AISI-SAE Numbers	UNS Numbers	AISI-SAE Numbers	UNS Numbers	AISI-SAE Numbers	UNS Numbers	AISI-SAE Numbers	UNS Numbers
Tool Steels (AISI and UNS Only) *Continued*							
M44	T11344	H41	T20841	S2	T41902	CD5	T90405
M46	T11346	H42	T20842	S4	T41904	CH12	T90812
M47	T11347	H43	T20843	S5	T41905	CH13	T90813
T1	T12001	A2	T30102	S6	T41906	CO1	T91501
T2	T12002	A3	T30103	S7	T41907	CS5	T91905
T4	T12004	A4	T30104	P2	T51602	…	…
T5	T12005	A5	T30105	P3	T51603	…	…

FIGURE 2.6.5 *continued.* Cross-reference of AISI/SAE and UNS numbers for steel classification.

Table 3: CLASSIFICATION OF TOOL STEELS

Category Designation	Letter Symbol	Group Designation
High-Speed Tool Steels	M	Molybdenum types
	T	Tungsten types
Hot-Work Tool Steels	H1–H19	Chromium types
	H20–H39	Tungsten types
	H40–H59	Molybdenum types
Cold-Work Tool Steels	D	High-carbon, high-chromium types
	A	Medium-alloy, air-hardening types
	O	Oil-hardening types
Shock-Resisting Tool Steels	S	
Mold Steels	P	
Special-Purpose Tool Steels	L	Low-alloy types
	F	Carbon-tungsten types
Water-Hardening Tool Steels	W	

From Machinery's Handbook 28, Industrial Press, New York, 2008

FIGURE 2.6.6 Categories of tool steels.

Table 6: MOLYBDENUM HIGH-SPEED STEELS

Identifying Chemical Composition and Typical Heat-Treatment Data

	AISI Type	M1	M2	M3 CL 1	M3 CL 2	M4	M6	M7	M10	M30	M33	M34	M36	M41	M42	M43	M44	M46	M47
Identifying Chemical Elements in Percent	C	0.80	0.85: 1.00	1.05	1.20	1.30	0.80	1.00	0.85: 1.00	0.80	0.90	0.90	0.80	1.10	1.10	1.20	1.15	1.25	1.10
	W	1.50	6.00	6.00	6.00	5.50	4.00	1.75	…	2.00	1.50	2.00	6.00	6.75	1.50	2.75	5.25	2.00	1.50
	Mo	8.00	5.00	5.00	5.00	4.50	5.00	8.75	8.00	8.00	9.50	8.00	5.00	3.75	9.50	8.00	6.25	8.25	9.50
	Cr	4.00	4.00	4.00	4.00	4.00	4.00	4.00	4.00	4.00	4.00	4.00	4.00	4.25	3.75	3.75	4.25	4.00	3.75
	V	1.00	2.00	2.40	3.00	4.00	1.50	2.00	2.00	1.25	1.15	2.00	2.00	2.00	1.15	1.60	2.25	3.20	1.25
	Co	…	…	…	…	…	12.00	…	…	5.00	8.00	8.00	8.00	5.00	8.00	8.25	12.00	8.25	5.00

From Machinery's Handbook 28, Industrial Press, New York, 2008

FIGURE 2.6.7 The compositions of "M" series tool steels.

UNS Designation of Tool Steels

UNS numbering for tool steels uses the prefix "T" followed by a five-digit number. The bottom section of **Figure 2.6.5 s**hows the UNS designations for tool steels cross-referenced to AISI numbers.

Cast Iron

When carbon content rises to 1.7–4.5 percent, the material is then known as **cast iron**. Cast iron also contains silicon and is recognized by its rough, scaly surface finish. **(See Figure 2.6.8.)**

The three basic types of cast iron are gray iron, malleable iron, and ductile iron. Gray cast iron is hard, stable, resistant to wear and heat, and breaks without flexing. Engine blocks and machine tool bases are often machined from gray cast iron. Malleable and ductile irons have the ability to flex and stretch before breaking. Machined parts include pipe fittings and brake rotors for automobiles.

Cast iron parts are cast by pouring molten iron into molds. The part that is produced has a rough outer scale. It can then be machined to finished sizes as required. The outer scale of cast iron is somewhat tough to penetrate but once it is removed, cast iron is readily machinable. **Figure 2.6.9** shows a rough iron casting and the same part after machining operations have been performed.

Designation of Cast Iron

The different types of cast irons can be classified by ASTM or UNS numbers. UNS designations for cast iron use the prefix "F" followed by a five-digit number to indicate the particular grade. **Figure 2.6.10** shows a cross-reference chart of some ASTM and UNS cast iron grades. HB is a measure of hardness, and higher numbers indicate higher hardness. T.S. stands for *tensile strength* measured in 1000 pounds per square inch (ksi). For example, a T.S. value of 20 ksi means the cast iron has a tensile strength of 20,000 pounds per square inch. That means it will withstand breakage when subjected to up to 20,000 pounds per square inch of pulling force.

Stainless Steels

Stainless steel refers to steels that have minimum chromium content of 10 percent. This makes them highly resistant to corrosion, or rust. Stainless steels can contain other alloying elements as well to give certain qualities. They have a slightly brighter, silver color than plain carbon and alloy steels. **(See Figure 2.6.11.)** Stainless steels are classified in three major types.

Austenitic Stainless Steels

Austenitic stainless steels are the most highly corrosion-resistant types. They contain nickel and are nonmagnetic. The food and chemical industries use machined

FIGURE 2.6.8 Cast iron is dull gray in color with a very rough surface.

FIGURE 2.6.9 A cast iron part before and after machining.

Table 4: UNS-ASTM Cast Iron Cross-Reference

UNS No	Description	ASTM Equivalent
UNS F10001	Cast Iron, Gray	ASTM A319(I)
UNS F10002	Cast Iron, Gray	ASTM A319(II)
UNS F10003	Cast Iron, Gray	ASTM A319(III)
UNS F10004	Cast Iron, Gray HB: 187 max	ASTM A159(G1800)
UNS F10005	Cast Iron, Gray HB: 170-229	ASTM A159(G2500)
UNS F10006	Cast Iron, Gray HB: 187-241	ASTM A159(G3000)
UNS F10007	Cast Iron, Gray HB: 207-255	ASTM A159(G3500)
UNS F10008	Cast Iron, Gray HB: 217-259	ASTM A159(G4000)
UNS F10009	Cast Iron, Gray HB: 170-229	ASTM A159(G2500a)
UNS F10010	Cast Iron, Gray HB: 207-255	ASTM A159(G3500b)
UNS F10011	Cast Iron, Gray HB: 207-255	ASTM A159(G3500c)
UNS F10012	Cast Iron, Gray HB: 241-321	ASTM A159(G4000d)
UNS F11401	Cast Iron, Gray. T.S.: 20ksi	ASTM A18 (20), ASTM A278 (20)
UNS F11501	Cast Iron, Gray. T.S.: 21ksi	ASTM A125 (A)
UNS F11601	Cast Iron, Gray. T.S.: 25ksi	ASTM A48 (25), ASTM A278 (25)
UNS F12101	Cast Iron, Gray. T.S.: 30ksi	ASTM A48 (30), ASTM A278 (30)
UNS F12102	Cast Iron, Gray. T.S.: 31ksi	ASTM A126 (B)
UNS F12401	Cast Iron, Gray. T.S.: 35ksi	ASTM A48 (35), A278 (35)
UNS F12801	Cast Iron, Gray. T.S.: 40ksi	ASTM A48 (40)
UNS F12802	Cast Iron, Gray. T.S.: 41ksi	ASTM A126 (C)
UNS F12803	Cast Iron, Gray. T.S.: 40ksi	ASTM A278 (40)

FIGURE 2.6.10 Cross-reference of ASTM and UNS classifications for cast iron.

FIGURE 2.6.11 Some pieces of stainless steel raw material. Notice that it is brighter than the CRS shown in Figure 2.6.3.

components made from austenitic stainless steels to prevent product contamination that would result from corrosion.

Ferritic Stainless Steels

Ferritic stainless steels have low carbon content and are less resistant to corrosion than austenitic grades. Automobile exhaust components are sometimes made from these magnetic stainless steels.

Martensitic Stainless Steels

Martensitic stainless steels have high carbon content and can be made harder by heat-treatment processes. Most are magnetic. Surgical equipment and knife blades are machined from martensitic stainless steels.

AISI Designation of Stainless Steel

In the AISI numbering system for stainless steels, as for tool steels, the numbers do not represent any specific types or amounts of alloying elements. Three-digit numbers are used, and generally austenitic grades start with 3, while ferritic and martensitic grades start with 4.

UNS Designation of Stainless Steel

UNS numbers for stainless steels use the prefix "S" followed by a five-digit number. **Figure 2.6.12** shows some UNS numbers for some stainless steels cross-referenced to AISI types and lists the elements that make up their compositions.

NONFERROUS METALS
Aluminum Alloys

Aluminum is a lightweight, silver-gray colored metal that is very common in the machining industry. Some examples of aluminum that may be used to produce machined components are shown in **Figure 2.6.13**. Aluminum is widely used in the aerospace industry because of its low

Table 6: STANDARD STAINLESS STEELS—TYPICAL COMPOSITIONS

AISI Type	Typical Composition (%)	AISI Type	Typical Composition (%)
Austenitic			
201 (S20100)	16–18Cr, 3.5–5.5Ni, 0.15C, 5.5–7.5Mn, 0.75Si, 0.060P, 0.030S, 0.25N	310 (S31000)	24–26Cr, 19–22Ni, 0.25C, 2.0Mn, 1.5Si, 0.045P, 0.030S
202 (S20200)	17–19Cr, 4–6Ni, 0.15C, 7.5–10.0Mn, 0.75Si, 0.060P, 0.030S, 0.25N	310S (S31008)	24–26Cr, 19–22Ni, 0.08C, 2.0Mn, 1.5Si, 0.045P, 0.30S
205 (S20500)	16.5–18Cr, 1–1.75Ni, 0.120.25C, 14–15.5 Mn, 0.75Si, 0.060P, 0.030S, 0.32–0.40N	314 (S31400)	23–26Cr, 19–22Ni, 0.25C, 2.0Mn, 1.5–3.0Si, 0.045P, 0.030S
301 (S30100)	16–18Cr, 6–8Ni, 0.15C, 2.0Mn, 0.75Si, 0.045P, 0.030S	316 (S31600)	16–18Cr, 10–14Ni, 0.08C, 2.0Mn, 0.75Si, 0.045P, 0.030S, 2.0–3.0Mo, 0.10N
302 (S30200)	17–19Cr, 8–10Ni, 0.15C, 2.0Mn, 0.75Si, 0.045P, 0.030S, 0.10N	316L (S31603)	16–18Cr, 10–14Ni, 0.03C, 2.0Mn, 0.75Si, 0.045P, 0.030S, 2.0–3.0Mo, 0.10N
302B (S30215)	17–19Cr, 8–10Ni, 0.15C, 2.0Mn, 2.0–3.0Si, 0.045P, 0.030S	316F (S31620)	16–18Cr, 10–14Ni, 0.08C, 2.0Mn, 1.0Si, 0.20P, 0.10S min, 1.75–2.50Mo
303 (S30300)	17–19Cr, 8–10Ni, 0.15C, 2.0Mn, 1.0Si, 0.20P, 0.015S min, 0.60Mo (optional)	316N (S31651)	16–18Cr, 10–14Ni, 0.08C, 2.0Mn, 0.75Si, 0.045P, 0.030S, 2–3Mo, 0.10–0.16N
303Se (S30323)	17–19Cr, 8–10Ni, 0.15C, 2.0Mn, 1.0Si, 0.20P, 0.060S, 0.15Se min	317 (S31700)	18–20Cr, 11–15Ni, 0.08C, 2.0Mn, 0.75Si, 0.045P, 0.030S, 3.0–4.0Mo, 0.10N max
304 (S30400)	18–20Cr, 8–10Ni, 0.08C, 2.0Mn, 0.75Si, 0.045P, 0.030S, 0.10N	317L (S31703)	18–20Cr, 11–15Ni, 0.03C, 2.0Mn, 0.75Si, 0.045P, 0.030S, 3–4Mo, 0.10N max
304L (S30403)	18–20Cr, 8–12Ni, 0.03C, 2.0Mn, 0.75Si, 0.045P, 0.030S, 0.10N	321 (S32100)	17–19Cr, 9–12Ni, 0.08C, 2.0Mn, 0.75Si, 0.045P, 0.030S [Ti, 5(C+N) min, 070 max], 0.10 max
304 Cu (S30430)	17–19Cr, 8–10Ni, 0.08C, 2.0Mn, 0.75Si, 0.045P, 0.030S, 3–4Cu	329 (S32900)	23–28Cr, 2.5–5Ni, 0.08C, 2.0Mn, 0.75Si, 0.040P, 0.030S, 1–2Mo
304N (S30451)	18–20Cr, 8–10.5Ni, 0.08C, 2.0Mn, 0.75Si, 0.045P, 0.030S, 0.10–0.16N	330 (N08330)	17–20Cr, 34–37Ni, 0.08C, 2.0Mn, 0.75–1.50Si, 0.040P, 0.030S
305 (S30500)	17–19Cr, 10.50–13Ni, 0.12C, 2.0Mn, 0.75Si, 0.045P, 0.030S	347 (S34700)	17–19Cr, 9–13Ni, 0.08C, 2.0Mn, 0.75Si, 0.045P, 0.030S (Nb+Ta, 10xC min, 1 max)
308 (S30800)	19–21Cr, 10–12Ni, 0.08C, 2.0Mn, 1.0Si, 0.045P, 0.030S	348 (S34800)	17–19Cr, 9–13Ni, 0.08C, 2.0Mn, 0.75Si, 0.045P, 0.030S (Nb+Ta, 10xC min, 1 max, but 0.10 Ta max), 0.20 Ca
309 (S30900)	22–24Cr, 12–15Ni, 0.20C, 2.0Mn, 1.0Si, 0.045P, 0.030S	384 (S38400)	15–17Cr, 17–19Ni, 0.08C, 2.0Mn, 1.0Si, 0.045P, 0.030S
309S (S30908)	22–24Cr, 12–15Ni, 0.08C, 2.0Mn, 1.0Si, 0.045P, 0.030S	…	…

FIGURE 2.6.12 Cross-reference of some stainless steel UNS numbers to AISI types and their major alloying elements. UNS numbers are in parentheses below the AISI numbers.

ASTM No.	UNS No.	Nominal Composition—Expressed as Percent by Weight
		Sand and Permanent Mold Castings
AZ92A	M11920	Al 8.3–9.7; Cu 0.25 max; Mg 89; Mn 0.1 min; Ni 0.01 max; Si 0.3 max; Zn 1.6–2.4; Other 0.3 max
EZ33A	M12330	Cu 0.1 max; Mg 93; Nd 2.5–4; Ni 0.01 max; Zn 2–3.1; Zr 0.5–1; Other 0.3 max
HK31A	M13310	Cu 0.1 max; Mg 96; Ni 0.01 max; Th 2.5–4; Zn 0.3 max; Zr 0.4–1; Other 0.3 max
HZ32A	M13320	Cu 0.1 max; Mg 94; Ni 0.01 max; Rare Earths 0.1 max; Th 2.5–4; Zn 1.7–2.5; Zr 0.5–1; Other 0.3 max
K1A	M18010	Mg 99; Zr 0.4–1; Other 0.3 max
QE22A	M18220	Ag 2–3; Cu 0.1 max; Mg 95; Nd 1.7–2.5; Ni 0.01 max; Zr 0.4–1; Other 0.03 max
QH21A	–	Ag 2–3; Cu 0.1 max; Mg 95; Nd 0.6–1.5; Ni 0.01 max; Th 0.6–1.6; Zn 0.2 max; Zr 0.4–1; other 0.03 max
ZE41A	M16410	Cu 0.1 max; Mg 94; Mn 0.15 max; Nd 0.75–1.75; Ni 0.01 max; Zn 3.5–5; Zr 0.4–1; Other 0.3 max
ZE63A	M16630	Cu 0.1 max; Mg 91; Nd 2.1–3; Ni 0.01 max; Zn 5.5–6; Zr 0.4–1; Other 0.3 max
ZH62A	M16620	Cu 0.1 max; Mg 92; Ni 0.01 max; Th 1.4–2.2; Zn 5.2–6.2; Zr 0.5–1; Other 0.3 max
ZK61A	M16610	Cu 0.1 max; Mg 93; Ni 0,01 max; Zn 5.5–6.5; Zr 0.6–1; Other 0.3 max
		Die Casting
AM60A	M10600	Al 5.5–6.5; Cu 0.35 max; Mg 94; Mn 0.13 min; Ni 0.03 max; Si 0.5 max; Zn 0.22 max
AM60B	M10603	Al 5.5–6.5; Cu 0.01 max; Fe 0.005 max; Mg 94; Mn 0.25 min; Ni 0.002 max; Si 0.1 max; Zn 0.22 max; Other 0.003 max
AS21X1	–	Al 1.7; Mg 97; Mn 0.4 min; Si 1.1
AS41XA	M10410	Al 3.5–5; Cu 0.06 max; Mg 94; Mn 0.2–0.5; Ni 0.03 max; Si 0.5–1.5; Zn 0.12 max; Other 0.3 max
AZ91A	M11910	Al 8.3–9.7; Cu0.1 max; Mg 90; Mn 0.13 min; Ni 0.03 max; Si 0.05 max; Zn 0.35–1 max; Other 0.3 max
AZ91B	M11912	Al 8.3–9.7; Cu 0.35 max; Mg 90; Mn 0.13 min; Ni 0.03 max; Si 0.5 max; Zn 0.35–1; Other 0.3 max

FIGURE 2.6.15 Some examples of cast magnesium alloys showing their ASTM and UNS designations and compositions.

FIGURE 2.6.16 Three copper alloys. From left to right: pure copper, brass, and bronze.

Designation of Copper Alloys

Historically, SAE designations of two or three digits had been assigned to different copper alloys starting with digits 4, 6, 7, or 8. These numbers did not indicate any specific alloying elements or amounts. Examples were SAE 41, SAE 70, or SAE 701. Though not widely used today, their descriptions can be found in *Machinery's Handbook*.

UNS is the most widely accepted method for classifying copper alloys. Copper alloys begin with the prefix "C" followed by a five-digit number. **Figure 2.6.17** shows UNS number ranges for groups of copper alloys and their compositions.

UNS Wrought Copper Alloys

UNS Range	Common Name	Major Alloying Elements
C10100-C15815	-	Minimum 99.3% Pure Copper
C16200-C19900	-	96%—less than 99.3% Pure Copper
C18000-C28000	Brasses	Zinc
C31200-C38500	Leaded Brasses	Zinc, Lead
C40400-C48600	Tin Brasses	Zinc, Tin, Lead
C50100-C52480	Phosphor Bronzes	Tin, Phosphorous
C53400 – C54400	Leaded Phosphor Bronzes	Lead, Tin, Phosphorous
C60800 – C64210	Aluminum Bronzes	Aluminum, Nickel, Iron, Silicon, Tin
C64700 – C67000	Silicon Bronzes	Tin, Silicon
C67200 – C69710	Miscellaneous Copper-Zinc Alloys	Zinc
C70100 – C72950	Copper-Nickel Alloys	Nickel, Iron
C73500 – C79830	Nickel Silvers	Nickel, Zinc

UNS Cast Copper Alloys

UNS Range	Common Name	Major Alloying Elements
C80100 – C81200	-	Minimum 99.3% Pure Copper
C81400 – C82800	-	94%—less than 99.3% Pure Copper
C83300 – C83810	Red Leaded Brasses	Zinc, Tin, Lead
C84200 – C84800	Semi-Red Leaded Brasses	Zinc, Tin, Lead
C86100 – C86800	Leaded Yellow Brasses	Zinc, Manganese, Iron, Lead
C87300 – C87900	Silicon Bronzes & Brasses	Zinc, Silicon
C89320 – C89940	High-Strength & Leaded High-Strength Yellow Brasses	Tin, Bismuth
C90200 – C91700	Tin Bronzes	Tin
C92200 – C92900	Leaded Tin Bronzes	Silicon, Zinc, Lead
C93100 – C94500	High Leaded Tin Bronzes	Silicon, Lead
C94700 – C94900	Nickel Bronzes	Silicon, Zinc, Nickel
C95200 – C95500	Aluminum Bronzes	Aluminum, Iron, Nickel
C96200 – C96950	Copper Nickels	Nickel, Iron
C97300 – C97800	Nickel Silvers	Nickel, Zinc, Lead, Tin
C98200 – C98840	Leaded Coppers	Lead
C99300 – C99750	Miscellaneous—Not Identified by Other Categories	

FIGURE 2.6.17 UNS designation numbers and major alloying elements for some wrought and cast copper alloys.

Titanium Alloys

Titanium alloys are metals that are very lightweight, very strong, and very expensive. Titanium has the best strength-to-weight ratio of any metal. Machined titanium components are used in the construction of aircraft, spacecraft, and motorsport racing to reduce weight and still maintain very high strength. Joint implants like hip and knee replacements may also be machined from titanium alloys. Titanium alloys are quite difficult to machine.

UNS Designation of Titanium Alloys

There are several designation systems for classifying titanium, including the Titanium Industry, SAE, ASTM, and U.S. military specifications (mil-specs) as well as common industry "trade names." Since these different systems can become very confusing, the UNS is the preferred method used to designate titanium alloys. UNS numbers for titanium alloys begin with the prefix "R" followed by a five-digit number. **Figure 2.6.18** shows some UNS numbers cross-referenced to trade name designations for some titanium alloys. It also lists the alloy compositions.

Superalloys

Superalloys are exotic nickel-based metals that have been created for use in very harsh conditions. They are very strong and can resist high temperatures. Superalloys can also be used in highly corrosive environments. Many have been engineered for very specific purposes.

Machined superalloy parts are used in jet engines and for fittings that carry hazardous chemicals. Superalloys are very difficult to machine.

UNS Designation of Superalloys

Many superalloys are identified by the name of the company that developed them. Several were created by Haynes International and are know by names like Haynes 230 or Haynes 242. Other Haynes International alloy names are Hastelloy® and Waspaloy®. Some superalloys have names like Inconel®, Incoloy®, and Monel®. UNS numbers for superalloys normally use the prefix "N," but some alloys are designated by "R" or "S." All prefixes are followed by a five-digit number. **Figure 2.6.19** shows some UNS numbers for superalloys.

Table 1: Chemical Composition of Titanium and Titanium Alloys		
Designation	**UNS No.**	**Nominal Composition—Expressed as Percent by Weight**
Unalloyed Grades		
Grade 1	R50250	C 0.1 max; Fe 0.2 max; H 0.015 max; N 0.03 max; O 0.18 max; Ti 99.5
Grade 2	R50400	C 0.1 max; Fe 0.3 max; H 0.015 max; N 0.03 max; O 0.25 max; Ti 99.2
Grade 3	R50550	C 0.1 max; Fe 0.3 max; H 0.015 max; N 0.05 max; O 0.35 max; Ti 99.1
Grade 4	R50700	C 0.1 max; Fe 0.5 max; H 0.015 max; N 0.05 max; O 0.4 max; Ti 99
Grade 7	R52400	C 0.1 max; Fe 0.3 max; H 0.015 max; N 0.03 max; O 0.25 max; Pd 0.2; Ti 99
Grade 11	R52250	Pd 0.2; Ti 99
Grade 12	R53400	Mo 0.3; Ni 0.8; Ti 99
Unalloyed Grades		
Ti-3Al-2.5V	R56320	Al 3; Ti 95; V2.5
Ti-5Al-2.5Sn	R54520	Al 5; Fe 0.5 max; O 0.2 max; Sn 2.5; Ti 92.5
Ti-5Al-2.5Sn,ELI	R54521	Al 5; Fe 0.25 max; O 0.12 max; Sn 2.5; Ti 92.5
Ti-6-Al-2Nb-1Ta-0.8Mo	R56210	Al 6; Mo 0.8; Nb 2; Sn 2.5; Ta 1; Ti 90
Ti-6Al-2Sn-4Zr-2Mo	R54620	Al 6; Mo 2; Ti 88; Zr 4
Ti-88Al-1Mo-1V	R54810	Al 8; Mo 1; Ti 90; V1
Alpha-Beta Alloys		
Ti-4Al-4Mo-2Sn-0.5Si	–	Al 4; Mo 4; Si 0.5; Sn 2; Ti 89
Ti-6Al-4V	R65400	Al 6; Fe 0.25 max; O 0.2 max; Ti 90; V 4
Ti-6Al-4V,ELI	R65400	Al 6; Fe 0.14 max; O 0.13 max; Ti 90; V 4
Ti-6Al-6V-2Sn	R56620	Al 6; Sn 2; Ti 86; V 6
Ti-6Al-2Sn-2Zr-2Mo-2Cr-.25Si	–	Al 6; Cr 2; Mo 2; Si 0.25; Sn 2; Ti 86; Zr 2
Beta Alloys		
Ti-8Mo-8V-2Fe-3Al	–	Al 3; Fe 2; Mo 8; Ti 79; V 8
Ti-15Mo-5Zr-3Al	–	Al 3; Mo 15; Ti 77; Zr 5
Ti-13V-2.7Al-7Sn-2Zr	–	Al 2.7; Sn 7; Ti 75; V 13; Zr 2
Beta III	R58030	Mo 11.5; Sn 4.5; Ti 78; Zr 6

FIGURE 2.6.18 Some UNS designations of titanium alloys and their compositions.

Table 1: Nominal Composition of Selected Wrought "Superalloys"

Material	UNS No.	Composition %
A-286	S66286	Ni 26; Cr 15; Ti 2, Mo 1.3, Al 0.2, B 0.015, Fe bal
Hastelloy G	No6007	Cr 22, Fe 19.5, Mo 6.5, co 2.5 max, Cb+Ta 2, Cu 2, Mn 1.5, Si 1 max, W 1 max, C 0.05 max, Ni bal
Hastelloy C-276	N10276	Mo 16, Cr 15.5, Fe 5, W 4, Co 2.5 max, Mn 1 max, V 0.35 max, Si 0.08 max, C 0.02 max, Ni bal
Hastelloy X	N06002	Cr 22, Fe 18.5, Mo 9, Co 1.5, Mn 1 max, Si 1 max, W 0.6, C 0.10, Ni bal
214	N07214	Ni 75, Cr 16, Al 4.5, Fe 3, Mn 0.2, Si 0.1, C 0.05, Other Y.01
230	N06230	Ni 57, Cr 22, W 14, Co 5 max, Mo 2, Mn 0.5, Si 0.4, Al 0.3, Fe 0.3 max, C 0.10, La 0.02
Haynes 188	R30188	Cr 22, Ni 22, W 14, Fe 3 max, Mn 1.25 max, Si 0.35, La 0.04, C 0.10, Co bal
IN-102	N06102	Cr 15, Fe 7, Mo 3, W 3, Cb 3, Ti 0.6, Al 0.4, Zr 0.03, B 0.005, Ni bal
Incoloy 800	N08800	Fe 45.5, Ni 32.5, Cr 21, Al 0.4, Ti 0.4, C 0.05
800HT	N08811	Fe 46, Ni 33, Cr 21, Mn 0.8, Si 0.5, Al 0.4, Ti 0.4, C 0.08
Incoloy 825	N08825	Ni 38-46, Cr 19.5-23.5, Fe 22 min, Mo 2.5-3.5, Cu 1.5-3, Ti 0.6-1.2, Mn 1 max, Si 0.5 max
907	N19907	Fe 42, Ni 38, Co 13, Nb 4.7, Ti 1.5, Si 0.15, Al 0.03
909	N19909	Fe 42, Ni 38, Co 13, Nb 4.7, Ti 1.5, Si 0.4, C 0.01, B 0.001
Inconel 600	N06600	Ni 76, Cr 15, Fe 8, Mn 0.5, Cu 0.2, Si 0.2, C 0.08
Inconel 625	N06625	Ni 61, Cr 21.5, Mo 9, Fe 4, Cb 3.6, Mn 0.2, Si 0.2, C0.05
Inconel 690	N06690	Ni 61, Cr 29, Fe 9, Cu 0.2, Mn 0.2, Si 0.2, C 0.02
Inconel 718	N07718	Ni 52.5, Cr 19, Fe 18.5, Cb 5.1, Mo 3.0, Ti 0.9, Al 0.5, Mn 0.2, Si 0.2, C 0.04
Inconel X750	N07750	Ni 73, Cr 15.5, Fe 7, Ti 2.5, Cb 1, Ti 0.9, Al 0.7, Mn 0.5, Si 0.2, C 0.04
L-605	R30605	Cr 20, W 15, Ni 10, Fe 3 max, Mn 1.5, Si 1 max, C 0.10, Co bal
M-252	N07252	Cr 20, Co 10, Mo10, Ti 2.6, Al 1, B 0.005, Ni bal
N-155	R30155	Cr 21, Co 20, Ni 20, Mo 3, W 2.5, Mn 1.5, Cb+Ta 1, Si 1 max, N 0.15, C 0.10, Fe bal
Nimonic 80A	N07080	Ni 74.5, Cr 20.5, Ti 2.4, Al 1.25, Fe 0.5, C 0.05
901	N09901	Ni 43, Fe 35, Cr 12, Mo 6, Ti 3, Co 1 max, Mn 0.5 max, Si 0.4 max, Al 0.2, C 0.05
Rene 41	N07041	Cr 19, Co 11, Mo 10, Ti 3.1, Fe 2, Al 1.5, B 0.005, Ni bal
Udimet 500	No7500	Co 18.5, Cr 18, Mo 4, Al 2.9, Ti 2.9, Zr 0.05, B 0.006, Ni bal
Ultimet	R31233	Co 54, Cr 26, Ni 9, Mo 5, Fe 3, W 2, Mn 0.8, Si 0.3, C 0.05, B 0.015
Waspaloy	N07001	Cr 19.5, Co 13.5, Mo 4.3, Ti 3, Fe 2, Al 1.3, Zr 0.04, B 0.006, Ni bal

Note: Y denotes Y_2O_3

FIGURE 2.6.19 Some UNS designations for superalloys. The composition percentage shows the elements that make up each superalloy.

SUMMARY

- There are nearly unlimited variations of metals in the world but they can be separated into the two basic categories of ferrous and nonferrous.
- Steel and cast iron are ferrous metals. Aluminum, copper, and magnesium are examples of nonferrous metals.
- An alloy is a combination of metals or metal mixed with other elements to produce desired characteristics.
- Plain carbon steels and alloy steels can be identified as low-, medium-, or high-carbon steels according to their carbon content.
- Cast iron is very hard but brittle because it has the highest carbon content of the ferrous metals.
- Stainless steels are a special type of corrosion-resistant steels due to the addition of at least 10 percent chromium.
- Tool steels are a group of steels used to make tools that "work" other metals by cutting, forming, bending, or other operations.
- AA, AISI, ASTM, SAE, IADS, and UNS are organizations that have played a role in creating systems for classifying metals.
- The AISI/SAE system is the most common system for identifying carbon and alloy steels.
- Aluminum alloys are lightweight metals and are frequently identified by the AA/IADS numbering systems.
- Magnesium alloys are also lightweight metals similar to, but stronger than, aluminum alloys and are most commonly identified by the UNS.
- Small particles of magnesium are flammable, so care must be taken when machining them.
- Copper and copper alloys are normally identified by the UNS.
- Titanium alloys are light, strong, expensive metals and are usually identified by the UNS.
- Superalloys are nickel-based metals engineered for use under extreme temperatures and corrosive conditions and are identified by the UNS.
- An understanding of the basic properties, uses, and identification of metals is important so that the proper material is chosen for any given application.

REVIEW QUESTIONS

1. What is the difference between ferrous and nonferrous metals?
2. Briefly describe an alloy.
3. Name three alloying elements added to steel.
4. What is cast iron?
5. What element is in stainless steel that makes it corrosion resistant?
6. Small particles of _____ alloys are flammable.
7. What are the two major benefits of titanium?
8. What does AISI stand for?
9. What does SAE stand for?
10. What is (are) the major alloying element(s) in 8730 steel?
11. What is the carbon content of 8730 steel?
12. What does IADS stand for?
13. What is the overall purity of 1030 aluminum?
14. What is (are) the major alloying element(s) of 6061 aluminum?
15. What does UNS stand for?

Heat Treatment of Metals UNIT 7

Learning Objectives

After completing this unit, the student should be able to:

- Demonstrate understanding of common heat treatment processes
- Demonstrate understanding of different types of heat-treating equipment
- Describe safety procedures and PPE for heat treating
- Demonstrate understanding of Rockwell and Brinell hardness scales
- Compare and contrast Rockwell and Brinell hardness testing methods

Key Terms

Annealing
Atmospheric furnace
Brinell hardness scale
Carburizing
Case hardening
Cyaniding
Direct hardening
Flame hardening

Hardening
Heat treatment
Induction hardening
Nitriding
Normalizing
Precipitation heat
treatment
Quenching

Rockwell hardness
scale
Solution heat
treatment
Surface hardening
Tempering (drawing)

INTRODUCTION

Heat treatment of metals is the controlled heating and cooling of metals to change their characteristics. It can be used to make the material harder, tougher, softer, more stable, or more easily machined. Most heat-treatment processes are performed on ferrous metals, but some are also performed on nonferrous metals.

Since heat treating can alter the relative hardness of metals, it is also important to be able to define and measure those levels of hardness.

This unit will explain the basics of both of these areas as they relate to the machining industry.

HARDENING

Hardening is most commonly performed on steels. The steel is heated to a hardening temperature in a special type of furnace and then cooled rapidly to transform the structure of the steel, resulting in an increase in material hardness.

Direct Hardening

Direct hardening can be performed on steel containing at least 0.3% of carbon. This amount of carbon is needed for the steel's structure to change and become hard when heated and then rapidly cooled. Direct hardening is sometimes called through hardening because the steel is hardened throughout. Direct hardening is usually performed by heating the steel in a special furnace. Different types of steels require different temperatures for hardening. Hardening temperatures for some alloy steels can be seen in the heat-treatment chart in **Figure 2.7.1** under the "Harden, Deg. F" column.

Preheating steel at lower temperatures prior to reaching the hardening temperature is a good practice. Rapidly heated steels can crack and deform more than when steels are heated more slowly.

In addition, heating to about 200°F below the hardening temperature and getting the entire part to that temperature will result in a more even hardening of the steel. After the preheating, the steel is then raised to the required hardening temperature and held there for a period of time so the entire thickness of the part reaches the hardening temperature. A good rule is to maintain temperature for about 1/2 hour per inch of part thickness.

Finally, the steel is rapidly cooled or **quenched**. Quenching can be done with different media, or substances. Submerging in water, brine (salt/water mix), or oil, and cooling with air blasts are common methods of quenching. Different steels require different quenching media. Refer to the "Quench" column in **Figure 2.7.1** to see the quenching media to be used when hardening some alloy steels. The key for the letter codes is listed below the table.

Surface Hardening

Surface hardening is generally performed on medium plain carbon or alloy steels. Only an outer layer is hardened, while the center or core remains in a softer, tougher condition. Sometimes this is desirable because a softer, tougher center is more resistant to shock and allows some part flexibility without its cracking or breaking while the hard outer layer, provides high wear resistance. **Figure 2.7.2** illustrates surface hardening.

Flame Hardening

Flame hardening is one of two common methods of surface hardening. The steel is heated to the hardening temperature using an open flame. Machine tool wear surfaces and large rolls and shafts are often flame hardened. Flame hardening is performed by heating the steel with a gas-fueled torch (such as propane or oxygen/acetylene mixture) and then quenching. **Figure 2.7.3** shows a flame-hardening operation. The depth of hardness is determined mostly by the heating time. A longer heating time produces a deeper hardened layer.

Induction Hardening

Induction hardening can also be used to surface harden steel. The steel is heated to the hardening temperature by running an electrical current through the material. Gear teeth and shafts can be hardened by induction processes, as shown in **Figure 2.7.4**.

Case Hardening

When steel contains less than 0.3% carbon, it is not directly hardenable because there is not enough carbon to transform the material structure. In these low-carbon steels, the outer layer of heated steel soaks up carbon from another source. Then the steel is hardened by heating and quenching to produce a hard shell on the outside of the part while the inside remains soft. This process is called **case hardening** because a hard, relatively thin outer case is created. The depth of this case is usually less than 0.050".

Case-hardened parts have similar benefits to those of surface-hardened medium-carbon steel parts and the base material is less expensive than medium carbon steels. Their soft inside makes them less likely to crack when subject to shock, but because of their lower carbon content, their inner cores are not as tough. There are three common methods of adding carbon to low-carbon steels.

Sometimes further machining operations are required after case hardening. In such situations, careful consideration must be given to machined dimensions before performing case hardening. If too much material is removed after case hardening, the hard outer surface may be removed and expose the softer, inner core.

SAE No.	Normalize Deg. F		Anneal Deg. F	Harden, Deg. F	Quench*	Temper, Deg. F
1330	⎰ ...	and/or	...	1525–1575	B	
	⎱ 1600–1700		1500–1600	1525–1575	B	
1335 & 1340	⎰ ...	and/or	...	1500–1550	E	
	⎱ 1600–1700		1500–1600	1525–1575	E	
2330	⎰ ...	and/or	...	1450–1500	E	
	⎱ 1600–1700		1400–1500	1450–1500	E	
2340 & 2345	⎰ ...	and/or	...	1425–1475	E	⎰ To desired hardness
	⎱ 1600–1700		1400–1500	1425–1475	E	
3130	1600–1700		...	1500–1550	B	
3135 to 3141	⎰ ...	and/or	...	1500–1550	E	
	⎱ 1600–1700		1450–1550	1500–1550	E	
3145 & 3150	...	and/or	...	1500–1550	E	
	1600–1700		1400–1500	1500–1550	E	
4037 & 4042	...		1525–1575	1500–1575	E	⎰ Gears, 350–450 ⎱ To desired hardness
4047 & 4053	...		1450–1550	1500–1575	E	
4063 & 4068	...		1450–1550	1475–1550	E	
4130	1600–1700	and/or	1450–1550	1600–1650	E	
4137 & 4140	1600–1700	and/or	1450–1550	1550–1600	E	⎰ To desired hardness
4145 & 4150	1600–1700	and/or	1450–1550	1500–1600	E	
4340	1600–1700	and draw	1450–1550	1475–1525	E	
4640	⎰ 1600–1700	and/or	1450–1550	1450–1500	E	To desired hardness
	⎱ 1600–1700	and/or	1450–1500	1450–1500	E	Gears, 350–450, 250–300
5045 & 5046	1600–1700	and/or	1450–1550	1475–1500	E	
5130 & 5132	1650–1750	and/or	1450–1550	1500–1550	G	To desired hardness
5135 to 5145	1650–1750	and/or	1450–1550	1500–1550	E	⎰ To desired hardness ⎱ Gears 350–400
5147 to 5152	1650–1750	and/or	1450–1550	1475–1550	E	⎰ To desired hardness ⎱ Gears 350–400
50100	⎰ ...		1350–1450	1425–1475	H	
51100	⎱					To desired hardness
52100	...		1350–1600	1500–1600	E	
6150	1650–1750	and/or	1550–1650	1600–1650	E	
9254 to 9262	...		...	1500–1650	E	
8627 to 8632	1600–1700	and/or	1450–1550	1550–1650	B	
8635 to 8641	1600–1700	and/or	1450–1550	1525–1575	E	
8642 to 8653	1600–1700	and/or	1450–1550	1500–1550	E	
8655 & 8660	1650–1750	and/or	1450–1550	1475–1550	E	⎰ To desired hardness
8735 & 8740	1600–1700	and/or	1450–1550	1525–1575	E	
8745 & 8750	1600–1700	and/or	1450–1550	1500–1550	E	
9437 & 9440	1600–1700	and/or	1450–1550	1550–1600	E	
9442 to 9747	1600–1700	and/or	1450–1550	1500–1600	E	
9840	1600–1700	and/or	1450–1550	1500–1550	E	
9845 to 9850	1600–1700	and/or	1450–1550	1500–1550	E	

Table 5b: Typical Heat Treatments for SAE Alloy Steels (Directly Hardenable Grades)

Symbols: B = water or oil; E = oil; G = water, caustic solution, or oil; H = water.

FIGURE 2.7.1 Heat-treating temperatures for some alloy steels.

From Machinery's Handbook 28, Industrial Press, New York, 2008

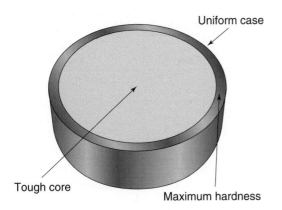

FIGURE 2.7.2 Surface hardening creates a hard outer surface while keeping the core softer and tougher.

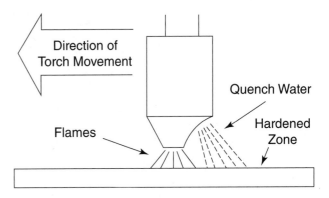

FIGURE 2.7.3 Flame hardening passes a torch over the surface of the steel. Water is applied after the torch to quench the steel.

Carburizing

Carburizing uses some type of material containing solid carbon to add carbon to the outer layer of the steel. One method involves heating the low-carbon steel to about 1600–1700°F in a heat-treating furnace. Then it is dipped or rolled in a carbon-rich powdered material until a layer of the powder coats the steel. Another process packs the low-carbon steel in a carbon-rich material and then heats it to the required temperature in a furnace. In both cases, the carbon from the carbon-rich material is absorbed by the heated steel. When the steel reaches the desired temperature, it is then removed from the furnace and quenched. **Figure 2.7.5** illustrates these carburizing methods.

Cyaniding

Cyaniding adds carbon to steel by placing it in a tank of heated liquid containing sodium cyanide. The outer surface of the steel absorbs carbon from the hot liquid, which creates a hard outer shell when the steel is quenched.

Nitriding

Nitriding heats steel in a sealed furnace containing a nitrogen-rich gas like ammonia at about 900°F. The nitrogen combines with the steel's other elements to create nitrides, which form a hard outer shell. No quenching is required to create this hard outer surface.

Courtesy of American Metal Treating Co.

FIGURE 2.7.4 Induction hardening of gear teeth. The glowing element heats teeth one at a time with electrical current. Note the water spray that quenches the teeth after hardening.

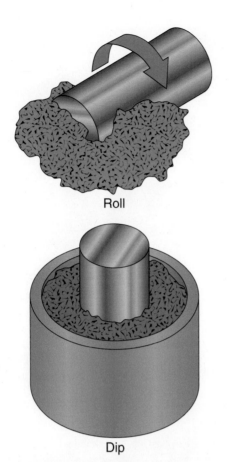

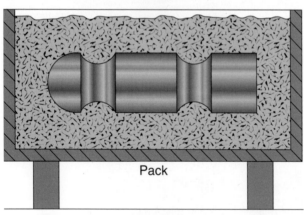

FIGURE 2.7.5 Rolling, dipping, and pack methods for carburizing low-carbon steels for case hardening.

First the steel is heated in a furnace to a temperature lower than the hardening temperature. The desired final degree of hardness and the alloy determine the temperature of the tempering operation. Consult the steel manufacturer's specifications for temperatures.

The steel is then held, or soaked, for a time recommended by the manufacturer. After soaking, most steels are allowed to cool in air at room temperature, but some steels require quenching to avoid a condition known as temper brittleness. Check the material manufacturer's specifications for the recommended method.

ANNEALING

Annealing is a heat-treating process used to return metals to their original pre-hardened condition. Hardened steels are often nearly impossible to machine by standard methods. Annealing softens them so that they can be more easily machined. This is accomplished by heating to a temperature slightly below the hardening temperature and then cooling very slowly instead of quenching. Sometimes the steel is heated in a furnace and then the furnace is shut down without the steel being removed to ensure very slow cooling. Steels can be hardened, tempered, and annealed repeatedly. The "Anneal Deg. F" column in Figure 2.7.1 shows the annealing temperatures for some alloy steels.

NORMALIZING

Normalizing is sometimes performed on medium- and high-carbon steels prior to hardening. It is also often applied to forged steel parts that have been "hammered" into shapes by large dies. Normalizing removes stresses from the steel and makes its structure more consistent so that better results will occur from other heat treating operations. Normalizing temperatures are very near annealing temperatures, but faster cooling provides a different result. In normalizing, the steel is removed from the furnace after heating and cooled at room temperature. The "Normalize Deg. F" column in Figure 2.7.1 shows normalizing temperatures for some alloy steels.

HEAT TREATMENT OF NONFERROUS METALS

Aluminum Alloys

Two common heat treatments are performed on aluminum. These are normally done at the material manufacturing stage prior to machining.

Solution Heat Treatment

Solution heat treatment is performed on wrought aluminum alloys. The material is heated to between 900°F

TEMPERING

After hardening, directly hardenable steel is very hard but too brittle to be used. **Tempering**, or **drawing**, will decrease the steel's hardness, increase toughness, and relieve some internal stress so it will be more durable and usable. Tempering of case-hardened low-carbon steels is not as common as with directly hardened steels, but it is performed in some instances.

Tempering should be done as soon as possible after hardening because some steels are unstable and can crack if left in the fully hardened condition.

and 1000°F for about an hour and then quenched. Unlike steel, this process makes the aluminum softer and more uniform in structure. Solution heat treatment of aluminum can be thought of as a cross between the annealing and normalizing operations that are performed on steel.

Precipitation Heat Treatment or Aging

When aluminum ages, it becomes tougher and more durable. **Precipitation heat treatment** is performed by heating aluminum alloys to around 300°F to artificially speed up the aging process.

HEAT-TREATING FURNACES

Most heat-treating operations are performed using specialized furnaces. They are usually powered by gas (bottled or natural) or electricity. Sizes and abilities of heat treating furnaces vary widely, but all perform the same basic task of heating metal to required temperatures.

Box Furnaces

The box furnace is very common in small- to medium-size shops. They are available in single- or dual-chamber versions with a wide variety of chamber sizes. Dual-chamber versions use one chamber for hardening and another for tempering. A typical dual-chamber box furnace is shown in **Figure 2.7.6**.

Production and Specialty Furnaces

Some facilities use furnaces with conveyors that move parts through the furnace continually. **(See Figure 2.7.7.)** These are generally used for production heat treatment

FIGURE 2.7.6 A dual-chamber box furnace. The top unit can reach temperatures of 2200°F and is used for hardening. The lower chamber can reach 1200°F and is used for tempering.

FIGURE 2.7.7 This conveyor-type furnace is used to heat treat springs. The springs are loaded on one end of the furnace and exit at the other end after being heat treated.

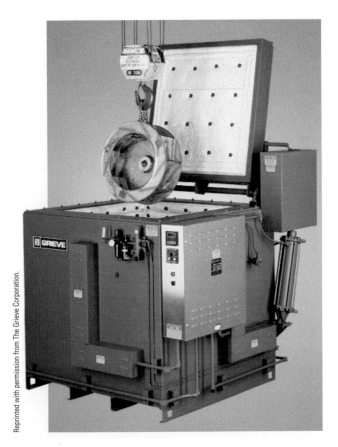

Reprinted with permission from The Grieve Corporation.

FIGURE 2.7.8 A pit-type furnace.

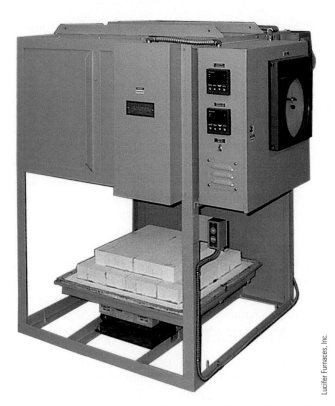

Lucifer Furnaces, Inc.

FIGURE 2.7.9 An elevator furnace. The bottom lowers for loading and unloading of parts.

of large numbers of parts. Some combine hardening, tempering, and quenching in one unit.

Another type of furnace that can heat treat parts continuously is the shaker furnace. Small parts are fed into a hopper and heated; they then pass through a bottom opening into a quenching tank.

Some larger furnaces are called pit furnaces and have a top lid that opens for loading parts, like the one shown in **Figure 2.7.8**. Other furnaces used for heavier parts have a bottom that lowers to receive the work and then is raised back into place. These are called elevator furnaces. **(See Figure 2.7.9.)**

Atmospheric Furnaces

Heat treating usually produces an oxidation, or scale, on the surface of the metal. To minimize or eliminate that oxidation, some furnaces can remove room air from the chamber and replace it with nitrogen, argon, or helium. These are called **atmospheric** or **atmospheric control furnaces**. All styles of furnaces are available with atmospheric control. Some models can preheat, harden, and quench within the controlled atmosphere. **Figure 2.7.10** shows an example of this type of furnace.

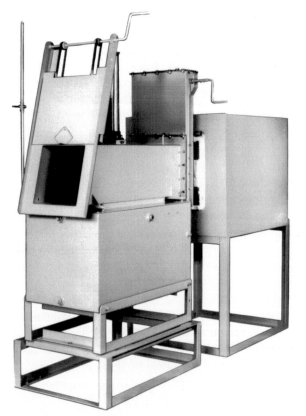

Lucifer Furnaces, Inc.

FIGURE 2.7.10 An atmospheric furnace can preheat, heat, and quench steel without any exposure to air to minimize oxidation.

To minimize oxidation in furnaces without atmospheric control, the steel can be wrapped in a 0.002"-thick stainless steel foil to seal it from contact with air.

Furnace Controls

Many modern heat-treating furnaces are equipped with microprocessor-powered controls. These can be used to program furnace temperatures and times. For example, a pre-heat temperature for hardening can be set to start at a certain time and stay at that temperature for a certain amount of time. Then the program can raise the temperature to the hardening temperature and hold it there for a particular amount of time. When annealing, the furnace can be programmed to reduce temperature slowly over several hours to ensure a very slow cooling and a successful annealing process. Controls like these allow unattended operation and greatly increase productivity.

HEAT-TREATMENT SAFETY

As with all machining processes, proper personal protection equipment (PPE) should be used when performing heat-treating operations. Because of the high-temperature exposure (2000°F plus), great care must be taken to avoid severe burns that can cause disfigurement and even death.

In addition to safety glasses, a heat-resistant face shield should be worn to protect the face from extreme temperatures. Fire-resistant gloves and a long-sleeved fire-resistant jacket should also be worn.

Always use some type of long-handled tongs to place metal in and remove it from a furnace. It is a good practice to preheat the tongs before using them to remove hot metal from a heat-treating furnace. **Figure 2.7.11** shows proper PPE for heat treating and use of tongs to handle hot metal.

If using a gas-fired furnace, always keep the door open when lighting to avoid gas buildup inside the furnace, which can lead to an explosion. When lighting, also stand to the side of the open door instead of directly in front of it to avoid severe burns from the possibility of flames extending past the furnace door. If the furnace does not light after two or three tries, wait several minutes before trying again to allow the gas to dissipate to avoid large, uncontrollable flames.

When working in a heat-treating area, always treat every piece of metal as if it is hot. Never assume it is cool enough to touch with bare hands. Steel at a temperature of 300°F looks exactly the same as steel at room temperature. A supply of small metal signs with the word "HOT" engraved or stamped on them can be used to remind others of potential danger.

When quenching, completely submerge the hot steel into the water, brine, or oil. Partial coverage when oil quenching can cause the oil to catch fire. If that happens, never use water or a water-based extinguisher on the fire.

FIGURE 2.7.11 In addition to standard PPE such as safety glasses, wear a heat-resistant face shield, long-sleeved fire-resistant jacket, and fire-resistant gloves when performing heat-treating operations, and always use tongs to handle hot metal.

A lid should be kept near an oil quench tank so that, in case of fire, it can be used to cover the tank and remove the oxygen so the fire will die.

HARDNESS SCALES AND TESTING

Since hardening, tempering, and annealing change the hardness of steels, there need to be methods for measuring and identifying the level of hardness. Two common measuring methods and scales are generally used to designate those hardness levels. Both make an indentation in a piece of steel and compare the size of the indentation to standardized scales.

Rockwell Hardness Scales

Rockwell hardness scales are usually used to designate the hardness of steels after heat-treating operations. Rockwell hardness testing uses a penetrator, or indenter, to make an impression in a piece of material and then uses the depth of the indentation to calculate a hardness value. There are several Rockwell scales that use different types of indenters and weights to perform testing. A 10-kg force, called the *minor load*, is first used to seat the

indenter into the material. Then a *major load*, determined by the scale being used, forces the indenter farther into the material. The difference between the depths of those two indentations determines the hardness. The principle of the Rockwell hardness test is shown in **Figure 2.7.12**.

There are several Rockwell hardness scales, but the A, B, and C scales are the most commonly used in the machining industry. The A scale uses a diamond-tipped penetrator called a *Brale* diamond penetrator and a 60-kg major load. The B scale uses a 1/16"-diameter hardened steel ball penetrator and a 100-kg major load. The C scale uses the Brale penetrator and a 150-kg major load. **Figure 2.7.13** shows some of the many different Rockwell scales as well as the 1/16"-diameter ball penetrator and the Brale diamond penetrator.

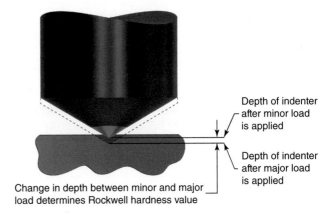

Change in depth between minor and major load determines Rockwell hardness value

FIGURE 2.7.12 The basic principle of the Rockwell hardness test.

Scale Symbol	Indenter	Major Load kgf	Typical Applications of Scales
HRA	Brale® Diamond	60	Cemented carbides, thin steel and shallow case hardened steel
HRB	1/16" ball	100	Copper alloys, soft steels, aluminum alloys, malleable iron
HRC	Brale® Diamond	150	Steel, hard cast irons, pearlitic malleable iron, titanium, deep case hardened steel and other materials harder than B100
HRD	Brale® Diamond	100	Thin steel and medium case hardened steel and pearlitic malleable iron
HRE	1/16" ball	100	Cast iron, aluminum and magnesium alloys, bearing metals
HRF	1/16" ball	60	Annealed copper alloys, thin soft sheet metals
HRG	1/16" ball	150	Phosphor bronze, beryllium copper, malleable irons. Upper limit G92 to avoid possible flattening of ball
HRH	1/8" ball	60	Aluminum, zinc, lead
HRK	1/8" ball	150	Bearing metals and other very soft or thin materials, including plastics (See ASTM D785). Use smallest ball and heaviest load that do not give anvil effect.
HRL	1/4" ball	60	
HRM	1/4" ball	100	
HRP	1/4" ball	150	
HRR	1/2" ball	60	
HRS	1/2" ball	100	
HRV	1/2" ball	150	

A

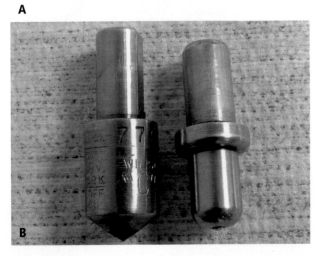

B

FIGURE 2.7.13 (A) Some of the Rockwell hardness scales showing the penetrator and major load used, and the materials those scales are used to measure. (B) The Brale diamond and the 1/16 ball indenter.

Rockwell hardness numbers must be connected to one of the scales. For example, a hardness of 50 on the C scale would be written as Rockwell C50, RC50, or HRC 50. A hardness of 90 on the B scale would be written as Rockwell B90, RB90, or HRB 90.

Performing Rockwell Hardness Testing

There are two basic types of Rockwell hardness testers, the open loop and the closed loop. The open loop uses a deadweight system that uses actual weights to drive the indenters into the test material. The second is a closed loop system that uses a microprocessor-controlled motor to apply the force of the minor and major loads to the penetrators. **Figure 2.7.14** shows an open loop Rockwell tester and labels its major parts. **Figure 2.7.15** shows a closed loop Rockwell hardness tester.

Before performing a Rockwell hardness test, first prepare the sample for testing by making sure it is clean and free of oxidation and scale. False readings can result if the sample is not prepared correctly.

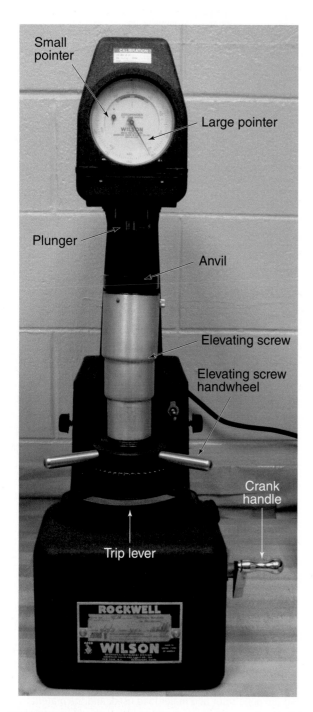

FIGURE 2.7.14 The major parts of an open loop or deadweight Rockwell hardness tester.

Wilson® Hardness Rockwell 2000 closed loop tester. Courtesy of Wilson Hardness

FIGURE 2.7.15 A closed loop tester uses a microprocessor-controlled motor to force the penetrator into the sample.

Follow these steps for performing a Rockwell hardness test on an open loop (dead weight) tester.

- Check that the crank handle is pulled forward (counterclockwise). **(See Figure 2.7.16.)**
- Select the appropriate anvil and install it on the elevating screw.
- Select the proper penetrator for the scale desired and install it in the plunger. **(See Figure 2.7.17.)**

- Select and install the proper weights according to the desired scale. Be sure they are correctly matched to the penetrator. **(See Figure 2.7.18.)**
- Place the test part on the anvil.
- Raise the elevating screw by turning the hand wheel clockwise. Continue until the small pointer is near the dot on the scale. Then continue until the large needle is vertical. This applies the minor load. Then adjust the dial face so that the "0" is aligned with the large needle. **(See Figure 2.7.19.)**
- Press the lever to trip the crank and release the major load. **(See Figure 2.7.20.)** The large needle will begin to rotate.
- When the needle stops, wait a few seconds and then return the crank handle to the forward starting position. This removes the major load, but the minor load is still applied.
- Record the reading using the proper scale on the dial. **(See Figure 2.7.21.)**
- Remove the minor load by turning the elevating screw handle counterclockwise until the penetrator clears the test part.

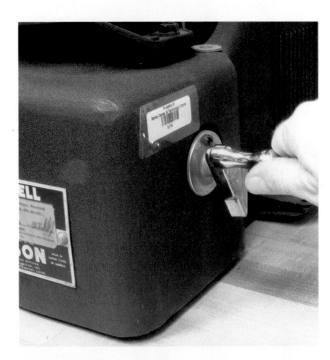

FIGURE 2.7.16 Be sure the crank handle is in the forward position before beginning the hardness test.

FIGURE 2.7.17 Installing an anvil and penetrator. Here the Brale point will be used for a C scale measurement.

FIGURE 2.7.18 Check for or install proper weights on the rear of the tester. All three weights equal the 150-kg major load for a C scale measurement.

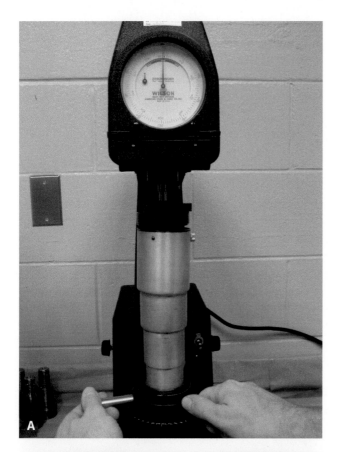

FIGURE 2.7.19 (A) Adjusting the elevating screw to set the minor load. (B) Set the small needle near the dot, the large needle at vertical, and adjust the dial face so the "0" aligns with the large needle.

FIGURE 2.7.20 Press the lever to trip the crank and apply the major load.

FIGURE 2.7.21 Record the hardness reading after releasing the major load. The result of this test is 50 on the C scale. This can be written as Rockwell C50, RC50, or HRC 50.

Brinell Hardness Scale

The **Brinell hardness scale** is most commonly used to designate hardness of nonferrous metals and steels before machining and heat treatment. Brinell hardness testers make an indentation on a piece of material with a 10-mm-diameter tungsten carbide ball. (Materials other than metals can be tested using a 5-mm ball.)

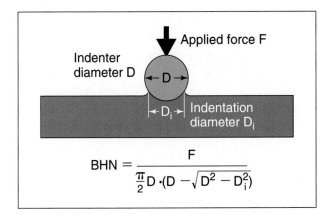

FIGURE 2.7.22 The principle of the Brinell hardness scale and formula for determining the HB number. "D" is the ball diameter, "Di" is the impression diameter, and "F" is the weight of the load applied.

A 500-kg, 1500-kg, or 3000-kg weight or load is used to press the ball into the material depending on the material type being tested. Harder materials require the use of heavier weights. After a wait of 10–30 seconds, the load is removed and the ball is retracted. The diameter of the impression is then input into a standard formula to calculate the Brinell hardness number. **Figure 2.7.22** shows the principle of the Brinell hardness system and the Brinell Hardness Number (BHN) formula.

Brinell hardness numbers are abbreviated as either HB (Hardness Brinell) or BHN. **Figure 2.7.23** shows the BHN ranges and the weight of the load used to perform the Brinell test for those ranges. Steels would generally be tested using the 3000-kg load. Brinell hardness numbers for steels typically range from around 100 for soft low-carbon steels to over 700 for fully hardened tool steels.

Brinell Hardness Number	Load (kg)
160–600	3000
80–300	1500
26–100	500

FIGURE 2.7.23 Brinell Hardness Number (BHN) ranges and the load sizes used to test those ranges.

Performing Brinell Hardness Testing

There are many variations of Brinell hardness testers, but two basic features can be used to classify them. Like Rockwell testers, the first difference is whether the tester is an open loop or closed loop type. This refers to how the load is applied. Open-loop-type testers use weights or hydraulic pressure to force the ball into test parts. Closed loop testers use a microprocessor-controlled motor to drive the ball into the material being tested.

Another difference relates to how the indentation is measured. The standard type of tester performs only the operation of making the indentation. After the indentation is made, it must then be measured with a Brinell microscope **(see Figure 2.7.24)** and the BHN calculated using the standard formula. The second type of tester performs the test, automatically measures the indentation, and outputs the HB number on a gage or readout.

Before the hardness test is performed, the sample to be tested must be clean and free of oxidation and scale. This can be done by machining, filing, or polishing with abrasive paper. If there is oxidation, surface marks, or bumps, false readings can result. A clean surface also makes it easier to measure the indentation if the test is done on a manual tester.

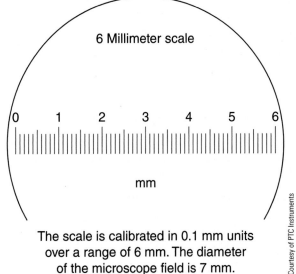

The scale is calibrated in 0.1 mm units over a range of 6 mm. The diameter of the microscope field is 7 mm.

Courtesy of PTC Instruments

FIGURE 2.7.24 A Brinell microscope. The microscope is placed over the impression made by the test and the scale in the viewfinder is used to measure the diameter of the impression.

Performing a Brinell hardness test is similar to performing a Rockwell test. Here are the basic steps. Refer to the manufacturer's instructions for details about a specific model.

- Load the test part on the anvil and preclamp by raising the elevating screw.

- Set the force of the major load. Depending on the model, this will be done by adjusting hydraulic or air pressure, installing the proper weights, or setting an electronic control.

- Apply the load. Depending on the model, this is done by pulling a lever to apply the load through hydraulic or air pressure, flipping a lever to release the weight, or pressing a button on the electronic control. Most

models will apply the load, pause for a preset amount of time, and then retract the load.

- Unclamp the sample part by lowering the elevating screw.

- On manual models, measure the indentation with a Brinell microscope, and then use the Brinell formula to calculate the BHN. On automatic models, simply read the BHN on the display of the control.

Cross-Reference of Brinell and Rockwell Hardness Values

A table of conversions between the Brinell 3000-kg scale and the Rockwell A, B, and C scales can be found in Appendix C.

SUMMARY

- Sometimes metals do not have the desired qualities in their raw state, so heat treatment is performed by controlled heating and cooling to change the characteristics of the material.

- Hardening is performed on steels by heating and rapidly cooling, or quenching.

- Direct hardening can be performed on steels containing at least 0.3% carbon and hardens the entire piece of the steel.

- Surface hardening is sometimes performed on medium-carbon steels by heating with a flame or electrical current so that only an outer layer becomes hard.

- Low-carbon steels with less than 0.3% carbon can be case hardened by adding carbon to the outer layer during or after heating to create a thin hard shell.

- Fully hardened steels are very brittle and are usually tempered, or drawn, to reduce hardness and increase toughness to a certain desired level.

- Annealing softens metals to their original state by heating and cooling very slowly.

- Normalizing relieves internal stresses from steels and makes them respond better to other heat treatment processes.

- Solution and precipitation heat treatment are sometimes performed on aluminum alloys to increase or decrease hardness.

- Heat-treating operations are performed in specialized furnaces that can be powered by either gas or electricity. There are many types and sizes of heat-treating furnaces depending on temperatures required, part size, volume of parts, and the need for atmosphere control. Always use proper PPE when performing heat treating operations to avoid injury from furnace or part temperatures. Lift materials with appropriate tongs and always treat all metal in a heat-treating area as if it is hot.

- A Rockwell tester is usually used to measure the hardness level of hardened steels by forcing an indenter into the material and measuring the depth of the impression. There are several different Rockwell scales and each requires the use of a different type of penetrator and load setting on the Rockwell tester.

- The Brinell hardness tester is usually used to measure hardness of nonferrous metals and steels in their soft state. It uses a ball-shaped penetrator to make an indentation and hardness is determined by the diameter of the impression made in the material.

- Both Rockwell and Brinell hardness testers can be either the open loop or closed loop design.

- By performing heat treating operations and verifying their results using standard hardness testing, proper material characteristics can be achieved for given specifications and applications.

REVIEW QUESTIONS

1. Briefly define the term *heat treatment*.

2. Direct hardening can be performed on steel containing at least _____ carbon.

3. _____ is the rapid cooling of metal during heat treatment.

4. What type of hardening operation that leaves the inner core in a softer condition is sometimes performed on medium-carbon steels?

5. In what two ways can the process described in the previous question be accomplished?

6. What method is used to harden low-carbon steels?

7. List two methods of performing the operation described in the previous question.

8. After hardening, steel is very hard and brittle. What operation is usually performed to reduce hardness and increase toughness so the steel is in a more usable condition?

9. Briefly define *annealing*.

10. What is the result in aluminum alloys after solution heat treatment?

11. A dual-chamber furnace allows the user to _____ in one chamber and _____ in the second chamber.

12. In addition to standard PPE, what specific PPE and safety precautions should be used during heat treating?

13. What hardness testing scale uses many different penetrators and loads?

14. What is a Brale penetrator and what hardness scale uses it?

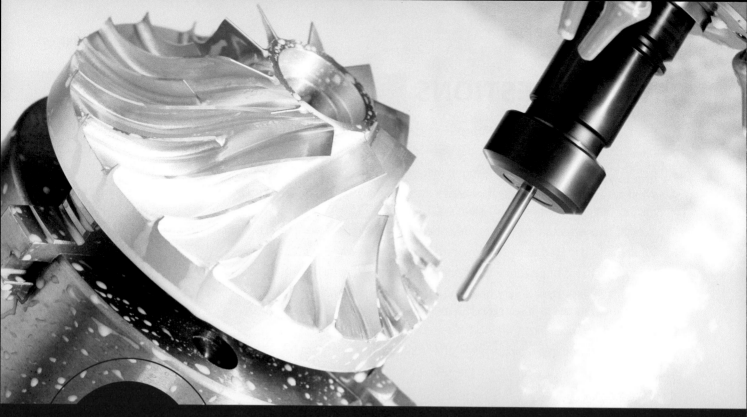

UNIT 8

Maintenance, Lubrication, and Cutting Fluid Overview

Learning Objectives

After completing this unit, the student should be able to:

- Demonstrate understanding of the importance of a routine maintenance program
- Identify different methods of machine tool lubrication
- Demonstrate understanding of routine machine tool maintenance inspection points
- Demonstrate understanding of the purpose of cutting fluids
- Demonstrate understanding of common types of cutting fluids
- Demonstrate understanding of methods of application of cutting fluids

Key Terms

Ball oiler
Chemical-based
  cutting fluid
Cold air gun
Cutting fluid
Flood system
Gib
Lubricant

Minimum Quantity
  Lubricant (MQL)
Mist system
Oil-based cutting fluid
Oil cup
Refractometer
Semi-solid and solid
  cutting compound

Semi-synthetic
Soluble oil (also
  known as
  emulsifiable or
  water-miscible oil)
Straight oil
Synthetic
Zerk

MAINTENANCE

Driving or owning a car requires periodic maintenance and repair. The same is true for equipment used in the machining field. Machine tools are complex pieces of equipment with mechanical, hydraulic, electrical, and even electronic or computerized systems. Some of those systems require routine maintenance so the machinery stays in proper working order and operates safely to prevent injury.

If a car owner never changes oil, rotates tires, or replaces brakes, the car will eventually develop many problems that require expensive repairs. If maintenance is ignored for too long, the car may not even be repairable. Machine tools can be much more expensive than cars, so they deserve the same care or better. If the machine does not run, both the company and the employee will lose money.

Several preventive maintenance tasks usually need to be done at specific intervals specified by the manufacturer. Following these recommended schedules found in manuals is the best way to keep equipment in good, safe operating condition so that it will provide long periods of reliable service. Every piece of equipment is unique, but a few common areas that apply to nearly every machine tool are briefly discussed here.

Lubrication

Frequent lubrication is one of the most important steps in maintaining machine tools. **Lubricants** (greases and oils) are vital to cool moving parts, minimize friction between them, and prevent their seizing. These lubricants are usually applied by four basic methods.

 CAUTION

There are many different types of greases and oils for very different uses. Always use the lubricant recommend by the manufacturer and consult the product's SDS before using.

Reservoirs

A reservoir is simply a tank that holds oil. Oil in these tanks may partially cover gear trains and coat the gears as they move against each other. Other times a pump circulates the oil through a gear box. On some reservoirs, a hand-operated pump is used to distribute oil to various machine lubrication points, as shown in **Figure 2.8.1**. This is often called a "one-shot" lubrication system because one single pump "shoots" or pressure feeds the oil to different lubrication points.

The locations of machine gear boxes are usually easily recognized because of their shape and size, a cap for

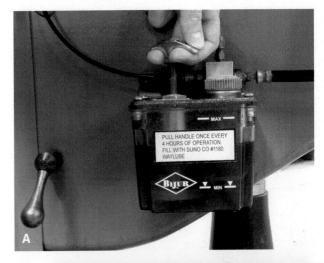

FIGURE 2.8.1 (A) Pulling the pump on a central reservoir, or "one shot," (B) can distribute oil to several different lubrication points through multiple oil lines.

filling, and perhaps a plug for draining. They often have a sight glass or a dipstick to check the oil level. **Figure 2.8.2** shows a typical machine tool oil gear box. These tanks are meant to be drained periodically and replaced with new oil. Follow the manufacturer's recommended schedule for changing oil.

Oil Cups

An **oil cup** is a small container mounted on a machine tool with a small lid that is opened to add lubricating oil. **(See Figure 2.8.3.)** It may fill a small reservoir, or slowly allow oil to flow to a lubrication point through a wick. Oil cups are usually to be checked and filled daily or as needed to maintain a particular level. A good level is usually verified by seeing oil in the cup.

Ball Oiler

A **ball oiler** contains a small, spring-loaded ball that acts as a valve. The valve keeps the oil in and debris out of

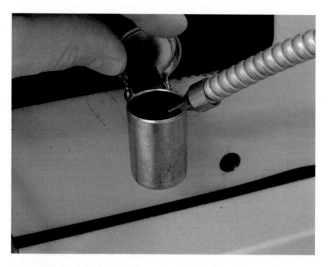

FIGURE 2.8.3 Oil cups like these may fill a small reservoir or allow oil to slowly flow to a specific lubrication point. Simply lift the cap and fill until oil is visible in the cup.

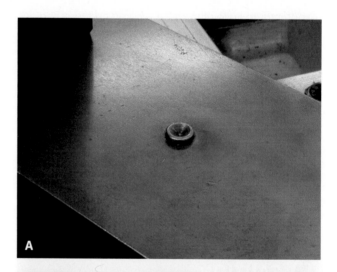

FIGURE 2.8.2 (A) The gear box of this lathe holds the oil that lubricates the gears that control the machine. The filler is at the upper right and the drain is at the lower right. (B) A sight glass verifies that oil is being circulated throughout the machine.

the lubrication system. The ball is depressed with an oil container nozzle tip to allow oil to flow to the lubrication point. After filling, the ball springs back and closes the hole. **Figure 2.8.4** shows a ball valve oil hole and how oil is added. Oil holes usually apply lubricants to sliding surfaces. They normally require oil to be applied on a

FIGURE 2.8.4 This ball oiler allows oil to be applied between two machine tool sliding surfaces that could not otherwise be reached. Depress the ball with an oil container nozzle and pump the container handle to apply oil.

FIGURE 2.8.5 A grease gun is pressed onto a zerk so that grease can be pumped into the required location.

FIGURE 2.8.6 The tapered gib can be adjusted by turning the screw to tighten the fit between the surfaces of a dovetail machine tool slide.

regular basis to keep the sliding surfaces lubricated and to flush debris out from between the mating surfaces.

Grease Zerks

A grease **zerk** is a fitting used to apply grease to a lubrication point. The tip of the grease zerk accepts a grease gun. Grease is then pumped from the gun, through the fitting, and to the required location. **(See Figure 2.8.5.)** Grease zerks have ball-type valves like ball oilers that maintain grease pressure and keep dirt out of the system.

Moving Parts and Wear Surfaces

Moving parts such as belts and gears should be checked periodically for wear and damage. Dry, cracked, or frayed belts should be replaced. Gears sometimes have adjustments to keep backlash at a minimum. Gears with broken or missing teeth should be replaced to prevent further damage to mating gears.

Sliding surfaces of machine tools should be checked for wear as well. Machine tools often use dovetail-shaped slides. These slides normally contain a tapered wedge, or **gib**, that can be used to tighten the slide as the dovetail surfaces wear. **Figure 2.8.6** shows a gib in a machine tool dovetail slide.

Machine tools also use threaded shafts called lead screws to transmit motion to slides. As these threads wear, backlash increases. Backlash, or play, is when a thread or gear has excess back-and-forth motion that does not cause any movement in the machine tool slide. Many machine tool screw threads contain an adjustable nut to minimize this backlash. After many years of heavy use, even adjustable nuts may wear to a point where replacement is necessary. Nonadjustable nuts will probably need replacement sooner.

 CAUTION

Always shut off power and tag out or lock out equipment before replacing or adjusting internal moving parts such as belts or gears.

CUTTING FLUIDS

Machining operations create heat as tools cut metals. **Cutting fluids** are substances applied to the cutting area where the tool and the workpiece make contact. They improve the efficiency and results of most machining operations. Cast iron is one material that is recommended to be machined "dry" or without any cutting fluid.

Cutting fluids lower the temperature of cutting tools and workpieces by cooling and reducing friction through lubrication of the tool/workpiece contact point. The ability of a cutting fluid to reduce friction is called its *lubricity*. A fluid with higher lubricity is better at reducing friction. The result of cooling and reducing friction is longer tool life, which means less frequent changing of cutting tools. Operations can also be performed at faster rates. The cooling effect of the fluid also helps prevent materials from expanding and affecting precision measurements.

Cutting fluids also wash chips away from the workpiece and prevent the buildup of material on the tool's cutting surfaces. Both of these actions improve the surface finish of the part being machined.

What follows is an introduction to the basics of cutting fluids and their applications. The two basic types are oil based and chemical based. Oil-based fluids range from straight oils to those with additives. Higher oil content provides greater lubricity but less cooling. Chemical-based fluids range from semi-synthetic to full-synthetic

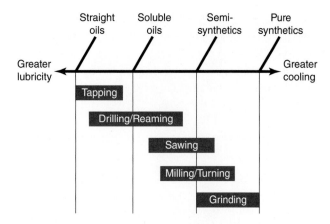

FIGURE 2.8.7 The lubricity/cooling relationship and ranges of general types of recommended cutting fluids for some common machining operations.

FIGURE 2.8.8 An example of a sulfurized, chlorinated cutting oil.

and are mixed with water. More water content provides greater cooling but less lubricity. **Figure 2.8.7** illustrates this lubricity versus cooling relationship and general recommended types of cutting fluids for some common machining operations. There are many variations of cutting fluids for specific applications that are available from many manufacturers. Those manufacturers can provide guidance in selecting the proper cutting fluid for a given operation and material.

 CAUTION

The large majority of today's cutting fluids are very safe, but be sure to always consult a product's SDS for proper handling and potential hazards.

Oil-Based Cutting Fluids

Oil-based cutting fluids consist primarily of petroleum or agricultural-based oils such as soy or vegetable. In the past, animal oils were used, but they are rarely used today because they tend to turn rancid, breed bacteria, and produce strong odors.

Straight Oils

Straight oils (commonly called cutting oils) are used mostly for light-duty, short-term operations and on nonferrous metals. They work very well on aluminum and magnesium alloys.

For heavier operations and when steels are machined, ingredients are added to improve the oil's performance. One common additive is called a *wetting agent*. Wetting agents help oil stick to the tool and workpiece. Ingredients called extreme pressure additives, such as sulfur and chlorine, improve lubricity to reduce friction. These sulfurized and chlorinated cutting oils are excellent

choices for operations such as drilling and tapping where there is significant surface contact between the cutting tool and workpiece.

Straight oils are very easy to maintain because they do not deteriorate, but they are fairly expensive. They can also create nuisance or harmful mist or smoke, so some controls may be needed to provide adequate ventilation. **Figure 2.8.8** shows an example of a straight cutting oil.

Soluble Oils

Soluble oils are similar to straight oils but contain additives that allow them to be combined with water. They are also called **emulsifiable** or **water-miscible** oils. The oil does not actually dissolve in water, but is emulsified. This means that very small oil droplets are suspended in the water. The resulting combination is used as the cutting fluid. Many different formulas of soluble oils are available. They are mostly used for light- to medium-duty mill and lathe operations on both ferrous and nonferrous metals. The soluble oil looks just like a straight cutting oil but, when mixed with water, it has a milky appearance. **(See Figure 2.8.9.)**

Soluble oils provide higher cooling rates than straight oils because of the water component. Because of the oil component, these cutting fluids still provide good lubricity and rust resistance for workpieces and machine

Photo courtesy of Master Chemical Company, a global supplier of industrial metal working products

FIGURE 2.8.9 A soluble oil is mixed with water to produce a milky white cutting fluid.

parts, but not as good as straight oils. Other additives help the oil to cling to tools and workpieces just like with straight oils.

Since water is a breeding ground for bacteria, these fluids must be maintained properly so they do not deteriorate and become rancid.

> ⊘ **CAUTION**
>
> Never use water-soluble oil as a cutting fluid for magnesium machining operations. Small magnesium particles are flammable, and water will only intensify a magnesium fire.

Chemical-Based Cutting Fluids

Chemical-based cutting fluids, as the name states, are based on chemicals and contain very little or no oil. Both of the two basic types are mixed with water like soluble oils. **Figure 2.8.10** shows a chemical-based cutting fluid concentrate and after combining with water.

> ⊘ **CAUTION**
>
> Since chemical-based fluids are combined with water, they should never be used for magnesium machining operations.

A

B

FIGURE 2.8.10 (A) A chemical-based cutting fluid concentrate, and (B) the product after mixing with water.

Synthetics

Synthetics contain no oil products and provide the highest level of cooling of all cutting fluids because of the water and their chemical makeup. Because they contain no oil, however, they do not provide the same levels of lubricity as oil-based fluids. Proper water/synthetic fluid concentration levels must be monitored closely. If the mixture is too lean (too much water in the mixture), corrosion can easily occur on workpieces and machinery. They are also easily contaminated by lubricating oils from machine tools, which makes them more difficult to maintain. Synthetics are also rather expensive. Many different formulas of synthetics are available for different operations on both ferrous and nonferrous metals, but they are particularly well suited for precision grinding operations because of the high temperatures created by those processes.

Semi-Synthetics

Semi-synthetics are crosses between soluble oils and synthetics. They provide better cooling than soluble oils, but not as high a level as pure synthetics. Because of their oil content, they lubricate better than pure synthetics, but not as well as straight or soluble oils.

Since semi-synthetics are mixed with water, concentration levels must be closely monitored, as with pure synthetics, to prevent corrosion. Different formulas are available for different operations involving both ferrous and nonferrous metals. Semi-synthetics are some of the most widely used cutting fluids for CNC lathe and mill operations.

Measuring Cutting Fluid Mixtures

Since some types of cutting fluids must be mixed with water, it is important to understand mixing ratios. Mixing ratios are usually given as parts of water to parts of the cutting fluid concentrate, such as 20:1, meaning 20 parts of water are to be combined with one part of the concentrate. Follow manufacturers' instructions on mixing concentrates with water.

Refractometer

When using soluble oils and synthetic and semi-synthetic cutting fluids, it is important to monitor and maintain proper fluid concentrations. A **refractometer** is a handheld tool that is commonly used to determine concentration levels. It measures the concentration level by how light is refracted, or "bent," as it goes through the liquid. A sample of the cutting fluid mixture is placed on a small glass area of the tool. The concentration level is then determined by reading a scale in the viewfinder. **Figure 2.8.11** shows the method for using a refractometer. **Figure 2.8.12** shows an illustration of a refractometer reading.

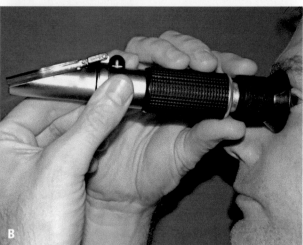

FIGURE 2.8.11 (A) The cutting fluid mixture is placed on the glass of the refractometer. (B) Then the user looks through the eyepiece to read the scale.

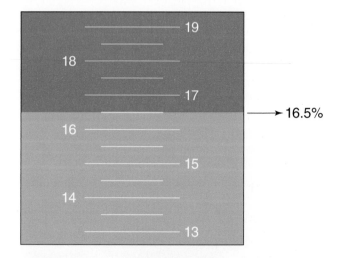

FIGURE 2.8.12 A sample refractometer reading. The scale is read where the two colors meet. This reading is 16.5, which means a 16.5% cutting fluid concentration.

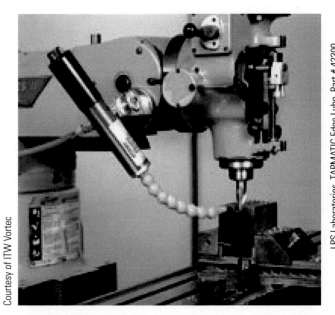

Courtesy of ITW Vortec

LPS Laboratories, TAPMATIC Edge Lube, Part # 42300. www.LPSLABS.com.

FIGURE 2.8.14 This stick cutting compound can be touched to a cutting tool before beginning a machining operation.

FIGURE 2.8.13 A cold air gun being used to provide cooling during a milling operation.

Cold Air Guns

Some machining applications use a **cold air gun** to cool the tool and workpiece. No lubrication is provided, only cooling. Compressed air flows through a spiral inside a canister that cools the air to temperatures below 0°F. The cold air is then directed at the cutting area. **Figure 2.8.13** shows a cold air gun in use.

Solid and Semi-Solid Cutting Compounds

The term *cutting fluid* usually means a liquid, but some substances—termed **solid and semi-solid cutting compounds**—used to cool and lubricate machining operations come in nonliquid forms such as wax sticks, bars, pastes, creams, and gels. Since they are manually applied to cutting tools before cutting operations, they are generally used for conventional, short-term operations, not for high-volume production. They are cleaner than liquids and provide very good lubricity because they stick to cutting tools very well. **Figure 2.8.14** shows an example of this type of cutting compound.

Methods of Application

All of the different types of cutting fluids or compounds must somehow be applied to cutting tools and workpieces. There are three basic methods used to accomplish that task.

Manual

Manual application is the simplest method of getting a cutting fluid to the point where cutting is taking place. The person performing the machining operation will use a brush, spray bottle, or squirt bottle to apply liquids. The semi-solid and solid compounds are almost always used this way. This method is usually used for short-term operations, not in high-volume production settings.

Flood System

Liquid cutting fluids may be applied by a **flood system**. A tank holds a certain volume of the cutting fluid that is pumped through a pipe or nozzle to the cutting area. The area is completely flooded by the cutting fluid. The fluid then returns to the tank, is filtered, and is recirculated through the system. **Figure 2.8.15** shows a flood system in use.

Mist System

Liquids can also be applied with a **mist system**. The cutting fluid is contained in a small tank. It is combined with compressed air (atomized) and sprayed on the cutting area. Unlike a flood system, the fluid in a mist system is generally not recovered or recirculated. **Figure 2.8.16** shows a mist system. The mist can quickly fill an area, so adequate ventilation must be provided to protect from inhalation of the mist.

Minimum Quantity Lubricant (MQL)

Another cutting fluid application process that is gaining popularity is **minimum quantity lubricant (MQL)**. This method is a mist system, but it uses lower air pressure to deliver the smallest fluid concentration required to provide adequate tool lubrication and cooling. The concentration must be closely controlled to provide proper results. There are three major benefits of MQL. First, lower amounts of a cutting fluid are used, so it is safer for workers and the environment. Second, cutting fluid costs are lower. Finally, machined parts, chips, tools, and equipment are kept nearly dry, so cleanup is much easier.

FIGURE 2.8.15 The flood coolant in this CNC machining operation cools, lubricates, and flushes chips away from the cutting area.

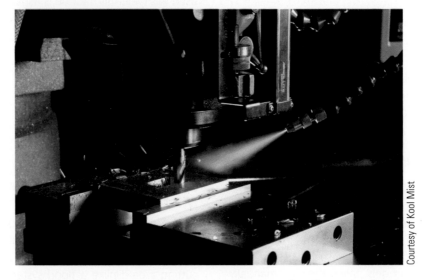

Courtesy of Kool Mist

FIGURE 2.8.16 A mist application for a milling operation. The cutting fluid is mixed with a stream of air and directed at the cutting area.

SUMMARY

- Machine tools and accessories are expensive and need to be properly maintained. One of the most important preventative maintenance tasks is to maintain proper machine lubrication.

- Periodic inspection of moving machine parts is another important part of machinery maintenance. Following a scheduled maintenance plan according to manufacturers' directions will keep machinery safe and reliable.

- Cutting fluids improve machining operations by allowing operations to be performed faster, improving surface finish, and reducing tool wear.

- Using proper cutting fluids improves efficiency during machining operations and proper maintenance keeps machinery in the best possible working condition. Both improve overall reliability and efficiency.

REVIEW QUESTIONS

1. Why is a routine maintenance plan important?

2. What is a lubricant?

3. List three methods of applying lubricants to machine tools.

4. List two components of machine tools that should be periodically inspected for wear.

5. Briefly describe a gib.

6. What are the purposes of cutting fluids?

7. What are the two major types of cutting fluids?

8. What is a wetting agent?

9. What two ingredients are often used as additives to improve lubrication qualities of cutting oils?

10. Oils that can be combined with water are called _____, _____, or _____ oils.

11. What is the major benefit of cutting fluids that contain water?

12. What is the difference between synthetic and semi-synthetic cutting fluids?

13. A _____ can be used to measure cutting fluid concentrations.

14. List three methods of applying cutting fluids.

15. Always review the _____ before using any lubricant or cutting fluid.

16. What type of cutting fluids should never be used when machining magnesium alloys?

SECTION 3 JOB PLANNING, BENCHWORK, AND LAYOUT

- # Unit 3
  ## Hand Tools
  Introduction
  Screwdrivers
  Pliers
  Hammers
  Wrenches
  Bench Vise
  Clamps
  Hacksaws
  Files
  Deburring
  Abrasives

- # Unit 4
  ## Saws and Cutoff Machines
  Introduction
  Power Hacksaws
  Band Sawing Machines
  Saw Blade Characteristics and Applications
  Band Saw Blade Welding
  Band Saw Blade Mounting/Removal
  Blade Speed
  The Abrasive Cutoff Saw
  Metal Cutting Circular (Cold) Saws

- # Unit 5
  ## Offhand Grinding
  Introduction
  Grinder Uses
  Abrasive Belt and Disc Machine Uses
  Grinding Wheels
  Pedestal Grinder Setup
  Grinding Procedures

- # Unit 6
  ## Drilling, Threading, Tapping, and Reaming
  Introduction
  Benchwork Holemaking Operations
  Threading and Tapping

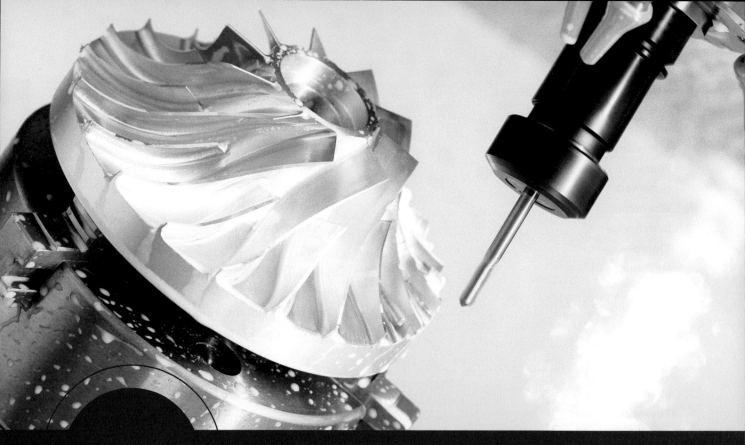

UNIT 1   Understanding Drawings

Learning Objectives

After completing this unit, the student should be able to:

- Identify and interpret title block information
- Identify line types and their uses
- Demonstrate understanding of the principle of orthographic projection
- Identify the three basic views frequently used on engineering drawings
- Identify isometric views
- Demonstrate understanding of basic symbols and notation used on engineering drawings
- Define *tolerance*
- Demonstrate understanding of unilateral, bilateral, and limit tolerances
- Demonstrate understanding of allowances and classes of fit for cylindrical components
- Define geometric dimensioning and tolerancing (GD&T) and explain its purpose
- Identify basic GD&T symbols
- Demonstrate understanding of basic GD&T feature control frames

Key Terms

Accumulated tolerance	Assembly drawing	Bill of materials
Allowance	Basic size	Center line
	Bilateral tolerance	Class of fit

Cutting plane line
Datum
Dimension line
Exploded isometric drawing
Extension line
Feature control frame
Feature of size
Form tolerance
Front view
Geometric dimensioning
  and tolerancing (GD&T)
Hidden line

Isometric view
Leader line
Least material condition (LMC)
Limit tolerance
Location tolerance
Maximum material condition
  (MMC)
Object (visible) line
Orientation tolerance
Orthographic projection
Phantom line
Profile tolerance

Reference dimension
Regardless of feature size (RFS)
Revision
Right side view
Runout tolerance
Scale
Section line
Section view
Title block
Top view
Typical dimension
Unilateral tolerance

IMPORTANCE OF ENGINEERING DRAWINGS

An important skill needed for success in the machining field is the ability to interpret engineering drawings, or prints. Engineering drawings show the sizes and shapes of components and their specific *features*, such as holes, slots, or surfaces. No matter how skilled you are at performing machining operations, if you are unable to properly interpret these drawings, you will not be able to produce machined components independently or efficiently within required specifications.

Engineering drawings can range from simple hand-drawn sketches to complex, computer-generated prints. They are produced in a standard format that enables machinists anywhere to understand them. Printed drawing sizes range from 8.5 × 11 up to 34 × 44 inches. Drawings for complex components may contain several sheets.

COMPONENTS OF ENGINEERING DRAWINGS

Engineering drawings are made up of several components. These elements make up a standard system of views, lines, and symbols that provide important information about required specifications for machined components.

Title Block

One major component of a drawing is the **title block**. The title block includes information such as the part name and number, tolerances, scale, material that the part should be made from, and any required heat treatment. Engineer and draftsperson names and drawing creation date are also usually included. A history of **revisions**, or changes, can be shown above the title block as well, in a box often called the *revision block*. Revisions include changes in dimensions, material, tolerances, or surface finish. Each revision should have a description, a date, and the initials of the person who approved the revision. A revision block may also be shown in the upper-right corner of a print.

It is good practice to check for revisions to ensure the print specifications are current.

Scale is the size of an actual object related to its size drawn on a print. Parts that are drawn to actual size on paper are *full scale*. This scale can be shown as 1:1. The 1:1 means that 1" on the drawing is equal to 1" on the actual part. The drawing is the same size as the part.

Large objects are frequently drawn smaller than actual size to make drawings easier to use. If an object is drawn half its actual size, the scale can be shown as 1/2, 1:2, or 1/2" = 1". These all mean that 1/2" on the drawing is equal to 1" on the actual part. The drawing is smaller than the actual part.

Small objects are often drawn larger than actual size to show details that might be too difficult to see if they were drawn to actual size. For example, if an object is drawn twice its actual size, the scale can be shown as 2/1, 2:1, 2X, or 2" = 1". These all mean that 2" on the drawing is equal to 1" on the actual part. The drawing is larger than the actual part.

In all cases, regardless of scale, drawing dimensions shown are actual part sizes.

Sometimes a print may have a **bill of materials**. A bill of materials can either list the raw materials used to make the machined part or list components that are assembled to produce the part specified on the print. A sample engineering drawing with the title block, revision block, and bill of materials labeled is shown in **Figure 3.1.1**.

Orthographic Projection

An engineering drawing is a two-dimensional representation of a three-dimensional object. For that reason, each side of the object is shown by a different *view*. Each view represents how the object would appear when looked at from a certain perspective or position. By studying these views, the part can be visualized in three dimensions. The number of views in a blueprint is determined by the shape of the part and how complex it is. A simple part may only require one view to show all of the information to machine it. A more complex part will require more views to aid in developing a mental image of its shape. This method of

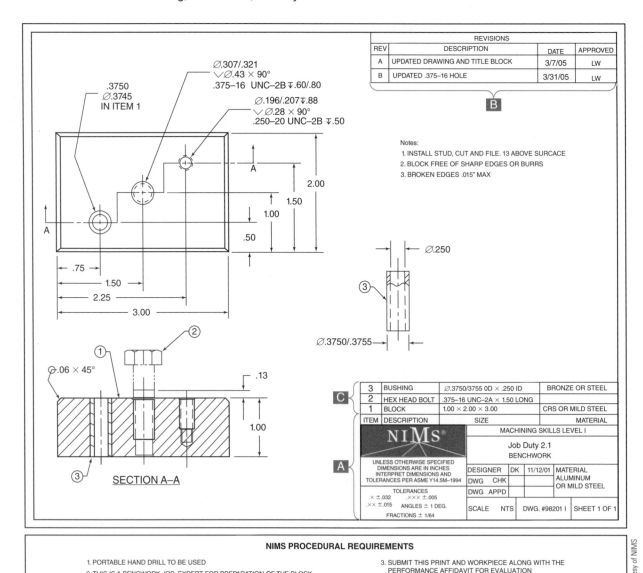

FIGURE 3.1.1 The title block (A) of an engineering drawing, or print, contains information including part name, tolerances, and scale. The revision block (B) shows a list of changes that have been made over time. The bill of materials (C) lists raw material or components used to produce the part.

representing a three-dimensional object in two dimensions using different views is called **orthographic projection**.

To understand orthographic projection, imagine an object inside a glass box with hinges where the sides meet. The top is hinged to open above the front, and the right side is hinged to open to the right of the front, as shown in **Figure 3.1.2**. Three-view drawings are common in machining and use the front view, right-side view, and top view. Three-view drawings provide sufficient information for many machined components.

The Front View

The **front view** is the view that normally shows the most details. The person creating the drawing decides which view will be the front view. It is not necessarily the front of

the object related to its use. For example, the front view of a motorcycle drawing may actually show its side because more details can be shown from that perspective. The front view is created by projecting the front of an object onto the front surface of the "glass box," as shown in **Figure 3.1.3**.

The Top View

The **top view** is created by projecting the top of an object onto the top surface of the "glass box," as shown in **Figure 3.1.4**.

The Right-Side View

The **right-side view** is created by projecting the right side of an object onto the right surface of the "glass box," as shown in **Figure 3.1.5**.

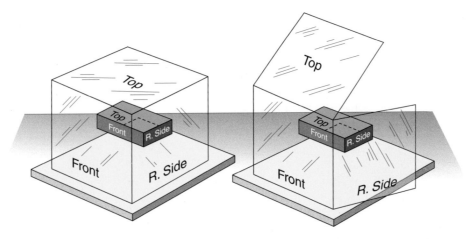

FIGURE 3.1.2 Orthographic projection uses different views to show three-dimensional objects in two dimensions. The "glass box" shows how different views are projected, or created.

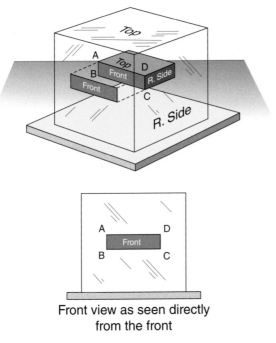

Front view as seen directly
from the front

FIGURE 3.1.3 The front view of a drawing is projected onto the front surface of the "glass box." Note the location of points A, B, C, and D.

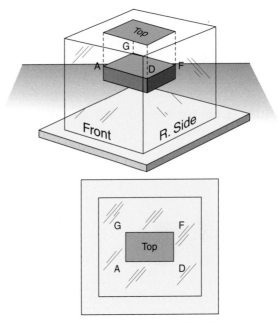

Top view as seen directly
from above

FIGURE 3.1.4 The top view of a drawing is projected onto the top surface of the "glass box." Note the location of points A, D, F, and G.

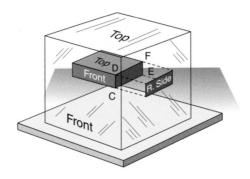

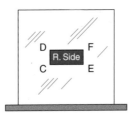

Right-side view as seen
directly from the right
side

FIGURE 3.1.5 The right-side view of a drawing is projected onto the right surface of the "glass box." Note the location of points C, D, E, and F.

Isometric View

Sometimes the views created through orthographic projection do not clearly show the shape of complex parts. To provide a better visualization of the part, a drawing may contain a three-dimensional view called an **isometric view**. **Figure 3.1.6** shows a simple isometric view.

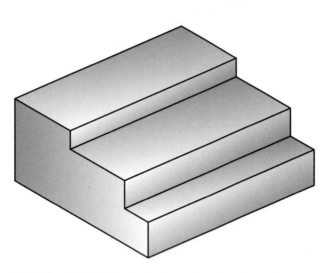

FIGURE 3.1.6 This simple isometric view shows how an object would appear in three dimensions. Isometric views are very helpful when trying to visualize complex objects.

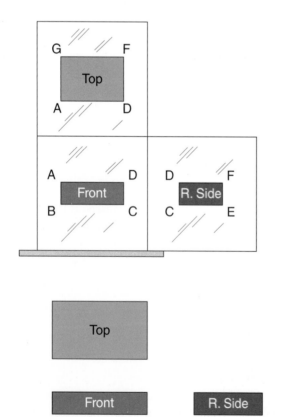

FIGURE 3.1.7 The positions of the three principal views in orthographic projection.

View Arrangement

After views are created through orthographic projection using the "glass box" method, the hinged box sides are opened completely to show the position of the three

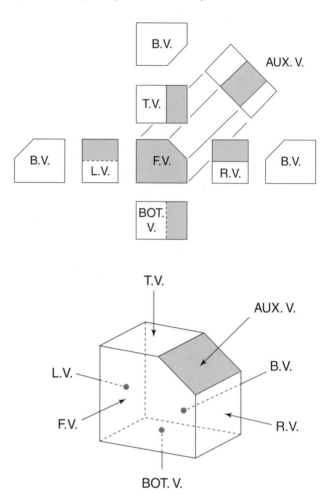

Name of view	Abbreviation
Front view	(F.V.)
Right side view	(R.V.)
Left side view	(L.V.)
Bottom view	(Bot. V.)
Back or rear view	(B.V.)
Auxiliary view	(Aux. V.)
Top view	(T.V.)

FIGURE 3.1.8 Arrangement and identification of projected views. Compare the surfaces from the isometric view to those in the projected views. Note that the back view can be placed in any of the three shown positions.

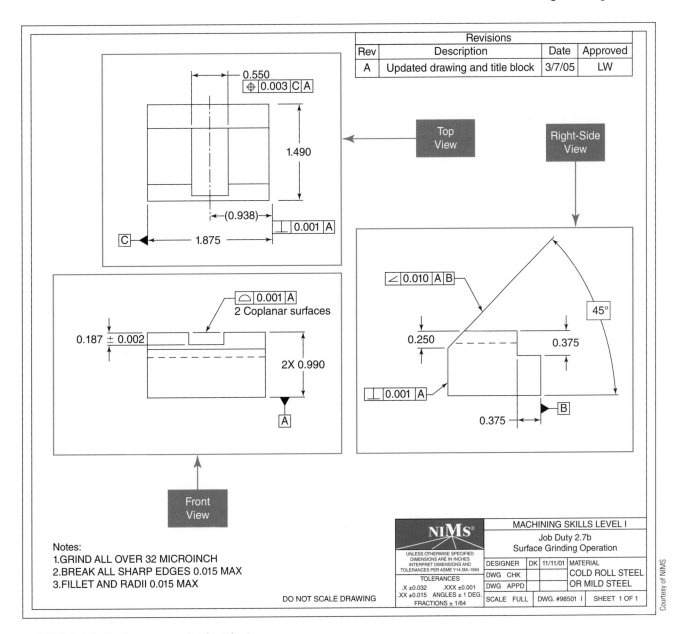

FIGURE 3.1.9 The front, top, and right-side views.

principal views. **(See Figure 3.1.7.)** Other surfaces of an object can be projected onto the "glass box" to create more views. **Figure 3.1.8** identifies these views and shows how they would be arranged on an engineering drawing. **Figure 3.1.9** labels the front view, top view, and side view in a typical three-view engineering drawing.

Line Types

Engineering drawings are made up of different styles of lines called *line types*. Each line type is used for a specific purpose. They are identified by the differences in their appearances. Line types are drawn as either thick or thin. Different types are also identified as continuous or broken, and by the size of the breaks. These different line types are used to form a drawing in the same way different letters of the alphabet are used to form words. Refer to **Figure 3.1.10** while reading about some of the commonly used line types explained in the following paragraphs. **Figure 3.1.11** shows an engineering drawing with some different line types labeled with letters. These examples will be referred to during the following discussion of line types.

Object (or Visible) Lines

Object (or **visible**) **lines** are used to show edges of an object that would be seen in any particular view. The outline of the object would be shown by object lines. They are thick and continuous. Object lines are labeled by the letter A in **Figure 3.1.11**.

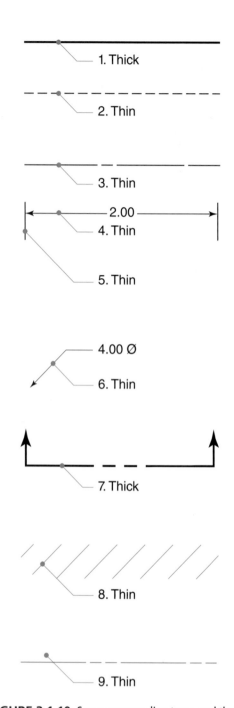

1. *Object or visible lines* are used to show part edges that are visible in a given view. They are thick and continuous.

2. *Hidden lines* show part edges that cannot be seen in a given view. They are thin and broken into small dashes.

3. *Center lines* show the center of a diameter of radius, or the center of a part. Hole locations are shown where two center lines cross. They are thin and broken into alternating long and short dashes.

4. *Dimension lines* have arrowheads at each end where they touch extension lines. A dimension is listed to show feature size. They are thin and have arrowheads at their ends.

5. *Extension lines* extend edges for dimensioning purposes. They are thin and continuous.

6. *Leader lines* are angled with an arrowhead at one end that touches a part feature. A dimension or note is placed at the other end of the leader. They are thin and continuous.

7. *Cutting plane lines* are used to make an imaginary cut through an object. The arrowheads indicate the viewing direction of the section view after the cut. They are thick and have one long and two short dashes alternately spaced.

8. *Section lines* show surfaces that have been "cut" in a section view created by a cutting plane line. They are thin and angled.

9. *Phantom lines* are used to show alternate positions of parts or outlines of adjacent parts. They are thin and have one long and two short dashes alternately spaced.

FIGURE 3.1.10 Some common line types and their descriptions.

Hidden Lines

Hidden lines are used to show edges that are not visible in a particular view. Hidden lines are thin and broken into a series of short dashes. Some examples of hidden lines are shown in **Figure 3.1.11** by the letter B.

Center Lines

A **center line** is used to show the center of a diameter or radius, or the center of a part. Center lines are thin and are broken into alternating long and short dashes. The center

point of a diameter or radius is located where the short dashes of two perpendicular center lines cross each other. Letter C in **Figure 3.1.11** shows some examples of center lines.

Extension Lines and Dimension Lines

Extension lines and dimension lines are used together. **Extension lines** extend from the edges of an object or feature of an object for dimensioning purposes. They are thin and continuous. Some extension lines are shown in **Figure 3.1.11** by the letter E.

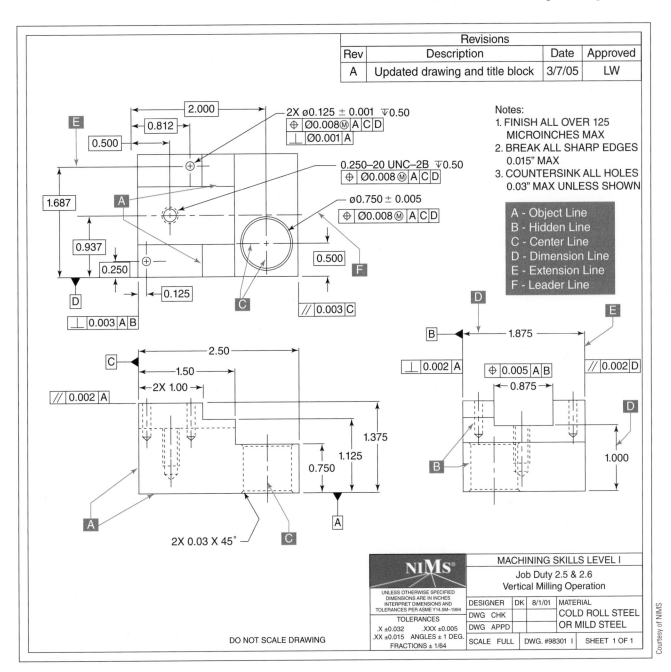

FIGURE 3.1.11 An example of an engineering drawing with some line types identified.

Dimension lines are used to specify sizes. They are thin and have arrowheads at the ends where they meet extension lines. There are a few different methods of placement of dimension lines and dimensions. When there is enough space between extension lines, the dimension line and the dimension are both placed between the extension lines. The dimension is located where the dimension line is broken. **(See Figure 3.1.12.)** Sometimes there is only room between extension lines for the dimension line, but not the dimension. In those cases, the dimension line is placed between the extension lines and is not broken, while the dimension is placed outside

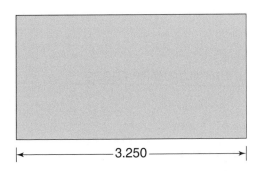

FIGURE 3.1.12 Dimension lines located inside extension lines with the dimension placed where the dimension line is broken.

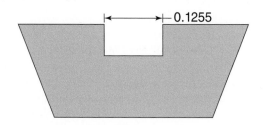

FIGURE 3.1.13 A dimension line located inside extension lines with the dimension placed outside the extension lines.

the extension lines, as shown in **Figure 3.1.13**. If there is room for the dimension but not the dimension line, the dimension can be placed between the extension lines, and the dimension line is broken and placed outside the extension lines. **(See Figure 3.1.14.)** Some dimension lines are identified by the letter D in **Figure 3.1.11**.

Leader Lines

A **leader line** is a thin angled line with an arrowhead on one end that points to a specific feature or detail.

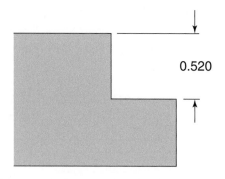

FIGURE 3.1.14 Dimension lines located outside extension lines with the dimension placed inside the extension lines.

The arrowhead touches the feature and a dimension or note is placed at the other end of the leader line. Some leader lines are identified by the letter F in **Figure 3.1.11**.

Phantom Lines

Phantom lines are used to show alternate positions of a part or outlines of adjacent parts. **Figure 3.1.15**

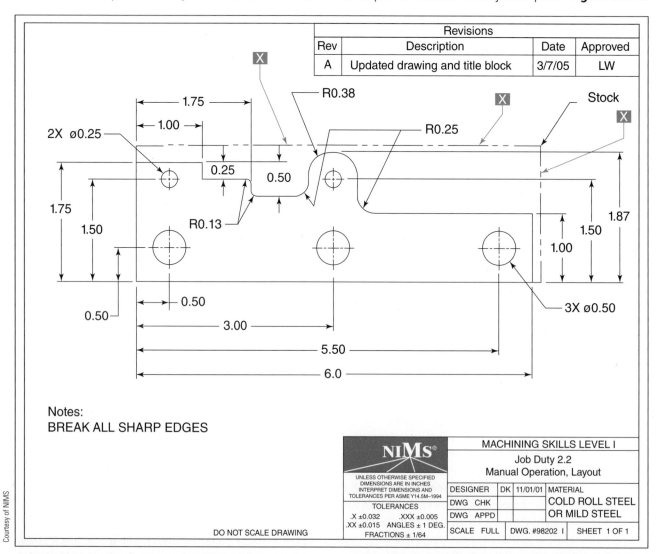

FIGURE 3.1.15 The phantom lines on this print shown at the X's show the outline of the material, or stock, used to make the part.

shows the use of a phantom line (letter X) to identify the outline of the material a part is to be made from.

Cutting Plane Lines, Section Views, and Section Lines

An object can be cut along an imaginary line by a **cutting plane line** to create a special type of view called a **section view**. A section view is used to show internal features more clearly. The cutting plane line is thick and

is drawn as one long and two short dashes alternately spaced. The ends of the cutting plane line make perpendicular turns and then end with arrowheads. The arrowheads show the viewing direction for the section view.

When the cutting plane line "cuts" an object and creates a section view, surfaces that have been "cut" are shown by **section lines**. Section lines are thin diagonal lines. **Figure 3.1.16** shows a cutting plane line and section lines in a section view (letters A and C).

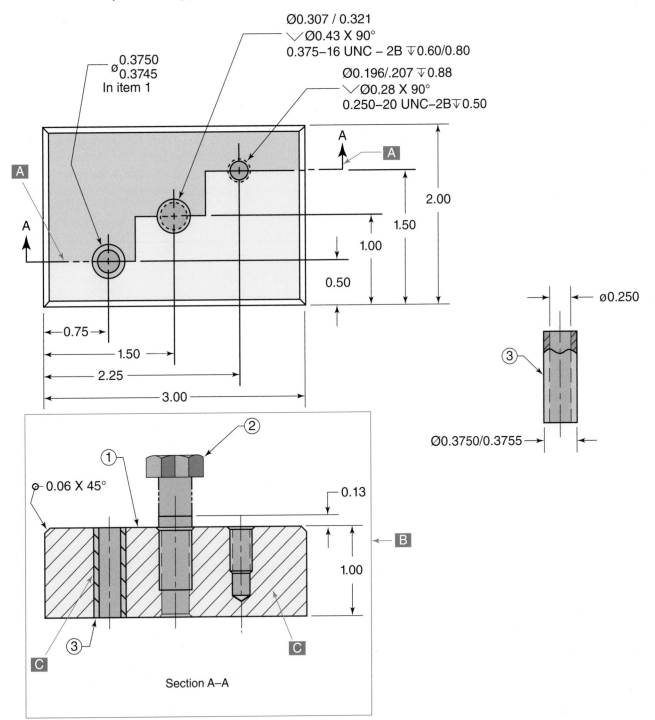

FIGURE 3.1.16 A cutting plane line (A) cuts the part along an imaginary line. The result is a section view (B) with section lines (C) showing surfaces exposed by the cut. Note that the ends of the cutting plane line are labeled "A" and that the section view is named Section A-A.

ASSEMBLY DRAWINGS

In the machining industry, several components are often produced as part of an assembly. An **assembly drawing** shows the components together as they would appear in the completed assembly (**Figure 3.1.17**). An **exploded isometric drawing** shows the components slightly separated to give more detail and direction about how they fit together. Parts are in line with their mating parts and a center line shows the direction of assembly (**Figure 3.1.18**).

BASIC SYMBOLS AND NOTATION

Interpreting engineering drawings also requires an understanding of the use of some basic symbols and notation. Diameters, radii, drilled holes, countersinks, counterbores, and depths are frequently specified with symbols or notations. The chart in **Figure 3.1.19** shows ways in which these features and specifications can be shown.

Radii on the corners of a workpiece are sometimes called fillets and rounds. A *fillet* is a radius on the inside corner and a *round* is a radius on the outside edge or corner of a part. Fillets and rounds are commonly found on stampings, forgings, and castings. They are sized in terms of their radius. Since many cast, forged, and stamped parts have several of these radii, their sizes are frequently found in the "Notes" section of the drawing. This helps keep the drawing from becoming cluttered with so many dimensions that it becomes hard to read. For example, a note might read "UNLESS OTHERWISE SPECIFIED ALL FILLETS AND ROUNDS R 1/4."

Surface finish requirements as described in Section 2.4 can be shown on prints by either symbols or notes. **Figure 3.1.20** lists a few examples of surface finish that can be specified.

Sometimes, there is more than one feature of the same size and shape on a part. For example, a block may contain four holes of the same size. Instead of using a separate leader line and dimension for all four holes, the drawing may contain *4X* before the hole dimension.

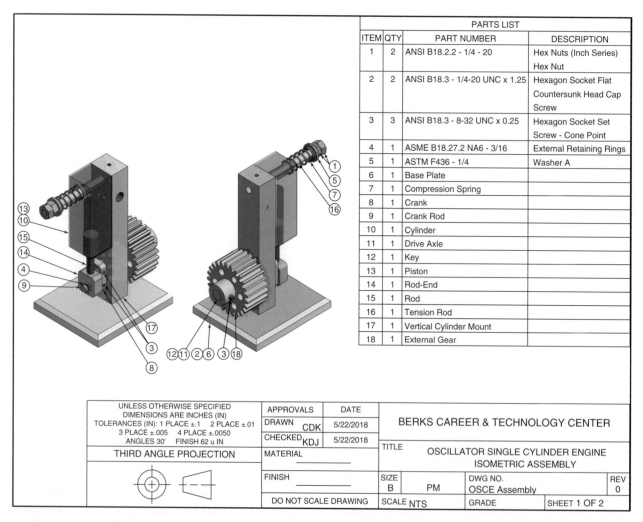

PARTS LIST			
ITEM	QTY	PART NUMBER	DESCRIPTION
1	2	ANSI B18.2.2 - 1/4 - 20	Hex Nuts (Inch Series) Hex Nut
2	2	ANSI B18.3 - 1/4-20 UNC x 1.25	Hexagon Socket Flat Countersunk Head Cap Screw
3	3	ANSI B18.3 - 8-32 UNC x 0.25	Hexagon Socket Set Screw - Cone Point
4	1	ASME B18.27.2 NA6 - 3/16	External Retaining Rings
5	1	ASTM F436 - 1/4	Washer A
6	1	Base Plate	
7	1	Compression Spring	
8	1	Crank	
9	1	Crank Rod	
10	1	Cylinder	
11	1	Drive Axle	
12	1	Key	
13	1	Piston	
14	1	Rod-End	
15	1	Rod	
16	1	Tension Rod	
17	1	Vertical Cylinder Mount	
18	1	External Gear	

UNLESS OTHERWISE SPECIFIED DIMENSIONS ARE INCHES (IN) TOLERANCES (IN): 1 PLACE ±.1 2 PLACE ±.01 3 PLACE ±.005 4 PLACE ±.0050 ANGLES 30' FINISH 62 u IN		APPROVALS	DATE	BERKS CAREER & TECHNOLOGY CENTER			
		DRAWN CDK	5/22/2018				
		CHECKED KDJ	5/22/2018	TITLE OSCILLATOR SINGLE CYLINDER ENGINE ISOMETRIC ASSEMBLY			
THIRD ANGLE PROJECTION		MATERIAL					
		FINISH		SIZE B	PM	DWG NO. OSCE Assembly	REV 0
		DO NOT SCALE DRAWING		SCALE NTS		GRADE	SHEET 1 OF 2

FIGURE 3.1.17 An assembly drawing shows an image of several components assembled as a final product.

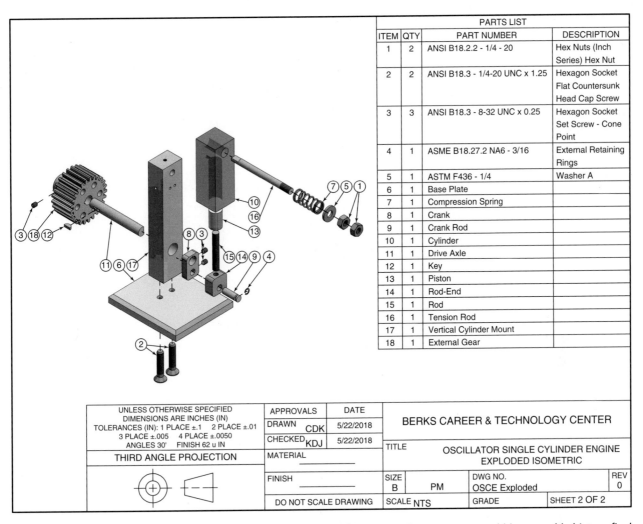

PARTS LIST			
ITEM	QTY	PART NUMBER	DESCRIPTION
1	2	ANSI B18.2.2 - 1/4 - 20	Hex Nuts (Inch Series) Hex Nut
2	2	ANSI B18.3 - 1/4-20 UNC x 1.25	Hexagon Socket Flat Countersunk Head Cap Screw
3	3	ANSI B18.3 - 8-32 UNC x 0.25	Hexagon Socket Set Screw - Cone Point
4	1	ASME B18.27.2 NA6 - 3/16	External Retaining Rings
5	1	ASTM F436 - 1/4	Washer A
6	1		Base Plate
7	1		Compression Spring
8	1		Crank
9	1		Crank Rod
10	1		Cylinder
11	1		Drive Axle
12	1		Key
13	1		Piston
14	1		Rod-End
15	1		Rod
16	1		Tension Rod
17	1		Vertical Cylinder Mount
18	1		External Gear

UNLESS OTHERWISE SPECIFIED DIMENSIONS ARE INCHES (IN) TOLERANCES (IN): 1 PLACE ±.1 2 PLACE ±.01 3 PLACE ±.005 4 PLACE ±.0050 ANGLES 30' FINISH 62 u IN	APPROVALS	DATE	BERKS CAREER & TECHNOLOGY CENTER			
	DRAWN CDK	5/22/2018				
	CHECKED KDJ	5/22/2018	TITLE	OSCILLATOR SINGLE CYLINDER ENGINE EXPLODED ISOMETRIC		
THIRD ANGLE PROJECTION	MATERIAL					
	FINISH		SIZE B	PM	DWG NO. OSCE Exploded	REV 0
	DO NOT SCALE DRAWING		SCALE NTS		GRADE	SHEET 2 OF 2

FIGURE 3.1.18 An exploded isometric drawing shows an image of how several components would be assembled into a final product. The center lines indicate where mating parts fit together and direction of assembly.

This means that there are four holes of the same size. Multiple, identical part features or equal dimensions can also be identified as **typical dimensions** by adding the abbreviation *TYP* after a dimension. The abbreviation *EQL SP* means that features are equally spaced. **Figure 3.1.21** shows examples of identifying multiple, identical features or dimensions using these methods.

TOLERANCE

Recall, from the measurement section, that a tolerance is an allowable variation from a given size. A dimension shown on a print is called a **basic size**. A tolerance is applied to the basic size to determine the largest and smallest acceptable size for a dimension. The largest acceptable size is often called the *high limit*, or *upper limit*. The smallest acceptable size is often called the *low limit*, or *lower limit*. The difference between the upper limit and the lower limit is the *total tolerance*. The amount of a tolerance can vary greatly depending on the intended

use of a machined component, but the principle of interpreting any tolerance is the same.

Bilateral Tolerances

A **bilateral tolerance** allows a dimension to vary both above and below basic size. A bilateral tolerance can take two forms. It can vary by equal amounts above and below basic size or by different amounts above and below basic size. The allowable amount above basic size is shown with a "+" symbol. The allowable amount below basic size is shown with a "−" symbol. When both amounts are equal, a "±" symbol is used.

Figure 3.1.22 shows a few examples of bilateral tolerances and how to determine upper limits, lower limits, and total tolerance.

Unilateral Tolerances

A **unilateral tolerance** allows a dimension to vary either above or below basic size, but not both. A unilateral tolerance is also shown using a "+" or "−" symbol, but

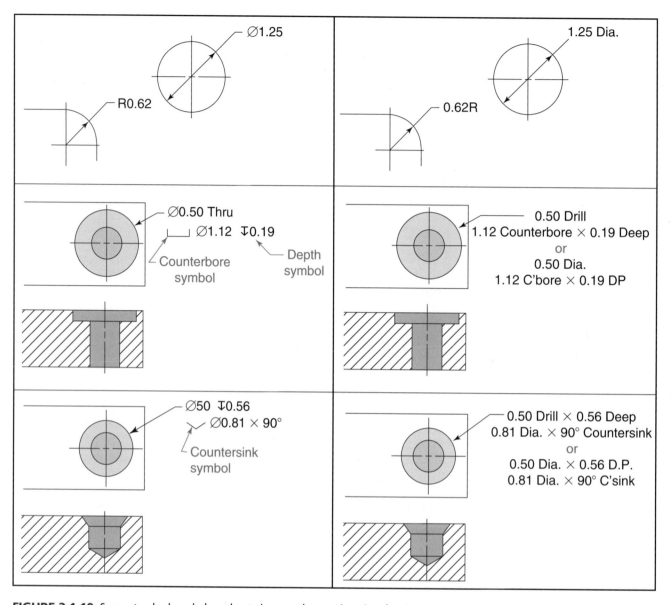

FIGURE 3.1.19 Some standard symbols and notations used on engineering drawings..

one amount is "0," indicating no variation is allowable in that direction.

Figure 3.1.23 shows two examples of unilateral tolerances and how to determine upper limits, lower limits, and total tolerance.

Limit Tolerances

A **limit tolerance** does not use the "+," "−," or "±" symbol. Instead of a basic dimension being listed, the upper and lower limits are shown. They are usually separated by a bar or a slash. To determine total tolerance from a limit tolerance, simply subtract the lower limit from the upper limit. **(See Figure 3.1.24.)**

Feature of Size, MMC, and LMC

Feature of Size

Recall that a feature is simply one characteristic of the part shown on the engineering drawing, such as a hole diameter, or a length, or a depth. A **feature of size** is a feature that is either cylindrical or has two opposing parallel surfaces. One way to determine a feature of size is this: If the feature can be measured with the external jaws or internal nibs of a dial caliper, it is a feature of size. Examples of features of size are the following:

• Hole diameters

• Shaft diameters

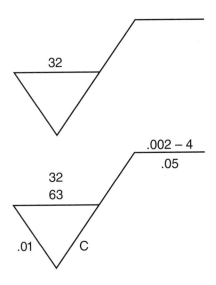

Grind all surfaces 32 microinch max

Finish all over 32 microinch max

Grind all surfaces $\overset{32}{\nabla}$

FIGURE 3.1.20 A few methods of specifying surface finish.

- Widths of slots
- Overall lengths of a parts

Examples of features that would not be features of size are the following:

- Depths of holes, steps, or slots
- Lengths of angles
- Lengths between two surfaces at angles to each other
- Radii, fillets, or rounds

Maximum Material Condition (MMC)

Maximum material condition (MMC) is a condition when a feature of size contains the most workpiece material within its given tolerance. Another way to remember MMC is to think about part weight. When is a part heavier? When it has more material: at its MMC. MMC can only be applied to a feature of size.

Consider the outside diameter of 1.500" shaft with a tolerance of ±0.005". The 1.500" is an external cylindrical dimension, so it is called an *external feature of size*. When is the most material present (and the object heavier)? When the diameter is at its upper limit of 1.505". The MMC of an external feature of size is always its upper limit.

Contrast that situation with a 1.500" hole diameter with a tolerance of ±0.005". In this case, the 1.500" is an internal cylindrical dimension, so it is called an *internal feature of size*. When is the most material present (and the object heavier)? When the diameter is at its lower limit of 1.495". The MMC of an internal feature of size is always its lower limit.

Least Material Condition (LMC)

Least material condition (LMC) is a condition when a feature of size contains the least material within its given tolerance. LMC can be remembered by thinking about part weight also. When is the part lighter? When it has less material: at its LMC. LMC can also only be applied to a feature of size.

Recall the outside diameter of 1.500" with a tolerance of ±0.005". It is an external feature of size. When is the least material present (and the object lighter)? When the diameter is at its lower limit of 1.495". The LMC of an external feature of size is always its lower limit.

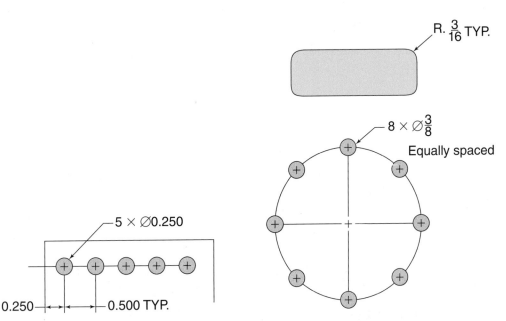

FIGURE 3.1.21 Multiple features can be identified by a few different methods.

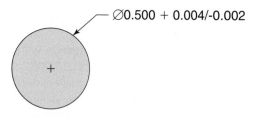

The 0.500 ± 0.002 shows an equal amount of allowable variation above and below basic size.

       Upper limit (largest allowable size) = 0.500 + 0.002 = 0.502
       Lower limit (smallest allowable size) = 0.500 − 0.002 = 0.498
       Total tolerance = 0.502 (upper limit) − 0.498 (Lower limit) = 0.004

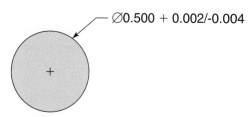

The 0.500 + 0.004/-0.002 shows different amounts of allowable variation above and below basic size.

       Upper limit (largest allowable size) = 0.500 + 0.004 = 0.504
       Lower limit (smallest allowable size) = 0.500 − 0.002 = 0.498
       Total tolerance = 0.504 (upper limit) − 0.498 (Lower limit) = 0.006

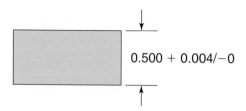

The 0.500 + 0.002/-0.004 also shows different amounts of variation above and below basic size.

       Upper limit (largest allowable size) = 0.500 + 0.002 = 0.502
       Lower limit (smallest allowable size) = 0.500 − 0.004 = 0.496
       Total tolerance = 0.502 (upper limit) − 0.496 (Lower limit) = 0.006

FIGURE 3.1.22 Bilateral tolerance examples.

       Upper limit (largest allowable size) = 0.500 + 0.004 = 0.504
       Lower limit (smallest allowable size) = 0.500 − 0 = 0.500
       Total tolerance = 0.504 (upper limit) − 0.500 (lower limit) = 0.004

FIGURE 3.1.23 Unilateral tolerance examples.

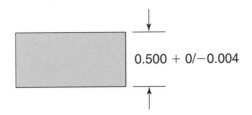

Upper limit (largest allowable size) = 0.500 + 0 = 0.500
Lower limit (smallest allowable size) = 0.500 − 0.004 = 0.496
Total tolerance = 0.500 (upper limit) − 0.496 (lower limit) = 0.004

FIGURE 3.1.23 *continued*

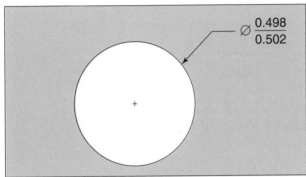

A limit dimension for a hole, or internal dimension, will usually show the lower limit first, or above the bar.

0.498/0.502 or $\frac{0.498}{0.502}$

0.498 is still the lower limit and 0.502 is the upper limit.

Total tolerance = 0.502−0.498 = 0.004.

FIGURE 3.1.24 Limit tolerance examples.

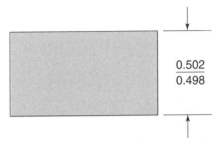

A limit dimension for an external dimension will usually show the upper limit first, or above the bar.

0.502/0.498 or $\frac{0.502}{0.498}$

0.498 is still the lower limit and 0.502 is still the upper limit.

Total tolerance = 0.502−0.498 = 0.004.

Notice that both of these limit tolerances have the same meaning as 0.500 ± 0.002.

Recall the 1.500" inside diameter with a tolerance of ±0.005". It is an internal feature of size. When is the least material present (and the object lighter)? When the diameter is at its upper limit of 1.505". The LMC of an internal feature of size is always its upper limit.

See **Figure 3.1.25** for an illustration of sizes at MMC and LMC and features that have no MMC and LMC because they are not features of size.

Tolerance Specifications

If a tolerance is bilateral or unilateral, it can be shown in two general ways. One way is to place the tolerance for the dimension with the dimension. That is called a *local tolerance*, or *specified tolerance*, because it is shown directly next to the dimension and applies only to that dimension. **Figure 3.1.26** identifies some specified tolerances with the letter A.

If there is no local tolerance, the tolerance is called *unspecified*. When a tolerance is unspecified, the tolerance for the dimension is determined by information in the *tolerance block* area of the title block. The tolerance block generally lists the tolerances for fractional, decimal, and angular dimensions. Dimensions shown with different amounts of decimal places are subject to different tolerances. For example, the 0.X in a tolerance block means the tolerance shown would be applied to one-place decimal dimensions such as 0.1 or 0.6. The 0.XX means the tolerance shown would be applied to two-place decimal dimensions such as 0.12 or 0.62, and so on. It is important to note that the values in the tolerance block do not apply to limit tolerances. For example, a .124/.125 limit tolerance cannot use the tolerance in the tolerance block in place of those limits, or in addition to those limits. **Figure 3.1.26** identifies some unspecified tolerances with the letter B.

Reference Dimensions

Sometimes a print may contain a dimension that is in parentheses. This **reference dimension** is provided to assist with producing the part but is not subject to standard tolerances and will not be inspected. For example, a reference dimension may indicate a raw material size that

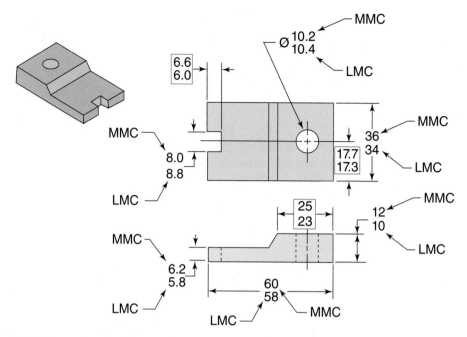

FIGURE 3.1.25 Identification of sizes at MMC and LMC for features of size. The dimensions in the red boxes are not features of size, so they do not have an MMC and LMC.

does require machining. That dimension is the nominal size, or ideal size, of the material. The material may actually vary from that size more than the standard tolerance but still be functional and acceptable. In other cases, several dimensions may add up to an overall length that is not critical. That overall length is helpful when cutting or preparing a piece of raw material for machining (**Figure 3.1.27**).

Even though a reference dimension is not subject to print tolerances, it is still important because it can have an impact on other dimensions. Refer back to **Figure 3.1.27** while reviewing this example.

- The four 1.00" dimensions shown are subject to a ±0.01" tolerance. Suppose that each 1.00" dimension is machined to 0.99". They are still in tolerance.

- If those four 0.99" dimensions are added together, the overall part length would only be 3.96". This is 0.04" under the 4.00" reference dimension, which is still acceptable because the 4.00" reference dimension is not subject to standard print tolerances.

- Now suppose that each 1.00" dimension is machined to 1.01". They are still in tolerance.

- If those four 1.01" dimensions are added together, the overall length would be 4.40". This is 0.04" over the 4.00" reference dimension, which is still acceptable because the 4.00" reference dimension is not subject to standard print tolerances.

- Since other dimensions add up to create the reference dimension, in reality the tolerances of those dimensions add up to create an acceptable variation for the reference dimension.

- Those four ±0.01" tolerances add up to an acceptable ±0.04" variation for the reference dimension. Since they add up, or accumulate, the result is often called an **accumulated tolerance.**

CLASSES OF FIT

Sometimes machining operations produce two mating parts, such as a shaft that fits inside a hub. If the print doesn't specifically call out the dimensions or tolerances, use of technical reference material may be required to determine the proper size ranges for those two mating parts. This relationship between the sizes of the two mating parts is called the **class of fit**.

Allowances

In machining, an **allowance** is the minimum amount of clearance, or the maximum amount of interference, between two mating parts that are features of size. Think of an allowance as the closest or tightest fit between the two mating parts. The allowance between any two mating parts can be found by subtracting the MMC of the external feature from the MMC of the internal feature.

A positive allowance provides clearance and specifies the minimum size difference between the mating parts.

- A shaft has an MMC of 1.998". It is the external feature of size.

- A hole in a pulley has an MMC of 2.000". It is the internal feature of size.

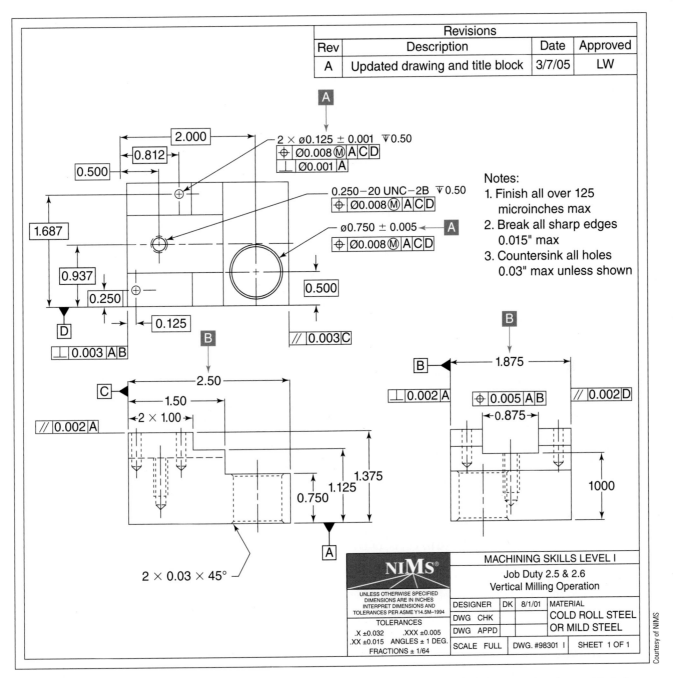

FIGURE 3.1.26 "A" shows dimensions with specified tolerances. "B" shows sizes with unspecified tolerances. The tolerance block shows that the 1.875 dimension has a tolerance of ±0.005 and that the 2.50 dimension has a tolerance of ±0.015.

- Subtracting 1.998" MMC of the external feature (shaft) from the 2.000" MMC of the internal feature (hole) results in 0.002". This is a positive allowance of 0.002" because the shaft is smaller than the hole.

A negative allowance creates interference and specifies the maximum interference between the surfaces of mating parts, where a shaft would be larger than a hole.

- A shaft has a MMC of 2.001". It is the external feature of size.
- A hole in a pulley has an MMC of 2.000". It is the internal feature of size.

- Subtracting the 2.001" MMC of the external feature (shaft) from the 2.000" MMC of the internal feature (hole) results in −0.001". This is a negative allowance of 0.001" because the shaft is larger than the hole.

The chart in **Figure 3.1.28** provides a summary of allowances.

Classifications of Fits

There are three general groups of fits specified by *ANSI (American National Standards Institute)*. One type is *running and sliding* fits, which are designated by the

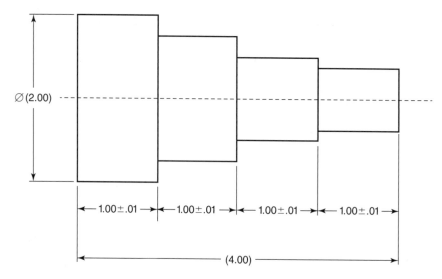

FIGURE 3.1.27 Reference dimensions are shown in parentheses and are not subject to standard print tolerances. The 2.00 reference dimension is a raw material size that does not require machining. The 4.00 is an overall reference length to consider when cutting a piece of the material to length in preparation for machining.

Positive allowance = minimum clearance	Negative allowance = maximum Interference
Clearance between mating parts	Interference between mating parts
Hole larger than shaft	Hole smaller than shaft
Shaft smaller than hole	Shaft larger than hole

FIGURE 3.1.28 Summary of allowances.

letters RC. *Locational fits* are designated by LC, LT, or LN. *Force fits* are designated by FN.

Figure 3.1.29 shows a portion of the RC class of fits tables. Refer to this table while reviewing the following example about determining sizes for a given class of fit.

EXAMPLE ONE: Steps to determine hole and shaft size for a Class RC2 Fit for a 1″ diameter.

1. Find the RC2 area in the table.

2. The far left column labeled "Nominal Size Range, Inches" shows the diameter ranges. (*Nominal* means "in name only." For example, 1/2 would be the nominal size for an object even if its actual size were slightly less than 0.500, like 0.498.) In this case, 1″ is the nominal size. Since 1″ falls between 0.71 and 1.19, that is the horizontal row where information will be located.

3. Follow that row across to the Class RC2 column.

4. The box in the "Clearance" column shows the range of clearance in thousandths of an inch. In this case those values are 0.3 and 1.2. To convert

to a proper decimal, multiply those values by 0.001. That gives a clearance range from 0.0003 to 0.0012. The difference between the two mating parts will be between these two values.

5. The "Standard Tolerance Limits" columns list the tolerances for the hole and the shaft.

6. In this case, the hole tolerance is shown as +0.5 and 0. Multiplication by 0.001 gives values of +0.00005 and 0. This means the 1″-diameter hole can be 0.0005 larger (+0.0005) but no smaller (0). The largest allowable size is 1 + 0.0005 = 1.0005″. The smallest allowable size is 1 + 0 = 1.00000.

7. The shaft tolerance is shown as −0.3 and −0.7. Multiplication by 0.001 gives values of −0.0003 and −0.0007. This means the 1″ shaft diameter can be from −0.0003 smaller to −0.0007 smaller. The largest allowable size is 1 − 0.0003 = 0.9997″. The smallest allowable size is 1 − 0.0007 + 0.9993″.

8. Subtracting the smallest shaft size from the largest hole size will reinforce the largest value listed in the clearance column: 1.0005 − 0.9993 = 0.0012″.

Table 8a: American National Standard Running and Sliding Fits

National Size Range, Inches	Class RC 1			Class RC 2			Class RC 3			Class RC 4		
	Clearance	Standard Tolerance Limits		Clearance	Standard Tolerance Limits		Clearance	Standard Tolerance Limits		Clearance	Standard Tolerance Limits	
		Hole H5	Shaft g4		Hole H6	Shaft g5		Hole H7	Shaft f6		Hole H8	Shaft f7
Over To	Values Shown below are in thousandths of an inch											
0–0.12	0.1	+0.2	–0.1	0.1	+0.25	–0.1	0.3	+0.4	–0.3	0.3	+0.6	–0.3
	0.45	0	–0.25	0.55	0	–0.3	0.95	0	–0.55	1.3	0	–0.7
0.12–0.24	0.15	+0.2	–0.15	0.15	+0.3	–0.15	0.4	+0.5	–0.4	0.4	+0.7	–0.4
	0.5	0	–0.3	0.65	0	–0.35	1.2	0	–0.7	1.6	0	–0.9
0.24–0.40	0.2	+0.25	–0.2	0.2	+0.4	–0.2	0.5	+0.6	–0.5	0.5	+0.9	–0.5
	0.6	0	–0.35	0.85	0	–0.45	1.5	0	–0.9	2.0	0	–1.1
0.40–0.71	0.25	+0.3	–0.25	0.25	+0.4	–0.25	0.6	+0.7	–0.6	0.6	+1.0	–0.6
	0.75	0	–0.45	0.95	0	–0.55	1.7	0	–1.0	2.3	0	–1.3
0.71–1.19	0.3	+0.4	–0.3	0.3	+0.5	–0.3	0.8	+0.8	–0.8	0.8	+1.2	–0.8
	0.95	0	–0.55	1.2	0	–0.7	2.1	0	–1.3	2.8	0	–1.6
1.19–1.97	0.4	+0.4	–0.4	0.4	+0.6	–0.4	1.0	+1.0	–1.0	1.0	+1.6	–1.0
	1.1	0	–0.7	1.4	0	–0.8	2.6	0	–1.6	3.6	0	–2.0
1.97–3.15	0.4	+0.5	–0.4	0.4	+0.7	–0.4	1.2	+1.2	–1.2	1.2	+1.8	–1.2
	1.2	0	–0.7	1.6	0	–0.9	3.1	0	–1.9	4.2	0	–2.4
3.15–4.73	0.5	+0.6	–0.5	0.5	+0.9	–0.5	1.4	+1.4	–1.4	1.4	+2.2	–1.4
	1.5	0	–0.9	2.0	0	–1.1	3.7	0	–2.3	5.0	0	–2.8
4.73–7.09	0.6	+0.7	–0.6	0.6	+1.0	–0.6	1.6	+1.6	–1.6	1.6	+2.5	–1.6
	1.8	0	–1.1	2.3	0	–1.3	4.2	0	–2.6	5.7	0	–3.2
7.09–9.85	0.6	+0.8	–0.6	0.6	+1.2	–0.6	2.0	+1.8	–2.0	2.0	+2.8	–2.0
	2.0	0	–1.2	2.6	0	–1.4	5.0	0	–3.2	6.6	0	–3.8
9.85–12.41	0.8	+0.9	–0.8	0.8	+1.2	–0.8	2.5	+2.0	–2.5	2.5	+3.0	–2.5
	2.3	0	–1.4	2.9	0	–1.7	5.7	0	–3.7	7.5	0	–4.5
12.41–15.75	1.0	+1.0	–1.0	1.0	+1.4	–1.0	3.0	+2.2	–3.0	3.0	+3.5	–3.0
	2.7	0	–1.7	3.4	0	–2.0	6.6	0	–4.4	8.7	0	–3.2
15.75–19.69	1.2	+1.0	–1.2	1.2	+1.6	–1.2	4.0	+2.5	–4.0	4.0	+4.0	–4.0
	3.0	0	–2.0	3.8	0	–2.2	8.1	0	–5.6	10.5	0	–6.5

Reprinted from ASME B4.1–1967 (2009), by permission of The American Society of Mechanical Engineers. American National Standards Institute (ANSI)

FIGURE 3.1.29 A portion of the RC class of fit table.

9. Subtracting the largest shaft size from the smallest hole size will reinforce the lowest value listed in the clearance column: 1.0000 − 0.9997 = 0.0003". This lowest clearance amount is the allowance, and it is a positive allowance because there is clearance.

Note that all of the hole tolerances in the RC table are either "0" or positive, meaning the hole can always be equal to or larger, but not smaller, than the nominal size of 1". In contrast, all shaft sizes in the RC table are negative, meaning the shaft can always be smaller, but not larger, than the nominal size of 1". Those combinations will always result in clearance between the shaft and the hole.

EXAMPLE TWO: Contrast Example One with a Class FN1 fit for a 1"-diameter hole and shaft.

Figure 3.1.30 shows a portion of the FN class of fits tables that will be used for this example.

1. Find the FN1 area in the table.

2. The far left column labeled "Nominal Size Range, Inches" again shows the diameter ranges. The nominal size of 1" falls between 0.95 and 1.19, so that is the horizontal row where information will be located.

3. Follow that row across to the Class FN1 column.

(Continued)

Table 11: ANSI Standard Force and Shrink Fits

Values shown below are in thousandths of an inch

Nominal Size Range, Inches Over	To	Class FN 1 Inter-ference	FN 1 Hole H6	FN 1 Shaft	Class FN 2 Inter-ference	FN 2 Hole H7	FN 2 Shaft s6	Class FN 3 Inter-ference	FN 3 Hole H7	FN 3 Shaft t6	Class FN 4 Inter-ference	FN 4 Hole H7	FN 4 Shaft u6	Class FN 5 Inter-ference	FN 5 Hole H8	FN 5 Shaft x7
0	0.12	0.05	+0.25	+0.5	0.2	+0.4	+0.85				0.3	+0.4	+0.95	0.3	+0.6	+1.3
		0.5	0	+0.3	0.85	0	+0.6				0.95	0	+0.7	1.3	0	+0.9
0.12	0.24	0.1	+0.3	+0.6	0.2	+0.5	+1.0				0.4	+0.5	+1.2	0.5	+0.7	+1.7
		0.6	0	+0.4	1.0	0	+0.7				1.2	0	+0.9	1.7	0	+1.2
0.24	0.40	0.1	+0.4	+0.75	0.4	+0.6	+1.4				0.6	+0.6	+1.6	0.5	+0.9	+2.0
		0.75	0	+0.5	1.4	0	+1.0				1.6	0	+1.2	2.0	0	+1.4
0.40	0.56	0.1	+0.4	+0.8	0.5	+0.7	+1.6				0.7	+0.7	+1.8	0.6	+1.0	+2.3
		0.8	0	+0.5	1.6	0	+1.2				1.8	0	+1.4	2.3	0	+1.6
0.56	0.71	0.2	+0.4	+0.9	0.5	+0.7	+1.6				0.7	+0.7	+1.8	0.8	+1.0	+2.5
		0.9	0	+0.6	1.6	0	+1.2				1.8	0	+1.4	2.5	0	+1.8
0.71	0.95	0.2	+0.5	+1.1	0.6	+0.8	+1.9				0.8	+0.8	+2.1	1.0	+1.2	+3.0
		1.1	0	+0.7	1.9	0	+1.4				2.1	0	+1.6	+3.0	0	+2.2
0.95	1.19	0.3	+0.5	+1.2	0.6	+0.8	+1.9	0.8	+0.8	+2.1	+1.0	+0.8	+2.3	1.3	+1.2	+3.3
		1.2	0	+0.8	1.9	0	+1.4	2.1	0	+1.6	2.3	0	+1.8	3.3	0	+2.5
1.19	1.58	0.3	+0.6	+1.3	0.8	+1.0	+2.4	1.0	+1.0	+2.6	1.5	+1.0	+3.1	1.4	+1.6	+4.0
		1.3	0	+0.9	2.4	0	+1.8	2.6	0	+2.0	3.1	0	+2.5	4.0	0	+3.0
1.58	1.97	0.4	+0.6	+1.4	0.8	+1.0	+2.4	1.2	+1.0	+2.8	1.8	+1.0	+3.4	2.4	+1.6	+5.0
		1.4	0	+1.0	2.4	0	+1.8	2.8	0	+2.2	3.4	0	+2.8	5.0	0	+4.0
1.97	2.56	0.6	+0.7	+1.8	0.8	+1.2	+2.7	1.3	+1.2	+3.2	2.3	+1.2	+4.2	3.2	+1.8	+6.2
		1.8	0	+1.3	2.7	0	+2.0	3.2	0	+2.5	4.2	0	+3.5	6.2	0	+5.0
2.56	3.15	0.7	+0.7	+1.7	1.0	+1.2	+2.9	1.8	+1.2	+3.7	2.8	+1.2	+4.7	4.2	+1.8	+7.2
		1.9	0	+1.4	2.9	0	+2.2	3.7	0	+3.0	4.7	0	+4.0	7.2	0	+6.0
3.15	3.94	0.9	+0.9	+2.4	1.4	+1.4	+3.7	2.1	+1.4	+4.4	3.6	+1.4	+5.9	4.8	+2.2	+8.4
		2.4	0	+1.8	3.7	0	+2.8	4.4	0	+3.5	5.9	0	+5.0	8.4	0	+7.0
3.94	4.73	1.1	+0.9	+2.6	1.6	+1.4	+3.9	2.6	+1.4	+4.9	4.6	+1.4	+6.9	5.8	+2.2	+9.4
		2.6	0	+2.0	3.9	0	+3.0	4.9	0	+4.0	6.9	0	+6.0	9.4	0	+8.0
4.73	5.52	1.2	+1.0	+2.9	1.9	+1.6	+4.5	3.4	+1.6	+6.0	5.4	+1.6	+8.0	7.5	+2.5	+11.6
		2.9	0	+2.2	4.5	0	+3.5	6.0	0	+5.0	8.0	0	+7.0	11.6	0	+10.0

FIGURE 3.1.30 A portion of the FN class of fit table.

4. The "Clearance" box is replaced with an "Interference" box, and this column shows the range of interference in thousandths of an inch. In this case those values are again 0.3 and 1.2. To convert to a proper decimal, multiply by 0.001. That gives an interference range from 0.0003 to 0.0012. The difference between the two mating parts will be between these two values, but in this case, keep in mind this is interference, not clearance.

5. The "Standard Tolerance Limits" columns list the tolerances for the hole and the shaft.

6. In this case, the hole tolerance is shown as +0.5 and 0. Multiplication by 0.001 gives values of +0.0005 and 0. This means the 10-diameter hole can be 0.0005 larger (+0.0005) but no smaller (0). The largest allowable size is 1 + 0.0005 = 1.0005". The smallest allowable size is 1 + 0 = 1.0000".

7. The shaft tolerance is shown as +0.8 and +1.2. Multiplication by 0.001 gives values of +0.0008 and +0.0012. This means the 1" shaft diameter can be from 0.0008 larger to 0.0012 larger. The largest allowable size is 1 + 0.0012 = 1.0012". The smallest allowable size is 1 + 0.0008 = 1.0008".

8. Subtracting the smallest shaft size from the largest hole size will reinforce the largest value listed in the interference column: 1.0005 − 1.0008 = −0.0003". The negative is just a reminder that the shaft is larger than the hole because this is an interference fit.

9. Subtracting the largest shaft size from the smallest hole size will reinforce the lowest value listed in the interference column: 1.0000 − 1.0012 = −0.0012". Again, the negative is just a reminder that the shaft is larger than the hole because this is an interference fit. This maximum interference is the allowance, and it is a negative allowance because there is interference.

Note that all the hole tolerances in the FN table are either "0" or positive, meaning the hole can always be equal to or larger, but not smaller, than the nominal size of 1". All shaft sizes in the FN table are also positive, meaning the shaft can always be larger, but not smaller, than the nominal size of 1". Those combinations will always result in interference between the shaft and the hole.

GEOMETRIC DIMENSIONING AND TOLERANCING (GD&T)

Geometric dimensioning and tolerancing (GD&T) uses symbols to show tolerances of *form*, *profile*, *orientation*, *location*, and *runout*. Use and interpretation of GD&T can be a very large and complex topic, so a basic overview is provided here. The overall purpose of GD&T is to define the largest tolerances that will still allow different parts to function together as designed. GD&T specifies larger tolerances than traditional coordinate tolerances that use unilateral, bilateral, and limit methods. Historically, many parts that did not meet traditional coordinate tolerances would be rejected when inspected, but would actually still function as designed. Larger tolerances make manufacturing easier. GD&T takes these situations and makes corrections in how a tolerance should be modified. This will become clearer as some examples are given. **Figure 3.1.31** shows a chart of some of the more commonly used GD&T symbols. Refer back to it while reading through the following GD&T topics.

Datum

In GD&T, dimensions are often related to a reference point called a **datum**. Only some features require a datum because they must be related to a reference point. Features that do not require the use of a datum are not related to any other feature or reference point. A datum is shown on a drawing as a capital letter inside a square with a line extending to the part feature. There is a triangle where this line contacts the feature. Sometimes a datum will be shown on the extension line of a surface feature. Several datums are identified on the print in **Figure 3.1.32**. More than one datum may be used to form positional relationships on a drawing. These datum planes are known as primary, secondary, or tertiary datums. There are three contact points on the primary datum plane, and not all of them are in the same line. The secondary datum plane is perpendicular to the primary datum plane. The secondary plane contains two points of contact. Tertiary datum planes are perpendicular to the primary and secondary datum plane. The tertiary datum plane has one point of contact. In **Figure 3.1.33**, an example of these datum planes may be seen in the section regarding the positioning of the holes. The primary datum plane is listed first, then the secondary, then the tertiary datum plane (A, B, C).

Feature Control Frame

A **feature control frame** is a rectangular box with sections for the geometric tolerancing symbol, the amount of the tolerance, and the reference to a specific datum (if required). It is used to show how to control the size and/or shape of a particular feature of a part. The tolerance amount is the total tolerance. **Figure 3.1.33** shows the parts of a feature control frame.

Interpretation of Geometric Tolerances

Correctly interpreting GD&T takes practice. The following basic explanations can help both beginning and

Category	Characteristic	Symbol	Uses a datum reference
Form	Straightness	—	Never
	Flatness	⟋⟍	
	Circularity (Roundness)	○	
	Cylindricity	/○/	
Profile	Profile of a line	⌒	Sometimes
	Profile of a surface	⌓	
Orientation	Angularity	∠	Always
	Perpendicularity	⊥	
	Parallelism	//	
Location	Position	⊕	
	Concentricity	◎	
	Symmetry	≡	
Runout	Circular runout	↗	
	Total runout	↗↗	

FIGURE 3.1.31 Geometric dimensioning and tolerancing (GD&T) symbols. Note datum use for the categories.

experienced print readers become more comfortable with understanding the more common GD&T symbols. The total tolerance amount shown in a feature control frame is called the *tolerance zone*.

Form Tolerances

Form tolerances are not related to other features, so their feature control frames will not contain datums.

Flatness means that an entire surface must be flat within a given tolerance zone. Keep in mind that flatness is not related to any other feature of a part. **Figure 3.1.34** shows a feature control frame specifying flatness and illustrates its meaning.

Circularity controls the diameter of any cross section of a cylinder. Imagine taking a slice of a cylindrical part and measuring how close to a true circle that slice is. Each of these cross sections needs to be within the specified tolerance zone, but they do not all need to be within the same tolerance zone. That means circularity can only be inspected at one point at a time around the periphery.

Cylindricity takes circularity one step further. It controls the surface of a cylinder across its entire length. The entire diameter must be within the specified tolerance zone, and it could be inspected at every point around the periphery. Thinking about the cross sections, as with circularity, every cross section of this cylinder needs to be within the same tolerance zone. **Figure 3.1.35** shows

feature control frames specifying circularity and cylindricity, and compares and contrasts their meanings.

Profile Tolerances

Profile tolerances are normally (but not always) related to another feature, so their feature control frames will frequently specify datums. They are applied just like the circularity and cylindricity specifications, but to noncylindrical surfaces.

The *profile of a line* is similar to circularity. All cross sections of the surface need to be within the specified tolerance zone, but they do not all need to be within the same tolerance zone. Only one cross section at a time can be inspected.

The *profile of a surface* is very similar to cylindricity. The entire surface must be within the specified tolerance zone. All cross sections of the surface need to be within the same tolerance zone. It could be inspected at every point on the surface. **Figure 3.1.36** shows feature control frames specifying profile of a line and profile of a surface, and it illustrates their meanings.

Orientation Tolerances

Orientation tolerances are related to other features, so their feature control frames will specify datums.

Parallelism means that every point on the surface must be an equal distance from the specified datum surface

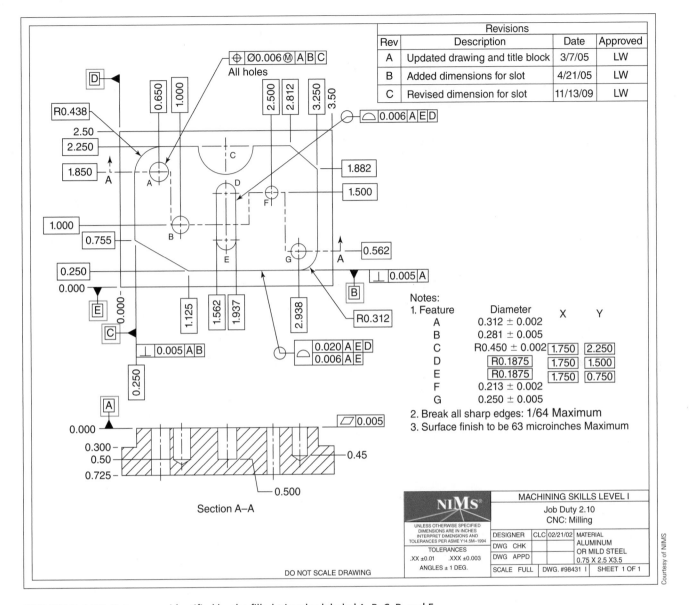

FIGURE 3.1.32 Datums are identified by the filled triangles labeled A, B, C, D, and E.

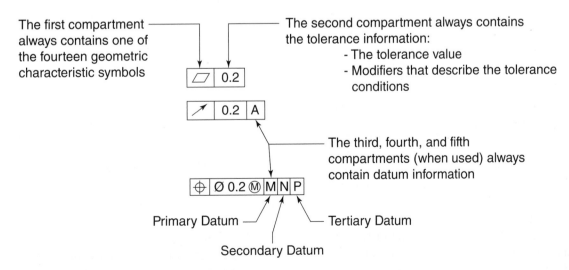

FIGURE 3.1.33 Parts of a feature control frame.

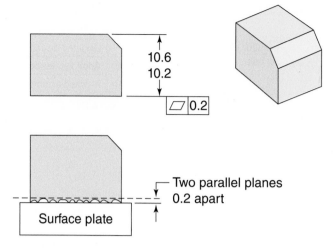

FIGURE 3.1.34 The flatness tolerance of 0.2 mm means that every point on the surface must be within a 0.2-mm tolerance zone.

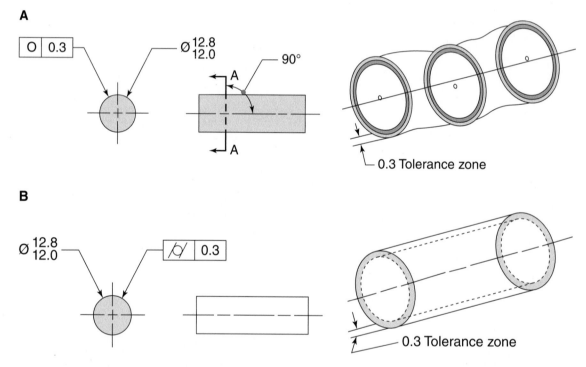

FIGURE 3.1.35 The circularity feature control frame in A establishes a 0.3-mm tolerance zone. All circular cross sections of the cylinder need to be within a 0.3-mm zone, but they do not need to be within the same zone. The cylindricity feature control frame in B establishes a 0.3-mm tolerance zone also, but the entire surface of the cylinder must be within that zone. (All circular cross sections need to be in the same tolerance zone.)

(parallel) within the given tolerance zone. **Figure 3.1.37** shows a feature control frame specifying parallelism and illustrates its meaning.

Perpendicularity means that a feature (frequently a surface) must be at a 90-degree angle to the specified datum (or datums) within the given tolerance zone. **Figure 3.1.38** shows a perpendicularity feature control frame and illustrates the perpendicularity tolerance.

Angularity is exactly like perpendicularity but is used for angular surfaces that are not 90 degrees.

Figure 3.1.39 shows a feature control frame specifying angularity and illustrates the angularity tolerance.

Location Tolerances

Location tolerances are used to specify the location of features related to other features, so their feature control frames will reference datums.

Concentricity compares relation of the axes of two or more cylindrical features. The entire feature must be within a cylindrical tolerance zone. Concentricity is not

A

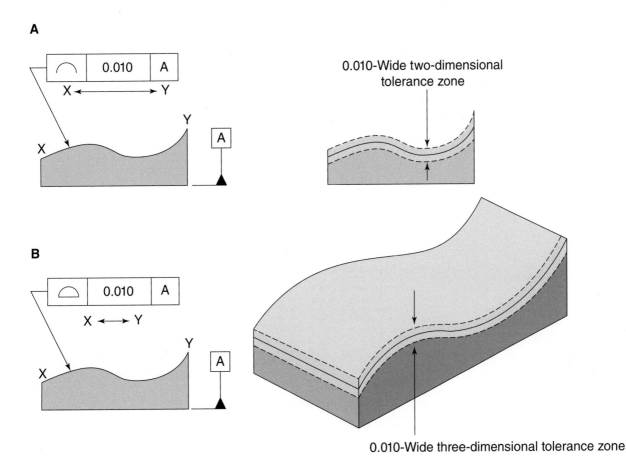

0.010-Wide two-dimensional tolerance zone

0.010-Wide three-dimensional tolerance zone

B

FIGURE 3.1.36 Consider the same shape with a profile of a line feature control frame (A) and then again with a profile of a surface feature control frame (B). In A, every cross section of the surface must be within a 0.010 two-dimensional tolerance zone, but every cross section does not need to be within the same 0.010 tolerance zone. In B, the entire surface needs to be within the 0.010 three-dimensional tolerance zone. (All cross sections need to be within the same 0.010 tolerance zone.)

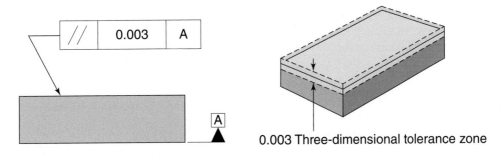

0.003 Three-dimensional tolerance zone

FIGURE 3.1.37 This parallelism feature control frame creates a 0.003 three-dimensional tolerance zone. All points on the surface must be within that tolerance zone.

used very often because it can be difficult, time consuming, and expensive to inspect.

Position (or *true position*) is used to specify the center of features such as holes or slots. For that reason, position feature control frames frequently contain more than one datum.

Remember that the major principle of GD&T is its ability to make it easier to manufacture acceptable mating parts because of greater tolerances. The position tolerance is a perfect example.

As shown in **Figure 3.1.40**, a traditional ±0.005" linear, or coordinate, tolerance on a hole location creates a 0.010" square tolerance zone. The center of the hole must be located inside that square zone. If the hole is centered at one of the extreme corners of the square, it is more than 0.005" from the specified print location. It is actually out of position by 0.007" and the tolerance zone across the corners of the square is 0.014". Since that position is acceptable, any location that is 0.007" from the print location should also be acceptable,

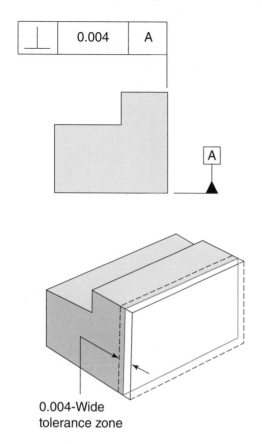

0.004-Wide
tolerance zone

FIGURE 3.1.38 This perpendicularity feature control frame creates a 0.004-wide tolerance zone. All points on the plane defining the surface must be within that tolerance zone.

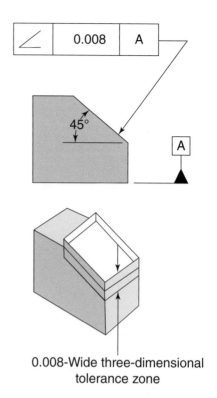

0.008-Wide three-dimensional
tolerance zone

FIGURE 3.1.39 This feature control frame for angularity creates a 0.008-wide three-dimensional tolerance zone. All points on the surface must be within that tolerance zone.

but by traditional tolerancing methods, parts would be rejected.

The position tolerance creates a circle that touches those four corners and results in a round tolerance zone. **Figure 3.1.41** shows some feature control frames for position tolerances.

Position Tolerance Modifiers and Bonus Tolerance

In some cases, as the size of a part feature changes within its tolerance, a position tolerance is permitted to change as well. The location of round features is one example. In some cases, when a diameter becomes larger, the location tolerance may increase. In other cases, when a diameter becomes smaller, the location tolerance may increase. This increase in tolerance is called the *bonus tolerance*. Three symbols called *position tolerance modifiers* are used to address situations like these.

MMC Modifier

A feature control frame that contains a circle with an "M" inside after the tolerance amount indicates that the specified tolerance applies at MMC. The second feature control frame in **Figure 3.1.41** contains an

MMC modifier. As the size of the feature moves away from MMC and toward LMC, the tolerance will increase.

Refer to the feature control frame and illustration in **Figure 3.1.42** for the following example. If the hole diameter is at MMC (smallest diameter of 0.498"), the tolerance zone is a 0.004"-diameter circle. If the diameter is at 0.499", the tolerance zone increases to a 0.005"-diameter circle. For every 0.001" of diameter increase, the diameter of the tolerance zone increases by 0.001". At LMC (largest diameter of 0.502"), the diameter of the tolerance zone would reach 0.008".

LMC Modifier

A feature control frame that contains a circle with an "L" inside after the tolerance amount indicates that the specified tolerance applies at LMC. The third feature control frame in **Figure 3.1.41** contains an LMC modifier. This is just the opposite of an MMC modifier. As the size of the feature moves away from LMC and toward MMC, the tolerance will increase.

Refer to the feature control frame and illustration in **Figure 3.1.43** for the following example. It is the opposite of the MMC example. If the hole diameter is at LMC (largest diameter of 0.502"), the tolerance zone is a 0.004"-diameter circle. If the diameter is at 0.501", the tolerance zone increases to a 0.005"-diameter circle. For every 0.001" of diameter decrease, the tolerance

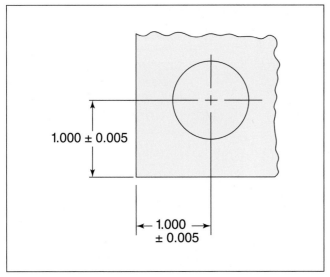

Coordinate tolerancing

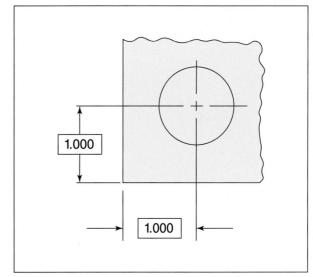

Positional tolerancing

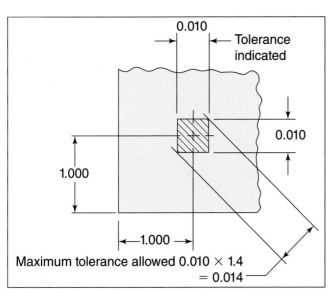

Coordinate tolerance zone (square)

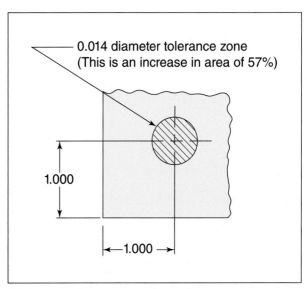

Positional tolerance zone (round)

FIGURE 3.1.40 The GD&T position tolerance increases the tolerance zone from the traditional method by 57% by creating a circular tolerance zone.

⊕	⌀0.008	A	B	C

⊕	⌀0.004 Ⓜ	A	B	C

⊕	⌀0.006 Ⓛ	A	B	C

⊕	⌀0.005 Ⓢ	A	B	C

FIGURE 3.1.41 Some position feature control frames. The diameter symbol and tolerance amount indicate the diameter of the tolerance zone. The letters A, B, and C reference datums.

Ø0.500 ± 0.002

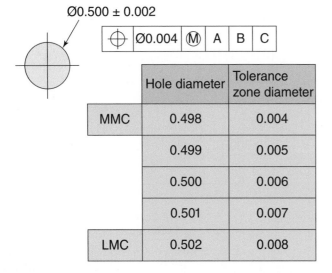

	Hole diameter	Tolerance zone diameter
MMC	0.498	0.004
	0.499	0.005
	0.500	0.006
	0.501	0.007
LMC	0.502	0.008

FIGURE 3.1.42 Effect of an MMC modifier on position and the bonus tolerance.

Ø0.500 ± 0.002

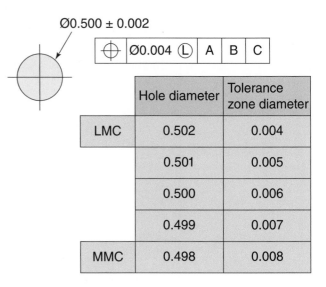

	Hole diameter	Tolerance zone diameter
LMC	0.502	0.004
	0.501	0.005
	0.500	0.006
	0.499	0.007
MMC	0.498	0.008

FIGURE 3.1.43 Effect of an LMC modifier on position and the bonus tolerance.

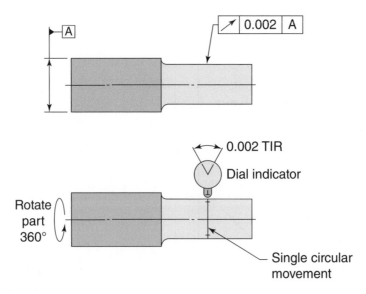

FIGURE 3.1.44 Circular runout requires the TIR (total indicator reading) when rotating the part to be within the specified tolerance at any one location. In this case, the datum diameter is first made to run true. Then the TIR is checked at one location on the surface noted by the feature control frame. Next, the indicator is moved to another point and the TIR is checked again. This is done at several locations along the part length. If the TIR is within the tolerance amount at each location, the part is acceptable.

increases by 0.001". At MMC (smallest diameter of 0.498"), the diameter of the tolerance zone would reach 0.008".

Regardless of Feature Size Modifier (RFS)

If a feature control frame contains the letter "S" inside a circle after the tolerance amount, the tolerance applies **regardless of feature size**, or **RFS**. This means that the tolerance is constant and cannot increase as the size of the feature varies from basic size. If the feature control frame has no modifier, GD&T standards state that RFS applies. The first and last feature control frames from **Figure 3.1.41** indicate that the tolerance applies RFS.

Runout Tolerances

Runout tolerances are used to check for runout of diameters related to a center axis, another diameter, or a

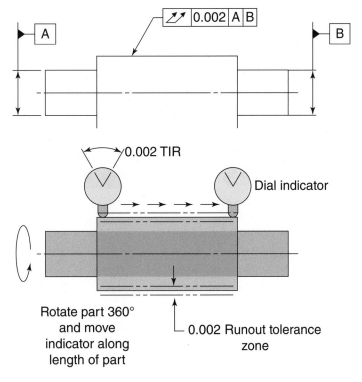

0.002 TIR

Dial indicator

Rotate part 360°
and move
indicator along
length of part

0.002 Runout tolerance
zone

FIGURE 3.1.45 Total runout requires the TIR when rotating the part to be within the specified tolerance at every location. In this case, the datum diameters are first made to run true. Then the TIR is checked at every location on the surface noted by the feature control frame by moving an indicator across the part while rotating it. If the TIR is within the tolerance amount across the entire surface, the part is acceptable.

perpendicular surface. They require use of one or more datums. The datums frequently used are the axis of a cylindrical part or other cylindrical features of the part.

Circular runout can be compared to circularity. Circularity checks only the diameter of the cross section of a cylinder and is not related or referenced to a datum. Circular runout checks the runout at one location around the periphery of a cylinder related to one or more datums.

Figure 3.1.44 shows a feature control frame specifying circular runout and illustrates the concept.

Total runout takes circular runout one step further. Total runout checks runout of the entire cylindrical surface related to one or more datums. The entire surface must be within the tolerance zone. **Figure 3.1.45** shows feature control frames specifying total runout and illustrates the concept.

SUMMARY

- The ability to read engineering drawings is an essential skill for success in machining careers.

- Engineering drawings use a system of views, line types, symbols, and notation to provide details about required specifications for machined components.

- Title blocks contain essential information such as the part name and number, tolerances, scale, and the material the part is to be made from.

- Orthographic projection and isometric views show different perspectives to help in the visualization of part shapes.

- Tolerances can be expressed in different formats and give the allowable limits of variation from basic sizes defined on prints.

(Continued)

- Class of fit and allowance are both used when describing the tolerances of and relationship between mating parts.
- Geometric dimensioning and tolerancing (GD&T) specifies tolerances of form, profile, orientation, location, and runout by using feature control frames. Some of these tolerances require reference to a datum, or reference, surface.

REVIEW QUESTIONS

1. Information such as tolerances and scale can be found in the _____ of an engineering drawing.
2. What view of a drawing usually shows the most detail?
3. Label the views shown in the sketch:

   a. _____
   b. _____
   c. _____
   d. _____
   e. _____
   f. _____

4. The line type used to show edges of an object that can be seen in a particular view is called a(n) _____ or _____ line.
5. What line type is used to show edges of an object that cannot be seen in a particular view?
6. What two line types work together to show sizes on an engineering drawing?
7. Identify the line types labeled in the print located on the following page:

   a. _____
   b. _____
   c. _____
   d. _____
   e. _____
   f. _____
   g. _____

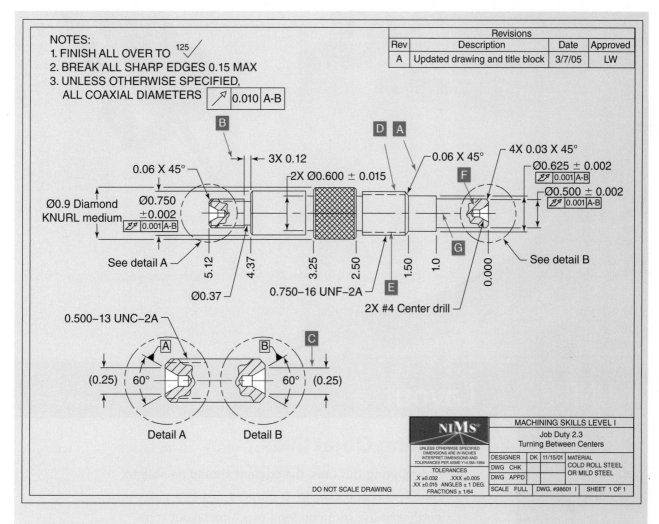

8. Define and briefly describe the following notations used on engineering drawings:

    a. TYP

    b. EQL SP

    c. 3X

9. A dimension listed on an engineering drawing is known as the _____ _____.

10. List and briefly describe the three types of tolerances used on prints.

11. Briefly describe the difference between a specified and an unspecified tolerance.

12. The relationship between sizes of mating parts is called _____ _____.

13. What does GD&T stand for?

14. A plane used as a reference for dimensions is called a(n) _____ _____.

15. A GD&T symbol and the amount of tolerance are shown in a(n) _____ _____.

16. What are the five categories of symbols used in GD&T?

17. What is a feature of size?

18. Briefly explain the benefit of a position tolerance over a traditional linear tolerance.

UNIT 2   Layout

Learning Objectives

After completing this unit, the student should be able to:

- Define *layout* and explain its purpose
- Identify and use common semi-precision layout tools
- Identify and use common precision layout tools
- Perform typical mathematical calculations required to perform layout
- Perform basic layout procedures

Key Terms

Angle plate	Layout	Rule holder
Center head	Layout fluid	Scriber
Center punch	(layout dye)	Surface gage
Divider	Layout fluid remover	Surface plate
Height gage	Parallel	Trammel
Hermaphrodite	Plain protractor	V-block
caliper	Prick punch	

INTRODUCTION

Layout is the term used to describe the process of locating and marking a workpiece as a visual reference to guide machining operations. Layout not only helps the machinist know where to machine the piece, but also, if multiple parts are to be made, layout can help to obtain the maximum number of parts from a piece of stock. Parts laid out and ready to machine are shown in **Figure 3.2.1**.

LAYOUT FLUID (LAYOUT DYE)

The surface finish on many materials may be rough, hard, or shiny. This can make it difficult to see layout lines. The solution to this problem is to coat the material with **layout fluid**, also called **layout dye**.

The purpose of layout dye is to provide contrast and make it easier to see layout lines. Layout dye is usually colored dark blue or red to give good visibility to the layout lines. The surface to be laid out should be clean and free of burrs before application. Dyes are available in aerosol spray, brush-cap bottles, or felt-tip applicators. Layout dye is shown in the layout in **Figure 3.2.2**. After application, the fluid dries quickly, so you will only need to wait a few moments to begin your layout.

 CAUTION

Layout fluid contains some hazardous materials, so review the SDS provided by the manufacturer for safety precautions. They will include working in an area with good ventilation, wearing safety glasses, and avoiding flame or ignition sources. Be sure to replace the lid on the container as soon as finished to avoid spills and to contain vapors.

Layout Fluid Remover

At some point, the layout dye will need to be removed from the part. The job may be finished, or a mistake may have been made when laying out the part and lines may need to be redrawn. **Layout fluid remover** is made

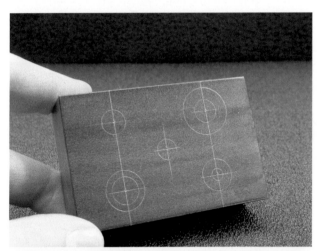

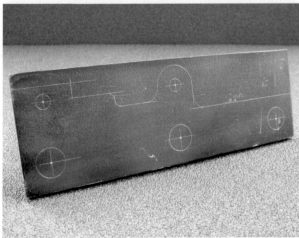

FIGURE 3.2.1 Layout lines provide guidelines for machining.

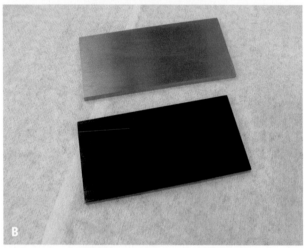

FIGURE 3.2.2 (A) Layout fluid is available in brush, spray aerosol, or felt-tip applicators. (B) A piece of steel before and after proper application of layout fluid.

specifically for that purpose, and is available in an aerosol spray can or as a liquid. It is applied to the surface and then quickly wiped off to remove the dye.

 CAUTION

Be sure to observe the same safety precautions using remover as when using the layout dye. An SDS should be kept on hand in case of an emergency such as inhalation of vapors, ingestion, or eye contact.

SEMI-PRECISION LAYOUT

Semi-precision layout is used to make simple layouts when extreme accuracy is not required. An example would be laying out a sheet of material to an accuracy of 1/64th inch before sawing it into parts to be finished on a milling machine. Hole locations with accuracy requirements of 1/64th of an inch are often laid out before drilling on a drill press. Several different tools are commonly used when performing semi-precision layout.

Scribers

The most common tool used to mark straight layout lines is the **scriber**. Scribers have a sharp, fine point on one or both ends made of hardened steel or tungsten carbide. Double-ended scribers have one point bent at a right angle to help mark hard-to-reach locations. To keep the point of a steel scriber sharp, hone it on a bench stone. Dull scribers do not make clear lines. When scribing a line, draw it once. Going over the line several times causes wide, inaccurate lines and dulls the scriber. The scriber should be tilted so that the point contacts the guiding edge of the measuring tool to scribe a neat, accurate line. Drag or pull the scriber instead of pushing it. Pushing a scriber can cause it to bounce and produce irregular lines. **Figure 3.2.3** shows the scriber in use.

CAUTION

Handle scribers carefully to prevent painful cuts and scratches. Do not carry a scriber in a pocket. Always wear safety glasses.

FIGURE 3.2.3 A scriber in use. Notice the tool is tilted so that the tip touches the guiding tool.

Layout with a Combination Set

The combination set can be used to perform a wide range of semi-precision layout operations. The different heads can be used to perform such operations as laying out perpendicular and angular lines and finding the center of round stock.

Combination Square

The square head and blade (or rule) of the combination set are called the combination square. It can be used in semi-precision layout operations to scribe straight lines and to lay out lines that are perpendicular, or square, to the edge of a workpiece or to other lines. Place the square head against the edge of a piece of material and use the blade (rule) to guide a scriber to scribe the line. The square head being used for layout is shown in **Figure 3.2.4**.

FIGURE 3.2.4 Layout of perpendicular lines using the combination square.

Angle Layout with the Combination Set

The combination set can be used to scribe angles when laying out a part. For 45-degree angles, the 45-degree side of the square head can be used. Place the 45-degree side of the square head against an edge and again use the blade as a guide for scribing a line. The protractor head can be adjusted and set for any angle up to 180 degrees. The head of the combination set is placed on the base line or a true edge and the line is scribed at the desired angle using the blade as a guide for the scriber. The square and protractor also have level vials, which can be used for machine tool setup purposes. **Figure 3.2.4** and **Figure 3.2.5** show angular line layout using the square head and the protractor head.

Center Head

The **center head** is used to find the center of a cylindrical part. The center head is assembled to the rule or blade and then laid across the end of the part. The rule will then cross the center of the cylindrical part. By scribing intersecting lines across the end of the part, the center point of the diameter can be found. The center head in use can be seen in **Figure 3.2.6**.

Divider

The **divider** is used to draw circles, radii, and arcs. Dividers have two legs with scribe points that are adjustable for different sizes. The easiest way to set a divider is to lightly place one point in the 1-inch graduation of a rule and then adjust the divider to the desired radius. Then, one point of the divider is placed at the center point of the arc or radius and the other point is used to scribe the circle or arc. Dividers are made in several sizes for diameters up to around 2 feet. **Figure 3.2.7** shows a divider being used to scribe an arc.

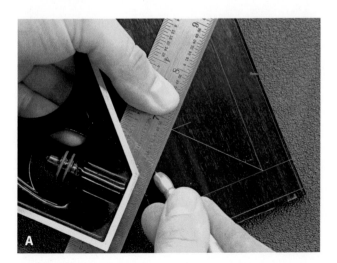
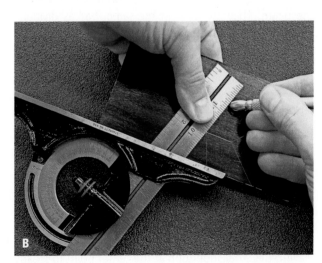

FIGURE 3.2.5 (A) Layout of angular lines using the square head and protractor head from the combination set. (B) The base of each head is held against a reference surface and the blade is used to guide the scriber.

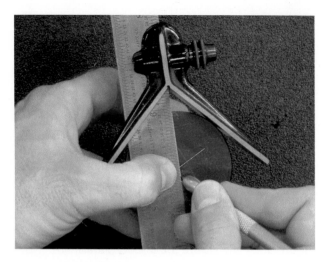

FIGURE 3.2.6 Finding the center of a cylindrical part using the center head and rule of the combination set.

A divider can also be used to *divide* lines or other shapes into sections. To lay out equal spaces, set the divider to the desired increment. Then place one point of the divider at a beginning point and "walk" the divider along the line to create divisions. **Figure 3.2.8** shows a divider being used to create equal spaces along a line.

 **CAUTION**

Handle dividers with the same care and precautions as scribers.

Trammel

Trammels are used to lay out circles or arcs that are too large for dividers. Trammels are composed of two sliding scribers, which are mounted on a long rod called the beam. They are used in the same way as dividers. **Figure 3.2.9** shows a beam trammel being used to lay out a large circle.

 **CAUTION**

Observe the same safety precautions when using a trammel as with a divider or scriber.

Prick and Center Punches

The **prick punch** can be used to mark the intersections of lines that locate the center points of circles or arcs. Having this small indention at the center point to steady one point of the divider or trammel makes it easier to scribe the desired circle or arc. The prick punch has an

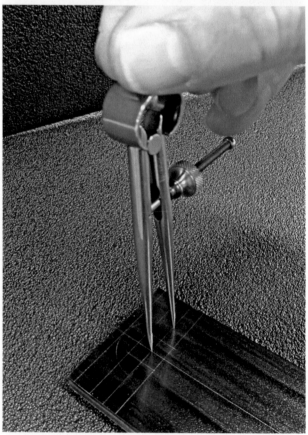

FIGURE 3.2.7 A divider is most frequently used to scribe circles and arcs.

FIGURE 3.2.8 Creating equal divisions along a line by "walking" a divider along the line.

included angle of 60 degrees so intersections to be marked can be readily seen. To use the prick punch, place the point where the center lines intersect and hold the punch perpendicular to the workpiece. Tap the punch lightly once with a ball peen hammer. After tapping the prick punch, check to be sure the

FIGURE 3.2.9 A trammel being used to scribe a large circle.

FIGURE 3.2.10 A prick and center punch. Note the different angles on the points.

indentation is precisely located. If not, lean the punch so the point faces the desired direction of movement and tap again to move the indentation. Keep the punch mark small. If the mark is too large, the divider can move around inside the mark and create an inaccurate circle or arc. After layout, the **center punch** can be used to enlarge the prick punch mark to aid in drilling operations. The center punch has an included angle of 90 degrees. When using the prick or center punch, strike it only once before checking the location. Slide the punch into the indentation before striking it again. Multiple strikes before checking can result in multiple indentations from the punch bouncing out of the original mark. The prick punch and center punch are shown in **Figure 3.2.10**.

 **CAUTION**

Always wear safety glasses when striking a punch with a hammer.

Hermaphrodite Caliper

Hermaphrodite calipers are used to scribe lines parallel to the edge of material. These calipers have one leg that is shaped like a divider and one that is shaped like an outside caliper. To set the hermaphrodite caliper, place the curved leg on the end of a steel rule and adjust the scriber point to the required dimension. To scribe a line, place the curved leg against the edge of the workpiece, and then scribe the line while keeping the curved leg against the edge. **Figure 3.2.11** shows how to set a hermaphrodite caliper, and **Figure 3.2.12** shows a line being scribed. Handle the caliper carefully to keep the legs from slipping while scribing the line.

FIGURE 3.2.11 Setting the hermaphrodite caliper with a rule.

 **CAUTION**

Remember that the scriber point is sharp. Handle this tool with the same precautions as a divider and scriber.

Plain Protractor

The **plain protractor** is another tool used to lay out angles in semi-precision layout work. The head of this protractor is graduated from 0 to 180 degrees like a combination set protractor. The advantage of the plain protractor is that it is smaller and flatter, and this can enable angle layout in places that a combination set protractor won't fit. **Figure 3.2.13** shows the use of a plain protractor.

Surface Plate

The **surface plate** provides a reference plane for layout just as it does for measurement tasks. Parts to be laid out can be placed directly on the surface plate, or they can be supported by workholding devices.

Surface Gage

The **surface gage** is made up of a base and a scriber mounted on a spindle that can be adjusted to different positions by a swivel bolt and fine adjustment screw. **Figure 3.2.14** shows some of the ways a surface gage can be set up for layout use. The surface gage can be placed on a surface plate and used to scribe parallel lines at a desired height from the surface plate. **Figure 3.2.15** shows a combination square being used to set the scribe to the desired dimension. A rule, or a combination set blade, can be clamped in a **rule holder** to perform this task as well. **Figure 3.2.16** shows setting a surface gage using these tools. After being set, the surface gage scribe is drawn carefully across the workpiece, making one thin line in the desired location. **(See Figure 3.2.17.)**

FIGURE 3.2.12 Scribing a line parallel to an edge with a hermaphrodite caliper. Keep the guiding leg at a consistent height along the edge so the scribed line is straight and drag the scriber point.

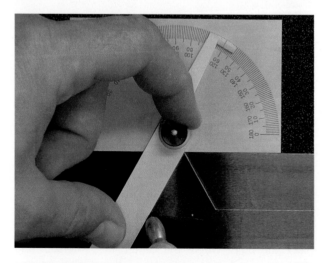

FIGURE 3.2.13 A plain protractor being used to lay out an angular line. The head is held against a reference edge and the scriber is guided by the blade.

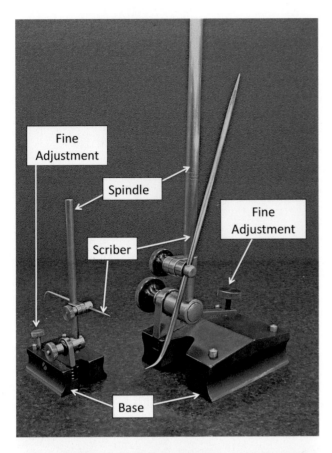

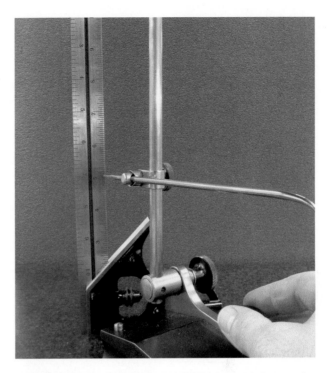

FIGURE 3.2.15 A combination square can be used to set the scriber height on a surface gage. Always keep the scriber point as parallel to the surface plate as possible and keep the spindle vertical.

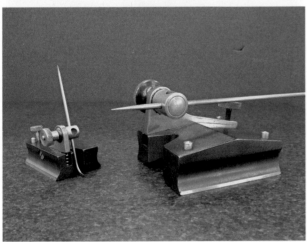

FIGURE 3.2.14 Surface gages are very versatile because of their ability to be adjusted to several different positions. The scriber can be mounted to the spindle with a clamp or directly mounted to the base when working close to the surface plate. The fine adjustment knob can be used for fine-tuning a setting after clamping the spindle and/or scriber.

CAUTION

Since the surface gage contains a sharp scriber point, always wear safety glasses when using and store it with the scriber in a position that will not cause injury.

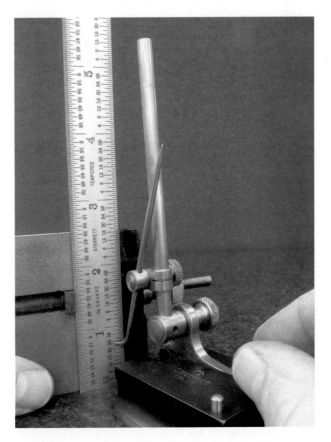

FIGURE 3.2.16 Setting the height of the surface gage scriber with a rule and rule holder. Note that the tip of the curved end of the scriber is parallel to the surface plate.

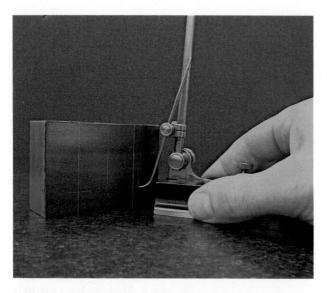

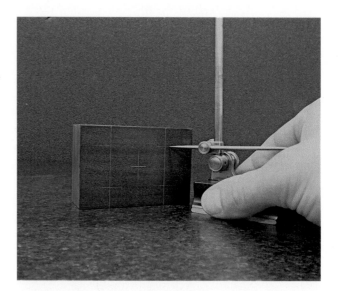

FIGURE 3.2.17 Scribing a line with a surface gage.

Workholding Accessories

Some thin or round parts are difficult to hold stationary on the surface plate while scribing them. In these cases, a method is needed to hold the workpiece steady.

Angle Plates

Angle plates are useful for holding parts square and steady during layout operations. Angle plates are made with all sides at 90-degree angles. The workpiece can be clamped to the angle plate or held against it and then positioned on the surface plate for layout operations. **(See Figure 3.2.18.)**

 CAUTION

Large angle plates can be heavy. Ask for help to move them and use proper lifting techniques. Use appropriate lifting tools when moving very large, heavy angle plates.

Parallels

Parallels are simply bars with opposite sides that are parallel within very close tolerances. They can be used to raise work above the surface plate and are available in a wide variety of shapes and sizes. Parallels can be solid or contain holes for clamping purposes. They can be made of steel or even granite like the surface plate. *Setup blocks* have parallel and square sides and can be used like parallels or angle plates. Some blocks with specific sizes are called 1-2-3 or 2-4-6 blocks because their outside dimensions are 1" × 2" × 3" or 2" × 4" × 6". **Figure 3.2.19** shows some examples of parallels and setup blocks.

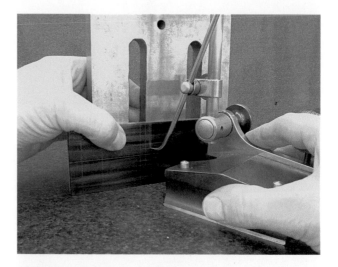

FIGURE 3.2.18 Scribing a line parallel to the surface plate using an angle plate.

FIGURE 3.2.19 Some different types of parallels and setup blocks that can be used to steady or secure workpieces during layout.

CAUTION

Some parallels and setup blocks can be large and heavy. Get help and use proper lifting techniques when moving these tools.

V-Blocks

V-blocks are square or rectangular blocks with one or more centrally located, 90-degree, V-shaped groove. V-blocks are useful for holding round work during layout operations. The workpiece is clamped with a U-shaped clamp to hold it steady. A V-block can also be used to position a square or rectangular workpiece at a 45-degree angle. A V-block in use is shown in **Figure 3.2.20**.

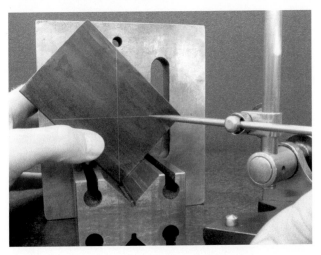

FIGURE 3.2.20 Some uses of a V-block when performing layout.

PRECISION LAYOUT

Precision layout is used when more accurate work than just sawing a part or drilling a hole on a drill press will be performed. Precision layout involves the use of more accurate tools.

Height Gage

The **height gage** is used with a surface plate to measure and mark off horizontal lines in precision layout work. It performs the same tasks as the surface gage. However, the scriber of this precision tool can be set directly without using any other measuring tools to set it. Height gages can use a vernier, dial, or digital scale. The scriber point is set to the desired dimension and then the height gage is carefully pulled along the part to scribe a layout line. Be sure to hold the base securely so the height gage does not tip on the surface plate. **(See Figure 3.2.21.)**

Precision Angular Layout

In precision angular layout, other tools will need to be used for angular lines that are more precise than the plain and combination set protractors.

Vernier Bevel Protractor

The vernier bevel protractor can be used to lay out angles with an accuracy of 5 minutes (1/12 of a degree). The base is generally held against a reference surface of the work and the blade is used for guiding a scriber. **(See Figure 3.2.22.)**

Sine Tools

Sine tools described in the precision measurement unit can also be used for layout of precision angles. They are set up for layout in the same way as when used for

FIGURE 3.2.21 Scribing lines with a vernier height gage.

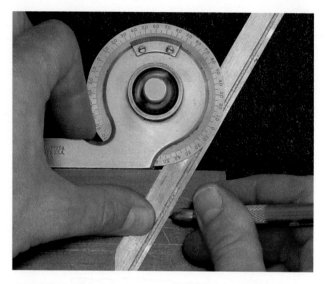

FIGURE 3.2.22 A vernier protractor being used for precision angular layout.

FIGURE 3.2.23 A sine bar holds the workpiece at the desired angle while a height gage scribes the angular line.

measurement by placing a gage block build under one of the rolls to raise the work to the desired angle. Then a height gage can be used to scribe the angular line. **Figure 3.2.23** shows a sine bar being used to lay out an angle.

BASIC LAYOUT CONSTRUCTION AND MATH

Some basic math concepts and relationships are frequently used while performing layout. They usually involve finding dimensions or locations that are not specified on a print. Here are a few examples of these situations.

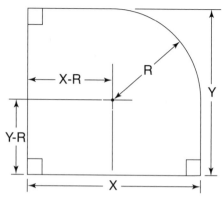

Dim 'X', 'Y', & 'R' (Radius) are given.

FIGURE 3.2.24 This sketch shows a typical layout situation.

EXAMPLE ONE: Suppose that a layout must be performed according to the sketch in **Figure 3.2.24**. Dimensions X, Y, and R (size of the radius) are given. Refer to **Figure 3.2.25** and use these steps to perform the layout to create a 90-degree arc with radius R that is tangent to the horizontal and vertical end lines.

1. Use reference edges to scribe lines at dimensions X and Y.
2. Use the formula X − R to find the horizontal distance from the left reference edge to the center point and scribe a line at that distance.
3. Use the formula Y − R to find the vertical distance from the bottom reference edge to the center point and scribe a line at that distance.
4. The intersection of those lines is the center of the radius. Construct the radius using that point.

EXAMPLE TWO: To lay out a 180-degree arc that is tangent to two lines or edges, refer to **Figure 3.2.26** and follow these steps.

1. Find the vertical center point of the radius with the formula Y − R and scribe a line at that distance from the bottom reference edge.
2. Find the center of the X dimension using X ÷ 2 or X/2 and scribe a line at that distance from the left reference edge.
3. The intersection of the lines created in steps 2 and 3 is the center point of the arc. Scribe the arc with radius R from that point.

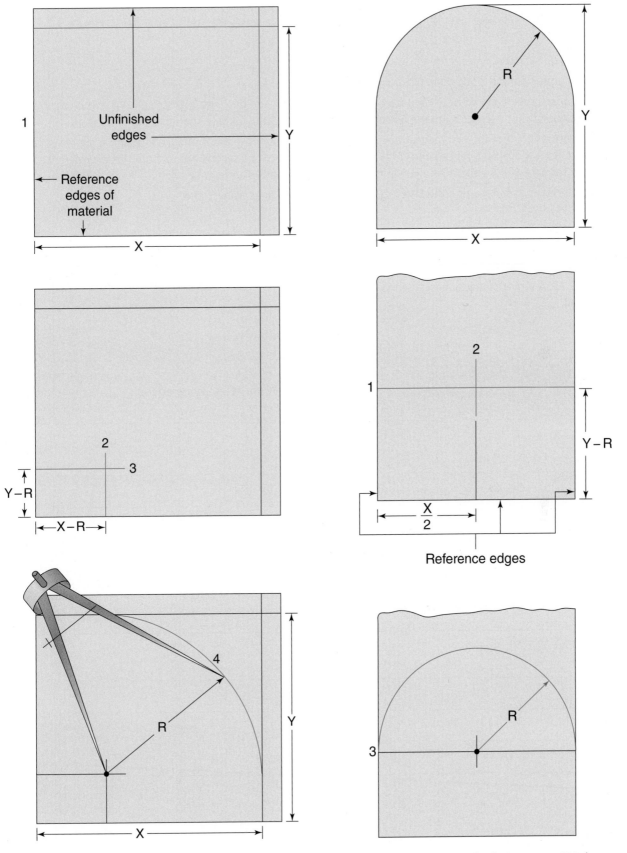

FIGURE 3.2.25 Steps for laying out a 90-degree arc tangent to two lines.

FIGURE 3.2.26 Procedure for laying out a 180-degree arc tangent to two parallel lines or edges.

EXAMPLE THREE: To lay out two tangent 90-degree arcs when only distances are given, refer to **Figure 3.2.27** and follow these steps:

1. Scribe heights Y_1 and Y_2 from the bottom reference edge.

2. Scribe length X from the left reference edge.

3. To find the center point of both radii from the bottom reference edge, use the formula $Y_1 + R$ or $Y_2 - R$ and scribe a line at that distance from the bottom reference edge. Both will give the same dimension.

4. To find center point of the first radius from the left reference edge, use the formula $X - R$ and scribe a line at that distance from the left reference edge. To find the center point of the second radius from the left reference edge, use the formula $X + R$ and scribe a line at that distance from the left reference edge.

5. Scribe arcs with radius R using the intersections created by the lines from step 4.

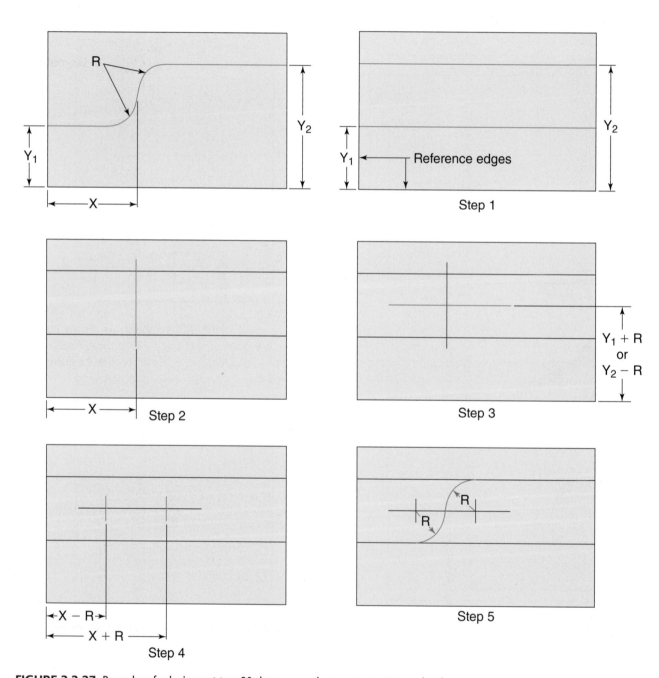

FIGURE 3.2.27 Procedure for laying out two 90-degree arcs that are tangent to each other.

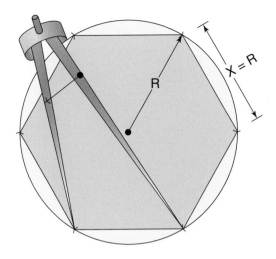

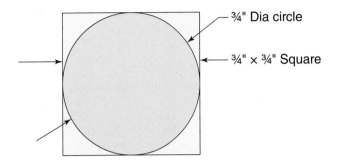

FIGURE 3.2.29 The diameter of a circle inscribed in a square is equal to the outside dimensions of the square. If a circle is to be cut from a square-shaped material, the outside dimension of the square must be equal or larger than the diameter of the circle.

FIGURE 3.2.28 A divider can be used to construct a hexagon inside a circle. After scribing the circle, keep the divider at the radius setting and walk around the circle. Then connect the intersections to form the sides of the hexagon.

When a circle is constructed, the radius R can be used to divide the circle into six parts. Start at any point on the periphery and walk a divider around the circle. Then the intersections on the periphery can be connected to create a hexagon. **(See Figure 3.2.28.)**

THE LAYOUT OF SQUARE SHAPES

Occasionally a machinist may need to lay out a circle from a square workpiece or a square from a circular workpiece. For this to happen, the machinist must determine the minimum size of the desired shape that will fit inside the available shape. Making a circle from a square is straightforward because the length and width dimensions must be equal to or greater than the diameter of the circle. For example, a 3/4" square can contain a 3/4" diameter circle, as shown in **Figure 3.2.29**. However, when a square is to be made from a circular workpiece, the math is a little more tricky.

This relationship may best be visualized by imagining a square contained within a circle with the corners touching the edge (inscribed), and a diagonal connecting the corners. This bisected square will quickly reveal two right triangles. In such a case, the diagonal's length will equal the diameter of the circle. That diagonal is also the hypotenuse of each triangle. **(See Figure 3.2.30.)** By solving one of the triangles for the length of its other sides, the largest square will be determined.

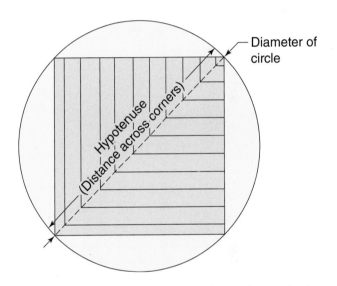

FIGURE 3.2.30 When a square is to be cut from a circular workpiece, the diagonal distance across the corners must be equal to or less than the diameter of the circle. Bisecting the square into a triangle can be used to determine the diameter of circle needed.

Since a square has all sides equal at 90-degree angles, and this triangle is constructed from a square, then it too must have two equal-length sides at 90 degrees to each other. This triangle then must have interior angles of 45, 45, and 90 degrees. **(See Figure 3.2.31.)**

There are several ways to go about solving this commonly appearing triangle, as shown in Section 2.2 of this text. One method of solving for the largest size inscribed square when the diameter is known is to multiply the diameter (which is the hypotenuse of the triangle) by a constant of 0.7071, because 0.7071:1 is the ratio of the diameter (the diagonal) to a side (leg) of the square. **(See Figure 3.2.32.)** Another method is to apply the Pythagorean theorem.

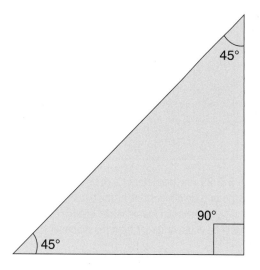

FIGURE 3.2.31 A bisected square results in a triangle with interior angles of 45, 45, and 90 degrees.

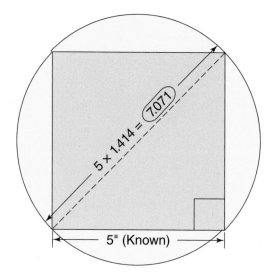

FIGURE 3.2.33 If the desired side is known, the largest-fitting diameter may be determined by multiplying a side by 1.414.

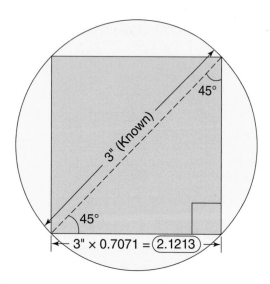

FIGURE 3.2.32 If the diameter is known, the largest-fitting square may be determined by multiplying the diameter by 0.7071.

If the size of the square is known, the diameter of the circle needed to produce it may be found by multiplying the length of the desired side by 1.414 (the square root of 2). **(See Figure 3.2.33.)**

Methods for Constructing Squares

Many methods may be used to construct a square for layout. Common methods for creating a square on a round workpiece include the following:

• Use a divider to construct each corner of the square by setting it to the width of the square as determined

by the triangle calculation shown earlier. The divider pivot is then placed at the position of the first corner of the desired square, while the marking point produces a mark at two adjacent corners. This process is repeated in both directions until each corner location is marked with two intersecting lines. **(See Figure 3.2.34.)**

• Use a center head from the combination set to establish one center line. Parallel lines may then be created on either side of the center line for the outer edges of the square. A square may then be used to create a center line perpendicular to the first center line, and the process is repeated. **(See Figure 3.2.35.)**

• Common methods for creating a square on a non-round workpiece include the following:

• Use a single straightedge on the workpiece, which serves as a baseline. A square may be used to construct two parallel lines from this edge and, finally, a parallel line opposite the baseline edge to close the square. **(See Figure 3.2.36.)**

• Use two acceptably square edges adjacent to each other. In this method, two parallel lines may be struck, at the desired distance from each edge. An alternate method is to strike two lines perpendicular to each edge. **(See Figure 3.2.37.)**

• Use two points. In this method, two points are placed where one of the desired edges of the square will be located. Next, a line is struck through the points, establishing the first edge of the square,

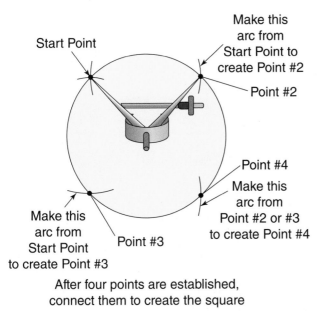

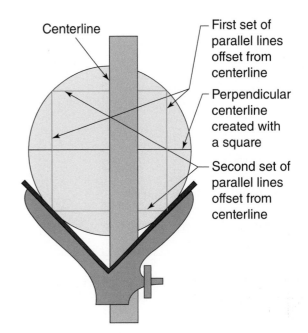

FIGURE 3.2.34 A divider being used to construct a square within a circle.

FIGURE 3.2.35 A square being constructed on a circular workpiece using center lines created with a center head.

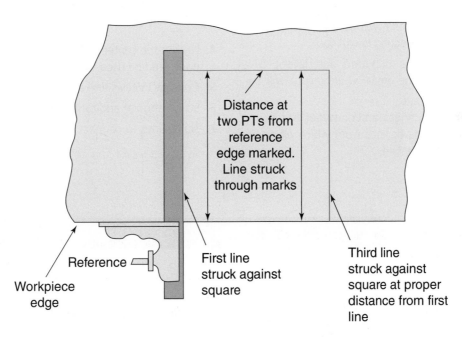

FIGURE 3.2.36 A square being used to create a line perpendicular to a straight workpiece edge. Parallel lines may then be created at the desired distance from these two lines.

which will also serve as a baseline for the rest of the construction. Then a flat square is placed against the line and a perpendicular line is struck. Finally, each opposite side of the square is struck parallel to the first two lines. **(See Figure 3.2.38.)**

LAYOUT PROCEDURE GUIDELINES

Every layout is unique, but there are a few general steps to follow for any layout. Keep in mind that these are

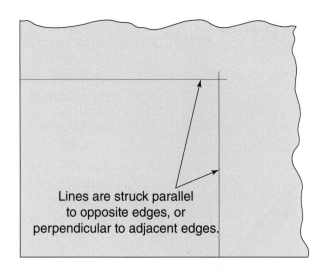

FIGURE 3.2.37 A workpiece with two already square sides may have parallel or perpendicular lines created directly from the edges.

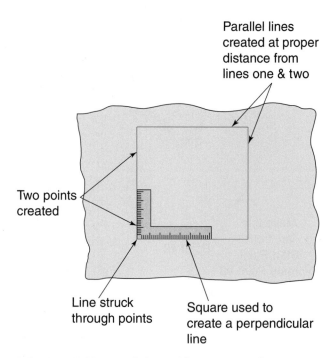

FIGURE 3.2.38 A workpiece with no square edges using a line as a reference and square to create an adjacent perpendicular line. These two lines may then be used to create parallel lines at the desired distance for the other sides of the square.

guidelines and that sometimes the order of these steps might be slightly different, depending on the situation.

1. Study the engineering drawing and develop a process plan of steps and tools that will be used. Keep tolerances in mind when selecting layout tools.
2. Make sure the part is deburred and layout dye has been applied in the areas where layout will be performed.
3. Decide on reference edges and/or layout baselines. These are the surfaces or lines from which all other measurements will be taken.

4. Lay out the center points of any circles or arcs. A prick punch can be used to lightly mark these center points.
5. Construct circles and arcs.
6. Scribe angular and tangent lines.
7. Scribe lines to connect any remaining points.

SUMMARY

- Proper layout can be a very helpful guide when performing machining operations, and it makes manufacturing easier.
- Layout errors can lead to serious machining errors that can result in great loss of time and money.
- Semi-precision layout can be used when print tolerances are 1/64" or greater. When tolerances are smaller than 1/64", precision layout tools are more appropriate.
- Always consult MSDS for layout chemicals and observe safety precautions when using sharp layout tools.
- Observe safe lifting and handling techniques when moving heavy objects such as angle plates and parallels.
- Layout often requires math skills when finding locations that are not shown on prints but are required for layout operations.
- By following basic guidelines and using the layout tools and concepts from this unit, you can produce a layout that will provide direction for machining operations.

REVIEW QUESTIONS

1. Define the term *layout* and briefly explain the purpose for layout.
2. What is the purpose of layout fluid (dye)?
3. Describe two safety precautions to observe when using layout dye.
4. Briefly define a scriber and its use.
5. What two angles can be laid out with the combination square?
6. What two tasks can a divider be used to perform?
7. What safety precautions should be observed when using layout tools that are used for scribing?
8. What would the divider setting be to scribe a 3/4"-diameter circle?
9. What would the divider setting be to scribe an arc with a 3/8" radius?
10. What semi-precision layout tool is used to scribe a line parallel to an edge at a set distance?
11. What tool can be used as a reference plane when performing layout?
12. Briefly describe the use of a surface gage.

UNIT 3 | Hand Tools

Learning Objectives

After completing this unit, the student should be able to:

- Identify common hand tools
- Describe the uses for common hand tools
- Describe hand tool safety precautions

Key Terms

Adjustable wrench	Hinged clamps	Rawhide
Arbor press	Hydraulic press	Set
Ball peen hammer	Kerf	Side cutting pliers
Box-end wrench	Loading	Slip joint pliers
C-clamp	Locking pliers	Socket wrench
Combination wrench	Needle nose pliers	Soft face hammer
Dead blow hammer	Nylon hammer	Spanner wrench
Deburring	Offset screwdrivers	Straight filing
Diagonal cutter	Open end wrench	Straight or slotted
Draw filing	Parallel clamp	screwdrivers
File card	Phillips screwdriver	Teeth Per Inch (TPI)
Hex key wrench	Pinning	Torx screwdrivers

INTRODUCTION

Proper hand tool use is an essential skill required in the machining industry. Hand tools are used for everything from setting up machine tools for operation to repairing them.

SCREWDRIVERS

Screwdrivers are simple hand tools that grip the heads of screws to loosen and tighten them. They are made with a wide variety of tip shapes. Each of these shapes is used on specific types of screw fasteners.

 CAUTION

Be sure to use the correct type and size of screwdriver on each type of screw to avoid injury to yourself or damage to equipment. It is also unsafe to carry screwdrivers in pockets or to use them around moving machinery. Screwdrivers are not safe to use as hammers or chisels or for prying. Be sure to keep the screwdriver lined up with the screw being driven to keep it from slipping and causing injury or damage to equipment.

Phillips

Phillips screwdrivers are used with Phillips head screws. The cross-shaped tip helps prevent slippage and comes in sizes 1 to 4. The size chosen should fit the screw snugly.

Straight

Straight or **slotted screwdrivers** have a broad, flat tip and are made in many sizes to fit the width and thickness of the screw slot. Use the biggest size that will fit the slot.

Offset

Offset screwdrivers are used when there is not enough room to use a screwdriver with a straight shank. These screwdrivers are angled at 90 degrees on one or both ends. They are made in a variety of sizes and tip shapes.

Torx

Torx screwdrivers are used in a wide variety of assembly applications such as automobile assemblies and cutting tool mounting applications. They are made in many sizes and have six splines on their tip. Torx screwdrivers should fit the screw snugly.

Examples of each of these screwdriver types are shown in **Figure 3.3.1**.

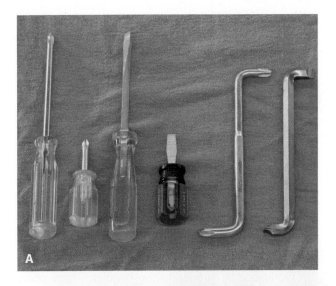

FIGURE 3.3.1 (A) Screwdrivers come in several sizes, lengths, and even offset styles. (B) Some of the most common tip styles: Phillips, straight, and Torx.

PLIERS

Pliers are used for a wide variety of holding and cutting tasks. There are several types, and they each have specific tasks they are used for. Pliers are not made to be used as wrenches. Some pliers come equipped with a cutting feature that can be used to cut wire.

Slip Joint Pliers

Slip joint pliers are used in many holding tasks. The slip joint enables the pliers to open wider in order to hold larger work.

Needle Nose Pliers

Needle nose pliers are also commonly used. The jaws on needle nose pliers taper toward the end to allow them to be used for holding small work. Some needle nose pliers have curved jaws to get into tight spaces.

Needle nose pliers may also be used for removing stringy chips from a lathe, but they should not be used around a machine in motion.

 **CAUTION**

Needle nose pliers should never be used around a machine in motion.

Locking Pliers

When pliers with a very high gripping power are needed, adjustable **locking pliers** are used. These pliers are adjusted to desired jaw opening size by rotating an adjusting screw in the handle. They may then be clamped on the workpiece by squeezing the handles. These pliers contain a mechanism that holds them in the clamped position so the user doesn't need to. They are released by pressing a lever on the handle.

Tongue-and-Groove Pliers

Sometimes pliers need to have the ability to hold larger-sized workpieces. These pliers have tongue-and-groove joints that enable them to be adjusted to various gripping ranges while the jaws are kept parallel to each other.

Side Cutting Pliers

Side cutting pliers, also known as linemen's pliers, have broad, flat jaws and are used for gripping as well as cutting wire and pins.

Diagonal Cutters

Diagonal cutters are used for light cutting of wire and pins. The cutting edge angle allows these cutters to cut nearly flush with a work surface. **Figure 3.3.2** shows examples of each of these pliers and cutters.

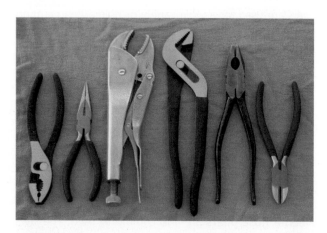

FIGURE 3.3.2 From left to right: Slip joint, needle nose, locking, tongue and groove, side cutting or lineman's pliers, and diagonal cutters.

HAMMERS

Hammers are frequently used to tap objects into alignment, drive pins, stamp objects, punch hole locations, and more. For general use, hard-headed steel hammers are used. When it is important not to mar a surface or damage a part, soft face hammers are used. Hammers should be checked for unsafe conditions such as loose or damaged heads or cracked handles. Many soft hammers have replaceable faces. Each type of hammer comes in several sizes or weights.

 **CAUTION**

Safety glasses should always be worn when using hammers. Hammers should not be struck against each other—fragments may break off causing injury.

Ball Peen

Ball peen hammers are made in a variety of sizes based on the weight of the head. These hammers are dual purpose, since they have two heads for two different functions. They have a striking face on one end and a rounded end that can be used for peening rivet heads or rough forming metal. The striking surface is used for light or heavy striking tasks, ranging from prick punch use during layout to striking chisels and punches. Ball peen hammers are shown in **Figure 3.3.3**.

Dead Blow

Dead blow hammers have sand or shot in the head to absorb the energy from striking and keep them from rebounding. The advantage of a dead blow hammer is that most of the energy from striking is transmitted to the object being struck instead of making the hammer rebound. Most hammer types are available as plain and dead blow styles.

Soft Face

Soft face hammers are used to strike surfaces that could easily be damaged by hard hammers. They are also used

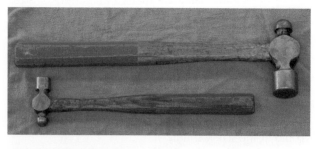

FIGURE 3.3.3 Ball peen hammers can be used for both light and heavy striking depending on size.

FIGURE 3.3.4 The head of a soft face dead blow hammer contains sand or shot to prevent rebounding.

for delicate positioning tasks such as alignment of parts on machine tools before final tightening and machining or for assembling precision components. Soft face dead blow hammers are frequently used to seat workpieces on parallels in machine vises. **Figure 3.3.4** shows a soft face dead blow hammer.

Nylon

Nylon hammers are used when a soft, yet more durable striking surface than most soft face hammers, is needed.

Rawhide

Rawhide is used for some hammers. The heads on these hammers are made from tightly rolled and shaped rawhide. Since rawhide is inherently a soft, compressible material, these hammers have little rebound. They are typically used when more force is required than plastic soft metal hammers, but the surface still needs to be protected from damage.

Soft Metal

Some hammer heads are made from soft metals like brass and copper. These hammers provide an option between the gentle non-marring nylon hammers and the unyielding steel-headed hammers. These can be used on heavier work, but there is more of a risk of damaging the workpiece. Another advantage of these non-ferrous hammers is that they will not produce sparks when used to strike other metal objects. Examples of some soft face hammers are shown in **Figure 3.3.5**.

WRENCHES

Wrenches are tools that are used to tighten and loosen threaded bolts and nuts. There are a wide variety of

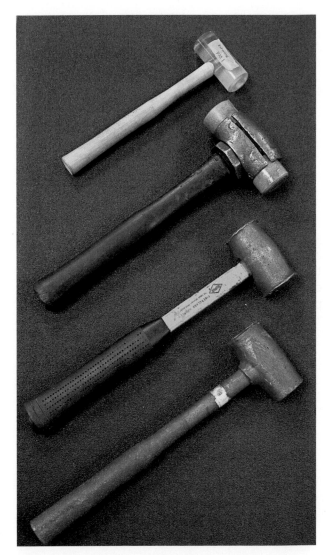

FIGURE 3.3.5 Examples of other soft face hammers.

wrenches to perform many tasks. Learning when and how to use each of these wrench types correctly will prevent personal injury and damage to equipment.

 CAUTION

Make sure the wrench fits the bolt or nut snugly. When possible, always pull the wrench toward yourself so that if it slips, you won't injure your hand. Don't put an extension on a wrench handle to increase leverage when the bolt or nut doesn't loosen. Don't hammer on a wrench. When pulling on a wrench, keep yourself in a balanced position so that you avoid straining your back, falling, or hitting anything if the bolt or nut loosens suddenly. You may find that a quick "snap" or "jerk" from a balanced position will shock the fastener and make it easier to loosen a bolt or nut.

Open-End Wrench

Open-end wrenches are light-duty wrenches with two parallel jaws that may be slid onto a hex or square drive surface. Most open-end wrenches position the gripping jaws at a 15-degree angle to permit use in tight spaces. When you use these wrenches in limited space, the wrench can be flipped over at the end of a stroke to enable another stroke. Open-end wrenches grip the fastener with only two surfaces and suffer from the tendency of the jaws to spread when great force is applied. **Figure 3.3.6** shows some open-end wrenches.

Box-End Wrench

Box-end wrenches are used when more torque must be applied to the fastener. The end of this wrench completely encircles the bolt or nut, giving the wrench more strength and eliminating the tendency of the jaws to spread. Since these wrenches encircle the fastener, they must be slid over the fastener from its end. These wrenches come in either 6 or 12-point varieties, as shown in **Figure 3.3.7**. The 6-point style has a better contact area with the fastener and is less likely to slip or round the corners of a fastener, but it can be harder to fit on the fastener when there is a small amount of

turning space. The 12-point style can be placed on the fastener at more positions than the 6 point. Twelve-point wrenches may also be used on fasteners with square or 12 pointed heads.

Double-Ended Wrench

There are several types of *double-ended wrenches*. Some have one box end and one open end of the same size for versatility and are called **combination wrenches**. Others have two box ends or two open ends of different sizes. **Figure 3.3.8** shows some double-ended wrenches.

Adjustable Wrench

The **adjustable wrench** is a versatile tool but must be properly adjusted to fit the fastener snugly in order to avoid damage to equipment and injury to you. **(See Figure 3.3.9.)** Adjustable wrenches come in many sizes, based on the length of the handle. Adjustable wrenches grip the fastener in the same manner as the open-end wrenches do, and therefore are susceptible to the jaws spreading. Adjustable wrenches are even more likely to have jaw flex due to the lack of a fixed jaw on one side.

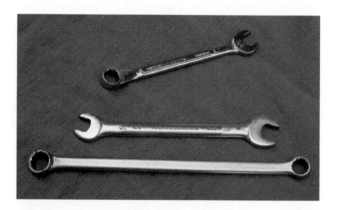

FIGURE 3.3.6 Open-end wrenches can be used when great force is not required. Most have heads set at a 15-degree angle for use in tight spaces.

FIGURE 3.3.8 Some double-ended wrenches. The combination wrench on top is a double-ended wrench with both a box and an open end of the same size.

FIGURE 3.3.7 Six- and 12-point box-end wrenches.

FIGURE 3.3.9 Adjustable wrenches come in different sizes and can be used when the specific size is not available.

It is important to adjust the wrench so it fits the bolt snugly. Pull the wrench toward you with the adjustable jaw in the direction of rotation to prevent injury if the wrench slips. **(See Figure 3.3.10.)**

Socket Wrench

Socket wrenches are made with a socket shaped like a box-end wrench in a hollow cylinder on one end and a square drive hole for attachment to a handle in the other end. A solid or ratcheting handle has a mating drive square and is used to transmit torque to the socket when the handle is pulled. Ratchet handles have the advantage of not having to be removed from the fastener each time a turn is made. Socket wrenches can have 6 or 12 contact points like box wrenches, as shown in **Figure 3.3.11**. Drive size is identified by the size of the square, such as 1/4, 3/8, 1/2, or 3/4". The drive size of the socket must match the drive size of the handle. Many accessories are available to allow a socket to reach into difficult-to-reach places. Socket handles and accessories are shown in **Figure 3.3.12**.

Spanner Wrench

Some threaded fasteners have holes or slots that are used for turning the fastener. **Spanner wrenches** fit these holes or slots. A *hook spanner* wrench has a hooked arm that fits into fasteners with slots, while *pin spanner* wrenches fit into fasteners with holes. Face spanner wrenches have pins that fit into the face of a fastener. Examples of these wrenches are shown in **Figure 3.3.13**.

Hex Key Wrench

Hex key wrenches are used on socket head cap and set screws. They are frequently used on cutting toolholders

FIGURE 3.3.10 When using an adjustable wrench, rotate it with the adjustable jaw in the direction of rotation.

Apply force in direction indicated.

FIGURE 3.3.11 Socket wrenches are available in 6-point and 12-point styles.

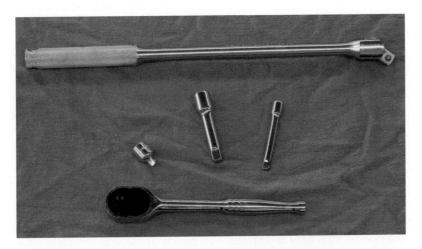

FIGURE 3.3.12 Solid and ratcheting socket handles and some socket wrench accessories used to reach fasteners in difficult-to-reach locations.

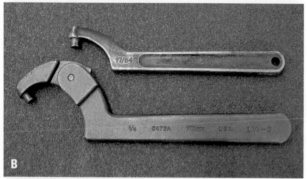

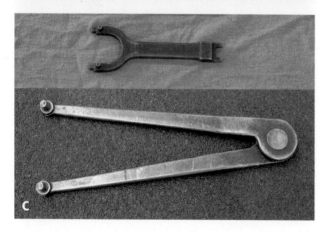

FIGURE 3.3.13 (A) Solid and adjustable hook spanner wrenches. (B) Solid and adjustable pin spanner wrenches. (C) Solid and adjustable face spanner wrenches. (D) Pin spanner and face spanner wrenches in use.

and fixtures. Some different styles of hex key wrenches can be seen in **Figure 3.3.14**.

BENCH VISE

A *bench vise* is a device used to hold workpieces securely to perform operations such as filing or hacksawing. Vises should be mounted securely on a stable workbench at a comfortable height to prevent worker fatigue. A vise handle should not extend into a walkway when it is clamped. Do not use a hammer or an extension on the handle to tighten the vise. Most bench vises feature a small anvil area behind the jaws for light hammering. The anvil is the only area of a vise that should be used for supporting hammering.

Bases

Swivel Base

Swivel-base vises have a base that can be swiveled (rotated) for better work positioning and clamped in the desired position. Do not hammer or use an extension to tighten the swivel clamp. **Figure 3.3.15** shows an example of this type of vise.

Fixed Base

Fixed-base vises are mounted in a fixed position on a bench. These vises usually are made from a solid casting and are directly fixed to the workbench with three bolts. **Figure 3.3.16** shows a fixed-base bench vise.

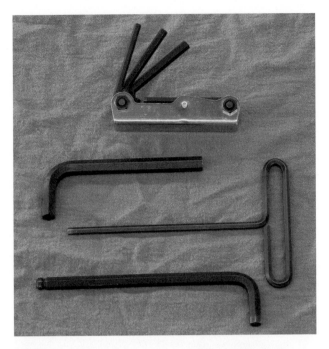

FIGURE 3.3.14 Hex key wrenches are available in fold-up sets, L-shaped types, and T-handle types. Some have ball ends to allow reaching and turning fasteners from an angle.

FIGURE 3.3.15 A swivel-base bench vise.

FIGURE 3.3.16 A fixed-base bench vise.

FIGURE 3.3.17 Smooth vise jaws are for lighter clamping while serrated jaws will bite into a workpiece and clamp work very securely.

Jaws

Vise jaws are gripping surfaces made of replaceable inserts held in place with pins or screws.

Hard Jaws

Hardened steel jaws are used when heavy clamping is needed to secure a workpiece. These jaws are available with straight grooves, serrated surfaces, or smooth surfaces. They will mar and possibly damage precision workpiece surfaces. **Figure 3.3.17** shows some bench vise jaws.

Soft Jaws

Soft jaws usually take the form of slip-on or magnetic caps that prevent the clamping action of the vise from damaging the workpiece. They may be made from aluminum, copper, rubber, plastic, or even wood. **Figure 3.3.18** shows some soft jaws on a bench vise.

FIGURE 3.3.18 Soft caps, such as this copper set, can be placed on vise jaws to reduce the chance of damage to parts when clamping.

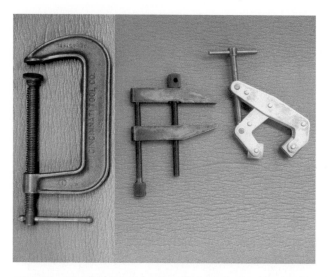

FIGURE 3.3.19 From left to right: C-clamp, parallel clamp, and hinged clamp.

CLAMPS

There are many types of clamps that may be used when performing benchwork. This may involve simply clamping parts together so that several can be machined at once.

C-Clamp

The **C-clamp** has a C-shaped frame and a screw to clamp the workpiece in place. This design is very strong and is useful for heavy-duty clamping. The size of the jaw opening and depth of the throat are used to determine the size of a clamp.

Parallel Clamp

Parallel clamps are useful for light-duty applications such as holding small and delicate parts. These clamps have two parallel jaws that use two parallel screws to adjust the jaw width and to clamp the parts.

Hinged Clamp

Hinged clamps feature two hinged clamping jaws that use a screw to force the jaws together for clamping. **Figure 3.3.19** shows some examples of these different types of clamps.

HACKSAWS

Hacksaws are simple, portable handsaws used for light-duty sawing operations. Sometimes using a hacksaw to cut work is quicker, easier, or more accurate than using a power saw. Hacksaws have three main parts: the blade, the handle, and the frame. Most hacksaw frames may be adjusted for different blade lengths, such as 8, 10, or 12 inch. **Figure 3.3.20** shows a hacksaw.

FIGURE 3.3.20 A hand hacksaw.

Hacksaw Blades

Most general-purpose hacksaw blades are 0.025 thick and are 1/2 inch wide. Blades are classified by their length and number of **teeth per inch (TPI)**. You should choose a blade that will have at least three teeth in contact with the work at all times while sawing to prevent them from snagging and breaking. To keep the blade from binding in the workpiece, the teeth are staggered from side to side in a pattern known as the **set**. Most hacksaw blades have a wavy set, in which the teeth progressively alternate in a gently repeating pattern to the right and then to the left. **Figure 3.3.21** shows a wavy set hacksaw blade. This results in the cut being slightly wider than the thickness of the blade. This narrow slot produced when sawing is called the **kerf**.

FIGURE 3.3.21 The wavy set of a hacksaw blade provides clearance for the blade during sawing.

Hacksaw Use

To use the hacksaw, first select the correct blade for the material being cut and mount it snugly in the saw frame with the teeth pointing away from the handle. If you need to produce a deeper cut than the frame will allow, mount the blade at a right angle to the frame by moving the mounts a quarter turn, as shown in **Figure 3.3.22**. The workpiece should be mounted in a vise close to the jaws to reduce vibration and chatter.

Use light pressure to start the cut. To keep the blade from sliding sideways when beginning, you may want to file a small notch or drag the saw lightly backwards while holding your thumbnail beside the cut to steady the blade. After the cut is started, it is best to use both hands to saw. Place one hand on the handle and the other hand on the other end of the frame. The saw cuts only on the forward stroke, so down pressure should be applied only on that stroke. Make about 40 to 50 forward strokes of the saw per minute, using the whole blade. Keep the blade positioned properly and avoid twisting. If you break or wear out a blade, the kerf of the old cut will be narrower than that of the replacement blade. A new cut should be started from the opposite side of the workpiece to avoid binding and destroying the set. **Figure 3.3.23** shows the proper hacksawing method.

Use caution so you do not make contact between your finger and the saw teeth.

Always wear safety glasses when using a hacksaw. Do not saw too fast or put too much pressure on the saw—it could break the blade and cause injury. Lighten up on the cutting pressure as you finish a cut and begin to break through the workpiece to avoid injury.

FILES

Files are used in benchwork for tasks such as shaping, smoothing, fitting, and deburring. It is essential to learn how to properly use files for times when filing is a more practical method than using a machine tool. Files are available in a multitude of shapes, sizes, and styles for different applications.

File Classification

Files can be classified according to length, shape of the cross section, cut or tooth type, and coarseness.

Length

The length of a file is measured from the heel to the tip of the file. **Figure 3.3.24** shows the basic parts of a file and how the length is determined.

Cross-Section Shapes

Another characteristic of files is the shape of their cross sections. Some frequently used shapes are square, flat, mill, pillar, knife, round, half round, and three square. **Figure 3.3.25** shows these cross-section shapes.

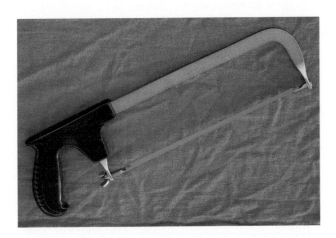

FIGURE 3.3.22 Some hacksaw frames allow blades to be mounted at 90-degree increments for sawing narrow strips.

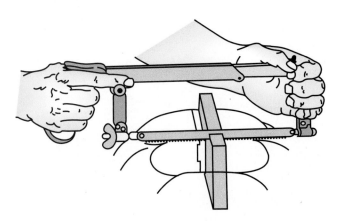

FIGURE 3.3.23 The proper way to hold a hacksaw. Hold the hacksaw level and apply pressure only to the forward stroke.

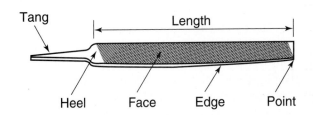

FIGURE 3.3.24 The basic parts of a file.

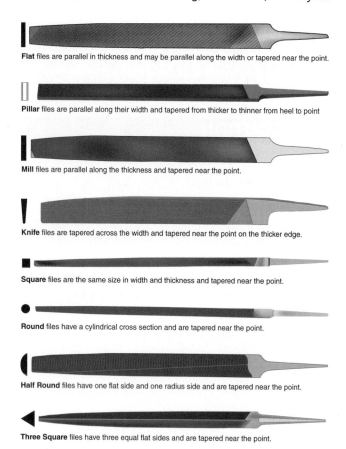

Flat files are parallel in thickness and may be parallel along the width or tapered near the point.

Pillar files are parallel along their width and tapered from thicker to thinner from heel to point

Mill files are parallel along the thickness and tapered near the point.

Knife files are tapered across the width and tapered near the point on the thicker edge.

Square files are the same size in width and thickness and tapered near the point.

Round files have a cylindrical cross section and are tapered near the point.

Half Round files have one flat side and one radius side and are tapered near the point.

Three Square files have three equal flat sides and are tapered near the point.

FIGURE 3.3.25 Shapes of some common file cross sections.

Cut or Tooth Classifications

Files come in different cut or tooth classifications. They are single cut, double cut, curved tooth, and rasp. Single-cut files have a single set of teeth. These files are used with light pressure to obtain a smooth surface and remove material slowly. Double-cut files have a second set of teeth perpendicular to the first set. They can be used with heavier pressure for faster metal removal and produce a rougher surface. The curved-tooth file can be used to file large flat surfaces but is not common in the machining field. The rasp-tooth file is also not very common in machining because it is designed for fast material removal on very soft materials such as wood or plastic. **Figure 3.3.26** shows examples of these cut classifications.

Coarseness

Files are also classified according to the coarseness of their teeth. From coarsest to smoothest, they are rough, coarse, bastard, second-cut, smooth, and dead smooth. Smoother files are used to produce finer surface finishes. Coarser files are used for more rapid material removal while leaving rougher surface finishes.

File coarseness is related to file length. Larger files will be coarser than smaller files of the same coarseness

FIGURE 3.3.26 From left to right: Rasp-tooth, curved-tooth, double-cut, and single-cut files.

classification. For example, a 12" second-cut file will be coarser than an 8" second-cut file.

Special Files

Files come in many different shapes, sizes, and styles that do not fit into the standard classifications. Very small files, called jeweler's or die maker's files, can be used for very intricate filing operations. They may have varying cross-section shapes, bends, or hooks to reach in small areas. **Figure 3.3.27** shows some examples of these files.

Files sometimes have an edge without teeth that is called a safe edge. A safe edge helps prevent filing into adjacent surfaces. **Figure 3.3.28** shows a file with a safe edge.

File Selection

When selecting files, longer, coarser, double-cut files are generally chosen to remove material quickly, while

FIGURE 3.3.27 Die maker's or jeweler's files come in many special shapes for intricate filing tasks.

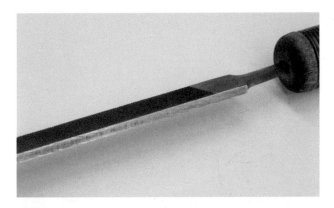

FIGURE 3.3.28 A safe edge on a file can help prevent cutting into adjacent surfaces when filing.

FIGURE 3.3.29 To mount a file in an unthreaded handle, place the tang in the handle and tap the end of the handle against a wooden workbench. Hold by the handle, not the file.

shorter, finer, single-cut files are generally chosen to create smooth surfaces. With practice and experience, you will begin to develop a sense of what type of file to use for different situations, since every job is unique.

General File Use

Gripping/Stance Techniques

Before filing, secure the workpiece at a comfortable height and position using clamps or a vise. Keep the line or surface you are filing close enough to the vise jaws to prevent it from chattering, but not so close that you file into the vise. When possible, position the surface to be filed horizontally, since this will allow the most natural hand motion.

When gripping a file for use, hold the file firmly by the handle with one hand and grasp the tip with the other hand for stability. Always use a file equipped with a handle to prevent injury from being stabbed by the tang. Some handles screw onto the file tang, while others just press onto the tang. **Figure 3.3.29** shows the proper method for mounting a file in an unthreaded handle.

When filing, stand with your feet shoulder-width apart and establish firm footing. Files cut only on the forward stroke, so use appropriate pressure only on the forward stroke when moving the file across the work surface. Be sure to lift the file during the return stroke, because dragging the file on the back stroke will only dull the file. Using too much or too little pressure when filing will cause the file to dull prematurely. You can rub chalk on a file to help prevent material from becoming lodged in the teeth of the file.

(!) **CAUTION**

Always use a file equipped with a handle to prevent injury from being stabbed by the tang.

(!) **CAUTION**

Don't use files for prying or hammering, because they are very hard and can shatter.

Filing

Straight filing is moving the file from tip to heel across the surface in a straight or angled motion, as shown in **Figure 3.3.30**. To finish a part to the final dimension and obtain a better finish, use a **draw filing** technique. Draw filing is moving the file in the direction of its width, as shown in **Figure 3.3.31**. Hold the file in both hands perpendicular to the workpiece and apply pressure when moving the file in the direction the teeth are pointed. When you use a single-cut file, the file will cut in only one direction. With the file handle in the right hand, a

FIGURE 3.3.30 Method for performing straight filing. Hold the file firmly and move across the work surface in the direction of the file's length. Apply pressure on the forward stroke and lift the file on the return stroke.

FIGURE 3.3.31 Method for performing draw filing. Hole the file lightly and move in the direction of the file's width. Move a single-cut file in the direction the teeth face. A double-cut file will cut moving both directions because the two sets of teeth point in opposite directions.

single cut file will cut on the forward stroke. With the file handle in the left hand, a single cut file will cut on the back stroke. When you draw file with a double-cut file, the file will cut in both directions of motion, because the two sets of teeth face opposite directions. When draw filing, use lighter pressure than when **straight filing**. Straight filing is used to remove material fairly quickly, while draw filing removes material much more slowly and provides a finer, smoother surface finish.

Cleaning and Care of Files

As you use files, their teeth will eventually become clogged with the material you are removing. This is called **loading**. Sometimes, particularly on softer metals such as aluminum, particles of the material being filed called *pins* can embed in the teeth. This is called **pinning**. Pinning and loading will occur more quickly if you use too much pressure. Clean your file frequently to keep pins from scratching the workpiece. If you notice scratches in the material, it's a sign that you need to clean your file. **File cards** are brushes used to clean files. A file card can have short soft or wire bristles. Many of them also have a small pick that can be used for removing pins. To remove the pins, use the pick and slide it between the teeth to push out the pins. Move the brush parallel with the angle of the teeth, as shown in **Figure 3.3.32**.

After use, files should be cleaned and stored so they do not come in contact with other files or tools that could dull them.

Do not clean the file with your hand or strike it on a workbench or other surface to remove debris.

FIGURE 3.3.32 (A) The soft bristle and wire bristle sides of a file card. (B) Clean files often during use by moving the bristles parallel with the angle of the teeth to remove filings.

Filing Tips

Every situation is unique, but here are some general guidelines to keep in mind when filing.

- Begin with coarser files and move to smoother cut files as you near the final size.
- As the part you are filing gets closer to final size, you need to check it more and more frequently for accuracy.

- When filing to a layout line, begin to check sizes before reaching the layout line.

- When filing a surface perpendicular to another surface, be sure to keep the file square to ensure perpendicularity.

- You can coat a surface with layout dye and then gently move a square or radius gage across the surface. Rubbing the layout dye indicates high spots on the surface. Don't force the measuring tool across the part; you will damage it.

- As the part nears final size, you might spend as much time checking it as you do filing it.

- Make sure you remove any burrs from the edge of the part before checking it, both for safety and to ensure accurate measurements.

DEBURRING

Deburring removes sharp raised edges from workpieces. Deburring is a critical task in the machining field because even small burrs or raised edges can prevent work from being accurately measured, located, or secured for hand and machining operations. Burrs can also cause serious cuts when you handle material and workpieces. A file can be used for deburring by moving the file along the edge of the part with a slight rolling motion. This cuts the burr off rather than pushing it back and forth. Special deburring tools are also available to fit a wide variety of situations. Some of these deburring tools are shown in **Figure 3.3.33**.

 CAUTION

Burrs can cause serious cuts when you handle material and workpieces.

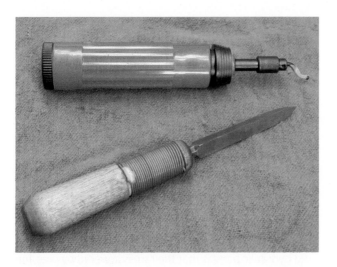

FIGURE 3.3.33 Some examples of tools used for deburring.

ABRASIVES

Abrasives are natural or synthetic materials that wear away material to produce a smooth surface or to remove burrs. They are generally used for light material removal after filing or during the final stages of workpiece machining. This final stage of light material removal is called *polishing*. Because of their hardness, abrasives are useful for polishing or deburring hardened steels that cannot be cut with files. Emery, garnet, aluminum oxide, and silicon carbide are examples of materials used as abrasives. Unlike files, abrasives are able to cut in any direction.

Cloth forms of abrasives have the abrasive grains glued to a flexible cloth backing. Paper forms have the abrasive grains glued to a paper backing. They are both available in different grit sizes. Grit refers to the size of the abrasive grains. Lower numbers refer to larger grit sizes, which are coarser. Higher numbers refer to finer grit sizes and are finer. Coarser grits remove material more quickly and leave a slightly rougher surface, while finer grits remove material more slowly and leave a smoother surface. **Figure 3.3.34** shows some examples of cloth and paper abrasives.

Stones have the abrasive grains bonded together in a solid form. They are available in many shapes, sizes, and grits as well. **Figure 3.3.35** shows some examples of abrasive stones.

PRESSES

Sometimes in the machining industry, two parts are designed to be assembled with an interference fit. Recall that Section 3.1 (Understanding Drawings) provided an example where a shaft would be larger than a hole. In cases like these, the two parts may require significant force to be assembled. A press is a device that can be used to create that force by using leverage or hydraulic pressure. Presses can also be used to disassemble parts that have an interference fit.

FIGURE 3.3.34 Cloth and paper abrasives are available in sheets and rolls of different sizes and grits.

FIGURE 3.3.35 Examples of abrasive stones used for deburring and polishing.

Arbor Press

An **arbor press** uses leverage to create force and can be used to assemble parts with light to medium interference fits. To assemble parts using an arbor press, follow these few steps.

- Align the two parts on the base of the press with the ram raised.
- Lower the ram to make contact with the top part. Keep fingers and hands out from under the ram.
- Pull the lever arm to press the two parts together. **(See Figure 3.3.36.)**

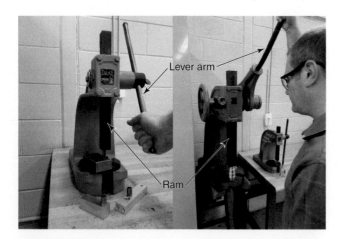

FIGURE 3.3.36 Using arbor presses to assemble parts with interference fits.

- Retract the ram and remove the assembly from the work area.
- Lower the ram to the lowest position to rest on the base when finished.

 CAUTION

Keep fingers and hands out from under the ram of an arbor press. The ram can be very heavy and cause injury if allowed to fall. Always lower the ram of an arbor press onto the base when not in use. Otherwise, the ram may unexpectedly fall and cause injury.

Hydraulic Press

A **hydraulic press** uses a hydraulic pump to move a cylinder to create force. It can generate a great deal more force than an arbor press and can be used to assemble and disassemble parts with higher interference fits. There are three general types of hydraulic presses. A manual model must be pumped using a handle. An electric model has a motor to operate the hydraulic pump. An air powered model uses an air powered motor to operate the hydraulic pump. Many models have a gage that shows the amount of pressure the press is creating. Small models typically can create 8 to 12 tons of pressure, while larger ones can range from 75 to 200 tons.

A typical hydraulic press has two upright frame pieces and a top section that holds the hydraulic mechanism. The work area is between the two uprights and is connected to the frame with removable pins. These pins can be removed and the work area raised and lowered as needed. They are often called H-frame presses because of this configuration. Maintain the level of hydraulic oil in the press, monitor the condition of oil/air lines and fittings, watch for oil and/or air leaks, and lubricate as needed according to the manufacturer's directions.

To use a manual hydraulic press, follow these general steps.

- Position the work table at the required height.
- Position the two parts to be assembled in the work area.
- Check to ensure that the hydraulic valve is closed.
- Lower the ram by pumping the handle to press the parts together. See **Figure 3.3.37**.
- Open the hydraulic valve so the ram will retract and the part can be removed.
- Close the hydraulic valve when finished.

The only difference when using an electric or air powered hydraulic press is that the ram is lowered by pressing a button or lever instead of pumping by hand.

FIGURE 3.3.37 (A) Using a manual hydraulic press to assemble parts with an interference fit. (B) The gage is showing pressure at about 7.5 tons.

SUMMARY

- The ability to use a variety of hand tools correctly and safely is an essential part of becoming a machinist.
- Choosing the wrong tool or misusing a hand tool can damage equipment and parts, or cause injury.
- Make sure that screwdrivers and wrenches correctly fit fasteners.
- Pliers come in various styles and sizes for different tasks. Some are adjustable and can be locked in the closed position. Others are also able to perform light-duty cutting tasks.
- Hammers are available in different types for tasks ranging from heavy driving to making light adjustments.
- Vises and clamps are often used to secure materials for benchwork.
- Hacksaws are compact, portable hand saws that can be used for many benchwork applications.
- Files are available in many different shapes and sizes and can be used for deburring and shaping workpieces by hand.
- Abrasives remove small amounts of material to smooth and deburr and can be used to cut hardened steel.
- Presses are used to assemble and disassemble parts with interference fits.

REVIEW QUESTIONS

1. What are three safety rules to observe when using screwdrivers?
2. List three types of screwdriver tips.
3. What is the advantage of using slip joint pliers?
4. What is an advantage of using locking pliers?
5. What are two uses for a ball peen hammer?
6. What is the advantage of a soft face hammer?
7. In what situation would a box-end wrench be chosen over an open-end wrench?

(Continued)

8. List two precautions to observe when using adjustable wrenches.

9. What is one method of preventing damage to work clamped in a vise?

10. List three safety precautions to be observed when using a hacksaw.

11. In which direction should hacksaw blade teeth point?

12. List two safety precautions that should be observed when using a file.

13. Will a single-cut or a double-cut file remove material more quickly?

14. Will a single-cut or a double-cut file produce a smoother surface?

15. _____ and _____ are two common filing problems that can be overcome by cleaning the file.

16. What tool is used to clean a file?

17. What are the two forms of abrasives used in benchwork?

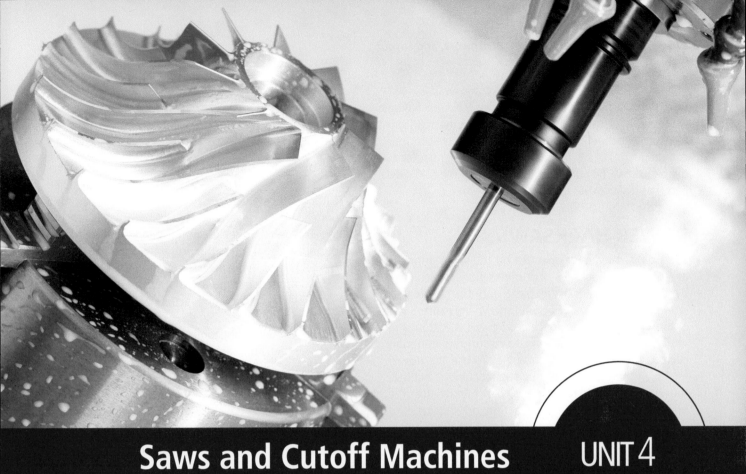

Saws and Cutoff Machines

Learning Objectives

After completing this unit, the student should be able to:

- Identify the various sawing machines used in the machine shop
- Operate band saws safely
- List the different band saw blade materials
- Define blade pitch
- Identify the three different tooth patterns and their uses
- Identify the three different blade sets and their uses
- Describe how to select proper band saw blade width
- Understand and be able to identify saw tooth geometry
- Explain the term *kerf*
- Calculate band saw blade length
- Describe the band welding procedure
- Describe blade mounting procedure for the vertical band saw

Key Terms

Alternate set	Hook tooth form	Skip tooth form
Bimetal	Horizontal band saw	Tooth form
Carbide tooth	Kerf	Tooth set
Carbon steel	Pitch	Teeth Per Inch (TPI)
Gauge	Power hacksaw	Variable-pitch blade
Gullet	Rake	Vertical band saw
High-Speed Steel	Raker set	Wavy set
(HSS)	Regular tooth form	

INTRODUCTION

One of the first operations performed in most machining processes is to saw raw material for a workpiece from bar, tube, or sheet material called *stock*. There are four common types of sawing machines: the power hacksaw, the band saw, the abrasive cutoff saw, and the cold saw. Each has distinct uses in the shop.

POWER HACKSAWS

The **power hacksaw** operates on the same principle as the hand hacksaw. Cutting action is achieved by drawing a saw blade back and forth across a workpiece. The power hacksaw shown in **Figure 3.4.1** is the earliest sawing machine we will mention and has been used for many years to cut off raw materials such as bar stock, plate, and pipe. To use the power hacksaw, the operator sets up the machine to achieve the number of blade strokes per minute for the material being cut. Cutting action on this type of saw occurs only on the forward stroke, making it very inefficient. For that reason, the power hacksaw has been largely replaced by newer band-style machines. However, the power hacksaw is still useful in some applications.

BAND SAWING MACHINES

Band sawing machines are the most commonly encountered saws. The blade is a continuous metal band with a series of teeth ground into one edge. Two large wheels are located at opposite ends of the saw that both support and drive the saw blade. When the power is turned on, the drive wheels begin to rotate, causing the saw blade to move. The continuous blade cuts

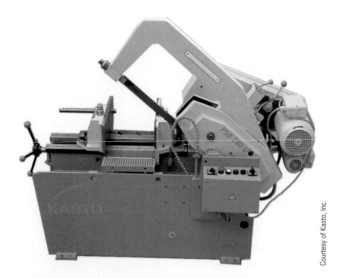

FIGURE 3.4.1 A power hacksaw only cuts in one direction.

Courtesy of Kasto, Inc.

constantly, which reduces sawing time. Blades on band saws cut more accurately than reciprocating hacksaws and build up less heat.

Horizontal Band Saws

The **horizontal band saw** gets its name from the fact that the blade is in a horizontal orientation. It is capable of cutting off large pieces of stock both quickly and accurately, holding tolerances to ±0.015 of an inch or better. This makes the horizontal band saw ideal for producing straight cuts through raw material. Most horizontal band saws use a vise to clamp material being cut. Most also incorporate a coolant system that floods the area where the cutting action is occurring with cutting fluid to help cool both the workpiece and the saw blade. Many modern horizontal band saws are equipped with power feeds that are controlled either hydraulically or by means of computer numerical control (CNC). **Figure 3.4.2** shows examples of manual, hydraulic, and CNC horizontal band saws.

Horizontal Band Saw Basic Operation

Here are some basic steps to follow when using a horizontal band saw to cut material. Keep in mind that because of the many different models, each machine will have its own specific controls.

- Adjust the arms that guide the blade as close together as possible for the size of the material being cut and lock them in place. **(See Figure 3.4.3.)**
- Measure the desired length and mount the material solidly in the saw's vise. Stock is normally cut about 1/16" to 1/8" longer than finished part requirements to allow for further material removal by other machining operations. Be sure the material is flat on the machine.
- If the piece of material is shorter than the width of the vise jaw, the moveable jaw may rotate and not hold it securely. Place another piece of the same size material near the far end of the vise to keep the clamping jaws parallel. **(See Figure 3.4.4.)**
- When cutting long bars, use a stand to support the stock to keep it flat and prevent material from falling. **(See Figure 3.4.5.)**
- Set the proper band speed for the material being cut.
- Lower the blade within about 1/4" of the material.
- Start the band.
- Lower the blade slowly into the material to begin the cut. Starting the cut too quickly can break teeth off the blade. Once several teeth are engaged in cutting, set an appropriate rate for the blade to cut through the material.
- After sawing, deburr the material, clean the saw, and return the excess material to a proper storage area.

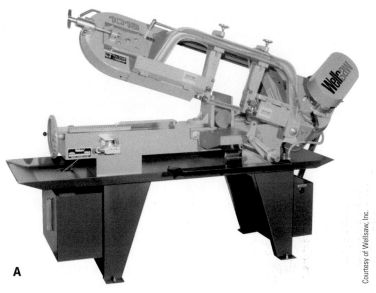

A

Courtesy of Wellsaw, Inc.

B

C

Courtesy of Behringer Saws, Inc.

FIGURE 3.4.2 (A) A manual horizontal band saw, (B) a hydraulic horizontal band saw, and (C) a CNC horizontal band saw.

FIGURE 3.4.3 Adjust the guide arms of a horizontal band saw as close as possible for the material being cut. This helps to keep the blade straight and results in straighter cuts.

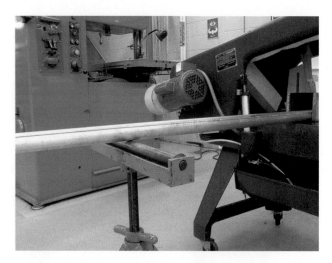

FIGURE 3.4.5 Use a stand to support long bars being cut on the horizontal band saw. This keeps them flat and prevents them from falling when the vise is unclamped.

FIGURE 3.4.4 When clamping short sections in the saw vise, use another piece of the same size material to keep the moveable jaw parallel to the solid jaw to ensure adequate clamping.

Horizontal Band Saw Safety

As with any machine tool, observing a few safety guidelines will keep everyone safe and prevent damage to equipment. Never attempt to hold material by hand when using a horizontal band saw, and keep all body parts clear of the moving saw blade and any other

moving machine parts. When moving long or heavy bars of material, get help to avoid injury, and always use proper lifting techniques. Always lock out or tag out the machine's power when adjusting guides, changing blades, or performing any machine maintenance.

Vertical Band Saws

The **vertical band saw** is common in almost every shop and is a very useful piece of equipment. It gets its name from the fact that the blade is in a vertical orientation supported by two wheels. The uppermost wheel is usually referred to as an idler wheel. The idler wheel has two main purposes: First, it supports the saw blade. Second, it can be raised and lowered by rotating the handle located just under the wheel to adjust the tension of the saw blade. The lower wheel supports the saw blade as well and provides the driving motion to the blade.

Vertical Band Saw Applications

The vertical band saw is often used to rough cut a part near finished size to remove any excess material before performing other machining operations. This can drastically reduce the time and tooling needed to produce a finished product. In **Figure 3.4.6**, a vertical band saw is being used to remove excess material from a part before beginning other machining operations.

The cuts produced on a band saw may be simple straight cuts or more complex contouring operations. This contouring ability is particularly useful, as it allows complex and intricate shapes to be cut both quickly and accurately. **(See Figure 3.4.7.)** Sometimes, a contour to be cut may be inside a closed boundary on a workpiece, with no place to enter the boundary with a saw cut. In these instances, it may be necessary to

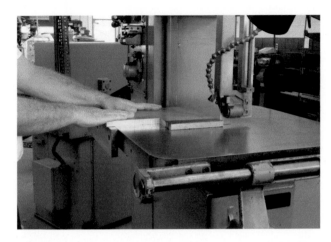

FIGURE 3.4.6 The vertical band saw is useful for removing large sections of material in preparation for other machining operations.

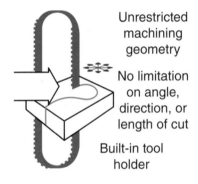

FIGURE 3.4.7 Contour sawing can be performed on the vertical band saw.

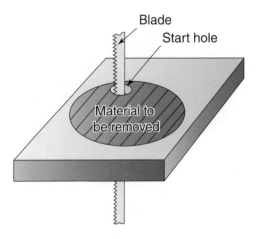

FIGURE 3.4.8 Method for sawing an enclosed internal contour. The blade is cut, placed through the start hole, and the ends are welded together. After sawing, the blade must be cut again to be removed.

FIGURE 3.4.9 A typical vertical band saw blade welder.

drill a hole in the inside of the boundary large enough to allow the blade to pass through. The blade may then be cut and inserted through the drilled hole. **(See Figure 3.4.8.)** Then its ends may be welded back together and installed in the machine with the workpiece surrounding it. Specialized band saw blade welders have been designed specifically for this task—some band saw machines even have them attached as standard equipment. **Figure 3.4.9** shows a blade

welder on a vertical band saw. Welding operations will be addressed later in this unit.

Often during contouring, it may be necessary to make tight turns on the inside corners of a contour or shape. It is a common practice to drill an appropriately

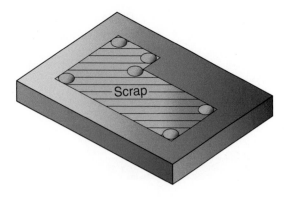

FIGURE 3.4.10 Drilled holes in internal corners provide clearance for the blade to make turns when contour sawing.

sized hole in the corner to allow the part to be pivoted around the blade once the blade enters the relief created by this hole. Often, if a radius is to reside in the inside corner on the finished part, the diameter of the drill will be chosen so that it closely matches the finished radius. **(See Figure 3.4.10.)**

Vertical Band Saw Safety

Certain safety precautions must always be followed when operating the vertical band saw. Extra caution must be observed when using a vertical band saw because the material is often held by hand, and hands are near the moving blade. Adjust the upper guide arm within 1/8" of the top of the work being cut, as shown in **Figure 3.4.11**, to limit blade exposure. Avoid placing fingers and hands directly in front of the blade while sawing. Do not use excessive force when pushing work into the blade. When sawing small parts, place a soft metal or wooden *push stick* behind the part and keep hands behind the push stick, as shown in **Figure 3.4.12**. Lock out or tag out the saw when changing blades or servicing. Do not force parts into the saw blade, but feed them consistently. Always deburr parts after sawing.

FIGURE 3.4.11 Keep the upper guides adjusted within 1/8" of the material being sawn to minimize exposure to the saw blade.

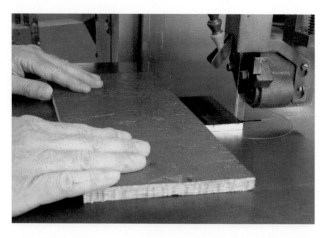

FIGURE 3.4.12 Use a soft metal or wooden push stick when sawing small parts to keep fingers and hands away from the blade.

SAW BLADE CHARACTERISTICS AND APPLICATIONS

Band saw and hacksaw blades, although simple in appearance, have carefully designed cutting geometry engineered to achieve durability and efficient cutting. An incorrect blade, or one whose critical features are worn, will cut slowly, require significantly more pressure to cut, and generate excessive heat. The many features of a saw blade are described in detail in the following paragraphs. **Figure 3.4.13** shows the parts of a saw blade.

These blade characteristics are important when the appropriate saw blade is selected for a job, as it will affect the quality of the cut as well as the life of the saw blade and time needed to complete the sawing process. Many band saws have a chart mounted on them to assist in choosing the correct blade.

Blade Material

Saw blades can be made of different types of materials. They are chosen according to the particular application and include factors such as material being cut, desired rate of cutting, and cost.

Carbon Steel

Carbon steel blades are the least expensive, but must be run at slower speeds than blades made of harder materials. Blades under 1/4" in width are usually only available in carbon steel and are often used for vertical contour sawing of small radii. Only the teeth of the *flex back* carbon blade are hardened. It is the least expensive blade material and is normally used when sawing non-ferrous metals. The entire blade of the *hard back* carbon blade is hardened, so it is stronger than the flex back. It can be used on non-ferrous metals and soft steels. It is more expensive than the flex back blade. **Figure 3.4.14** shows a carbon steel blade.

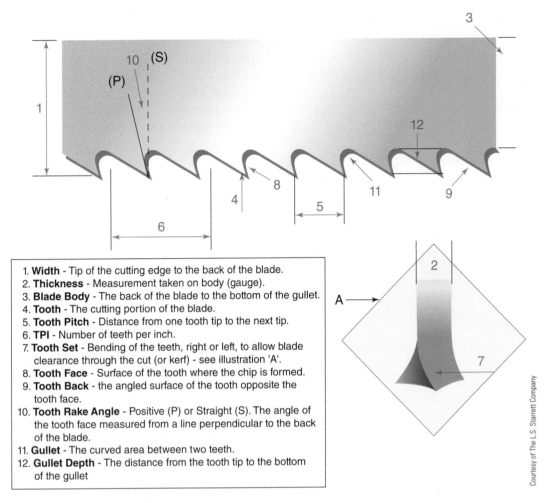

1. **Width** - Tip of the cutting edge to the back of the blade.
2. **Thickness** - Measurement taken on body (gauge).
3. **Blade Body** - The back of the blade to the bottom of the gullet.
4. **Tooth** - The cutting portion of the blade.
5. **Tooth Pitch** - Distance from one tooth tip to the next tip.
6. **TPI** - Number of teeth per inch.
7. **Tooth Set** - Bending of the teeth, right or left, to allow blade clearance through the cut (or kerf) - see illustration 'A'.
8. **Tooth Face** - Surface of the tooth where the chip is formed.
9. **Tooth Back** - the angled surface of the tooth opposite the tooth face.
10. **Tooth Rake Angle** - Positive (P) or Straight (S). The angle of the tooth face measured from a line perpendicular to the back of the blade.
11. **Gullet** - The curved area between two teeth.
12. **Gullet Depth** - The distance from the tooth tip to the bottom of the gullet

Courtesy of The L.S. Starrett Company

FIGURE 3.4.13 Parts of a saw blade.

Courtesy of Lenox

FIGURE 3.4.14 A typical carbon steel saw blade.

Courtesy of Lenox

FIGURE 3.4.15 A bimetal band saw blade.

Bimetal

A **bimetal** blade has a carbon steel body with a strip of **high-speed steel (HSS)** welded to the one edge. The teeth are cut into the high-speed steel strip. Bimetal blades can be used for horizontal and vertical band sawing operations. They are more expensive than carbon steel blades, but can be run at higher speeds. **Figure 3.4.15** shows an example of a bimetal blade.

Carbide Tooth

Carbide tooth blades have tungsten carbide brazed to a carbon steel body. They can be operated at very high speeds and cut very tough materials that even bimetal blades cannot cut. Carbide tooth blades are very expensive and are most often used for high-speed, high-production raw material cutting on horizontal band saws. **Figure 3.4.16** shows a carbide tooth blade.

Tooth Set

Tooth set refers to the staggered arrangement of the saw teeth. Tooth set is necessary to provide clearance for the saw blade body as it travels through a workpiece. Item 7 in **Figure 3.4.13** shows the tooth set.

FIGURE 3.4.16 A carbide-tooth band saw blade.

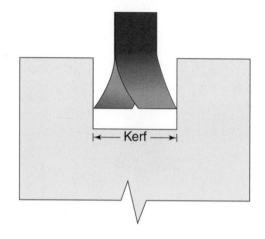

FIGURE 3.4.17 The kerf is the slot produced by sawing.

The width of the slot produced by the saw blade in the workpiece is called the **kerf**. It is important to remember that the actual kerf will always be slightly wider than the saw blade itself because of the tooth set. **Figure 3.4.17** illustrates a saw kerf.

There are three common types of tooth sets that can be selected depending on the sawing application. They are shown in **Figure 3.4.18** and explained in the following paragraphs.

Alternate Set

An alternate tooth setting pattern, or **alternate set**, has teeth set in a pattern in which every other tooth switches the side of the blade it is set to. The pattern simply sets the teeth left, right, left, right, and so on. This set is the most aggressive and is recommended for softer, non-ferrous materials such as aluminum alloys.

Raker Set

A raker tooth setting pattern, or **raker set**, has small groups of teeth with an alternate tooth set, but places a tooth with a neutral set in between them. The raker set is best suited where the workpiece material being cut consists of large round stock or thick steel. It is also a good choice when performing contouring on a vertical band saw.

Wavy Set

A **wavy set** appears as a wave of tooth set with groups of teeth set to one side and with the next group gradually changing to the other side. The wavy set saw blade is best suited for sawing operations where the cross-sectional area consists of varying thickness common to many structural materials, such as I-beams, channels, and tubing.

Blade Pitch or TPI

Blade **pitch** is the spacing of teeth from one tooth to the next, as shown in Item 5 of **Figure 3.4.13**. Generally the thinner the workpiece, the finer the pitch of the saw blade should be. A commonly accepted rule of thumb is that there should always be at least three saw teeth engaged in the workpiece at all times if at all possible. This prevents teeth from straddling the work across its thickness and being broken off. **(See Figure 3.4.19.)** **Variable-pitch blades** have been developed with a constantly changing tooth pitch to help alleviate this

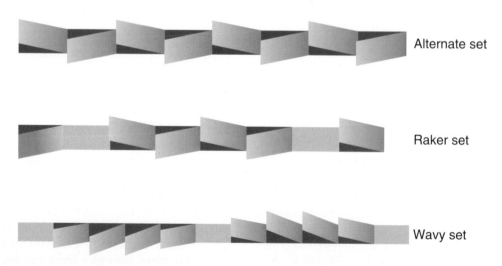

Alternate set

Raker set

Wavy set

FIGURE 3.4.18 The alternate, raker, and wavy blade sets.

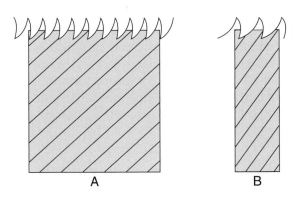

FIGURE 3.4.19 (A) Always keep a minimum of three teeth cutting in the material at all times. Six to 12 teeth is ideal. (B) If less than three teeth are in contact with the material being sawn, teeth can straddle the work and be broken and stripped from the blade.

FIGURE 3.4.20 A variable-pitch saw blade has teeth of different sizes to help reduce teeth stripping and minimize vibration during cutting.

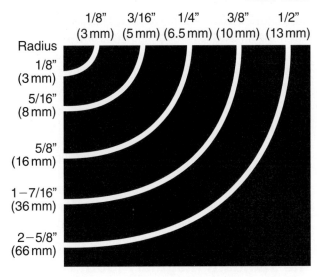

FIGURE 3.4.21 A chart like this shows the minimum radii that can be sawn with particular blade widths. If the radius fall between two widths, select the narrower blade.

problem on thin sections and reduce vibration during cutting. **Figure 3.4.20** shows a variable-pitch blade.

Blade **teeth per inch (TPI)** is another way to quantify pitch by determining the number of teeth per inch that have been ground into the saw blade. To determine the minimum TPI for sawing a given material thickness, use the formula $TPI = 3/t$, where t = material thickness. If the answer is fractional, always round up to choose the minimum pitch.

Blade Width

Saw blade *width* is the distance from the back of the blade to the tip of the cutting tooth and is shown as item 1 in **Figure 3.4.13**. Blades are normally available in 1/8" to 3" widths. Narrower blades (1/8–1/2") are generally used for contouring on vertical band saws, while wider blades (3/4–3") are more common on horizontal band saws for making straight cuts.

A blade's width limits its ability to make tight turns while sawing a contour. For this reason, a narrow blade will be able to cut a smaller radius (make a tighter turn) during contour cutting than a wider blade. Blade wear (particularly the tooth set) and saw blade guide wear will result if a blade is forced to cut a tighter radius than it has been designed for. Charts like the one shown in **Figure 3.4.21** can be used to select the maximum blade width for a given radius to be cut. If the radius falls between two blade widths on the chart, select the narrower blade because the wider blade will not be able to turn sharply enough to cut the desired radius.

Blade Thickness or Gauge

The thickness of a saw blade is often referred to as the gauge, shown as item 2 in **Figure 3.4.13**. Blade gauge sizes range from 0.014" to 0.063". Common sizes include 0.020", 0.025", 0.032", 0.035", 0.042", and 0.050".

Correct gauge size depends mostly on the size of the band saw wheels. Thinner gauges have more flexibility and will normally be used on machines with smaller diameter wheels, while less flexible thicker gauges will normally be used with larger diameter wheels. Using thicker gauge blades on machines with smaller wheels puts greater strain on the blade and can cause it to weaken and break more easily. The best option is to consult the manual for the machine and use the gauge recommended by the manufacturer.

Rake

The **rake** is the angle of the saw tooth's cutting face (item 10 in **Figure 3.4.13**.). A zero- or straight-rake angle tooth will be exactly perpendicular to the path of motion. Zero-rake blades are well suited for sawing structural materials such as angles, I-beams, and tubing because the teeth will not cut as aggressively and will not straddle thinner-walled areas (which would cause teeth to break).

Positive-rake geometry is also commonly used and will have more of a shearing action on the metal due to its steep inclination. Positive-rake teeth are generally thinner and weaker. Positive-rake blades work well when sawing solids because there are no thin-walled areas where the more aggressive teeth can straddle material and be broken.

Gullet

The **gullet** is the curved area at the root of a saw tooth in which the metal chips are formed into curls. This curved shape provides strength for adjacent teeth and a void for chips to be held in as they pass through the work being cut. The gullet and gullet depth are identified as items 11 and 12 in **Figure 3.4.13**.

Tooth Patterns

Tooth forms, or patterns, refer to the shape and pattern of the saw teeth. Each pattern has advantages and disadvantages, and which one is best depends on the job being done. **Figure 3.4.22** shows the different types of tooth forms. All tooth types can be used in both vertical and horizontal band sawing applications.

The Standard (Regular) Tooth Form

The standard blade or **regular tooth form** has large radii in the gullet area of the saw teeth and a zero rake angle. This blade is well suited for general cutting of many types of steel and will give an accurate cut and good surface finish.

Skip Tooth Form

The **skip tooth form** also has zero rake. The main difference between the skip tooth blade and the standard tooth blade is that every other tooth on the skip tooth blade has been omitted. This opens up the area between the adjoining saw teeth, allowing some additional room for chip clearance. The skip tooth blade is mainly used to cut softer materials such as aluminum and brass where higher blade speeds are often used that also create a higher volume of chips.

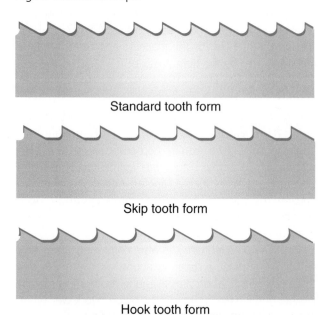

Standard tooth form

Skip tooth form

Hook tooth form

FIGURE 3.4.22 The standard, skip, and hook tooth forms (or patterns).

Hook Tooth Form

The **hook tooth form** has a positive rake angle (shaped like a hook) and large gullet area. The hook tooth blade has a built-in chip-breaker design that makes it ideal for cutting softer materials that commonly cause chips to stick in the saw blade teeth, such as aluminum and copper.

BAND SAW BLADE WELDING

Blades for band saws can be purchased prewelded to length. This is normally the case for wide blades used for horizontal band saws and those with carbide teeth, but blades for vertical band sawing often need to be cut from a coil of blade stock and welded to make a continuous band. After selecting the proper blade material type, set, pitch (or TPI), width, and tooth form, it is time to begin the blade welding process.

Band Length

Band length can be found in the saw manufacturer's manual, but it can also be calculated if that information is not available. Refer to **Figure 3.4.23** while following these steps:

- First, disconnect power, and lock out or tag out the machine before removing guards.
- Adjust the idler wheel to its midpoint position.

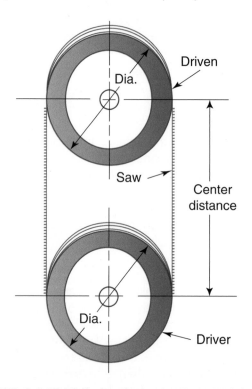

FIGURE 3.4.23 Blade length can be determined by first measuring the wheel diameter and distance between the wheels. Then those numbers are input into the blade length formula: $2 \times$ Center Distance $+ \pi \times$ Wheel Diameter.

- Measure the diameter of one of the saw's wheels. (The upper and lower wheels are the same size.)
- Measure the center distance between the wheels.
- Calculate the circumference of the wheel diameter.
- Multiply the center distance by 2 and add to the circumference. The answer is the required band length. (**Figure 3.4.23** shows a formula that can be used for this calculation.)
- Measure and cut the length from the blade stock coil using the machine's blade shear or snips.

Band Welding

Many vertical band saws have a built-in resistance-type butt welder for welding blades. Follow these general steps to weld the blade. Consult the manufacturer's documentation for specific welders and procedures.

- Grind the ends of the blade square using the built-in grinder. **(See Figure 3.4.24.)**
- Set the welding force according to the width of the blade. **(See Figure 3.4.25.)**
- Clean the welder's clamping jaws and clamp the blade. Make sure the ends are touching and that there are no gaps. **(See Figure 3.4.26.)**
- Press and HOLD the welding lever. Look away from the welder when performing the welding operation to prevent eye injury from welding flash. **(See Figure 3.4.27.)** During welding, the moveable jaw pushes the blade ends together while heating the blade for welding.
- Since the moveable jaw has pushed the blade ends together, unclamp the moveable jaw before releasing the weld lever. **(See Figure 3.4.28.)** If this is not done, the weld can separate if it has not cooled. The jaws can also be damaged as they are pulled across the teeth of the saw blade.
- Check for uniform flash on both sides of the weld. **(See Figure 3.4.29.)**
- Reclamp the blade and press the anneal button until the blade turns a dull red color. **(See Figure 3.4.30.)** This annealing is needed to soften the blade because the welding process makes it brittle. If a blade is not annealed, it will break when mounted on the machine's wheels.
- Flex the blade near the weld to check weld strength.
- Use the machine's grinder to remove the weld flash or the blade will not fit through the machine's blade guides. Do not grind the teeth, and avoid grinding the blade too thin. Check for thickness using the machine's thickness gage. **(See Figure 3.4.31.)**

FIGURE 3.4.24 (A) Flip the ends so the teeth are opposite before grinding. When flipped back for welding, the ends will align even if the ground ends are not exactly square. (B) Using the grinder to square the blade ends.

 **CAUTION**

Look away from the welder during welding.

FIGURE 3.4.25 Selecting the proper weld setting according to blade width.

FIGURE 3.4.26 Clamp the blade in the welder's jaws with the ends touching.

FIGURE 3.4.27 Press and hold the weld lever to weld the blade.

FIGURE 3.4.28 Unclamp the moveable jaw before releasing the weld lever to avoid breaking the weld and damaging the clamping jaws.

FIGURE 3.4.29 A properly welded blade should have equal flash on both sides.

FIGURE 3.4.30 Heat the blade to a dull red color to anneal it.

FIGURE 3.4.32 Loosen the blade by turning the tension crank.

FIGURE 3.4.31 Check the blade thickness at the weld using the thickness gage.

FIGURE 3.4.33 This gage shows when blade tension is correct. Notice the two scales. The inner scale is for carbon blades and the outer scale is for bimetal blades. Here the tension is set for a ½"-wide carbon steel blade.

BAND SAW BLADE MOUNTING/REMOVAL

After the band saw blade has been welded, follow these general steps for removing an installed blade and installing the new blade:

- Turn off the main power, unplug the saw, or lock out or tag out the machine.
- Loosen the upper wheel to release tension on the band. **(See Figure 3.4.32.)**
- Remove the band and store in a proper location. Use gloves to prevent cuts.
- Remove the blade guides.
- Install the new blade loosely on the wheels. Be sure that teeth are pointing downward toward the machine table and that the blade is just touching the backup bearing.
- Tighten the band to the proper tension. Some saws have a gage that shows proper tension. **(See Figure 3.4.33.)**
- Turn on the machine power and jog the machine (start and stop quickly a few times) to be sure that the blade is tracking properly on the wheels.
- Disconnect the power again.

(a)   (b)

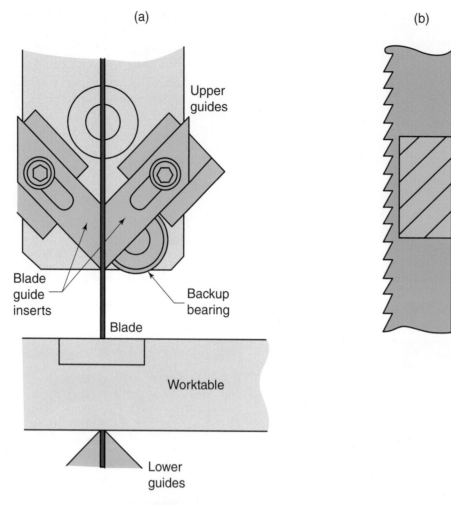

FIGURE 3.4.34 Proper guide installation.

- Install blade guides. Be sure to use the correct guides for the blade width. The teeth should stick out past the guides. If not, the set will be ruined and the blade will not cut. If teeth stick out too far past the guides, it will move from side to side during cutting. There should only be about 0.001" to 0.002" clearance between the guides and the blade. When installing the guides, be sure that the blade is not pushed to either side but remains straight. If the blade is moved, crooked cuts will result. **Figure 3.4.34** shows proper guide installation.

BLADE SPEED

Before beginning a sawing operation, it is necessary to know how fast the saw should be moving. Operating a saw at the wrong speed is not only unsafe, but it can result in inaccurate cuts and reduced blade life and possible machine damage as well. Band saws usually have a chart on the machine that contains important information regarding blade speed. This chart will contain information that will help determine the number of teeth per inch the blade should have for the material thickness and how fast the saw should run to cut that material. Materials such as aluminum will usually run at a higher speed than harder materials such as steel. Material hardness and blade material are factors in selecting band speed. Speeds for band sawing machines are given in surface feet per minute (SFPM) or just feet per minute (FPM). That is the distance that 1 tooth on the blade would travel in 1 minute. Set the speed according to the manufacturer's machine manual. Some saws may have only a few pulley settings, while others may have a variable speed control. **(See Figure 3.4.35.)**

THE ABRASIVE CUTOFF SAW

A metal saw blade is not always the best option for cutting off a piece of material, particularly when the material has been hardened. In these instances, an abrasive cutoff saw, or "chop saw" like the one shown in **Figure 3.4.36** may be used. The main part of a chop saw consists of a high-speed motor that has an abrasive cutting disc mounted to its shaft. This disc is very similar to a grinding wheel and is made of bonded abrasive particles. Each of the particles acts as a small cutting edge and can produce cuts very quickly. Abrasive saws create a great deal of dust and sparks, which can be harmful to other equipment in the machine shop. Chop saws also can heat up the workpiece and can change heat treatment of the same metals near the cut. Chop saws are occasionally used in the machining environment, but they see most of their use in the welding and fabrication fields.

⊘ CAUTION

Wear a dust mask for protection from nuisance dust when using an abrasive saw.

FIGURE 3.4.35 (A) The speed controls of a typical vertical band saw. The desired range is selected by shifting the lever. Here the transmission is in the medium speed range. (B) The speed control wheel is then turned until the desired speed is shown on the speed indicator. The speed on this saw is set at about 450 feet per minute because the transmission is in the medium speed range.

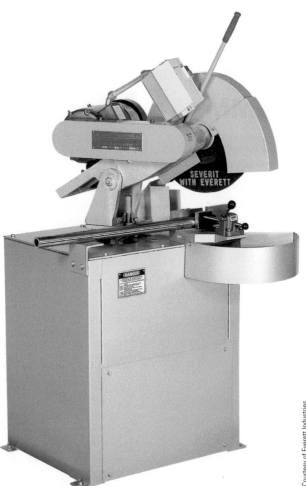

Courtesy of Everett Industries

FIGURE 3.4.36 An abrasive "chop saw" can cut through hardened steel.

METAL CUTTING CIRCULAR (COLD) SAWS

Cold saws are also circular-type saws and at first glance appear to be heavy-duty versions of abrasive chop saws. Cold saws remedy many of the drawbacks of abrasive chop saws, since they use a metallic toothed blade instead of a high-friction abrasive-type blade. Because cold saws have a much lower blade speed and produce curled metal chips instead of fine dust, they cut cooler, do not affect the heat treatment of the work, and are more accurate. Cold saws even offer some advantages over band saws for certain applications, such as producing smooth surface finishes, very little blade "walking" or flexing, and the ability to cut thin sections with less tooth damage. Many cold saws also achieve cut-length accuracies and squareness around 0.005". Cold saws may use a high-speed steel (HSS) blade or one with carbide teeth. **Figure 3.4.37** shows a cold saw.

Courtesy of Dake Corporation

FIGURE 3.4.37 A cold saw can cut parts to within .005" of desired sizes without generating heat like the chop saw.

SUMMARY

- Various types of sawing machines can be used to perform simple operations such as cutting off stock before other machining operations begin.

- Saws can be used to rough cut excess material from a workpiece, which can drastically reduce the time needed to produce a finished product and wear on expensive tooling that would otherwise be used in cutting away this excess stock, both of which offer significant economic advantages.

- Proper blade selection is one of the most important aspects to consider when preparing to perform sawing operations.

- Welding blades and installing them on a band saw are important skills that need to be performed when preparing for sawing operations.

- Selecting and setting correct sawing speeds allows for efficient and safe sawing operations.

- The abrasive cutoff saw works well for cutting hardened material.

- The cold saw can make accurate cuts at low speeds and generates little heat.

REVIEW QUESTIONS

1. Sawing machines can be divided into roughly four categories. They are _____, _____, _____, and the _____.

2. The vertical band saw is particularly useful, as it can be used to make straight cuts as well as _____.

3. The horizontal band saw is ideal for cutting _____ _____.

4. Cutting action on the power hacksaw is very similar to the _____ _____.

5. List three safety precautions to observe when operating a horizontal band saw.

6. Briefly describe the process to prepare for internal contour sawing on the vertical band saw.

7. List three safety precautions to observe when operating a vertical band saw.

8. What type of band saw blade has HSS teeth welded to a carbon steel body?

9. How many saw teeth should be engaged in the workpiece at any given time?

10. Name the three different types of tooth patterns.

11. What are the three types of tooth set and why is tooth set necessary?

12. The slot created in a workpiece by the saw blade is known as the _____.

13. Explain how to use a push stick.

14. Saws should be _____ _____ when a part is mounted or the machine is serviced.

15. Saw guides should be mounted _____ above the part on a vertical saw.

16. Why must a band saw blade be annealed after welding?

17. Why does a band saw blade need to be ground after welding?

18. Clearance between the vertical band saw guides and the blade should be _____.

19. Band saw cutting speeds are given in _____.

UNIT 5 Offhand Grinding

Learning Objectives

After completing this unit, the student should be able to:

- Explain the purpose for offhand grinding
- Select the correct grinding wheel for the operation to be performed
- Identify different types of offhand grinding machines
- Describe the process of mounting a grinding wheel
- Explain how to set up a pedestal grinder for safe operation
- Demonstrate safe offhand grinding procedures

Key Terms

Aluminum oxide	Loading	Tool rest
Blotter	Ring test	Wheel dresser
Glazing	Silicon carbide	Wheel flange
Grit	Spark breaker	

INTRODUCTION

Grinding machines cut metal by passing an abrasive cutting surface over the workpiece under pressure. Grinding machines usually use abrasives in the shape of a wheel, but disc and belt machines are also sometimes used. Abrasives can cut hardened steels that cannot be cut by other machining methods.

Parts ground by abrasive machines are held by hand, and for that reason, these operations are often referred to as "offhand" grinding. Because the material removal is controlled almost exclusively by hand, very keen hand-eye coordination and manual dexterity are necessary. Because hands and fingers are in such close proximity of the rotating grinding wheel, the operator must also be extremely safety conscious when operating this machine tool.

GRINDER USES

The pedestal grinder and bench grinder **(Figure 3.5.1)** are used for a wide variety of operations in a machine shop. The only difference between the two is that the bench grinder is somewhat smaller and is mounted on a workbench. The pedestal grinder is larger and has its own base, which is normally bolted to the floor. Sharpening high-speed steel lathe tool bits and drills are common pedestal and bench grinder tasks. Punches and chisels are also resharpened, and mushrooming can be ground off the striking end of these tools to keep them safe. Worn straight screwdriver tips can be reground to fit screw slots correctly. Heavy deburring can also be performed. Pedestal grinders usually have a water cup mounted in the front for cooling parts as they heat up during grinding. Only ferrous metals should be ground on a pedestal grinder. Grinding non-ferrous metals on a pedestal grinder will clog the grinder wheels, causing them to cut very inefficiently.

Another type of grinder is the tool grinder. **(See Figure 3.5.2.)** It has adjustable tool rests that can be tilted to various angles. Tool grinders are generally used to sharpen lathe cutting tools. Only ferrous metals or carbide tools should be ground on a tool grinder, depending on the type of wheel installed. Some tool grinders have a coolant system that keeps the tool and wheel cool during grinding.

FIGURE 3.5.1 (A) A pedestal grinder. (B) A bench grinder.

FIGURE 3.5.2 A tool grinder has adjustable tool rests and is commonly used to grind lathe cutting tools. It can be bench or pedestal mounted.

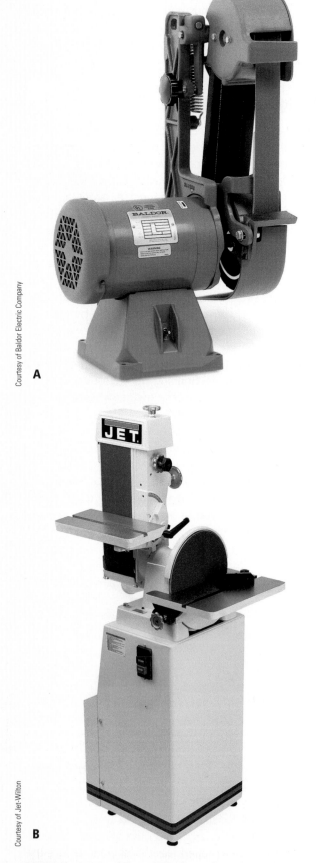

Courtesy of Baldor Electric Company

A

Courtesy of Jet-Wilton

B

FIGURE 3.5.3 (A) A light-duty abrasive belt machine. (B) A heavier-duty belt and disc machine.

ABRASIVE BELT AND DISC MACHINE USES

Abrasive belts and discs normally remove material at slower rates than pedestal grinders, so they are often used for lighter deburring tasks. Both ferrous and non-ferrous metals can be ground with abrasive discs and belts. **Figure 3.5.3** shows some belt and disc machines.

GRINDING WHEELS

Many varieties of grinding wheels are available for various applications. The characteristics are usually labeled on the side of the wheel on the blotter near the mounting hole. The **blotter** is a round disc of heavy paper surrounding the center area of a grinding wheel. It provides a cushion when the wheel is mounted and also provides information about the wheel. Grinding wheel selection for precision grinding can involve many characteristics, but for offhand grinding, the wheel selection is usually based on four main variables: abrasive type, **grit** (coarseness), wheel size (diameter, hole size, and width), and maximum wheel speed. **Figure 3.5.4** shows a wheel blotter and identifies those four important pieces of information, which are explained in the following paragraphs.

ABRASIVE TYPE

Grinding wheels are constructed with different abrasive types and are selected based on the types of materials they will be used for. The two most common pedestal grinder abrasive types are **aluminum oxide** and **silicon carbide**. Aluminum oxide is quite possibly the most common and is used for general-purpose grinding of ferrous metals. Aluminum oxide wheels are gray in color and can be used on pedestal, bench, and tool grinders.

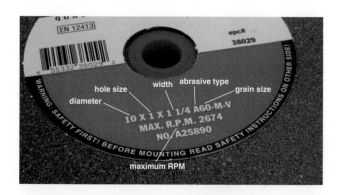

FIGURE 3.5.4 A grinding wheel blotter shows wheel information such as abrasive type, grain size, wheel size, and maximum wheel RPM.

FIGURE 3.5.5 An aluminum oxide wheel.

Figure **3.5.5** shows an aluminum oxide wheel. Silicon carbide wheels are used for grinding the extremely hard carbides used in cutting tools as well as non-ferrous metals. Silicon carbide wheels are green in color and are usually used on tool grinders. **Figure 3.5.6** shows a silicon carbide wheel. Tool grinders can also be fitted with diamond-impregnated wheels for fine grinding of carbide cutting tools. Carbide is the only material that should be ground on a diamond-impregnated wheel. **Figure 3.5.7** shows a diamond-impregnated wheel. Refer to **Figure 3.5.4** for the abrasive type identified on the wheel blotter.

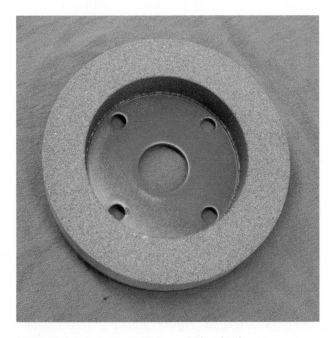

FIGURE 3.5.6 A green silicon carbide wheel.

FIGURE 3.5.7 A diamond-impregnated wheel.

WHEEL GRIT (ABRASIVE GRAIN SIZE)

Pedestal grinders are usually capable of holding two wheels, one on either end of the machine. Often, a coarser grit wheel is mounted on one side for rough grinding (heavy material removal, resulting in a poor surface finish) and a finer grit wheel on the other side for finish work (light material removal, resulting in a finer surface finish and more accuracy). Wheel manufacturers determine grit size by screening the abrasive grains through a screen of a particular standard-sized mesh. Common wheel grits are 24, 36, 60, 80, and 120. Smaller grit numbers refer to coarser grit wheels, and larger grit numbers refer to finer grit wheels. Select these wheels based on the task to be performed. Coarser wheels on bench and pedestal grinders are commonly 36 grit wheels. Finer wheels on these machines are commonly 60 or 80 grit wheels. Refer to **Figure 3.5.4** for the grit size specified on the blotter.

WHEEL SIZE

One of the factors to consider when selecting a grinding wheel is its size. There are three considerations when determining wheel size: outside wheel diameter, wheel width, and the mounting hole diameter. The grinder manufacturer's manual will list appropriate wheel sizes for a given machine. The recommended wheel size must match dimensions labeled on the side of the wheel. Refer to **Figure 3.5.4** for size specifications on the blotter. If the mounting hole is larger than the spindle, it may be permissible to use a properly sized bushing to reduce the opening to the correct size. **Figure 3.5.8** shows a wheel and a reducing bushing.

FIGURE 3.5.8 A grinding wheel manufacturer may allow use of a reducing bushing to mount a wheel on a slightly smaller machine arbor.

MAXIMUM WHEEL SPEED

After the appropriate wheel dimensions have been determined, the next consideration is to make sure that the wheel is rated to run at the speed of the grinder. Grinding wheels rotating at high speed are under an immense amount of centrifugal force, and an underrated wheel could fracture and fly apart. Again, either the machine manual or the machine's information plate should be consulted to determine the machine's operating RPM. The machine's speed should never exceed the maximum speed printed on the blotter. Refer to **Figure 3.5.4** for the maximum speed listed on the blotter.

 **CAUTION**

Make sure that the maximum safe speed for the wheel is not less than the speed of the grinder. The maximum speed should be printed on the blotter.

Grinding Wheel Storage

It is important to remember that grinding wheels are made of brittle materials and are relatively fragile. The wheels selected should be stored with padding between them to prevent damage. In addition, ANSI B7.1-2000 Section 2.3 requires that grinding wheels are never stored where they could be subject to:

• Exposure to water or other solvents

• Any temperature or humidity condition that causes water or condensation on the wheels

• Freezing temperatures

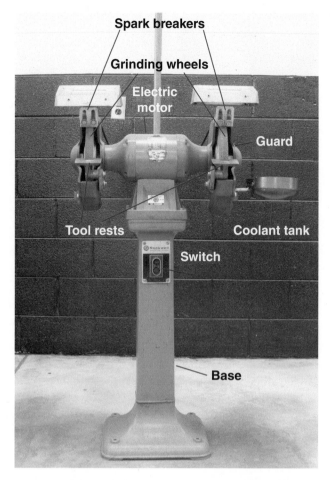

FIGURE 3.5.9 The major parts of a typical pedestal grinder.

PEDESTAL GRINDER SETUP

Once a suitable wheel, or wheels, has been chosen, it, or they, must be inspected and then mounted to the grinder. Then some adjustments must be made before beginning grinding. Become familiar with the major parts of the pedestal grinder shown in **Figure 3.5.9**, because they will be discussed during the setup described next.

Grinding Wheel Ring Testing

To be sure that a grinding wheel did not sustain dangerous cracks during removal, storage, or transport, always perform a **ring test** prior to mounting a wheel. A ring test is performed by suspending the wheel through its hole on a finger or other object and then tapping it lightly with something dense but non-metallic such as a screwdriver handle. A clear ringing sound should be heard (instead of a dull thud). A wheel that passes this test is called "sound." Failure to produce a ringing sound means the wheel is probably cracked and unsafe. **Figure 3.5.10** shows a wheel being ring tested.

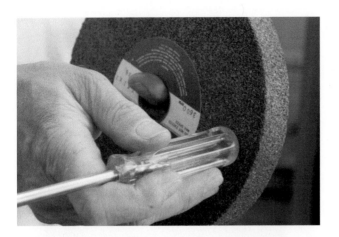

FIGURE 3.5.10 Suspend a grinding wheel and lightly tap with a non-metallic object to perform a ring test. A clear ring indicates that the wheel is "sound." A dull sound means a wheel is damaged and should not be mounted on a grinder.

 CAUTION

Never mount a wheel that has been dropped. Damaged and cracked wheels should be destroyed and discarded.

Grinding Wheel Mounting

The grinding wheel must have a blotter on both sides. Although seemingly unimportant, blotters are a critical safety item and are necessary since they serve as cushions to conform to the minor surface irregularities on the side of the wheel when the wheel is tightened. This reduces the chance of cracking the wheel when it is tightened.

Most pedestal grinders use two thick, precision, washer-like discs called **wheel flanges (Figure 3.5.11)** to distribute the clamping force during tightening. Flanges should be the correct size, totally seated against

Wheel flange

FIGURE 3.5.11 The wheel flanges of a pedestal grinder distribute the clamping pressure evenly on the wheel. Always use a blotter on each side of the wheel between the flange and the wheel.

FIGURE 3.5.12 A piece of wood may be needed to keep wheels from spinning when loosening or tightening the nuts that secure the wheel flanges.

the blotter material, and used on each side of the wheel. When removing and installing the wheels, it is important to know that most left-side machine spindles have left-handed threads, and right-side machine spindles have right-hand threads so that mounting nuts do not come loose during use. Since a wrench must be used to tighten and remove the nut from the machine spindle, it may be necessary to apply friction with a piece of wood to keep the old wheel from rotating while removing it and for tightening the new wheel. **(See Figure 3.5.12.)** Be careful when tightening the new wheel to apply only enough torque to securely seat the wheel. Consult the machine manual for a torque specification. Be sure to replace and adjust all guards, tool rests, and spark breakers.

 CAUTION

Always disconnect power from the machine before removing or installing grinding wheels.

 CAUTION

When starting a grinder, stand to the side in case the wheel comes apart under the sudden stress of startup, and let the wheel run for one full minute before using to be sure the wheel is safe and does not break.

Tool Rest and Adjustment

Since pedestal grinders are offhand machines, their designers have included a small platform at the face of the wheel to rest the workpiece against during grinding. This device is called the **tool rest**. Tool rests help to support the work during grinding and may be set at any

FIGURE 3.5.13 Keep the clearance between the tool rest and the wheel within 1/16". Always wait until the wheel has completely stopped before making adjustments.

desired angle to aid in grinding uniform angled surfaces on the work. After installing the wheel or wheels, adjust the tool rest within 1/16" of the wheel face, as shown in **Figure 3.5.13**.

 CAUTION

The space between the tool rest and the wheel should never exceed 1/16" to prevent parts from being pulled in between the rest and the wheel. As the wheel wears, the space will increase and the rest will need to be adjusted. Always stop the grinder before adjusting the tool rest.

Spark Breaker and Adjustment

Because of the friction created during grinding, the very small metal chips have a very high heat concentration. They are usually orange in color due to their temperature and are therefore referred to as "sparks." Grinding sparks will be extremely hot in the first second or so after their creation and then fizzle out to the natural color of the metal. As the sparks are created, they tend to wrap around the periphery of the rotating wheel and are often ejected down from the top of the wheel into the area where the operator is holding the workpiece. A device called a **spark breaker** (sometimes called a *spark arrestor*) is used to catch and control most of the sparks and to keep them from contacting the operator's hand.

The spark breaker should be adjusted within 1/16" of the wheel just like the tool rest. Having the spark breaker properly adjusted can prevent sparks from hitting the user's hands, and also help contain the wheel in the machine in the event it breaks. A properly adjusted spark breaker is shown in **Figure 3.5.14**.

FIGURE 3.5.14 The spark breaker should also be adjusted within 1/16" of the grinding wheel. Always wait until the wheel has completely stopped before making adjustments.

Grinding Wheel Dressing

A new wheel should be dressed because its periphery does not always run true with its mounting hole. This can cause uneven grinding and machine vibration. Hand dressing will not completely true the wheel, but it will improve the concentricity of the periphery of the wheel and provide a more even grinding surface, as well as prevent machine vibration. **Wheel dressers**, sometimes called star dressers because of the shape of the dressing wheels, are used to dress aluminum oxide wheels.

To dress the wheel, hold the wheel dresser securely on the tool rest and apply enough pressure so the dresser removes material from the surface of the wheel. Then move the dresser back and forth across the wheel until contact is made around the entire periphery of the wheel. Keep about half of the dresser in contact with the wheel at each edge. Moving the dresser completely off either side of the wheel can round the edge of the wheel instead of keeping the surface straight. **Figure 3.5.15** shows how to position a wheel dresser when dressing a wheel.

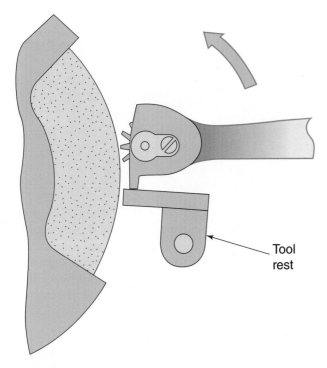

Tool
rest

FIGURE 3.5.15 To dress a grinding wheel, place the wheel dresser on the tool rest and lift the handle upward to make contact with the wheel. Then slide the dresser from side to side on the tool rest to dress the periphery of the wheel. Always wear a face shield when dressing grinding wheels.

 CAUTION

Always wear a face shield when dressing the wheel. After dressing, the tool rest and spark breaker distance should be rechecked to be sure they are both within 1/16" of the wheel face. If they are not, stop the grinder and adjust them.

Wheel Wear

Grinding wheels are designed to wear as they are used. As the wheel wears, dull grains should be released from the wheel, exposing new, sharp grains. Sometimes the wheel face will wear unevenly and develop grooves when grinding is done in one spot on the wheel. Dressing the wheel removes the grooves and makes a flatter surface for grinding.

Wheel Loading

When soft metals are ground on the pedestal grinder, small pieces of material can become lodged in the wheel instead of the wheel releasing dull grains. The wheel will then cut inefficiently. This condition is called **loading**. **Figure 3.5.16** shows a loaded wheel. Dressing the wheel will remove the loading and make it cut more efficiently.

FIGURE 3.5.16 A loaded wheel results from grinding soft metals. Note the pieces of material embedded in the wheel.

FIGURE 3.5.17 A glazed wheel results from grinding very hard materials. Note the shiny appearance of the wheel compared to a sharp wheel.

Continual use of a loaded wheel requires extreme pressure and creates excessive heat that can lead to thermal discoloration of the workpiece and discomfort as the operator holds the hot workpiece.

Wheel Glazing

When very hard metals are ground on the pedestal grinder, the grains of the wheel become dull more quickly than normal, so again, new, sharp grains are not exposed. This also will cause it to cut inefficiently. This condition is called **glazing**. **Figure 3.5.17** shows a glazed wheel. Dressing the wheel will remove the glazing and make it cut more efficiently. Continual use of a glazed wheel (just like a loaded wheel) requires extreme pressure and creates excessive heat that can lead to burns.

GRINDING PROCEDURES

Before using a pedestal grinder, first check the tool rest and spark breaker to be sure they are within 1/16" of the wheel and that they are securely tightened.

Upon machine startup, always stand to one side of the grinder and wait about a minute for it to come up to speed and to ensure that the wheel is safe. Stand to the side of the grinder when grinding. Use the tool rest to help support the workpiece, and use moderate pressure when grinding. Too much pressure will cause heat to build up quickly. When parts start to heat up, dip them in the water cup for cooling. If grinding tools such as lathe cutting tools, drills, and chisels, avoid heating them to the point where they discolor. This can affect their hardness and reduce their durability as cutting tools.

Grind only on the face of the wheel, not on the sides. The wheels are not designed to withstand side pressure. Move across the face of the wheel to avoid uneven wear and creating grooves in the wheel face. Dress the wheel when necessary if it loads, glazes, or wears unevenly. **Figure 3.5.18** shows proper grinder use.

Always wait until the wheel comes to a complete stop before making any adjustments to the grinder.

FIGURE 3.5.18 The proper method for performing grinding on a pedestal grinder. Stand to the side of the wheel, keep fingers and hands away from the wheel, and use the tool rest to help support the work.

Never hold tools so that the tip can be pulled between the wheel and tool rest. Screwdrivers, chisels, and punches should be held up rather than down to avoid accidents. Do not wear gloves when using a pedestal grinder. Never use rags near a grinder.

SUMMARY

- Offhand abrasive machines such as grinders and disc and belt machines can cut hardened steel, which makes them useful for sharpening some types of cutting and striking tools.
- Belt and disc machines can be used for both ferrous and non-ferrous metals, but grinders should be used only for ferrous metals.
- The tool grinder is a type of pedestal grinder that is useful for grinding lathe tools.
- Grinding wheels are made with different abrasive types for different applications, and come in many different sizes and grits.
- Grinding wheels should never be mounted on a machine that exceeds their recommended RPM rating, and they should be stored properly to prevent damage.
- Always ring test a grinding wheel before installing on a grinding machine, and destroy and discard any wheels that are not sound.
- Observe safety precautions such as standing to one side of a grinder when starting and allowing a new wheel to run for about a minute before beginning grinding.
- Keep the grinder's tool rest and spark breaker adjusted within 1/16" of the wheel.
- Never wear gloves when grinding or use rags near a running grinding wheel.
- Dress worn, loaded, and glazed wheels to keep them cutting efficiently.
- When grinding, use moderate pressure and grind only on the face of the wheel. Move across the entire face of the wheel to avoid wearing grooves in the wheel.
- When parts begin to heat up, cool them by dipping them in water.

REVIEW QUESTIONS

1. What is the main benefit of offhand abrasive machining or grinding?

2. What type of metals should not be ground on a pedestal grinder?

3. Which wheel is finer, a 60 grit or a 36 grit?

4. If a grinder runs at 3400 RPM and a wheel is rated to run at 3000 RPM, is the wheel safe to use on that grinder?

5. How is a ring test performed?

6. Why is it necessary to have blotters on both sides of a wheel?

7. The maximum distance that a spark breaker and tool rest should be from a wheel is _____.

8. When should a grinding wheel be dressed?

9. Where is the best place to stand when using a grinder?

UNIT 6

Drilling, Threading, Tapping, and Reaming

Learning Objectives

After completing this unit, the student should be able to:

- Demonstrate understanding of benchwork drilling operations
- Demonstrate understanding of countersinking, spotfacing, and counterboring
- Identify various reamer types and explain their uses
- Demonstrate understanding of standardized thread systems and their designations
- Identify various tap types and explain their uses
- Demonstrate understanding of tap drill selection
- Identify various thread-cutting die types and explain their uses
- Calculate the number of turns of a tap and die to produce a given depth and length of thread
- Demonstrate understanding of tap removal techniques

Key Terms

Adjustable reamers	**Counterboring**	**Fine pitch**
Bottoming chamfer tap	**Countersinking**	**Flutes**
Center drill	**Crest**	**General-purpose**
Chamfer	**Die stock**	**reamers**
Class of fit	**Drilling**	**Hand reamers**
Coarse pitch	**Electric drill**	**Hand tap**
Combination drill and	**Expansion reamers**	**Internal thread**
countersink	**External thread**	**ISO Metric**

Jobber's reamers
Lead
Major diameter
Minor diameter
National Pipe Thread (NPT)
Nominal
Percentage of thread
Pilot
Pitch
Pitch diameter
Plug chamfer tap

Reamer
Reaming
Root
Spiral-flute tap
Spiral-point (gun) tap
Spot drill
Spotfacing
straight tap wrench
Tap
Tap drill
Tap extractor

Taper chamfer tap
Taper pin reamers
Thread
Thread-cutting die
Thread depth
Thread forming tap
Thread series
Threads Per Inch (TPI)
Twist drill
T-handle tap wrench
Unified Thread Standard

INTRODUCTION

Producing and modifying holes may be performed with portable hand tools at a workbench. These operations create and modify holes to accept fasteners, pins, or shafts.

BENCHWORK HOLEMAKING OPERATIONS

Drilling is the process of using a cylindrical rotating cutting tool that is sharpened on its end to create a hole. Drilling is one of the most efficient ways of producing a hole in a workpiece. Drill bits may be mounted in and driven by a handheld electric drill motor, often simply called an **electric drill**.

Twist Drilling

General-purpose hand drilling is usually done with a **twist drill** bit. Twist drill bits have spiral grooves in their sides called **flutes** that give them the appearance of having been twisted. **(See Figure 3.6.1.)** These spiral flutes evacuate chips from the hole while the drill is rotating. The detailed nomenclature of twist drills will be covered in Section 5, Unit 2.

Twist drill bits are sized according to diameter by four methods: letter, wire gage (number), fractional, and metric:

- Letter size drill bits range from "A" (0.2340") to "Z" (0.4130").
- Wire gage or number drill bits range from #1 (0.2280") to #80 (0.0135").
- Fractional drill bits range from 1/64" to 2.5" in 1/64" increments.
- Metric drill bits range from 0.050 mm to 32 mm.

It should be noted that even though a drill bit may be a particular size, twist drill bits always cut a diameter slightly larger than their own diameter. The amount a hole will be over its specific drill size depends on many variables, including drill tip condition, workpiece material, and cutting methods. Generally, a properly sharpened drill bit under 1/2" in diameter will produce a hole that is no

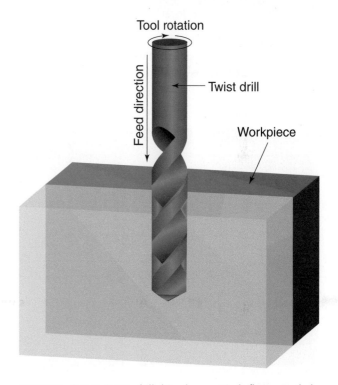

FIGURE 3.6.1 Twist drill bits have spiral flutes to help evacuate chips during drilling.

more than 0.004" larger than its stated size. This must be taken into consideration when selecting drill bit sizes.

Drilling Procedure

Before beginning to drill a hole, it is recommended that the location be center punched to prevent the drill from "walking" off location. The indentation from the center punch serves to hold the rotating drill bit in that position until the entire point penetrates the workpiece. If a more precise hole position is desired, the hole may be started (after center punching) with a shorter drill bit with a specially designed point. Two main types of these starter drill bits are available. One is a **spot drill** bit and the other is a **combination drill and countersink**. **(See Figure 3.6.2.)** The combination drill and countersink is sometimes referred to as a **center drill** because some styles can be used to create a hole designed to mount a workpiece

FIGURE 3.6.2 A combination drill and countersink (upper) and a spot drill (lower). Either of these tools may be used to spot-drill hole locations to prevent the twist drill from walking off location.

FIGURE 3.6.3 A spot-drilled hole location ready for twist drilling.

between center points on a lathe. (Refer to Section 5.2 for more details.) Once the hole has been spot drilled or "spotted," the hole may be completed with the twist drill bit. **(See Figure 3.6.3.)** Rotating twist drill bits do not pull themselves through the material and therefore must be pushed with moderate hand pressure. It is always important to hold the drill at the appropriate angle to the surface being drilled for the duration of the operation. When chips become long, relieve pressure to allow them to break and fall off the drill so they do not wrap around the tool and cause injury.

When drilling a hole completely through a workpiece, you should be aware that the drill bit may have a tendency to "grab" as it breaks through the other side. When a rotating drill bit grabs, it will attempt to either spin the workpiece or the drill motor itself. The thinner and softer the material, the more the drill bit may tend

FIGURE 3.6.4 A hole being drilled in a properly clamped workpiece.

 CAUTION

Never hold a workpiece with your hand. Always clamp the workpiece to a workbench with appropriate clamps or in a vise. **(See Figure 3.6.4.)** As you drill and the drill nears the breakthrough point, lighten up on the pressure to help reduce "grabbing."

to grab as it breaks through. To reduce this grabbing, reduce pressure when breaking through the workpiece.

Counterboring, Countersinking, and Spotfacing

In addition to plain straight holes being drilled, holes may also need to be modified. **Counterboring** increases the diameter of a hole to a certain depth in order to allow a screw head or nut to be positioned flush with or below the workpiece surface. **(See Figure 3.6.5.)** Counterbore bits have a guide on their end called a **pilot** that keeps the counterbore bit aligned in the existing hole. The pilot diameter should be about 0.003" to 0.005" smaller than the existing hole diameter. Less clearance may cause the pilot to bind or seize in the hole. More clearance will allow the counterbore bit to walk off location. The pilot can be an integral part of the counterbore, or it may be interchangeable so different diameter pilots can be used with a single counterbore. Some counterbore bits are shown in **Figure 3.6.6**.

Spotfacing is the process of machining a flat spot on a rough surface surrounding a hole opening so that bolts, nuts, and washers will be properly seated when tightened. Counterbore bits are usually used to perform this operation. **Figure 3.6.7** shows a counterbored hole and a spotfaced hole.

Countersinking is the process of cutting a tapered opening in a hole so that a flathead screw can be installed flush with the workpiece surface. Countersinking is

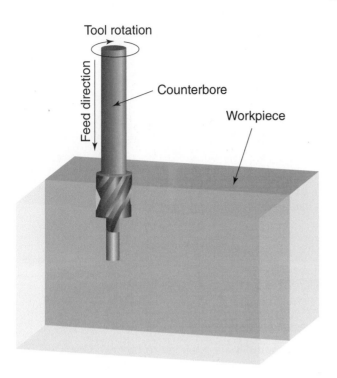

FIGURE 3.6.5 A counterbore being used to machine a hole for accommodating a bolt head or nut.

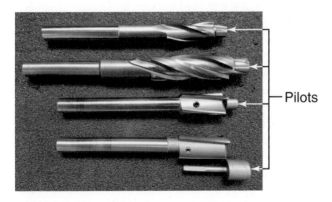

FIGURE 3.6.6 Examples of counterbore bits. The top two have integral pilots. The bottom two have interchangeable pilots that are secured with a small screw.

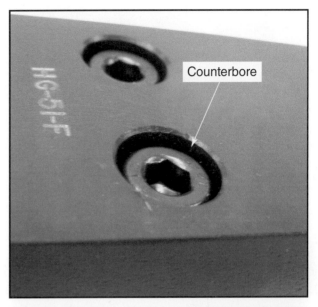

FIGURE 3.6.7 A workpiece with a counterbored hole and a spotfaced hole. Counterboring fully recesses a screw head while spotfacing simply provides a flat surface. Both features were created with the same cutting tool.

also frequently performed to create a **chamfer** at the opening of a hole to allow easier entry of pins, aid in starting threads and for deburring. Countersinks are commonly found with included angles of 60, 82, 90, and 100 degrees. **Figure 3.6.8** shows several different types of countersinks. Countersinking with an 82-degree countersink bit will accommodate a standard flathead screw. **(See Figure 3.6.9.)**

Reaming

Reaming is the process of finishing a hole to a precise size with a smooth finished surface. The cutting tools used for reaming are called **reamers**. A hole slightly smaller

FIGURE 3.6.8 Various sizes of countersinks.

FIGURE 3.6.9 Cross section of a countersunk hole accommodating the 82-degree head of a flathead screw.

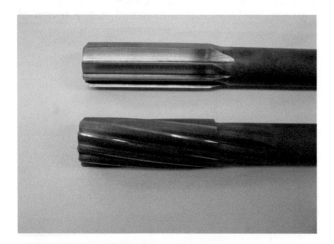

FIGURE 3.6.10 Examples of chucking reamers, with straight flutes and spiral flutes.

than the desired size must first be drilled. Reamers may be driven with a drill motor, but since they remove only small amounts of material, they may also be driven by hand power. Reamers driven with power tools are called machine reamers or chucking reamers, and reamers driven by hand are called hand reamers.

Chucking Reamers

There are several different types of machine reamers. Chucking reamers used for benchwork have straight shanks that may be mounted in an electric drill. General-purpose chucking reamers have straight flutes. Spiral flute reamers are available for use when reaming holes with interruptions such as keyways or grooves. **Jobber's reamers** have longer flutes and are useful when reaming deep holes. The cutting edges of these reamers are at the end, so cutting only takes place at the end of the tool, not on the periphery. **Figure 3.6.10** shows some examples of chucking reamers.

FIGURE 3.6.11 A hand reamer in use.

Hand Reamers

Hand reamers are intended to be driven by hand and are fitted with a square driving surface for mounting a wrench. Special two-handled wrenches designed for driving taps are typically used to turn hand reamers. The cutting edges of hand reamers are on the periphery instead of at the end like chucking and jobber's reamers. **Figure 3.6.11** shows a hand reamer in use. Hand reamers are available with either straight flutes or spiral flutes for holes with interruptions.

General-purpose reamers are used for finishing holes to a particular size. They are tapered slightly for about one-third to one-half of the flute length to make starting the reamer easier.

Expansion reamers have slots cut into the body that allow the reamer to expand to the exact size needed when the adjusting screw in the end of the reamer is tightened.

Adjustable reamers are similar to expansion reamers, but they have a different construction, which allows a much greater range of adjustability. The bodies are threaded and have tapered slots to hold cutting blades. Two adjusting nuts hold the blades in place. To adjust the reamer's diameter, one nut is loosened and the other tightened to slide the blades in the tapered slots. Moving the blades toward the shank increases the diameter, and moving the blades toward the tip decreases the diameter.

Taper pin reamers are used to create precise tapers within straight holes that allow them to receive tapered pins. **Figure 3.6.12** shows some examples of hand reamers.

Reamer Use

In order to create a reamed hole, a hole is first drilled, leaving a small amount of material to be removed by the reamer. With power tool reaming with machine reamers,

FIGURE 3.6.12 Examples of hand reamers, including general-purpose spiral-flute, expansion, adjustable, and taper pin. Note the square end of the shank for driving.

the drilled holes should be produced undersize by the following amounts. (Drill accordingly, remembering that twist drills commonly produce slightly oversized holes.)

- 0.010" for a hole up to 1/4"
- 0.015" for a hole between 1/4" and 1/2"
- 0.025" for a hole between 1/2" and 1-1/2"

Be sure to apply cutting fluid of a type appropriate for the material being reamed. The reamer should be driven at about one-half the RPM and advanced at about twice the rate that was used when the hole was drilled.

When hand reaming, it is usually desirable for the reamer to remove as little material as possible, since the reamer is powered by hand. Therefore it is customary to leave about 0.001" to 0.008" of material in a hole. Use a tap wrench to turn the reamer in a clockwise direction, and check to make sure the reamer starts squarely. A combination or solid square can be used to check perpendicularity. A block with perpendicular surfaces can also be used as a guide. Advance the reamer smoothly through the hole. Continue to turn the reamer clockwise when removing it from the hole. Turning reamers backwards will only dull their cutting edges. **Figure 3.6.13** shows the procedure for hand reaming.

Reamers are precision finishing tools and must be handled with care. Never allow reamers to contact each other during storage, since their edges are hard and keen and tiny dings can cause significant damage to the reamer. Most reamers come shipped in cardboard or plastic tubes that are ideal for storage.

THREADING AND TAPPING

A **thread** is a spiral groove made on a round external or internal diameter. An **external thread** can be cut on the outside diameter of a workpiece with a cutting tool called a **thread-cutting die**. An **internal thread** can be produced in the inside diameter of a workpiece with a tool called a **tap**. **(See Figure 3.6.14.)**

Threads allow mating parts to be screwed together, are used to transmit power, and are used in the mechanisms of measuring instruments. Most tools and equipment in use today are assembled with many threaded fasteners. Dealing with threads occurs almost daily in the machining industry, so it is essential to have an understanding about them and how they are created.

Basic Thread Terminology

It is important that a part produced in one region of the world will be able to interchange with another part made in a different region of the world. To ensure this interchangeability, threads have been standardized by form, size, tolerance, allowance, classification, and class of fit. For inch series threads, the **Unified Thread Standard** has established the form (shape) and dimensional standards for the consistent interchange of threaded parts. Metric threads are standardized also, by the **ISO Metric** screw thread system, sometimes simply called M-series threads. Both of these thread systems are considered 60-degree V-threads because of the shape and included angle of their form.

Unified National and M-series threads have several elements, as shown in **Figure 3.6.15**. The definitions are as follows:

- The **major diameter** is the largest diameter of a thread.
- The **minor diameter** is the smallest diameter of a thread. This diameter is measured from the base of a thread on one side to the base of the thread on the opposite side.
- The **crest**
  - On an external thread, the crest is the peak of one thread and creates the points where the major diameter is measured.
  - On an internal thread, the crest is the peak of one thread and creates the points where the minor diameter is measured.
- The **root**
  - On an external thread, the root is the valley between threads and creates the points where the minor diameter is measured.
  - On an internal thread, the root is the valley between the threads and creates the points where the major diameter is measured.

Select the desired reamer and mount in a tap wrench.

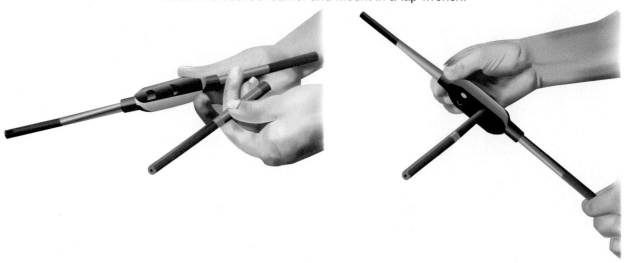

Turn the reamer clockwise in the hole to start and check for perpendicularity with a square. Apply light downward pressure while rotating clockwise to ream the hole. Check for perpendicularity frequently.

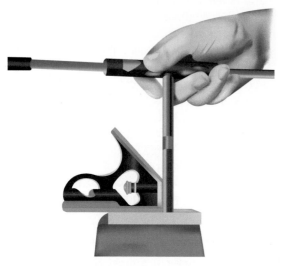

When the reamer passes through the workpiece and begins to turn easily, the hole is complete. Continue to turn clockwise when removing the reamer from the hole.

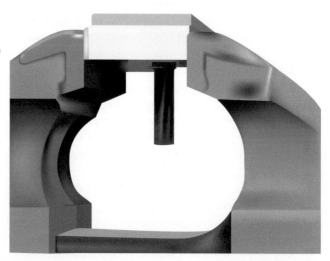

FIGURE 3.6.13 Steps for performing a hand reaming operation.

FIGURE 3.6.14 (A) An external thread being produced with a thread-cutting die. (B) An internal thread being produced with a tap.

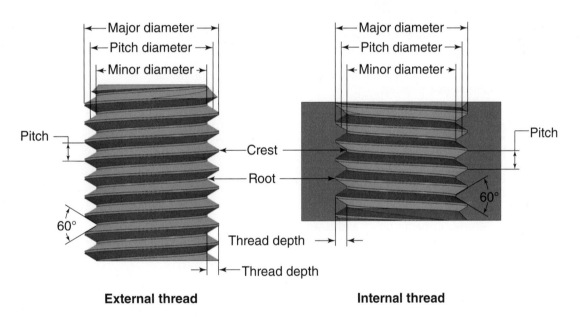

FIGURE 3.6.15 The elements of a 60-degree V-form thread.

- The **threads per inch (TPI)** is the number of threads contained within 1 inch of the workpiece length. (INCH SERIES THREADS ONLY)
- **Pitch** is the distance from a point on one thread to the same point on the adjacent thread. For inch series threads, the pitch may be calculated by dividing 1 by the TPI (1/TPI = pitch).
- The **lead** is the distance a thread will advance in one revolution.
- **Pitch diameter** is the imaginary diameter measured where the thickness of the thread and space within the groove are equal. The pitch diameter of a thread is responsible for the fit between two mating threads.
- **Thread depth** is the depth of a single thread measured from the crest to the root.

Threads may be produced to fit more tightly or more loosely with their mating parts depending on their intended use. This relationship is called **class of fit**. Inch thread fit classes range from 1 to 3, with 3 achieving the closest fit. It is the pitch diameter of mating threads that determines the fit.

Within a thread system, various diameters and pitch combinations are defined as standard. These combinations are called **thread series**. For each nominal major diameter, the thread series defines one **coarse pitch** and one **fine pitch** combination (some extra-fine combinations exist as well). Coarse pitch threads have larger spacing between threads (fewer TPI), while fine pitch threads have smaller spacing (more TPI). **(See Figure 3.6.16.)**

FIGURE 3.6.16 Comparison of a coarse-pitch thread and a fine-pitch thread of the same nominal diameter.

Thread Designations

Inch Series Thread Designations

The designation system used to show thread size, type, and fit is also standardized. For inch series threads, the first number shown is the **nominal** major diameter of the thread. Nominal means "in name only" and is an approximation of a targeted size. Most external major diameters actually measure less (while internal major diameters measure more) than their nominal size, to allow for clearance when assembling. For example, if the major diameter of a thread is between 0.495" and 0.498", its nominal major diameter is 0.500", or 1/2". The second designated number is the number of threads per inch. Next are the initials "UN," that designate the Unified National form. Also shown are the thread series initials of C (coarse) or F (fine), the class of fit from 1 (loosest) to 3 (closest) and whether the thread is external (A) or internal (B). **(See Figure 3.6.17.)** For example, **Figure 3.6.18** shows a typical ½-20 bolt and its thread designation. The major diameter of this bolt is nominally 1/2" and it has 20 threads per inch. It is a Unified National Fine series and has a class 2 fit, external thread.

NOTE:

Unified National threads use a numbering system to designate the major diameter on threads smaller than 1/4" major diameter. These number threads range from #0 to #12. Those sizes and their nominal major diameters are listed in **Figure 3.6.19**.

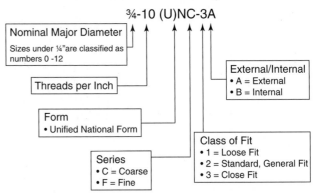

Unified Screw Thread Designation

FIGURE 3.6.17 Thread designation diagram for Unified National inch series threads.

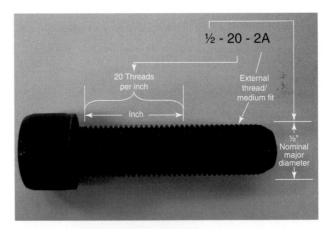

FIGURE 3.6.18 A ½-20 bolt with thread dimensions and its corresponding thread designation.

Thread Size	Nominal Major Diameter
0	0.060
1	0.073
2	0.086
3	0.099
4	0.112
5	0.125
6	0.138
8	0.164
10	0.190
12	0.216

FIGURE 3.6.19 Nominal major diameters of number designated thread sizes

Metric Series Thread Designations

Metric thread designations first begin with the letter "M" to indicate that they are metric. The number following the "M" identifies the nominal major diameter and the second number is the pitch of the thread in millimeters (mm). **(See Figure 3.6.20.)** For example, **Figure 3.6.21** shows a typical M8x1.5 and its thread designation. The major diameter of this bolt is nominally 8 mm and it has a pitch of 1.5 mm.

Tap Drills

Prior to tapping a hole, a hole must be produced with a drill bit that is smaller than the major diameter of tap. This operation will provide enough material so that the tap will produce the proper thread form and size.

If a drill bit is too small for a given thread, the tap will be very difficult to turn and could cause the tap to break. If the drill bit is too large for a given thread, the tap may still produce a thread, but the thread depth may be too shallow and lack strength. A thread with 100% thread depth offers very little strength advantage over 75% thread depth. Further, the force required to drive a tap to produce a 100% thread depth is much, much greater than the force needed to produce a 75% thread depth. This full thread depth compared to a partial thread depth is called **percentage of thread**. **(See Figure 3.6.22.)**

The recommended drill bit size for a hole prior to tapping is called the **tap drill** size. Charts like the ones shown in **Figure 3.6.23** list tap drill sizes for common thread sizes. Most tap drill charts provide drill sizes to produce about 75% thread depth. Many threads can be created with

ISO Metric Thread Designation

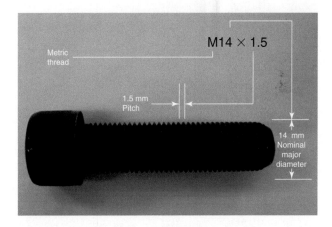

FIGURE 3.6.20 Thread designation diagram for M-series threads.

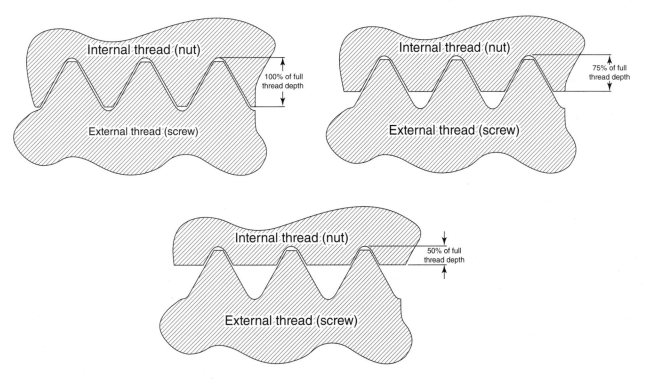

FIGURE 3.6.21 An M14x1.5 bolt with thread dimensions and its corresponding thread designation.

FIGURE 3.6.22 Diagram depicting percentage of thread height.

Starrett®

Precision, Quality and Innovation...
Since 1880

INCH/METRIC TAP DRILL SIZES & DECIMAL EQUIVALENTS

DRILL SIZE	DECIMAL EQUIVALENT	TAP SIZE
80	.0135	
79	.0145	
1/64	.0156	
78	.0160	
77	.0180	
76	.0200	
75	.0210	
74	.0225	
73	.0240	
72	.0250	
71	.0260	
70	.0280	
69	.0292	
68	.0310	
1/32	.0312	
67	.0320	
66	.0330	
65	.0350	
64	.0360	
63	.0370	
62	.0380	
61	.0390	
60	.0400	
59	.0410	
58	.0420	
57	.0430	
56	.0465	
3/64	.0469	0 - 80
55	.0520	
54	.0550	
53	.0595	1 - 64, 72
1/16	.0625	
52	.0635	
51	.0670	
50	.0700	2 - 56, 64
49	.0730	
48	.0760	
5/64	.0781	
47	.0785	3 - 48
46	.0810	
45	.0820	3 - 56
44	.0860	
43	.0890	4 - 40
42	.0935	4 - 48
3/32	.0938	
41	.0960	
40	.0980	
39	.0995	
38	.1015	5 - 40
37	.1040	5 - 44
36	.1065	6 - 32
7/64	.1094	
35	.1100	
34	.1110	
33	.1130	6 - 40
32	.1160	
31	.1200	
1/8	.1250	
30	.1285	
29	.1360	8 - 32, 36
28	.1405	
9/64	.1406	
27	.1440	
26	.1470	
25	.1495	10 - 24
24	.1520	
23	.1540	
5/32	.1562	
22	.1570	
21	.1590	10 - 32
20	.1610	
19	.1660	
18	.1695	
11/64	.1719	
17	.1730	
16	.1770	12 - 24
15	.1800	
14	.1820	12 - 28
13	.1850	
3/16	.1875	
12	.1890	
11	.1910	

DRILL SIZE	DECIMAL EQUIVALENT	TAP SIZE
10	.1935	
9	.1960	
8	.1990	
7	.2010	1/4 - 20
13/64	.2031	
6	.2040	
5	.2055	
4	.2090	
3	.2130	1/4 - 28
7/32	.2188	
2	.2210	
1	.2280	
A	.2340	
15/64	.2344	
B	.2380	
C	.2420	
D	.2460	
1/4 E	.2500	
F	.2570	5/16 - 18
G	.2610	
17/64	.2656	
H	.2660	
I	.2720	5/16 - 24
J	.2770	
K	.2810	
9/32	.2812	
L	.2900	
M	.2950	
19/64	.2969	
N	.3020	
5/16	.3125	3/8 - 16
O	.3160	
P	.3230	
21/64	.3281	
Q	.3320	3/8 - 24
R	.3390	
11/32	.3438	
S	.3480	
T	.3580	
23/64	.3594	
U	.3680	7/16 - 14
3/8	.3750	
V	.3770	
W	.3860	
25/64	.3906	7/16 - 20
X	.3970	
Y	.4040	
13/32	.4062	
Z	.4130	
27/64	.4219	1/2 - 13
7/16	.4375	
29/64	.4531	1/2 - 20
15/32	.4688	
31/64	.4844	9/16 - 12
1/2	.5000	
33/64	.5156	9/16 - 18
17/32	.5312	5/8 - 11
9/32? 35/64	.5469	
9/16	.5625	
37/64	.5781	5/8 - 18
19/32	.5938	
39/64	.6094	
5/8	.6250	
41/64	.6406	
21/32	.6562	3/4 - 10
43/64	.6719	
11/16	.6875	3/4 - 16
45/64	.7031	
23/32	.7188	
47/64	.7344	
3/4	.7500	
49/64	.7656	7/8 - 9
25/32	.7812	
51/64	.7969	
13/16	.8125	7/8 - 14
53/64	.8281	
27/32	.8438	
55/64	.8594	
7/8	.8750	1 - 8
57/64	.8906	
29/32	.9062	

DRILL SIZE	DECIMAL EQUIVALENT	TAP SIZE
59/64	.9219	1 - 12
15/16	.9375	1 - 14
61/64	.9531	
31/32	.9688	
63/64	.9844	1 1/8 - 7
1	1.0000	
1 3/64	1.0469	1 1/8 - 12
1 7/64	1.1094	1 1/4 - 7
1 1/8	1.1250	
1 11/64	1.1719	1 1/4 - 12
1 7/32	1.2188	1 3/8 - 6
1 1/4	1.2500	
1 19/64	1.2969	1 3/8 - 12
1 11/32	1.3438	1 1/2 - 6
1 3/8	1.3750	
1 27/64	1.4219	1 1/2 - 12
1 1/2	1.5000	

METRIC TAP DRILL SIZES

METRIC TAP	TAP DRILL (mm)	DECIMAL (Inch)
M1.6 x 0.35	1.25	.0492
M1.8 x 0.35	1.45	.0571
M2 x 0.4	1.60	.0630
M2.2 x 0.45	1.75	.0689
M2.5 x 0.45	2.05	.0807
M3 x 0.5	2.50	.0984
M3.5 x 0.6	2.90	.1142
M4 x 0.7	3.30	.1299
M4.5 x 0.75	3.70	.1457
M5 x 0.8	4.20	.1654
M6 x 1	5.00	.1968
M7 x 1	6.00	.2362
M8 x 1.25	6.70	.2638
M8 x 1	7.00	.2756
M10 x 1.5	8.50	.3346
M10 x 1.25	8.70	.3425
M12 x 1.75	10.20	.4016
M12 x 1.25	10.80	.4252
M14 x 2	12.00	.4724
M14 x 1.5	12.50	.4921
M16 x 2	14.00	.5512
M16 x 1.5	14.50	.5709
M18 x 2.5	15.50	.6102
M18 x 1.5	16.50	.6496
M20 x 2.5	17.50	.6890
M20 x 1.5	18.50	.7283
M22 x 2.5	19.50	.7677
M22 x 1.5	20.50	.8071
M24 x 3	21.00	.8268
M24 x 2	22.00	.8661
M27 x 3	24.00	.9449
M27 x 2	25.00	.9843
M30 x 3.5	26.50	1.0433
M30 x 2	28.00	1.1024
M33 x 3.5	29.50	1.1614
M33 x 2	31.00	1.2205
M36 x 4	32.00	1.2598
M36 x 3	33.00	1.2992
M39 x 4	35.00	1.3780
M39 x 3	36.00	1.4173

PIPE THREAD SIZES (NPSC)

THREAD	DRILL	THREAD	DRILL
1/8 - 27	11/32	1 1/2 - 11 1/2	1 3/4
1/4 - 18	7/16	2 - 11 1/2	2 7/32
3/8 - 18	37/64	2 1/2 - 8	2 21/32
1/2 - 14	23/32	3 - 8	3 1/4
3/4 - 14	59/64	3 1/2 - 8	3 3/4
1 - 11 1/2	15/32	4 - 8	4 1/4
1 1/4 - 11 1/2	1 1/2		

Bulletin 1214-5M/S 05/10   starrett.com

FIGURE 3.6.23A This chart lists the correct tap drill to the left of the tap size. For example, a 5/16" drill is used for a 3/8-16 tap.

Tap/Drill Sizes

Tap Size	Cut Taps Drill Size	Dec. Equiv.	Form Taps Drill Size	Dec. Equiv.
0–80	3/64	.0469	54	.0550
M1.6 × 0.35	1.25	.0492	1.45	.0571
M1.8 × 0.35	1.45	.0571	1.65	.0650
1–64	53	.0595	51	.0670
1–72	53	.0595	51	.0670
M2 × 0.40	1.60	.0630	1.80	.0709
2–56	50	.0700	5/64	.0781
2–64	50	.0700	47	.0785
M2.2 × 0.45	1.75	.0689	2.00	.0787
M2.5 × 0.45	2.05	.0807	2.30	.0906
3–48	47	.0785	43	.0890
3–56	46	.0810	2.30	.0905
4–40	43	.0890	38	.1015
4–48	42	.0935	2.60	.1024
M3 × 0.50	2.50	.0984	7/64	.1094
5–40	38	.1015	33	.1130
5–44	37	.1040	2.90	.1142
M3.5 × 0.60	2.90	.1142	3.20	.1260
6–32	36	.1065	1/8	.1250
6–40	33	.1130	3.25	.1280
M4 × 0.70	3.30	.1299	3.70	.1476
8–32	29	.1360	25	.1495
8–36	29	.1360	24	.1520
M4.5 × 0.75	3.70	.1476	4.10	.1614
10–24	26	.1470	11/64	.1719
10–32	21	.1590	16	.1770
M5 × 0.80	4.20	.1654	14	.1820
12–24	16	.1770	8	.1990
12–28	15	.1800	7	.2010
M6 × 1.00	5.00	.1969	7/32	.2188
1/4–20	7	.2010	1	.2280
1/4–28	3	.2130	15/64	.2340
M7 × 1.00	6.00	.2362	F	.2570
5/16–18	F	.2570	L	.2900
5/16–24	I	.2720	M	.2950
M8 × 1.25	6.70	.2638	7.40	.2913
M8 × 1.00	7.00	.2756	19/64	.2969
3/8–16	5/16	.3125	S	.3480
3/8–24	Q	.3320	T	.3580
M10 × 1.50	8.50	.3346	U	.3680
M10 × 1.25	8.70	.3425	9.40	.3701
7/16–14	U	.3680	Y	.4040
7/16–20	25/64	.3906	Z	.4130

Tap Size	Cut Taps Drill Size	Dec. Equiv.	Form Taps Drill Size	Dec. Equiv.
M12 × 1.75	10.20	.4016	11.20	.4409
M12 × 1.25	10.80	.4252	11.50	.4528
1/2–13	27/64	.4219	15/32	.4682
1/2–20	29/64	.4531	12.25	.4823
M14 × 2.00	12.00	.4224	33/64	.5156
9/16–12	31/64	.4844	17/32	.5312
9/16–18	33/64	.5156	13.50	.5315
5/8–11	17/32	.5312	14.75	.5807
5/8–18	37/64	.5781	15.25	.6004
M16 × 2.00	14.00	.5512	19/32	.5938
M16 × 1.50	14.50	.5209	15.25	.6004
M18 × 2.50	15.50	.6102	39/64	.6094
M18 × 1.50	16.50	.6496	17.25	.6791
3/4–10	21/32	.6562	45/64	.7031
3/4–16	11/16	.6875	23/32	.7188
M20 × 2.50	17.50	.6890		
M20 × 1.50	18.50	.7283		
M22 × 2.50	19.50	.7677		
M22 × 1.50	20.50	.8071		
7/8–9	49/64	.7656		
7/8–14	13/16	.8125		
M24 × 3.00	21.00	.8268		
M24 × 2.00	22.00	.8661		
1–8	7/8	.8750		
1–12	59/64	.9219		
M27 × 3.00	24.00	.9449		
M27 × 2.00	25.00	.9843		
1–1/8–7	63/64	.9844		
1–1/8–12	1–3/64	1.0469		
M30 × 3.50	26.50	1.0433		
M30 × 2.00	28.00	1.1024		
1–1/4–7	1–7/64	1.1094		
1–1/4–12	1–11/64	1.1719		
M33 × 3.50	29.50	1.1614		
M33 × 2.00	31.00	1.2205		
1–3/8–6	1–7/32	1.2188		
1–3/8–12	1–19/64	1.2969		
M36 × 4.00	32.00	1.2598		
M36 × 3.00	33.00	1.2992		
1–1/2–6	1–11/32	1.3438		
1–1/2–12	1–27/64	1.4219		
M39 × 4.00	35.00	1.3780		
M39 × 3.00	36.00	1.4173		

FIGURE 3.6.23B This tap drill chart shows drill sizes for inch and metric taps. Note that there are two columns of drill sizes. One is for cutting taps and one is for forming taps.

Pipe thread sizes (NPT)			
Thread	Drill	Thread	Drill
$1/8 - 27$	$11/32$	$1 1/2 - 11 1/2$	$1 3/4$
$1/4 - 18$	$7/16$	$2 - 11 1/2$	$2 7/32$
$3/8 - 18$	$37/64$	$2 1/2 - 8$	$2 21/32$
$1/2 - 14$	$23/32$	$3 - 8$	$3 1/4$
$3/4 - 14$	$59/64$	$3 1/2 - 8$	$3 3/4$
$1 - 11 1/2$	$15/32$	$4 - 8$	$4 1/4$
$1 1/4 - 11 1/2$	$1 1/2$		

FIGURE 3.6.24 Most tap drill charts will also list pipe tap drill sizes.

60% thread depth with very little loss of thread strength. Note that on chart 3.6.23B, two drill sizes are given for each thread size. One size is for cutting taps and the other for forming taps. The differences between cutting taps and forming taps will be described later in this unit.

A common error is to use a drill bit that is the size of the major diameter of the tap. If the hole is drilled to the major diameter, the tap will simply fall through the hole and not produce any threads.

Tapered Pipe Threads

NPT is the abbreviation for **National Pipe Thread**. These threads are tapered 3/4 inch per foot to help seal them as they tighten together. NPT threads are sized according to the nominal inside pipe diameter rather than the major thread diameter. Most tap drill charts also list pipe thread tap drill sizes. **(See Figure 3.6.24.)**

Tap Styles

Taps are available in many styles and chamfer types for various needs. These two characteristics create dramatically different results when using a tap.

Tap Style

The tap style has a profound impact on the way the chips are formed when the tap is in use. The most common are the following:

- **Hand Tap**—Hand taps cut threads and have straight flutes from the tip toward the shank. These taps curl the chips tightly and break them into small pieces. The chips are mostly stored in the flutes while the tap is in use. **(See Figure 3.6.25.)**

- **Spiral-Point (Gun) Tap**—These taps also cut threads and have straight flutes with a special angle ground on their ends. The angle on the end serves to create stringy chips that are projected forward as the tap is in use. Since the chips are pushed forward, these taps are normally used on through holes where the chips can escape out the bottom of the hole. **(See Figure 3.6.26.)**

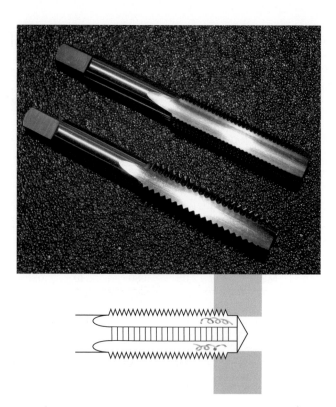

FIGURE 3.6.25 With a standard hand tap, chips are formed into small pieces that accumulate in the flute space.

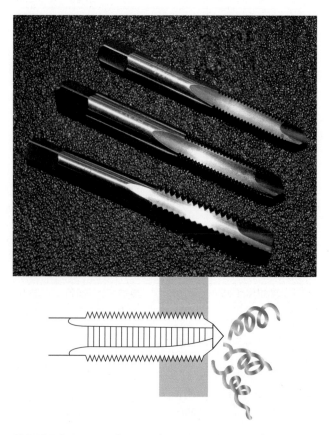

FIGURE 3.6.26 With a spiral-point tap, chips are formed into stringy pieces that are projected forward so that they may exit the bottom of a through hole.

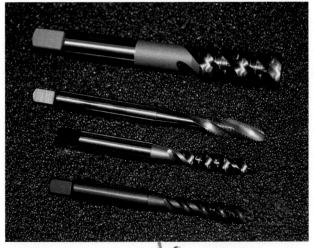

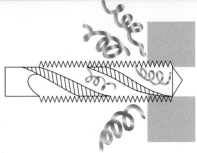

FIGURE 3.6.27 With a spiral-flute tap, chips are formed into stringy pieces that are guided back through the flutes of the tap and ejected out the opening of the hole in the same manner as a twist drill.

- **Spiral-Flute Tap**—These taps also cut threads and have a spiral flute much like the flutes on a twist drill. Like a twist drill, these spiral grooves propel the chips backward out of the hole while the tap is in use. **(See Figure 3.6.27.)**
- **Thread Forming Tap**—These taps do not cut, but form threads by displacing material into the shape of the threads. They are often used with ductile materials such as aluminum. No chips are produced by a thread forming tap. **(See Figure 3.6.28.)**

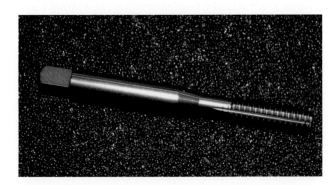

FIGURE 3.6.28 A form tap can be used to tap ductile metals. They do not cut and produce chips, but form the thread by displacing material into the shape of the threads of the tap.

Tap Chamfer Type

When taps are made, short chamfers are ground onto their ends. This results in each leading thread being progressively larger than the thread before it. This treatment makes the tap easier to start in the drilled hole. There are three main chamfer types available and they can be purchased as a set. They are as follows:

- **Taper Chamfer Tap**—These taps are sometimes called starter taps, since they may be used to start a tapped hole that will be finished with another tap. The threads on these taps have 7 to 10 threads chamfered to make the tap easier to start into the hole. Taper taps are usually used for tapping through holes.

NOTE:

It is a common mistake to think that these taps produce tapered threads like those used for pipe fittings. However, these taps are only chamfered on the end of the tap to make starting easier. These taps DO produce straight threads. **(See Figure 3.6.29.)**

- **Plug Chamfer Taps**—These taps are the most general purpose of the three and they are either used by themselves or after a taper tap when it is necessary to thread deeper into a hole. Plug taps have three to five threads chamfered. These taps may be used in through holes or in blind holes (when a hole doesn't go all the way through a workpiece) when there is adequate clearance in the bottom of the hole. **(See Figure 3.6.30.)**
- **Bottoming Chamfer Taps**—These taps are used when a hole is blind, since they can cut full threads nearly all the way to the bottom of a hole. These taps only chamfer the first one or two threads. **(See Figure 3.6.31.)**

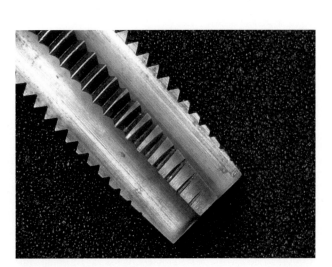

FIGURE 3.6.29 Taper chamfer taps have a chamfer of 7 to 10 threads to make them easy to start.

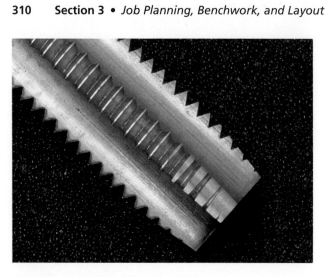

FIGURE 3.6.30 Plug chamfer taps have a chamfer of three to five threads for relatively easy starting and for producing full threads fairly close to the bottom of blind holes.

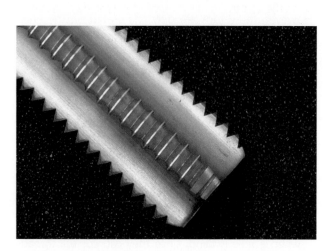

FIGURE 3.6.31 Bottoming chamfer taps have a chamfer on one to two threads for finishing previously tapped holes and for producing full threads very close to the bottom of blind holes.

Tap Use

Before tapping a hole, be sure the hole was drilled with the correct tap drill size and that it is clear of chips and debris. It is a good idea to countersink the hole to be tapped to the major diameter of the thread before starting the tapping operation. This has several benefits, such as helping to prevent a burr at the top of the tapped hole and aiding in starting the tap (and the fastener that will be threaded into the hole).

Once the hole has been prepared, the workpiece should be securely mounted in a vise with the hole positioned vertically. Taps may be driven with either a **T-handle tap wrench** or a **straight tap wrench.** T-handle wrenches are typically used for smaller tap sizes and afford better feel and sensitivity for how much torque is being applied to help avoid breaking the tap.

FIGURE 3.6.32 A T-handle tap wrench and a straight tap wrench provide leverage for turning taps by hand.

Straight tap handles usually provide more leverage, which helps you to apply the higher torque needed for larger tap sizes. These tap wrenches are shown in **Figure 3.6.32**.

Check to be sure the cutting edges of the tap are sharp. Worn, dull taps may require excessive torque to cut, resulting in tap breakage or threads fitting too tightly. Apply cutting fluid to the tap and start turning it into the hole while assisting with downward pressure. Once the tap starts cutting, it will pull itself into the work. Reverse slightly every half to full turn to avoid chips clogging the tap flutes and causing breakage. Look at the tap from several different angles as you use it to ensure the hole is being tapped straight. A combination or solid square can be used to check perpendicularity. A block with perpendicular surfaces can also be used as a guide. If the tap isn't straight when you check, reverse it slightly and gently straighten as the tap is advanced. **(See Figure 3.6.33.)**

Sometimes chips will bind within the flutes and cause breakage when the tap is backed off. If resistance is felt when you back a tap off, turn it in the forward direction slightly and try backing off again. Smaller tapped holes have more of a tendency to clog, since their flutes are smaller and have less area to contain chips. When you feel the tap contact the end of a blind hole, back it out and remove all debris still in the hole.

After the hole has been tapped and cleaned, check the fit and depth with a mating part or a gage. If the tapped hole is too tight, it is likely that the tap is dull and worn. Using a new tap usually solves this problem. Loose-fitting holes can result if the hole is drilled too large or if the tap isn't kept straight during tapping.

It is common for a hole to be threaded to a specified depth, especially threaded blind holes. In these cases, it is convenient to determine how many turns of the tap are needed to reach the required depth. When using inch-based taps, the number of turns needed can

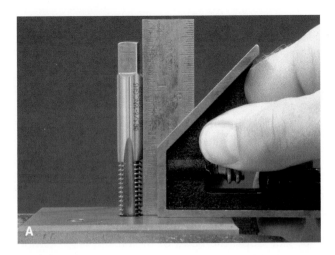

FIGURE 3.6.33 A square or block can be used to keep a tap straight during hand tapping. Check at two places, each 90 degrees apart.

be found by multiplying depth by TPI. Here are a few examples:

- Tap 3/8-16 to a depth of 0.5".
  - Depth is 0.5".
  - TPI is 16.
  - $0.5 \times 16 = 8$.
  - 8 turns are needed.
- Tap 5-40 to a depth of 0.28".
  - Depth is 0.28".
  - TPI is 40.
  - $0.28 \times 40 = 11.2$
  - 11.2, or about 11-1/4 turns are needed.

When using metric taps, the number of turns needed can be found by dividing depth by pitch. Here are a few examples:

- Tap M5 $\times$ 0.8 to a depth of 7 mm.
  - Depth is 7 mm.
  - Pitch is 0.8.
  - $7 \div .8 = 8.75$ (8-3/4).
  - 8-3/4 turns are needed.
- Tap M12 $\times$ 1.75 to a depth of 15 mm.
  - Depth is 15 mm.
  - Pitch is 1.75.
  - $15 \div 1.75 = 8.5174$.
  - About 8.5 (8-1/2) turns are needed.

Depth of a thread in a hole refers to the usable amount of thread. Remember that a tap has a chamfer on the end, so those few tapered threads at the chamfer will not produce full usable threads. These formulas are estimates because the number of incomplete threads on a tap varies according to the type of tap chamfer. For that reason, threads will most often not be deep enough. To check the depth of the thread after tapping, clean the chips/debris from the hole, then follow these steps.

- Measure the entire length of a screw that is the size of the threaded hole, or the entire length of a threaded plug gage.
- Thread the screw, or gage, into the tapped hole until it stops.
- Use a caliper or depth micrometer to measure the exposed length of the screw, or gage **(Figure 3.6.34)**.
- Subtract that measurement from the entire length of the screw, or gage.

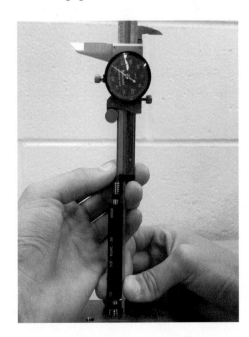

FIGURE 3.6.34 Measuring depth of thread in a tapped hole using a plug gage and the depth rod of a caliper. Depth is found by subtracting the exposed length of the gage from the entire length of the gage.

- The result is the depth of the threaded portion of the hole.
- If the threads are not deep enough, subtract the actual depth from the required depth.
- For inch-based threads, multiply that difference by the TPI to determine how many more turns of the tap are needed.
- For metric threads, divide that difference by the pitch to determine how many more turns of the tap are needed.
- Recheck to verify correct depth.

Broken Tap Removal

Since taps are so hard and brittle, occasionally a tap may break inside a hole. Tap removal is tricky, but there are few techniques that may be attempted. If the tap is in pieces in the hole, it may be able to be removed by using a hammer and punch to break it into smaller pieces. Then a scriber or tweezers may work to turn and remove the pieces.

 CAUTION

When using this method, cover the tap with a rag when striking with a punch. This will help contain hard, sharp fragments of the tap that can be thrown and cause injury.

If the tap is somewhat intact, a **tap extractor** may be used to remove a broken tap. Tap extractors have fingers that slide down the flutes of the broken tap, allowing it to be reversed out of the hole with a tap wrench. **(See Figure 3.6.35.)** If none of the other methods is successful, an electrical discharge machine (EDM) must be used to erode the broken tap from the hole.

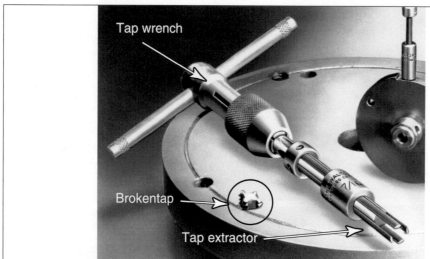

	**Step 1:** Thoroughly remove all chips of the broken tap. Insert the extractor fingers into the flutes of the broken tap, pushing them gently, but firmly, into position.
	**Step 2:** Push the holder down until it touches the broken tap. Slide the sleeve down until it touches the work.
	**Step 3:** Apply a tap wrench to the square end of the holder. Twist forward and backward a few times to loosen, then back out the broken tap.

FIGURE 3.6.35 A tap extractor engaging with the flutes of a broken tap.

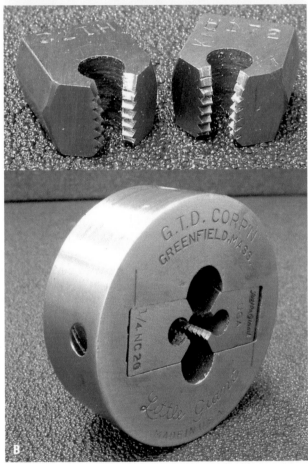

FIGURE 3.6.36 (A) A split round die. (B) The cutters of a two-piece die and the cutters mounted in a collar. (C) A solid hexagon die. (D) A solid round die.

Die Use

When it is necessary to cut external threads on a round workpiece, a threading die can be used. Threading dies can be made of high carbon steel or high speed steel (HSS). High carbon steel dies are mostly used for repairing existing damaged threads. They may wear quickly if used to produce new threads, especially when used on harder materials. High speed steel dies are harder and more durable, and can be used to create new threads. Some dies are slightly adjustable to adjust the fit of the thread,

while others are solid and cannot be adjusted. The split round die has a screw that can be used to make slight adjustments. The two-piece die has two cutters that are mounted in a collar. When using a two-piece die, it is important that the chamfered side of both cutters faces the same direction. Screws on each side of the collar allow for more adjustment than with the split round die. Solid dies are available in round and hexagon shapes. **Figure 3.6.36** shows examples of these types of thread-cutting dies. The workpiece diameter should never be

FIGURE 3.6.37 Creating a chamfer on the end of a workpiece helps a thread-cutting die to start more easily.

FIGURE 3.6.39 Cutting external threads with a hand threading die. Always start with the chamfered side of the die.

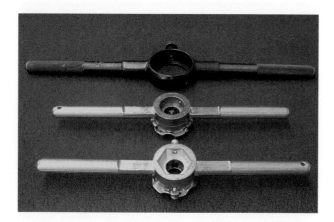

FIGURE 3.6.38 Die stocks used to hold thread-cutting dies.

larger than the nominal major diameter of the thread to be cut, and dies cut best when the workpiece diameter is actually a few thousandths smaller. Creating a chamfer on the end of the workpiece with a file, grinder, or belt/disc sander as shown in **Figure 3.6.37** will assist in starting the die.

One side of the thread-cutting teeth will have a chamfer for starting the cut. Some dies are labeled "start this side" to help the user identify this feature. **Die stocks** are used to hold threading dies when cutting threads. **(See Figure 3.6.38.)**

If using an adjustable die, first adjust the die to fit over a finished part or screw so it is slightly loose. Then mount the die in a die stock and place the chamfered side of the cutting teeth over the workpiece and apply downward pressure as the die and stock are rotated. Dies easily walk when advanced down a workpiece, causing deeper threads to be cut on one side than the other. Pay special attention to this and frequently check that the die is square on the workpiece and that the thread depth is uniform. Always use a generous supply of cutting fluid and back off the die frequently as you would a tap. **(See Figure 3.6.39.)** The last few threads will be incomplete because of the chamfer on the die. Flip the die so that the unchamfered side leads, and run it over the work to recut the last few incomplete threads to full depth. Check the thread with a mating part or a go/no-go ring gage. If the go gage does not fit, adjust the die slightly and repeat the process until the go gage threads onto the work freely.

If a thread length is given, the required number of turns of a threading die can be determined the same way as when tapping. For inch-based threads, multiply length by TPI. For metric threads, divide length by pitch.

SUMMARY

- Accurately locating and producing drilled holes is essential to performing benchwork operations.
- Drilled holes may be modified by counterboring, spotfacing, or countersinking to properly accommodate the head of a fastener.
- Holes that must have especially smooth, straight, and precise diameters may be drilled smaller than their intended size and finished with a reamer.
- Threading systems have been standardized so that all threads made to those specifications are sure to interchange.

- The main threading systems used in benchwork operations are Unified National (inch) and ISO Metric V-thread systems.
- Threads may be created internally with a tap or externally with a die.
- There are three main tap types in a set based on their chamfer type: taper, plug, and bottoming.
- Threading dies are used to repair existing damaged external threads or produce new external threads. They can be solid or adjustable and are made of either high carbon steel or high speed steel.

REVIEW QUESTIONS

1. Define *drilling*.
2. What factors might determine when a hole must be finished by reaming?
3. What are two reasons why a hole may need to be countersunk?
4. Explain the purpose of a counterbore.
5. What is the purpose of the pilot on a counterbore?
6. What should be done as a drill nears the "breakthrough" point during drilling?
7. Define *thread* as it relates to benchwork.
8. Explain the major diameter of a thread.
9. What is the TPI of a ½-20 thread?
10. Name two types of tap wrenches.
11. A 3/8-16 threaded hole needs to be tapped ¼ " deeper. How many turns of the tap are needed to gain that ¼ " of depth?

SECTION 4 DRILL PRESS

Introduction to the Drill Press — UNIT 1

Learning Objectives

After completing this unit, the student should be able to:

- Identify types of drill presses
- Identify the major components of the drill press and their functions

Key Terms

Feed
Gang drill press
Micro drill press

RPM (Revolutions
per minute)
Radial-arm drill press
(radial drill press)

Sensitive drill press
Spindle
Upright drill press

INTRODUCTION

The drill press is one of the most basic machine tools used in the machining field. The drill press performs holemaking operations by pushing a rotating cutting tool into the workpiece. **(See Figure 4.1.1.)** The **spindle** is the part of the drill press that holds and rotates the cutting tool. *Spindle speed* is the term used in machining to refer to the **RPM (revolutions per minute)** of the machine's rotating spindle. Advancement of the cutting tool into a workpiece is called **feed**.

Common operations performed on the drill press include drilling, reaming, countersinking, counterboring, spotfacing, and tapping. **(See Figure 4.1.2.)** Drill presses provide greater power and control than hand-held electric drills when performing holemaking operations and are available in different styles and sizes depending on the type of work to be performed.

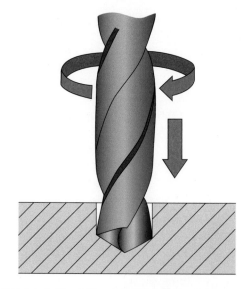

FIGURE 4.1.1 The drill press advances a rotating cutting tool into a workpiece to perform holemaking operations.

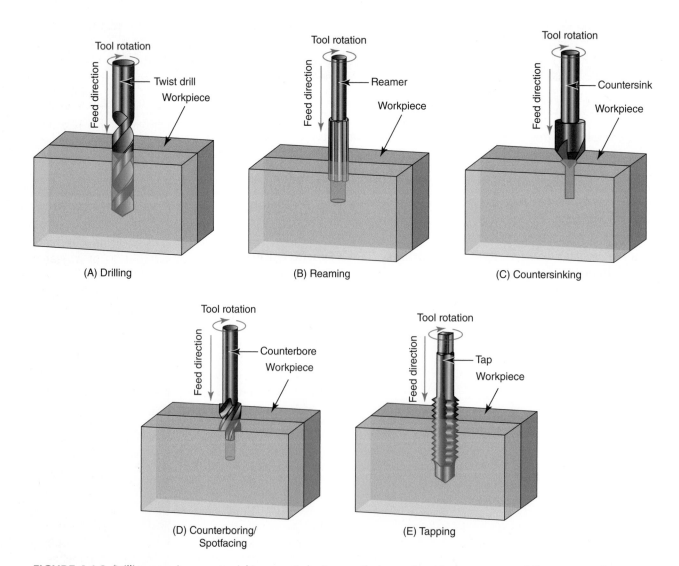

(A) Drilling

(B) Reaming

(C) Countersinking

(D) Counterboring/ Spotfacing

(E) Tapping

FIGURE 4.1.2 Drilling, reaming, countersinking, counterboring, spotfacing, and tapping are common drill press operations.

UPRIGHT DRILL PRESS

The **upright drill press** consists of a column mounted at 90 degrees to a base. The column supports the head which contains the various mechanisms used to power the spindle and feed cutting tools into the workpiece. An adjustable worktable is also mounted to the column which will support the workpiece during holemaking operations. Upright drill presses are available in bench-top or floor models, as shown in **Figure 4.1.3**. When using an upright drill press, work is positioned and secured beneath the spindle to establish the location for the holemaking operation.

Upright drill presses are available with either belt-driven or gear-driven heads. Belt-driven heads sometimes

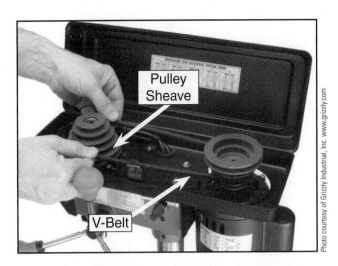

Photo courtesy of Grizzly Industrial, Inc. www.grizzly.com

FIGURE 4.1.4 The belts on a step pulley drill press need to be manually moved between the different pulley combinations to set spindle speed. Always turn off the spindle and the main power switch before moving the belts.

utilize a step pulley system for changing spindle speeds. This requires stopping the machine spindle, and then manually changing belts between different sets of pulleys. **(See Figure 4.1.4.)** Other belt-driven machines make use of a variable-speed pulley system that allows the setting of any spindle speed within the machine's range. The spindle speed of this type of drill press is adjusted while the spindle is rotating by turning a hand wheel. **(See Figure 4.1.5.)** A drill press with a geared head contains a gear box with levers that are moved to various positions to produce different spindle speeds. The speed is set on a gear-driven drill press while the spindle is off. Geared-head drill presses provide greater torque to the spindle than belt-driven machines. **Figure 4.1.6** shows the speed controls of a geared-head drill press.

The size of an upright drill press is identified by the largest-diameter workpiece that can be drilled on center. That size would be double the distance from the column to the center of the spindle. For example, a 20" drill press would be able to drill a hole in the center of a 20"-diameter workpiece, therefore the distance from the column to the center of the spindle would be 10". **(See Figure 4.1.7.)**

An upright drill press may contain a power feed mechanism that can automatically feed the cutting tool into the work by engaging a clutch drive mechanism. Tools can also be fed into the work by manually applying pressure on the feed handle. A drill press without a power-feed mechanism is sometimes called a **sensitive drill press** because the operator can "feel" the cutting action of the tool and adjust the rate of feed for optimal cutting pressure by increasing or decreasing pressure on the feed handle.

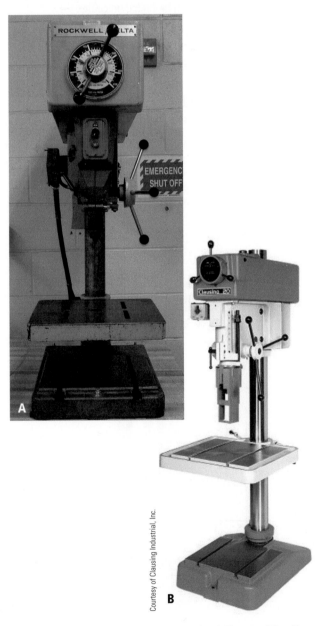

Courtesy of Clausing Industrial, Inc.

FIGURE 4.1.3 (A) A bench-top upright drill press. (B) A floor model upright drill press.

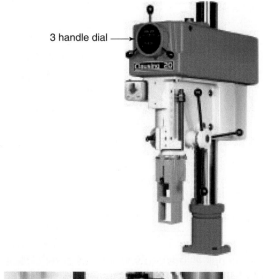

3 handle dial

FIGURE 4.1.5 A speed-change hand wheel is rotated while the spindle is running to change spindle speed on a drill press with a variable-speed pulley system.

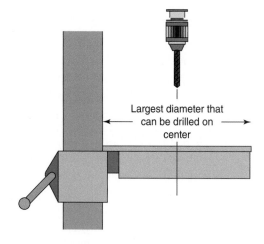

FIGURE 4.1.7 Drill press size is determined by the largest-diameter workpiece that can be drilled on center.

Largest diameter that can be drilled on center

Drill Press Controls

The controls of every upright drill press are slightly different, but every drill press has the same basic parts. **Figure 4.1.8** labels the parts of a typical upright drill press. Refer to it while reading the following discussion of functions of the parts.

The *base* of the drill press provides a solid foundation for the entire machine. The top of the base is a flat surface that is perpendicular to the spindle. Large

FIGURE 4.1.6 Speed selector levers are set at various positions to set spindle speed on a geared-head drill press. Only move these levers when the spindle is stopped.

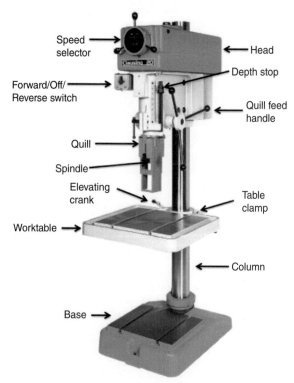

Speed selector — Head
Forward/Off/ Reverse switch — Depth stop
Quill feed handle
Quill
Spindle
Elevating crank — Table clamp
Worktable
Column
Base

FIGURE 4.1.8 The parts of an upright drill press.

workpieces can be mounted directly to the base using the clamping slots.

The *column* is mounted at 90 degrees to the base and supports the *worktable* and the *head* of the drill press. The worktable can be rotated around the column to any position. It can also be raised and lowered with the *elevating crank*. The worktable on some drill presses can also be tilted. When in position, the table is locked in place using the *table clamp*. The table has slots machined into its face. These slots are used to clamp the work piece into position.

The spindle holds the cutting tools for holemaking operations. Belts or gears rotate the spindle at the desired RPM. The *forward/off/reverse switch* starts the motor to rotate the spindle. Most drill presses have both forward and reverse spindle directions. Some may also have a low and high range. The *speed selector* is used to adjust the spindle speed of the drill press. Some models use a variable-speed drive while other machines may require manually changing belt positions on pulleys or positioning levers to change gear settings.

The *quill* is moved down with the *quill feed handle* to feed cutting tools into the work. The quill itself does not rotate however, the spindle rotates inside the quill. Some drill presses are equipped with a *power-feed mechanism* that will automatically feed the quill when a clutch lever is engaged. The *depth stop* can be set to limit the travel of the quill to feed to a desired depth. On a drill press with power feed, the depth stop will automatically disengage the feed clutch when the quill reaches the depth set with the stop.

Gang Drill Press

A **gang drill press** contains multiple drilling heads attached to a single base and worktable. Its construction and operation is very similar to the standard upright drill press. The multiple heads allow numerous cutting tools to be mounted, reducing the need to repeatedly change tools in and out of the spindle when performing different holemaking operations on multiple workpieces. It can also be used when more than one type of cutting tool is needed to produce a certain type of hole. **Figure 4.1.9** shows a gang drill press.

Courtesy of Clausing Industrial, Inc.

FIGURE 4.1.9 A gang drill press.

RADIAL-ARM DRILL PRESS

The **radial-arm drill press** (often called a **radial drill press**) is the largest of the drilling machines. It is designed for machining large-diameter holes or large workpieces. The radial drill press consists of a heavy base and column. An arm is fitted around the column that can be raised and lowered as well as rotated 360 degrees around the column. An assembly that houses the spindle and feed controls is attached to this arm and can be moved back and forth along the length of the arm. The size of a radial-arm drill press is given by the maximum distance from the column to the center of the spindle. When using a radial drill press, work is secured to the worktable and the spindle is positioned over the workpiece in the desired location for the holemaking operation. **Figure 4.1.10** shows a radial-arm drill press and labels its major parts. Refer to this figure while reading the following discussion about the functions of the parts.

The *base* of the radial drill press is the foundation, just like the base of the upright drill press. Clamping slots can be used to mount workpieces to the base. The table is mounted to the base and is used for mounting workpieces using the clamping slots.

The *column* is mounted at 90 degrees to the base and supports the radial arm of the drill press. The *radial arm* is mounted to the column. It can be manually rotated around the column with a handle at the end of the arm. Moving the *positioning lever* up or down activates a motor that moves the arm up or down on the column. After the arm is moved to the desired location, it is locked in place by either a manual or hydraulic clamp.

The *drill head* is mounted to the radial arm. It can be positioned at any location along the arm by turning the *hand wheel* and then locked in place with either a manual or hydraulic clamp. The spindle holds and rotates holemaking tools, just like the spindle of an upright drill press. *Speed selection* controls set the spindle RPM and *feed selection* controls set the feed rate. The *spindle power lever* starts the machine spindle. The quill is fed manually by rotating the *feed handle* or under power by engaging the *feed clutch*.

Micro Drill Press

A **micro drill press** (sometimes called a *precision drill press*) operates at very high spindle speeds and is used to produce very small holes. Some models can drill holes as small as 0.002″ and achieve spindle speeds up to 20,000 RPM. These high-speed spindles produce very little torque, so they can only produce enough power to machine hole diameters up to about ¼″. Many micro drill presses are equipped with dial gages to closely monitor quill movement to reduce tool breakage. **Figure 4.1.11** shows a micro drill press.

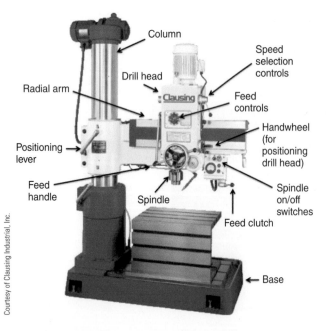

Courtesy of Clausing Industrial, Inc.

FIGURE 4.1.10 The parts of a radial-arm drill press.

Courtesy of Servo Products Co., Eastlake, OH, USA

FIGURE 4.1.11 This micro drill can be used to drill hole diameters as small as 0.002″ at speeds up to 20,000 RPM.

SUMMARY

- Drill presses are machine tools that are widely used to perform holemaking operations such as drilling, reaming, countersinking, counterboring, and tapping. They are available in different styles and sizes to meet different needs.

- The upright drill press is available in both bench-top and floor models. Its spindle may be driven by a step pulley system, a variable-speed drive, or a gear-driven mechanism.

- With the upright drill press, cutting tools are fed into the work by applying pressure to a handle or by power feed on models equipped with that option.

- The size of an upright drill press is identified by the largest diameter that can be drilled on center.

- The radial-arm drill press is designed for machining large holes and large workpieces. The large-diameter column supports an arm that is positioned at the desired hole location. The arm contains a drill head that houses the machine spindle and the speed and feed controls.

- The size of a radial drill press is determined by the maximum distance from the column to the center of the machine spindle.

- The micro drill press can be used to drill holes as small as 0.002". It can achieve very high spindle speeds and is equipped with a gage to closely monitor feed distance to reduce tool breakage.

REVIEW QUESTIONS

1. Upright drill presses are available in _____ and _____ models.
2. List two general types of upright drill press heads.
3. Explain how the size of an upright drill press is determined.
4. Briefly describe the term *sensitive drill press.*
5. Briefly describe a gang drill press and its benefit.
6. When would a radial-arm drill press most likely be used?
7. Briefly describe a micro drill press and its purpose.

Tools, Toolholding, and Workholding for the Drill Press

Learning Objectives

After completing this unit, the student should be able to:

- Identify the major parts of the twist drill
- Explain the function of each part of the twist drill
- Explain the various toolholding and workholding devices used on the drill press
- Identify which type of toolholding and workholding device should be used in various situations

Key Terms

Angle plate	**Flutes**	**Morse taper sleeve**
Body clearance	**Heel**	**Pilot**
Chisel edge	**Helix angle**	**Pin vise chuck**
Dead center	**Lip clearance**	**Shank**
Drill body	**Lips**	**T-nut**
Drill chuck	**Margin**	**V-block**
Drill drift	**Morse taper**	**Web**
Drill point	**Morse taper**	
Drill point gage	**extension socket**	

INTRODUCTION

All of the holemaking tools that were covered in Section 3.6 for handheld electric drill applications may also be used in the drill press with the added benefit of increased power and control. Since there is considerably more torque available, a workpiece being machined in a drill press must be securely mounted to prevent injury to the operator. Before any machining process can begin on the drill press, the cutting tools must be properly mounted and the workpiece must be properly secured.

TYPES OF CUTTING-TOOL MATERIALS

High-speed steel (HSS) is very popular due to its low cost, and ability to flex under impact without breaking or chipping. HSS can also be combined with other alloying ingredients such as cobalt. Some cutting tools may have up to 8% cobalt added to them. These HSS tools are often labeled as simply "cobalt." This variety of HSS offers the same advantages as standard HSS, but can operate at up to 10% higher speeds and feeds due to its slightly higher hardness, better toughness, wear resistance, and heat resistance.

Tungsten carbide is also a cutting-tool material used for high-performance and high-production operations. Tungsten offers superior tool life due to its extreme hardness, wear resistance, and ability to withstand heat. While carbide is good at many things, it is also very brittle, and easily chipped under impact. While some tooling is made of solid carbide, more economical tooling is available that uses small pieces of carbide as the tool's cutting edges. The carbide pieces can either be mounted to the steel tool body by brazing or by screws. Tooling using replaceable carbide held in place with screws is

known as inserted or indexable tooling. Solid carbide tools can be distinguished from HSS cutting tools by weight. Because carbide is a very dense material carbide tools may be up to twice the weight of similar HSS cutting tools. Many tools in use today have surface coatings applied to them such as Titanium Nitride (TiN). A tool that has been TiN coated will have a shiny gold appearance. These coatings can further extend the life of the tool. In its uncoated state carbide will appear as a darker, grayish color as opposed to high-speed steel's bright, shiny luster or deep black oxide finish. **(See Figure 4.2.1.)**

DRILL BITS

The purpose and function of a drill bit was covered lightly in Section 3.6; however, there are some additional details about a drill bit that are important to understand when the machining processes become more demanding.

Twist Drills

A twist drill can be divided into three main parts: the drill point, drill body, and shank. **(See Figure 4.2.2.)** Each of these parts serves a vital role.

The Drill Point

The **drill point** is the cone-shaped area at the very tip of the drill bit. The drill point is made of the following subparts shown in **Figure 4.2.3**:

- The **lips** are the angled cutting edges that shear the metal into chips as the drill rotates. The lips are the only part of the drill that actually cuts. Each twist drill has two lips. The included angle of the lips on a general-purpose twist drill is 118 degrees. Other point angles may perform better when drilling certain types of materials. If the drill tip is improperly sharpened so that one lip is longer than the other, the drill will cut larger than its intended size.

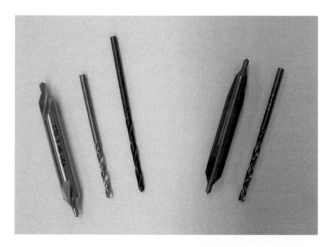

FIGURE 4.2.1 High-speed steel (HSS) and carbide drills. The HSS drills on the left have a bright or black finish versus the dull gray finish of the carbide drills on the right.

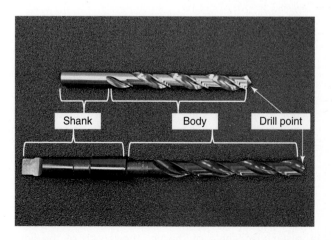

FIGURE 4.2.2 The parts of a twist drill showing the point, body, and shank.

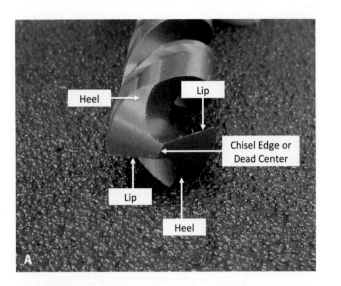

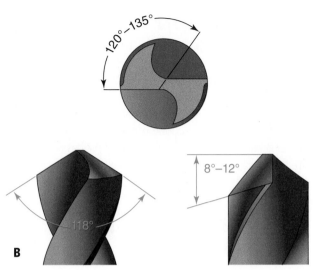

FIGURE 4.2.3 (A) The dead center, lips, and heel of a twist drill. (B) The included angle, lip clearance, and dead center angle of a general-purpose twist drill point.

FIGURE 4.2.4 Checking the angle of the drill point and the length of the cutting lips using a drill point gage. Both lips must be the same angle and the same length.

- The **chisel edge** is sometimes also called the **dead center** and lies between the lips at the center of the drill point. The chisel edge does no actual cutting but is instead forced into the material when drilling. The rotating chisel edge is the pivot point of the rotating drill as it contacts the workpiece. The pressure from the chisel edge as it rotates on the workpiece forces the material outward into the lips to be cut into chips.

Lip clearance is the relief angle from the lips back to the **heel**. A general purpose twist drill has lip clearance between 8 and 12 degrees. The angle from the lips to the dead center should be between 120 and 135 degrees. The included angle of the tip, the length of the lips, lip clearance, and angle of the dead center should be verified before using a twist drill. Check the angle of the

drill point and the length of the lips with a **drill point gage**, as shown in **Figure 4.2.4**. The angle of the dead center and the lip clearance can be checked using a protractor as shown in **Figure 4.2.5**.

The Drill Body

The **drill body** extends from the tip to the beginning of the shank and makes up the majority of the drill bit. The drill body is made of the following main parts shown in **Figure 4.2.6**:

- The **flutes** run the length of the body section and create the cutting lips. They also provide a pathway for chips to flow out of the hole during drilling operations and for cutting fluid to flow into the hole to help cool and lubricate the cutting lips (or point). A twist drill is so named because its flutes are spiraled and appear twisted.

- The **margin** is a thin raised strip that runs the length of the drill body along the edge of the flutes and gives the drill bit its diameter. The margin fits closely in the hole being produced by the lips and stabilizes the drill.

- The **body clearance** is the area just behind the margin and is slightly smaller in diameter so that the body does not rub on the walls of the hole during drilling. If there were no body clearance, the entire diameter of the drill would be in contact with the walls of the drilled hole, creating excessive friction and heat from the additional surface contact against the walls of the hole being drilled.

- The **helix angle** is the angle of a drill's spiral relative to the center axis of the drill.

- The **web** connects the flutes and makes up the centermost part of the bit. The web is slightly tapered and becomes thicker toward the shank.

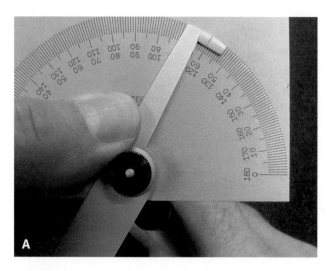

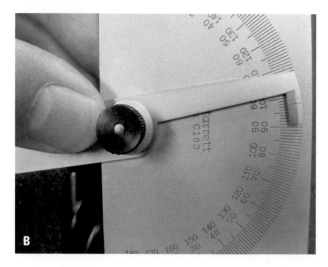

FIGURE 4.2.5 (A) The angle of the dead center should be between 120 and 135 degrees. (B) Lip clearance should be between 8 and 12 degrees.

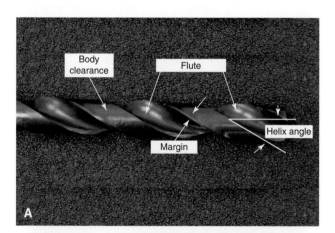

FIGURE 4.2.6 (A) The drill body, showing the flutes, margin, body clearance, and helix angle and (B) the blue shows the web of a twist drill.

The Drill Shank

The **shank** of a drill bit provides an area for mounting the drill bit into some type of holder, frequently called a toolholding device. Many types of drills can be purchased with either a straight shank that is generally held in a standard **drill chuck** or a tapered shank that can be inserted directly into a machine spindle. Tapered shanks are most commonly found on drills larger than 1/2-inch in diameter. The **Morse taper** is a popular standardized style of taper used in drill press spindles and for holemaking tools. Morse tapers are numbered from smallest to largest by the numbers 0 through 7.

Each size has a slightly different amount of taper, but all are approximately a 5/8" change in diameter per 12 inches of length. Taper shank tools have a flat section at the end of the shank called the tang. This is used when removing the tool from the machine spindle.

High-speed steel twist drills are often made with their shanks intentionally softer than the rest of the drill. This allows the toolholding device to obtain a better grip. **Figure 4.2.7** shows straight and Morse taper drill shanks.

Drill Sharpening

Twist drill points will wear as they are used, and may become damaged, so it is important to know how to resharpen drills by hand to keep them sharp and reduce tool replacement cost. Twist drills can be resharpened many, many times before they need to be replaced. Follow these basic steps to sharpen a drill point using a bench or pedestal grinder.

- Hold the drill with the cutting lip up, at 59 degrees to the wheel face, and with the tip slightly higher than the shank. **(See Figure 4.2.8.)**
- Lightly touch the lip against the wheel and lower the shank end of the drill to move from the lip toward the heel. Apply more pressure when moving toward the heel to create the 8- to 12-degree lip clearance. **(See Figure 4.2.9.)** Back the drill away from the wheel and repeat the process two to three times.
- Repeat with the other lip and as needed to remove any wear or chips at the cutting edge.
- Check the drill point angle and lip length with a drill point gage and grind as needed to create the proper angle and make both lips the same length.
- Check the angle of the dead center and the lip clearance with a protractor.

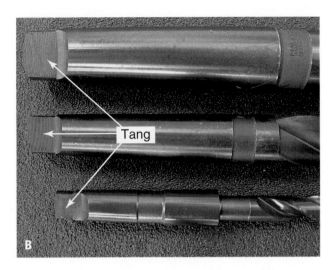

FIGURE 4.2.7 (A) Straight shank bits and (B) taper shank bits.

FIGURE 4.2.8 Hold the drill at 59° to the wheel face with the tip slightly higher than the shank.

After sharpening, the twist drill can be tested by drilling a sample hole. The drill should create equal size and shape chips from both flutes, and the hole should be within 0.003" to 0.005" of the drill size.

Spotting Drills and Combination Drills and Countersinks

Recall from Section 3.6 that special drills are available to create a shallow starting hole so that the twist drill does not "walk" when it initially starts to penetrate the workpiece. Two main types of these starter drills are available, the combination drill and countersink (often called a center drill), and the spot drill. **(See Figure 4.2.10.)** Once the hole has been spot drilled or "spotted," it may be completed with the twist drill.

A combination drill and countersink has a small drill pilot that transitions into a countersink. The included

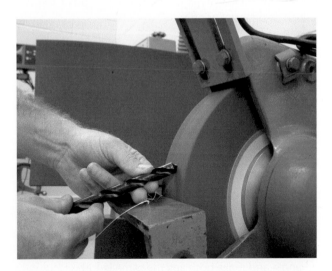

FIGURE 4.2.9 Lower the shank and increase pressure to move from the lip toward the heel.

FIGURE 4.2.10 A combination drill and countersink and a spot drill.

FIGURE 4.2.11 A combination drill and countersink (often called a center drill) properly mounted in a drill press chuck.

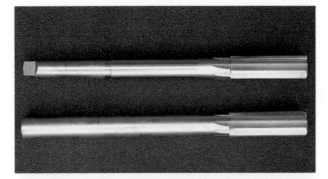

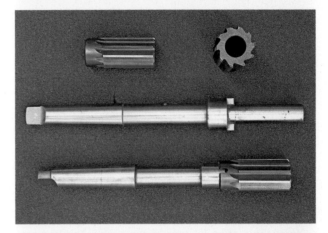

angle of the tip of the pilot portion is 118 degrees. The included angle of the countersink portion can be 60 degrees, 82 degrees, 90 degrees, or 100 degrees. A spotting drill resembles a short twist drill with short flutes. These drills are very stable due to their solid design and can withstand heavier feed rates than combination drills and countersinks. Because of their short length, they do not deflect like twist drills and will not wander off location during use like a twist drill might. The included angle of the tip of a spot drill can be 60 degrees, 82 degrees, 90 degrees, 118 degrees, 120 degrees, 142 degrees, or 145 degrees. Both cutting tools are usually held in drill chucks mounted in the drill press spindle. **(See Figure 4.2.11.)**

REAMERS

Recall from Section 3.6 that reaming is the process of finishing a previously drilled hole to a more precise size and smoother-finished surface than is possible with a drill bit. To properly ream a hole, a twist drill is selected that is slightly smaller than the reamer. Follow these guidelines for drilling a hole that will be reamed.

- 0.010" undersize for a hole up to 1/4"
- 0.015" undersize for a hole between 1/4" and 1/2"
- 0.025" undersize for a hole between 1/2" and 1-1/2"

This will leave a small amount of stock in the hole that can then be cut away by the reamer. Section 3.6 covered

FIGURE 4.2.12 From top to bottom: chucking, shell, and expansion reamers.

the styles and use of hand-fed reamers, but reaming is also commonly done on the drill press. Machine reamers for drill press use are usually made of high-speed steel, cobalt, or carbide. Variations of these machine-driven reamers include general-purpose chucking reamers, shell reamers, and expansion reamers. All are available as either straight or taper shank tools. Chucking reamers are a set size. So are shell reamers, but different diameters can be mounted, removed, and remounted on a driving arbor. Expansion reamers have a screw in the end to enlarge the tool diameter slightly. **(See Figure 4.2.12.)**

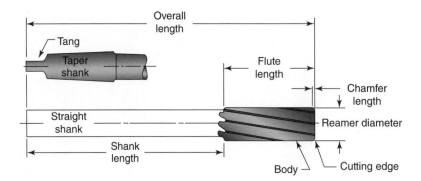

FIGURE 4.2.13 Parts of a straight machine reamer.

Reamer Parts

Machine reamers can be divided into three parts, as shown in **Figure 4.2.13**:

- The *cutting tip* of a reamer is typically produced with a 45-degree chamfered cutting edge. On a straight reamer, the chamfer is the only place where cutting action occurs when producing a reamed hole.

- Straight machine reamers have a very slight back taper, meaning the tool diameter decreases moving away from the cutting tip toward the shank. This prevents excess rubbing and wear during cutting. Taper reamers cut on the periphery to produce the desired angle.

- The flutes on a reamer help form the cutting edge geometry and act as a way to remove chips from the hole. General-purpose chucking reamers have straight flutes. Spiral flute reamers have helical flutes that, depending on the direction of spiral, either cause the chips to be projected forward or pulled out the opening of the hole during cutting. Spiral flute reamers are also ideal for use when reaming holes with interruptions such as keyways or grooves, since the spiral cutting edge can span the interruption and minimize tool vibration. **(See Figure 4.2.14.)**

- The shank is the portion of the reamer that is used to grip and rotate the reamer while machining. Reamer shanks are available as Morse taper or straight style. The taper shank reamer fits directly into the internal taper in the drill press spindle. A straight shank is typically held with a drill chuck.

Reamer Sizes

Reamers are available in a wide variety of different sizes. Inch unit reamers are available in common fractional increments and countless decimal increments. Like drill bits, reamers also come in number (wire gauge) sizes and letter sizes. Metric-sized reamers are also available. Reamers are available in "over/under" sizes for standard dowel pins. When assembling and disassembling mating parts aligned with dowel pins, it is usually ideal to have the pin stay in one of the parts while sliding freely in and out of the mating part. This is accomplished by making one hole slightly smaller by using the undersized reamer (interference fit) and one hole slightly larger by using the oversized (clearance fit) by about 0.001". Reamers are also available for producing tapered holes, such as the Morse taper used in machine tool spindles or tapered dowel pins. **(See Figure 4.2.15.)**

FIGURE 4.2.14 Spiral flute reamers.

FIGURE 4.2.15 Taper reamers.

FIGURE 4.2.16 Zero-flute countersinks.

COUNTERSINKS AND COUNTERBORES

Recall from Section 3.6 that the countersink is a cutting tool used after drilling to chamfer the opening of a hole to eliminate any sharp edges or burrs, or to create a conical recess for accommodating the head of a flathead screw. Countersinks come in a variety of diameters and are available with a 60-degree, 82-degree, 90-degree, 100-degree, or 120-degree included angle. The appropriate countersink angle is usually determined by specifications on the drawing or by the head of the desired fastener. Countersinks are available with either one flute or mulitple flutes. A hybrid countersink called a zero flute is also available.

- Zero-flute countersinks have an angled hole through the tool body that forms a cutting edge. These countersinks are particularly well suited for avoiding chatter and for deburring holes. Each tool has a relatively small working diameter range. **(See Figure 4.2.16.)**

- Single-flute countersinks cut quickly and are likely to resist chatter and vibration.

- Multiple flute countersinks with three or more flutes provide enhanced tool life by distributing wear across more cutting edges. **(See Figure 4.2.17.)**

Recall also that counterbores are used to create a smooth, flat enlargement on the opening of an existing hole. When done properly, a screw or bolt head will fit flush with or slightly below the top of the workpiece surface. Counterbore tools are also used for spotfacing operations which provide a shallow flat surface for a fastener head to seat against, as shown in **Figure 4.2.18**. Spotfacing is most often used on parts

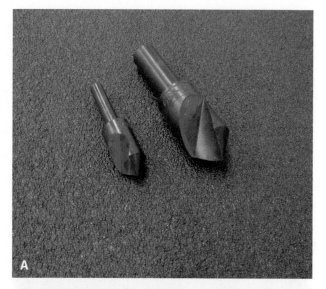

FIGURE 4.2.17 Single flute countersinks (A) and multiple flute countersinks (B).

that have been manufactured from a casting process. Examples of this include spotfaced holes that can be found on many automotive parts, such as engine intake manifolds and cylinder heads.

A counterbore has a guide on its end called a **pilot** that keeps the counterbore aligned in the existing hole. The pilot diameter should be about 0.003" to 0.005" smaller than the existing hole diameter. Less clearance may cause the pilot to bind or seize in the hole while more clearance will allow the counterbore to walk off location. The pilot can be an integral part of the counterbore or interchangeable so that different-sized pilots can be used with the same cutting diameter. **(See Figure 4.2.19.)** Counterbores for drill press operations are available with either straight shanks or tapered shanks.

FIGURE 4.2.18 (A) Counterbored holes and (B) a spotfaced hole.

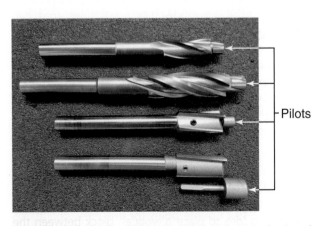

FIGURE 4.2.19 Counterbores with integral pilots (top) and interchangeable pilots (bottom).

TOOLHOLDING

Morse Taper-Shank Toolholding

Many holemaking cutting tools over 1/2" in diameter have taper shanks instead of straight shanks. The internal Morse taper in the drill press spindle can be used for rugged, fast, and accurate direct mounting of taper-shank cutting tools. Since many sizes of Morse tapers exist, adapters are available to convert the cutting-tool shank to match the drill press spindle taper size. **Morse taper sleeves** increase the size of a tool's shank taper and **Morse taper extension sockets** can be used to reduce the size of a tool's shank taper. **(See Figure 4.2.20.)**

When assembling tapers, be sure that the bore and shank are both clean and burr free. The end of Morse tapers used on holemaking tools have a tang that must be aligned with a mating slot in the receiving bore. **(See Figure 4.2.21.)** The tapers are self-holding and are assembled using a quick, forceful motion. Pressure from the drilling operation further secures the tapers together.

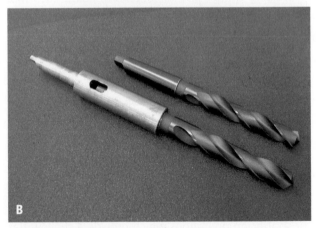

FIGURE 4.2.20 (A) Morse taper sleeves increase a tool's shank size. (B) Morse taper sockets reduce a tool's shank size.

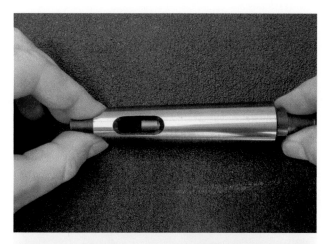

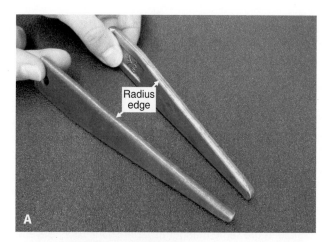

FIGURE 4.2.21 The Morse taper tang must be aligned with the slot in the mating bore.

 **CAUTION**

Hold cutting tools by the shank, not by the flutes, when mounting them in a drill press spindle, sleeve, or socket to prevent cuts from sharp edges.

To remove a taper-shank tool from the drill press spindle, a wedge-shaped **drill drift** is inserted between the tapers through an opening in the spindle and lightly tapped with a hammer. The radius edge of the drift

FIGURE 4.2.22 (A) Two types of drifts. One has the radius on the angled edge and the other has the radius on the straight edge. When using, place the flat edge against the tang of the tool. (B) A drill drift being used to remove a drill from the drill press. Note the wooden block to prevent the tool from dropping onto the machine table.

should face up and the flat edge should be against the tang of the tool. Always hold the mounted tooling by hand when separating tapers so the tool and machine table are not damaged when it releases from the spindle. It is a good idea to place a wooden block between the tool and the worktable to stop the tool from falling. **(See Figure 4.2.22.)**

CAUTION

After removing tools from a drill press spindle, immediately remove the drift from the spindle. Starting the machine spindle will violently throw the drift and can cause serious injury.

Straight-Shank Toolholding

Straight-shank tooling must be mounted in a drill chuck. Drill chucks are often equipped with Morse taper shanks and are mounted in and removed from the drill press spindle in the same manner as taper-shank tools. Several different types of chucks exist for different applications.

Keyed drill chucks are often called Jacobs-type chucks and are available in different sizes with various gripping capacities. Some models have the ability to grip as small as a few thousandths of an inch, while others can hold straight-shank tools up to about 1" in diameter. Most drill press chucks have three jaws that must be tightened with a special chuck key. **(See Figure 4.2.23.)**

FIGURE 4.2.23 A Jacobs-type drill chuck and chuck key.

 CAUTION

Remove the chuck key from the drill chuck immediately after mounting tools in or removing them from the chuck. Starting the machine spindle will violently throw the key and can cause serious injury.

Drill chucks are also available in keyless styles that can be tightened by rotating a knurled collar by hand without the use of any tools. **(See Figure 4.2.24.)** Keyless chucks enable tools to be changed more quickly.

Specialty chucks are also available for small-hole drilling. The **pin vise chuck** is a micro drill chuck that has a much slimmer design, has a smaller capacity, and will enable very small drills to run truer than a larger chuck. **(See Figure 4.2.25.)** Sensitive micro drilling adapters are also available for small-hole drilling. The advantage with this type of drill is that pressure can be gently applied to the drill bit under fine finger pressure. These spring-loaded adapters allow high levels of sensitivity when drilling with fragile small diameter drill bits. **(See Figure 4.2.26.)**

WORKHOLDING

As with any machine tool, one of the first things to consider when setting up to run the drill press is to select a way to safely secure the part for machining. There are numerous workholding devices available. Many of these

FIGURE 4.2.24 A keyless drill chuck.

FIGURE 4.2.25 A pin vise chuck.

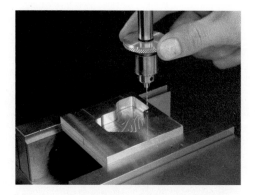

FIGURE 4.2.26 A spring-loaded micro drilling adapter.

devices are commercially available and others may be custom made for a particular job.

Drill Press Vise

The drill press vise offers ease of use and fast setup times. A drill press vise is similar to the bench vise, but has a lower profile, a precise flat bottom bed, and smooth jaws. **(See Figure 4.2.27.)** A workpiece with perpendicular sides is set in the vise on a pair of parallels to ensure that the part rests parallel to the machine table surface. The moveable jaw is then tightened to secure the workpiece. As the drill press vise is tightened, the work should be tapped down with a soft faced hammer to seat it on the parallels. When both parallels are snug, the work is properly seated. **(See Figure 4.2.28.)** When clamping a round workpiece, it is better to use one wider parallel and position it so contact occurs at the 6 o'clock position, or lowest point on the work, instead of on the edges of a pair of parallels, as shown in **Figure 4.2.29**.

Drill press vises can be purchased with a V-shaped slot ground into the jaw faces that allows for the direct clamping of round workpieces. **Figure 4.2.30** shows some different types of drill press vises.

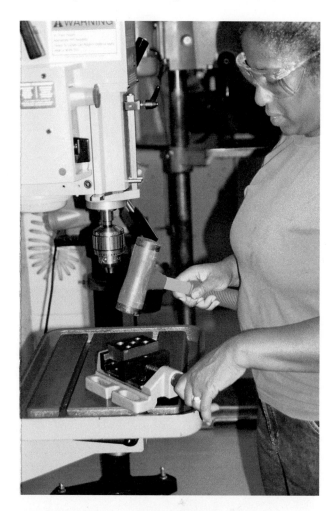

FIGURE 4.2.28 The workpiece should be tapped down with a soft face hammer to properly seat it on the parallels.

⊘ CAUTION

Always clamp a round workpiece so that the center-line of the workpiece is below the top of the vise jaws, to prevent the part from "popping" out of the vise. **(See Figure 4.2.31.)**

V-Block

A **V-block** is another workholding device that can be used to secure a round workpiece. The V-block consists of a steel or cast iron block with a "V" shape machined into it. The V cradles the work and provides two points of contact between it and the workpiece. The part is also clamped so that pressure is applied to the workpiece, forcing it against the V. These three points of contact offer more locating accuracy and gripping power than would be supplied by a vise. The V-block itself can then be clamped into a vise or clamped directly to the table. **(See Figure 4.2.32.) Figure 4.2.33** shows some different styles of V-blocks.

FIGURE 4.2.27 A drill press vise holding a workpiece for a drilling operation.

FIGURE 4.2.29 Correct use (A) and incorrect use (B) of parallels when clamping round work in a vise.

FIGURE 4.2.31 Always clamp round parts so that their centerline is below the top of the vise jaws as shown in (A), never like in (B), or the work will not be secure and "pop" out when tightened or during machining.

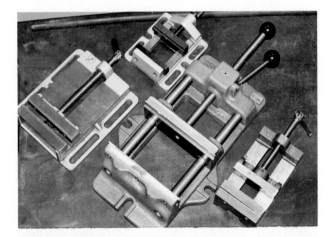

FIGURE 4.2.30 There are several types of drill press vises for holding different types of parts.

FIGURE 4.2.32 A V-block holding round work for drilling.

FIGURE 4.2.33 There are several different types of V-blocks for holding parts of different shapes and sizes.

FIGURE 4.2.35 An angle plate used to hold a part in a drill press.

FIGURE 4.2.34 A typical angle plate.

FIGURE 4.2.36 A T-nut (circled) fits into T-slots of a drill press table.

Angle Plate

An **angle plate** is a simple yet very useful workholding device. **(See Figure 4.2.34.)** Angle plates are usually made of hardened tool steel or cast iron with working surfaces that form 90-degree angles to each other. An angle plate is most often used when a hole is to be drilled parallel to a large flat surface on the workpiece. To use an angle plate, the workpiece is first secured to the angle plate by some type of clamping device such as C-clamps or tool maker's clamps. The angle plate is then positioned on the drill press table and it too is secured with clamps, as shown in **Figure 4.2.35**.

Hold-Down Clamps

When a workpiece does not lend itself well to being held in a vise or clamped to an angle plate or V-block, clamps may be used to secure the work directly to the machine table. Most drill presses either have T-shaped slots, which accept special nuts called **T-nuts**, or through-slots for clamping purposes. **(See Figure 4.2.36.)** Many styles of clamps are available, as shown in **Figure 4.2.37**. Always use at least two clamps when securing work to the drill press table. Using only one clamp can allow the work to pivot at the single clamping point. To use hold-down clamps, the workpiece must be oriented on the table in a position that will allow the operation to be completed. When drilling a hole completely through a workpiece, it must also be raised high enough off the table with parallels or blocks so that the drill bit will not damage the worktable when it breaks through the bottom of the part. The clamps must be positioned on top of the workpiece in such a manner that will not interfere with the drilling operation. The opposite end of the clamp is supported by riser or step blocks. Always

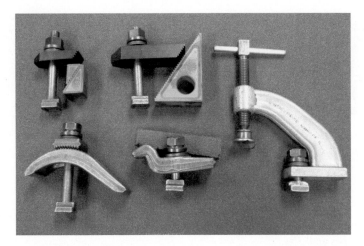

FIGURE 4.2.37 Some of the many types of clamps used for drill press work.

FIGURE 4.2.38 When drilling completely through the work, the work must be raised so that the drill will not damage the drill press table.

raise the rear of the clamp slightly so that all clamp contact with the part occurs on the clamp's tip and position the tip of the clamp over the parallels or blocks used to raise the work. **(See Figure 4.2.38.)** Uses of some other types of clamps and methods are shown in **Figure 4.2.39**.

 CAUTION

When directly clamping to the drill press table, use at least two clamps. When using clamp styles that require studs, select a stud long enough to engage fully into the T-nut in the table and pass fully through the hex nut on top of the clamp. It is also important to remember to position the stud or bolt as close to the workpiece as possible, so that the clamping force is directed toward the workpiece and not the step block.

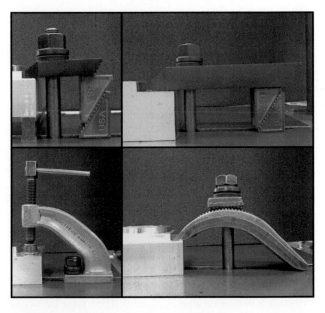

FIGURE 4.2.39 Some methods of clamping work directly to the table.

SUMMARY

- Drill press cutting tools are available in various types of materials.

- Carbide is often used for high-production or high-performance operations, while HSS is more economical and able to withstand more impact during use.

- Both the twist drill bit and reamer can be broken down into three distinct areas. The first is the point, which is the area where the cutting action takes place. The second is the body section. The body contains the flutes, which provide an area for chips to travel out of the hole and help guide the tool. The third area is the shank. The shank can be straight or tapered. The shank provides an area for mounting to the spindle (taper shank) or drill chuck (straight shank).

- As twist drill points wear, they can be sharpened using a bench or pedestal grinder.

- Reamers remove a small amount of material from a drilled hole. Reamers produce a better surface finish, more precise diameter, and better roundness than a twist drill bit.

- Often a countersink operation is necessary to break the sharp edges on the top and bottom of a drilled hole. The countersink is also used to provide a conical recess in a workpiece to accept a flathead screw.

- Counterbores may be used to create a recess to accept a bolt head or to create a flat spotface so that a bolt head or nut may seat flatly and squarely.

- Sleeves and sockets are used to adapt different-size Morse tapers to each other.

- Drill chucks are available in many styles and are used to hold straight-shank tooling.

- Some devices, such as hold-down clamps, can be used to clamp work or other workholding devices to the drill press table, while other devices such as the drill press vise offer ease of use and fast setup times.

- V-blocks can be used to securely hold round workpieces while performing holemaking operations.

- The angle plate can be used to securely hold a workpiece to perform holemaking operations at a right angle to a surface on the part.

REVIEW QUESTIONS

1. What are the two main materials used to make holemaking cutting tools?

2. What two types of shanks are commonly found on twist drills?

3. Name three functions of the flutes on a twist drill. _____ _____ _____

4. Name two types of flute styles for machine reamers. _____

5. How might an interruption such as a keyway influence the selection of a reamer?

6. What is the major difference between counterboring and spotfacing?

7. What must be used to secure a workpiece to an angle plate?

8. What devices are used to elevate a workpiece to prevent drilling into the machine table or vise bed?

9. If 50 parts measuring 1" × 1" × 3" each needed a hole drilled in a side, which workholding device would be best?

10. Which would be the best workholding device for holding a 1/2" × 8" × 8" flat plate for drilling a hole in its edge?

11. Which would be the best workholding device for holding a 1"-diameter × 6"-long shaft so that a cross hole for a grease fitting could be drilled?

12. Which would be the best workholding device for holding a 3/4" × 12" × 14" flat plate for drilling a series of holes through the large surface of the plate?

Learning Objectives

After completing this unit, the student should be able to:

- Describe drill press safety procedures
- Define cutting speed and perform speed and feed calculations for holemaking operations
- Demonstrate understanding of drilling operations
- Demonstrate understanding of reaming operations
- Demonstrate understanding of countersinking operations and calculate countersink feed depth
- Demonstrate understanding of counterboring/spotfacing operations
- Demonstrate understanding of tapping operations and estimate number of tap turns to achieve a given thread depth

Key Terms

Blind hole	Inches per revolution (IPR)	Surface feet per minute (SFPM)
Center finder (wiggler)	Machinability	Tapping head
Cutting speed	Pecking (Peck drilling)	Through hole (thru hole)
Feed	Pilot hole	
Feed per revolution (FPR)	Spotting	

INTRODUCTION

After becoming familiar with drill press safety, and selecting proper tooling and workholding, the next step is to determine proper speeds and feeds so the workpiece can be machined.

GENERAL DRILL PRESS SAFETY

Safe drill press operation can be ensured by observing some basic precautions. Specific safety notes are shown throughout this unit, but here are some precautions that should be observed during any drill press operation.

CAUTION

- Always wear safety glasses and observe all general shop safety rules when operating a drill press.
- Always remove drifts from drill press spindles immediately after use.
- Always remove keys from drill chucks immediately after loosening or tightening chucks.
- Always secure the workpiece before beginning any holemaking operation. Never attempt to hold a workpiece by hand when drilling.
- Do not touch rotating drill press spindles, chucks, and cutting tools.
- Always shut off the spindle and let it come to a complete stop before adjusting tools or the workpiece. Do not use hands or rags to slow the spindle. Allow it to stop on its own.
- Keep all items and body parts away from the rotating drill press spindle. Remove chips from the work area using a brush or rag only after the spindle has come to a complete stop. Never remove chips by hand.
- Never use compressed air to clean chips, debris, and cutting fluids from the drill press, cutting tool, or workpiece.
- Observe good housekeeping rules. Do not pile rags, tools, and parts on the drill press. Remove chips and cutting fluids from the drill press after use.

SPEED AND FEED

Spindle speeds are set on the drill press (and most machine tools) in RPM. Four basic factors will influence the RPM required to perform any holemaking operation. They are: cutting-tool material (HSS or carbide), material being machined, type of cutting tool, and diameter of the cutting tool. It is very important that the proper spindle RPM be determined before any drilling operation begins. Speeds that are too slow result in inefficiency,

which equals lost time and money. Speeds that are too fast can create excessive heat and cause severe damage to cutting tools and workpieces.

Feed refers to the movement, or advancement, of the cutting tool into the workpiece, and feed rate is the rate the tool is advanced into the workpiece. Proper feed rates are also very important for drill press operations (and all machining operations). Feed rates that are too slow, like speeds that are too slow, result in inefficient operations. Feed rates that are too great can cause rapid tool wear and even tool breakage.

A relative measure of how easily a material can be machined is called its **machinability**. SAE-AISI 1212 steel is used as the reference material and is given a machinability rating of 100%. Generally, softer materials will have higher ratings and harder materials will have lower ratings. For example, a 70% rating means that material can be removed at a rate that is 70% of the rate used to machine SAE-AISI 1212. **Figure 4.3.1** shows a portion of a machinability rating chart.

CAUTION

Improper speed and feed can create unsafe conditions. Extreme speeds cause overheating and tool dulling and failure. Extreme feeds can cause drill breakage. Both can lead to sharp tool fragments or workpieces being thrown from the work area.

Cutting Speed and RPM Calculation

Before choosing spindle speeds, it is important to gain an understanding of cutting speed. **Cutting speed** is the distance that a point on the circumference of a rotating cutting tool travels in 1 minute. It is stated in **surface feet per minute (SFPM)** and may sometimes be shown as just SFM or FPM. An illustration of cutting speed is shown in **Figure 4.3.2**.

There are many resources available for selecting the proper cutting speed for any given material and operation. The *Machinery's Handbook* is one very common resource. Cutting-tool manufacturers often also make cutting-speed charts for their particular tools. **Figure 4.3.3** shows an example of a chart showing recommended cutting speeds for drilling and reaming operations for some carbon steels using HSS cutting tools. Many charts will also list cutting speeds for carbide cutting tools. Because carbide is very heat resistant, cutting speeds for carbide tools can be at least two to three times faster than speeds used for HSS tools.

SAE-AISI Designation	Rating %	SAE-AISI Designation	Rating %	SAE-AISI Designation	Rating %	SAE-AISI Designation	Rating %	SAE-AISI Designation	Rating %
Chromium Steel									
5015	78	5130	57	5140A	70	5150A	64	5160A	60
5060A	60	5132A	72	5145A	66	5152A	64	E51100A	40
5120	76	5135A	72	5147A	66	5155A	60	E52100A	40
Chromium Vanadium Steel									
6102	57	6118	66	6145	66	6150A	60	6152A	60
Alloy Steel - Boron									
50B44A	70	50B50A	70	51B60A	60	94B17	66	94B30A	72
50B46A	70	50B60A	64	81B45A	66	–	–	–	–
Stainless Steel									
301	55	308	27	317	35	403	55	418	40
302	50	309	28	321	36	405	60	420	45
303	65	310	30	330	30	410	55	430F	65
304	40	314	32	347	40	416	90	440	50
Tool Steel									
A2, A3, A4	16	D5, D7	16	H24, H25	15	O1, O2, O7	16	S1, S2, S5	20
A6, A8, A9	16	H10, H11	20	H26, H42	15	O6	38	T1	14
A7	11	H13, H14	20	M2	14	P2, P3, P4	25	T4	11
A10	27	H19	20	M3	11	P5, P6	25	Ta5	8
D2, D3, D4	11	H21, H22	15	M15	8	P20, P21	22	W (AB)	30

"A" Indicates annealed. Carbon Steel 1212 (100% rating) is the comparison material for ratings. The information on this table was supplied by DoAll Company and Texaco.

FIGURE 4.3.1 A machinability rating chart tells how easy or difficult it will be to machine a given material.

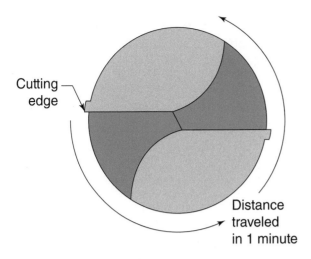

Cutting edge

Distance traveled in 1 minute

FIGURE 4.3.2 Cutting speed is the distance a point on the cutting edge (circumference) of a rotating cutting tool will travel in 1 minute.

Once a cutting speed has been selected from a chart, an appropriate spindle RPM must be calculated for the given cutting tool and operation. There are two basic formulas that can be used to calculate RPM. The first is:

$$RPM = \frac{12 \cdot CS}{\pi \cdot D}$$

CS = cutting speed
D = tool diameter
π is the value of Pi, approximately 3.14159.

The second formula is a simplification of the first. Since $12 \div \pi$ is approximately 3.82, that formula is:

$$RPM = \frac{3.82 \cdot CS}{D}$$

3.82 is a constant
CS = cutting speed
D = tool diameter.

The 3.82 in the second formula is also sometimes rounded to 4, changing it to $\frac{4 \cdot CS}{D}$. All of these formulas are commonly used in the machining industry and results may vary slightly depending on which formula is used. All the examples in this book use $\frac{3.82 \cdot CS}{D}$. It is important to *remember that cutting speed and RPM are not the same, but that cutting speed is used to calculate RPM.* Following are some examples of how to find cutting speeds from the chart in **Figure 4.3.3** and then calculate spindle RPM.

Material		Brinell Hardness	Drilling	Reaming
			HSS	
			Speed (fpm)	
Free-machining plain carbon steels (Resulfurized): 1212, 1213, 1215	{	100–150	120	80
		150–200	125	80
(Resulfurized): 1108, 1109, 1115, 1117, 1118, 1120, 1126, 1211	{	100–150	110	75
		150–200	120	80
		175–225	100	65
(Resulfurized): 1132, 1137, 1139, 1140, 1144, 1146, 1151	{	275–325	70	45
		325–375	45	30
		375–425	35	20
(Leaded): 11L17, 11L18, 12L13, 12L14	{	100–150	130	85
		150–200	120	80
		200–250	90	60
Plain carbon steels: 1006, 1008, 1009, 1010, 1012, 1015, 1016, 1017, 1018, 1019, 1020, 1021, 1022, 1023, 1024, 1025, 1026, 1513, 1514	{	100–125	100	65
		125–175	90	60
		175–225	70	45
		225–275	60	40
Plain carbon steels: 1027, 1030, 1033, 1035, 1036, 1037, 1038, 1039, 1040, 1041, 1042, 1043, 1045, 1046, 1048, 1049, 1050, 1052, 1524, 1526, 1527, 1541	{	125–175	90	60
		175–225	75	50
		225–275	60	40
		275–325	50	30
		325–375	35	20
		375–425	25	15

FIGURE 4.3.3 An example of a chart showing recommended cutting speeds for drilling and reaming operations for some carbon steels using HSS cutting tools.

EXAMPLE ONE: *Determine cutting speed and calculate spindle RPM for drilling a 1/4"-diameter hole in AISI-SAE 1020 steel with a Brinell hardness of 125 to 175.*

First find 1020 in the left column of the chart and then locate the correct Brinell hardness row. Following that row across to the drilling column shows a cutting speed of 90 surface feet per minute. Insert 90 into the RPM formula for *CS* and 1/4 or 0.25 into the formula for *D*.

$$RPM = \frac{3.82 \cdot CS}{D}$$

$$RPM = \frac{3.82 \cdot 90}{0.25}$$

$$RPM = \frac{343.8}{0.25}$$

$$RPM = 1375$$

Contrast Example One with Example Two.

EXAMPLE TWO: *Determine cutting speed and calculate spindle RPM for drilling a 1.25"-diameter hole in AISI-SAE 1020 steel with a Brinell hardness of 125 to 175.*

Since this is the same material, the cutting speed is still 90 surface feet per minute. Again insert 90 into the RPM formula for *CS*, but this time *D* = 1.25.

$$RPM = \frac{3.82 \cdot CS}{D}$$

$$RPM = \frac{3.82 \cdot 90}{1.25}$$

$$RPM = \frac{343.8}{1.25}$$

$$RPM = 275$$

Note that a larger tool diameter used to machine the same material at the same cutting speed will produce a slower spindle speed. Conversely, smaller tool diameters will yield faster spindle speeds.

EXAMPLE THREE: *Determine cutting speed and calculate spindle RPM for drilling a 1/4"-diameter hole in AISI-SAE 1050 with a Brinell hardness of 325 to 375.*

First find 1050 in the left column of the chart and then locate the correct Brinell hardness row. Following that row across to the drilling column shows a cutting

speed of 35 surface feet per minute. Insert 35 into the RPM formula for *CS* and 0.25 into the formula for *D*.

$$RPM = \frac{3.82 \cdot CS}{D}$$

$$RPM = \frac{3.82 \cdot 35}{0.25}$$

$$RPM = \frac{133.7}{0.25}$$

$$RPM = 535$$

Contrast this answer with Example One. The same-diameter drill is used, but the harder material requiring a lower cutting speed results in a much slower spindle speed. Conversely, when the same-diameter tool is used to machine softer materials, faster spindle speeds will result from the RPM calculation formula.

Feed Rates for Drill Press Operations

Feed rates for drill press operations are specified in **IPR (inches per revolution)** or **FPR (feed per revolution)**. IPR and FPR refer to the distance that a cutting tool advances into a workpiece for each revolution of the cutting tool. For example, 0.003 IPR (or FPR) means that the cutting tool would advance into the work 0.003" every time the tool makes one complete revolution. **(See Figure 4.3.4.)**

Smaller-diameter holemaking tools require lighter feed rates to prevent tool breakage, while larger-diameter tools can be fed at heavier rates because the larger tools are much more solid and can withstand more pressure. **Figure 4.3.5** shows a chart of recommended feed rates for twist drills. The lower values in each range should be used for machining harder metals with lower machinability ratings such as tool steels and superalloys. Higher values can be used for softer metals such as aluminum and brass with higher machinability ratings.

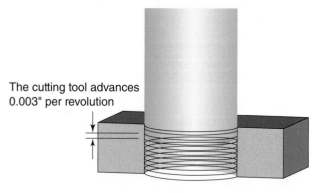

The cutting tool advances 0.003" per revolution

FIGURE 4.3.4 A cutting tool is advanced into the workpiece at a rate of 0.003" per revolution.

Diameter Range	FPR—Light Conditions	FPR—Medium Conditions	FPR—Heavy
inch	inch	inch	inch
1/16–1/8	0.0005–0.001	0.001–0.002	0.002–0.004
1/8–1/4	0.001–0.003	0.003–0.005	0.004–0.006
1/4–3/8	0.003–0.005	0.005–0.007	0.006–0.010
3/8–1/2	0.004–0.006	0.005–0.008	0.008–0.012
1/2–3/4	0.005–0.007	0.007–0.010	0.009–0.014
3/4–1.0	0.007–0.010	0.009–0.014	0.014–0.020

FIGURE 4.3.5 A chart provides guidelines for appropriate twist drilling feed rates.

LOCATING HOLES ON THE DRILL PRESS

Holes produced on a drill press are frequently located according to a layout, so it is important that the layout is accurate enough to meet required tolerances. Scribing a circle equal to the desired hole diameter is helpful to provide a visual reference to monitor the actual location of the hole being created. If a prick punch mark was created during layout, it can be slightly enlarged with a center punch to help keep the drilling tool on the correct location.

If the workpiece is to be held in a vise, V-block, or clamped to an angle plate, it should be secured in the workholding device. This workholding device with the mounted workpiece can then be positioned so the hole to be drilled is aligned with the center of the drill press spindle. If the workpiece is to be directly clamped to the table, position it initially so the desired hole location is aligned with the center of the drill press spindle. Clamps can then be positioned and tightened down to securely hold the part. When using a vise, position it on the table with the handle end at approximately a 10 o'clock position. In this position, if clamps fail, the vise would only rotate slightly before hitting the column and stopping instead of rotating into the operator. **(See Figure 4.3.6.)**

A **center finder**, or **wiggler**, is sometimes used to position the workpiece under the center of the drill press spindle. It contains a small chuck with several different attachments. **(See Figure 4.3.7.)** Follow these steps to locate a hole using the center finder.

- Mount the pointed attachment, or probe, snugly in the center finder's chuck. The probe will still be able to be moved in its chuck.
- Mount the center finder in the drill chuck of the drill press.
- Visually align the pointed probe vertically.
- Turn on the spindle and set RPM between 500 and 1000.

FIGURE 4.3.6 A vise positioned on the table, so that if the clamps were to fail, the vise would hit the column instead of the operator.

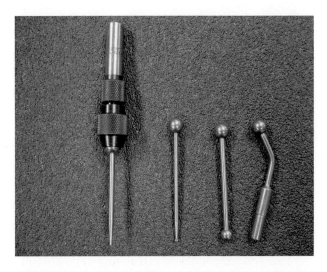

FIGURE 4.3.7 A center finder set with its various attachments. The pointed probe is normally used for establishing a hole location on the drill press.

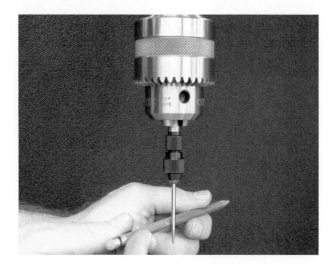

FIGURE 4.3.8 A pencil being used to true a center finder for use.

- Using a small-diameter rod or pencil to press against the side of the probe, gently apply pressure in toward the center of the spindle until the point runs true. **(See Figure 4.3.8.)**
- Stop the machine spindle.
- Position the workpiece can by bringing the quill down near a punch mark or scribed intersection on the workpiece, as shown in **Figure 4.3.9**. The work can be positioned by tapping it or the workholding device with a soft hammer to adjust location.
- When location is established, tighten clamps to secure the workpiece to the machine table.

A *pointed edge finder* (**Figure 4.3.10**) can be used much the same way as the wiggler to position a workpiece.

- Mount the pointed edge finder in the drill press spindle.
- Turn on the spindle and set RPM between 500 and 1000.
- Align the tip to run true in the same manner used to true the wiggler.
- Position the workpiece so that the desired hole location is aligned with the point. **(See Figure 4.3.11.)**
- The pointed edge finder tip can also be brought into very light contact with a punch mark while rotating.
- When the tip is running true, the punch mark is aligned, and the location is established.
- If the tip is not running true, adjust the workpiece until the point runs true.
- Tighten clamps to secure the workpiece or workholding device.

Another method for aligning a workpiece when a punch mark is present is to use a center drill (combination drill and countersink) mounted in the drill press spindle.

FIGURE 4.3.9 Prior to clamping the vise to the table, the work is positioned by aligning the center finder point with a punch mark.

FIGURE 4.3.10 A cone-pointed edge finder.

- Approximately locate the workpiece beneath the center drill.
- Tighten the work holding clamps with light pressure so that the vise or workpiece can still "float" or move beneath the clamps.
- Start the spindle and bring the center drill into light contact with the punch mark. The punch mark will cause the workpiece to move slightly so that the punch mark is aligned with the rotating drill.
- Retract the quill.
- Stop the spindle.
- Secure clamps.

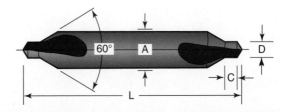

Size	Body Dia (A)	Drill Dia (D)	Drill Length (C)	OAL (L)
00	1/8	0.025	0.030	1 1/8
0	1/8	1/32	0.038	1 1/8
1	1/8	3/64	3/64	1 1/4
2	3/16	5/64	5/64	1 7/8
3	1/4	7/64	7/64	2
4	5/16	1/8	1/8	2 1/8
5	7/16	3/16	3/16	2 3/4
6	1/2	7/32	7/32	3
7	5/8	1/4	1/4	3 1/4
8	3/4	5/16	5/16	3 1/2

FIGURE 4.3.12 A chart shows the dimensions of standard center drills.

FIGURE 4.3.11 Prior to clamping the vise to the table, the work is positioned by aligning the pointed edge finder point with a punch mark.

 **CAUTION**

When using this method for establishing location, shut off the spindle and allow it to come to a stop before tightening clamps. Do not reach around the rotating drill press spindle to tighten the clamps.

SPOT DRILLING

After the workpiece is positioned and securely clamped, a center drill (combination drill and countersink) or spot drill is usually used to create a more positive starting point for a twist drill. This operation is sometimes called **spotting**, and the mark created can be called a *spot*. Select a drill with a pilot diameter that is slightly larger than the dead center of the twist drill to be used. The chart in **Figure 4.3.12** shows sizes of center drills and their pilots. If using a spot drill, the spot drill's diameter is not as critical, but if using a spot drill larger than the desired hole diameter, be sure to avoid drilling too deep and creating a diameter larger than the desired hole size. When calculating spindle speed for a spot drill, use the drill diameter. Theoretically, when calculating spindle speed for a center drill, the pilot diameter should be used. However, this often results in

an impractical spindle speed for smaller center drills, so some judgment must be used to set a realistic speed.

A good rule for spot depth is to drill to a depth that is equal to about 1/2 the length of the point of the twist drill to be used. At this depth, if there is a location error, steps can be taken to adjust the hole location. If the depth of the spot is equal to or greater than the length of the twist drill point, a location error is very difficult to correct. The length of a drill point depends on the included angle of the tip. **Figure 4.3.13** shows how to calculate drill tip length for some common drill point angles. Only use these quick formulas as estimates. Since they do no account for the flat on the drill point created by its dead center, the length of the point will actually be slightly less than the quick calculation. This results in a slightly deeper result than expected. If a more accurate drill point length is desired, one way to measure its length is by using an optical comparator.

- 90° included angle tip: $0.5 \cdot$ drill Ø = Tip length
- 118° included angle tip: $0.3 \cdot$ drill Ø = Tip length
- 135° included angle tip: $0.207 \cdot$ drill Ø = Tip length

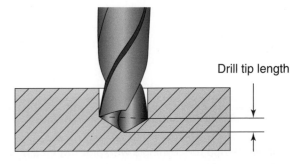

FIGURE 4.3.13 Drill tip length can be estimated by using a formula.

FIGURE 4.3.14 A hole being spotted with a center drill.

The depth stop on the drill press quill can be used to set the depth of the spot.

- Adjust the depth stop so that the tip of the spot drill nearly touches the surface of the workpiece. A piece of paper can be used as a feeler gauge to establish the distance between the workpiece and the drill tip if there is no punch mark on the workpiece.

- Retract the quill.

- Move the stop the amount of the desired spot depth.

- Start the spindle and set the appropriate speed.

- Use the feed handle to feed the drill to the desired depth. Apply an ample amount of an appropriate cutting fluid while drilling and feed center drills lightly to prevent the pilot from being broken off in the work.

- Retract the quill frequently to clear chips from the point. This alternating drilling and retracting motion is called **pecking**, or **peck drilling**. **Figure 4.3.14** shows spotting with a center drill.

⚠ CAUTION

Do not feed a spot drill past its point because there is no body clearance on a spot drill. If the body is fed into the workpiece it may heat up and break, throwing fragments that can cause injury.

DRILLING

After spotting, the next step in the process is to select and mount the desired twist drill in the machine spindle. When using a straight-shank drill, follow these steps.

- Open the drill chuck and insert the shank portion of the drill into the drill chuck.

- The drill bit should be inserted far enough into the chuck that the entire shank is firmly held, however, the drill bit should never be so far in the chuck that the chuck jaws will tighten onto the flutes. This may cause damage to the drill or chuck jaws.

- If possible, insert the drill so that the end of the shank bottoms in the drill chuck. This prevents the drill from being pushed back into the chuck when drilling.

- When the desired hole is larger than 1/2" in diameter, a smaller hole called a **pilot hole** is often drilled before the larger drill bit is used. For the pilot hole, select a drill that is slightly larger than the width of the dead center of the larger drill size. **(See Figure 4.3.15.)** The pilot hole creates relief for the dead center and reduces the pressure required to feed the drill into the work.

- Secure the workpiece.

- Calculate and set the proper spindle speed using one of the formulas discussed previously.

- On a drill press with a power-feed option, set the power feed to an appropriate rate.

- Start the spindle and begin to feed the drill into the workpiece manually using the feed handle. Manually feeding the drill press gives the operator more control.

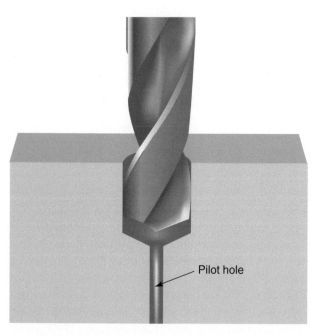

Pilot hole

FIGURE 4.3.15 A pilot hole is drilled to a diameter slightly larger than the width of the larger drill's dead center.

- Before the drill cuts to its full diameter, stop the spindle and check the distance between the edge of the cone-shaped depression being cut by the drill point and the edge of the laid-out circle.
- If the hole is out of position, the drilling operation can be stopped and the hole relocated by creating a mark with a chisel as shown in **Figure 4.3.16**. It is

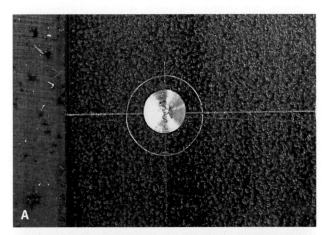

FIGURE 4.3.16 Steps can be taken using a chisel to pull a drill point back on center. (A) The drill has started off location. (B) Make a small mark with a chisel along the cone in the direction the hole needs to be moved. (C) The lips of the drill will be pulled by the chisel mark and bring the drill to the proper location.

important to perform this visual inspection because once the drill reaches its full diameter, the location can no longer be adjusted using this method because the body is contained by the walls of the hole.

- After the hole location is confirmed or corrected, apply a liberal amount of an appropriate cutting fluid to the drill and/or on the workpiece.
- Feed the quill using the feed handle, or engage the power-feed mechanism if available, to drill to the desired depth.

Chips should curl out of the drill flutes and be of equal size. This indicates a properly sharpened drill. Peck drilling will break chips into short pieces so they fall away from the drill. It also clears the chips out of the hole and allows cutting fluid to reach the tip of the drill. If the drill is not retracted frequently when drilling deep holes, chips can accumulate in the flutes causing the drill to bind and break inside the hole. A high-pitch squealing noise indicates that spindle speed is too high, or that the feed pressure is too light. When drilling ferrous metals with HSS drills, brown or blue chips indicate that the feed rate is too high.

ⓘ CAUTION

Peck drill to avoid long stringy chips that can wrap around the spindle or whip around the work area.

Through and Blind Holes

Holes that are drilled entirely through a workpiece are called **through holes** (or **thru holes**). Holes that are only drilled partially through a workpiece are called **blind holes**. The depth of a blind hole is specified as full-diameter depth. This is the depth of the hole not including the angled bottom portion created by the drill point. **(See Figure 4.3.17.)**

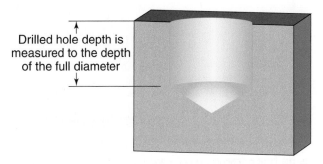

Drilled hole depth is measured to the depth of the full diameter

FIGURE 4.3.17 A blind hole is dimensioned to its full-diameter depth, not the depth to its point.

To drill a blind hole, the depth stop on the drill press can be used to limit the distance of the quill feed. There are a few ways to set and check the depth of a blind hole.

- If depth is not critical, the quill can be set against the depth stop just as the drill begins cutting its full diameter. Then the stop can be moved a distance equal to the desired depth. **(See Figure 4.3.18.)**

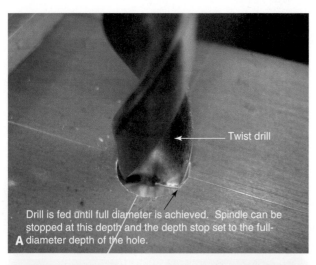

Twist drill

Drill is fed until full diameter is achieved. Spindle can be stopped at this depth and the depth stop set to the full-diameter depth of the hole.

A

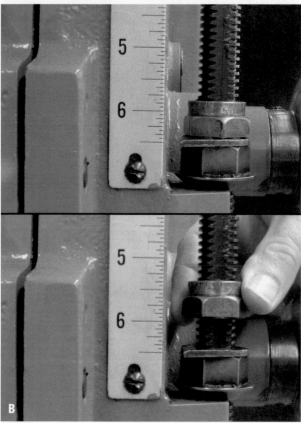

B

FIGURE 4.3.18 (A) Once a drill is engaged with the work to its full diameter with the quill against the depth stop, (B) the depth stop can be adjusted for the desired full-diameter depth.

- The drill tip can also be brought into light contact against a spacer on the top surface of the workpiece with the quill against the depth stop. The length of the drill point and the spacer thickness can be added to the desired depth and then the stop moved the total amount. **(See Figure 4.3.19.)** Then start the spindle and drill the hole until the quill reaches the depth stop.

- If depth is critical, use this same method but set the depth stop so the hole depth is shallow. After reaching the stop, depth can be checked by placing a pin gage in the hole and measuring the exposed length of the pin. Subtracting that value from the overall pin length will give the depth of the drilled hole. **(See Figure 4.3.20.)** Then adjust the depth stop until the desired depth is reached.

On drill presses with a power-feed mechanism, the depth stop will usually disengage the power feed slightly before reaching the full depth. The quill can be fed manually to reach the full depth.

It is good practice to set the depth stop even when drilling through holes to avoid accidentally drilling into the workholding device or worktable, or hitting the

FIGURE 4.3.19 Another method of setting the depth stop requires touching the drill tip on a spacer of known thickness on the work surface when the quill is against the depth stop and then adjusting the stop to the desired full-diameter depth plus the drill tip length and the spacer thickness.

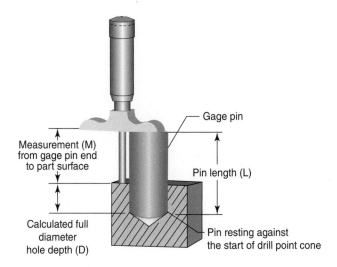

FIGURE 4.3.20 A gage pin can be used to check the full-diameter depth of a hole.

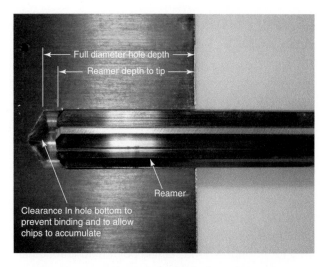

FIGURE 4.3.21 A reamer should have a small amount of clearance in the bottom of a drilled hole so it doesn't bottom out.

top surface of the workpiece with a drill chuck, sleeve, extension socket, or the face of the spindle. When manually feeding while drilling a through hole, decrease feed pressure when breaking through the bottom of the workpiece to avoid binding or breaking the drill.

REAMING

When a hole must be machined to a more accurate diameter or surface finish than can be produced by drilling, a reaming operation may be selected to meet requirements. Reaming slightly enlarges an existing hole and provides a very smooth surface, so a hole must first be drilled with a standard twist drill before reaming. Drill size for creating the initial hole should be below the final size by the following amounts.

- 0.010" for a hole up to 1/4"
- 0.015" for a hole between 1/4" and 1/2"
- 0.025" for a hole between 1/2" and 1-1/2"

When too little material is left in a hole, the reamer may only rub against the side of the hole instead of actually cutting. If too much material is left in a hole, material will extend beyond the reamer's cutting edges. This can cause the reamer to bind and break.

 CAUTION

Too much material in the hole can cause the reamer to wear prematurely or to break and throw sharp fragments, leading to injury.

After drilling, it is a good idea to ream the hole without moving the position of the workpiece to ensure accurate alignment. To ream a hole, follow these steps.

- Mount the desired reamer in the drill press.
- Calculate and set an appropriate spindle speed for the reamer. If reaming cutting speeds are not listed on an available cutting-speed chart, use a speed about 50% to 60% of that used for drilling the same material.
- When reaming, IPR should be about twice that for the same size drill and material type, so set the feed rate accordingly if using a drill press with a power-feed mechanism.
- Apply an ample amount of appropriate cutting fluid to the reamer and start the machine spindle.
- If feeding the reamer manually, maintain a steady rate to produce a consistent surface finish. Move quickly enough so that the reamer cuts freely, but not too quickly that it is forced. Pecking is generally not needed except when reaming deeper than the length of the flutes. In this case, the pecks should be about the length of the flute to keep chips from packing in the flutes and scoring or gouging the walls of the hole.

Blind hole depth for reamed holes is specified in the same manner as drilled holes. The depth can be set and checked similar to the methods used when drilling blind holes. When reaming a blind hole, keep in mind that the flattened end of the reamer prevents it from traveling as deep as the drill. The drilled hole will need to be slightly deeper than the required reamer depth. This will keep the reamer from bottoming out in the drilled hole. **(See Figure 4.3.21.)**

 **CAUTION**

Forcing a reamer against the bottom of a drilled hole can result in reamer breakage.

COUNTERBORING AND SPOTFACING

Counterboring and spotfacing can be performed on the drill press using the depth stop to machine the desired depth in the same manner as when drilling and reaming. Counterbores have small pilot diameters on the cutting end. These pilots are used to align the counterbore with an existing diameter. Be sure to use a pilot with a diameter that is about 0.003" to 0.005" smaller than the existing hole to prevent binding. Counterboring and spotfacing speeds, like reaming, should be about 50% to 60% of those used for drilling the same material. Feed rates for counterboring are usually lighter than those used for drilling. A good starting point is about one-half of the rate used with a similar-sized drill.

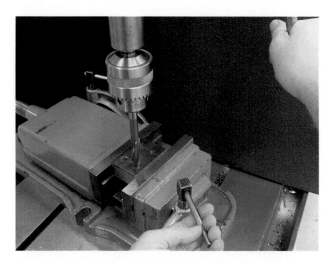

FIGURE 4.3.22 The pilot of the counterbore is used to check alignment with the hole. When the pilot slips into the hole without binding or flexing the counterbore tool, tighten the clamps.

 CAUTION

If the pilot is too large, the tool may seize in the hole, break, and throw sharp fragments that can cause injury.

It is usually easier to perform counterboring or spotfacing immediately after drilling so the workpiece does not need to be repositioned. Follow these steps to perform counterboring.

- Mount the cutting tool in the spindle. When possible, bottom the shank of a straight-shank tool in the drill chuck.

- If the work was moved after drilling, use the pilot to align the workpiece. Adjust the location of the work until the pilot slips inside the existing hole without binding or flexing. Then clamp the work securely. **(See Figure 4.3.22.)**

- Use the feed handle to bring the cutting edge of the counterbore within about 1/32" of the workpiece surface and set the depth stop at this location.

- Calculate and set spindle RPM, then start the spindle.

- Make small adjustments of about 0.001" to 0.003" to the depth stop and bring the counterbore against the stop after each adjustment. When the tool begins to touch the surface of the part, a reference point has been set.

- Adjust the depth stop to the required counterbore or spotface depth.

- Apply cutting fluid to the tool's cutting edges and feed the quill until it reaches the desired depth set by the stop. **Figure 4.3.23** shows this method of setting depth.

FIGURE 4.3.23 (A) Adjust the depth stop in small increments until the counterbore lightly touches the surface of the work. (B) Then the depth stop can be adjusted to counterbore to the desired depth.

- If depth is critical, set the depth stop short of the desired depth, then stop and measure the depth before making adjustments to reach final depth. When counterboring and spotfacing, chips should form much like when drilling.

 CAUTION

Peck and retract the tool occasionally to release chips before they become long and hazardous.

CHAMFERING AND COUNTERSINKING

A countersink is used to perform countersinking or chamfering operations on existing holes. Speeds for countersinking are normally about 25% of those used when drilling the same material. Feeds for countersinking are usually light, just as when counterboring or spotfacing, because of the large amount of surface contact between the tool and the workpiece.

Feed depth amount for a countersink depends on the angle of the tool, the existing hole size, and the desired countersink (or chamfer) diameter. When using a 90-degree countersink, feed depth can be found using the following formula:

$$\frac{D - d}{2},$$

D = desired countersink or chamfer diameter
d = existing hole diameter.

Figure 4.3.24 illustrates this concept and shows variations of this formula for different countersink angles.

A countersink will center itself in an existing hole, so the workpiece or vise is normally allowed to float on

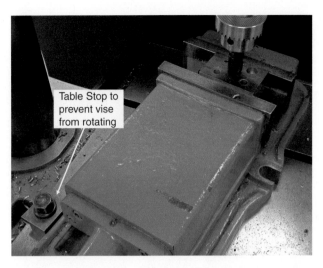

FIGURE 4.3.25 A countersinking operation and proper position of a table stop to prevent vise movement.

the drill press table instead of being tightly clamped. Instead, a positive stop is clamped to the worktable to prevent the vise or workpiece from rotating. To chamfer or countersink a hole, follow these steps.

- Mount the tool in the spindle.
- Calculate and set spindle RPM.
- Set the depth stop so the tool is within about 1/32" of touching the edge of the existing hole.
- Adjust the depth stop in small increments of about 0.001" to 0.003" until the countersink touches the edge of the existing hole.
- Adjust the depth stop to achieve the required depth.
- While countersinking, the vise (or workpiece if large enough to be placed directly on the table) will move slightly in a small circular motion. This is the tool continually centering itself in the hole and is normal.

Figure 4.3.25 shows a countersinking operation and proper position of a table stop to prevent vise movement.

TAPPING

The drill press can also be used to properly align taps, eliminating the need for using a square as when hand tapping. If the workpiece has been moved between the drilling and tapping operation, the hole can be realigned with the spindle using a center point mounted in the spindle, as shown in **Figure 4.3.26**. Clamp the vise or workpiece securely to the table before tapping to prevent the work from moving out of position.

Countersink Included Angle	Formula for F (Amount of Feed)
60	$F = \left(\dfrac{D - d}{2}\right) \times 1.732$
82	$F = \left(\dfrac{D - d}{2}\right) \times 1.15$
90	$F = \left(\dfrac{D - d}{2}\right)$
100	$F = \left(\dfrac{D - d}{2}\right) \times 0.839$

FIGURE 4.3.24 Calculating depth for a desired countersink diameter.

FIGURE 4.3.26 A hole to be tapped can be realigned with the spindle by bringing a spindle-mounted center point inside a hole and against the work. The vise "floats" into position and is then clamped securely to the table before tapping.

After securing the workpiece, follow these steps.

• Mount the center point in a spindle-mounted chuck. This center point fits in a hole in the top of the tap or in the back of a tap wrench/handle.

• Mount the tap in a tap wrench/handle.

• Place the tap in the top of the drilled hole.

• Lower the quill so the tip of the tap center is in the hole in the tap or tap wrench/handle.

• If using a spring-loaded center, bring the center into contact with the tap or wrench/handle by lowering the quill until the spring is compressed. Lock the quill in position with the quill lock. **(See Figure 4.3.27.)**

• If using a solid tap center, one hand must apply light pressure to the quill feed handle to keep the tap

FIGURE 4.3.27 A spring-loaded tap center maintains alignment and applies pressure downward when starting a tap.

FIGURE 4.3.28 A solid tap center is used to maintain alignment when starting a tap. Quill pressure is applied with one hand to help the tap engage while turning the tap wrench with the other hand.

aligned while turning the tap with the other hand, as shown in **Figure 4.3.28**.

• Turn the tap about one half to one full turn at a time.

- After each forward motion, back the tap out one half turn before continuing. This helps break the chip being formed as it cuts. Not backing the tap up will often cause the tap to bind in the hole and break.

⊗ CAUTION

Mounting a tap in a drill chuck and running the spindle to drive a tap into a workpiece is not recommended. This can cause taps to shatter.

Sometimes a thread depth is specified for a threaded hole. Remember that thread depth is like drill depth. The specified depth is the amount of full, usable threads, not just the distance the tap was fed into the hole. It is useful to estimate the number of turns of a tap to achieve a certain depth. As a review of some information from Section 3.6, the number of turns for an inch-based thread can be determined by multiplying the desired depth by the number of threads per inch (TPI). For example, if a 3/8-16 thread must be 5/8" deep, estimate the number of turns by multiplying 5/8 by 16. Ten turns would be needed. When using metric taps,

the number of turns can be determined by dividing the desired depth by the pitch. For example, if a M8 × 1.25 thread must be 12 mm deep, estimate the number of turns by dividing 12 by 1.25. That answer is 9.6, so just over 9 ½ turns would be needed. Remember that these are estimates because of the different tap chamfers available. Depth will most likely not be deep enough. To check depth after tapping the estimated number of turns, first remove the tap and clean the hole. Then follow these steps.

- Measure the entire length of a screw that is the size of the threaded hole. (A plug gage could also be used.)

- Thread the screw (or gage) into the tapped hole until it stops.

- Use a caliper or depth micrometer to measure the exposed length of the screw (or gage).

- Subtract that measurement from the entire length of the screw (or gage).

- The result is the usable depth of the thread. **(See Figure 4.3.29.)**

- If the threads are not deep enough, subtract that actual depth from the required depth.

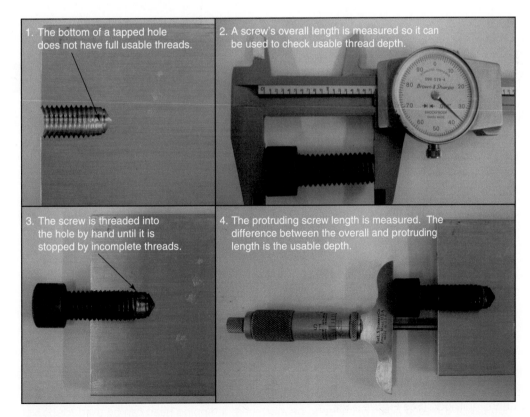

FIGURE 4.3.29 The usable depth of thread can be checked using a screw. The screw is turned into the hole until it bottoms. The protruding length is measured and is subtracted from the overall screw length.

- For inch-based threads, multiply that difference by the TPI to determine how many more turns of the tap are needed. For example, suppose the 3/8-16 thread is only 1/2" deep instead of 5/8". An additional 1/8" of thread is needed. Since $1/8 \times 16 = 2$, two additional turns are needed to reach the minimum depth of 5/8".

- For metric threads, divide that difference by the pitch to determine how many more turns of the tap are needed. For example, suppose the M8 × 1.25 thread is only 9.5 mm instead of 12 mm. An additional 2.5 mm of thread is needed. Since $2.5 \div 1.25 = 2$, two additional turns are needed to reach the minimum depth of 12 mm.

- Recheck to verify correct depth.

When a large number of holes needs to be tapped using a drill press, turning the tap by hand can be a slow, tiresome process. A **tapping head** is a specially designed tool that allows the spindle to be run to drive taps. A tapping head contains an adjustable clutch that limits torque to prevent tap breakage. It also contains a mechanism that reverses tap direction when the quill is raised to retract the tap from the hole. **Figure 4.3.30** shows a tapping head.

Courtesy of Tapmatic Corporation

FIGURE 4.3.30 A reversible tapping head is used to advance and retract a tap using machine power.

SUMMARY

- The drill press can be used to safely perform several types of holemaking operations by observing some basic safety guidelines.

- Different cutting speeds are required for different materials and are used to calculate spindle RPM for the given type of cutting tool and operation. Application of appropriate speeds and feed rates will result in safe, efficient drill press operations.

- After mounting a workpiece on the drill press, hole locations can then be established so that spotting can be performed to create a positive starting point for a twist drill.

- Drilling can be used to create through or blind holes.

- Reaming can be performed if a hole requires a more accurate size and smoother surface finish than can be created by drilling alone.

- Counterboring and spotfacing can be performed on existing holes using the drill press depth stop to machine to a desired depth.

- The depth stop can also be used when performing countersinking or chamfering operations on the drill press.

- The drill press is useful when tapping holes by using the spindle to keep taps perpendicular to the top surface of a workpiece.

- If a large number of threaded holes will need to be machined, a tapping head can be used for power tapping.

REVIEW QUESTIONS

1. List five drill press safety guidelines.
2. Briefly define the term *cutting speed*.
3. What unit of measure is used to define cutting speeds?
4. What is *feed,* and how is feed stated for drill press operations?
5. Briefly describe two methods for locating hole positions on the drill press.
6. What is *spotting* and why is it performed?
7. What is pecking, or peck drilling?
8. Briefly contrast the speed and feed differences between drilling and reaming operations.
9. What drill press component can be used to control counterbore depth?
10. Calculate feed depth for a 90-degree countersink used to machine a 0.40"-diameter chamfer on an existing 1/4"-diameter hole.
11. Calculate feed depth for an 82-degree countersink used to machine a 1/2"-diameter countersink on an existing 9/32"-diameter hole.
12. What benefit does tapping on the drill press offer over hand tapping?
13. How many turns must a 1/4-20 tap be turned into a hole to produce 3/8" of threads in a hole? (Disregard any chamfer on the tap.)
14. How many turns must a 5/16-18 tap be turned into a hole to produce 9/16" of threads in a hole? (Disregard any chamfer on the tap.)

SECTION 5 TURNING

Introduction to the Lathe

UNIT 1

Learning Objectives

After completing this unit, the student should be able to:

- Explain the principal operation of a lathe
- Identify and explain the functions of the parts of the lathe
- Explain how lathe size is specified

Key Terms

Apron	Feed rod	Spindle nose
Bed	Gap bed	Step cone pulley
Carriage	Gear box	Swing
Compound rest	Gib	Tailstock
Cross slide	Half nut lever	Thread dial
Feed change lever (feed change knob)	Headstock	Ways
	Leadscrew	Workholding devices
Feed control lever	Longitudinal feed	
Feed rack	Saddle	
Feed reverse lever (feed reverse knob)	Spindle	
	Spindle clutch lever	

INTRODUCTION

The engine lathe is one of the most versatile and oldest machine tools in the machining field. The principal operation of the engine lathe is to rigidly hold and rotate a workpiece against a cutting tool. The tool travels along the outside of the workpiece to shave off material and produce cylindrical parts. **(See Figure 5.1.1.)** The lathe is used to produce a wide variety of parts. **(See Figure 5.1.2.)**

The four main sections of the lathe are the bed, the headstock, the tailstock, and the carriage.

FIGURE 5.1.1 The lathe holds and rotates work against a cutting tool to machine cylindrical components. Movement of the cutting tool determines the shape of the cylindrical part.

FIGURE 5.1.2 Some examples of components machined on lathes.

 CAUTION

The engine lathe, like all industrial machinery, is very powerful and potentially dangerous. It is important to always wear the proper PPE when operating the lathe. This consists of safety glasses with the ANSI Z87 rating and some type of heavy work boot, preferably with safety toes. To avoid becoming entangled in the moving parts of the machinery, shirttails should always be tucked in and shirtsleeves should be rolled up above the elbow. Jewelry such as rings, necklaces, and bracelets or watches should also be removed. Long hair should be safely secured so it does not become tangled in moving machine parts.

HEADSTOCK

The **headstock** is located at the upper-left side of the engine lathe. It is comprised of a casting that contains the mechanisms used to hold and rotate the workpiece and control the rate of tool movement. **Figure 5.1.3** identifies the lathe headstock.

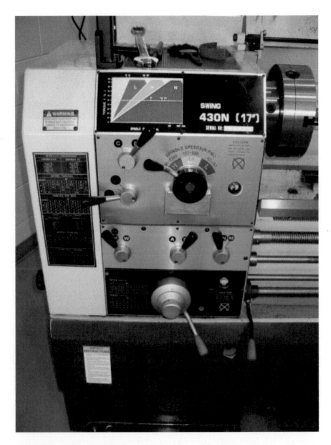

FIGURE 5.1.3 The headstock of a lathe contains the spindle and the belts or gears that power the spindle and transmit power to guide the cutting tool.

Spindle

One of the most important parts of the headstock and lathe itself is the **spindle**. The spindle is the part of the machine that holds and rotates work during machining. It is a precisely ground hollow tube supported by precision bearings. Long workpieces can be placed through the center hole of the spindle. The front of the spindle hole contains an internal taper that can be used to align and secure tapered workholding accessories. A knockout bar with a soft metal end is inserted through the hollow spindle to remove these accessories by forcing the mating tapers loose from each other. **Figure 5.1.4** shows an illustration of a lathe spindle. The spindle is driven by a heavy-duty electric motor through a series of either pulleys or gears. The spindle is usually started and stopped by a lever-controlled clutch, but some lathes use on/off buttons to start and stop the spindle. There is also a control to operate the spindle in either a forward or reverse direction.

CAUTION

Do not use a hard steel knockout bar, or the spindle bore and workholding accessory can be damaged.

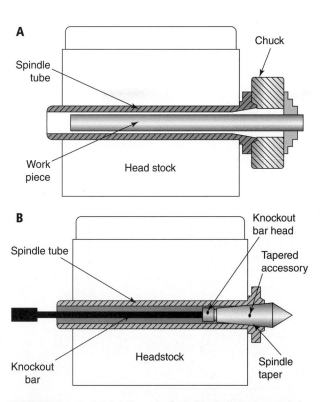

FIGURE 5.1.4 (A) A lathe spindle holds and rotates work for machining. Its hollow construction allows long workpieces to be placed through it. **NOTE: Material should never extend out of the left side of spindle bore, or imbalance can cause whipping and ultimately injury or machine damage.** (B) The tapered front section aligns and secures accessories that can be removed with a knockout bar.

Spindle Noses

The headstock **spindle nose** is used to attach various **workholding devices** to the spindle. Workholding devices are the accessories mounted to the spindle used to secure material for machining operations. Many types of spindle noses are used by lathe manufacturers.

The threaded spindle nose (**Figure 5.1.5**) is common in older lathes but is seldom used today. Workholding devices are simply threaded onto the spindle nose and tightened.

The tapered spindle nose (sometimes called "L-taper") has a long taper with a key and a threaded collar. The taper and key align with a mating taper and keyway on a workholding device for alignment. The threaded collar is tightened onto the threads of the workholding device. **Figure 5.1.6** shows a tapered spindle nose.

The cam-lock spindle nose has a short taper that fits into a mating taper on a workholding device. The

FIGURE 5.1.5 The threaded spindle nose is not very common in lathes today.

FIGURE 5.1.6 A tapered spindle nose, sometimes called an L-taper.

FIGURE 5.1.7 A cam-lock spindle and some mating work-holding devices.

workholding device has pins that fit into the holes in the spindle nose. A special wrench is then used to tighten the cam-lock pins on the spindle nose to secure the device. **Figure 5.1.7** shows a cam-lock spindle nose.

 CAUTION

Never attempt to mount a workholding device to a threaded spindle nose by running the spindle under power and holding the chuck. Always thread it on by hand with the spindle stopped.

Belt Drive Lathes

A belt drive lathe transmits power from the motor to the spindle by means of belts and what is known as a **step cone pulley**. The step cone pulley has diameters of different sizes that change ratios to achieve different spindle speeds. To change the spindle speed of belt-driven machines, first the tension on the belt must be released. This is usually done by lifting the motor itself upward with

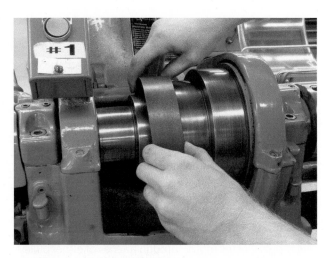

FIGURE 5.1.8 Changing the belt position on a lathe spindle with a step cone pulley to change spindle RPM.

a lever. The belt is then moved to the desired pulley and the belt tension reapplied. A chart is usually located on the machine that will indicate which step to use for a specific RPM setting. **Figure 5.1.8** shows a step cone pulley.

Geared Head Lathe

As its name implies, a geared head lathe uses a series of gears to transmit power from the motor to the spindle. Larger lathes are usually driven in this manner because the gears can transmit considerable amounts of power without slipping, as is needed during heavy machining operations. To set the speed on a geared head lathe, levers or knobs located on the front of the headstock are set to various positions. A chart on the headstock will show the positions of these levers needed to set specific spindle speeds. **Figure 5.1.9** shows how these levers are used to set speeds on a typical geared head lathe.

FIGURE 5.1.9 Setting spindle RPM on a geared head lathe is done by placing levers in the positions indicated by a chart on the lathe's headstock.

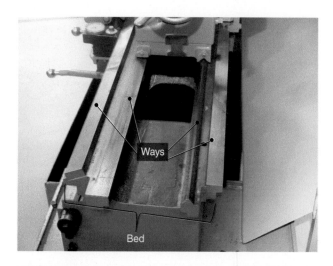

FIGURE 5.1.11 The bed is the foundation of the lathe. The ways are precision-ground rails that are frequently flame hardened for durability. To prevent excessive wear and damage, do not lay tools on the ways.

LATHE BED

The lathe **bed** is located on the right side of the headstock and is the foundation of the entire machine. Lathe beds are heavy castings designed to both be strong enough to handle heavy machining forces and to ensure smooth, precise cutting motion. The top of the bed contains precision-ground flat and V-shaped rails known as the **ways**. **Figure 5.1.11** identifies the bed and the ways of a lathe. Most lathe ways are flame hardened for strength and wear resistance. Since the ways are so hard, it is important to protect them from sudden shocks that can cause damage and affect machine accuracy. Be careful not to drop accessories or tools on the ways. It is a good habit to avoid laying tools, especially hammers, wrenches, and files, on the ways. A wooden board can be placed on the ways to serve as a place to lay tools while also protecting the ways.

CARRIAGE

The **carriage** supports the lathe cutting tool and provides the tool with the movement needed to perform machining operations. It slides along the ways and contains two primary sections called the **saddle** and the **apron**. **Figure 5.1.12** shows the carriage of a lathe and identifies the saddle and apron, described next.

Saddle

The saddle is an H-shaped casting that slides back and forth on the ways. The apron is suspended from the saddle. The sliding motion of the saddle is parallel to the ways and is called **longitudinal feed**. The saddle supports both the **cross slide** and the **compound rest**.

CAUTION

Operating the lathe at an excessively high RPM can damage equipment and can be very dangerous. Always check to see if a workholding device is labeled with a maximum RPM rating. Be sure to never exceed that maximum speed or a workpiece may be ejected from the machine, causing serious injury or even death.

Quick-Change Gear Box

Just under the headstock is another gear train called the quick-change **gear box**. The quick-change gear box controls the rate of movement of the cutting tool. This tool movement is called *feed* and the rate of the movement is called the *feed rate*. Feed rates on the lathe are measured by how far the cutting tool advances each time the spindle turns one revolution. This is called *feed per revolution (FPR)*. Available feed rates on the lathe range from about 0.001" to 0.120" per revolution. A chart will show settings for knobs or levers required to set desired feed rates. **Figure 5.1.10** shows how feed rates are set on a lathe using the quick-change gear box.

CAUTION

Excessive feed rates can cause the cutting tool to shatter and/or the workpiece to be pulled out of the spindle workholding device.

FIGURE 5.1.10 The quick-change gear box changes the amount of feed per revolution by changing lever positions.

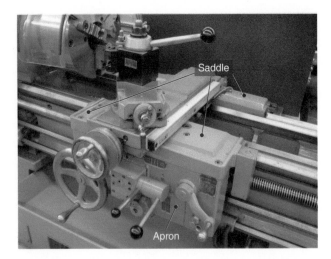

FIGURE 5.1.12 The carriage supports cutting tools and moves along the ways to provide tool movement. The top section that rests on the ways is the saddle and the apron is suspended from the saddle and hangs down in front of the bed.

Cross Slide

The cross slide is mounted on top of the saddle and provides cutting-tool movement perpendicular to the ways. The cross slide is attached to the saddle with a dovetail-shaped slide for smooth movement. The cross-slide movement is controlled by the cross-slide hand wheel. The hand wheel has a graduated micrometer collar that can be used to accurately control the amount of movement. Movement of the cross slide is called cross feed. **Figure 5.1.13** shows the cross slide.

Compound Rest

The compound rest is mounted on top of the cross slide and provides and allows angular tool movement. Just like the cross slide, the compound rest has a dovetail slide precisely controlled by a small hand wheel with a graduated micrometer collar. The compound rest can be rotated 360 degrees and set at any angular position to permit angular tool movements. First, locking screws are loosened. Then the compound rest is set to the desired angle using the angular graduations. Finally, the locking screws are tightened. **Figure 5.1.14** shows the compound rest and how it can be adjusted to different angles.

Gibs

Since both the compound rest and cross slide are used to control the tool motion while machining, it is important that both move smoothly and accurately. Over time, the sliding motion of both of these devices can cause wear on the dovetail slides. This wear can have an effect on machine accuracy and rigidity. A small wedge-shaped piece of steel or iron known as a **gib** is used to compensate for this wear. Gibs

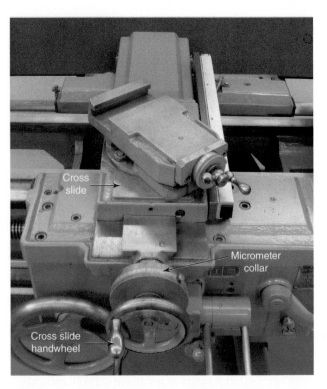

FIGURE 5.1.13 The cross slide provides tool movement perpendicular to the ways. A graduated micrometer dial on the hand wheel allows precise movements to be made.

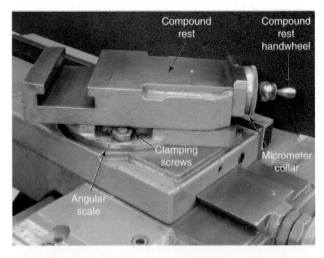

FIGURE 5.1.14 The compound rest can be adjusted to any angle and is used for tool movement other than longitudinal and cross feed.

are located near both the cross-slide and compound-rest hand wheels behind a small adjusting screw. When the motion of the compound or cross slide becomes too loose, the adjusting screw is tightened slightly to push the gib forward. The tapered shape of the gib causes it to become tighter as it moves forward, reducing clearance caused by wear. **Figure 5.1.15** shows a sample gib and location of the cross slide gib.

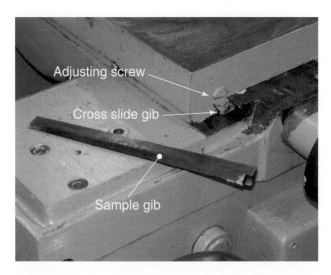

FIGURE 5.1.15 Adjusting the gib on the compound rest tightens the dovetail slide and reduces sloppiness caused by wear.

FIGURE 5.1.16 The feed rack, leadscrew, feed rod, and spindle clutch lever.

Also on the apron is the **feed control lever**. The feed control clutch causes machine-powered movement of either the carriage or the cross slide when engaged, depending on other machine settings.

The apron also contains the **feed change lever** or **feed change knob** that is used to switch between longitudinal and cross feed. A **feed reverse lever** or **feed reverse knob** will reverse the direction of the carriage or cross slide movement. The feed reverse control for some lathes is located on the headstock instead of the apron.

The **half-nut lever** is also located on the apron. It controls a split nut that can be engaged directly to the lathe's leadscrew to perform thread-cutting operations. The **thread dial** is used to determine the correct time to engage the half-nut lever when cutting threads.

There is also a second spindle clutch lever located on the apron on most lathes. **Figure 5.1.17** shows the apron and labels these parts.

⚠ CAUTION

Proper machine maintenance is very important. A machine in disrepair is not only unable to produce quality parts, it is also hazardous to the operator. Any faulty or broken equipment should be reported to the proper person immediately.

Leadscrew and Feed Rod

The **leadscrew** is a very long threaded rod supported by bearings at both ends. This screw is used to transmit motion to the carriage for thread-cutting operations.

The **feed rod** is a long shaft, either round or hexagonal, that transmits power to the carriage apron gear train. The gear train then uses this power to provide motion to either the cross slide or the carriage. The **feed rack** is a bar that spans the length of the bed and contains gear teeth. This rack meshes with a gear in the carriage to create longitudinal movement. A **spindle clutch lever** is also often located near the headstock where the leadscrew and feed rod enter the headstock. The leadscrew, feed rod, feed rack, and spindle clutch of a lathe are shown in **Figure 5.1.16**.

Apron

The apron is attached to the bottom of the saddle and hangs down in front of the bed. The apron contains a hand wheel that engages a gear with the feed rack to move the carriage along the ways. Some apron hand wheels have a graduated micrometer dial to allow precise control of the longitudinal carriage movement.

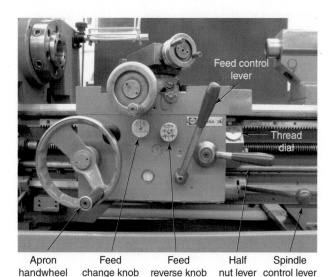

FIGURE 5.1.17 The apron contains several controls for the lathe.

TAILSTOCK

The **tailstock** can be used to secure workholding accessories to help support the workpiece in many operations. It can also hold cutting tools for performing standard holemaking operations. The tailstock, much like the carriage, slides longitudinally on the lathe's ways. After positioning, it can be locked to the ways with a locking lever or bolt(s). The tailstock has a hand wheel with a graduated micrometer collar and a precision-ground *quill* that fits inside the tailstock. The quill also moves longitudinally when the tailstock hand wheel is rotated. The quill has graduations, like on a rule, to show the amount of its movement. The graduations on the quill and the micrometer collar are very useful for controlling depth when performing holemaking operations. A locking lever is provided to lock the quill in any position.

The quill has a tapered hole through its center. The great majority of lathes use the Morse taper. This allows taper-shank tools such as chucks, drills, reamers, and counterbores to be used for holemaking operations. Workholding accessories used to help support workpieces can also be mounted in this Morse taper. The lathe tailstock body is constructed from two pieces so its alignment with the headstock can be changed by two adjusting screws (one on each side). **Figure 5.1.18** shows a lathe tailstock and identifies its parts.

LATHE SIZE

The size of a lathe is specified by measurements known as the swing and bed length. Other measurements are often specified as well. Refer to **Figure 5.1.19** while reading descriptions about lathe size.

Swing

The **swing** of the lathe is determined by the biggest-diameter workpiece that can be mounted in the spindle without touching the ways. Swing over the carriage is also often specified. That is the largest diameter that can be mounted in the spindle without touching the carriage. Some lathes are equipped with what is known as a **gap bed**. On a gap-bed lathe, a small portion of the lathe bed can be removed to allow larger-diameter workpieces to be machined. That portion is sometimes called the gap. **Figure 5.1.20** shows a gap-bed lathe.

Bed Length

The bed length is another factor that determines lathe size. Bed length is measured from the headstock to the end of the bed. Bed length is often mistaken to mean the maximum length of workpiece that can be machined on the lathe. However, this is not the case. The maximum workpiece length is determined by distance between workholding devices called *centers*. This measurement is called the distance between centers.

FIGURE 5.1.18 The parts of a lathe tailstock. Note the graduation on the quill and the micrometer collar on the hand wheel. The inside of the quill contains a Morse taper that can hold workholding accessories and holemaking tools.

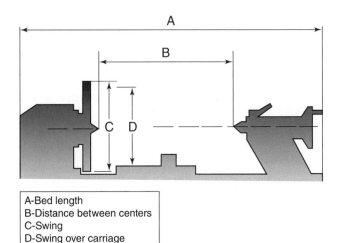

A-Bed length
B-Distance between centers
C-Swing
D-Swing over carriage

FIGURE 5.1.19 Dimensions used to specify lathe size.

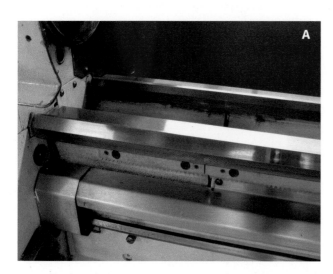

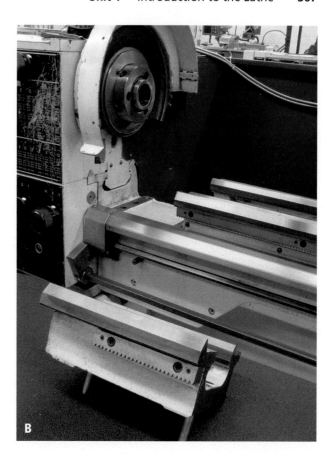

FIGURE 5.1.20 (A) A lathe with a removable gap. Notice the split line at the feed rack and the bed below the ways. (B) The same lathe with the gap removed to allow machining of diameters larger than the swing.

SUMMARY

- The engine lathe is used to machine cylindrical workpieces.

- The headstock is located at the left end of the machine and contains the spindle as well as the gears or belts that power the spindle. Controls for setting spindle speed are also located on the lathe headstock.

- Different types of spindle noses are used for mounting different workholding devices to the lathe spindle. These workholding devices hold and rotate workpieces during machining operations.

- The bottom section of the headstock holds the quick-change gear box that controls the rate of cutting-tool movement known as feed.

- The ways are precision-ground rails located on the top of the lathe bed that hold and guide the carriage and the tailstock.

- The saddle and the apron are the main parts of the carriage. The carriage travels along the ways with longitudinal movement. Several feed and threading controls are located on the apron.

- The cross slide provides tool movement perpendicular to the ways, while the compound rest can be set to any angle to provide angular tool movement. Both the cross slide and the compound rest have graduated dials to allow for precise amounts of tool movement.

- The feed rod runs the entire length of the lathe and transmits motion from the gear box to the carriage and cross slide.

- The leadscrew also runs the length of the lathe and transmits motion to the carriage for threading operations.

- The lathe tailstock can be positioned along the ways and clamped to the ways at any location. It contains a Morse taper that can secure workholding accessories and holemaking tools.

REVIEW QUESTIONS

1. List the four main parts of the engine lathe.
2. What are the two main purposes of the lathe spindle?
3. What part of the lathe is used to set the feed rate of the cutting tool?
4. What are the two main components of the carriage?
5. What function does the compound rest serve?
6. What is the purpose of the leadscrew of a lathe?
7. What two functions can the lathe tailstock perform?
8. The standard taper in most lathe tailstocks is the _____ taper.
9. Define the swing and the bed length of a lathe.

Tools, Toolholding, and Workholding for the Lathe

Learning Objectives

After completing this unit, the student should be able to:

- Explain the differences between universal-type and independent-type chucks
- Explain the function and application of a three-jaw universal chuck
- Explain the function and application of a four-jaw independent chuck
- Explain the function and application of collets
- Demonstrate understanding of various types of lathe centers
- Demonstrate understanding of mandrels
- Identify and explain the applications of a steady rest and follower rest
- Demonstrate understanding of various toolholding devices
- Demonstrate understanding of basic cutting tools and cutting-tool geometry
- Demonstrate understanding of carbide inserts and toolholders

Key Terms

Chuck key	Follower rest	Rocker-type toolholder
Collet	Four-jaw chuck	Rocker-type tool post
Dead center	Independent chuck	Spring collet
Drawtube	Indexable tool post	Steady rest
Drive dog (lathe dog)	Insert	Three-jaw chuck
Drive plate	Jaw-type chuck	Toolholding device
Emergency collet	Live center	Turning
Faceplate	Mandrel	Universal chuck
Facing	Quick-change toolholder	Workholding device
Flex collet	Quick-change tool post	

INTRODUCTION

Before any machining processes can begin on a lathe, there must be a means of "holding," or mounting, the workpiece securely in the machine. The devices used to mount work in a machine are referred to as **workholding devices**. Workholding devices must accurately locate and rigidly secure the work to the machine to withstand the significant forces created during machining operations. These devices include chucks, collets, lathe centers, and mandrels.

Of equal importance is securely mounting the cutting tools for the various operations that will be performed. The devices used to make tool mounting possible are called **toolholding devices**. These devices include tool posts, toolholders, and Morse taper accessories.

WORKHOLDING

Jaw-Type Chucks

A **jaw-type chuck** is a device used to clamp a workpiece in the lathe spindle by applying pressure on multiple sides. Jaw-type chucks consist of a chuck body that contains sliding work-gripping jaws for securing the workpiece. Since the jaws slide in the body of the chuck, they are able to accommodate a broad range of workpiece sizes. Jaw-type chucks can be used to hold work for machining external and internal diameters. **Figure 5.2.1** shows some jaw-type chucks.

The jaws of **independent chucks** are advanced and retracted independently of each other and allow for fine-tuning the position of the workpiece for maximum accuracy. The jaws of **universal chucks** advance and retract simultaneously. A scroll mechanism contained in the chuck body moves the jaws in this manner. **(See Figure 5.2.2.)** They offer little fine adjustment part location, but they are quicker and easier to use.

Jaw-type chucks are usually equipped with reversible, stepped jaws. These jaws may be used to clamp on the outside surface of a workpiece by applying pressure inward toward the center of the workpiece. Also, since the jaws have steps that form ledges, a part with a center hole may be placed over the jaws, pressed back against the face of the steps, and clamped by moving the jaws outward. **(See Figure 5.2.3.)** Some chucks use jaws that may also be removed, flipped, and reinstalled to increase the size capacity of the chuck. Others have different sets of jaws that can be installed to increase capacity. Caution should be used when removing and installing chuck jaws. Each jaw may be numbered and dedicated to one position in the chuck body, so it is important that the jaws are installed in the correct positions.

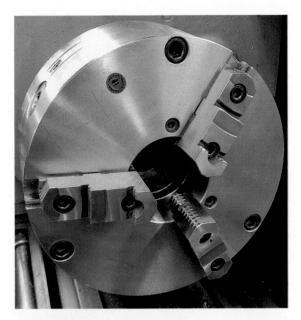

FIGURE 5.2.1 Various styles of jaw-type lathe chucks, including a three-jaw, four-jaw, and six-jaw chuck.

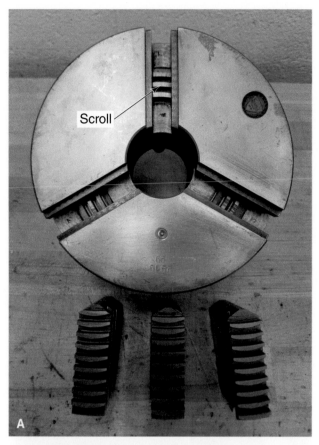

FIGURE 5.2.3 A chuck being used to grip a workpiece by its inside diameter.

 **CAUTION**

A maximum RPM rating is listed on every jaw-type chuck. Never exceed the manufacturer's maximum rating when using any chuck.

Three-Jaw Chuck

One of the most commonly used lathe workholding devices is the universal **three-jaw chuck**, shown in **Figure 5.2.4**. This chuck is so named because it has three work-gripping jaws that secure the workpiece. The jaws are closed and opened by inserting a special wrench called a **chuck key** into a socket on the chuck. Since this chuck is a universal type, when the chuck key is turned, all of the jaws move in unison. Since the movement of all three jaws is synchronized, the universal three-jaw chuck is often referred to as a self-centering chuck. In this way, the universal three-jaw chuck is similar to the drill chuck.

The three-jaw chuck is ideal for holding workpieces that are round or have a number of flat sides that is divisible by three. Examples of these types of workpieces are those that are triangular or hexagonal in shape.

FIGURE 5.2.2 (A) The grooves on the backs of universal chuck jaws mate with the scroll. Turning the chuck wrench rotates the scroll, which moves all the jaws simultaneously. (B) An independent chuck uses an individual screw thread to move each jaw separately.

FIGURE 5.2.4 A universal three-jaw chuck.

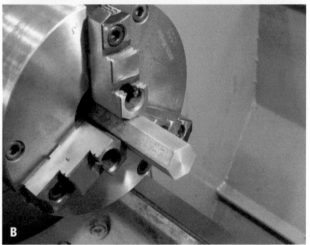

FIGURE 5.2.5 (A) Round and (B) hexagonal workpieces can be properly secured in a three-jaw chuck.

Square or octagonal material cannot be held properly in a three-jaw chuck. **Figure 5.2.5** shows round and hexagonal workpieces clamped in a three-jaw chuck.

Installing a workpiece in a three-jaw chuck is accomplished by placing the workpiece into the chuck and tightening the jaws with the chuck key. The chuck must be tight enough so that the part remains secure under heavy machining, quick spindle starting and stopping, and high speeds. High speeds could pull the jaws outward due to centrifugal force. Quick spindle stops can cause the scroll to rotate and loosen the jaws. However, it is also important that the jaws are not over-tightened, as this may mar the workpiece surfaces, distort the shape of hollow parts, and even permanently damage the scroll.

While common and easy to use, three-jaw chucks are not ideal for all types of work. They may not repeat (allow accurate removal and replacement to the same position) precisely enough for all applications. For this reason, when possible, work requiring high accuracy (more precise than 0.003 TIR) should not be removed from this type of chuck until all machining operations have been completed. This is because once removed, the workpiece cannot be placed back into the chuck jaws in the exact same position. Therefore it may no longer run true enough for some requirements.

Four-Jaw Chuck

Four-jaw chucks are also often used on the lathe. Unlike the three-jaw chuck, the jaws of most four-jaw chucks move independently of each other, and each has its own adjusting screw. Four-jaw chucks are usually chosen when the necessary part accuracy cannot be

achieved with a three-jaw chuck or when the number of sides of the workpiece is divisible by four. Four-jaw chucks are also capable of clamping many workpiece shapes eccentrically, or off-center. **Figure 5.2.6** shows some uses of a four-jaw chuck.

 CAUTION

Never leave a chuck key in a chuck. This can be extremely dangerous. A chuck key will easily fly out of the chuck if the machine is turned on and cause severe damage to the machine and serious injury. Always remove the chuck key immediately after mounting work in or removing it from a chuck.

Since a four-jaw chuck is an independent-type chuck, each jaw must be adjusted and tightened individually to align a workpiece in the chuck. Here are some general

FIGURE 5.2.6 (A) A four-jaw independent chuck can be used to hold square work. (B) It can also hold a workpiece eccentrically, or off-center.

steps for mounting a cylindrical workpiece in a four-jaw chuck.

- Lightly secure and visually align the workpiece by using the concentric index rings on the face of the chuck.
- Most four-jaw chuck jaws are numbered 1 to 4, so position the chuck with jaw #1 facing up.
- Mount a dial indicator on the lathe carriage, bring the indicator contact against the top of the workpiece, and apply some preload.
- Set a "0" on the indicator.
- Rotate the chuck 180 degrees to jaw #3 and observe the direction and amount of indicator movement. Make sure the indicator doesn't run out of travel or come out of contact with the workpiece. If this happens, readjust the rough part alignment before recording precise indicator readings.
- Adjust jaws #1 and #3 until the desired accuracy is achieved.
- Repeat the adjustment for jaws #2 and #4 until the desired accuracy is achieved.
- Tighten jaws #1 and #3 securely and equally while maintaining the desired TIR (total indicator reading).
- Tighten jaws #2 and #4 securely and equally while maintaining the desired TIR.
- Rotate the chuck and verify the TIR around the entire workpiece (**Figure 5.2.7**).

Chuck Installation/Removal

The first step to removing and installing a chuck on a lathe is to place a protective barrier on the machine ways to prevent damage in case the chuck is dropped. A wooden board or block is commonly used to protect the ways. **(See Figure 5.2.8.)**

FIGURE 5.2.7 A dial indicator being used to center work in a four-jaw independent chuck.

FIGURE 5.2.8 A wooden block can serve as a protector for preventing damage to the lathe bed ways during chuck removal and installation.

! CAUTION

When mounting and removing chucks, use a wooden board to support the chuck and protect the ways. Get assistance or use appropriate lifting equipment when moving heavy lathe chucks. If using a piece of overhead lifting equipment such as a hoist or crane, wear a hard hat as required by OSHA. Never stand under a chuck suspended from a hoist or crane.

The fastening procedure for chuck mounting is dependent on the spindle nose type of the lathe. On cam-lock spindle noses, the chuck and spindle tapers will often remain locked together after loosening the locks. Striking the back of the chuck with a dead blow hammer might be necessary to release the tapers. Chucks should be handled carefully and laid on their backs for storage. Before installing another chuck, remove any dirt, dust, and debris from both the spindle nose and the mounting surfaces of the chuck. If the chuck is mounted onto a dirty spindle nose, it will most likely wobble when in motion, causing runout on the workpiece.

Drill Chucks

A Morse taper-shank drill chuck can be mounted in the headstock spindle by using a reducing sleeve. The sleeve's external taper mates with the spindle bore. The internal taper of the sleeve accepts the Morse taper of the drill chuck. This workholding method can be used to hold diameters that are too small to be gripped in three-jaw and four-jaw chucks. **(See Figure 5.2.9.)** The chuck and the sleeve can be removed from the headstock using a knockout bar.

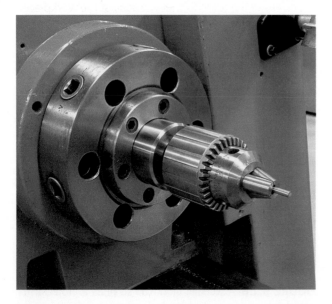

FIGURE 5.2.9 A drill chuck mounted in a lathe spindle being used to grip a small-diameter workpiece.

Collets

Collets are another way of holding workpieces on the lathe. They have holes in their centers to match the size and shape of the workpiece to be held. An external taper on the collet allows it to be drawn against a mating taper to provide gripping force. The most common types of collets are spring collets and flex collets.

Collets have some advantages over other workholding devices. They don't have the mass of jaw-chuck bodies so they are not greatly affected by centrifugal force at high RPM. Collets also cause very little distortion to hollow workpieces and are less likely to mar surfaces because of their larger surface contact and more even clamping pressure. Collets can also achieve the accuracy and repeatability of a four-jaw chuck, but they are much quicker and easier to use. Collets can be used to hold work for machining external and internal diameters like jaw-type chucks.

Spring Collets

Spring collets are very accurately ground cylindrical sleeves. The spring collet has slits that begin at its front and continue about three-quarters of the way toward the back. The back end of the collet is threaded, which allows it to be pulled, or drawn, against a mating tapered surface. The slits allow the collet to constrict and grip the workpiece when the taper of the collet is drawn against the mating taper.

Spring collets come in a variety of sizes and shapes to accommodate many types, shapes, and sizes of workpieces. Collets are most commonly found with straight round bores but are also available with square- and hex-shaped holes. Spring collets are purchased for particular standard sizes and have a gripping range of only a few thousandths of an inch. If many different workpiece sizes must be machined, many different collets are needed.

Since nonstandard sizes often cannot be adequately secured in standard collets, **emergency collets** made from brass or mild steel are available that can be easily machined to match any desired size.

Step collets are used to grip on short workpieces whose diameters are too large to fit within the envelope of standard spring collets. These collets may be designed for one size or have a soft head that can be machined to accommodate a variety of part sizes and shapes, much like an emergency collet.

An *expanding collet* includes an integral arbor instead of a hole. When it is drawn against the taper, the arbor expands and can grip on an existing inside diameter of a workpiece. Most expanding collets are machinable like emergency collets, so they can be customized for a particular size. **Figure 5.2.10** shows some examples of spring collets.

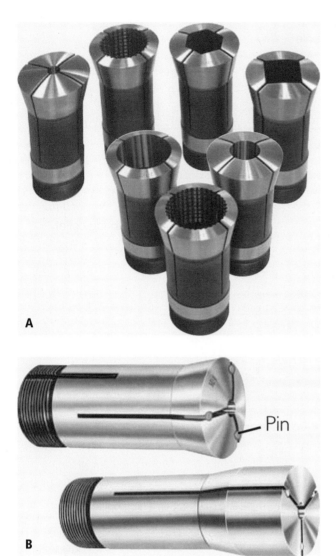

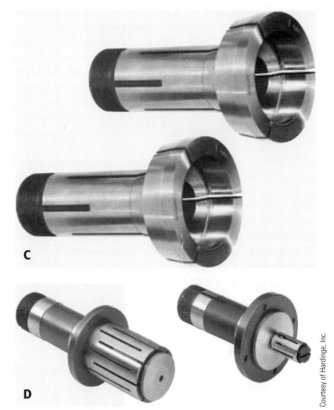

Courtesy of Hardinge, Inc.

FIGURE 5.2.10 (A) Smooth and serrated round collets, square collets, and hex collets. (B) The hole size of emergency collets can be machined to any desired size. The pins keep the collet in the clamped position during machining of the hole. Then they are removed to allow the collet to constrict on the work when tightened. (C) Step collets can be used to hold diameters larger than the shank portion of the collet. (D) Expanding collets can hold work by an existing bore for external machining.

One method of using spring collets for lathe operations uses a spindle collet adapter and a **drawtube**, which draws (pulls) the collet against a mating taper so it squeezes against the workpiece. Follow these general steps to mount and remove a workpiece using a collet.

- Clean the spindle bore and the appropriate collet adapter, or sleeve.
- Insert the adapter in the tapered bore of the headstock spindle. This adapter has an external taper that mates with the spindle bore taper to lock it in place.
- Clean the bore of the adapter and the collet. The internal taper of the adapter matches the external taper of the collet.
- Place the collet in the sleeve. The sleeve usually has a small pin, or key, that aligns with a slot on the collet.
- Insert the drawtube through the back of the headstock spindle bore and thread onto the back end of the collet 2 to 3 turns, but do not tighten.
- Place the workpiece in the collet.

- Tighten the drawtube to pull the collet into the sleeve and cause the collet to squeeze against the workpiece. **See Figure 5.2.11** illustrates how a collet functions.

FIGURE 5.2.11 (A) A workpiece held in a collet.

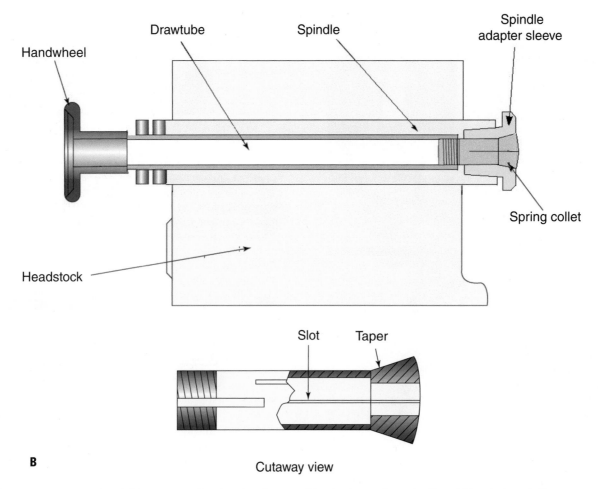

B   Cutaway view

FIGURE 5.2.11 *continued.* (B) A cross section showing the assembly of a draw-tube and collet within a lathe spindle.

- To remove the workpiece, loosen the drawtube about one full turn.

- If the part does not come loose by itself, tap the handle of the drawtube with your palm to release the collet from the taper. Do not completely unthread the drawtube from the collet and use the drawtube to strike the collet. This will damage the threads of the drawtube and the collet.

- Remove the workpiece from the collet.

- Unthread the drawtube from the collet, and remove the collet and drawtube from the spindle.

- Use a knockout bar with a soft metal end to remove the collet adapter from the headstock. **Figure 5.2.12** illustrates this step.

Another method uses a collet chuck mounted on the spindle of the lathe. These chucks mount to the spindle nose just like other chucks. The collet is then placed inside the bore of the chuck. One chuck style uses a key to rotate a threaded ring inside the chuck to draw the collet closed on the work. Another style tightens the collet on the work by turning a hand wheel. **Figure 5.2.13** shows collet chucks.

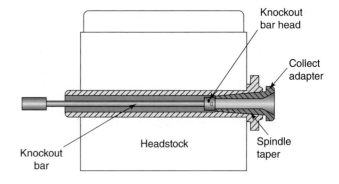

FIGURE 5.2.12 A knockout bar being used to remove a collet adapter from a lathe spindle.

Flex Collets

A variation of the spring collet is the **flex collet**. Unlike the solid spring collet, the flex collet consists of steel segments attached to a hub. This gives the flex collet a size range of up to 1/8". Flex collets can contain steel sections molded into a rubber body or attached to a slotted steel hub with a spring-loaded clamping mechanism. **(See Figure 5.2.14.)**

Collet chucks are needed when using flex collets. Follow these general steps for using a flex collet.

- Open the chuck completely to remove the front retaining cap.
- Clean the collet and the inside of the chuck.
- Place the collet inside the chuck.
- Reattach the cap by tightening the chuck with the key or hand wheel.
- Place the workpiece inside the collet.
- Tighten the chuck with the hand wheel or key to force the collet against the taper of the chuck. This causes the steel segments to close down onto the workpiece. **Figure 5.2.15** shows work held in a flex collet.

Courtesy of Hardinge, Inc.

FIGURE 5.2.13 Examples of spring collet chucks.

FIGURE 5.2.14 Some examples of flex collets.

FIGURE 5.2.15 A workpiece held in a key-type flex collet chuck.

Faceplate

Sometimes an irregularly shaped workpiece needs to be secured in the lathe for machining, and neither a chuck nor collet can be used. In these situations, a faceplate can be used. A **faceplate** is mounted to the headstock spindle and is generally made of cast iron with a series of slots machined into its face. The work is positioned against the faceplate and secured using clamping devices anchored through the machined slots. **Figure 5.2.16** shows an application of a faceplate. Although faceplates are seldom used today, they are occasionally still useful for securing irregularly shaped workpieces such as castings.

 CAUTION

When using faceplates there are some unique safety considerations.

When mounting and removing faceplates, take the same safety precautions as when working with lathe chucks. Get assistance or use appropriate lifting equipment when moving heavy faceplates. If using a piece of overhead lifting equipment such as a hoist or crane, wear a hard hat as required by OSHA. Never stand under a faceplate suspended from a hoist or crane.

Great care should be taken when using a faceplate to avoid the workpiece coming loose due to insufficient clamping pressure. Unlike a jaw-type chuck or a collet, where a loose workpiece is surrounded by the chuck jaws or collet body and is unlikely to fly out of the workholding device before the operator can react, faceplate clamps can come completely off, causing immediate hazards from the workpiece and the clamps.

Since oddly shaped, large work is commonly secured to faceplates, be aware of workpiece imbalance and use conservative spindle speeds. A severely imbalanced workpiece running at high spindle speeds can cause extreme vibration that can damage the machine bearings or create a dangerous situation if clamps fail. Always use a conservative RPM when starting an unbalanced workpiece.

Workholding Between Centers

Holding work between *centers* is another method of holding workpieces on the lathe. Since the work is supported at each end, only external diameters can be machined using this workholding method. **(See Figure 5.2.17.)**

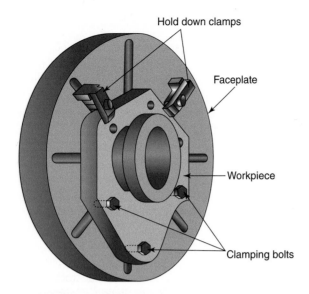

FIGURE **5.2.16** Faceplates can be used for holding large or irregular-shaped workpieces.

FIGURE **5.2.17** A workpiece held between lathe centers.

A lathe center is a cylindrical steel device with a 60-degree included angled point on one end and a Morse taper on the other end. One center is mounted in the headstock spindle using a reducing sleeve and another is mounted in the quill of the tailstock. Be sure that all mating surfaces are clean and burr free before installation. Holes are drilled in the ends of the workpiece for mounting between the lathe centers.

The main advantage of working between centers is that work can be easily removed, turned end for end, and reinstalled with a high level of repeatability, or accuracy. Diameters machined from each end will maintain concentricity and have very little runout because the centers accurately locate the work in the same position each time.

FIGURE **5.2.18** Some examples of live centers that can be mounted in the lathe tailstock.

FIGURE **5.2.19** A dead center is solid and is usually mounted in the lathe headstock.

Lathe Centers

Lathe centers are available in two main varieties: live and dead. A **live center** is mounted in a bearing cartridge and freely rotates with the workpiece while the mounting shank stays stationary. Live centers are used in the tailstock end of the lathe and serve only to align and support the tailstock end of the workpiece. **Figure 5.2.18** shows some examples of live centers.

A **dead center (Figure 5.2.19)** is usually used in the headstock spindle and has no rotating parts. Occasionally a dead center may be used in place of a live center in a tailstock. This is often the case when performing operations such as knurling where there may be extreme amounts of pressure exerted on the workpiece that could damage live center bearings. This practice is becoming less widely used since the quality and design of modern live centers have the ability to handle the extreme force with no damage. If a dead center is used in the tailstock,

FIGURE 5.2.20 When using a dead center in the lathe tailstock, apply grease or anti-seize compound in the center hold to reduce friction between the center and the rotating workpiece.

it is important that a lubricant such as grease or anti-seize compound is applied between the work and the center to reduce friction. **(See Figure 5.2.20.)** Spindle speeds should also be kept low to minimize friction and wear on the center and the center hole in the workpiece.

NOTE:

In addition to being used when mounting work between centers, a tailstock-mounted center can also be used to help support long workpieces held in chucks and collets. When the length of the workpiece extends past the chuck or collet more than 3 to 4 times its diameter, it is a good idea to use a tailstock center for support. **(See Figure 5.2.21.)**

FIGURE 5.2.21 A tailstock-mounted center can also be used to support long workpieces held in chucks or collets.

Center and Tailstock Alignment

When working between centers, it is important that the centers run true and that tailstock is properly aligned with the headstock. A center can be inspected after installation by placing a test indicator on its workholding surface and rotating the center to check for runout. **(See Figure 5.2.22.)** Any amount of runout shown on the indicator will show up as misaligned surfaces on machined parts. A headstock center may be able to be machined by turning or grinding so that it will run true.

The tailstock can be purposely offset for machining tapers, which will be described in Section 5.5, "Taper Turning." Tailstock alignment for producing straight diameters can be accomplished by a few different methods.

Approximate tailstock alignment can be performed by bringing the points of the tailstock and headstock centers near each other and visually inspecting alignment. Be careful not to bump the centers together and damage or deform their points. Most tailstocks also have witness lines on the body below the hand wheel that indicate the aligned position. A screw on each side of the tailstock is used to adjust its position. **(See Figure 5.2.23.)**

Precise alignment can be done by following these steps.

- Mount a *test bar* between the centers. A test bar is a precision straight cylindrical bar with concentric center holes in each end.

- Mount a dial indicator on the cross slide or compound rest and orient as shown in Figure 5.2.24. Make sure the indicator is set up to measure movement perpendicular to the ways. Also center the indicator contact vertically on the diameter of the test bar.

- Bring the indicator contact against the side of the test bar near one end and preload the indicator.

FIGURE 5.2.22 Checking runout of a headstock center using a test indicator.

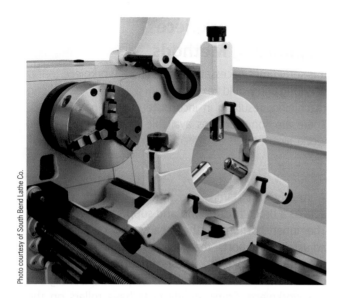

Photo courtesy of South Bend Lathe Co.

FIGURE 5.2.31 A steady rest can provide extra support for long workpieces to prevent them from flexing away from the cutting tool.

a center-drilled hole, use a tailstock center to support the other end.

- Close and lock the top section of the steady rest, then adjust the supports so they contact the work with light pressure. The supports are then locked into position to prevent them from moving during operation.
- Unlock and open the top of the steady rest and then remove the workpiece from the lathe.
- Unclamp the rest from the bed, move it to the desired location on the ways, and reclamp it to the bed.
- Remount the workpiece in the lathe, then close and lock the top of the rest. Be careful not to drop the workpiece on the supports or they may be moved out of adjustment.
- Apply lubricant to the area where the supports contact the work.
- Check the adjustment and contact of the supports frequently during use to maintain light contact between the supports and the workpiece. Keep the supports well lubricated during use as well.

Follower Rest

The **follower rest** is similar to the steady rest in that it provides additional support to the workpiece during machining operations by contacting and stabilizing the outside diameter. The major difference is that the follower rest is attached to the carriage and moves along the length of the workpiece with the tool during machining. It only has two supports and is usually placed directly on the opposite side of the work from the cutting tool. The cutting tool acts as the third support as it pushes the workpiece against the two rest supports during machining. **(See Figure 5.2.32.)**

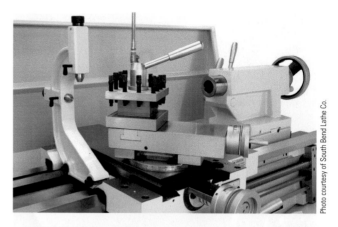

Photo courtesy of South Bend Lathe Co.

FIGURE 5.2.32 A follower rest attaches to the carriage and moves with the cutting tool for support when machining long workpieces.

Follow these general steps to set up a follower rest:

- Mount the workpiece in the headstock and support it with a tailstock-mounted center.
- Retract the follower rest supports.
- Bolt the follower rest to the carriage.
- Adjust the two supports until they lightly contact the workpiece.
- Apply ample lubrication to the surface in contact with the supports to prevent wear and damage to the supports and the work.
- Since the supports of the follower rest contact the surface being machined, they must be readjusted after each pass as the diameter changes.

TOOLHOLDING

Properly securing cutting tools is another important aspect of operating the lathe. Toolholding devices are used to mount cutting tools on the lathe. To be effective, lathe toolholding devices must be able to rigidly hold the cutting tool to produce consistent and accurate results. Chatter is a word used to describe the excessive noise and vibration that will often occur when a tool is not adequately supported. Chatter results in a poor surface finish on the part and can lead to excessive tool wear or breakage. Another important attribute of toolholding is the ease and speed of changing between the wide varieties of cutting tools.

Rocker-Type Toolholding

One method of holding a tool in a lathe is the **rocker-type tool post**. The name for this assembly comes from the way the toolholder height is adjusted up and down by loosening a clamping bolt and rocking the holder on top of a rounded shim. Rocker-type tool posts and holders have been largely replaced since they do not offer as much rigidity as newer toolholding methods. Changing from one tool to another is also not as efficient as with the more modern methods. Although rocker-arm tool posts and holders are not widely used today, there are still many attachments available in this style.

Rocker-Type Toolholders

Rocker-type toolholders are available in left-hand, neutral (or straight), and right-hand styles, as shown in **Figure 5.2.33**. They are also available for holding many different types of cutting tools, as shown in **Figure 5.2.34**.

Rocker-Type Tool Post

The rocker-type tool post secures the rocker-type holder to the lathe. This device consists of a tool post with clamping screw, tool post ring, base, and tool post rocker (or wedge), as shown in **Figure 5.2.35**. The main post is sometimes referred to as a "lantern" post because of its shape. Follow these steps to mount a tool using this type of holder.

- Place the tool post through the base and slide the base into the T slot on top of the compound rest.
- Slide the tool post ring over the tool post.
- Place the rocker through the tool post.
- Place a rocker-type holder through the tool post on top of the rocker.
- Tighten the clamping bolt to secure both the toolholder and the tool post to the compound rest.
- Mount the tool bit by placing it in the toolholder and tightening the toolholder clamping screw. **Figure 5.2.36** shows the tool post and holder mounted to the compound rest.

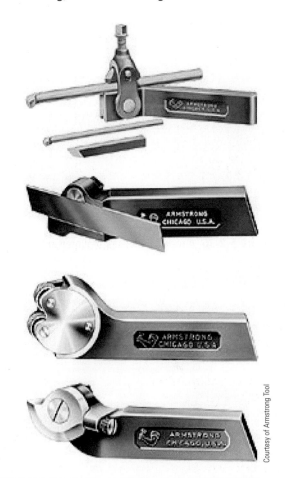

Courtesy of Armstrong Tool

FIGURE 5.2.34 Rocker-type holders are available in different styles for holding different types of cutting tools.

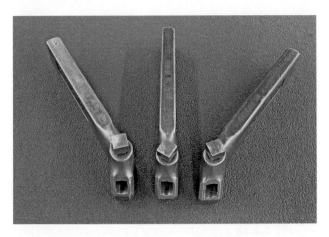

FIGURE 5.2.33 From left to right: Left-hand, straight, and right-hand rocker-type toolholders.

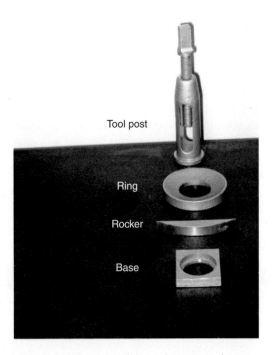

FIGURE 5.2.35 The parts of the rocker-type tool post.

FIGURE 5.2.36 A rocker toolholder mounted in the tool post.

FIGURE 5.2.37 Adjust the rocker in the tool post to raise or lower the tip of the cutting tool to obtain the proper height, and then tighten the clamping screw.

After mounting the tool bit, the tip of the tool must be set to a height that aligns it to the center of the lathe spindle, which will also be the center of the workpiece. A headstock or tailstock center can be used as a reference point. To adjust this height, adjust the rocker and holder to move the tool tip up or down, then tighten the tool post clamping screw. **Figure 5.2.37** shows how to make this adjustment.

⊘ **CAUTION**

Do not hold the toolholder in your hand to tighten the tool bit clamping screw. The wrench, tool bit, and holder can slip and cause serious injury from the sharp tool bit.

Quick-Change Tool Holding
Quick-Change Tool Post

The **quick-change tool post** is much more convenient to use than the rocker-type post. The quick-change tool post consists of a T-nut, clamping stud, and tool post. Quick-change tool posts are usually equipped with a dovetail that mates with a dovetail on a toolholder. The tool post is clamped to the compound rest using the T-nut and clamping stud. The handle moves a clamping mechanism that locks and unlocks toolholders mounted on the post. A holder is slid onto the dovetail and the handle is pulled to lock the holder to the post. Reversing the handle unlocks the post so the holder can be removed. **Figure 5.2.38** shows a quick-change tool post. The quick-change tool post allows several tools to be preset in different holders. They can be quickly and easily changed without the need to reset their positions as with the rocker-type post and holder.

Quick-Change Toolholders

The dovetail-shaped **quick-change toolholders** slide onto the matching dovetail on the quick-change posts. Several styles of holders are available. Some holders allow cutting tools to be clamped by tightening screws, and others feature integrated tools. **Figure 5.2.39** shows some examples of quick-change holders. Tool height is set by moving the adjusting nut to raise or lower the position of the holder on the tool post, as shown in **Figure 5.2.40**.

Indexable Tool Posts

An **indexable tool post** consists of a T-nut, clamping stud, and a multi-sided tool block with either removable or integral toolholders. **(See Figure 5.2.41.)** This block

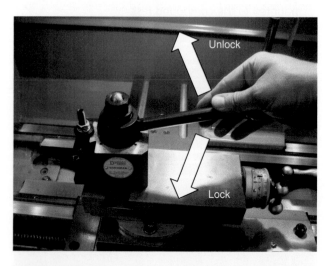

FIGURE 5.2.38 Pulling the lever of the quick change tool post locks holders to the dovetail on the post. Pushing the lever unlocks the holders.

FIGURE 5.2.39 Some of the different styles of quick-change toolholders.

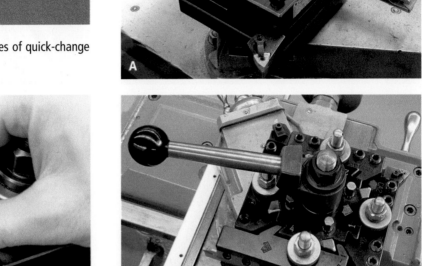

FIGURE 5.2.41 (A) An indexable holder can have integrated toolholding slots or (B) accept quick-change holders.

FIGURE 5.2.40 Tool height is set by adjusting the nut on top of the quick-change toolholder.

can be loaded with multiple tools simultaneously, then easily rotated and secured to position the desired tool for machining. These tool posts typically have detents, or notches, that allow them to be rotated and then repeatedly and accurately positioned in the same locations. An indexable tool post can hold multiple cutting tools, which can greatly reduce setup time. Those with integral toolholders may only hold four tools and would require resetting of tools if more than four different cutting tools were needed. Others utilize the quick-change system, so even though only a certain number of tools may be mounted to the post at one time, many tools can be preset so they can be quickly and easily removed and installed back in the same position.

To install, a T-nut is placed into the T-slot machined into the compound rest. The tool post is clamped to the compound rest by tightening the stud into the T-nut. A lever located on top of the tool block is used to unlock the tool post, index it to the desired position, and then lock it in place for machining. **(See Figure 5.2.42.)**

LATHE CUTTING TOOLS

Facing and turning are the two most common lathe operations. **Facing** cuts a flat surface on the end of a workpiece. **Turning** cuts the diameter of a workpiece. Cutting tools made of high-speed tool steel (HSS) have been used for many years for turning and facing on the lathe. A square or rectangular tool bit blank can be ground on a pedestal grinder to produce the desired tool shape, or geometry. Some standard tool types can also be purchased with the correct geometry already existing. **(See Figure 5.2.43.)** That geometry includes angles that allow the tool to cut material properly during machining operations. When HSS tools wear, they can be resharpened by grinding, but then they need to be reset in their toolholders because the location of the cutting tip changes during resharpening.

Brazed carbide tools consist of a carbon steel shank with a small piece of carbide brazed on the end where cutting action takes place. **(See Figure 5.2.44.)** These can be

FIGURE 5.2.42 (A) The lever of the indexable tool post is used to unlock the tool post. (B) It can then be rotated to the desired position and locked in place for machining.

FIGURE 5.2.43 (A) An HSS tool bit blank such as the one shown on the left can be shaped on a pedestal grinder to create cutting tools with the required geometry to perform turning and facing. (B) A few samples of the countless shapes of cutting tool that can be made from HSS blanks. (C) Some standard HSS tools for specific operations that can be purchased ready to use.

FIGURE 5.2.44 Some examples of brazed carbide cutting tools.

resharpened by grinding with a silicon carbide or diamond wheel, but they also need to be reset after sharpening.

Inserted (or indexable) tools use interchangeable tips called **inserts** that are fastened on a steel shank. Inserts can be made from very hard materials such as carbide, ceramic, or diamond. Carbide is most common, and will be used for examples in this text. These inserts are manufactured in different grades, or compositions, for different applications and for machining different materials. The different grades can be compared to alloy steels. Different alloy steels contain varying amounts of elements to give them characteristics such as toughness, hardness, and shock resistance. The same is true of carbide cutting tools. When these inserts wear, they are simply replaced. **(See Figure 5.2.45.)** The new insert places the cutting tip in the same original location, so these tools require little or no resetting. For that reason, carbide inserted tools are becoming very common, but HSS and brazed carbide tools are still found in many shops.

General turning and facing tools can first be divided into the three basic categories of left-hand, right-hand, and neutral tools. The cutting direction of a left-hand tool is from left to right and the cutting direction of a right-hand tool is from right to left. A neutral tool is able to cut in either direction. **(See Figure 5.2.46.)** Keep in mind that these are basic styles and that there are many variations of these styles.

FIGURE 5.2.45 Inserted (or indexable) cutting tools use carbide tips called inserts. They can be easily replaced instead of being resharpened like HSS and brazed carbide tools.

FIGURE 5.2.46 (A) A right-hand tool cuts from right to left. (B) A left-hand tool cuts from left to right. (C) Neutral tools can cut in either direction.

Basic Tool Geometry

Facing and turning tools share some characteristics regardless of whether they are HSS, brazed carbide, or inserted carbide.

The *lead angle,* or *side cutting angle,* is the angle of the cutting edge of the tool relative to its shank. The lead angle can be either positive, negative, or zero. This surface can also be called the leading edge. **Figure 5.2.47** shows examples of these different lead angles.

The *end cutting-edge angle* is the angle between the surface being machined and the front edge of the tool. Without this angle the front of the tool would rub across the machined surface. **(See Figure 5.2.48.)**

The *side clearance angle* is needed for the tool to be able to cut. Without it, the tool will only rub. **Figure 5.2.49** illustrates the side clearance angle.

Back rake is the term used to describe the angle of the top of the cutting tool relative to a horizontal line through the center of the workpiece. It can be positive, negative, or neutral, as shown in **Figure 5.2.50**. *Side rake* is the angle of the top of the cutting tool to a vertical line. It can also be positive, negative, or neutral, as shown in **Figure 5.2.51**. Positive-rake angles make chips thinner and require less force when cutting, but as they become steeper the tool becomes smaller and weaker.

Tool nose radius is the radius at the tip of the tool where the leading edge and the end cutting edge meet. **(See Figure 5.2.52.)** The tool nose radius directly affects the surface finish of the machined surface. Generally, larger nose radii generate better surface finish and create a stronger cutting tip. However, if the radius becomes too large, chatter and deflection may develop because of the large amount of surface contact (pressure) between the tool and the workpiece. In contrast, as a tool's nose radius becomes smaller, the tip becomes weaker, less tolerant of heat, and will wear more rapidly, requiring frequent resharpening or insert replacement. Nose radii of 0.015" or 0.031" are common for general purpose turning and facing.

If using HSS tools, these angles must all be ground on a tool bit blank using the pedestal grinder and an aluminum oxide wheel. A radius is normally created on the tool nose by hand using an abrasive stone. Brazed carbide tools can be purchased with varying options and can be resharpened on a grinder using a silicon carbide or a diamond wheel. Carbide inserts come ready for use in a multitude of different combinations of these angles and with different tool nose radius sizes.

Carbide Inserts

Tooling manufacturers invest an extraordinary amount of research and development into making a carbide insert for every possible application. Insert material, shape, tolerance, edge geometry, chipformer geometry,

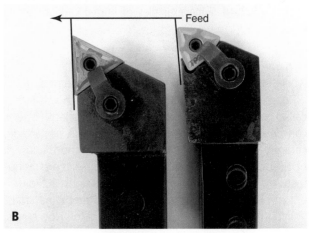

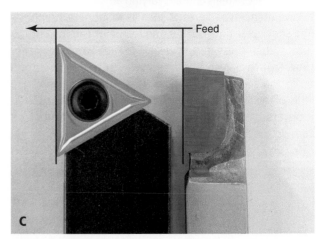

FIGURE 5.2.47 Examples of right-hand tool lead angles: (A) Positive. (B) Negative. (C) Zero.

and coatings are tailored to produce high-performance cutting characteristics for any machining situation. The depth of this information goes far beyond what can be covered in this book. In fact, it is common for a tooling manufacturer to have a 500-page catalog for each of their product lines. However, the basic identification system for general purpose carbide inserts has been standardized by

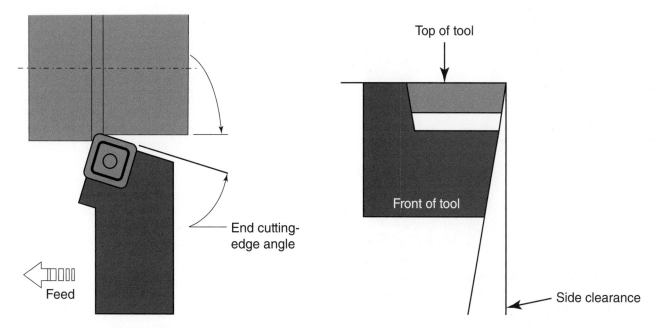

FIGURE 5.2.48 The end cutting-edge angle provides clearance so the front of the cutting tool does not rub across the surface being machined.

FIGURE 5.2.49 The side clearance of a lathe cutting tool prevents rubbing so the tool can cut.

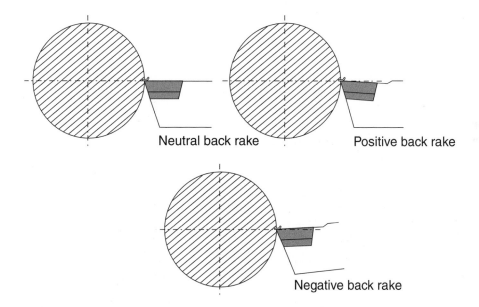

FIGURE 5.2.50 Back rake, or top rake, angles.

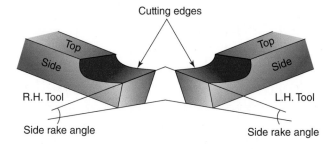

FIGURE 5.2.51 Side rake angles for a right-hand tool and left-hand tool.

the American National Standards Institute (ANSI). That system identifies inserts based on their shape, clearance angle, tolerance during manufacture, insert features (related to how the insert is mounted on a holder), size, thickness, and nose radius/cutting point. Additional information may be included to identify handedness of cutting direction, cutting-edge preparation, facet size, and chipformer style. **Figure 5.2.53** shows a helpful chart for decoding any insert identified using this system. A sample insert designation is listed at the top of the chart for reference.

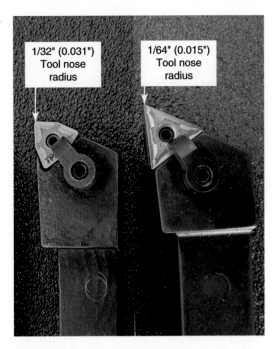

1/32" (0.031") Tool nose radius

1/64" (0.015") Tool nose radius

FIGURE 5.2.52 Tool nose radius is the size of the radius at the tip of the cutting tool.

Shape

The first identifying characteristic of the insert is the shape. This is designated by a letter in the first position of the identification in **Figure 5.2.53**. An insert's shape determines not only the type of part features that the tool can access but also the strength of the insert and the amount of cutting force between the work and the tool. Notice the effect of insert shape on strength and machining forces in **Figure 5.2.54.**

The most common shapes are round, square, 80-degree rhombus, trigon, triangular, 55-degree rhombus, and 35-degree rhombus. The insert identification system uses a letter to denote each insert shape. **Figure 5.2.55** shows these shapes, their identification letters, and the common applications for each.

Clearance Angle

The clearance angle refers to the amount of relief built into the insert under the cutting edge. This is designated by a letter and the angle is measured as shown in the second position of **Figure 5.2.53**. The clearance angle on the insert determines the rake angle at which the insert needs to be mounted in a toolholder. It is important to

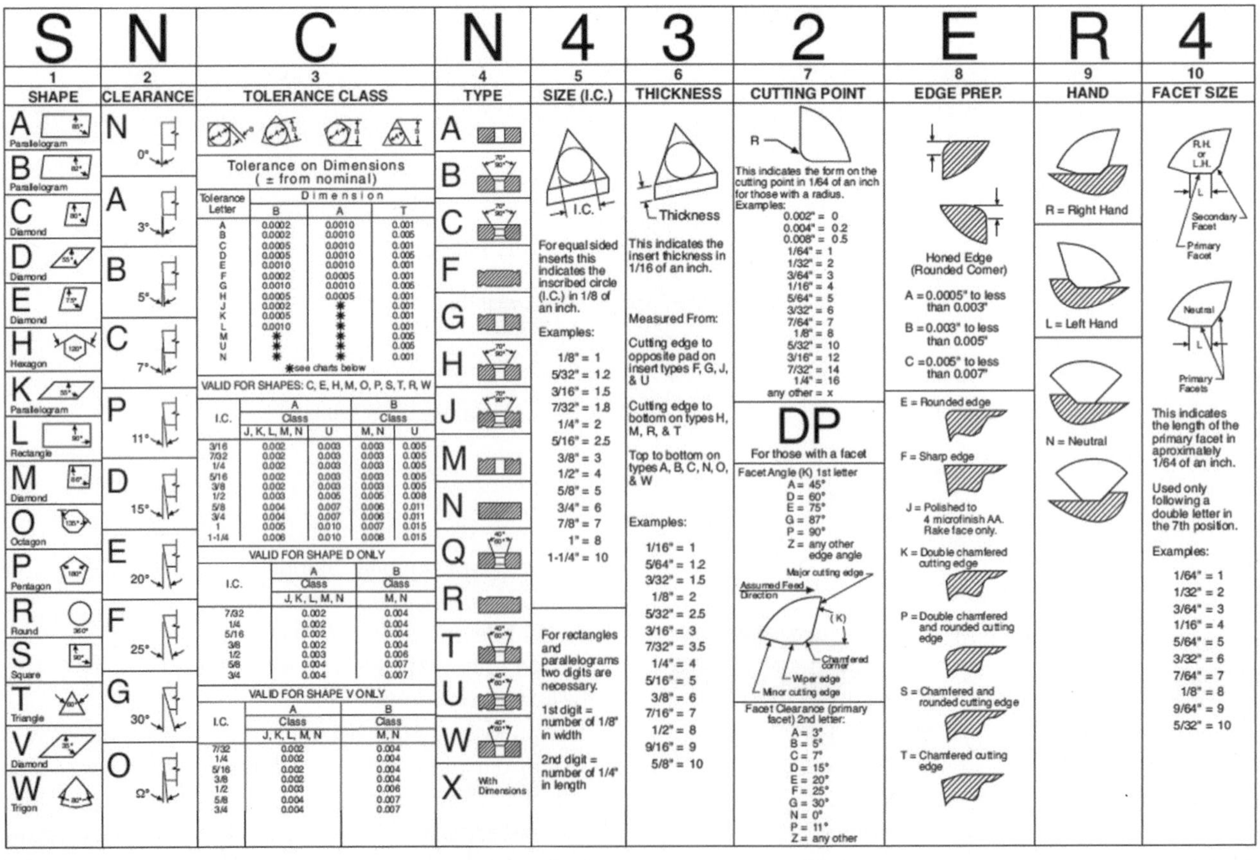

S N C N 4 3 2 E R 4

1	2	3	4	5	6	7	8	9	10
SHAPE	CLEARANCE	TOLERANCE CLASS	TYPE	SIZE (I.C.)	THICKNESS	CUTTING POINT	EDGE PREP.	HAND	FACET SIZE

SHAPE
- A — Parallelogram (85°)
- B — Parallelogram (82°)
- C — Diamond (80°)
- D — Diamond (55°)
- E — Diamond (75°)
- H — Hexagon (120°)
- K — Parallelogram (55°)
- L — Rectangle
- M — Diamond (86°)
- O — Octagon
- P — Pentagon
- R — Round (360°)
- S — Square (90°)
- T — Triangle (60°)
- V — Diamond (35°)
- W — Trigon (80°)

CLEARANCE
- N — 0°
- A — 3°
- B — 5°
- C — 7°
- P — 11°
- D — 15°
- E — 20°
- F — 25°
- G — 30°
- O — Ω°

TOLERANCE CLASS

Tolerance on Dimensions (± from nominal)

Tolerance Letter	Dimension B	A	T
A	0.0002	0.0010	0.001
B	0.0002	0.0010	0.005
C	0.0005	0.0010	0.001
D	0.0005	0.0010	0.005
E	0.0010	0.0010	0.001
F	0.0002	0.0005	0.001
G	0.0010	0.0010	0.005
H	0.0005	0.0005	0.001
J	*	*	0.001
K	*	*	0.001
L	*	*	0.001
M	0.0010	*	0.005
N	*	*	0.001

*see charts below

VALID FOR SHAPES: C, E, H, M, O, P, S, T, R, W

I.C.	Class A (J,K,L,M,N)	(U)	Class B (M,N)	(U)
3/16	0.002	0.003	0.003	0.005
7/32	0.002	0.003	0.003	0.005
1/4	0.002	0.003	0.003	0.005
5/16	0.002	0.003	0.003	0.005
3/8	0.003	0.005	0.005	0.008
1/2	0.003	0.005	0.005	0.008
5/8	0.004	0.007	0.006	0.011
3/4	0.004	0.007	0.006	0.011
1	0.005	0.010	0.007	0.015
1-1/4	0.006	0.010	0.008	0.015

VALID FOR SHAPE D ONLY

I.C.	A Class (J,K,L,M,N)	B Class (M,N)
7/32	0.002	0.004
1/4	0.002	0.004
5/16	0.002	0.004
3/8	0.002	0.006
1/2	0.003	0.006
5/8	0.004	0.007
3/4	0.004	0.007

VALID FOR SHAPE V ONLY

I.C.	A Class (J,K,L,M,N)	B Class (M,N)
7/32	0.002	0.004
1/4	0.002	0.004
5/16	0.002	0.004
3/8	0.002	0.004
1/2	0.003	0.006
5/8	0.004	0.007
3/4	0.004	0.007

TYPE
A, B, C, F, G, H, J, M, N, Q, R, T, U, W, X (With Dimensions)

SIZE (I.C.)

For equal sided inserts this indicates the inscribed circle (I.C.) in 1/8 of an inch.

Examples:
- 1/8" = 1
- 5/32" = 1.25
- 3/16" = 1.5
- 7/32" = 1.8
- 1/4" = 2
- 5/16" = 2.5
- 3/8" = 3
- 1/2" = 4
- 5/8" = 5
- 3/4" = 6
- 7/8" = 7
- 1" = 8
- 1-1/4" = 10

For rectangles and parallelograms two digits are necessary.

1st digit = number of 1/8" in width
2nd digit = number of 1/4" in length

THICKNESS

Thickness

This indicates the insert thickness in 1/16 of an inch.

Measured From:
Cutting edge to opposite pad on insert types F, G, J, & U
Cutting edge to bottom on types H, M, R, & T
Top to bottom on types A, B, C, N, O, & W

Examples:
- 1/16" = 1
- 5/64" = 1.2
- 3/32" = 1.5
- 1/8" = 2
- 5/32" = 2.5
- 3/16" = 3
- 7/32" = 3.5
- 1/4" = 4
- 5/16" = 5
- 3/8" = 6
- 7/16" = 7
- 1/2" = 8
- 9/16" = 9
- 5/8" = 10

CUTTING POINT

R — This indicates the form on the cutting point in 1/64 of an inch for those with a radius.

Examples:
- 0.002" = 0
- 0.004" = 0.2
- 0.008" = 0.5
- 1/64" = 1
- 1/32" = 2
- 3/64" = 3
- 1/16" = 4
- 5/64" = 5
- 3/32" = 6
- 7/64" = 7
- 1/8" = 8
- 5/32" = 10
- 3/16" = 12
- 7/32" = 14
- 1/4" = 16
- any other = x

DP — For those with a facet.

Facet Angle (K) 1st letter
- A = 45°
- D = 60°
- E = 75°
- G = 87°
- P = 90°
- Z = any other edge angle

[diagram: Major cutting edge, Assumed Feed Direction, (K), Chamfered Wiper edge, Minor cutting edge]

Facet Clearance (primary facet) 2nd letter:
- A = 3°
- B = 5°
- C = 7°
- D = 15°
- E = 20°
- F = 25°
- G = 30°
- N = 0°
- P = 11°
- Z = any other

EDGE PREP.

Honed Edge (Rounded Corner)
- A = 0.0005" to less than 0.003"
- B = 0.003" to less than 0.005"
- C = 0.005" to less than 0.007"
- E = Rounded edge
- F = Sharp edge
- J = Polished to 4 microfinish AA. Rake face only.
- K = Double chamfered cutting edge
- P = Double chamfered and rounded cutting edge
- S = Chamfered and rounded cutting edge
- T = Chamfered cutting edge

HAND
- R = Right Hand
- L = Left Hand
- N = Neutral

FACET SIZE

R.H. or L.H.
Secondary Facet
Primary Facet
Neutral
Primary Facets

This indicates the length of the primary facet in approximately 1/64 of an inch.

Used only following a double letter in the 7th position.

Examples:
- 1/64" = 1
- 1/32" = 2
- 3/64" = 3
- 1/16" = 4
- 5/64" = 5
- 3/32" = 6
- 7/64" = 7
- 1/8" = 8
- 9/64" = 9
- 5/32" = 10

FIGURE 5.2.53 The ANSI carbide turning insert identification system.

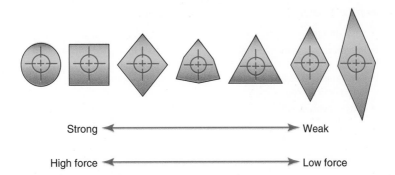

Strong ⟵⟶ Weak

High force ⟵⟶ Low force

FIGURE 5.2.54 Various insert shapes are required to cut different types of features. The shape of an insert also has an effect on its strength and the forces created during machining.

	Round	Square	80° Rhombus	Trigon	Triangular	55° Rhombus	35° Rhombus
Insert Shape	R ●	S ■	C ◇80°	W ▲80°	T ▲	D ◇55	V ◇35
Facing	FAIR	GOOD	GOOD	GOOD	FAIR	FAIR	N/A
Longitudinal turning	N/A	N/A	GOOD	FAIR	FAIR	GOOD	GOOD
Intricate profiling	N/A	N/A	FAIR	FAIR	FAIR	GOOD	GOOD
Interrupted cutting	GOOD	GOOD	FAIR	FAIR	FAIR	N/A	N/A
Roughing	GOOD	GOOD	GOOD	FAIR	FAIR	N/A	N/A
Light roughing/semi-finishing	N/A	FAIR	GOOD	GOOD	GOOD	GOOD	N/A
Finishing	N/A	N/A	FAIR	FAIR	GOOD	GOOD	GOOD
Low power consumption	N/A	N/A	FAIR	FAIR	GOOD	GOOD	GOOD
Low chatter/vibration	N/A	N/A	FAIR	FAIR	GOOD	GOOD	GOOD
Hard material cutting	GOOD	GOOD	N/A	N/A	N/A	N/A	N/A

FIGURE 5.2.55 The common carbide insert shapes, their identification letters, and the common uses for each.

note that the overall clearance while machining may be either built into the insert, provided by the toolholder, or a combination of both. Zero degree inserts must be used in holders that tilt the insert in a negative rake direction, so that clearance is provided between the cutting tool and the workpiece. Positive clearance inserts (generally those with angles greater than 3 degrees) may be held flat (in a neutral toolholder) or tilted back (in a positive toolholder). **Figure 5.2.56** shows the inclination of the insert in a holder so that proper clearance can be achieved. Inserts with a zero-degree angle are stronger and have the added benefit of having double the number of cutting edges, since they can be flipped and used on either side. Zero-degree inserts generally consume more power and create more cutting force.

Tolerance

Tolerance is designated by a letter in the third position of **Figure 5.2.53**. Inserts are produced to dimensional tolerances like any other manufactured item. These tolerances ensure that the user can expect a certain amount of repeatability when the insert is indexed or it is replaced with another of the same type. Inserts made to tighter tolerances are more carefully produced and therefore cost more.

Insert Features

The fourth position of **Figure 5.2.53** specifies the way that the insert is fastened to the toolholder and the placement of the chipformer. Some inserts are fastened through a

Negative inserts

- Very strong
- Preferred choice for external turning
- Can withstand heavy and interrupted cutting
- Often double sided

Positive inserts

- Less cutting force
- Used for boring and external turning on non-rigid parts
- Single sided

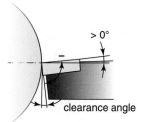

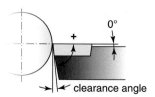

FIGURE 5.2.56 Notice how the holder must hold the insert on an incline to give a zero-degree insert clearance. The insert must be tilted in two directions so that clearance is provided on the front as well as the side of the cutting edge. Notice also how a holder with no incline (or a positive incline) could not use an insert having a zero-degree clearance angle, since it would rub the workpiece.

hole in their center, while others are clamped into place. Many inserts use both a hole fastener and a clamp. Chipformers are carefully designed geometry formed into the surfaces of the insert. Their intent is to create a specific chip formation so that the chip curls tightly and breaks into small manageable pieces. Inserts are available with a chipformer on one side, both sides, or not at all.

Insert Size

The fifth position of **Figure 5.2.53** specifies one factor of the insert size. Since inserts come in a variety of shapes, a universal method was developed to define their size regardless of shape. By determining the largest circle that can fit within the edges of the insert, most insert shapes can be sized. This size is called the *inscribed circle (I.C.)*. **Figure 5.2.57** shows an inscribed circle on two different

insert shapes. As shown, inch series inscribed circle diameters are denoted as the number of 1/8ths of an inch. For example, an insert with a size designation of "3" can fit a 3/8" diameter inscribed circle within its edges.

Insert Thickness

The sixth position of **Figure 5.2.53** specifies another aspect of the insert size, its thickness. The thickness of an insert is noted in terms of the number of 1/16ths of an inch. For example, an insert with a size designation of "2" measures 2/16" or 1/8" (reduced) thick.

Tool Nose Radius/Cutting Point

The seventh position of **Figure 5.2.53** specifies the cutting point. For turning inserts, the cutting point is normally a radius. As discussed previously, the radius on the tool tip helps provide strength to the insert and affects the surface finish and cutting force. Insert nose radii are noted in the system as the number of 1/64ths of an inch. For example, a "1" means that the tool has a nose radius of 1/64".

Manufacturer's Option

Positions eight through ten of **Figure 5.2.53** provide optional space for the manufacturer to make special notes about additional information.

Chip Former

To further help with chip control, carbide manufacturers invest a great deal of research and development into creating a working surface on inserts to further refine

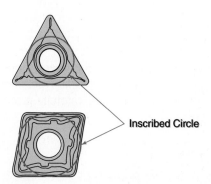

FIGURE 5.2.57 The red circles represent inscribed circles on two different insert shapes.

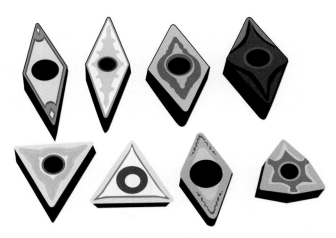

FIGURE 5.2.58 Notice the different chipformer geometries on these inserts.

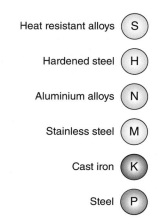

Heat resistant alloys	S
Hardened steel	H
Aluminium alloys	N
Stainless steel	M
Cast iron	K
Steel	P

FIGURE 5.2.59 The standardized ISO identifications for a various workpiece material types are usually referred to when selecting carbide insert grades.

the formation of chips. These chipformer geometries may sometimes appear like ornate pieces of art, but they are scientifically refined patterns defined for specific cutting conditions. Inserts are available with chipformers designed for specific material types, hardness conditions, and whether roughing or finishing is to be performed. Many manufacturers include designations for their various styles of chipformers at the end of the ANSI identification. **Figure 5.2.58** shows three different chipformer geometries and their designations.

Insert Grade

Generally an insert made of hard carbide will sacrifice toughness, and a tough insert will sacrifice hardness. These properties make up the insert *grade* and must be matched to each application. Workpiece material, the operation type (finishing, medium roughing, or heavy roughing), and machining conditions (heat treatment, rigidity, interruptions, chip evacuation, cutting fluid supply, etc.) all influence the selection of the proper carbide grade. **Figure 5.2.59** shows the standardized ISO identifications for various workpiece material types. Within each ISO workpiece material code, many grades are offered ranging from very tough (softer) to wear resistant (harder). Tougher grades can handle more impact. Wear resistant grades last longer and stand up better to harsh materials. It is important to note that most carbide manufacturers have devised their own proprietary grade designations, and it is usually best to refer to the applications catalog for each brand.

External Toolholders for Carbide Inserts

A toolholder is available to suit every insert option. The standardized system shown in **Figure 5.2.60** is used to identify the various configurations of toolholders.

Insert Mounting

The first position of the designation in **Figure 5.2.60** identifies the method used for fastening the insert. Methods include screws, cams, clamps, and tapers. **Figure 5.2.61** shows some common methods of fastening.

Insert Shape

The second position in **Figure 5.2.60** identifies the shape of the insert. The same identification letters used in the insert classification system are also used for toolholders.

Toolholder Style

The third position in **Figure 5.2.60** identifies the toolholder style. This specifies whether the shank is straight or offset. The end or side cutting angle created when the insert is mounted in the holder is also specified here.

Insert Clearance Angle

The fourth position in **Figure 5.2.60** identifies the insert clearance angle that the holder will accept. The insert and the holder must have compatible angles, or the insert will not seat properly in the toolholder pocket.

Hand of Tool

The fifth position in **Figure 5.2.60** identifies the hand of the tool. The toolholder positions the insert to make the desired direction of cut possible. As mentioned earlier in the unit, a right-hand tool cuts from right to left, a left-hand tool cuts from left to right, and a neutral tool can cut in either direction.

Pocket Style

Position six in **Figure 5.2.60** specifies information about the pocket in which the insert will mount. The pocket of an insert provides support to the insert and helps to accurately position it in the holder. If the holder did not

External Toolholders Identification System

MWLNR-16-3C

M	W	L	N	R		-16-	3	C	-	
1	2	3	4	5	6	7	8	9	10	

1. Method of holding

C = Clamp lock assembly
(PC toolholders)
D = Top clamp using center hole
M = Multiple lock assembly
(Pin and clamp lock)
(M-Type toolholders)
P = Pin lock assembly
(NL/PL toolholders)
S = Screw lock
T* = Taper stem
(Tee-lock toolholders)

2. Insert shape

C = 80° Diamond
D = 55° Diamond
R = Round
S = Square
T = Triangle
V = 35° Diamond
W = Trigon

3. Toolholder style

A = Straight shank with 0° side cutting edge angle.

B = Straight shank with 15° side cutting edge angle.

C = Shank with 0° end cutting edge angle.

D = Straight shank with 45° side cutting edge angle.

E = Straight shank with 30° side cutting edge angle.

F = Offset shank with 0° end cutting edge angle.

G = Offset shank with 0° side cutting edge angle.

J = Offset shank with −3° side cutting edge angle.

K = Offset with 15° end cutting edge angle.

L = Offset shank with −5° end or side cutting edge angle.

M = Straight shank with 50° side cutting edge angle.

O* = Offset shank with centrally located round insert.

P = Straight shank with 27 1/2° side cutting edge angle.

Q = Offset shank with −17 1/2° cutting edge angle.

R = Offset shank with 15° side cutting edge angle.

S = Offset shank with 45° side cutting edge angle.

T* = offset shank with −30° side or end cutting edge angle.

V = Straight shank with 17 1/2° side cutting edge angle.

W = Offset shank with 10° side cutting edge angle.

4. Insert clearance angle

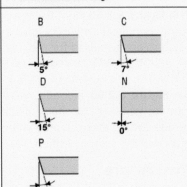

B

C

5°

7°

D

N

15°

0°

P

11°

5. Hand of tool

L = Left
N = Neutral
R = Right

8. Insert size I.C.

Number of eights of I.C.

6. Pocket style

S = Single wall pocket construction. Full pocket construction when letter position is vacant.

9. Qualified surface & length

A = Qualified back and end. 4" long.
B = Qualified back and end. 4.5" long.
C = Qualified back and end. 5" long.
D = Qualified back and end. 6" long.
E = Qualified back and end. 7" long.
F = Qualified back and end. 8" long.
J = Qualified back and end. 3.5" long.
M = Qualified front and end. 4" long.
N = Qualified front and end. 4.5" long.
P = Qualified front and end. 5" long.
R = Qualified front and end. 6" long.
S = Qualified front and end. 7" long.
T = Qualified front and end. 8" long.

7. Toolholder shank size

For square shanks, the number represents the number of sixteenths of width and height. For rectangular shanks, the first digit represents the number of eighths of width and the second digit represents the number of quarters of height.

10. Manufacturers option

* Seco standard

FIGURE 5.2.60 The standardized identification system for external turning toolholders.

Toolholder systems, external turning

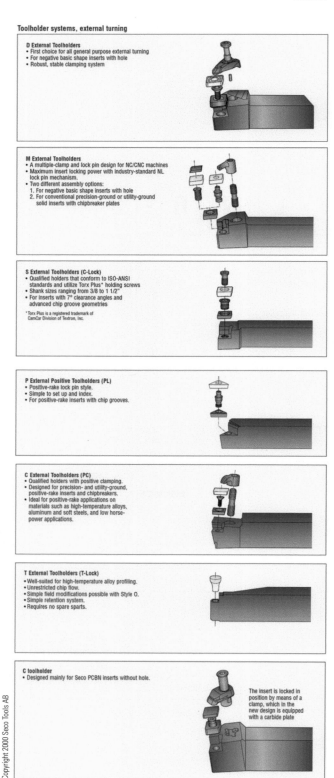

D External Toolholders
- First choice for all general purpose external turning
- For negative basic shape inserts with hole
- Robust, stable clamping system

M External Toolholders
- A multiple-clamp and lock pin design for NC/CNC machines
- Maximum insert locking power with industry-standard NL lock pin mechanism.
- Two different assembly options:
  1. For negative basic shape inserts with hole
  2. For conventional precision-ground or utility-ground solid inserts with chipbreaker plates

S External Toolholders (C-Lock)
- Qualified holders that conform to ISO-ANSI standards and utilize Torx Plus* holding screws
- Shank sizes ranging from 3/8 to 1 1/2"
- For inserts with 7° clearance angles and advanced chip groove geometries
  * Torx Plus is a registered trademark of CamCar Division of Textron, Inc.

P External Positive Toolholders (PL)
- Positive-rake lock pin style.
- Simple to set up and index.
- For positive-rake inserts with chip grooves.

C External Toolholders (PC)
- Qualified holders with positive clamping.
- Designed for precision- and utility-ground, positive-rake inserts and chipbreakers.
- Ideal for positive-rake applications on materials such as high-temperature alloys, aluminum and soft steels, and low horse-power applications.

T External Toolholders (T-Lock)
- Well-suited for high-temperature alloy profiling.
- Unrestricted chip flow.
- Simple field modifications possible with Style O.
- Simple retention system.
- Requires no spare sparts.

C toolholder
- Designed mainly for Seco PCBN inserts without hole.

The insert is locked in position by means of a clamp, which in the new design is equipped with a carbide plate.

FIGURE 5.2.61 Several different methods for mounting inserts on toolholders for external turning.

have a pocket, the mounting screw or clamp would be under great force during machining, and the insert could slip out of position or break the screw. Some holder pockets have just one wall, which only provides support to the insert in one direction. Other holders have a full pocket that has at least two walls to provide greater support and positioning of the insert.

Toolhoolder Shank Size

The seventh position in **Figure 5.2.60** identifies the shank. Shanks can be either square or rectangular. The number listed on a square shank holder will be the number of 1/16ths of an inch of height and of width. Rectangular shanks will be identified by two numbers, the first for the width in 1/8ths of an inch and the second for the number of 1/4ths of an inch of height.

Insert Size

The size of insert that the toolholder pocket will accept is specified in position eight of **Figure 5.2.60**. This value represents the number of 1/8ths of an inch of the insert inscribed circle.

Qualified Dimensions

The ninth position in **Figure 5.2.60** identifies the *qualified dimensions* of the holder. The total length of the holder plus the insert is called the *qualified end length*. Both toolholders and inserts are manufactured to consistent, accurate dimensions. This prevents machine setup changes and adjustments when an insert is indexed or replaced or a holder is exchanged.

Manufacturer's Option

Position ten of the identification system provides optional space for the manufacturer to make a special note about a unique feature of their product.

Internal Toolholders for Carbide Inserts

Internal toolholders use an identification system similar to external holders. This standardized system and its components are shown in **Figure 5.2.62**. Three common insert mounting methods for internal holders are shown in **Figure 5.2.63**.

Holemaking Tools

Standard holemaking operations including drilling, reaming, countersinking, and counterboring can be performed by mounting cutting tools in the lathe tailstock.

Taper-Shank Tools

The Morse taper in the lathe tailstock quill can be used for mounting taper-shank holemaking tools or chucks that can hold straight shank tools. Morse taper sleeves and sockets can be used to adapt taper shanks to fit the taper size in the tailstock. Follow these steps to mount and remove a Morse taper cutting tool or chuck in the tailstock.

Internal Toolholders Identification System

1. Type of bar

S = Steel
A = Steel with
    coolant hole
C = Carbide
E = Carbide with
    coolant hole
H = Heavy metal
J = Heavy metal with
    coolant hole

2. Shank diameter (Inch)

Indicates bar diameter in sixteenths of an inch.

Stepped bar shows smallest diameter first.

3. Boring bar length (Inch)

F = 3"
G = 3.5"
H = 4"
J = 4.5"
K = 5"
L = 5.5"
M = 6"
N = 6.5"
P = 6.75"
Q = 7"
R = 8"
S = 10"
T = 12"
U = 14"
V = 16"
W = 18"
Y = 20"
X = Special Length

4. Method of holding insert

C = Clamp lock assembly
    (PC toolholders)
M = Multiple lock assembly
    (Pin and clamp lock)
    (M-Type toolholders)
P = Pin lock assembly
    (NL/PL toolholders)
S = Screw lock

5. Insert shape

C = 80° Diamond
D = 55° Diamond
R = Round (inch)
S = Square
T = Triangle
V = 35° Diamond
W = Trigon

6. Bar style

U = Offset shank with negative
    3° (93°) end or side cutting
    edge angle.
F = Offset shank with 0° (90°) end
    cutting edge angle.
G = Offset shank side/end cutting
    (round insert)
K = Offset shank with 15° (75°) end
    cutting edge angle.
L = Offset shank with negative 5°
    (95°) end or side cutting
    edge angle.
Q = Offset shank with negative 17.5°
    (107.5°) cutting edge angle.
P = Offset shank with 27° 30' side
    and end cutting edge angle.

7. Insert clearance angle

B   5°
C   7°
D   15°
N   0°
P   11°

8. Hand of bar

R = Right
L = Left

9. Insert size I.C.

Number of eights of I.C.

FIGURE 5.2.62 The standardized identification system for internal turning toolholders.

Toolholder systems, internal turning

M Internal Boring Bars
- Proven lock pin for negative rake geometry inserts.
- Ideal for unground, negative rake inserts or utility, and precision ground inserts with a chipbreaker plate.
- Available with integral coolant delivery.

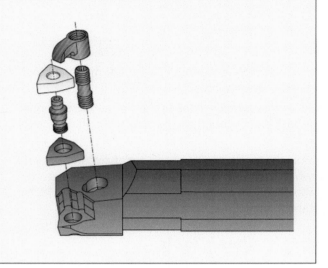

S Internal Boring Bars (C-Lock)
- For bores as small as .180"
- Available with steel or carbide shanks, ranging in size from 3/16" to 1".
- Designed to ISO-ANSI standards.

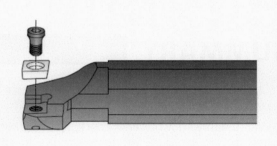

C Internal Boring Bars (PCBN)
- Insert is held down with clamp.
- For rougher machining, clamp is equipped with a carbide pressure plate that reduces clamp wear and distributes the clamping force onto the insert surface.

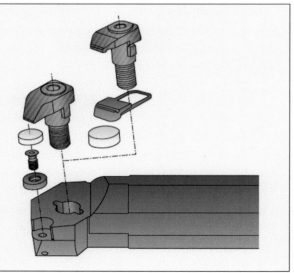

FIGURE 5.2.63 Several different methods for mounting inserts on toolholders for internal turning.

- Extend the tailstock quill about 2" out of the tailstock body so the shank does not bottom in the bore.
- Check that the tailstock bore and tool (or chuck) shank are both clean and burr free.
- Align the tang with the slot in the bore.
- Push the tools in the tailstock just like mounting tools in a drill press spindle.
- To remove a taper-shank tool from the tailstock quill, retract the quill with the hand wheel until resistance is felt.
- Hold the tool in one hand and forcefully turn the hand wheel a bit farther to release the taper lock between the shank and bore. **Figure 5.2.64** shows a Morse taper drill mounted in a lathe tailstock.

Straight Shank Tools

Holemaking cutting tools with straight shanks can simply be mounted in a chuck that is installed in the lathe tailstock, just like those used in drill presses.

FIGURE 5.2.64 A Morse taper-shank drill mounted in a lathe tailstock. To remove the drill, retract the quill with the hand wheel until the drill is released from the tapered bore.

SUMMARY

- Choosing the proper lathe workholding device for any situation is crucial to safely, accurately, and efficiently perform machining operations.
- With the three-jaw universal chuck, all three jaws move simultaneously on a scroll mechanism when the chuck is tightened or loosened with the chuck key. It can secure round work or work with a number of sides divisible by three. The three-jaw universal chuck offers ease of use and can center work within about 0.003" accuracy.
- When higher accuracy is required, a four-jaw independent chuck can be selected for workholding. All four jaws move independently of each other to allow more control over workpiece position. The four-jaw independent chuck requires more time to set up for use than the three-jaw chuck, but it is required for work with a number of sides that is divisible by four. It can also be used to offset a workpiece to machine eccentric diameters.
- Drill chucks can also be mounted in the spindle for machining of smaller-diameter workpieces.
- Spring collets can quickly secure work and can be used with a headstock sleeve and drawtube or in a collet chuck, while flex collets need to be used with collet chucks. Collets can be as precise as four-jaw chucks and work well with higher spindle speeds because they have less mass than jaw-type chucks and are not as likely to cause vibration.
- A lathe faceplate is useful for holding an odd-shaped workpiece that could not be held by other methods. The major drawbacks of the faceplate are the large amount of required setup time and the unique potential safety hazards.
- Mounting work between centers allows parts to be flipped end for end while maintaining concentricity, but care must be taken to be sure that lathe centers run true and that the tailstock is accurately aligned with the headstock. However, work held between centers is limited to machining of outside diameters.
- If an outside diameter must be machined to be concentric with an existing bore, mandrels can be used to locate and secure the workpiece by its inside diameter. Mandrels are then usually held between centers for machining.
- Long parts present problems of flex while being machined, so a steady rest can be used for extra support to machine middle sections of long workpieces or to perform machining on the ends of long workpieces.

- The follower rest is bolted to the carriage and travels along the part with the cutting tool to provide support when machining long, slender diameters.

- As the need for faster and more efficient machining operations has developed over time, different types of toolholding devices have been developed to meet those demands.

- One toolholding device is the rocker-style toolholder and tool post. The tool post can contain up to four separate pieces. These posts and toolholders have an initial low cost but require resetting different cutting tools every time they are used, which can be very time consuming.

- The quick-change tool post and holder allow many tools to be preset, and quickly and easily removed from and reinstalled on the tool post.

- Indexable tool posts can be rotated to and secured at different positions by a locking lever. They may have integral toolholders or use quick-change holders. The integral holders limit the user to a certain number of tools, while the quick-change types allow the user to preset any number of tools required for a specific operation or group of operations.

- Specific geometry of cutting tools is needed for the tools to cut properly.

- ANSI provides a standard system for classification and identification of carbide inserts and tool holders.

- The Morse taper in the lathe tailstock is useful for mounting taper-shank tools for performing holemaking operations.

- Drill chucks with Morse taper shanks can also be mounted in the tailstock to perform holemaking operations using straight-shank tools.

REVIEW QUESTIONS

1. What is the special name for the type of jaw-type chuck that has a mechanism to move the jaws simultaneously?
2. The most common variation of the above chuck has how many jaws?
3. Name two material shapes that can be properly held in a three-jaw chuck.
4. List two advantages of using a self-centering chuck.
5. Name two material shapes that can be properly held in a four-jaw chuck.
6. List three benefits of holding a workpiece between centers.
7. List three potential advantages of using an independent chuck for a workholding situation.
8. List three characteristics of a workpiece that would make a mandrel the ideal workholding device.
9. What type of mandrel would be ideal for gripping a workpiece on an accurately machined but odd-sized hole?
10. Name the type of tailstock center that would be selected to support a workpiece for most lathe operations.
11. What two auxiliary devices can be used to stabilize long, slender workpieces for turning operations?
12. Explain the differences between the two auxiliary devices in the previous question.
13. Name the device that is used to transmit the spindle power to the workpiece when held between centers.
14. Which two tool posts are the most efficient if several tools are to be used?
15. What is the major advantage of using inserted/indexable carbide cutting tools over brazed carbide tools?
16. What direction is a right-hand turning tool designed to cut?
17. What is an inscribed circle related to carbide inserts?
18. Would a 0.010" or 0.031" tool nose radius be stronger?
19. Which device may be used for either toolholding or workholding?

Machining Operations on the Lathe

Learning Objectives

After completing this unit, the student should be able to:

- Explain the relationship between depth of cut and diameter reduction
- Compare and contrast roughing and finishing operations
- Explain lathe speed and feed terms, and calculate spindle speeds and machining time
- Demonstrate understanding of lathe safety precautions
- Demonstrate understanding of the purpose of facing, turning, and shouldering operations
- Demonstrate understanding of lathe holemaking operations
- Explain how to use taps and dies to cut threads on the lathe
- Demonstrate understanding of form cutting
- Demonstrate understanding of grooving and cutoff operations
- Demonstrate understanding of the purpose and process of knurling

Key Terms

Boring	**Finishing**	**Knurling**
Cutoff (parting)	**Form cutting**	**Roughing**
Depth of cut	**Grooving**	**Shouldering**
Feed per revolution (FPR)	**Inches per revolution (IPR)**	

INTRODUCTION

Once appropriate lathe toolholding and workholding devices are selected to machine a workpiece, it is time to move from the planning stage to the processing stage. In this phase, the lathe can be used to machine many different types of external and internal cylindrical part features. The first step in performing any lathe operation is to determine cutting rates, including calculation of spindle speeds and feed rates. Then appropriate cutting tools are selected for the given operation. Once the workpiece and tools are mounted, setup is complete and machining can begin.

DEPTH OF CUT, SPEED, FEED, AND TIME CALCULATION

Before beginning lathe operations, appropriate speed and feed rates need to be determined. Speeds and feeds that are too slow result in inefficiency and lost time. Speeds and feeds that are too fast will result in rapid tool wear and can damage workpieces. Recall that speed on the lathe is set in RPM (revolutions per minute) of the spindle and that feed is measured in distance traveled per revolution of the spindle.

Depth of cut, speed, feed rate, and machining time are important related factors that need to be understood and considered to safely and efficiently perform lathe operations.

Depth of Cut

Depth of cut refers to the distance that the cutting tool is engaged in the workpiece. When cutting a diameter, the diameter is reduced by two times the depth of cut. **(See Figure 5.3.1.)**

Depth of Cut and Cross-Slide Micrometer Collars

A lathe's cross-slide micrometer collar can be graduated in one of two ways: diametrically or radially. Radial graduations are less common but do appear on some machines. If a machine is equipped with a radially graduated micrometer dial, the graduations indicate the actual amount of tool movement, which is equal to the depth of cut. For example, suppose a radially graduated collar is zeroed after touching a cutting tool to a workpiece diameter. If the cross slide is advanced by a 0.050" collar reading, the cutting tool will remove 0.050" from the part's radius (one side of the part). That 0.050" depth of cut would reduce the diameter by 0.100".

Conversely, a diametric, or diameter-reading, collar is graduated to show the amount of material removed from the part's overall diameter (both sides). For example, suppose a diametric collar is zeroed after touching a cutting tool to a workpiece diameter. If the cross slide is advanced by a 0.050" collar reading, the cutting tool will remove only 0.025" per side. That 0.025" depth of cut would reduce the diameter by 0.050".

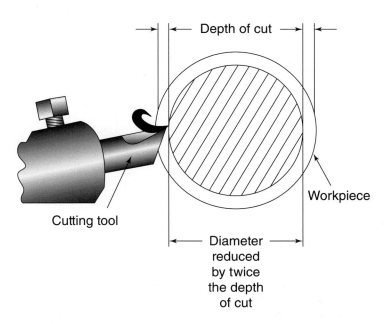

FIGURE 5.3.1 Depth of cut is the distance that the tool is fed into a diameter. The diameter is reduced by twice the depth of the cut amount.

Speed

Recall the standard RPM formula used for drill press operations: $RPM = \dfrac{3.82 \cdot CS}{D}$, where CS = cutting speed in surface feet per minute and D = diameter of the cutting tool. The same formula is used to calculate RPM for lathe operations, except that D = diameter of the workpiece, instead of the tool.

EXAMPLE: Determine spindle RPM for machining a 1.5"-diameter piece of 1020 CRS at 90 SFPM.

$CS = 90$ and $D = 1.5$. Insert these values into the RPM formula and solve.

$$RPM = \frac{3.82 \cdot 90}{1.5}$$
$$= \frac{343.8}{1.5}$$
$$= 229.2$$

The spindle speed would be set to the available speed closest to 229 RPM.

Remember that the proper cutting speed for any given material and tool type (HSS or carbide) needs to be selected using charts such as those in the *Machinery's Handbook* or other reference materials. **Figure 5.3.2** shows an example of a cutting speed table.

Feed

Feed rates for lathe operations are similar to those for drill press operations. They are given in **IPR (inches per revolution)** or **FPR (feed per revolution in inches)**.

For example, a feed rate of 0.002 IPR (or FPR) means that every time the spindle makes one revolution, the cutting tool advances 0.002" across the surface of the work. A feed rate of 0.015 IPR (or FPR) means that every time the spindle makes one revolution, the cutting tool advances 0.015" across the surface of the work.

When setting lathe feed rates, keep in mind that on most lathes, the cross-slide feed rate is one-half of the longitudinal feed rate.

Roughing and Finishing

In lathe operations, and all machining, there is rough cutting and finish cutting. The purpose of rough cutting, or **roughing**, is to remove as much material as quickly as possible to get close to the desired size. There is little concern for creating a smooth surface. Roughing uses slower cutting speeds, deeper cuts, and higher feed rates.

After roughing, finish cutting, or **finishing**, is used to produce a smoother surface and bring the part to the final desired size. Finishing uses higher cutting speeds, lighter cuts, and lower feed rates. **Figure 5.3.3**

Material	Brinell Hardness	Cutting Speed in SFPM	
		HSS	Carbide
Plain Carbon Steel: AISI/SAE 1006–1026, 1513, 1514	100–125	120	350–600
	125–175	110	300–550
	175–225	90	275–450
	225–275	70	225–350
Alloy Steel: AISI/SAE 1330–1345, 4032–4047, 4130–4161, 4337–4340, 5130–5160, 8630–8660, 8740, 9254–9262	175–225	85	250–375
	225–275	70	225–350
	275–325	60	180–300
	325–375	40	125–200
	375–425	30	90–150
Tool steel: AISI 01, 02, 06, 07	175–225	70	225–350
Tool steel: AISI A2, A3, A8, A9, A10	200–250	70	225–350
Tool steel: AISI A4, A6	200–250	55	175–275
Tool steel: AISI A7	225–275	45	125–225
Stainless steel: AISI 405, 409, 429, 430, 434, 436, 442, 446, 502	135–185	90	325–450
Stainless steel: AISI 301, 302, 303, 304, 305, 308, 309, 310, 314, 316, 317, 330	135–185	75	275–400
	225–275	65	225–350
Gray cast iron: ASTM A18, A278 (20 KSI TS)	120–150	120	240–600
Gray cast iron: ASTM A18, A278 (25 KSI TS)	160–200	90	200–450
Gray cast iron: ASTM A48, A278 (30, 35, & 40 KSI TS), A	190–200	80	175–400

FIGURE 5.3.2 Sample chart showing cutting speeds in SFPM for turning some materials using HSS and carbide tools.

	Depth of Cut	Speed	Feed	Surface Finish
Roughing Operations	Deeper, 0.050–0.250	Slower	Higher, 0.010–0.040 IPR	Rougher
Finishing Operations	Shallower, 0.010–0.050	Faster	Lighter, 0.001–0.010 IPR	Smoother

FIGURE 5.3.3 The relationship of speed, feed, and depth of cut to roughing and finishing operations.

summarizes roughing and finishing and the relationship between roughing and finishing and depth of cut, speed, and feed.

Machining Time Calculation

It may be important to estimate the time required to perform a lathe operation or operations. This is vital when machining large parts or large numbers of parts. Machining time can be calculated using the following formula.

$$\text{Time (in minutes)} = \frac{L}{\text{RPM} \cdot \text{Feed Rate}},$$

where L = length of the cut, RPM = spindle speed in RPM, and Feed Rate = IPR (or FPR).

The following examples are extreme and may not be common, but they show the importance of being able to calculate estimated machining time.

EXAMPLE ONE: Machine a 30" length of 16"-diameter AISI/SAE 4140 steel using a cutting speed of 110 SFPM and a feed rate of 0.005 IPR.

First calculate RPM = $\frac{3.82 \cdot 110}{16}$

$= \frac{420.2}{16} = 26.3 \approx 26$ RPM

Then calculate Time = $\frac{30}{20 \cdot 0.005}$

$= \frac{30}{0.13} = 230.8$ Minutes

To convert to hours:

$= \frac{230.8 \text{ Minutes}}{1} \cdot \frac{1 \text{ Hour}}{60 \text{ Minutes}} \approx 3.85$ Hours

If a workpiece requires many depth cuts, a time calculation can be even more important.

EXAMPLE TWO: Begin with a 20"-diameter piece of cast iron 80" long. Machine a 16" diameter 72" long at 200 SFPM, using 0.008 IPR, and taking depth cuts of 0.200". Estimate machining time.

First, determine how many cuts will be required. Subtract the finish diameter (16") from the starting diameter (20") to find the total diameter change of 4". Since each 0.200"-deep cut reduces the diameter by 0.400", divide the total diameter change (4") by the diameter change of each cut (0.400"). That answer is the number of cuts required. In this case, 10 cuts would be needed.

Then calculate spindle RPM using the average of the beginning and ending sizes. This works well for estimating, but during actual machining, spindle RPM should be calculated and adjusted as material is removed. The average diameter in this case would be (20 + 16) ÷ 2 = 18". Average spindle speed would be

$$\frac{3.82 \cdot 200}{18} = \frac{764}{18} = 42 \text{ RPM}$$

Then calculate time for each cut. L = 72, RPM = 42, and Feed Rate = 0.008.

$$\text{Time} = \frac{72}{42 \cdot 0.008}$$

$$= \frac{72}{0.336} = 214 \text{ Minutes}$$

Calculate time for all 10 cuts. Time = 214 · 10 = 2140 Minutes.

To convert to hours:

$$\frac{2140 \text{ Minutes}}{1} \cdot \frac{1 \text{ Hour}}{60 \text{ Minutes}} \approx 35.6 \text{ Hours}$$

Examples like these show the importance of being able to calculate machining time to estimate costs and completion dates.

GENERAL LATHE SAFETY

Like any machine tool, the lathe can be very dangerous, but by observing a few basic precautions, safe operation can be ensured. Specific safety notes are shown throughout this unit, but following are some precautions that should be observed during any lathe operation.

 CAUTION

- Always wear ANSI Z87 rated safety glasses when operating a lathe.
- Wear appropriate hard-soled work shoes.
- Wear short sleeves or roll up long sleeves past the elbows.
- Do not wear any loose clothing that can become caught in moving machine parts.
- Remove watches, rings, and other jewelry.
- Secure long hair so it cannot become tangled in moving machine parts.
- Make sure all machine guards and covers are in place before operating any lathe.
- Avoid extending long workpieces beyond the left end of the headstock.
- Never operate a lathe that is locked out or tagged out or remove another person's lock or tag.
- When operating a lathe, stay focused on the machine. Do not become distracted by other activities or talk to others.
- Never walk away from the lathe while it is running.
- Do not let others adjust work, tool, or machine settings and do not adjust other people's setups.
- Avoid rapidly and forcefully moving cutting tools into the workpiece. This can break tools and throw sharp fragments toward the operator. Small-diameter and short workpieces can also be pulled from the rotating workholding device.
- Never touch a workpiece or workholding device that is rotating or attempt to stop a lathe spindle by hand or with a rag. Allow the spindle to stop on its own.
- Always shut off the spindle and let it come to a complete stop before adjusting workholding or toolholding devices, to take measurements, or to clean the lathe.
- Remove chips from the workpiece and tool using a brush, pliers, or a chip hook only after the spindle has come to a complete stop. Never remove chips by hand.
- Never use compressed air to clean chips, debris, and cutting fluids from the lathe.

FACING AND TURNING OPERATIONS

Facing and turning are two of the most common lathe operations. Recall that facing is cutting across the end of a workpiece to machine the end flat, while turning

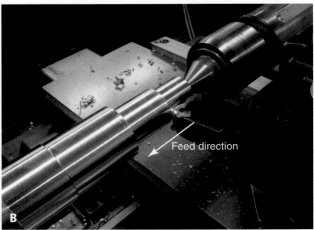

FIGURE 5.3.4 (A) Facing machines the end of a workpiece and lengths, (B) while turning machines workpiece diameters.

is reducing the outside diameter of a workpiece. **(See Figure 5.3.4.)**

Tool Nose Radius and Depth of Cut

For most situations, it is preferred that the chips produced by the lathe cutting tool break into small pieces and fall in a shower formation to the chip pan. The lead angle, depth of cut, tool nose radius (TNR), feed rate, and tool chipformer must all be working together to make this happen. By forcing a properly formed chip curl against a solid object such as the work or the tool body, the brittle chip can break into small manageable pieces. The tool lead angle acts to direct the chip curl during cutting as shown in **Figure 5.3.5**.

Figure 5.3.6 shows the formation of chips created by two different nose radii sizes while being used at equal depths of cut and equal feed rates. For proper chip breaking, the depth of cut should be at least 2/3 of the tool nose radius. For example, if this rule is followed, a 1/32" tool nose radius should take a minimum depth of cut of .021". Also, surface texture will often appear

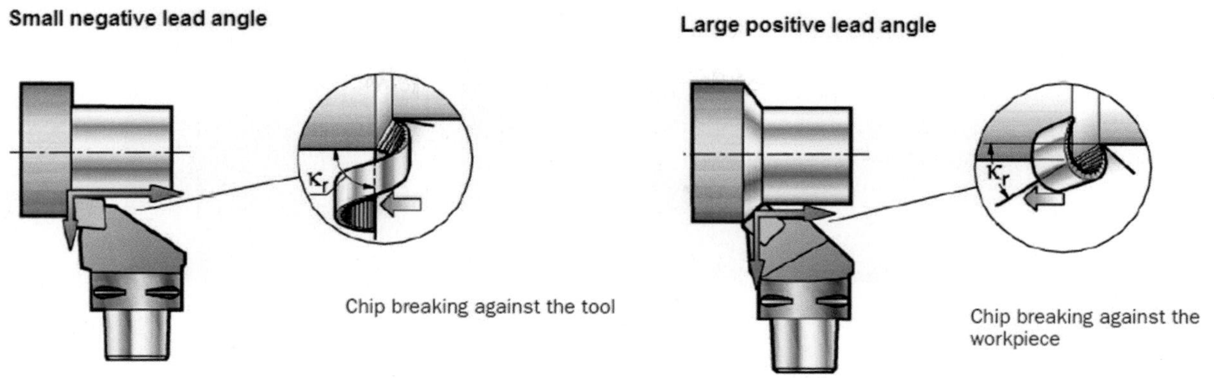

Small negative lead angle

Large positive lead angle

Chip breaking against the tool

Chip breaking against the workpiece

Copyright © Sandvik AB

FIGURE 5.3.5 The cutting tool lead angle directs the chip curl during cutting so the chip breaks as it comes in contact with either the tool or the workpiece.

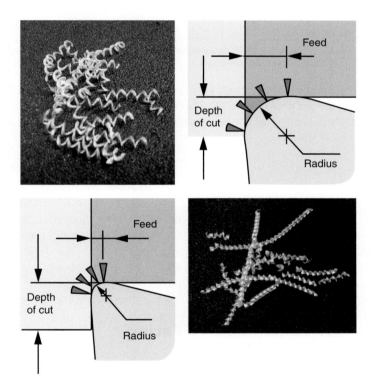

Feed

Depth of cut

Radius

Feed

Depth of cut

Radius

FIGURE 5.3.6 The different formation of chips created by two different nose radii sizes while being used at equal depths of cut.

better when this rule is followed. **Figure 5.3.7** shows the chip formation created by three different cut depths with the same nose radius.

By increasing the feed rate, the chip becomes thicker. As chip thickness increases, chips are more likely to break into pieces instead of forming continuous strings. So it is good practice to use the highest feed rate that produces the required surface finish. Another benefit of using higher feed rates is reduced machining time.

Multiple Turning Passes

There are some other variables to consider that can affect results during turning operations. These include depth of cut, feed rate, heat that can create expansion, and the relationship of the workpiece diameter to its length. Because of the pressure between the cutting tool and workpiece, different depths of cut at a particular feed rate may create slightly different size diameters. The same is true of using different feed rates with the

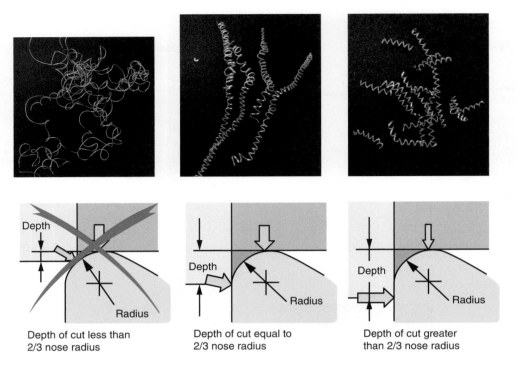

Depth of cut less than
2/3 nose radius

Depth of cut equal to
2/3 nose radius

Depth of cut greater
than 2/3 nose radius

FIGURE 5.3.7 The image on the left shows the stringy chip formation created when the depth of cut is less than 2/3 of the tool nose radius (TNR). The center image shows a chip formation when the depth of cut is equal to 2/3 of the TNR. The right image shows the chip formation when the depth of cut is greater than 2/3 of the TNR.

same depth of cut. Also, as the length of the workpiece becomes longer related to its diameter, there is the possibility of the workpiece "flexing" away from the cutting tool. This can also create some slight variation in diameter. For these reasons, it is important to use a strategy of taking multiple final turning passes using the same spindle RPM, depth of cut, and feed rate to be able to accurately predict an outcome.

Consider this example.

- Suppose a diameter is roughed within 0.060" of final size.

- A tool with a 1/64" (0.015") nose radius is used for finishing.

- One finishing pass would be taken to reduce the diameter by 0.020". This sets a 0.010" depth of cut and follows the rule of cutting a minimum of 2/3 of the tool nose radius.

- The result is measured.

- A second pass of the same depth is taken and the workpiece measured again to verify that the desired diameter change was achieved. There would still be 0.020" on the workpiece diameter for a final pass to reach the desired finish diameter. By doing this, that final cutting pass using the same settings will produce a predictable result.

This multiple pass strategy makes achieving desired size easier because it eliminates variables in the cutting operation. The first pass sets the parameters of the operation. The second pass verifies those parameters and their results. The final pass achieves the desired end result. If machining multiple workpieces, two final passes will be sufficient for subsequent parts, because the parameters are known and predictable.

Facing

Since cylindrical material is often sawed to a rough length for lathe work, facing is typically the first operation performed to clean up saw-cut end surfaces and create the desired overall length of the workpiece. To perform facing, the workpiece is usually held in a chuck or collet. A left-hand tool works well for facing operations since the cutting tool is normally fed from the outside of the work toward the center. **(See Figure 5.3.8.)**

When facing, consider that roughing cut depths for general machining can range from about 0.050" to 0.100" and feed rates about 0.010" to 0.020" IPR. Finishing cut depths generally range from about 0.010" to 0.030" and feed rates about 0.003" to 0.010" IPR. These are only general guidelines. Use tables from cutting tool manufacturers or *Machinery's Handbook* to determine appropriate values.

FIGURE 5.3.8 Most facing is done from the outside diameter toward the center of the workpiece using a left-hand cutting tool.

FIGURE 5.3.9 (A) Setting the height of the cutting tool tip using a tailstock center. (B) The toolpost is then rotated to orient the tool properly for facing.

Here are general steps to follow to perform a facing operation.

- Mount the cutting tool in a tool holder and tool post.
- Adjust the tool tip to the proper height. For all lathe operations, the tip of the cutting tool must be in line with the center line, or axis, of the workpiece. A mounted lathe center can be used as a reference for setting tool height, as shown in **Figure 5.3.9**.
- Rotate the toolholder or tool post to set the tool, as shown in **Figure 5.3.10**. This is for safety reasons. If cutting forces move the toolholder, it will push away from instead of being pushed into the workpiece, which can gouge the workpiece or break the cutting tool. This position also creates a positive lead angle that allows the tool to cut more freely.
- Avoid excessive tool overhang to create a rigid setup and eliminate chatter or vibration. **(See Figure 5.3.11.)**
- Calculate and set an appropriate spindle speed and feed rate.
- Set the feed-change lever for cross feed.
- Set the feed direction lever so the cross slide moves from the outside of the work toward the center.
- Position the tip of the cutting tool within about 1/16" of the right end of the workpiece.
- Start the spindle.
- Gently bring the tool tip into contact with the rotating workpiece using either the apron hand wheel or the compound rest. This process is called *touching off*. **(See Figure 5.3.12.)**
- Move the tool clear of the workpiece by retracting the cross slide.

- Move the carriage toward the headstock to set the desired depth of cut. This can be done either with the apron hand wheel or with the compound rest. Remember the rule of using a depth of cut that is at least 2/3 of the tool nose radius.
- If the apron hand wheel is not graduated, a micrometer stop or long travel dial indicator can be used to measure the amount of carriage movement. Some lathes are equipped with a digital readout (DRO) that shows carriage movement on a display. **Figure 5.3.13** shows each of these methods.
- Lock the carriage to the ways to ensure a straight cut. If left unlocked, the carriage may drift during facing, resulting in an uneven surface.
- Engage the feed control lever to feed the tool across the end of the work with the cross slide. **Figure 5.3.14** shows a facing cut.
- Disengage the feed control lever to stop the tool before it passes the center of the work. Running the tool

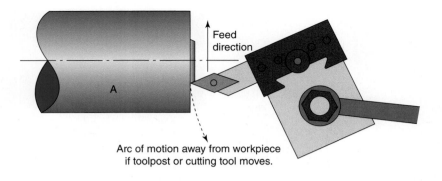

Feed
direction

A

Arc of motion away from workpiece
if toolpost or cutting tool moves.

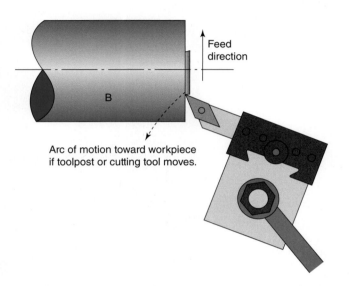

Feed
direction

B

Arc of motion toward workpiece
if toolpost or cutting tool moves.

FIGURE 5.3.10 Preferred orientation of the cutting tool when facing so that if the cutting pressure moves the tool, it will be (A) pushed away from the workpiece instead of (B) digging being pushed into the work, which can gouge the work or break the tool.

past center can cause excessive tool wear and may break the tip.

• Move the carriage toward the tailstock to move the tool away from the workpiece.

• If a nub is left on the workpiece, the tool tip is either above or below center. This can be determined by the shape of the nub, as shown in **Figure 5.3.15**.

• Adjust the tool height if needed.

• Repeat the process until either the end of the work is cleaned up or the desired part length is achieved.

Turning

Turning may be used to produce a constant diameter, a tapered diameter, or a contoured diameter (sphere, radius, or other nonstandard shape). *Straight turning* of a constant diameter is a common operation performed on the lathe. During straight turning operations, the workpiece can be held in a chuck, collet, between centers,

or on a mandrel. A steady or follower rest may also be used for extra support of very long work.

When performing straight turning, the turning tool must always be set on center just as when facing. Most turning is done from the tailstock toward the headstock using a right-hand tool. To turn from the headstock toward the tailstock, use a left-hand tool. **(See Figure 5.3.16.)** Here are general steps to follow to perform a straight turning operation.

• Mount the workpiece in the spindle using the desired workholding method.

• As with facing, it is good practice to set the tool during roughing for safety in case cutting forces push the tool away from the workpiece. **(See Figure 5.3.17.)**

• Keep the tool short for rigidity.

• Set the tool height by adjusting the tip on center just like the facing tool.

• Position the tool tip near the workpiece diameter.

FIGURE 5.3.11 (A) Mount cutting tools short and rigidly. (B) Excessive extension or overhang of the tool, toolholder, or compound rest will cause vibration during machining.

- Move the carriage toward the headstock to the farthest required tool location.
- Rotate the spindle by hand to check for clearance between the toolholder and compound rest and the workholding device to make sure that there will be no collisions during machining.
- Repeat this at the beginning point of the cut as well, especially if a center or steady rest is supporting the work. **(See Figure 5.3.18.)**
- Calculate and set an appropriate spindle speed and feed rate for the material being machined.
- Set the lathe to the proper longitudinal feed direction using the feed-change lever.
- Position the tool tip near the beginning of the cut.
- Start the spindle.
- Advance the tool with the cross slide to touch off on the diameter of the workpiece. **(See Figure 5.3.19.)**
- Use the apron hand wheel to move the tool toward the tailstock to clear the workpiece.
- Advance the tool with the cross slide to the desired depth of cut. Roughing cut depths for general turning

FIGURE 5.3.12 Touching off. Move slowly to avoid jamming the tool into the workpiece and causing breakage.

range from about 0.050" to 0.100" and feed rates about 0.010" to 0.020" IPR. These depend on the material type, cutting tooling, rigidity, and machine power, so use tables from cutting tool manufacturers or *Machinery's Handbook* as a guide.

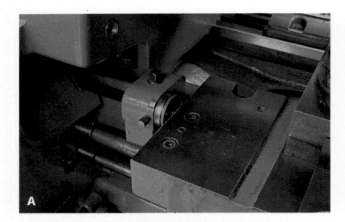

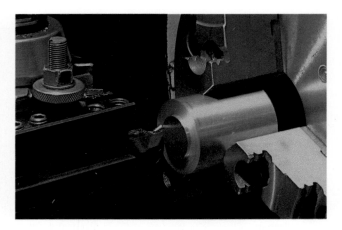

FIGURE 5.3.14 After locking the carriage to the ways, feed the cutting tool across the part to take a facing cut.

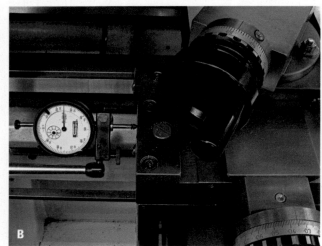

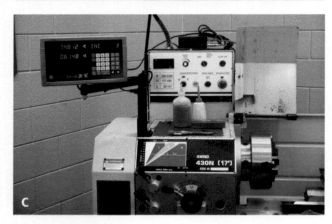

FIGURE 5.3.13 Three methods to accurately monitor carriage movement. (A) A micrometer stop. (B) A travel dial indicator. (C) A digital readout.

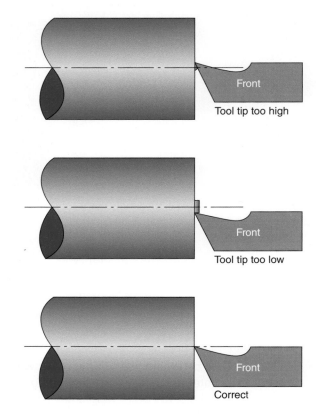

Tool tip too high

Tool tip too low

Correct

FIGURE 5.3.15 (A) Notice the cone-shaped nub, indicating the tool tip is above center. (B) Notice the cylindrical nub, indicating the tool tip is below center. (C) No nub remains when the tool tip is on center.

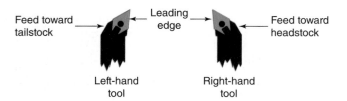

FIGURE 5.3.16 A left-hand tool is used to turn toward the tailstock, while a right-hand tool is used to turn toward the headstock.

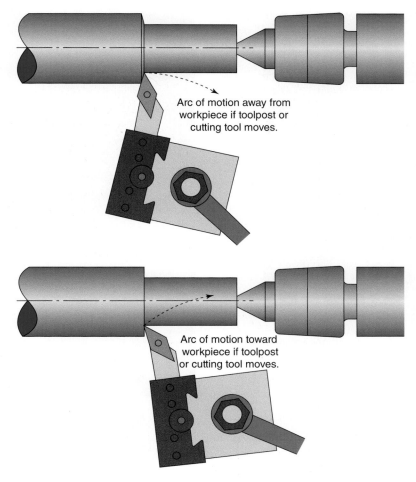

FIGURE 5.3.17 Preferred orientation of the cutting tool when turning so that if the cutting pressure moves the tool, it will be (A) pushed away from the workpiece instead of (B) digging into the work.

FIGURE 5.3.18 (A) Rotate the spindle by hand with the tool near the beginning of the cut (A) and (B) at end of the cut to check for clearance between the tool and the workholding device.

FIGURE 5.3.19 Touching off on the workpiece diameter to set a reference point for taking depth cuts.

- Engage the feed control lever to begin longitudinal feed. **Figure 5.3.20** shows a straight turning operation on a lathe.

- To stop the carriage at the correct location to produce the desired length of cut, a micrometer stop, travel dial indicator, or digital readout can be used.

- Use the apron hand wheel to return the carriage to the beginning position to prepare for the next cutting pass.

- Before continuing, determine finishing cut depth values. Finishing cut depths for general turning range from about 0.010" to 0.030" and feed rates about 0.003" to 0.010" IPR, but again use tables from cutting tool manufacturers or *Machinery's Handbook* to determine appropriate values. Also keep in mind the rule of setting minimum cut depth to 2/3 of the tool nose radius.

- Once a finishing cut depth is determined, rough the diameter to a size so that three finishing passes can be taken to reach the final size.

- Adjust the feed rate to the finishing feed rate.

- Set the depth of cut and machine a pass.

- Measure workpiece diameter.

Always shut off the spindle and let it come to a complete stop before taking part measurements.

FIGURE 5.3.20 A straight turning cut being taken on the lathe.

- Set the tool to take another pass with the same depth of cut and measure the workpiece diameter again.

- Taking these two equal depth cut passes will show what can be expected at the given RPM, depth of cut, and feed rate. This makes the result of the final pass predictable.

- Set the depth of cut for a final pass to finish the diameter to the final desired size.

Shouldering

Shouldering combines turning and facing to create a step where two different diameters meet. Three common shoulder types are square, filleted, and angular. **(See Figure 5.3.21.)** To machine shoulders, first rough the smaller diameter and shoulder length with the cutting tool set in the position recommended for turning. Be sure to allow enough material for finishing the shoulder to the desired shape.

Square Shouldering

To finish a square shoulder, follow these steps.

- Position the tool so only the tip will contact the shoulder when the cross slide is fed across the shoulder. **(See Figure 5.3.22.)**

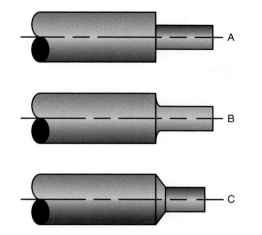

FIGURE 5.3.21 (A) Square, (B) filleted, and (C) angular shoulders can be machined as a transition between two diameters.

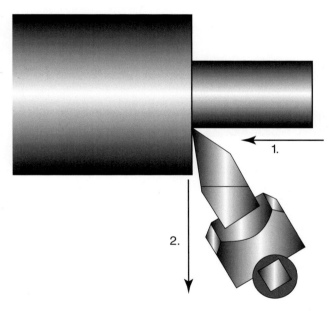

FIGURE 5.3.22 Tool position for finish machining of a square shoulder.

Figure 5.3.23 illustrates these following remaining steps.

- Touch off the tool against the roughed diameter and set the cross slide dial (or digital readout) to zero.
- Take small depth cuts across the shoulder to establish a flat surface, using the cross-slide zero to avoid touching the diameter.
- Machine the shoulder until it is within about 0.005" to 0.010" of the desired length.

- Machine a turning pass to finish the diameter to size, and at the end of the turning pass, gently feed the tool by hand to the finish length position.
- Lock the carriage to the ways and use the cross slide to feed the tool across the shoulder. A micrometer stop, dial indicator, or digital readout can be used to monitor shoulder length.

Filleted Shouldering

To finish a filleted shoulder, use a tool with the desired radius on the tip. When roughing the length of the shoulder, be sure to leave material on the shoulder equal to the size of the fillet radius. See **Figure 5.3.24** for an illustration of this process.

Angular Shouldering

The compound rest can be used to machine an angular shoulder. **Figure 5.3.25** illustrates these general steps for finishing an angular shoulder.

- Turn the diameter to leave enough material for a finishing pass.
- Turn the shoulder length within about 0.005" to 0.010" of the dimension where the angle meets the smaller diameter.
- Finish the smaller diameter and bring the tool to a position that will create the desired length where the angle begins.
- Use the compound rest to machine the angular shoulder.

FILING AND POLISHING

Filing and polishing can be performed on the lathe to deburr edges and create very smooth surfaces.

🛈 **CAUTION**

Always use a file with a handle. Use of a file without a handle can lead to serious injuries from the tang.

Facing and turning, like all machining operations, will create burrs on the edges of a workpiece. These burrs can be easily removed by filing the rotating workpiece following these steps.

- Use spindle speeds equal to turning speeds when filing, but do not exceed safety ratings for any work-holding device.
- Since no carriage or cross-slide movement is required, place the feed rod and leadscrew in neutral.
- Move the carriage away from the area to be filed.

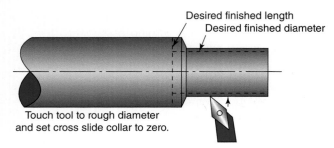

Step 1

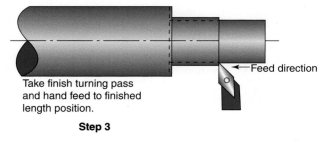

Step 3

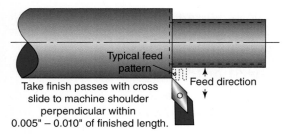

Step 2

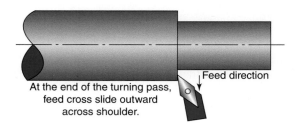

Step 4

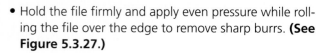

FIGURE 5.3.23 Steps for finishing a square shoulder.

⚠ CAUTION

Always file left-handed on the lathe, as shown in **Figure 5.3.26**, to avoid reaching over the rotating workholding device, Stay clear of chucks, lathe dogs, and steady rests. If the file hits any of these rotating parts it can be thrown out of the hands or pull hands into the machine, causing serious injury. If work is supported by a tailstock center, be careful not to hit the center with the file.

- Hold the file firmly and apply even pressure while rolling the file over the edge to remove sharp burrs. **(See Figure 5.3.27.)**

Abrasive paper or cloth can be used to polish a workpiece to remove small amounts of material and produce a smooth surface finish. Smaller numbers represent coarser grits that will remove material quickly, while higher numbers are finer grits that remove material slowly but produce higher surface finishes.

- Use spindle speeds equal to turning for polishing, but again do not exceed recommended speeds for any workholding device.
- Move the carriage away from the polishing area.
- Place the leadscrew and feed rod in neutral.

⚠ CAUTION

Use strips of paper or cloth to polish and be sure to remove any stray strings so they do not wrap around the rotating workpiece and entangle fingers. Never wrap the cloth or paper completely around the workpiece with your fingers or hands or the ends can grab and pull you into the rotating work. See **Figure 5.3.28** for the method to polish safely and for unsafe conditions to avoid.

CENTER AND SPOT DRILLING

Once a facing operation is complete, center drilling frequently follows. Center drilling or spot drilling is performed for two main reasons. One is to create 60-degree bearing surfaces for lathe centers. The other is to create a starting point that will prevent a twist drill from walking off center when it begins to cut.

Center Drills

Center drills (combination drill and countersinks) are usually held in drill chucks mounted in the lathe tailstock.

- Position the tailstock with the center drill within about 1/16" of the workpiece.
- Lock the tailstock to the ways.

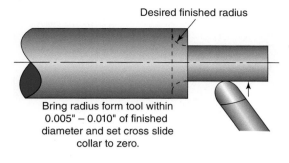

Desired finished radius

Bring radius form tool within
0.005" – 0.010" of finished
diameter and set cross slide
collar to zero.

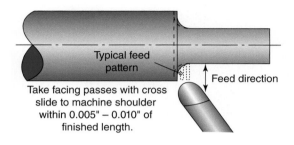

Typical feed
pattern

Feed direction

Take facing passes with cross
slide to machine shoulder
within 0.005" – 0.010" of
finished length.

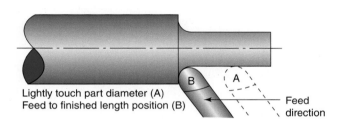

Lightly touch part diameter (A)
Feed to finished length position (B)

Feed
direction

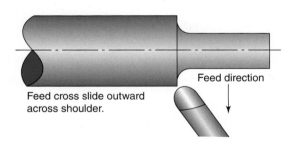

Feed direction

Feed cross slide outward
across shoulder.

FIGURE 5.3.24 Steps for finishing a filleted shoulder.

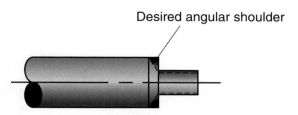

Desired angular shoulder

Rough shoulder within 0.005"–0.010"
of length at smaller diameter. Leave
material on small diameter for finish
turning pass.

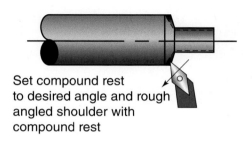

Set compound rest
to desired angle and rough
angled shoulder with
compound rest

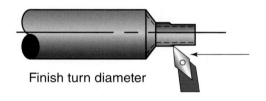

Finish turn diameter

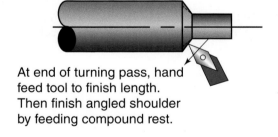

At end of turning pass, hand
feed tool to finish length.
Then finish angled shoulder
by feeding compound rest.

FIGURE 5.3.25 Steps for finishing an angular shoulder.

- Set the spindle RPM and start the spindle.
- Feed the center drill into the work using the tailstock hand wheel. The small pilots of center drills are fragile, so apply adequate cutting fluid, use light feed, and peck (retract frequently) to remove chips from the hole to prevent breakage. **Figure 5.3.29** shows center drilling on a lathe and proper center drill depth for mounting work between lathe centers.

Spotting Drills

A spotting drill can also be used to create a starting point for a twist drill. A spotting drill resembles a short twist drill with short flutes. These drills are very stable due to their solid design and can withstand heavier feed rates than center drills. The tips of spot drills are usually available with either a 90-degree or 118-degree included

FIGURE 5.3.26 File left-handed on the lathe to avoid reaching over a rotating workholding device. Always use a file with a handle.

FIGURE 5.3.27 Deburr sharp edges of a workpiece by rolling the file over the edge as it is being pushed across the work.

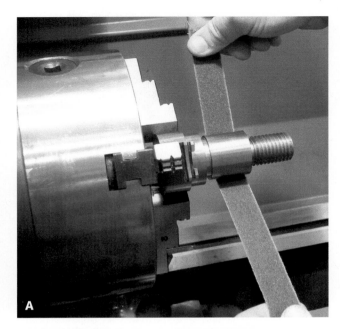

FIGURE 5.3.28 (A) Hold a strip of abrasive cloth or paper between the thumb and forefinger, keeping fingers and hands away from rotating work. Never wrap cloth or paper around fingers (B) or completely around the work (C).

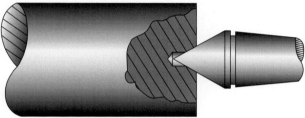

Correctly drilled hole

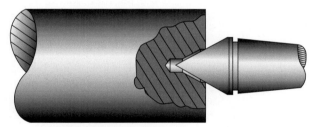

Hole drilled too deep

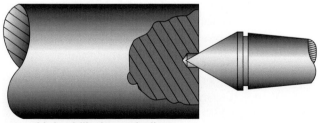

B Hole drilled too shallow

FIGURE 5.3.29 (A) Center drilling a workpiece held in a three-jaw chuck using a tailstock-mounted drill chuck. (B) Depth of the center drilled hole is important when the work is to be mounted on lathe centers.

angle. Since they do not have a smaller pilot tip, a larger diameter can be achieved with a shallower depth.

Spotting drills are also normally held in drill chucks mounted in the lathe tailstock and fed using the tailstock hand wheel. **Figure 5.3.30** shows some spotting drills.

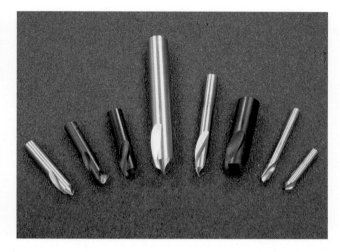

FIGURE 5.3.30 Spotting drills come in different sizes and with different tip angles. Note that there is no margin or body clearance. They are more solid than center drills and can withstand heavier feed rates.

⊘ CAUTION

Spotting drills have no body clearance so they should not be fed into workpieces beyond the cutting lips or the drill may heat up and cause damage to workpiece or tool.

HOLEMAKING ON THE LATHE

It is often necessary to machine holes on the lathe. There are a few different methods of holemaking, and the method chosen depends on factors including size, surface finish, and accuracy.

Drilling

Drilling on the lathe is similar to drilling on a drill press except that the work rotates instead of the drill. Straight-shank drills are held in drill chucks mounted in the tailstock, while taper-shank drills can be mounted directly in the tailstock. Drill sleeves and sockets can be used to adapt different taper sizes to fit the tailstock taper. When drilling holes larger than 1/2 " in diameter, drilling with a pilot drill first will greatly reduce the feed pressure required to drill the larger size. Use a pilot drill diameter that is slightly larger than the width of the dead center of the larger drill.

Follow these steps for drilling on the lathe.

- Calculate and set an appropriate spindle speed. When calculating spindle speed, use the diameter of the drill.
- Mount the drill in the tailstock.
- Position the tailstock with the drill near the face of the workpiece and lock the tailstock to the ways.

FIGURE 5.3.31 Drilling a hole with a Morse taper-shank drill held directly in the lathe tailstock.

- Start the spindle.
- Apply an appropriate cutting fluid to the tip and in the flutes of the drill. The cutting fluid applied to the cutting edges lubricates and cools the cutting zone. The fluid in the flutes helps the chips slide through the flutes and out of the hole.
- Since very few lathes have power feed on the tailstock, advance the drill with the quill-feed hand wheel. When full diameter is reached, note the graduations on the tailstock quill or hand wheel. The hole depth will be referenced from this position. (Some hand wheels have a collar that can be set to "0".)
- Drill using the tailstock hand wheel to advance the drill. If the drill produces a high-pitched squealing sound and/or fine stringy chips, the feed rate should be increased. If the chips are purple or blue and the spindle is laboring, the feed should be reduced.
- Frequently retract the drill to clear chips out of the hole and to keep cutting fluid on the drill.
- Use graduations on the quill and the hand wheel to control depth. **Figure 5.3.31** shows a drilling operation on the lathe.

(!) CAUTION

Improper speed or feeds and/or inadequate cutting fluid can cause the drill tip to become very hot and "weld" itself to the workpiece. This can lead to drill breakage and flying debris that can cause injury.

Reaming

Reaming is performed on the lathe to slightly enlarge a drilled hole to a size more accurate than a drill can produce and to produce a smooth surface finish. The

size of the hole that should be drilled before reaming depends on the desired finished hole size. Use these guidelines for drilling or prepare for reaming:

- Drill about 0.010" undersize for reaming holes 1/4" diameter and under.
- Drill about 0.015" undersize for reaming holes between 1/4" and 1/2" diameter.
- Drill about 0.020" undersize for reaming holes between 1/2" and 1" diameter.
- Drill about 0.030" undersize for reaming holes over 1" in diameter.

Reamers may also have straight or Morse taper shanks like twist drills and are mounted by the same methods used for mounting drills.

- Mount the reamer in the tailstock and position the reamer near the hole opening.
- Lock the tailstock to the ways.
- Set a spindle speed about one-half the speed as would be used for a drill of the same size.
- Apply ample cutting fluid to the reamer.
- Advance the reamer until it just touches the hole.
- Note the graduations on the tailstock quill or hand wheel. (Some hand wheels have a collar that can be set to "0".)
- Feed the reamer into the work at about twice the rate used for a similar-sized drill bit. It is important to keep a reamer moving at this constant rate so that each cutting edge is penetrating the material and a consistent surface finish is achieved.
- As with drilling, use the quill and hand wheel graduations to monitor depth. **Figure 5.3.32** shows a reaming operation being performed on the lathe.

FIGURE 5.3.32 Reaming a hole with a straight-shank reamer held in a tailstock-mounted drill chuck.

Counterboring and Countersinking

Counterboring and countersinking operations can be performed on the lathe with the tailstock using the same methods used for drilling and reaming.

Boring

Boring is using a single-point cutting tool to enlarge an existing hole. Boring is used to machine internal diameters similar to the way turning machines external diameters. A *boring bar* is the name of the cutting tool used for performing boring operations. A boring bar can be made from a solid piece of HSS or carbide, or it can be a steel bar that holds an HSS or carbide cutting tool bit. Inserted boring bars using carbide inserts similar to those used for turning and facing tools are also available. Boring tools are available in left-hand and right-hand styles. **Figure 5.3.33** illustrates the principle of boring and **Figure 5.3.34** shows a few styles of boring bars. One advantage of boring over other holemaking operations is that any desired hole size can be machined. Boring is also used to produce large-diameter holes that are beyond the range of drills and reamers. Unlike reaming, boring does not follow the path of the existing hole, so it can be used to correct misaligned holes or create eccentric diameters.

To perform boring operations, follow these steps.

- Mount the workpiece in a chuck or a collet. A steady rest can also be used for extra support when boring a long workpiece.

- Select the largest available boring bar diameter that will safely fit into the existing hole to avoid vibration.

- Mount the boring bar in a tool post parallel to the ways and set the tip of the cutting tool on center as would be done for turning or facing operations. Keep the length of the bar as short as possible to minimize vibration, but be sure it is long enough to reach the desired depth without chance of collision between the toolholder and the workpiece. **(See Figure 5.3.35.)**

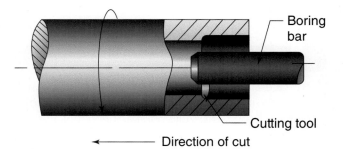

FIGURE 5.3.33 Boring uses a single-point cutting tool to machine internal diameters, much like turning operations machine external diameters.

FIGURE 5.3.34 (A) Boring bars used for small-diameter hole machining are often made from solid HSS, solid carbide, or brazed carbide. (B) Boring bar used to hold HSS or carbide tool bits. (C) Some carbide inserted boring bars.

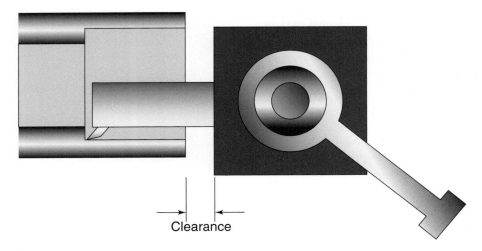

FIGURE 5.3.35 Mount a boring bar as short as possible to minimize chatter, but be sure it is long enough so there is no collision between the toolholder and the workpiece.

- Calculate and set an appropriate spindle speed and feed rate. Start with speeds and feeds equal to those used for turning, but since a boring bar is not as rigid as a turning tool, expect that speed may need to be reduced to eliminate vibration.
- Set the lathe for the correct direction of longitudinal feed.
- Start the spindle and carefully move the boring bar about 1/16" inside the existing hole.
- Touch off the existing diameter using the cross slide. **(See Figure 5.3.36.)**
- Retract the bar from the hole by moving the carriage away from the headstock.
- Set depth of cut with the cross slide. Keep in mind the 2/3 nose radius rule when setting cut depth.

FIGURE 5.3.36 Touching off against an existing inside diameter with a boring tool.

- Engage the feed control lever to machine a boring pass. As with turning, cut length can be monitored by setting a micrometer stop, or using a travel indicator or digital readout.
- Keep in mind the multiple pass strategy when approaching the desired final diameter.

Internal Shouldering

Internal shouldering can be performed in a method similar to shouldering outside diameters and lengths.

- During roughing, leave a small amount of material on both the shoulder and diameter for finishing. A typical amount would be about 0.005" to 0.010" on the shoulder and enough on the diameter to allow a depth cut equal to 2/3 of the tool nose radius.
- Machine the finish pass and gently feed the tool to the final shoulder depth.
- Use the cross slide to face the internal shoulder, as shown in **Figure 5.3.37**.

 **CAUTION**

Extra care should be taken when performing boring operations since the cutting tool is inside a hole and often out of direct sight. Collision of a boring tool against the bottom of a blind hole can damage the workpiece, tooling, and machine and create a hazardous situation where broken tools or workpieces can be violently thrown from the lathe and cause serious injury.

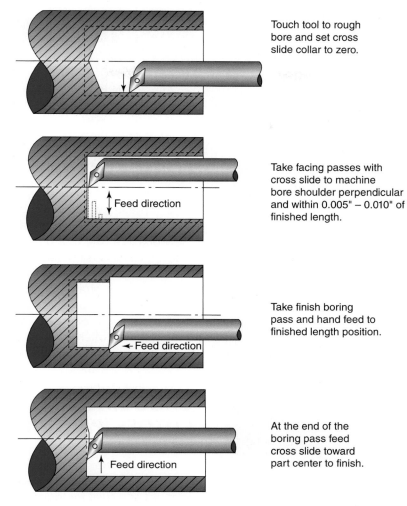

Touch tool to rough bore and set cross slide collar to zero.

Take facing passes with cross slide to machine bore shoulder perpendicular and within 0.005" − 0.010" of finished length.

Feed direction

Take finish boring pass and hand feed to finished length position.

← Feed direction

At the end of the boring pass feed cross slide toward part center to finish.

Feed direction

FIGURE 5.3.37 Method for creating an internal square shoulder with a boring bar.

THREAD CUTTING WITH TAPS AND DIES

When perform tapping on the lathe, the tailstock can be used for tap alignment. After first drilling with the appropriate tap drill size, follow these steps.

- Chamfering the hole with a countersink is a good practice and often specified on a print.

- Mount a spring-loaded tap center in a tailstock drill chuck to keep the tap aligned with the hole. A dead center may also be used to align the tap.

- The tap can be fed into the hole by two different methods.

    1. Set the spindle to a very low speed so it will not turn, and turn the tap by hand with a tap wrench.

    2. Place the spindle in neutral, hold the tap handle stationary, and rotate the lathe spindle by hand.

FIGURE 5.3.38 Using the tailstock for alignment to tap a hole on the lathe.

(See Figure 5.3.38.) The handle of the tap wrench can be placed against the compound rest to keep it from rotating.

External threads can be cut with a threading die on the lathe using methods similar to tapping.

- Place the spindle in neutral.
- Place the die mounted in a diestock against the end of the workpiece.
- Remove tools from the tailstock quill and position the face of the quill near the diestock.
- Lock the tailstock to the ways.
- Use the tailstock hand wheel to lightly bring the face of the quill against the diestock to keep the die square.
- Another method of alignment is to use a tailstock mounted drill chuck. Open the jaws so they do not extend past the chuck body and bring the face of the chuck body lightly against the diestock.
- Apply plenty of cutting fluid to the die.
- Rotate the headstock spindle by hand to cut the thread with the handle of the diestock against the saddle or compound rest to keep it from rotating. Apply light pressure with the tailstock hand wheel to keep the quill or chuck against the die for alignment. **Figure 5.3.39** shows a thread being cut on the lathe with a die.

FIGURE 5.3.39 Cutting external threads with a threading die on the lathe. The face of a tailstock mounted chuck or the tailstock quill will help keep the die square.

 **CAUTION**

Do not attempt to run the spindle under power while threading using these methods. The tap or die can shatter and/or be thrown from the lathe, causing serious injury.

FORM CUTTING

Form cutting is used to produce contoured surfaces on a conventional lathe. The cutting tool contains the reverse form of the desired part shape. When fed into the workpiece, the tool produces the desired shape in the part. Form tools can take many different shapes. They are often custom ground from HSS or carbide tool bits. **Figure 5.3.40** illustrates some form tools and a workpiece with some surfaces produced by form cutting. Because of the large amount of tool contact during form cutting, spindle speeds often need to be drastically reduced to eliminate vibration and chatter.

GROOVING AND CUTOFF (PARTING)

The lathe will often be used to perform **grooving** to machine grooves or recesses on workpieces. These grooves may be needed to accept O-rings or retaining rings, act as a relief for threading, or to provide clearance on shoulders for mating parts. They may be straight or require a special cutting tool profile to create a contour within the groove, similar to form cutting. **Figure 5.3.41** illustrates some groove shapes that can be machined on the lathe.

It is also quite common to perform a **cutoff** (or **parting**) operation on the lathe that uses a special narrow cutting tool to cut off the end workpiece to a desired length. This operation is often called *parting-off* or simply *parting*. Grooving and cutoff operations can be performed on work held in chucks or collets.

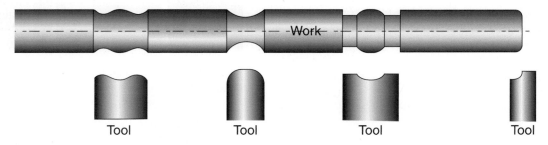

| Tool | Tool | Tool | Tool |

FIGURE 5.3.40 Form tools can be used to machine contoured surfaces.

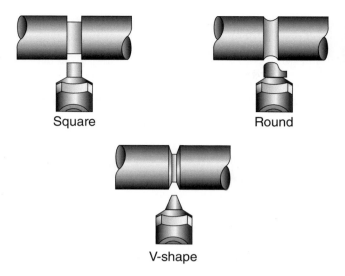

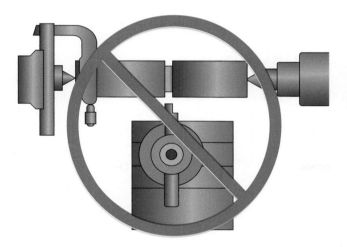

FIGURE 5.3.41 Some common groove shapes machined on the lathe.

FIGURE 5.3.42 Cutoff operations should never be performed on work mounted between centers. Damage to equipment or personal injury can result.

① CAUTION

Never perform cutoff operations on work that is mounted between centers. **(See Figure 5.3.42.)**

Sometimes the same tool can be used for both standard grooving and cutoff operations. Specialized grooving tools are available in many shapes and sizes, including square (90 degrees), partial radius, full radius, or custom form. Cutoff tools are available with a cutting face that is angled either to the left or right to minimize the amount of uncut material left on the part that is cut off. Both grooving and cutoff tools are available in HSS or inserted carbide types. **Figure 5.3.43** shows some examples of grooving and cutoff tools.

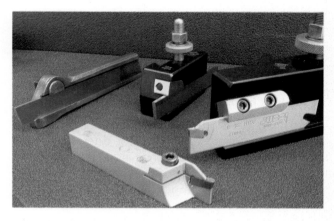

FIGURE 5.3.43 Examples of the many different grooving and cutoff tools and holders available in both HSS and inserted carbide styles.

FIGURE 5.3.44 When performing grooving or parting, be sure the tool is set 90 degrees to the ways.

Both grooving and parting operations require special care because of the substantial width of the cutting tool and large amount of contact between the tool and the workpiece. Grooving and cutoff operations require extremely rigid setups. Be sure the tool tip is on center and keep the tool as short as possible to minimize vibration and chatter. Here are general steps for grooving and cutting off.

- Set the tool at 90 degrees to the lathe's ways so that the sides of the tool will not rub the sidewalls of the groove and create excessive heat. **(See Figure 5.3.44.)**
- Position the tool at the desired location with the carriage.
- Lock the carriage to the ways to prevent unwanted carriage movement.
- Set spindle speeds to about one-fourth to one-third of those used for turning. If vibration occurs during cutting, lower spindle speed until it stops.

- Start the spindle.
- Feed the tool with the cross slide to the desired depth or until the part is completely cut off. Pausing periodically will break the chip so it falls away from the work. Continually apply a liberal amount of cutting fluid to the tool. **Figure 5.3.45** shows a grooving operation.

Internal grooves can be machined with boring bars that hold tools similar to those used for external grooving operations. The tool is positioned inside the hole at the desired location and fed straight into the workpiece as when performing external grooving. **(See Figure 5.3.46.)** Extra care must be taken when machining internal grooves, as when boring, because the cutting tool often cannot be seen. The tool may also need to be removed from the hole frequently to clear chips so it does not bind during the grooving operation.

When cutting both external and internal grooves, sometimes the groove to be machined may be wider than the width of the available tool. In cases like this, the groove can be cut in steps, as shown in **Figure 5.3.47**.

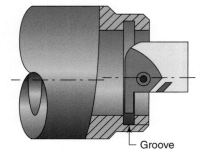

FIGURE 5.3.45 Machining an external groove with an HSS tool. The chip should curl into a circle on top of the tool.

Groove

FIGURE 5.3.46 A specially designed boring bar can machine internal grooves.

The large amount of tool contact during grooving and parting generates substantial heat that can cause expansion of the workpiece. To prevent this expansion from pinching the blade of the cutting tool on the sidewalls of the cut, constantly apply a generous amount of cutting fluid when performing grooving and parting.

 CAUTION

Failure to take these precautions can result in the tool binding, breaking, and broken pieces being thrown out of the lathe, causing serious injury. When cutting off, allow the part to drop freely. Do not attempt to hold the part by hand. When machining deep grooves or cutting off large diameters, the chance of tool binding can be greatly reduced by cutting to a partial depth and then making the groove wider to provide more tool clearance. **(See Figure 5.3.48.)**

KNURLING

Knurling is producing a raised pattern on the circumference of a workpiece. Knurling is accomplished by pressing a knurling tool with two wheels, called rolls, against a rotating workpiece. Basic knurl patterns are straight and diamond, with both available in fine, medium, and coarse types. **(See Figure 5.3.49.)** The pattern is raised because instead of the material being cut as in most machining operations, the knurling tool forms the material by applying pressure. This raised pattern causes the workpiece diameter to increase after knurling. A medium diamond knurl will increase workpiece diameter by about 0.015" to 0.020".

The most common reason for producing a knurl is to provide a gripping surface on handles, knobs, or levers. Since knurling increases workpiece diameter, sometimes knurling is used to repair worn cylindrical parts. For example, a worn shaft can be knurled to increase its diameter and then machined back to its original size. Knurling can also be used for decorative purposes or to create serrations that lock cylindrical components in place after being pressed in a hole. **(See Figure 5.3.50.)**

The knurling tool contains two hardened steel rolls of the desired pattern. Knurling tools are available in different styles for use with different toolholding devices, but the two basic types are the bump type and the clamp type. The bump-type knurling tool is pressed against the workpiece from one side. The clamp-type knurling tool clamps the workpiece between the rolls to minimize part distortion and flex. It is especially useful for knurling small diameters and long sections. Some examples of knurling tools are shown in **Figure 5.3.51**.

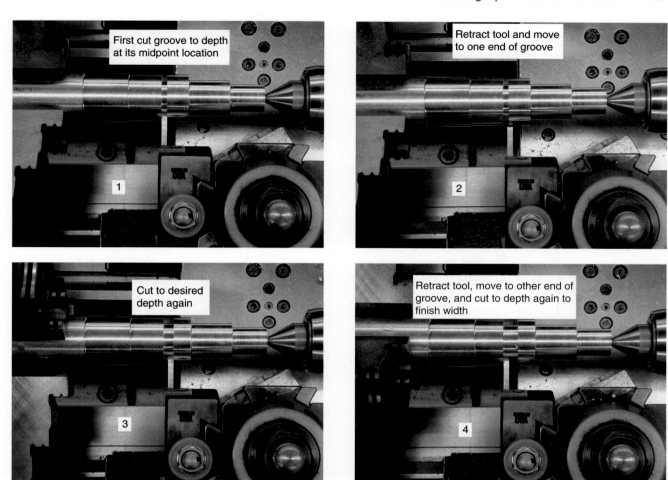

FIGURE 5.3.47 These steps for machining a groove wider than the width of the grooving tool can be used when cutting both external and internal grooves.

FIGURE 5.3.48 Method for machining deep grooves or parting off large diameters. Cut the groove to a partial depth, then make the groove wider before cutting to final depth.

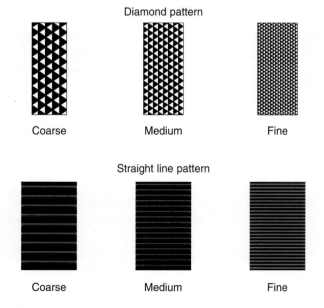

FIGURE 5.3.49 Coarse, medium, and fine diamond and straight knurls.

FIGURE 5.3.50 Knurling created the serrations on this insert that will lock it into place after being pressed into a hole.

FIGURE 5.3.51 (A) Bump-type knurling tools. The upper style has a rotating head with fine, medium, and coarse rolls. (B) A clamp-type knurling tool is ideal for knurling small diameters and long workpieces.

Knurling can be done when holding work in a chuck, collet, or between centers by following these steps.

- Mount the knurling tool 90 degrees to the ways (parallel with the cross slide).
- Adjust the height so that both rolls will contact the workpiece with equal pressure when brought against the workpiece.
- Set a spindle speed about ¼ of turning speed.
- Set a heavy longitudinal feed rate of about 0.020" to 0.040" IPR.
- Start the spindle.
- Engage the longitudinal feed.

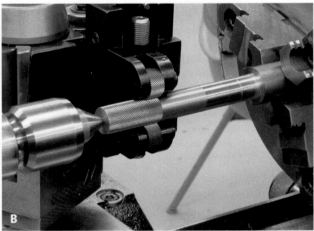

FIGURE 5.3.52 (A) Knurling with a bump-type and (B) clamp-type tool.

- Quickly bring the rolls in contact with the rotating workpiece. The cross slide is used to apply pressure when using a bump-type tool. A clamp-type tool is tightened against the part with an adjusting screw or handle. **Figure 5.3.52** shows knurling using both types of tools.
- Apply plenty of cutting fluid to lubricate the rolls and workpiece, and to flush debris out of the forming area. Use a flood system or a hand held squirt bottle to apply cutting fluid.

 CAUTION

Do not apply cutting fluid with a brush while the spindle and rolls are rotating. The brush can easily be pulled between the rolls and the work, causing damage to both the workpiece and the knurling tool, and may cause the workpiece to be pulled from the work-holding device, leading to injury.

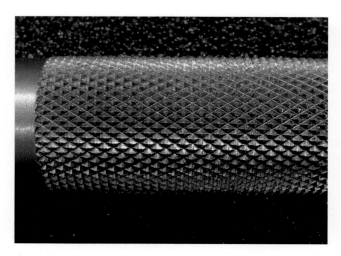

FIGURE 5.3.53 The pattern on a properly knurled surface will come to points.

FIGURE 5.3.54 When knurling, keep the tool on the workpiece to keep the rolls in the pattern.

- When the tool reaches the end of the section to be knurled, stop the spindle and check the depth of the pattern. Do not disengage the feed control lever.
- The pattern should come to a point, as shown in **Figure 5.3.53**. If it does not, reverse the spindle or the feed direction, and allow the tool to feed back across the workpiece.
- Do not allow the knurling tool to feed off the workpiece and keep at least half of the width of the rolls on the part. **(See Figure 5.3.54.)**
- Continue this process until the pattern is fully formed. Apply more pressure if needed to reach the required

depth. Try to finish the knurl in as few passes as possible because as more passes are made, the material has a tendency to work harden, or become brittle. If this occurs, small particles of the material will begin to flake off the workpiece and a poor-quality knurl will result.
- When the pattern is finished, retract the rolls from the workpiece. If using a bump-type knurling tool, simply retract the cross slide. If using a clamp-type knurling tool, unclamp the tool before retracting the cross slide or the knurl will be damaged as the tool is dragged across the knurled surface.

SUMMARY

- Many different machining operations can be performed on the lathe, and each operation must be performed using appropriate speeds and feeds to ensure safety, produce desired results, and prevent damage to work and equipment.
- Depth of cut, spindle speed, and feed rate are vital factors to consider when performing both roughing and finishing operations, and the ability to calculate machining time is also important to estimate cost and time needed for completion of operations.
- Facing and turning are common operations used to machine lengths and diameters of workpieces.
- When two different diameters meet, different types of shoulders can be machined where the smaller diameter transitions to the larger diameter.
- Burrs created by facing, turning, and shouldering operations can be safely and quickly removed by filing so long as a few precautions are observed.
- Abrasive paper and cloth can be used to remove small amounts of material and polish workpiece surfaces to very smooth surface finishes.

(Continued)

- A center drill held in a drill chuck can be used to create a start point for a twist drill or to produce holes for mounting work on lathe centers.
- Spotting drills are also used to create start points for twist drills.
- Drilling, reaming, countersinking, and counterboring can be performed by mounting tools in a drill chuck or in the tailstock's Morse taper bore. The tailstock can also be used to align taps and threading dies used to cut internal and external threads.
- Boring uses a single-point tool mounted on a bar to produce nonstandard-size holes or diameters beyond the range of standard drills or reamers. Boring is similar to turning but machines internal diameters instead of external diameters.
- When contoured cylindrical shapes are desired, form cutting tools can be custom ground to create the desired workpiece shape. The shape of the form tool is exactly opposite of the shape created in the part when the tool is fed into the workpiece.
- To cut a groove into a workpiece, the tool is fed straight into the diameter with the cross slide to produce a specified groove depth or workpiece diameter.
- In parting, a cutoff tool is fed into the work just as when grooving, except that instead of stopping at a specific depth, the tool is fed in until the workpiece is completely cut off.
- Knurling is a process that displaces material by force, instead of cutting as in other lathe operations, to produce a raised pattern on an external diameter. Because of this displacement, the diameter of a workpiece will increase during knurling. Knurling is performed to improve part appearance, provide a gripping surface, increase diameter for remachining, or create serrations on the workpiece.

REVIEW QUESTIONS

1. If a 0.050" depth of cut is taken on the diameter of a workpiece, by how much will the diameter be reduced?
2. A lathe cross slide uses a diameter-reading micrometer collar. If the cross slide is advanced by 0.150", what depth of cut would result?
3. In what units are feed rates measured for lathe operations?
4. Are deeper cuts used for roughing or finishing operations?
5. Calculate spindle RPM and machining time for cutting a 1.5" diameter 4" long at 225 SFPM using a feed rate of 0.004".
6. List three safety precautions related to clothing that should be observed during lathe operation.
7. What two materials are most commonly used for lathe cutting tools?
8. What feature of a lathe cutting tool has a direct effect on surface finish?
9. Is a left-hand or right-hand tool normally used for facing?
10. What part of the lathe is used to feed the tool during facing?
11. When facing, why should the tool not be fed past the center of the workpiece?
12. Should a left-hand or right-hand tool be used when turning toward the headstock?
13. When and how should chips be removed from the work and cutting tool?
14. What are two reasons for center drilling on the lathe?
15. When drilling and reaming on the lathe, how are the tools usually fed into the workpiece?
16. How can hole depth be controlled during drilling operations?

17. What are two reasons boring may be selected to produce a hole instead of drilling and reaming?

18. Why must extra care be taken when performing boring operations?

19. How can a tap be aligned when threading a hole on the lathe?

20. Briefly define form cutting.

21. How do grooving and cutoff speeds compare to turning speeds?

22. How can tool binding be overcome when cutting deep grooves or cutting off large diameters?

23. List the two basic knurl patterns.

24. How is knurling different from other lathe operations?

25. List and briefly describe the two different types of knurling tools.

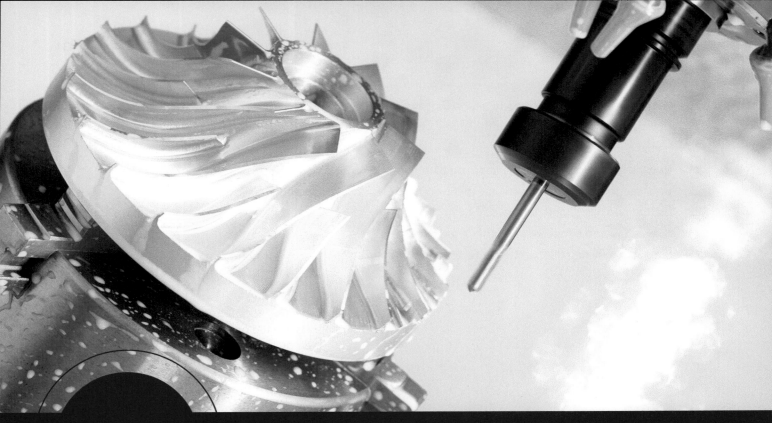

UNIT 4 — Manual Lathe Threading

Learning Objectives

After completing this unit, the student should be able to:

- Identify the parts of a thread and define thread terminology
- Describe the difference between left-hand and right-hand threads
- Identify and describe the different classes of fit
- Locate appropriate thread reference data from charts
- Perform calculations required for thread cutting
- Demonstrate understanding of workpiece and tooling setup for thread cutting on the lathe
- Demonstrate understanding of the lathe thread cutting process
- Demonstrate understanding of various methods of thread measurement

Key Terms

Acme thread	Lead	Thread measuring
Buttress thread	Left-hand threads	wires (Thread wires)
Center gage	Length of	Three-wire method
Flank	engagement	
Helix	Right-hand threads	
Helix angle	Tapered pipe threads	

INTRODUCTION

Unit 6 of Section 3 provided an introduction to the many uses of screw threads and the characteristics of threads. The focus of Section 3.6 was the production of threads with hand-powered taps and dies. Threading may also be performed on the lathe while the workpiece rotates and a single-point cutting tool cuts the spiral groove.

Threading with a lathe offers much greater control over fit, finish, and form than threading performed with a tap or die. To cut quality threads it is important to understand thread calculations, how to properly set up and operate the machine, and how to measure threads.

SCREW THREAD TERMINOLOGY

Section 3 covered the basic characteristics of threads relevant to hand threading, but since the lathe will provide more control over form and fit, it is essential to understand more terms and details. Following are threading terms that were not covered in Section 3. Refer to **Figure 5.4.1** for an illustration of all the major parts of a 60-degree V thread.

- The **flank** is the thread surface that joins the crest to the root. On V-threads the flanks create the shape of the "V." The shape and angle of the flanks are directly determined by the form of the cutting tool.
- The **helix** of a thread refers to the spiral, or helical, shape created by the thread grooves.

- The **helix angle** is the angle of a thread's spiral. Coarser threads have a greater helix angle than finer threads.
- Crest clearance or root clearance is the distance between the crest and root of two mating threads.

NOTE:

This area of clearance does not determine the fit of a thread. **(See Figure 5.4.2.)**

- **Lead** is how far a thread will move in relationship to its mating thread in one revolution. A good example of lead can be found in the micrometer. When the thimble is rotated one complete turn the micrometer will open or close by 0.025 of an inch. Threads are also manufactured with multiple leads, such as a double lead and triple lead. A double-lead thread has two separate grooves and a triple-lead thread has three separate grooves. In a single-lead thread the lead will be equal to the pitch. On a double-lead screw the movement between the two mating threads will be double the distance of the pitch with every rotation. If a double-lead thread were used in the previous example, the micrometer would open or close 0.050 of an inch for every rotation of the thimble. The helix angle is much more dramatic on a double-lead thread versus a single-lead thread, as shown in **Figure 5.4.3**.

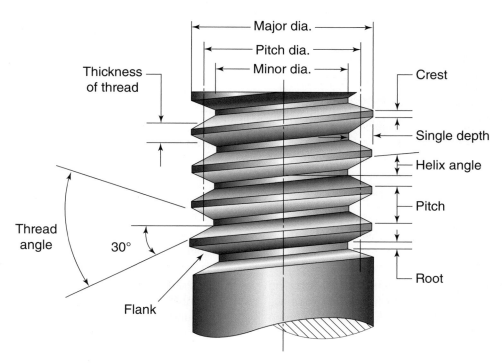

FIGURE 5.4.1 The major parts of a 60-degree V thread. The flanks are the surfaces where the two mating threads make actual contact.

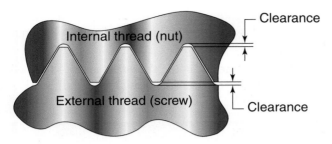

FIGURE 5.4.2 The clearance between the crest and root of mating threads. There is no contact at this point and this area does not determine fit between mating threads.

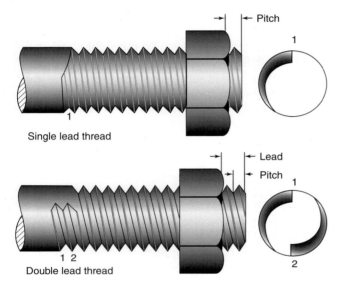

Single lead thread

Double lead thread

FIGURE 5.4.3 A double lead thread has a much more steep helix angle than that of a single lead thread.

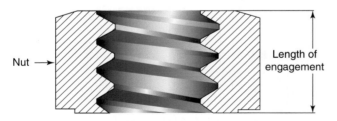

FIGURE 5.4.4 The length of engagement determines how many threads will be engaged with the mating part.

- **Length of engagement** is the distance an externally threaded part threads into an internally threaded part. **(See Figure 5.4.4.)**

- Screw threads can be produced with the helix spiraling in one of two directions. **Right-hand threads** have a helix that causes assembled fasteners to tighten when rotated clockwise. **Left-hand threads** have a helix that leans the opposite of a right-hand thread, causing the fasteners to become tightened when rotated counterclockwise. **(See Figure 5.4.5.)**

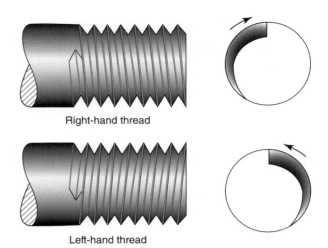

Right-hand thread

Left-hand thread

FIGURE 5.4.5 A right-hand and left-hand thread have opposite thread helix directions.

CLASS OF FIT

To maintain the standardization of threads and ensure consistent levels of fit (looseness or tightness), standard classes of fit have been defined. The class of fit for a thread is determined by the clearance between the mating thread flanks. The pitch diameter determines that amount of clearance. The closer the pitch diameters of the external and internal threads, the less clearance there is between the flanks, and the closer the fit. When there is greater difference between the external and internal pitch diameters, there is more clearance between the flanks, and the fit is looser. It is a common misconception that the major or minor diameters affect fit. There should always be clearance between the root and crests of mating threads. Proper threads will only make contact on their flanks.

There are six common classes of fit. Three of these classes are for external threads and are indicated by the letter A, and three are for internal threads and are indicated by the letter B.

- **Class 1A and 1B:** Class 1A and 1B are intended to achieve a quite loose fit between mating threads. This class is ideal for quick assembly and disassembly, where speed is more important than precision. Items that must be assembled "in the field" or in dirty, extreme environments may sometimes be made with this type of thread fit. This class is seldom used.

- **Class 2A and 2B:** This class is ideal for general assembly fasteners and is the most common class of fit found on general-purpose nuts and bolts. Class 2A and 2B are manufactured with enough clearance to make them easy to assemble and still offer enough thread engagement to achieve considerable strength.

• **Class 3A and 3B:** This class has little or no clearance between mating threads. Class 3 fit is used when a very accurate or high-strength assembly is required. This class of fit is more expensive to achieve since its production must be monitored closely to ensure accuracy.

Since it is the pitch diameters of the two parts that determine this clearance and, ultimately, the fit, allowable size limits for the pitch diameter of each standard thread size are specified for each class of fit. When a desired class of fit is to be achieved, the pitch diameter must be produced between these size limits for *both* the internal and external mating threads.

DETERMINING THREAD DATA

Prior to performing any thread cutting, all data for the desired thread need to be referenced from a chart like the one shown in **Figure 5.4.6**, or calculated where necessary. These data will be needed for machine setup and thread measurement. The data of particular concern for lathe thread cutting are the following:

• Major diameter (external thread)
• Minor diameter (internal thread)
• Pitch diameter
• Compound-rest in-feed

Major Diameter for External Threads

After cutting an external thread on a workpiece, the only surface that did not get cut by the threading tool should be a small flat at the crest of the thread. This surface is a remnant of the original diameter and achieves the dimension for the major diameter. It is important that this diameter be cut to the correct dimension prior to threading. The major diameters vary by thread fit class, and most are slightly smaller than the nominal major thread diameter.

EXAMPLE: Determine the major diameter limits for a ¾-10 UNC 2A thread.

Refer to the chart in **Figure 5.4.6** and locate ¾-10 UNC in the left-most "Nominal Size" column. Then find the appropriate class of fit designation for that thread series. Follow the row across to the "Major Diameter" column under the "External" heading and determine the upper and lower limits for that thread.

Upper limit = 0.7482

Lower limit = 0.7353

The major diameter of the threaded section should be produced to a dimension between these limits before beginning external thread cutting.

Minor Diameter for Internal Threads

Like external threads, after cutting an internal thread on a workpiece, the only surface that did not get cut by the threading tool should be a small flat at the crest of the thread. This surface is a remnant of the original diameter and achieves the dimension for the minor diameter. It is again important that this diameter be cut to the correct dimension prior to threading.

EXAMPLE: Determine the minor diameter limits for a ¾-10 UNC 2B thread.

Refer to the chart in **Figure 5.4.6** and again locate ¾-10 UNC in the leftmost "Nominal Size" column. Then again find the appropriate class of fit designation for that thread series. This time, follow the row across to the "Minor Diameter" column under the "Internal" heading and determine the upper and lower limits for that thread.

Upper limit = 0.6630

Lower limit = 0.6420

The minor diameter of the threaded section should be produced to a dimension between these limits before beginning internal thread cutting.

Pitch Diameter

The pitch diameter is important for achieving a desired thread fit and is a standardized dimension directly referenced from a chart. This is the main feature that will be measured when cutting external threads.

EXAMPLE: Determine the pitch diameter limits for a ¾-10 UNC 2A thread.

Refer to the chart in **Figure 5.4.6** and again locate ¾-10 UNC in the leftmost "Nominal Size" column, then find the appropriate class of fit designation for that thread series. Follow the row across to the "Pitch Diameter" column under the "External" heading and determine the upper and lower limits for that thread.

Upper limit = 0.6832

Lower limit = 0.6773

Compound-Rest In-Feed

As will be discussed later, the compound rest will be the only machine slide used to set the depth of cut when threading. When cutting 60-degree V-threads, this slide will be set to advance the tool on an approximate

Standard Series and Selected Combinations—Unified Screw Threads

Nominal Size, Threads per Inch, and Series Designation	External									Internal					
	Class	Allowance	Major Diameter			Pitch Diameter		UNR Minor Dia., Max (Ref.)		Class	Minor Diameter		Pitch Diameter		Major Diameter
			Max	Min	Min	Max	Min				Min	Max	Min	Max	Min
5/8–32 UN	2A	0.0011	0.6239	0.6179	—	0.6036	0.6000	0.5867		2B	0.591	0.599	0.6047	0.6093	0.6250
	3A	0.0000	0.6250	0.6190	—	0.6047	0.6020	0.5878		3B	0.5910	0.5969	0.6047	0.6082	0.6250
11/16–12 UN	2A	0.0016	0.6859	0.6745	—	0.6318	0.6264	0.5867		2B	0.597	0.615	0.6334	0.6405	0.6875
	3A	0.0000	0.6875	0.6761	—	0.6334	0.6293	0.5883		3B	0.5970	0.6085	0.6334	0.6387	0.6875
11/16–16 UN	2A	0.0014	0.6861	0.6767	—	0.6455	0.6407	0.6116		2B	0.620	0.634	0.6469	0.6531	0.6875
	3A	0.0000	0.6875	0.6781	—	0.6469	0.6433	0.6130		3B	0.6200	0.6284	0.6469	0.6515	0.6875
11/16–20 UN	2A	0.0013	0.6862	0.6781	—	0.6537	0.6494	0.6267		2B	0.633	0.645	0.6550	0.6606	0.6875
	3A	0.0000	0.6875	0.6794	—	0.6550	0.6518	0.6280		3B	0.6330	0.6412	0.6550	0.6592	0.6875
11/16–24 UNEF	2A	0.0012	0.6863	0.6791	—	0.6592	0.6552	0.6367		2B	0.642	0.652	0.6604	0.6656	0.6875
	3A	0.0000	0.6875	0.6803	—	0.6604	0.6574	0.6379		3B	0.6420	0.6494	0.6604	0.6643	0.6875
11/16–28 UN	2A	0.0011	0.6864	0.6799	—	0.6632	0.6594	0.6438		2B	0.649	0.657	0.6643	0.6692	0.6875
	3A	0.0000	0.6875	0.6810	—	0.6643	0.6615	0.6449		3B	0.6490	0.6551	0.6643	0.6680	0.6875
11/16–32 UN	2A	0.0011	0.6864	0.6804	—	0.6661	0.6625	0.6492		2B	0.654	0.661	0.6672	0.6718	0.6875
	3A	0.0000	0.6875	0.6815	—	0.6672	0.6645	0.6503		3B	0.6540	0.6594	0.6672	0.6707	0.6875
3/4–10 UNC	1A	0.0018	0.7482	0.7288	—	0.6832	0.6744	0.6291		1B	0.642	0.663	0.6850	0.6965	0.7500
	2A	0.0018	0.7482	0.7353	0.7288	0.6832	0.6773	0.6291		2B	0.642	0.663	0.6850	0.6927	0.7500
	3A	0.0000	0.7500	0.7371	—	0.6850	0.6806	0.6309		3B	0.6420	0.6545	0.6850	0.6907	0.7500
3/4–12 UN	2A	0.0017	0.7483	0.7369	—	0.6942	0.6887	0.6491		2B	0.660	0.678	0.6959	0.7031	0.7500
	3A	0.0000	0.7500	0.7386	—	0.6959	0.6918	0.6508		3B	0.6600	0.6707	0.6959	0.7013	0.7500
3/4–14 UNS	2A	0.0015	0.7485	0.7382	—	0.7021	0.6970	0.6635		2B	0.673	0.688	0.7036	0.7103	0.7500
3/4–16 UNF	1A	0.0015	0.7485	0.7343	—	0.7079	0.7004	0.6740		1B	0.682	0.696	0.7094	0.7192	0.7500
	2A	0.0015	0.7485	0.7391	—	0.7079	0.7029	0.6740		2B	0.682	0.696	0.7094	0.7159	0.7500
	3A	0.0000	0.7500	0.7406	—	0.7094	0.7056	0.6755		3B	0.6820	0.6908	0.7094	0.7143	0.7500
3/4–18 UNS	2A	0.0014	0.7486	0.7399	—	0.7125	0.7079	0.6825		2B	0.690	0.703	0.7139	0.7199	0.7500
3/4–20 UNEF	2A	0.0013	0.7487	0.7406	—	0.7162	0.7118	0.6892		2B	0.696	0.707	0.7175	0.7232	0.7500
	3A	0.0000	0.7500	0.7419	—	0.7175	0.7142	0.6905		3B	0.6960	0.7037	0.7175	0.7218	0.7500
3/4–24 UNS	2A	0.0012	0.7488	0.7416	—	0.7217	0.7176	0.6992		2B	0.705	0.715	0.7229	0.7282	0.7500
3/4–27 UNS	2A	0.0012	0.7488	0.7421	—	0.7247	0.7208	0.7047		3B	0.710	0.719	0.7259	0.7310	0.7500

FIGURE 5.4.6 A thread-cutting data chart specifies important thread dimensions.

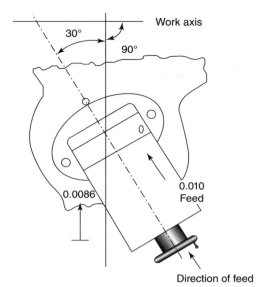

At a 30-degree setting, 0.010" of compound rest feed advances the tool approximately 0.0086" radially into the workpiece.

FIGURE 5.4.7 The distance a compound rest must be advanced to produce a full thread must be calculated since the tool will advance at an angle.

30-degree angle (**Figure 5.4.7**). The following formula can be used to estimate the amount of in-feed for the compound rest for a 60-degree external thread:

$$\text{Compound in-feed} = 0.7 \times \text{Pitch}$$

OR

$$\text{Compound in-feed} = \frac{0.7}{N}$$

EXAMPLE: Determine the compound in-feed for a ¾-10 UNC 2A thread.

Since the thread designation states that there are to be 10 threads per inch:

$$\text{Compound in-feed} = 0.7 \times \text{Pitch}$$
$$= 0.7 \times 0.1 = 0.07$$

OR

$$\text{Compound in-feed} = \frac{0.7}{N} = \frac{0.7}{10} = 0.07$$

PRODUCING THREADS ON THE LATHE

Thread cutting is accomplished on the lathe by taking several successive depth cuts along the diameter of a workpiece with a single-point cutting tool. As the thread depth becomes deeper during successive cuts, the cutting tool will create more contact with the work. This contact can result in higher cutting pressures, which can lead to chatter, excessive heat, and deflection of the workpiece or cutting tool. In an effort to minimize the cutting-tool contact (and pressure) during threading, it is customary to advance a threading tool on an angle approximately equal to half the included angle of the thread form (**Figure 5.4.8**). Advancing the tool on this angle ensures that only the leading edge of the tool cuts the majority of the material. It is often advantageous to then set the compound angle slightly less than this angle (29 to 29-1/2 degrees instead of 30 degrees for a 60-degree form) so the adjacent cutting edge cuts lightly during each pass.

LATHE SETUP

To properly accomplish single-point threading, many steps of machine setup must be performed prior to any cutting:

- The quick-change gear box and carriage feed direction will need to be properly set up so that the cutting tool feeds appropriately during cutting.
- The threading dial gear must be meshed with the leadscrew.
- The workpiece must be mounted securely so that it cannot deflect or slip during cutting.
- The compound rest must be set at the angle of desired compound in-feed.
- The cutting tool must be mounted in a sturdy holder.
- The cutting tool's form must be properly positioned.

Installing the Workpiece

It is good practice to machine the diameter to be threaded and cut the thread without removing and remounting the workpiece. If this is not possible, be sure the workpiece does not run out more than 0.001" TIR when remounted. If this is not done, thread depth will be inconsistent and mating parts may not be concentric to each other when assembled. Greater runout can also affect the fit between mating parts.

Workholding for External Threading

The same devices used to mount workpieces for outside turning operations can be used when threading. When cutting external threads on a workpiece extending from a chuck or collet more than about three times its diameter, a live center mounted in the tailstock should be used to support the work and prevent deflection. Remember that a fair amount of torque is required to turn the workpiece during thread cutting. It is very important to ensure that the workpiece is securely gripped since any slippage will cause the threads to be damaged.

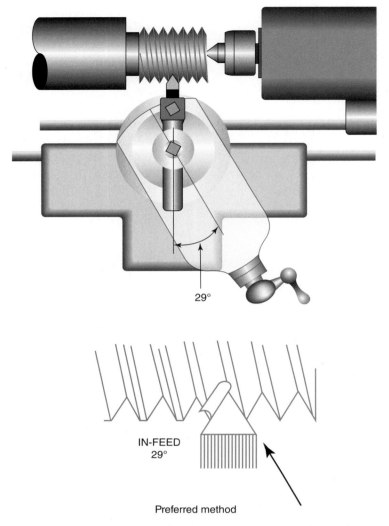

29°

IN-FEED
29°

Preferred method

FIGURE 5.4.8 The compound rest is set at approximately 1/2 the included angle of the thread so that one cutting edge is dominant when removing material.

Workholding for Internal Threading

When mounting a workpiece for internal threading, the same principles apply. Work may not be held between centers, however, since the hole in the end of the part will need to be open in order to accept the internal threading tool.

Setting the Quick-Change Gear Box

To set the quick-change gear box, first determine the required number of threads per inch. Most lathes have a chart located on the gear box labeled with lever settings for a desired TPI. Be sure to carefully read the labeling and to not confuse these numbers with the feed per revolution values. The whole numbers represent the number of threads per inch available. **(See Figure 5.4.9.)** The chart lists the required positioning of the gear-box levers for a given TPI.

The quick-change gear box directly transmits power to the leadscrew. It is necessary to ensure any necessary couplings or levers are set to make the leadscrew rotate. The leadscrew rotation direction must also be set on the gear box to ensure the carriage advances in the correct direction. If right-hand threads are to be produced, the lathe spindle should rotate forward while the carriage advances longitudinally toward the headstock. For left-hand threads the spindle should also rotate forward, but the carriage should advance toward the tailstock.

Setting the Compound Rest

The compound rest should be used to produce the angular in-feed of the cutting tool during the threading operation. For 60-degree V-threads, it should be set between 29 and 29-1/2 degrees, moving from a position where the compound is parallel with the cross slide. This is usually accomplished by loosening locking

MM				1 IN		
.2	LT1Z	1.2	LR6Z	6.5	HS7Y	72	LA6R	22	LB4S	7½ HA3S

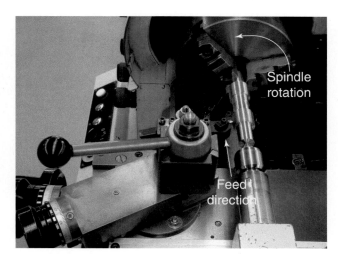

FIGURE 5.4.9 A thread-cutting chart affixed to the quick-change gear box shows the lever settings for cutting threads. The left half is for metric threads with the numbers on the left showing the pitch in millimeters and those on the right showing the machine lever settings. The right half of the chart is for inch based threads with the left numbers showing TPI and the right numbers showing the lever settings.

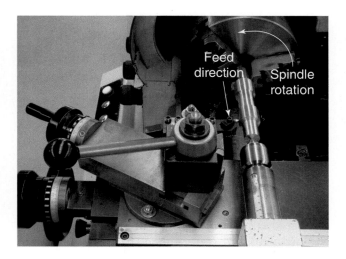

FIGURE 5.4.11 Spindle rotation, compound-rest orientation, and feed direction for cutting external right-hand threads.

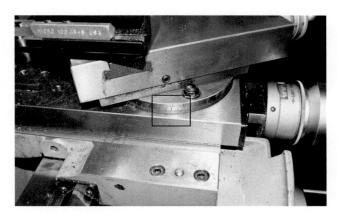

FIGURE 5.4.10 The graduations around the compound-rest base are set between 29 and 29-1/2 degrees for cutting 60-degree V-threads.

FIGURE 5.4.12 Spindle rotation, compound-rest orientation, and feed direction for cutting external left-hand threads.

screws located at the base of the compound rest. Once unclamped, the compound rest may be rotated until the 29- or 29.5-degree mark on the compound lines up with the reference mark on the cross slide. The locking screws must then be retightened. **(See Figure 5.4.10.)**

When cutting external right-hand threads the compound is positioned as shown in **Figure 5.4.11**. As the spindle rotates forward, the carriage advances toward the headstock.

When cutting external left-hand threads the compound is positioned as shown in **Figure 5.4.12**. As the spindle rotates forward, the carriage advances toward the tailstock.

When cutting internal right-hand threads the compound is positioned as shown in **Figure 5.4.13**. As the spindle rotates forward, the carriage advances toward the headstock.

FIGURE 5.4.13 Spindle rotation, compound-rest orientation, and feed direction for cutting internal right-hand threads.

FIGURE 5.4.14 Spindle rotation, compound-rest orientation, and feed direction for cutting internal left-hand threads.

When cutting internal left-hand threads the compound is positioned as shown in **Figure 5.4.14**. As the spindle rotates forward, the carriage advances toward the tailstock.

Setting the Spindle Speed

Threading on a lathe is done with a lower spindle speed than would be used for turning similar-sized diameters. This is done for two reasons. First, threading requires a large amount of tool contact with the workpiece and chatter may be prevented by a lower spindle speed. Second, threading requires a great deal of operator attention since the tool advances so quickly and the half-nut lever must be engaged and disengaged at precisely the correct time. Since the carriage feed is directly linked to the spindle rotation, running the spindle at a lower speed will provide the operator with more time to engage and disengage the half-nut lever. A spindle speed about one-fourth of a turning speed for the same diameter is a good starting point.

 **CAUTION**

Always make a trial pass with the tool far away from the work, headstock, and tailstock to check the rate of carriage motion. If the rate is too fast to safely engage and disengage the half-nut lever, reduce spindle speed accordingly.

Installing and Aligning the Cutting Tool

There are a few different styles of cutting tools for machining external threads. An HSS tool bit can be ground on a pedestal grinder to the proper thread form.

Its sides must have side clearance angles to cut and prevent rubbing. Brazed carbide tools can be purchased with the desired thread form and clearance angles. Carbide inserted threading tools are also available from many cutting-tool manufacturers. **Figure 5.4.15** shows some examples of external thread-cutting tools. When cutting internal threads, toolholders similar to those used for boring are used to hold the cutting tool so that it can be extended into the hole in the workpiece. **Figure 5.4.16** shows some different types of internal threading tools.

With both external and internal tools, always be sure to mount the tool with the least amount of overhang possible to ensure maximum rigidity. Internal threading tools are especially subject to rigidity issues because of the slender shank necessary to extend into a hole. Always select the largest-diameter internal threading

FIGURE 5.4.15 Various thread-cutting tools for external threads.

FIGURE 5.4.16 Various thread-cutting tools for internal threads.

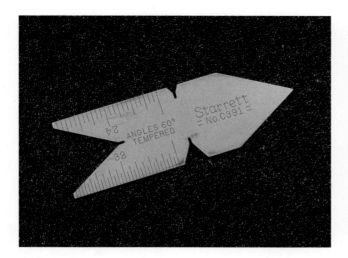

FIGURE 5.4.17 A center gage.

FIGURE 5.4.18 A center gage being used to align the "V" form of a thread-cutting tool.

tool bar that will fit into the hole and still provide enough clearance to retract the tool from the thread groove at the end of the pass.

The alignment of the thread-cutting tool's form to the workpiece is very important.

- Place the threading tool in a holder and adjust the tool to the correct height. The height of the cutting tool needs to be set at the part's center line just as for turning and boring operations.
- A small gage called a **center gage** (sometimes referred to as a **fishtail gage**) is used to properly align the threading tool. **Figure 5.4.17** shows a center gage.
- Hold the center gage against the side of the workpiece surface.
- Bring the tool close to one of the side "V"-grooves in the center gage by moving the cross slide.
- Adjust the tool position until the tip fits properly into the "V"-groove in the center gage, as shown in **Figure 5.4.18**. This will locate the cutting edges of the threading tool perpendicular to the workpiece.
- Tighten the toolholding device and recheck with the center gage to ensure the tool did not move during tightening.

LATHE THREADING OPERATION

Referencing the Cutting Tool

Once the workpiece and tool are mounted, one final step must be performed: the cutting tool must be referenced. This zero reference point serves as a baseline from which the thread depth can be monitored as successive cuts are produced.

- Bring the tool tip near the workpiece diameter to be threaded.
- Start the spindle.
- Bring the tool tip against the diameter to touch off. Applying a coating of layout fluid to the workpiece diameter can make it easier to make a light touch.
- Set the cross-slide and compound-rest hand-wheel collars to "0" while in this position. **(See Figure 5.4.19.)**
- It is important to make sure the backlash is removed from both the compound rest and the cross slide before setting the collars to "0." For external threading, move both away from the operator position prior to setting "0." For internal threading, move both toward the operator position prior to setting "0."

FIGURE 5.4.19 The hand-wheel micrometer's collars are set to zero when the tool is in contact with the diameter to be threaded.

Threading Dial and Half-Nuts

When performing turning operations on the lathe, the feed lever is used to supply power to the carriage. However, when threading, the half-nut lever is used instead to engage the carriage directly to the leadscrew. **(See Figure 5.4.20.)** The half-nuts actually resemble a nut that has been split into two pieces. When the half-nut lever is engaged, the two halves of the nut clamp down around the leadscrew, engaging the carriage with the leadscrew. **(See Figure 5.4.21.)**

A thread cannot be cut to full depth in one pass, so successive passes must be taken to produce full depth. Each pass must occur within exactly the same groove. The timing of the half-nut engagement is monitored by the rotating marks on the thread dial. This dial allows the half-nut to be engaged with the leadscrew in the same location every time. The thread dial rotates when the spindle is turning and half-nut is not engaged (carriage is stationary). When the appropriate line on the thread dial lines up with the reference mark, the half-nut is engaged with the leadscrew and the thread dial will stop rotating as the carriage moves to perform the threading pass.

Typically, a thread dial will have equally spaced main graduations marked 1, 2, 3, and 4 and additional graduations in between those numbers. **Figure 5.4.22** shows when to engage the half-nut for a given TPI using this type of thread dial. Since there is some variation in gear-box and thread-dial design among machine manufacturers, it is best to refer to the lathe's manual for the proper numbers of engagement on the thread dial.

Threading Tool In-Feed and Positioning

All tool in-feed for V-threading will be done with the compound rest so that one cutting edge of the tool performs most of the cutting. The cross slide is only used for retracting the tool at the end of the threading cut and for repositioning at the start of the next cut. The compound-rest is progressively advanced for each pass and the total cutting tool in-feed is monitored by its micrometer collar.

To start the threading process, position the tool at the start of the thread after touching off and setting the compound-rest and cross-slide collars to "0." Then follow these steps to cut the thread.

- Start the spindle.
- Use the compound-rest to advance the cutting tool only 0.001" to 0.002" for the first pass. **(See Figure 5.4.23.)** This light cut will reveal a tracing of the thread's helix and allow the TPI to be checked with a screw pitch gage before taking further cuts.
- Engage the half-nut lever at the proper time as indicated by the threading dial to begin the cutting pass. **(See Figure 5.4.24.)**

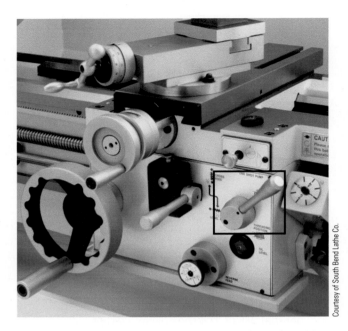

Courtesy of South Bend Lathe Co.

FIGURE 5.4.20 The lever used for engaging the half-nuts during threading.

FIGURE 5.4.21 A split half-nut assembly is used for carriage motion during threading by closing the two halves on the lead screw.

TPI	Thread Dial Half-nut Engagement Points
EVEN NUMBER	Any graduation
ODD NUMBER	Any numbered graduation OR Any un-numbered graduation (Once one option is selected, use only that option throughout the threading process.)
ANY ½ NUMBER	1 and 3 numbered graduations OR 2 and 4 numbered graduations (Once one option is selected, use only that option throughout the threading process.)
ANY OTHER FRACTIONAL NUMBER	Any graduation, but once selected, use only that graduation throughout the threading process

FIGURE 5.4.22 Half-nut engagement points for given TPI values with a typical thread dial.

FIGURE 5.4.23 The cross slide is set to zero and the compound is advanced 0.001 to 0.002″ for a trial cut.

FIGURE 5.4.24 The threading dial must be watched closely so that the half-nut lever is engaged at the appropriate time.

FIGURE 5.4.25 A screw pitch gage is used to verify that the correct number of threads per inch is being produced.

- When the tool reaches the end of the threaded section, disengage the half-nut lever.
- Rotate the cross slide hand wheel one full turn to retract the tool out of the thread groove.
- Move the carriage back to the starting position of the thread. Since the tool has been retracted, there will be no damage to the threads.
- Return the cross slide to the reference "0" position to prepare for the next pass.
- Stop the spindle.
- Use a screw pitch gage to ensure that the proper number of threads per inch is being cut before continuing. **(See Figure 5.4.25.)**
- Advance the compound-rest about 0.010" to 0.015" for the first actual threading pass.
- Take another pass by engaging the half-nut at the proper time and disengaging at the end of the thread.
- Retract the tool with the cross slide and return to the starting point again.
- Return the cross slide to the reference "0" to prepare for the next pass.
- Repeat the process by taking depth cuts with the compound rest for subsequent passes. Decrease the amount of compound rest in-feed with each pass until depth is about 0.002" per pass. This will help chip formation and tool life as tool contact and depth increase. When nearing completion, take finishing passes around 0.0005" to 0.001" in depth.
- Surface finish can often be improved by taking the last few passes by feeding the tool radially into the work with the cross slide at 0.0005" to 0.001" (instead of with the compound rest).
- Because of the large amount of tool contact and potential for workpiece and tool flex, it is also a good idea to take a few "spring" passes without advancing

FIGURE 5.4.26 Some threads extend all the way across a diameter, and the half-nut lever may be disengaged when the cutting tool exits the cut.

the tool when nearing the final depth. This will often remove small amounts of material (around 0.0005") and also improve surface finish.

Methods for Terminating a Thread

Sometimes a drawing may call for threads to be cut all the way across the entire length of a diameter. In these cases, the half-nut lever may simply be disengaged after the threading tool exits the work, as shown in **Figure 5.4.26**. Other times, a diameter must be threaded up to a shoulder or other feature. In these instances, two methods may be used to terminate the thread groove.

A thread relief or undercut may be produced if the drawing allows. This undercut is a narrow groove that provides a space for the threading tool to stop after the half-nut lever is disengaged without damaging the threads. Thread relief grooves may also be used as spaces to start the threading tool when cutting threads. **(See Figure 5.4.27.)**

FIGURE 5.4.27 A thread relief groove serves as an area to disengage the half-nut lever without causing damage to the thread or adjacent surfaces.

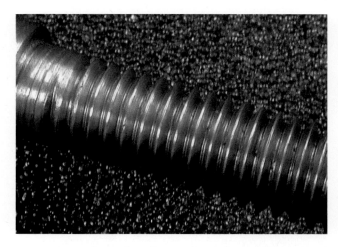

FIGURE 5.4.28 A threading tool may be quickly retracted at the end of each pass to terminate the thread, resulting in a "vanishing" thread groove.

If a narrow relief groove is not permitted, the operator must use the cross slide to rapidly retract the cutting tool at the desired thread length. **(See Figure 5.4.28.)** Care must be taken to ensure that the tool is retracted at the same point at each pass, so the tool does not gouge into uncut material. When using this method to cut external threads, it is a good idea to set the cross-slide hand-wheel lever at the 9 o'clock position when cutting external threads (3 o'clock when cutting internal threads). This makes it easier to retract the tool by simply pushing the lever in a downward motion and can help to prevent accidentally moving the cross slide in the wrong direction when retracting the tool. **(See Figure 5.4.29.)**

THREAD MEASUREMENT

The size, shape, and accuracy of screw threads produced on the lathe can be measured in a variety of ways. The type of measuring tool used is determined by the tolerances stated on the print and what details of the thread will be checked. The most common inspection performed on a thread is measurement of the pitch diameter, but inspection of the thread form may also be desired.

Thread Ring and Plug Gages

Go and no-go thread ring gages are often used to inspect the pitch diameter of external threads. They cannot give an actual pitch diameter measurement, but they will determine if the pitch diameter is within the proper limits. These gages are produced to inspect a particular class of fit, so be sure the class of fit identified on the gage matches the desired thread specification. The go gage should turn freely on the thread, while the no-go gage should not fit on the thread. If the go ring

FIGURE 5.4.29 (A) A cross-slide dial set to 9 o'clock for easy tool retraction at the end of an external thread-cutting pass. (B) Cross-slide dial set to 3 o'clock for easier retraction at the end of an internal threading pass.

gage does not fit, the thread needs to be cut deeper. If the no-go gage fits, the thread is too deep, is out of tolerance, and will be rejected. **Figure 5.4.30** shows go and no-go thread ring gages in use.

Go and no-go thread plug gages are used to inspect internal threads. They also cannot measure actual pitch diameter, but they will determine if the pitch diameter is within proper limits. They are also produced to inspect a particular class of fit. If the go gage does not fit in the threaded hole, the thread must be cut deeper. If the no-go gage fits, the thread is too deep, is out of tolerance, and will be rejected. **Figure 5.4.31** shows a go/no-go thread plug gage in use.

Thread Micrometer

If the actual measurement of the pitch diameter of an external thread is desired, or if ring gages are not available, a thread micrometer can be used. The thread

FIGURE 5.4.30 Checking an external thread with go and no-go thread ring gages.

FIGURE 5.4.32 A thread micrometer being used to measure the pitch diameter of an external thread.

FIGURE 5.4.31 Checking an internal thread with a double end go/no-go thread plug gage.

FIGURE 5.4.33 Three thread wires carefully placed in a thread groove can be measured with a micrometer to determine pitch diameter.

micrometer is similar in appearance to the standard outside-diameter micrometer except that the thread micrometer has a 60-degree conical point on its spindle and a 60-degree "V" on its anvil. The thread micrometer provides a quick and accurate means of measuring the pitch diameter of a thread. To use the thread micrometer, first find the correct pitch diameter for the thread using reference materials. Since each thread micrometer can only measure a specific range of threads per inch, it is important to select a micrometer that will correctly measure the thread being machined. **Figure 5.4.32** shows a thread micrometer in use.

Three-Wire Method

If ring gages or a thread micrometer are not available, the **three-wire method** may be used to measure pitch diameter. Precise pin gages called **thread measuring wires** (often just called **thread wires**) are used with a standard micrometer. Three of these **thread wires** of equal diameter are placed in the thread groove and the micrometer is used to measure across those wires, as shown in **Figure 5.4.33**. A formula is used to determine the acceptable maximum and minimum measurements over wires for known pitch diameter limits. Thread measuring wires are available in sets. The set contains

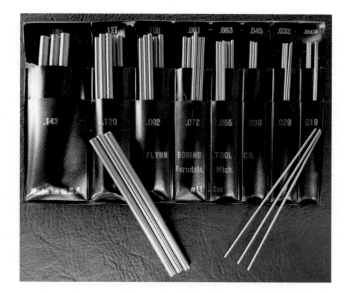

FIGURE 5.4.34 A thread wire measuring set contains several sets of three matched wire diameters.

Threads per Inch	Pitch, Inch	Wire Diameters for American Standard Threads		Pitch-Line Contact
		Max.	Min.	
4	0.2500	0.2250	0.1400	0.1443
4½	0.2222	0.2000	0.1244	0.1283
5	0.2000	0.1800	0.1120	0.1155
5½	0.1818	0.1636	0.1018	0.1050
6	0.1667	0.1500	0.0933	0.0962
7	0.1428	0.1283	0.0800	0.0825
8	0.1250	0.1125	0.0700	0.0722
9	0.1111	0.1000	0.0622	0.0641
10	0.1000	0.0900	0.0560	0.0577
11	0.0909	0.0818	0.0509	0.0525
12	0.0833	0.0750	0.0467	0.0481
13	0.0769	0.0692	0.0431	0.0444
14	0.0714	0.0643	0.0400	0.0412
16	0.0625	0.0562	0.0350	0.0361
18	0.0555	0.0500	0.0311	0.0321
20	0.0500	0.0450	0.0280	0.0289
22	0.0454	0.0409	0.0254	0.0262
24	0.0417	0.0375	0.0233	0.0240
28	0.0357	0.0321	0.0200	0.0206
32	0.0312	0.0281	0.0175	0.0180
36	0.0278	0.0250	0.0156	0.0160
40	0.0250	0.0225	0.0140	0.0144

FIGURE 5.4.35 A table may be used to determine acceptable wire diameters for using the three-wire method on a particular TPI.

several size increments of three equal size diameter wires. **(See Figure 5.4.34.)**

To measure threads by the three-wire method, the pitch diameter must first be determined as explained earlier. Then the ideal wire size for measurement is located using reference tables, and measurement over wires is calculated.

Determining Wire Size

A reference such as the *Machinery's Handbook* can be used to first determine the correct wire diameter for a given thread size according to the TPI.

EXAMPLE: Determine the maximum, minimum, and best wire size for a ¾-10 UNC 2A thread.

Since the thread to be measured has 10 threads per inch, find the row labeled 10 TPI in the Wire Diameters for American Standard Threads table shown in **Figure 5.4.35**. Follow the 10 TPI row across to the "max" and "min" columns.

Maximum wire size = 0.0900

Minimum wire size = 0.0560

Any diameter wire between these two sizes can be used, but the best wire size, which will make contact at the pitch line, can be found using the formula .57735 × P, where P is the pitch. Those values are in shown in the far right column of Figure 5.4.35. A chart from a thread wire set shown in **Figure 5.4.36** lists 0.063" wires to be used for 10 TPI. These are in the required range and also very close to the best wire size.

Wire Measurement Calculation

Once the wire size has been chosen, the desired or target measurements for that correct class of thread may be determined using the formula:

$$\text{Measurement} = \text{Pitch Diameter} - (0.86603 \times \text{pitch}) + (3 \times \text{wire diameter})$$

EXAMPLE: Determine the maximum and minimum acceptable measurement over wires for a ¾-10 UNC 2A thread. Use 0.063" diameter thread wires and the previously determined pitch diameter limits:

Pitch diameter upper limit = 0.6832

Pitch diameter lower limit = 0.6773

(Continued)

EXAMPLE: (continued)

Upper-Limit Measurement = 0.6832 − (0.86603 × 0.100) + (3 × 0.063)

= 0.6832 − 0.086603 + .189

= .7856

Lower-Limit Measurement = 0.6773 − (0.86603 × 0.100) + (3 × 0.063)

= 0.6773 − 0.086603 + .189

= 0.7797

The thread is then cut until the measurement over wires is between the upper and lower calculated measurements.

Note: The far right column of **Figure 5.4.36** lists a value to add to the major diameter for the measurement over wires. That result applies only to the measurement over wires for the upper limit of a Class 3A thread.

Thread Measuring Wire Set

From chart below, select wire size opposite number of threads to be checked.

To obtain correct measurement over wires, add decimal in third column to major diameter of thread to be measured.

For 60° Threads — U.S.S., S.A.E., N.C., N.F.

THREADS per Inch	WIRE SIZE Inches	ADD TO MAJOR Diameter, Inches
40	0.019	0.0191
36	0.019	0.0149
32	0.025	0.0277
28	0.029	0.0329
24	0.029	0.0239
20	0.032	0.0203
18	0.0395	0.0343
16	0.0395	0.0238
14	0.045	0.0268
13	0.045	0.0185
12	0.055	0.0387
11	0.055	0.0273
10	0.063	0.0375
9	0.072	0.0476
8	0.081	0.0536
7	0.092	0.0595
6	0.108	0.0714
5½	0.120	0.0845
5	0.127	0.0779
4½	0.143	0.0923
4	0.143	0.0502
3½	0.185	0.1220

FIGURE 5.4.36 A chart from a thread wire measuring set shows which wire size to use for a given TPI. (The far right column lists a value to add to the major diameter for the measurement over wires. That result only applies only to the measurement over wires for the upper limit of a Class 3A thread.)

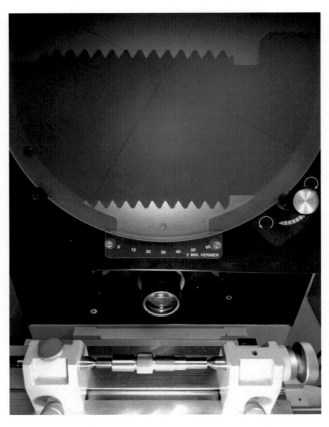

FIGURE 5.4.37 An optical comparator being used to project a magnified image of a thread for inspection of its form.

Thread Form Measurement

The form of an external thread may also be inspected if desired. The optical comparator is usually used to magnify the thread so that measurements, including the flank angles, crest shape, and root shape, can be clearly seen. **Figure 5.4.37** shows a comparator being used to inspect thread form.

OTHER THREAD FORMS

While the 60-degree thread forms discussed up to this point may be the most common, other forms are produced to meet various engineering and design needs. These threads are cut using similar methods as those used to cut Unified threads (60-degree V-threads). The cutting tool must match the shape of the thread form, and the compound rest is fed at an angle slightly less than half of the included thread angle.

Acme Thread

The **Acme thread** has a very distinct appearance since the included angle of the thread form is 29 degrees and the threads are thicker and somewhat square. Acme threads are strong and commonly found in power

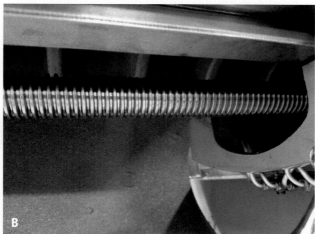

FIGURE 5.4.38 Lathe and milling machine leadscrews are commonly made with Acme threads.

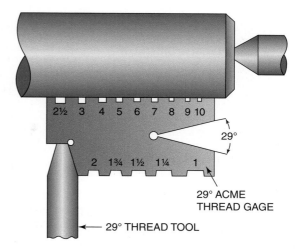

FIGURE 5.4.39 A special gage is used to size the flat on the tip of an Acme thread-cutting tool and to align the tool with the diameter to be threaded.

FIGURE 5.4.40 A tapered pipe thread allows mating parts to wedge together, helping them to seal.

transmission applications where strength is of the utmost importance. They are used for applications such as jacks, large valves, and lathe and milling machine leadscrews. **(See Figure 5.4.38.)** Since each size of Acme thread has a different-size flat at the root, a different-size cutting tool is needed to machine each size. The compound rest would be set to feed the tool at 14 degrees. **Figure 5.4.39** shows an Acme threading tool and the gage used to size and align Acme threading tools.

Tapered Pipe Threads

Tapered pipe threads change in diameter from one end to the other. The diameter change on this type of thread is 3/4 of an inch per every foot of length. The taper allows for two mating tapered pipe threads to be wedged tightly together as they are assembled and tightened. This wedging action makes a nearly leak-proof assembly. A pipe-sealing compound may still be needed to protect against leaks of fluids or gasses as

they pass through the joint. **Figure 5.4.40** shows a tapered pipe thread.

The included angle of the tapered pipe thread form is also 60 degrees, so the tool and compound rest settings are the same as when cutting a standard 60-degree V-shaped thread. The tool is aligned with the center gage on a straight section of the workpiece diameter, not on the tapered section. **(See Figure 5.4.41.)**

Buttress Threads

The **buttress thread** is easily identified since the thread form is asymmetrical. On this thread, one flank is nearly perpendicular to the center line of the screw and is called

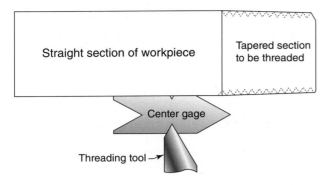

FIGURE 5.4.41 A threading tool for cutting tapered pipe threads must be aligned by using a center gage on the straight diameter and not the taper.

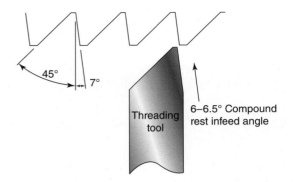

FIGURE 5.4.42 The profile of a buttress thread, threading tool, and tool in-feed angle.

the pressure flank. This thread form is often used in applications requiring high thrust in one direction. The process of cutting a buttress thread is very similar to that of cutting a 60-degree V thread. The shape of the cutting tool is made to match the asymmetrical shape of the thread groove. The compound rest would be set to feed the tool between 1/2 and 1 degree less than the pressure flank angle. The buttress thread is shown in **Figure 5.4.42**.

SUMMARY

- Single-point thread cutting on the lathe offers greater control over fit, form, and finish than thread cutting with a tap and die.
- Many terms and mathematical calculations are associated with both cutting and measuring threads, and a thorough understanding of each is essential.
- The major diameter, pitch diameter, and minor diameter limits for a certain class of fit are found by referencing a thread data chart.
- The fit of a thread is determined by the pitch diameter and is the basis for measurement when checking the size of a thread.
- The quick-change gear box on a lathe contains a gear train that rotates the leadscrew at the appropriate speed to produce the selected number of threads per inch.
- The leadscrew is directly coupled to the carriage by a pair of lever-activated half-nuts.
- The half-nuts are engaged at the appropriate time by observing the threading dial so that each threading pass cuts in exactly the same groove.
- Since threading typically generates a lot of force between the tool and workpiece, all in-feed of the tool is performed by the angled compound rest.
- The angle of the compound rest allows one cutting-edge of the thread-cutting tool to perform the majority of the cutting and greatly reduces cutting forces.
- Spring passes may be performed near the completion of the thread cutting to take any material not cut due to deflection from excessive cutting forces.
- The thread may be terminated at the appropriate length by quickly retracting the tool at the end of the pass or by allowing the tool to enter a relief groove.
- As the thread approaches its final size, the pitch diameter may be checked with either a ring gage (external), a plug gage (internal), a thread micrometer (external), or by using the three-wire method (external).
- Thread form may be visually inspected with an optical comparator.

REVIEW QUESTIONS

1. The distance of actual contact of two mating threads measured along the length of the two threads is the _____.

2. What feature of mating threads determines the class of fit?

3. How many classes of fit are there in the Unified National system?

4. Determine the major diameter limits for the following threads using the *Machinery's Handbook*:

   a. ½-20 UNF 2A

   upper: _____

   lower: _____

   b. ¾-16 UNF 3A

   upper: _____

   lower: _____

   c. 1-8 UNC 2A

   upper: _____

   lower: _____

   d. 1 ¼-12 UNF 3A

   upper: _____

   lower: _____

5. Determine the minor diameter limits for the following threads using the *Machinery's Handbook*:

   a. ½-20 UNF 2B

   upper: _____

   lower: _____

   b. ¾-16 UNF 3B

   upper: _____

   lower: _____

   c. 1-8 UNC 2B

   upper: _____

   lower: _____

   d. 1 ¼-12 UNF 3B

   upper: _____

   lower: _____

6. Determine the pitch diameter limits for the following threads using the *Machinery's Handbook*:

   a. ½-20 UNF 2B

   upper: _____

   lower: _____

   b. ¾-16 UNF 3A

   upper: _____

   lower: _____

   c. 1-8 UNC 2B

   upper: _____

   lower: _____

   d. 1 ¼-12 UNF 3A

   upper: _____

   lower: _____

7. Determine the approximate compound-rest in-feed for cutting a 1½-12 UNF 2A thread.

8. What is the name for the rotating device that transmits power to the carriage for thread-cutting motion?

9. When threading, what is the reason for feeding the cutting tool at an angle for each depth cut?

10. Name the device used to track the timing of half-nut lever engagement.

11. Why should the depth of cut be reduced for each pass as the thread is cut deeper?

12. What dimension of the thread is measured by using a thread micrometer or the three-wire method?

13. What measuring tool is used to visually inspect the form, or shape, of an external thread?

14. List two applications of Acme threads.

15. What type of thread is machined on a tapered diameter and is used to create leak-proof joints?

UNIT 5 Taper Turning

Learning Objectives

After completing this unit, the student should be able to:

• Define a taper
• Demonstrate understanding of taper specification methods
• Perform taper calculations
• List methods of turning tapers and their benefits and drawbacks
• Demonstrate understanding of setup procedures for taper turning methods

Key Terms

Offset tailstock
  method
Radius type center
  drill

Tailstock offset or
  setover
Taper
Taper attachment

Taper per foot (TPF)
Taper per inch (TPI)
Tool bit method

INTRODUCTION

A **taper** is a constant change in diameter of a cylindrical part. Bell-shaped cylindrical surfaces are not considered tapers. **(See Figure 5.5.1.)** Tapers are common in machine tools to provide alignment and securing of workholding and toolholding devices. Tapers can be either self-holding or self-releasing. Self-holding tapers can align *and* secure mating parts. Examples of self-holding tapers are the Morse and Jacobs tapers, commonly found in drill presses, lathes, and drill chucks and shanks. Self-releasing tapers only provide alignment and must be secured by additional methods, usually threaded drawbars. The taper used in most modern CNC milling machines is an example of a self-releasing taper and is based on the NMTB taper (National Machine Tool Builders' Association) or AMT taper (Association for Manufacturing Technology).

Tapers can also be machined for appearance, weight reduction, or clearance. Examples might include a taper on a pool cue or a machine or tool handle. Short tapers might create clearance between the shoulders of mating cylindrical parts. Regardless of the applications, turning tapers on the lathe is a skill that is often required in the machining field.

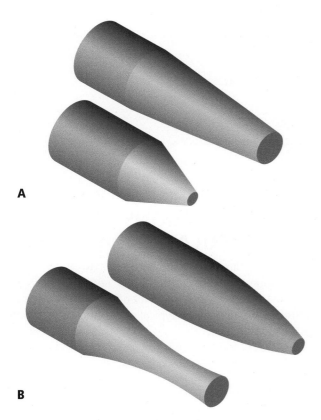

FIGURE 5.5.1 The parts in A are tapered because they each have a constant rate of diameter change. The parts in B are bell-shaped and not considered tapers.

This unit will address terms and calculations involving tapers and the methods for machining tapers on the lathe.

TYPICAL TAPER SPECIFICATIONS

Tapers can be specified on prints by two basic methods. The first is by an angular dimension. Since a taper is a constant diameter change, the second method is a rate of diameter change over a given length.

Angular Specification

Angular specifications of tapers are simply stated by the angle in degrees, but there are two ways of expressing a taper in the angular format.

The included angle of a taper is the measure of the entire angle from one side of the tapered cylinder to the other side. An example of an included angle would be the measurement of the full angle of the point of prick and center punches (60 and 90 degrees, respectively).

A taper designated by a centerline angle is measured from one side of the tapered cylinder to the centerline of the cylinder. The centerline measurement of the center punches mentioned earlier would be 30 and 45 degrees, respectively. See **Figure 5.5.2** for an illustration of the difference between included and centerline angles.

Rate-of-Change Specification

The rate-of-change specifications for tapers are ratios stated by the amount of diameter change over a given length.

Taper per inch (TPI) is the diameter change in 1 inch of length. For example, 1/2" TPI would mean that in

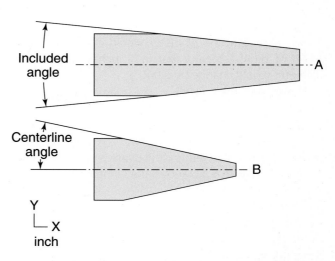

FIGURE 5.5.2 (A) A taper specified by an included angle, and (B) by a centerline angle.

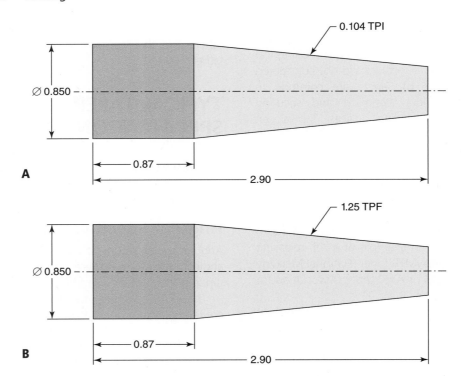

FIGURE 5.5.3 (A) Taper specified by TPI (taper per inch). (B) Taper specified by TPF (taper per foot). Both of these specifications would produce parts with the same dimensions.

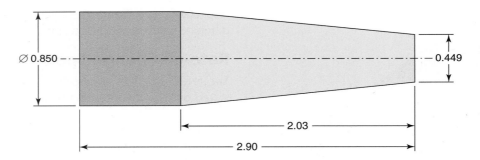

FIGURE 5.5.4 A taper specified by end diameters and length.

1 inch of length, the diameter would change by 1/2". A TPI of 0.045" means that in 1 inch of length, the diameter would change by 0.045".

Taper per foot (TPF) is the diameter change in 1 foot, or 12", of length. Contrasted with the previous examples, 1/2" TPF would mean that in 12" of length, the diameter would change by 1/2". A TPF of 0.045" means that in 12" of length, the diameter would change by 0.045". **Figure 5.5.3** shows the comparison between TPI and TPF specifications.

A taper can also be designated by the diameter of the large end, the diameter of the small end, and the length of the taper, as shown in **Figure 5.5.4**.

Standard tapers such as the Morse taper can be specified on a print by a leader or note identifying the type of taper.

TAPER DIMENSIONS AND CALCULATIONS

Several taper dimensions are important to know when turning tapers on the lathe, and a few formulas can be used to determine taper dimensions. The sketch in **Figure 5.5.5** shows some typical dimensions used when working with tapers.

One basic formula that is used frequently when working with tapers is shown below.

$$TPI = \frac{D - d}{l}$$

• D is the diameter of the large end of the taper
• d is the diameter of the small end of the taper
• l is the length of the taper

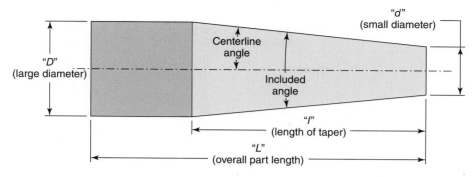

FIGURE 5.5.5 Summary of commonly used taper dimensions.

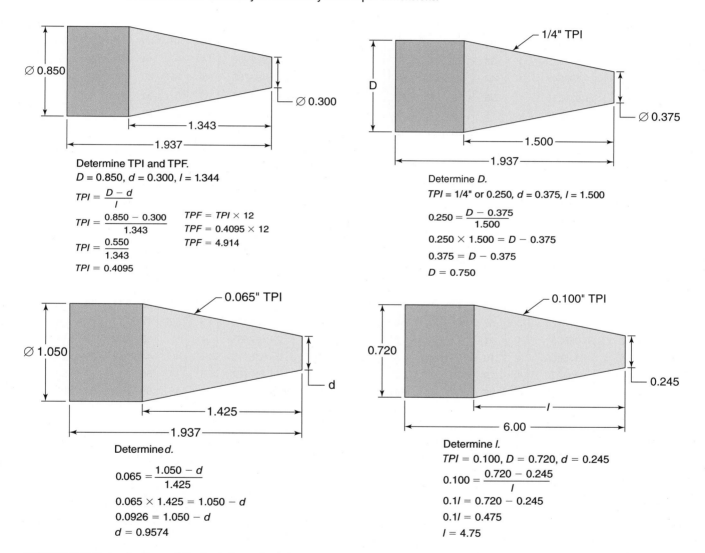

Determine TPI and TPF.

$D = 0.850$, $d = 0.300$, $l = 1.344$

$$TPI = \frac{D - d}{l}$$

$$TPI = \frac{0.850 - 0.300}{1.343}$$

$$TPI = \frac{0.550}{1.343}$$

$$TPI = 0.4095$$

$TPF = TPI \times 12$

$TPF = 0.4095 \times 12$

$TPF = 4.914$

Determine D.

$TPI = 1/4"$ or 0.250, $d = 0.375$, $l = 1.500$

$$0.250 = \frac{D - 0.375}{1.500}$$

$0.250 \times 1.500 = D - 0.375$

$0.375 = D - 0.375$

$D = 0.750$

Determine d.

$$0.065 = \frac{1.050 - d}{1.425}$$

$0.065 \times 1.425 = 1.050 - d$

$0.0926 = 1.050 - d$

$d = 0.9574$

Determine l.

$TPI = 0.100$, $D = 0.720$, $d = 0.245$

$$0.100 = \frac{0.720 - 0.245}{l}$$

$0.1l = 0.720 - 0.245$

$0.1l = 0.475$

$l = 4.75$

FIGURE 5.5.6 Applications of the basic taper formula.

Always remember that l is the length of the tapered section, not the overall length of the part (another taper formula that will be discussed later uses L as the overall length of the part). Become familiar with this formula because it is often used when working with tapers.

This formula is commonly used when dealing with rate-of-change specifications. With this formula, any one of the variables can be found if the others are available. For example, if the end diameters and length are given, TPI can be found. If one end diameter, length, and TPI are given, the other end diameter can be found. Further, TPF can be found by multiplying TPI by 12. See **Figure 5.5.6** for examples of uses of this formula.

Tapers per Foot and Corresponding Angles

Taper per Foot	Included Angle			Angle with Centerline			Taper per Foot	Included Angle			Angle with Centerline		
1/64	0.074604°	0°	4'	20″	0°	2'	14″	1 7/8	8.934318°	8°	56'	4″	4°	28'	2″
1/32	0.140208	0	8	57	0	4	29	1 15/16	9.230863	9	13	51	4	36	56
1/16	0.298415	0	17	54	0	8	57	2	9.527283	9	31	38	4	45	59
1/32	0.447621	0	26	51	0	13	26	2 1/8	10.119738	10	7	11	5	3	36
1/8	0.596826	0	35	49	0	17	54	2 1/4	10.711650	10	42	42	5	21	21
3/32	0.746028	0	44	46	0	22	23	2 3/8	11.302990	11	18	11	5	39	5
3/16	0.895228	0	53	43	0	26	51	2 1/2	11.893726	11	53	37	5	56	49
3/32	1.044425	1	2	40	0	31	20	2 5/8	12.483829	12	29	2	6	14	31
1/4	1.193619	1	11	37	0	35	49	2 3/4	13.073267	13	4	24	6	32	12
8/32	1.342808	1	20	34	0	40	17	2 7/8	13.662012	13	39	43	6	49	52
5/16	1.491993	1	29	31	0	44	46	3	14.250033	14	15	0	7	7	30
11/32	1.641173	1	38	28	0	49	14	3 1/8	14.837300	14	50	14	7	25	7
3/8	1.790347	1	47	25	0	53	43	3 1/4	15.423785	15	25	26	7	42	43
13/32	1.939516	1	56	22	0	58	11	3 3/8	16.009458	16	0	34	8	0	17
7/16	2.088677	2	5	19	1	2	40	3 1/2	16.594290	16	35	39	8	17	50
15/32	2.237832	2	14	16	1	7	8	3 5/8	17.178253	17	10	42	8	35	21
1/2	2.386979	2	23	13	1	11	37	3 3/4	17.761318	17	45	41	8	52	50
17/32	2.536118	2	32	10	1	16	5	3 7/8	18.343458	18	20	36	9	10	18
7/16	2.685248	2	41	7	1	20	33	4	18.924644	18	55	29	9	27	44
19/32	2.834369	2	50	4	1	25	2	4 1/8	19.504850	19	30	17	9	45	9
5/8	2.983481	2	59	1	1	29	30	4 1/4	20.084047	20	5	3	10	2	31
21/32	3.132582	3	7	57	1	33	59	4 3/8	20.662210	20	39	44	10	19	52
11/16	3.281673	3	16	54	1	38	27	4 1/2	21.239311	21	14	22	10	37	11
23/32	3.430753	3	25	51	1	42	55	4 5/8	21.815324	21	48	55	10	54	28
3/4	3.579821	3	34	47	1	47	24	4 3/4	22.390223	22	23	25	11	11	42
25/32	3.728877	3	43	44	1	51	52	4 7/8	22.963983	22	57	50	11	28	55
13/16	3.877921	3	52	41	1	56	20	5	23.536578	23	32	12	11	46	6
27/32	4.026951	4	1	37	2	0	49	5 1/8	24.107983	24	6	29	12	3	14
7/8	4.175968	4	10	33	2	5	17	5 1/4	24.678175	24	40	41	12	20	21
30/32	4.324970	4	19	30	2	9	45	5 3/8	25.247127	25	14	50	12	37	25
15/16	4.473958	4	28	26	2	14	13	5 1/2	25.814817	25	48	53	12	54	27
31/32	4.622931	4	37	23	2	18	41	5 5/8	26.381221	26	22	52	13	11	26
1	4.771888	4	46	19	2	23	9	5 3/4	26.946316	26	56	47	13	28	23
1 1/16	5.069753	5	4	11	2	32	6	5 7/8	27.510079	27	30	36	13	45	18
1 1/8	5.367550	5	22	3	2	41	2	6	28.072487	28	4	21	14	2	10
1 3/16	5.665275	5	39	55	2	49	57	6 1/8	28.633518	28	38	1	14	19	0
1 1/4	5.962922	5	57	47	2	58	53	6 1/4	29.193151	29	11	35	14	35	48
1 5/16	6.260400	6	15	38	3	7	49	6 3/8	29.751364	29	45	5	14	52	32
1 3/8	6.557973	6	33	29	3	16	44	6 1/2	30.308136	30	18	29	15	9	15
1 7/16	6.855367	6	51	19	3	25	40	6 5/8	30.863447	30	51	48	15	25	54
1 1/2	7.152669	7	9	10	3	34	35	6 3/4	31.417276	31	25	2	15	42	31
1 9/16	7.449874	7	27	0	3	43	30	6 7/8	31.969603	31	58	11	15	59	5
1 5/8	7.746979	7	44	49	3	52	25	7	32.520409	32	31	13	16	15	37
1 11/16	8.043980	8	2	38	4	1	19	7 1/8	33.069676	33	4	11	16	32	5
1 3/4	8.340873	8	20	27	4	10	14	7 1/4	33.617383	33	37	3	16	48	31
1 13/16	8.637654	8	38	16	4	19	8	7 3/8	34.163514	34	9	49	17	4	54

FIGURE 5.5.7 This chart gives the corresponding centerline and included angles for TPF values in 1/64″ increments.

When an angle is given on a print, it is sometimes desirable to convert that angle to TPI or TPF. However, when TPI or TPF are known, it is sometimes desirable to find the corresponding included or centerline angle. If the TPF is a 1/64" fractional value, the angle can be found by using a chart like the one in **Figure 5.5.7**. If not, some formulas based on right-angle trigonometry can be used.

Converting TPI or TPF to an Angular Dimension

When TPI or TPF is known, the following formula can be used to find the centerline angle:

Centerline Angle = Arctan (TPF ÷ 24)

All that means is that the centerline angle has a tangent value that is (*TPF* ÷ 24). If TPI is known, first multiply TPI by 12 to get TPF, then divide TPF by 24. That is the tangent of the centerline angle. Look at this example:

EXAMPLE: What is the centerline angle if TPF is 3/4"?

$$Tangent\ of\ centerline\ angle = \frac{3}{4} \div 24$$

$$= \frac{3}{4} \cdot \frac{1}{24}$$

$$= \frac{1}{32} = 0.03125$$

On a scientific calculator, press "2nd," "TAN," "0.03125," "=," and the answer is 1.7899. That is the angle in degrees. To convert to degrees, minutes, and seconds, use the methods discussed in Section 2.2 under "Angular Measurement and Conversion."

If the included angle is preferred, adjust the original formula to:

Included Angle = Arctan (TPF ÷ 12)

Converting an Angular Dimension to TPI or TPF

A similar method can be used to find the TPI or TPF corresponding to a given angle. When given a centerline angle, use this formula to calculate TPF:

TPF = 24 (Tan X).

EXAMPLE: Find the TPF and TPI given a centerline angle of 8 degrees.

TPF = 24 (Tan 8)

= 24(0.14054)

= 3.37298

To find TPI, divide the TPF by 12. So TPI = 3.37298 ÷ 12 = 0.2811. TPI can also be calculated by adjusting the original formula to *TPI* = 2 (Tan X).

A summary table of several taper calculation formulas is given in **Figure 5.5.8**.

TAPER TURNING METHODS

Different taper turning methods require different dimensions, and each has its benefits and drawbacks. See **Figure 5.5.9** for a summary of the different methods, their benefits/drawbacks, and the dimensional information needed for each method. When turning tapers, generally the same speeds and feeds used for straight turning are used. Calculate spindle RPM using the largest diameter being machined.

Tool Bit Method

The **tool bit method** of machining uses a flat edge of a cutting tool to cut the taper. It can cut only short tapers and either the included or centerline angle must be known. The flat edge of the cutting tool is set at the proper angle using a protractor. Then the tool is fed against the work to the proper depth or length depending on the print specification.

Taper length is limited because of the size of the cutting tool. For short lengths such as 0.050", use speeds the same as for straight turning. As lengths increase and the amount of tool contact increases, spindle speeds may need to be lowered considerably to eliminate chatter. **Figure 5.5.10** illustrates the tool bit taper cutting method.

Compound-Rest Method

Since the compound rest can be rotated to any angle, it can be used to machine a taper when the included or centerline angle is known. This method can be used to machine external and internal tapers, but taper length is limited by the travel of the compound rest.

- If an included angle is given, first divide by two to get the centerline angle.

- Set the compound rest to the centerline angle. Be sure to set the angle from a position parallel to the ways.

- If the angle is given from the face of the part, find its complement. Then set the compound rest to that angle from a position parallel to the ways.

$TPI = \dfrac{D-d}{l}$ $TPI = \dfrac{TPF}{12}$	To determine taper per inch (TPI), subtract the small end diameter (*d*) from the large end diameter (*D*) and divide by the length of the taper (*l*). If taper per foot is known, determine taper per inch by dividing taper per foot (TPF) by 12.
$TPF = \left(\dfrac{D-d}{l}\right) \times 12$ $TPF = TPI \times 12$	To determine taper per foot, subtract the small end diameter (*d*) from the large diameter (*D*), divide by the length (*l*) of taper in inches, and then multiply by 12. If TPI is known, multiply the TPI by 12.
$D = (TPI \times l) + d$	To determine large end diameter of a part if taper per inch, small end diameter, and length of taper are known, multiply the taper per inch (TPI) by the length of the taper (*l*), and add the small diameter (*d*).
$d = D - (TPI \times l)$	To determine small end diameter of a part if taper per inch, large end diameter, and length of taper are known, multiply the taper per inch (TPI) by the length (*l*) of the taper, and subtract that answer from the large diameter (*D*).
Centerline angle = *Arctan* ($TPF \div 24$)	To determine angle to centerline given taper per foot, first divide taper per foot (TPF) by 24. The answer is the tangent of the angle.
Included angle = $2 \times Arctan\left(\dfrac{TPF}{24}\right)$	To determine included angle given taper per foot, first divide taper per foot (TPF) by 12. The answer is the tangent of the angle.
$TPF = 24\,(TanX)$	To determine taper per foot given centerline angle, multiply the tangent of the centerline angle by 24.
$TPF = 24\left(Tan\left(\dfrac{x}{2}\right)\right)$ when x = included angle	To determine taper per foot given included angle, multiply the tangent of the included angle by 12.

FIGURE 5.5.8 Summary of frequently used taper formulas.

Taper Turning Method	Required Information	Benefits	Drawbacks
Tool Bit Method	• The centerline angle of the taper	• Quick setup time • Any angle can be machined	• Taper length limited by tool size. • Custom ground form tool may be needed
Compound Rest Method	• The centerline angle of the taper	• Quick setup time. • Any angle can be machined • External and internal tapers can be produced	• Taper length limited by compound rest travel • Hand feed only
Taper Attachment Method	• TPF or angle (Some attachments have centerline angle graduations and some have included angle graduations.)	• Longer tapers can be machined • External and internal tapers can be produced. • Power feed can be used	• Longer set up time • Taper limited to approximately 10° centerline angle • Length is limited by taper attachment travel • Taper attachment backlash
Offset Tailstock Method	• TPI or TPF	• Very long tapers can be machined. (Only limited by distance between centers.) • No backlash concerns	• Taper amount limited by tailstock offset adjustment distance and binding of lathe dog in drive plate slot • Long setup time • Only external tapers can be machined • Use of bell-type center drills and/or ball centers recommended

FIGURE 5.5.9 Taper turning methods, required information, and their benefits and drawbacks.

FIGURE 5.5.10 The tool bit method for machining short tapers.

- The gib of the compound rest may need to be loosened slightly so the compound rest can be moved freely.
- Set an appropriate spindle speed. To begin cutting the taper, position the compound rest with the tool tip near the corner where the taper begins.
- Alternately move either the cross slide or the carriage toward the work in 0.001" increments and move the compound rest forward and backward until the tool touches the corner.
- Move the cross slide or carriage (depending on print specifications) to set cut depths and feed with the compound rest to cut the taper until the proper size is reached. **Figure 5.5.11** illustrates this method.

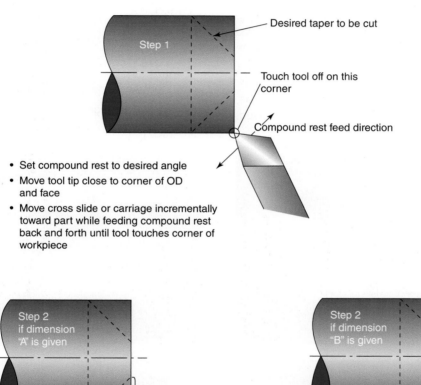

- Set compound rest to desired angle
- Move tool tip close to corner of OD and face
- Move cross slide or carriage incrementally toward part while feeding compound rest back and forth until tool touches corner of workpiece

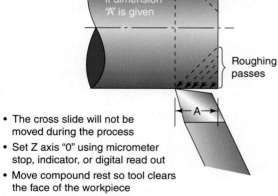

- The cross slide will not be moved during the process
- Set Z axis "0" using micrometer stop, indicator, or digital read out
- Move compound rest so tool clears the face of the workpiece
- Move carriage toward headstock incrementally and take roughing passes in direction of arrows using compound rest
- Take final pass with compound rest when carriage has been advanced to total distance of dimension "A"

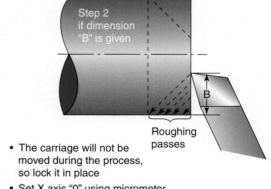

- The carriage will not be moved during the process, so lock it in place
- Set X axis "0" using micrometer collar or digital read out
- Move compound rest so tool clears the face of the workpiece
- Move cross slide incrementally toward part center and take roughing passes in direction of arrows using compound rest
- Take final pass with compound rest when cross slide has been advanced to total distance of dimension "B"

FIGURE 5.5.11 Methods for machining tapers using the compound rest.

Taper Attachment Method

A **taper attachment** is a lathe accessory that can move the cross slide either in or out as the carriage moves longitudinally along the ways. It can be set up using either the centerline angle or TPF value, but knowing the TPF or TPI is helpful when checking the setup. Both external and internal tapers can be cut. Another benefit is that the taper can be cut using the carriage power feed. Longer tapers can be cut than with the compound rest, but length is still limited by the length of the taper attachment.

Setup

To set up the taper attachment, follow these steps.

- Lock the attachment to the ways and position the carriage near the middle of the attachment.
- Loosen the guide rail locking screws and adjust the rail to the desired angular or TPF setting. (One end is graduated in degrees and the other end in TPF.) Then reclamp the locking screws. **(See Figure 5.5.12.)**
- Position the compound rest parallel to the cross slide. Because the taper attachment moves the cross slide, whenever the direction of the carriage is reversed,

there will be backlash from the reversal of the cross slide. Using the compound rest to advance the tool will eliminate potential problems caused by backlash. **(See Figure 5.5.13.)**

Check the taper attachment setting using dial indicators or the lathe's digital readout. Follow these steps if using indicators.

- Mount a test bar or a straight part in the machine.
- Position the taper attachment so that the length of the part or the test bar is within its travel. **(See Figure 5.5.14.)**
- Move the carriage far enough past the end of the part or test bar to remove backlash, then move it at least 1" in the direction that cutting will occur.
- Mount a plunge-type indicator on the compound rest and use the compound rest to bring the indicator into contact with the part or test bar to preload the indicator. Set the indicator on zero.
- Mount another indicator to measure longitudinal movement of the carriage. Set the indicator on zero.
- Move the carriage 1" in the cutting direction.

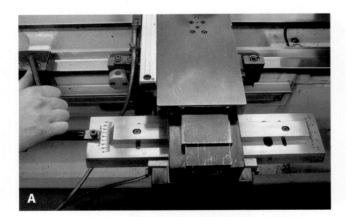
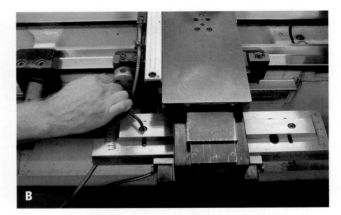

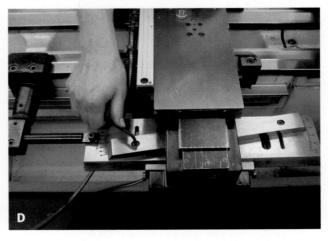

FIGURE 5.5.12 Adjusting the taper attachment. (A) Clamp the attachment to the ways. (B) Loosen the locking screws. (C) Adjust the guide rail. (D) Retighten the locking screws.

FIGURE 5.5.13 Positioning the compound rest parallel with the cross slide for setting cut depth can aid in eliminating backlash when using the taper attachment.

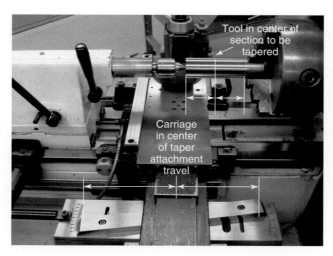

FIGURE 5.5.14 Make sure that the section of the work to be tapered is within the travel of the taper attachment.

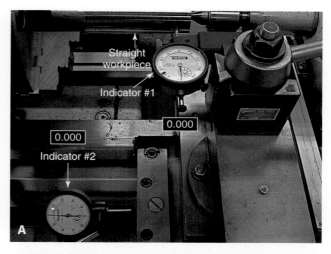

FIGURE 5.5.15 Using dial indicators to check the setting of the taper attachment. (A) Zero indicator #1 on the straight part or test bar and indicator #2 against the carriage. (B) Move the carriage 1″ in the cutting direction (measured with indicator #2) and check reading of indicator #1. Reading should be 1/2 of the TPI.

- Record the reading on the compound rest mounted dial indicator. It should be 1/2 of the TPI because only one side of the taper setting is being measured.
- Adjust the taper attachment and repeat the check as needed until the indicator shows the correct reading. **(See Figure 5.5.15.)**

Follow these steps to use a digital readout to check the setup of a taper attachment.

- Move the carriage in the cutting direction until cross-slide movement is shown on the digital readout. This ensures backlash has been removed from the cross-slide.
- Zero both the cross-slide (X-axis) and longitudinal (Z-axis) fields of the digital readout.
- Move the carriage another 1″ in the cutting direction.
- Record the movement of the cross slide displayed on the digital readout. In the diameter setting, the digital

readout should display the TPI. In the radius setting, the digital readout should display 1/2 of the TPI.

- Adjust the taper attachment and repeat the check as needed until the X-axis of the digital readout displays the proper value. **(See Figure 5.5.16.)**

Machining

After the taper attachment setting is verified, machining can begin.

- Set an appropriate spindle speed and feed rate.
- Move the carriage far enough toward the tailstock past the start point to remove backlash.
- Move in the cutting direction to position the tool near the beginning of the cut.
- Use the compound rest to set a cut depth.

FIGURE 5.5.16 Using the lathe's digital readout to check the taper attachment setting. (A) Zero both cross slide and longitudinal feed fields. (B) Then move the carriage 1" in the cutting direction and check the cross-slide movement reading. In diameter mode, it should equal TPI. (C) In radius mode, it should be 1/2 of the TPI.

- Engage the longitudinal feed to begin cutting. **(See Figure 5.5.17.)**
- After the cutting pass, return the carriage to the starting position. Because of the backlash when reversing direction, when the carriage is returned to the starting position, the tool will move away from the work slightly. This is one benefit of the backlash, because the tool will not be dragged across the surface of the part as it would be in a straight turning operation. **(See Figure 5.5.18.)**

FIGURE 5.5.17 Taking a cut with the taper attachment. Remember to bring the carriage far enough past the start point before taking a cut to eliminate backlash.

FIGURE 5.5.18 One benefit of the backlash in the taper attachment. When returning the carriage to the starting position the tool will not be in contact with the work.

- Continue taking passes by setting cut depth with the compound rest until the desired taper length is reached.
- Be sure to move the carriage far enough toward the tailstock past the start point to remove backlash after each pass.

The taper attachment can be simply unclamped from the ways at any time to perform straight turning, then repositioned and reclamped to perform taper turning when needed without having to reset the attachment.

Offset Tailstock Method

Another method for turning tapers is available that requires no special attachment. In the **offset tailstock method,** the tailstock center is offset from the headstock

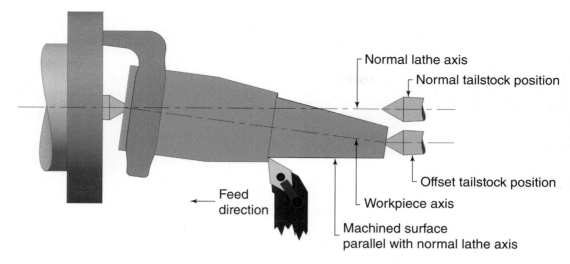

FIGURE 5.5.19 The principle of the offset tailstock method for taper turning.

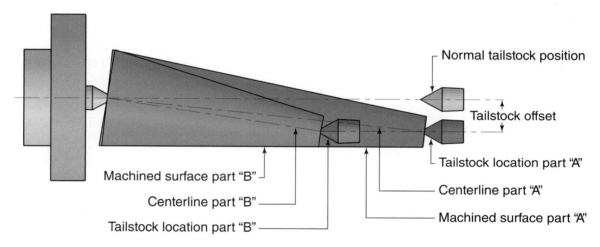

FIGURE 5.5.20 Variation in part lengths, or even different depths of center drilled holes, can produce different tapers from the same tailstock offset amount because the tailstock location changes between parts.

spindle centerline, as shown in **Figure 5.5.19**. This method can only be used for external tapers because the work must be held between centers. Its main benefit is that very long tapers can be machined. Length is only limited by the distance between lathe centers. There is also no concern about backlash as with the taper attachment.

There are some drawbacks to the offset tailstock method, and that is why it is usually only used if other methods will not produce desired results. First, the tailstock will have to be moved before the operation begins and then reset to zero after machining is complete and before any other operations can begin. Second, steep tapers often cannot be machined because the tailstock only has a certain amount of adjustment. Further, the offset amount needs to be different if part lengths are

different. Even if part lengths are the same, variation in depths of the center-drilled holes will likely require adjustment of the tailstock offset amount. The centers will contact the work at different depths, so the tailstock will be at a different location creating the same issue as when machining different length parts. **Figure 5.5.20** illustrates this. So unlike the tool bit, compound-rest, and taper attachment methods, which can be used for several parts once the setup is complete, the offset tailstock method may require frequent adjustment if multiple parts are to be machined. Finally, several trial cuts and adjustments are almost always needed to produce an accurate taper.

Since the offset tailstock method shifts the alignment of the lathe's centers, they do not make good contact with standard center-drilled holes as in straight turning

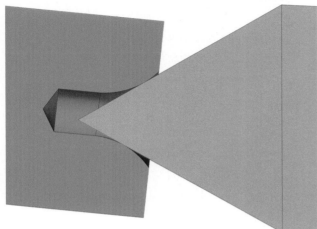

FIGURE 5.5.21 Drilling with a radius-type center drill will allow the workpiece to pivot more easily on the centers.

operations. This can cause binding between the center-drilled holes and the lathe centers, and it can deform the center-drilled holes. For those reasons, drilling with a **radius-type center drill** is recommended when using the offset tailstock method. The radius-type center drill creates a radius inside the hole instead of the standard 60 degree tapered hole. This shape allows the workpiece to pivot more easily on the lathe centers. **(See Figure 5.5.21.)** Center drilling must be performed before offsetting the tailstock or on a lathe with the tailstock in the standard position.

Setup

Before setting up the lathe for the offset tailstock method, the **tailstock offset** or **setover** must be calculated. That is the distance that the tailstock must be moved from its alignment with the centerline of the lathe spindle. This calculation is approximate and will probably require some further adjustment due to the variables of center-drill sizes and drilled-hole depths.

To calculate setover, first determine TPI using methods discussed earlier. Then use the following formula:

$$Setover = \frac{L \cdot TPI}{2}$$

Note that here the overall length of the part *L* is used, not the length of the tapered section of the work. Here is an example:

EXAMPLE: Calculate tailstock setover if TPI is 0.050 and the overall length of the part is 11″.

$$Setover = \frac{L \cdot TPI}{2}$$

$$= \frac{11 \cdot 0.050}{2}$$

$$= \frac{0.55}{2} = 0.275$$

Once setover is calculated, the tailstock of the lathe must be moved. There are a few different methods of setting the offset amount, but using a plunge-type dial indicator is probably the most widely used.

- Mount a dial indicator on the cross slide with its travel direction parallel to cross-slide movement.
- Position the indicator near the tailstock quill.
- Move the cross slide to contact the tailstock quill and preload the indicator with about 1/4 turn of pressure.
- Zero the indicator.
- Adjust the tailstock until the indicator shows movement equal to the setover amount. **(See Figure 5.5.22.)**
- Tighten the tailstock adjusting screws and recheck the indicator reading. It may take a few minor adjustments while locking the adjusting screws to maintain the correct position.

Machining

Once tailstock setover is complete, the work can be mounted between centers. When turning long parts, spindle speeds often need to be reduced drastically to eliminate vibration and chatter because of the long, thin, unsupported workpiece.

- Position the cutting tool near the beginning of the cut.
- Advance the tool with the cross slide to set cut depth.
- Engage the power feed to take a cutting pass. Remove just enough material so that the taper can be checked for accuracy.
- Additional adjustment is often needed to the tailstock offset and several trial cuts may be required to produce the correct taper. Expect that this can be a time-consuming process.

FIGURE 5.5.22 Setting tailstock offset with a dial indicator. (A) Zero the indicator against the tailstock quill. (B) Then adjust the tailstock until the indicator displays the offset amount.

SUMMARY

- A taper is a constant rate of diameter change on a cylindrical part.
- Common standard tapers in the machining industry are used for aligning and securing mating parts on machine tools and can be self-holding or self-releasing.
- Tapers can be specified by several different methods.
- Becoming familiar with the use of formulas and math calculations is important to understand taper specifications and how to machine tapers.
- Taper turning can be performed by some different methods depending on the configuration of the workpiece and available machines, tools, and accessories. Each of these methods has its benefits and drawbacks that need to be considered in order to efficiently produce a taper within required specifications.
- Machining of tapers can be a time-consuming and involved process and requires patience and practice to become proficient at performing this important machining task.

REVIEW QUESTIONS

1. Briefly define a taper.
2. What is the difference between an included angle and a centerline angle?
3. What does TPI stand for in relation to tapers?
4. What are the TPI and TPF of a part with end diameters of 3/4" and 1/2" and a taper length of 4.25"?
5. What are the corresponding centerline and included angles to 5/8" TPF?
6. What is the corresponding centerline angle of a part with TPI of 0.070"? Round your answer to the nearest minute.
7. What is the limitation of the tool bit taper turning method?

(Continued)

8. What must be known to use the compound-rest taper turning method?

9. What taper turning methods allow use of the lathe's power feed?

10. What two steps can be taken to eliminate backlash when using the taper attachment?

11. The TPI specified on a print is 0.030". If checking a taper attachment setup with a cross-slide-mounted dial indicator, how much movement should register on the indicator if the carriage is moved 1" in the cutting direction?

12. If TPF is 0.42", how much movement should register on the cross slide's digital readout in radius mode if the carriage is moved 1" in the cutting direction?

13. What is the benefit of using the offset tailstock method for turning tapers?

14. What are two ways to reduce uneven pressure on lathe centers when using the offset tailstock method?

15. Calculate tailstock setover for a 13.5" part with a TPF of 0.27.

MILLING SECTION 6

- Unit 3
  Vertical Milling Machine Operations
  Introduction
  General Milling Machine Safety
  Tramming the Vertical Milling
  Machine Head
  Aligning Workholding Devices
  Speeds and Feeds for Milling
  Operations
  Holemaking Operations
  Milling Basics
  Squaring a Block
  Angular Milling
  Milling Steps, Slots, and Keyseats
  Milling Radii
  Pocket Milling

- Unit 4
  Indexing and Rotary Table Operations
  Introduction
  Parts of the Rotary Table
  Rotary Table Setup
  Rotary Table Operations
  The Indexing Head
  Indexing Head Operations

Introduction to the Vertical Milling Machine

UNIT 1

Learning Objectives

After completing this unit, the student should be able to:

• Identify the components of the vertical milling machine

• Explain the function of the components of the vertical milling machine

Key Terms

Drawbar	**R-8 taper**	**X-axis**
Knee	**Ram**	**Y-axis**
Leadscrew	**Saddle**	**Z-axis**
Quill	**Turret**	

INTRODUCTION

Conventional, or manual, milling machines are primarily used to machine flat and angled surfaces by feeding a workpiece into a rotating cutting tool to remove material. They are also commonly used to position work more accurately for the same type of holemaking operations than can be accomplished with a drill press. By combining these operations, components can be machined to countless desired shapes. **Figure 6.1.1** shows some examples of parts produced on milling machines.

The vertical spindle milling machine (frequently called the *knee mill* or just the *mill*) is widely used in many machining careers in many different industries. Movements of the vertical milling machine are often identified by the Cartesian coordinate system using **X-, Y-,** and **Z-axes**, as shown in **Figure 6.1.2**.

FIGURE 6.1.1 Some sample parts produced by milling operations.

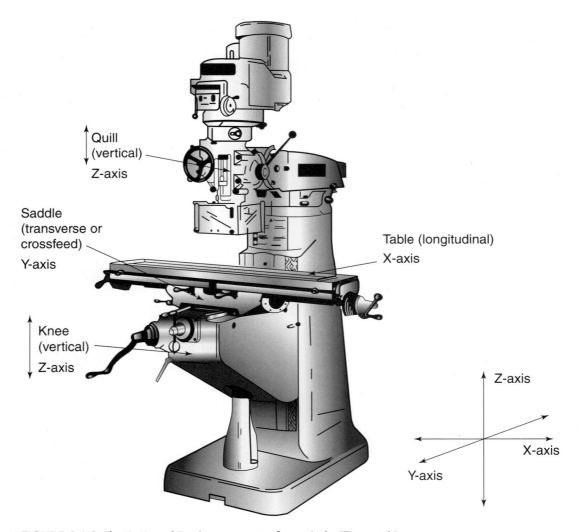

Quill
(vertical)
Z-axis

Saddle
(transverse or
crossfeed)
Y-axis

Table (longitudinal)
X-axis

Knee
(vertical)
Z-axis

Z-axis

X-axis

Y-axis

FIGURE 6.1.2 The X-, Y-, and Z-axis movements of a vertical milling machine.

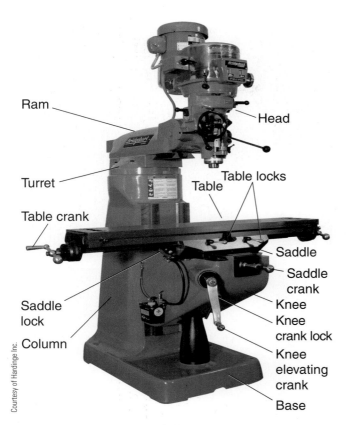

Courtesy of Hardinge Inc.

FIGURE 6.1.3 The major parts of a typical vertical milling machine.

Figure 6.1.3 shows an example of a typical vertical milling machine and labels its major parts. Refer to this figure while reading about the various parts of the vertical mill. This unit will provide an explanation of the parts of the vertical milling machine and their functions—the first step in learning to safely operate the vertical mill to efficiently perform machining operations.

BASE AND COLUMN

The *base and column* portion of the vertical mill is a single cast-iron unit that provides a heavy, solid base for the machine. Most modern-day vertical milling machine bases and columns are manufactured by a process known as meehanite casting. This process creates a cast-iron base that has a very uniform composition and is highly wear resistant. The *turret* and *ram* are mounted on top of the column. The *knee* is fitted to a vertically oriented dovetail machined into the front face of the column.

KNEE

The **knee** is heavy casting with a dovetail slot at its rear. This slot is used to attach the knee to the column portion of the mill. The dovetail slide also allows the knee to be raised and lowered as needed. The knee is both supported

FIGURE 6.1.4 The adjustable micrometer collar of the knee-elevating crank allows accurate movements of the knee.

and moved by a heavy *elevating screw* mounted inside the knee. The elevating screw is attached to the *elevating crank* by a geared mechanism, so when the crank handle is turned, the elevating screw rotates and raises or lowers the knee. Turning the handle clockwise raises the knee, while turning it counterclockwise lowers the knee.

The elevating crank contains an adjustable micrometer collar (**Figure 6.1.4**) so that the amount of knee movement can be accurately controlled. The locking collar can be loosened and the micrometer collar rotated to set a "0" reference. Then the locking collar can be retightened and the knee moved the desired amount. On most vertical milling machines, each graduation on the micrometer collar equals 0.001″ and one full turn of the crank moves the knee 0.100″. The knee of the vertical mill provides movement along the Z-axis in the Cartesian coordinate system.

A *clamping lever* locks the elevating crank in place after positioning the knee, and two additional clamps can be used to more rigidly secure the knee to the column. **(See Figure 6.1.5.)** Always be sure that all of these clamps are

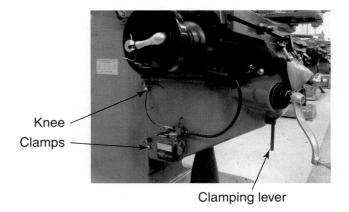

FIGURE 6.1.5 Clamps for securing the knee to the column.

released before raising or lowering the knee to prevent damage to the elevating crank mechanism and dovetail slide.

Saddle

The **saddle** is mounted on another machined dovetail on top of the knee. The saddle permits movement toward and away from the column along the Y-axis. A nut inside the saddle is attached to another heavy screw called a **leadscrew**. The saddle crank handle located on the front of the knee turns this leadscrew. Turning this handle clockwise moves the saddle toward the column, while turning it counterclockwise moves the saddle away from the column. This handle also has an adjustable micrometer collar so that the amount of saddle movement can be accurately controlled. On most vertical mills, each collar graduation is 0.001" and one rotation of this handle moves the saddle 0.200". After positioning the saddle, the *saddle lock* can be used to secure it in place to prevent unwanted movement. **(See Figure 6.1.6.)** Be sure to release this lock before moving the saddle to prevent damage to the dovetail slide or leadscrew mechanism.

Table

The *table* is mounted on another machined dovetail on the top of the saddle. The table allows movement from left to right along the X-axis. As with the knee and saddle a leadscrew provides the table movement. This leadscrew is turned by one of two crank handles located at each end of the table to move the table to the left or right. Turning either handle clockwise moves the table away from the operator position, while turning them counterclockwise moves the table toward the operator position. **(See Figure 6.1.7.)** These handles also have adjustable micrometer collars like the saddle handle to accurately control the amount of table movement. They will also normally have 0.001" graduations and move the table 0.200" for each rotation of the handle. Two locks on the front of the saddle can be used to lock the table in place. **(See Figure 6.1.8.)** Be sure to release these locks before moving the table to prevent damage to the dovetail slide and screw mechanism.

The table provides a flat reference surface used for locating workpieces for machining operations. T-shaped slots machined in the table accommodate clamping equipment used to secure workpieces or workholding devices for machining. Every effort should be made to protect the surface of the table from damage. When installing heavy devices, such as mill vises, they should always be gently placed on the table. Do not place cutting tools, hammers, files, wrenches, or other rough tools on the machine table. Always use some type of protector, such as a plastic tray or wooden board, to hold

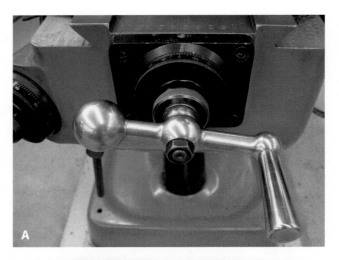

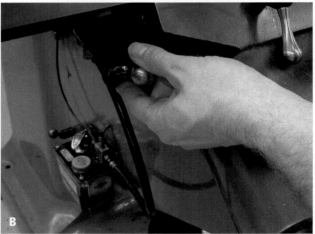

FIGURE 6.1.6 (A) The saddle crank moves the saddle back and forth. Note the adjustable micrometer collar on the handle. (B) After positioning, the saddle lock secures the saddle to prevent unwanted movement.

FIGURE 6.1.7 Table crank handles move the table left and right. Note the adjustable micrometer collar on the handle.

FIGURE 6.1.8 Table locks secure the table to prevent unwanted movement.

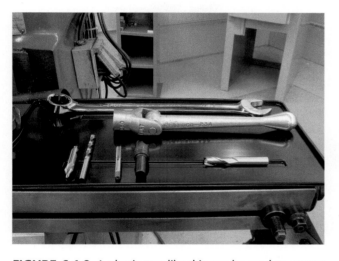

FIGURE 6.1.9 A plastic tray like this can be used to protect the surface of the machine table from rough tools and parts.

tools and parts while machining. This will help protect the table from being scarred, chipped, and damaged. **(See Figure 6.1.9.)**

TURRET

The top of the column casting is a machined flat surface. The **turret** rests on this surface and allows the entire machine head to be swiveled 360 degrees. A protractor on the turret is graduated in degrees and contains a zero mark to position the head in the center of the column. Loosening the clamping bolts allows the turret to be swiveled by pushing on the head. When in position, the clamps are tightened. **(See Figure 6.1.10.)**

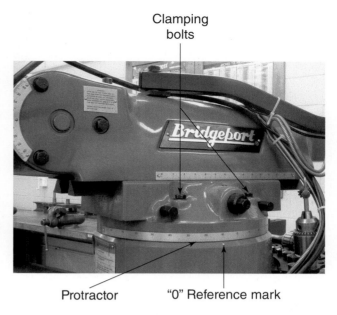

Clamping bolts

Protractor   "0" Reference mark

FIGURE 6.1.10 The turret of the vertical milling machine can be rotated. Note the protractor and clamping bolts.

 CAUTION

Always be sure the turret clamping bolts are tightened before beginning any machining operation.

The turret should never be moved during a machining operation. The force of a rotating cutting tool can cause a loose turret to move and violently pull workpieces from the machine and shatter cutting tools, causing serious injury.

RAM

The **ram** allows the entire head to be moved forward and backward and then locked in position. This movement is provided to increase the workpiece capacity of the machine. The bottom of the ram contains a dovetail that is fitted to a mating dovetail slot in the top of the turret. The dovetail ensures that the head moves back and forth accurately in a straight line. The ram also contains a rack and pinion gear system. Rotating the *ram adjusting lever* or *nut* moves the ram forward or backward. To move the ram, first loosen the clamping bolts, then rotate the adjusting lever. When in position, retighten the clamps. **(See Figure 6.1.11.)**

Clamping bolts

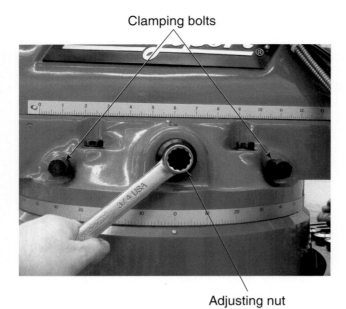

Adjusting nut

FIGURE 6.1.11 The ram adjusting nut and clamping bolts.

CAUTION

As with the turret, always be sure the ram clamping bolts are tightened before beginning any machining operation, or serious injury can result. The ram should never be moved as a machine slide during machining operations.

HEAD

The head of the vertical mill contains the mechanisms for holding and driving cutting tools. Its basic construction and components are similar to the head of a drill press but with added features. **Figure 6.1.12** shows a typical vertical milling machine head and labels its parts.

Spindle

The *spindle* is a precisely ground shaft. A hole passes through its middle to accommodate various cutting-tool-holding devices and the **drawbar**. The threaded drawbar passes through the spindle from the top of the head. These threads provide a means of securing tools such as drill chucks and collets to the spindle. The top of the drawbar contains a hex for tightening the drawbar into the toolholding device with a wrench. The inside diameter of the lower end of the spindle is tapered for accurately centering toolholding devices in the spindle. Most modern vertical milling machines utilize a standard taper known as an **R-8 taper**. A small key inside the spindle acts as a way to both align toolholding devices

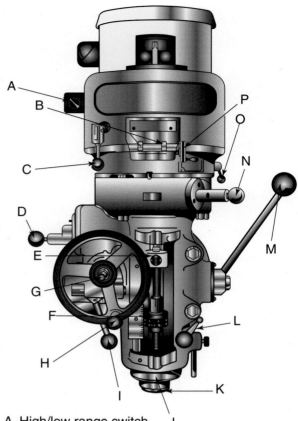

A. High/low range switch
B. Variable speed dial
C. Spindle brake
D. Quill feed selector knob
E. Quill stop
F. Micrometer adjusting nut
G. Feed reverse knob
H. Manual feed handwheel
I. Feed control lever
J. Quill
K. Spindle
L. Quill lock
M. Quill feed handle
N. Power feed transmission engagement crank
O. High/neutral/low lever
P. Speed change handwheel

FIGURE 6.1.12 The parts of a typical vertical milling machine head.

inside the spindle and aid in driving them. **Figure 6.1.13** illustrates the main parts of the spindle.

Setting Spindle Speed

A motor on top of the head rotates the spindle. Most vertical mills have a *high/neutral/low range lever* (**Figure 6.1.14**) that controls a small transmission gear train inside the head. This lever should be placed in the range that corresponds with the desired RPM. Most vertical milling machine speed ranges are 60 to 500 RPM in low range and 500 to 4000 RPM in high range. The neutral position allows the spindle to be rotated freely by hand, which can be helpful during machine setup. It is very important to remember that this lever should only be moved when the spindle is not rotating. Adjusting this lever with the spindle running will cause damage to the machine head's powertrain.

Many older milling machines are equipped with a step cone belt pulley system. On mills of this type the spindle RPM is set by positioning the belt in different locations on the pulley.

! CAUTION

When manually changing belt positions on a machine with step cone pulleys, turn off the main power to the vertical mill. Never attempt to change these belt positions with the spindle rotating.

•

Modern mills are equipped with a variable-speed belt drive. To set the RPM on this type of mill, first select the desired range (high or low) and turn on the spindle. Then a dial located on the front of the head is rotated until the desired RPM is reached. **(See Figure 6.1.15.)** It is important to remember that the variable speed adjustment dial only be turned when the machine is running.

Spindle Brake

The *spindle brake lever* can be slightly rotated forward or back to quickly stop the spindle. It can also be pulled out to lock the spindle in place. It is important to remember to release the lock before starting the spindle to avoid excessive wear on the spindle brake mechanism. This brake will only function when the spindle is in low or high range, not when in neutral. **Figure 6.1.16** shows the use of the spindle brake.

! CAUTION

If an emergency requires the spindle to be stopped immediately, the brake can be engaged prior to shutting off the spindle power. This does cause extra strain on the brake and motor, but that extra wear is more acceptable than the result from not stopping the spindle when an emergency arises.

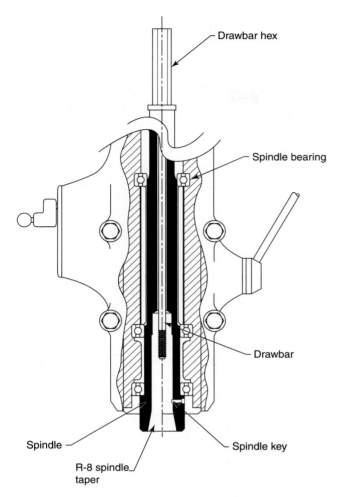

FIGURE 6.1.13 The parts of a typical milling machine spindle. Note the drawbar, which tightens the toolholding device.

Drawbar hex

Spindle bearing

Drawbar

Spindle

Spindle key

R-8 spindle taper

FIGURE 6.1.14 Vertical mill high/neutral/low gear-change lever.

SPEED RANGE
HI   NEUTRAL   LO

FIGURE 6.1.16 The milling machine spindle brake can be used to quickly stop spindle rotation or to lock the spindle.

FIGURE 6.1.15 The vertical mill variable-speed control. (A) Rotating the hand wheel changes spindle RPM, (B) which is shown in the "window" of the variable-speed dial.

Quill

The **quill** on the vertical milling machine is much like the quill on the drill press. The *quill feed handle* can be used when performing holemaking operations or to position cutting tools for milling operations. A spring inside the head balances the weight of the quill and helps it to stay where positioned with the feed handle. The *micrometer adjusting nut* can be set to limit quill travel by moving it so the *quill stop* comes in contact with the nut. Graduations on the adjusting nut along with those on the front of the head allow for accurate positioning.

FIGURE 6.1.17 The micrometer adjusting nut can be set so that the quill stop limits quill travel.

(See Figure 6.1.17.) The *quill lock* can be used to lock the quill in an extended position by pulling down on the lock as shown in **Figure 6.1.18**. The quill of the vertical mill provides additional movement along the Z-axis (in addition to the knee).

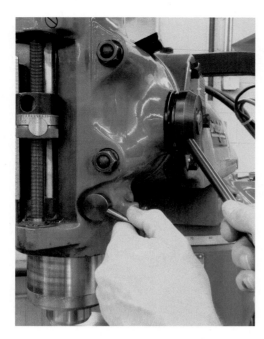

FIGURE 6.1.18 The quill lock secures the quill in an extended position.

Power Quill Feed

When performing holemaking operations the quill can be set to feed under power. First, the *power feed transmission engagement crank* located near the upper-right portion of the head must be engaged. **(See Figure 6.1.19.)** Only engage and disengage the crank when the spindle is not running. To move the crank, pull the knob out slightly, then rotate it to the engage or disengage position. Most vertical milling machines offer three feed rate settings for quill feed. The *quill feed selector knob* on the left side of the head can be positioned to select 0.0015", 0.003", or 0.006" IPR (inches per revolution). **(See Figure 6.1.20.)** These numbers indicate how far the quill will advance every time the spindle makes one revolution. Changing the quill feed selector knob should also only be done when the spindle is not running.

Figure 6.1.21 shows the *feed reversing knob* that is used to set quill feed direction for either upward or downward motion. Pushing the small knurled sleeve all the way in will set the quill to feed in a downward direction. Pulling the sleeve all the way out will set the quill to feed in an upward direction. The middle setting is a neutral position. This knob may be repositioned with the spindle running so that the slow-turning gears will mesh.

The *feed control lever* is located on the left side of the head and is used to start the quill feed. Before beginning quill feed, the spindle must be turned on. The quill will begin to feed when the clutch is engaged by pulling the handle out. **(See Figure 6.1.22.)**

FIGURE 6.1.19 The power feed transmission engagement crank engages the quill power feed gear-train mechanism.

When using downward power quill feed, the micrometer adjusting nut can be used to automatically disengage the feed. When the quill stop makes contact with the micrometer adjusting nut, the feed control lever will automatically disengage and quill feed will stop. When feeding upward, the feed control lever will automatically disengage when the quill stop makes contact with a small web on the head casting. **(See Figure 6.1.23.)**

FIGURE 6.1.20 The quill feed selector knob is used to set the quill feed rate in IPR (inches per revolution).

FIGURE 6.1.21 The feed reversing knob sets the direction of quill feed. Pushing in sets downward feed and pulling out sets upward feed. There is also a neutral position between the two settings.

Head Movements

The vertical mill head can be rotated to machine angular surfaces or to produce angled holes. It can be rotated a full 360 degrees in a clockwise or counterclockwise direction when viewed from the front of the machine. Four clamping bolts on the front of the head are loosened, and then an adjusting screw is turned to rotate the head in the desired direction. A protractor aids in positioning the head at the desired angle. The four clamping bolts are then retightened. **(See Figure 6.1.24.)**

The head can also be tilted 45 degrees forward and 45 degrees backward. To tilt the head, first loosen the

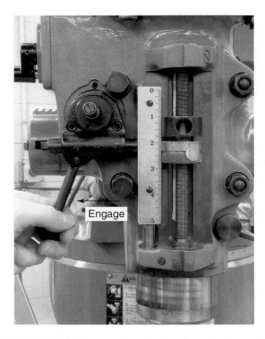

FIGURE 6.1.22 Pulling out on the feed control lever begins power quill feed.

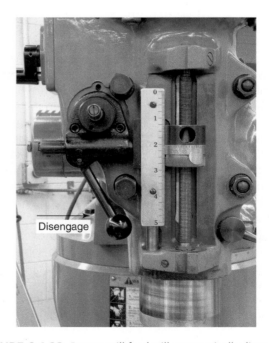

FIGURE 6.1.23 Power quill feed will automatically disengage when the quill stop reaches the micrometer adjusting nut.

three clamping bolts on the side of the head. Then turn an adjusting screw to tilt the head in the desired direction. Another protractor aids in positioning the head at the desired angle in this direction. After the head is positioned to the desired angle retighten the three clamping bolts. **(See Figure 6.1.25.)**

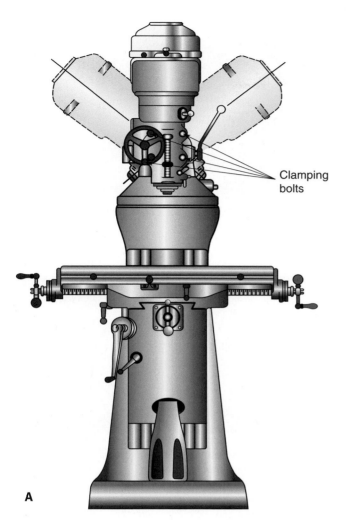

Clamping
bolts

OPTIONAL FEATURES

In addition to the standard features previously described, vertical milling machines may be equipped with other devices designed to increase efficiency. One such item is a digital readout, or DRO, as shown in **Figure 6.1.26**. The digital readout is frequently used in place of the micrometer collars.

When table or saddle direction is reversed, backlash can cause errors in positioning when using only micrometer collars. For example, suppose the saddle micrometer collar is set to "0" after being rotated clockwise. So long as all motion is in the same direction, movement can be accurately determined using the micrometer collar. When the direction of travel is reversed, backlash between the leadscrew and nut allows the saddle handle to be moved slightly before there is any actual table movement. Since the micrometer collar turns with the handle, this causes errors in measuring movement. A digital readout is not affected by backlash since it displays only the true amount of movement on the screen. The digital readout has become an indispensable piece of equipment on the milling machines used today. Many DROs can be programmed to aid in creating bolt hole circles and calculating trigonometric function values.

A

Clamping bolts

Adjusting   "0" Reference
screw   mark   Protractor

B

C

FIGURE 6.1.24 (A) The milling machine head can be tilted both right and left. (B) After loosening four clamping bolts on the front of the head, it can be (C) tilted by turning the adjusting screw. A protractor helps to obtain the desired setting.

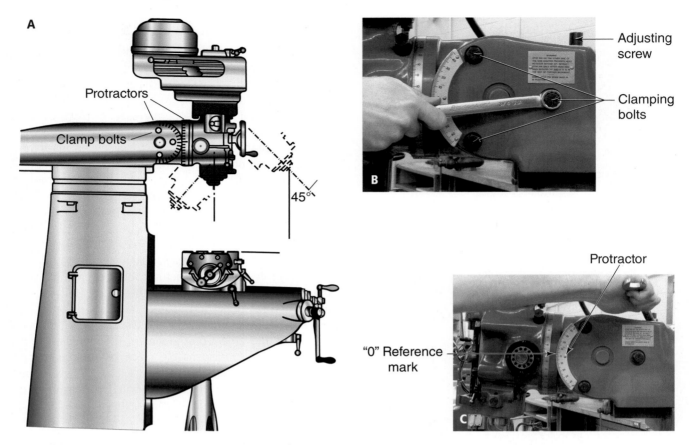

FIGURE 6.1.25   (A) The head can also be tilted forward or back. (B) After loosening three clamping bolts on the side of the ram, the head is tilted by turning the adjusting screw. (C) The protractor helps to obtain the desired setting.

FIGURE 6.1.26 A digital readout (DRO) displays only actual machine movement, making positioning easier and more accurate.

A power feed unit is another device that is often added to a vertical mill. The power feed devices make it easier and faster to move the machine table compared to manually turning handles and cranks. **Figure 6.1.27** shows a power feed unit installed on the table movement

FIGURE 6.1.27 A power feed unit attached to the table movement of a vertical mill.

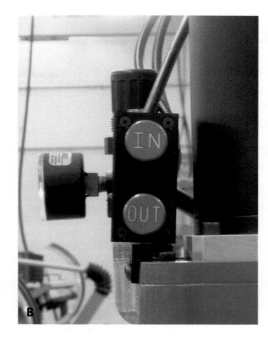

FIGURE 6.1.28 The power drawbar on a vertical milling machine allows fast mounting and removal of cutting tools by simply pressing the IN and OUT buttons.

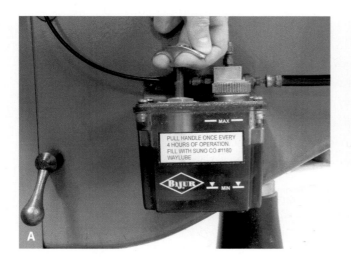

FIGURE 6.1.29 (A) Pulling the handle of a one-shot oiling system (B) pumps oil through a series of lines to lubricate moving machine parts.

(X-axis) of a vertical mill. It is important to remember to loosen the table locks before engaging power feed units. This will help to prevent excessive wear to the machine screws and dovetail slides.

A power drawbar can be used to mount and remove toolholding devices much more quickly than manually tightening and loosening the drawbar. A button located on the head automatically activates an air driven mechanism to tighten or loosen the drawbar. **Figure 6.1.28** shows a power drawbar unit.

A one-shot lubrication system is another feature that is included on most milling machines. **(See Figure 6.1.29.)** The one shot system contains a reservoir that holds a special lubricant commonly called way oil. When the handle on the one-shot system is pulled, a measured amount of oil is pumped from the reservoir through several oil lines to lubricate the various slides and working components of the machine. The one-shot system lever should be pulled at the beginning of each day to adequately lubricate machine surfaces.

SUMMARY

- Milling machines are primarily used to produce flat and angled surfaces and to perform holemaking operations.
- The vertical milling machine offers several options for positioning the head and workpiece to perform these machining operations.
- Workpieces can be mounted to the table and positioned by moving the knee, saddle, and table.
- The milling machine head can be tilted and rotated to many different angular positions to suit different situations.
- The milling machine spindle utilizes an internal R-8 taper and drawbar to align, hold, and rotate cutting tools required to perform different machining operations.
- The quill can be used to position cutting tools for milling or to perform holemaking operations using manual or power feed.
- Optional accessories such as DROs, power feed units, and power drawbars can increase efficiency when operating a vertical mill.
- A one-shot lubrication system provides oil to several points on the milling machine.

REVIEW QUESTIONS

1. What part of the vertical milling machine allows workpieces to be raised and lowered?
2. Briefly describe the direction of movement provided by the saddle.
3. When a table handle is turned clockwise the table moves _____ the operator's position.
4. What distance does the saddle or table usually move when a handle is rotated one full turn?
5. List three items that should not be placed on a milling machine table.
6. Briefly describe the function of the turret and the ram of a vertical milling machine.
7. What is the name of the taper found in most modern vertical milling machines?
8. What part of the vertical mill can be raised and lowered to position cutting tools or used to feed tools during holemaking operations?
9. How can power quill feed be automatically stopped when feeding in a downward direction?
10. What is a benefit of using a DRO instead of micrometer collars for positioning on a vertical milling machine?

Tools, Toolholding, and Workholding for the Vertical Milling Machine

UNIT 2

Learning Objectives

After completing this unit, the student should be able to:

- Identify and demonstrate understanding of various cutting tools used on the milling machine
- Demonstrate understanding of carbide inserts
- Identify and demonstrate understanding of various toolholding devices used on the milling machine
- Identify and demonstrate understanding of various workholding devices used on the milling machine

Key Terms

Arbor	**Endmill**	**Fly cutter**
Ballnose endmill	**Face mill**	**Shell endmill**
collet block	**Fixture**	**Stub arbor**

INTRODUCTION

Before any machining process may begin on the milling machine, the cutting tools must be properly selected and mounted and the workpiece must be properly secured. All of the holemaking tools that were covered in previous sections may also be used in the vertical milling machine. The workpiece and cutting tool must be very securely mounted in a milling machine to prevent injury to the operator during heavy machining and so that desired specifications can be achieved.

CUTTER SHANKS AND ARBORS

Milling cutting tools may be mounted either with **arbors** or directly by their shanks. Some tools have a shank similar to that of a straight-shank drill bit that can be directly clamped in the spindle with a toolholder or collet. Other tools have a mounting hole so they can be mounted on an arbor to secure the cutting tool. **(See Figure 6.2.1.)** The arbor is then mounted in the spindle of the mill. Remember from Unit 1 that most manual vertical milling machines have a special spindle taper called an R-8 taper. **Figure 6.2.2** shows an R-8 taper shank. The toolholding devices must have this taper shank in order to be mounted into the machine spindle.

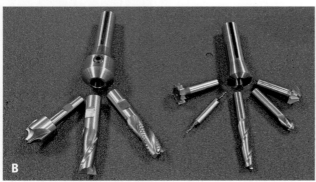

FIGURE 6.2.1 (A) Some arbor-type cutters and mating arbors for use in a vertical milling machine. (B) Some shank-type cutters and holders for use in a vertical milling machine.

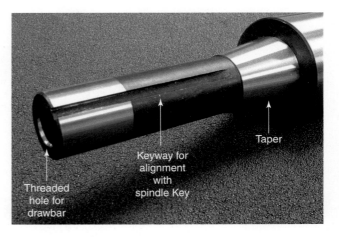

FIGURE 6.2.2 An R-8 taper shank.

CUTTING-TOOL MATERIALS

Milling cutters may be made from one solid piece of HSS or carbide, or from a steel shank with replaceable cutting edges called inserts. Inserted cutters come in a variety of shapes and sizes and consist of a cutter body that is usually constructed of mild steel and disposable inserts. The cutter body has a series of pockets machined around its periphery. Inserts fit in these pockets and are secured with mounting screws. Since all the cutting action is performed by the very hard insert portion of the tool, high cutting speeds can be used, making inserted cutters very efficient cutting tools. Inserts can be made from very hard materials such as carbide, ceramic, or diamond. Carbide is most common, and will be used for examples in this text. **(See Figure 6.2.3.)**

When a solid endmill wears as it is used, it must be reground with a special grinder to produce sharp new cutting edges. Worn or damaged inserted endmills may be renewed by simply replacing the inserts or rotating them to a new cutting edge.

Carbide Inserts

Insert characteristics are tailored to produce high performance results in any given machining situation. The basic identification system for general purpose carbide inserts has been standardized by American National Standards Institute (ANSI). That system identifies inserts based on their shape, clearance angle, tolerance during manufacture, insert features (related to how the insert is mounted on a cutter body), size, thickness, and nose radius/cutting point. Additional information may be included to identify handedness of cutter rotation cutting edge preparation, facet size, and chipformer style. **Figure 6.2.4** shows a helpful chart for decoding any insert identified using this system. A sample insert designation is listed at the top of the chart for reference.

FIGURE 6.2.3 Solid carbide and inserted carbide cutting tools for milling.

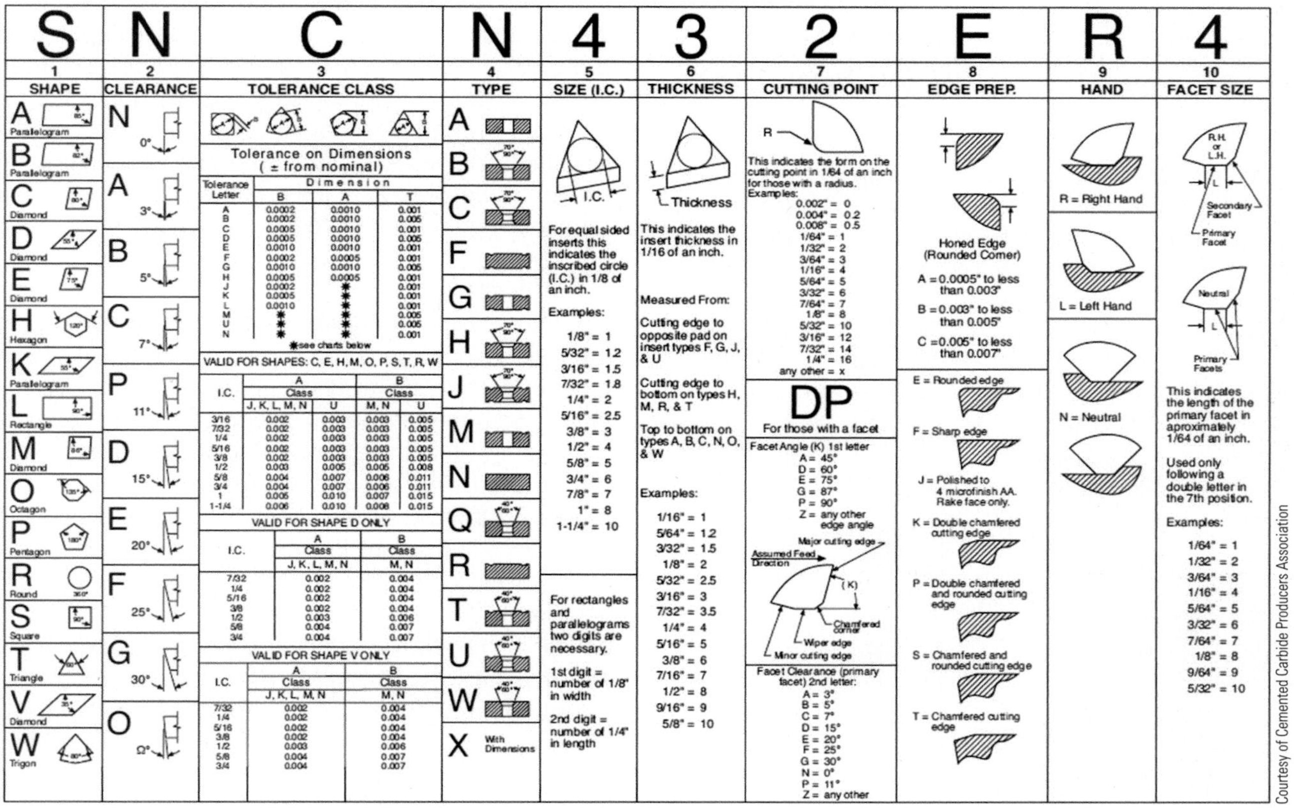

FIGURE 6.2.4 The ANSI carbide milling insert identification system.

Courtesy of Kennametal, Inc.

Courtesy of Cemented Carbide Producers Association

Shape

The first identifying characteristic of the insert is the shape. This is designated by a letter in the first position of the identification in **Figure 6.2.4**. An insert's shape determines not only the type of part features that the tool can access but also the strength of the insert and the amount of cutting force between the work and the tool. Notice the effect of insert shape on strength and machining forces

CARBIDE INSERT SHAPES

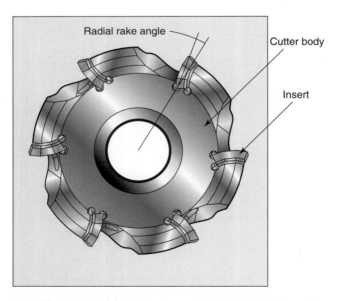

Strong ← — — — — — — — — → Weak

High force ← — — — — — — — → Low force

FIGURE 6.2.5 Various insert shapes are available. The shape of an insert has an effect on its strength and the forces created during machining.

in **Figure 6.2.5**. The most common shapes for milling are round, square, rectangular, octagonal, hexagon, trigon, and triangular. The insert identification system uses a letter to denote each insert shape.

Side Clearance Angle

The side clearance angle designated in position two of **Figure 6.2.4** refers to the amount of relief built into the insert behind the cutting edge. This clearance angle allows radial rake to be built into the cutter without the side of the insert rubbing as it is cutting. **(See Figure 6.2.6.)** It is important to note that the overall clearance may be either built into the insert, provided by the cutter body, or a combination of both. The clearance angle on the insert determines the angle that the insert needs to be mounted in a cutter body. Zero-degree inserts must be used in cutters that tilt the insert in a negative radial rake direction, so that clearance is provided between the cutting tool and the workpiece. Positive clearance inserts (generally those with

FIGURE 6.2.6 The radial rake angle is built into the cutter body. A positive radial rake is shown. Notice that the insert must have an adequate clearance angle to prevent rubbing.

angles greater than 3 degrees) may be held in-line with the tool axis (in a neutral tool body) or tilted at a positive radial rake direction. Inserts with a zero-degree angle are stronger and have the added benefit of having double the number of cutting edges, since they may be flipped and used on either side. Zero-degree inserts generally consume more power and create more cutting force, requiring more spindle horsepower and more rigidity.

Tolerance

Tolerance is designated by a letter in the third position of **Figure 6.2.4**. Inserts are produced to dimensional tolerances like any other manufactured item. These tolerances ensure that the user can expect a certain amount of repeatability when the insert is indexed or it is replaced with another of the same type. Inserts made to tighter tolerances are more carefully produced and cost more.

Insert Features

The fourth position of **Figure 6.2.4** specifies the way that the insert is fastened to the toolholder and the placement of the chipformer. Inserts may be fastened through a hole in their center, retained by a screw head, clamped with a wedge, or held in place under spring tension. **Figure 6.2.7** shows illustrations of various fastening systems. Chipformers are carefully designed geometry formed into the surfaces of the insert. Their intent is to create a specific chip formation so that the chip curls tightly and breaks into small manageable pieces. Inserts are available with a chipformer on one side, both sides, or not at all.

Insert Size

The fifth position of **Figure 6.2.4** specifies one factor of insert size. Since inserts come in such a variety of shapes, a universal method was developed to define their size no matter what their shape. By determining the largest circle that can fit within the edges of the insert, all insert shapes can be sized. This size is called the *inscribed circle (I.C.)*. **Figure 6.2.8** shows an *inscribed circle* on two different insert shapes. Inch series inscribed circle diameters are denoted as the number of 1/8ths of an inch. For example, an insert with a size designation of "3" can fit a 3/8" diameter inscribed circle within its edges. Area number 5 in **Figure 6.2.4** shows the size of the inscribed circle. A circle would not inscribe properly in rectangular inserts, so two digits are given for these. The first number represents the number of 1/8ths of an inch of width, and the second number represents the number of 1/4ths of an inch in length.

Insert Thickness

The thickness of an insert is noted in position six of **Figure 6.2.4.** The thickness of an insert is expressed in terms of the number of 1/16ths of an inch. For example, an insert with a size designation of "2" measures 2/16" or 1/8" (2/16 fraction reduced to simplest terms) thick.

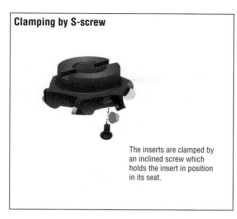

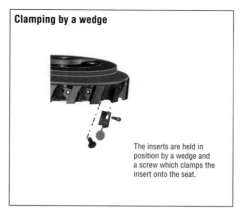

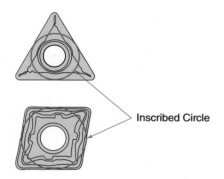

FIGURE 6.2.7 Several different methods for mounting inserts to milling cutter bodies.

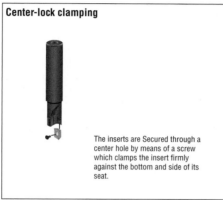

FIGURE 6.2.8 The inscribed circle is shown on two different insert shapes.

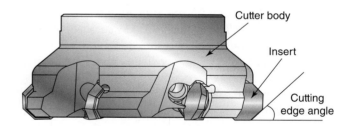

FIGURE 6.2.9 The cutting edge angle of an insert is shown.

Tool Nose Radius/Cutting Point

The radius on the tool tip helps provide strength to the insert and affects the surface finish and cutting force. The seventh position of **Figure 6.2.4** shows the codes for either the nose radius or cutting angle on the side of the insert. Optional notations for axial or end-clearance angles can also be listed here. Cutting angle is the angle the leading edge of the insert forms with the workpiece as shown in **Figure 6.2.9**. Axial insert clearance allows the insert to be tilted in the cutter body to provide a positive rake angle as shown in **Figure 6.2.10**.

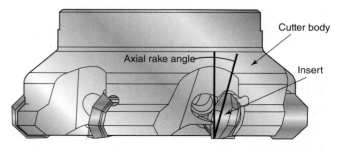

FIGURE 6.2.10 A positive axial rake angle is shown.

Manufacturer's Options

Positions eight through ten of the identification system provide optional space for the manufacturer to make special notes about additional information. **(See Figure 6.2.4.)**

Chip Formation

For most situations, it is strongly preferred that the chips produced by the cutting tool break into small, manageable pieces. The cutting angle, depth of cut, tool nose radius, feed rate, and insert chipformer must all be working together to control chip size and formation.

For proper chip breaking, surface texture, and tool life, the depth of cut should always be at least 2/3rds of the tool nose radius. For example, if this rule is followed, a 1/32" tool nose radius should not be used to take a depth of cut less than 0.021". By increasing the feed rate, the chip becomes thicker. As chip thickness increases, chips are more likely to break into pieces instead of forming continuous strings.

To further help with chip control, carbide manufacturers invest a great deal of research and development into creating a working surface on inserts to further refine the formation of chips. These chipformer geometries may sometimes appear more like ornate pieces of art, but they are scientifically refined patterns defined for specific cutting conditions. Inserts are available with chipformers designed for specific material types, hardness conditions, and whether roughing or finishing is to be performed.

Insert Grade

Generally an insert made of hard carbide will sacrifice toughness, and a tough insert will sacrifice hardness. These properties make up the insert *grade* and must be matched to each application. Workpiece material, the operation type (finishing, medium roughing, or heavy roughing), and machining conditions (heat treatment, rigidity, interruptions, chip evacuation, cutting fluid supply, etc.) all influence the selection of the proper carbide grade. **Figure 6.2.11** shows the standardized ISO identifications for a various workpiece material types. Within each ISO workpiece material code, many grades are offered ranging from tough (softer) to wear resistant (harder). Tougher grades can handle more impact from interruptions in the cutting action. Wear-resistant grades last longer and stand up better to harsh materials. It is important to note that most carbide manufacturers have devised their own proprietary grade designations, and it is usually best to refer to the applications catalog for each brand.

PROPER CUTTING-TOOL STORAGE

Milling cutters should always be stored in a manner that prevents them from contacting each other. This keeps the sharp cutting edges from becoming dull or chipped.

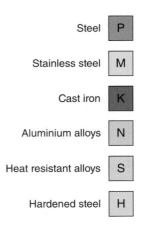

Steel	P
Stainless steel	M
Cast iron	K
Aluminium alloys	N
Heat resistant alloys	S
Hardened steel	H

FIGURE 6.2.11 The standardized ISO identifications for a various workpiece materials are shown.

If stored in a tool chest drawer, the drawer should be lined with a plastic, rubber, felt, or wood bottom, and dividers should be used to keep the tools from rolling into each other. The plastic or cardboard containers used to ship many cutters also provide good protection.

ENDMILLS

At first glance, a solid **endmill** with helical flutes resembles a twist drill bit. Endmills have cutting edges both on their end and on their periphery for milling. Endmills are useful when machining a wide variety of features such as pockets, slots, keyways, and steps. Flat general-purpose endmills have a straight outside diameter and a flat face on their ends. Specially designed endmills are also available for performing milling functions such as *roughing* (removing large amounts of material) and creating convex radii, concave radii, T-slots, woodruff keyseats, and dovetails.

Endmills are available in standard fractional inch and metric diameters and come in a variety of flute counts and styles. Since the flutes form the cutting edges, the number of flutes is the same as the number of cutting edges. The most common flute counts are two and four, although many others are available for special purposes. **(See Figure 6.2.12.)** Selection of the cutting-tool flute

FIGURE 6.2.12 Endmills with two, three, four, and six flutes.

number depends on the material to be cut and the cutting conditions. Since each cutting edge removes a certain volume of material, feed rates and cut depth can usually be increased with a higher flute count. Soft materials such as aluminum generally allow higher feed rates and produce larger chips as a result. Larger chips require the additional flute space of a two-flute endmill to prevent the flutes from packing with chips. The increased strength and rigidity of a four-flute tool can be beneficial when machining harder materials such as tool steel and stainless steel. The additional cutting edges of a four flute tool can also help to create better surface finishes.

Endmills can also be classified as either center-cutting endmills or non-center-cutting endmills. The ability to cut on its end is determined by the way the cutting edges come together on the bottom face of the endmill. On center-cutting endmills, each of the cutting edges extends slightly past the center of the tool, leaving no material uncut when fed straight down into the workpiece (plunge cutting). The cutting edges on a non-center-cutting endmill do not touch in the center and make it unable to plunge cut. **(See Figure 6.2.13.)** Almost all two-flute endmills are center-cutting, while only certain endmills with more than two flutes are center-cutting.

FIGURE 6.2.13 Center-cutting endmills with two, three, and four flutes, and a non-center-cutting four-flute endmill.

Roughing Endmills

Roughing endmills are identified by their serrated cutting edges. The serrations are specially designed to aggressively remove material without the chatter, heat, and horsepower consumption normally associated with heavy material removal. These serrations will produce an uneven surface finish that is usually remachined with a standard endmill. **Figure 6.2.14** shows several roughing endmills.

Ballnose Endmills

The **ballnose endmill** (sometimes called a ball endmill or ballmill) is an endmill with a half-round sphere ground on its end. The radius of the sphere is proportional to the diameter of the outside diameter of the endmill. For example, a 1"-diameter ballnose endmill will have a ball end radius of 1/2". A ballmill may be used to mill a concave (inside) radius in a workpiece. **(See Figure 6.2.15.)**

Radius Endmills

A *radius endmill* (sometimes called a bullnose endmill) is a flat endmill with a special radiused edge on the corners of the cutting edges. **(See Figure 6.2.16.)** Endmills with a corner radius provide longer tool life since the shape of the corner is stronger and more durable. They are also used when a fillet radius is required between two perpendicular surfaces. Endmills for producing fillets may be custom ordered to suit the necessary radius.

Corner-Rounding Cutters

Corner-rounding cutters are ground with the form of a concave radius on each cutting edge. These are used to create convex (outside) radii on the corners of workpieces. Corner-rounding cutters are shown in **Figure 6.2.17**.

FIGURE 6.2.14 Roughing endmills.

A

B

FIGURE 6.2.15 (A) A two-flute and a four-flute ballnose end-mill. (B) A carbide inserted ballnose endmill.

FIGURE 6.2.16 Radius, or bullnose, endmills.

FIGURE 6.2.17 (A) A set of shank-type corner-rounding endmills. (B) An arbor-type corner-rounding cutter.

Chamfer Endmills

Chamfer endmills are used to produce a bevel on the workpiece edge. They are available in 60-, 82-, and 90-degree included angles. A tool with a 90-degree included angle will produce a 45-degree chamfer. Chamfer endmills are shown in **Figure 6.2.18**.

Tapered Endmills

Tapered endmills are used to mill angled surfaces without tilting the machine head. They are available in various degrees of taper up to 45 degrees. **Figure 6.2.19** shows a tapered endmill.

B

FIGURE 6.2.18 (A) An HSS chamfer endmill. (B) A carbide insert chamfer mill.

FIGURE 6.2.19 A tapered endmill.

FLAT-SURFACE MILLING CUTTERS

It is often important to create a flat surface on the top of a workpiece (called the part face). This operation is accomplished with one of three types of milling cutters.

The **fly cutter** is a simple cutting head that holds a single-edged cutting tool, as shown in **Figure 6.2.20**. Fly cutters are capable of creating fine surface finishes and are easily sharpened by removing the tool bit and grinding with a pedestal grinder or honing with an abrasive stone. Fly cutters generally should not be used to take heavy roughing passes. Fly cutters can be dangerous if care is not taken and should only be used when the operator is protected by a guard that can withstand the impact of a cutter if it should come out of the cutter body.

A **shell endmill** is a multiple-flute hollow cutter that is mounted onto an arbor. **(See Figure 6.2.21.)** They are usually made of solid HSS, and flute counts of 8, 10, 12, or 14 are common. Although technically an endmill since it is capable of cutting on its end and side, the relatively large diameter of a shell mill also makes it ideal for machining flat surfaces. Shell endmills are capable of machining up to and against square shoulders.

A **face mill** is a multiple-cutting-edge milling cutter that has replaceable inserted cutting edges. Face mills are available either with a center hole for mounting on a shell mill arbor or with an integral shank. **(See Figure 6.2.22.)** The multiple cutting tooth design makes face mills ideal for high-performance roughing and finishing operations. Face mills are not normally designed to machine up to or against square shoulders. The cutting edges are easily renewed by replacing or rotating the inserts.

FIGURE 6.2.21 An HSS shell endmill, a carbide inserted face mill, and the arbor used for mounting them.

FIGURE 6.2.20 A flycutter uses a single cutting edge. The cutting edge of this flycutter was ground from an HSS lathe tool bit blank.

FIGURE 6.2.22 A face mill with an integrated R-8 shank.

SPECIALTY MILLING CUTTERS

T-Slot Cutters

The T-slots like those in vertical milling machine and drill press tables can be machined with a special milling tool called a *T-slot cutter*. To machine a T-slot, an endmill of the appropriate size is first used to machine the straight opening of the slot. Then a T-slot cutter is passed through the slot to form the wider opening at the bottom. **(See Figure 6.2.23.)**

Dovetail Cutter

To mill the dovetails used in the slides of machine tools, a special tool called a *dovetail cutter* is used. Because of their fragile angled tips, a dovetail is first milled near the finished size with a standard endmill. Then the dovetail cutter is used only to machine the angled surface. **(See Figure 6.2.24.)**

Woodruff Keyseat Cutter

A woodruff key is a half-round key often used to drive a flywheel, pulley, or gear attached to a shaft such as the flywheel on a small engine crankshaft. When this type of key is needed, a half-round slot must be machined in the shaft to accept the woodruff key. A specialty milling cutter called a *woodruff keyseat cutter* is available for this operation. **(See Figure 6.2.25.)**

FIGURE 6.2.23 A solid HSS T-slot cutter.

Courtesy of Weldon Tool, Millersburg, PA.

FIGURE 6.2.24 Dovetail cutters.

FIGURE 6.2.25 Woodruff keyseat cutters.

Slitting Saws

Slitting saws are available to produce narrow slits in the workpiece. These cutters closely resemble the circular saw blades used in woodworking. Slitting saws range from about 0.006" to 1/8" thick. **Figure 6.2.26** shows examples of slitting saws.

A

B

FIGURE 6.2.26 (A) Some HSS slitting saws. (B) A carbide inserted slitting saw.

FIGURE 6.2.27 An arbor-type concave milling cutter.

FIGURE 6.2.28 An arbor-type convex milling cutter.

Form Milling Cutters

Milling cutters are also available with special profiles ground into their cutting edges. These "forms" are directly transferred onto the workpiece when the cutter is in use. The most common of these forms is the concave and convex radius. **Figure 6.2.27** shows a *concave cutter* that is used to make a convex radius on the workpiece. **Figure 6.2.28** shows a *convex cutter* that is used to make a concave radius on the workpiece. These cutters are available with a straight shank or a center hole for arbor mounting.

TOOLHOLDING

After the cutting tool has been selected it must be mounted properly into the milling machine spindle. Vertical milling machines are usually equipped with an R-8 taper inside of the spindle instead of the Morse taper found in the drill press. **Figure 6.2.29** shows an R-8 taper next to a Morse taper. Some cutting tools and most arbors have integrated R-8 shanks to allow them to be directly mounted in the machine spindle. The R-8 taper is unlike the Morse taper used in the drill press and is not sufficiently self-holding for heavy milling. The drawbar of the milling machine is inserted into the top of the machine through the spindle and is threaded into the back of the toolholder to secure the mating tapers.

Follow these steps to install an R-8 taper shank device:

• Place the machine spindle in either low or high range.
• Clean the machine spindle taper and the shank of the toolholder.

FIGURE 6.2.29 An R-8 taper (top) and a Morse taper (bottom).

• Partially insert the holder gently into the spindle until contact is felt with the spindle key.
• Rotate the toolholder until its slot aligns with the spindle key, and then insert the toolholder the rest of the way into the spindle.
• Thread the drawbar into the toolholder by hand.
• Apply the spindle brake to prevent the spindle from rotating and tighten the drawbar using an appropriate wrench.

Once fully assembled, the mating tapers will lock together from the tension of the drawbar. Follow these steps to remove an R-8 taper shank device.

• Place the spindle in either low or high range.
• Hold the spindle brake to prevent the spindle from turning.
• Loosen the drawbar 1 to 2 turns with an appropriate wrench.
• Lightly strike the top of the drawbar with a brass hammer to release the taper shank holder from the spindle taper. Most vertical milling machines have a tool that has a drawbar wrench on one end and a brass hammer on the other end.
• Do not completely unthread the drawbar before tapping with a hammer. This can damage the threads on both the drawbar and holder, and it can also cause the holder to fall out of the spindle.
• After the holder is loose, unthread the drawbar the rest of the way to remove the holder.

Endmill Toolholders

Endmill toolholders have an R-8 shank and a hole for standard-size straight tool shanks. There are also one or two threaded holes and set screws on the side for securing the cutting tool. **(See Figure 6.2.30.)** This screw must be tightened onto a flat machined onto the tool's shank to keep the tool securely clamped.

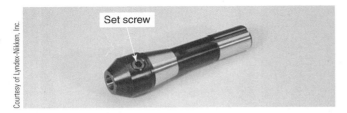

FIGURE 6.2.30 An endmill toolholder with an R-8 shank.

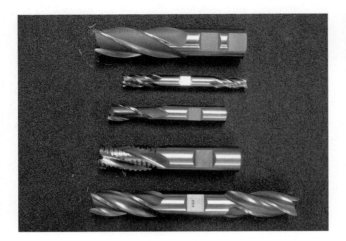

FIGURE 6.2.31 Weldon shank endmills. Note the flats on the shanks.

These holders are mostly used for holding "Weldon shank" endmills, which have a special shank with the necessary flat. **(See Figure 6.2.31.)** Endmill holders are inexpensive, rugged, simple, and capable of transmitting high torque. A different-size holder is necessary for each endmill diameter. Endmill holders often cause the endmill to run out slightly since the endmill shank must have clearance within the holder's bore to allow hand assembly. When the set screw is tightened, the endmill is forced to one side of the bore, causing runout. For general machining applications, this is not an issue, but for high-speed, high-precision operations, these holders can cause tools to cut oversize, cause some vibration, and cause premature tool wear.

Drill Chucks

Drill chucks for the milling machine are identical to those used in the drill press, but with shanks designed to fit the R-8 spindle taper. **(See Figure 6.2.32.)** Drill chucks have the greatest "one size fits all" size range capability of all the toolholding devices. However, these are intended to hold low-torque tooling such as straight-shank drills, reamers, countersinks, and counterbores. It is not a good practice to hold other milling tools such as endmills or taps because drill chucks are not designed to withstand side pressure caused by milling due to the relatively high amount of runout and minimal gripping force.

FIGURE 6.2.32 Drill chucks with R-8 shanks.

FIGURE 6.2.33 An R-8 shank Morse taper adapter.

Morse Taper Adapters

Taper-shank holemaking tools may be used in a milling machine with the use of a Morse taper adapter. This adapter has a female Morse taper in one end to receive drill shanks and an R-8 taper in the other to mount into the milling machine spindle. **(See Figure 6.2.33.)**

Shell Mill Arbors

Shell mill arbors are simple holders onto which shell mills and face mills are mounted. The arbor consists of a round pilot diameter for accurately locating the center of the mounted tool and two opposing drive keys to prevent slippage of the cutting tool on the arbor. **(See Figure 6.2.34.)** Cutters slide over the pilot and the drive keys by hand and the tool is retained by a bolt or cap screw.

FIGURE 6.2.34 An R-8 shell mill arbor.

FIGURE 6.2.35 An R-8 stub arbor.

FIGURE 6.2.36 An R-8 collet.

Stub Arbors

Arbor type milling cutters such as slitting saws and form cutters have a straight hole through their center with a single keyway. These cutters must be secured onto a **stub arbor** for mounting in the machine spindle. Most vertical mill stub arbors come with an R-8 shank. They contain a series of spacers to accommodate different-width cutters. A locking nut or screw is located on the end to secure the cutting tool. **(See Figure 6.2.35.)** Use spacers on each side of the cutter and be sure to securely tighten the clamping nut or screw.

R-8 Collets

Another option for holding straight shank cutting tools in the vertical mill is the R-8 collet shown in **Figure 6.2.36**. R-8 collets have an internal thread in their back ends to accept the drawbar threads. When the drawbar is tightened, the collet is pulled up into the spindle taper, causing the collet to contract and hold the cutting tool. Collets offer superior runout control and the shortest projection from the spindle nose of all of the other tool-holding devices. R-8 collets come in increments of 1/32" and can only expand or contract about 0.005 to 0.010" above and below their marked size.

Follow these steps to install a straight shank tool using an R-8 collet:

- Place the machine spindle in either low or high range.
- Clean the machine spindle taper and the outside of the collet.
- Partially insert the collet gently into the spindle until contact is felt with the spindle key.

- Rotate the collet until its slot aligns with the spindle key, and then insert the collet the rest of the way into the spindle.
- Thread the drawbar into the collet by hand leaving about 1/16" to 1/8" of vertical collet movement.
- Insert the straight shank tool into the collet.
- Apply the spindle brake to prevent the spindle from rotating and tighten the drawbar using an appropriate wrench.

Follow these steps to remove a tool mounted in an R-8 collet.

- Place the spindle in either low or high range.
- Hold the spindle brake to prevent the spindle from turning.
- Loosen the drawbar 1 to 2 turns with an appropriate wrench.
- Hold the cutting tool with one hand and lightly strike the top of the drawbar with a brass hammer to release the taper shank holder from the spindle taper. Most vertical milling machines have a tool that has a drawbar wrench on one end and a brass hammer on the other end.
- Do not completely unthread the drawbar before tapping with a hammer. This can damage the threads on both the drawbar and holder, and it can also cause the collet and/or tool to fall out of the spindle.
- Remove the cutting tool from the collet.
- Unthread the drawbar the rest of the way to remove the collet.

WORKHOLDING

Machine tools are very expensive investments. The more time used to actually machine the product, the more profitable the investment is. Using the best workholding device for the job can save much time. There are numerous workholding devices available for milling that range from simple hold-down clamps and machine vises to elaborate and expensive custom fixtures specifically designed to hold one particular workpiece. Many of the same workholding devices discussed in the drill press unit, such as V-blocks and angle plates, are also used to hold work on the milling machine. However, there are some more sophisticated and precise alternatives available specifically for milling. These devices can aid in producing a workpiece to precise tolerances and drastically reduce setup time.

Hold-Down Clamps

Often the size or shape of a part makes it difficult to hold in a vise but the volume of work does not justify the expense of creating a custom fixture. Clamps are an

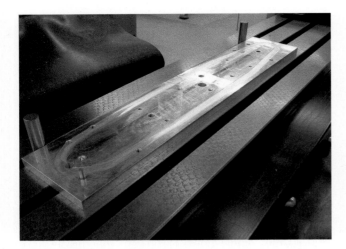

FIGURE 6.2.37 Pins placed in the table T-slots can be used to align a workpiece or workholding device.

extremely universal method for securing a workpiece to a machine table. Unfortunately, holding work with hold-down clamps provides no accurate provision for repeated locating from part to part. Therefore, every time a workpiece is clamped it must be properly aligned and located. One time-saving method is to use two pins that fit snugly into the table T-slots as a "backstop." A straight edge of a workpiece can be placed against these two pins and the part will be quickly aligned parallel to the machine table's travel. **(See Figure 6.2.37.)**

> ## ⊘ CAUTION
>
> Clamps require special attention since they always (with the exception of toe clamps) protrude above the top of the work surface, and the potential for collision with cutting tools is very high.

Clamps are available in many different variations. One type of clamping system is the *step block clamp* style, which allows a stud or bolt to be anchored to the machine table T-slot and a strap to be drawn down onto a part's surface to secure it to the table. **(See Figure 6.2.38.)** Step clamps are generally purchased in a set that consists of the following: T-nuts, studs of various lengths, clamps of various lengths, step/riser blocks of various heights, and necessary nuts and washers. **(See Figure 6.2.39.)**

Here are general steps for mounting a workpiece using step clamps.

- Position the workpiece on the machine table so that there is enough room and table travel to finish the job in one setup (if possible).
- Insert T-nuts into the table's T-slots as close to the workpiece as possible so that the maximum clamping

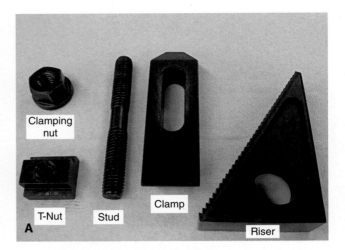

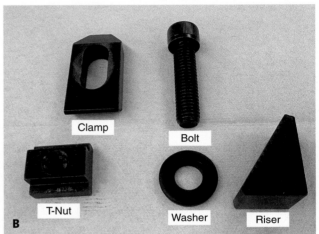

FIGURE 6.2.38 (A) A step clamp assembly usually contains a T-nut, stud, clamp, riser block, and clamping nut. (B) A bolt and washer may be used in place of the stud and nut.

FIGURE 6.2.39 A typical step clamp set with T-nuts, clamping nuts, and assorted sizes of studs, clamps, and riser blocks.

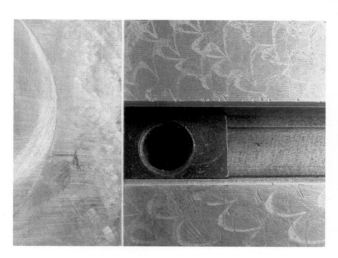

FIGURE 6.2.40 After sliding the T-nut in the T-slot, it should be positioned as close to the work as possible.

pressure will be applied to the workpiece when clamps are tightened. **(See Figure 6.2.40.)**

- Select a stud from the set with enough height to provide full thread engagement in both the T-nut and clamping nut.
- Thread the stud fully into the T-nut.
- Select an appropriate size clamp. Choose the shortest clamp possible for maximum holding power.
- Select a step block that is tall enough to slightly elevate the rear portion of the clamp when it is tightened. This directs clamping force toward the tip of the clamp and onto the workpiece.
- Thread a clamping nut onto the stud against the clamp and tighten securely. **(See Figure 6.2.41.)**

Toe Clamps

Toe clamps bolt into T-nuts and allow a part to be held down against the table by gripping only on its edges. These clamps use special gripping jaws to grab into the material and pull the work tightly against the table. **(See Figure 6.2.42.)** Toe clamps are ideal for machining large flat surfaces on plates in one setup since they do not protrude above the workpiece. These clamps usually leave marks on the workpiece where they grip.

Toggle Clamps

A toggle clamp has a lever-actuated clamping arm. Force from these clamps is applied and released by flipping a handle. **(See Figure 6.2.43.)** Since wrenches and other tools are not required to release and tighten toggle clamps, they are ideal for use in setups that will be used to produce multiples of the same part.

FIGURE 6.2.41 (A) Shows a proper step clamp setup. The T-nut and stud are close to the work. The back of the clamp is slightly raised by the riser block to direct clamping pressure toward the work. The stud length provides adequate thread engagement in both the T-nut and clamping nut. (B) Avoid situations where the back of the clamp is lower than the front and the stud is closer to the riser than the work, directing more clamping pressure toward the riser than the work.

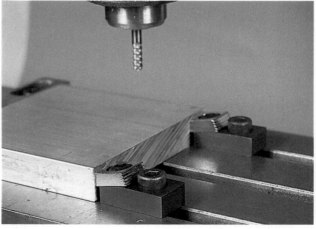

Courtesy of Mitee-Bite Products

FIGURE 6.2.42 Toe clamps are first secured to the table with T-nuts. Then the adjusting screws are tightened to grip the work by its sides.

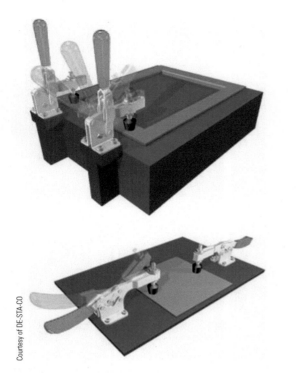

Courtesy of DE-STA-CO

FIGURE 6.2.43 Toggle clamps can be quickly locked and released without the use of tools.

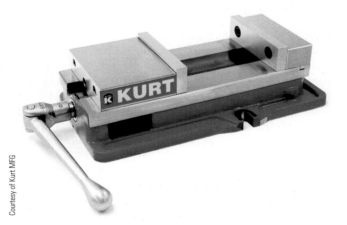

Courtesy of Kurt MFG

FIGURE 6.2.44 A typical milling machine vise.

Milling Vises

Milling vises are similar to drill press vises but are made to a higher degree of precision. **(See Figure 6.2.44.)** Vises are commonly used workholding devices for milling since they are highly versatile, precise, and repeatable. By clamping work between two jaws, they are ideal for securing materials with two parallel edges or workpieces with opposing convex surfaces (round or D-shaped). **(See Figure 6.2.45.)** It is important when considering a vise for holding thin material that the workpiece has substantial thickness to resist

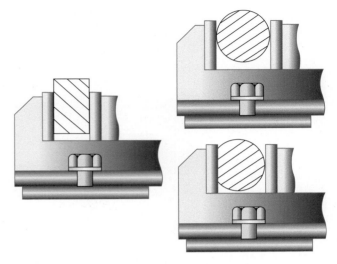

FIGURE 6.2.45 The milling vise works well for holding parts with parallel sides, round work, and D-shaped work.

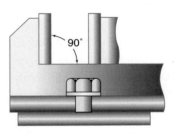

FIGURE 6.2.46 The solid jaw of the milling vise is perpendicular to the bottom for machining perpendicular surfaces.

bowing under the clamping pressure. Since the vise jaws are parallel to each other and perpendicular to the machine table, they can also accurately position a workpiece perpendicular to the surfaces clamped by the jaws. **(See Figure 6.2.46.)**

The milling vise consists of a heavy platform section called a bed. The solid jaw is at the rear of the bed and does not move. A solid jaw is either cast as one piece with the bed or secured to the bed with socket head cap screws and keys or pins to prevent any movement. The top of the bed has two flat precision guideways called the bearing surface. When the vise handle is rotated, the moveable jaw slides back and forth on this bearing surface to clamp and unclamp the workpiece. The bed of the vise also contains a screw that passes through a nut on the underside of the moveable jaw. **Figure 6.2.47** shows the various components of the milling vise.

Parallels like those mentioned in the drill press unit are also essential for mounting work in a milling vise. These precision pieces of steel are used to elevate the workpiece while still maintaining precision parallel alignment with the bearing surface of a vise. Parallels may also be used as "feeler gauges" to ensure that the workpiece is properly seated. When tightening a vise on

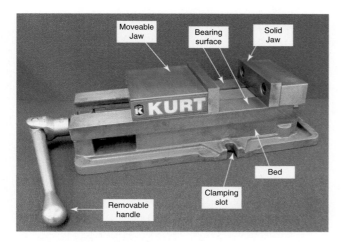

FIGURE 6.2.47 The parts of a milling machine vise.

Courtesy of Palmgren

FIGURE 6.2.49 An angle vise can be moved on a hinge and then locked to position work at nearly any desired angle.

precision workpiece surfaces, a properly seated part will be snug against the parallels so that the parallels cannot be easily moved by hand. Often, the workpiece must be lightly tapped on each corner with a soft-faced hammer as the vise is tightened to ensure the work is seated.

Modern mill vises have a mechanism on the movable jaw that pulls the jaw down tight against the bearing surface to maintain perpendicularity with the bed surface. When the vise jaws stay perpendicular this prevents the workpiece from lifting off of the parallels. Earlier vises did not have this feature, and the clearance required for the jaw to slide allowed the jaw to tip upwards when tightened on the workpiece often caused the work to lift.

Milling vises may either be directly mounted to the machine table or mounted on a swivel base. The swivel base is graduated in 1-degree increments and allows the entire vise to be rotated to a desired angle and locked in that position with the vise bed clamping nuts. This feature is convenient for performing operations such as milling angles on the end of a workpiece or drilling holes at an angle to a part's centerline. **(See Figure 6.2.48.)**

Specialty Vises

Special vises are available for angular workholding. The angle vise has a base plate that supports a hinged bed. This vise may be hinged and locked in a desired angle for milling operations. Many angle vises have a graduated support arm that can be used to visually set the vise at a desired angle. **(See Figure 6.2.49.)** The *sine vise* is a specialty vice for milling angles of higher precision. The sine vise is similar to the angle vise but has a special base that resembles a sine bar. This vise may be set to precision angles using gage blocks in the same manner as a sine bar is set. **(See Figure 6.2.50.)**

Courtesy of Kurt MFG

FIGURE 6.2.48 A milling vise mounted on a swivel base vise allows angular positioning of workpieces.

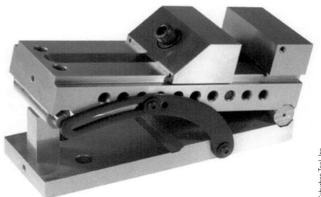

Suburban Tool, Inc.

FIGURE 6.2.50 The sine vise can be set using gage blocks for very precise angular positioning.

FIGURE 6.2.51 A three-jaw chuck fixture for milling.

Chucks/Collet Fixtures

A jaw-type chuck, similar to the chucks used on lathes, or a **collet block** may be used on the milling machine to hold and locate workpieces. These devices may be mounted so that parts are positioned either vertically or horizontally. The chucks or collets may also be part of what is called an indexer. An indexer has the ability to incrementally rotate the workpiece to allow features to be machined around a part's periphery in angular increments or patterns such as hole circles. **Figure 6.2.51** shows an image of a mounted three-jaw chuck for milling. **Figure 6.2.52** shows an image of collet blocks. **Figure 6.2.53** shows an image of an indexer.

Vacuum Plates, Magnetic, and Adhesive-Based Workholding

Workpieces are often difficult to hold using the methods just described. An example might be a part that requires four sides and the top surface of the part to be machined in one setup. Other parts may be too thin or fragile to handle the force of mechanical clamping. For these situations selecting a workholding device that holds by attracting, or sticking to, the workpiece may be the answer.

Vacuum plates are tooling plates that have a pattern of vacuum ports or slots through which vacuum is applied. When a workpiece is placed over these ports and the vacuum pump is turned on, the workpiece is securely pulled down onto the face of the vacuum plate. This is not ideal for heavy milling, but it is very easy to change parts and works well for light duty machining operations. This method also ensures flatness when machining thin, flexible pieces.

Magnetic workholding is another method sometimes used. This method uses a magnetic chuck to hold a workpiece in position. The magnet can be turned off

FIGURE 6.2.52 Square- and hex-shaped collet blocks.

Courtesy of Yuasa International Inc.

FIGURE 6.2.53 This indexing fixture can use collets or a three-jaw chuck. It also tilts for angular positioning.

for part changing and then magnetized for machining. Using this device leaves fewer surfaces obstructed for machining, and thin parts can be held at a consistent height by pulling them against the flat surface of the chuck. **(See Figure 6.2.54.)**

FIGURE 6.2.54 Work secured on a magnetic chuck for milling.

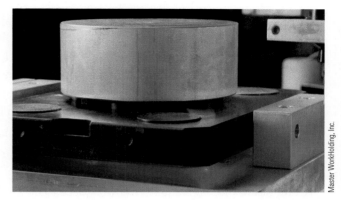

FIGURE 6.2.55 This adhesive-based workholding process cures the adhesive with ultraviolet light.

FIGURE 6.2.56 Custom milling fixtures are sometimes created to hold unusually shaped parts.

Adhesives may be used to "stick" workpieces to a flat plate or contact surface. This is usually accomplished by applying double-sided tape between the workpiece and the tooling plate or machine table. Some technologies even use a glue-based adhesive that cures when exposed to ultraviolet light. The major advantage of adhesive workholding (in addition to speed and accessibility to machine part surfaces) is the low amount of workpiece distortion. This type of workholding is primarily being used today in the aerospace industry to hold down composite materials with complex shapes. **(See Figure 6.2.55.)**

Fixtures

Sometimes a part has a unique shape and needs precise location along with a highly secure fastening system. For this, a **fixture** must be used. A fixture is a custom workholding device specifically designed to accommodate a particular workpiece. This method can be valuable for unusual parts and allows consistent referencing from critical dimensional surfaces. Fixtures usually are expensive due to their totally custom nature, which requires much thought, planning, design, custom machining, and materials. They are, however, the ideal technique for holding unusual parts that will see high production numbers. **(See Figure 6.2.56.)**

SUMMARY

- There is a wide variety of cutting-tool shapes, sizes, and materials to match almost any cutting condition required when performing vertical milling machine operations.
- ANSI provides a standard system for classification and identification of carbide inserts for milling.

(Continued)

- Endmills have cutting edges on their ends and sides and can be used to cut with both of these surfaces. Endmills are commonly used for machining features such as slots, pockets, steps, and flat surfaces. They are available in several different styles and numbers of flutes.

- Center-cutting endmills have cutting edges extending all the way across their end and have the ability to be used for plunge cutting, much like drills.

- Surface milling cutters are used mostly for machining large, flat surfaces.

- Specialty cutters are available for specific purposes, including cutting T-slots or dovetails.

- Most vertical milling machines have an R-8 spindle taper, so cutting tools require either an integrated R-8 shank, an R-8 toolholder, adapter, or arbor.

- The milling machine requires the workpiece to be rigidly secured before any machining process begins.

- The accuracy and time needed to complete a job can be drastically affected by the manner in which the workpiece is secured.

- There are numerous workholding devices available to the machinist, ranging from the milling vise to the collet fixture, and an appropriate method needs to be selected based on factors including workpiece shape, size, and number of parts to be machined. Specialty fixtures are usually built with a specific workpiece and operation in mind, but they can be expensive to produce.

- To perform milling operations, it is important to become familiar with the many different types of cutting tools, toolholding devices, and workholding devices available.

REVIEW QUESTIONS

1. Many of the cutters used in machining are made of either _____ or _____.

2. What is the major advantage of using inserted/indexable carbide cutting tools?

3. What feature of a carbide insert controls the shape of the material shavings as they are removed during machining?

4. What type of cutting tool would most likely be used to machine the special long slots in a milling machine table?

5. Which type of cutter would most likely be used to machine a half-moon-shaped slot into a shaft to accommodate a key to secure a flywheel?

6. What workpiece factors might cause a four-flute endmill to be selected over a two-flute endmill for an application?

7. List three types of milling cutters that are used to machine the top surface on the workpiece flat:

    a. _____

    b. _____

    c. _____

8. Most manual vertical milling machine spindles are equipped with what style of taper?

9. A _____ is used to retain the toolholder in the milling machine spindle.

10. What type of toolholder might be selected for mounting a slitting saw?

11. Sketch a corner-rounding endmill.

12. What toolholding device uses two drive keys and a pilot diameter to mount large milling cutters?

13. List two workholding applications where hold-down clamps would be best:

    a. _____

    b. _____

14. List the four basic pieces of a step clamp set:

   a. _____

   b. _____

   c. _____

   d. _____

15. When using step clamps, how should the clamp straps be positioned for best clamping?

16. Irregularly shaped work may be held in a custom workholding device called a(n) _____.

UNIT 3

Vertical Milling Machine Operations

Learning Objectives

After completing this unit, the student should be able to:

- Demonstrate understanding of vertical milling machine safety practices
- Perform milling machine head tramming
- Calculate speeds and feeds for milling operations
- Use an edge finder to establish a reference location
- Use an indicator to locate the center of a part feature
- Perform boring operations on the milling machine
- Demonstrate understanding of conventional and climb milling
- Demonstrate understanding of the process of squaring a block on the milling machine
- Demonstrate understanding of the basic steps of milling rectangular pockets

Key Terms

Boring head
Chip load
Climb milling
Conventional milling
Edge finder

Face milling
Feed per tooth (FPT)
Inches per minute (IPM)
Inches per tooth (IPT)

Peripheral milling
Tramming

INTRODUCTION

After appropriate cutting tools, toolholding devices, and workholding devices are selected to perform desired milling machine operations on a workpiece, speeds and feeds must be determined before beginning to set up the machine. The vertical mill must then be correctly aligned, workholding devices properly and securely mounted to the machine table, and cutting tools mounted before machining operations can begin.

GENERAL MILLING MACHINE SAFETY

The milling machine, like any machine tool, can be very dangerous, but by following some basic guidelines, safe operation can be ensured. Specific safety notes will be discussed throughout this unit, but the following are some precautions that should be observed during any milling machine operation.

⊘ CAUTION

- Always wear safety glasses when operating a milling machine.
- Wear appropriate hard-soled work shoes.
- Wear short sleeves or roll up long sleeves past the elbows.
- Do not wear any loose clothing that can become caught in moving machine parts.
- Remove watches, rings, and other jewelry.
- Secure long hair so it cannot become tangled in moving machine parts.
- Make sure all machine guards and covers are in place before operating any milling machine.
- Never operate a milling machine that is locked out or tagged out, and never remove another person's lock or tag.
- Be sure all cutting tools and workpieces are secure before beginning any machining.
- After tightening or loosening a drawbar, remove the drawbar wrench immediately.
- Get help when moving heavy workpieces or workholding devices and use proper lifting techniques.
- When operating a milling machine, stay focused on the machine. Do not become distracted by other activities or talk to others.
- Never walk away from the mill while it is running.
- Do not let others adjust work, tool, or machine settings, and do not adjust other people's setups.
- Keep the machining area clear of all items including rags and tools to ensure that nothing comes in contact with rotating cutting tools. Items can become entangled in cutting tools or be violently thrown from the work area if they touch a rotating cutting tool.

- Always shut off the spindle and let it come to a complete stop before adjusting workholding or toolholding devices or to take measurements or clean the machine.
- Use caution when handling milling cutters so sharp edges do not cause cuts.
- Remove chips from the workpiece and tool using a brush only after the spindle has come to a complete stop. Never remove chips by hand.
- Never use compressed air to clean chips, debris, and cutting fluids from the mill.

TRAMMING THE VERTICAL MILLING MACHINE HEAD

Tramming is the process of adjusting the head so that the spindle is perpendicular to the top surface of the machine table. This is required to produce perpendicular machined surfaces when milling and to ensure that machined holes are perpendicular to the table. To make tramming easier, it is best to remove any workholding devices from the machine table. As an example, suppose that the mill head is tilted in both directions as shown in **Figure 6.3.1**.

- First loosen both sets of clamping bolts.
- Turn the adjusting screws to align each reference mark with its "0" on each protractor. **(See Figure 6.3.2.)**
- Snug the clamping bolts just beyond finger-tight with a wrench.
- Move the quill to its lowest position and lock it in place using the quill lock.

FIGURE 6.3.1 This tilted milling machine head must be trammed before machining square and parallel surfaces.

Reference   Reference

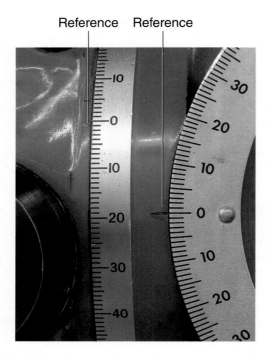

FIGURE 6.3.2 After loosening the head clamping bolts, move the head to set each protractor to its "0" mark. Then tighten the clamping bolts just beyond finger-tight.

- Place the beam of a solid square on the table and slide the blade against one side of the quill. Be sure the table and square are free of dirt and burrs.
- Use the adjusting screw to more closely align the quill with the square.
- Repeat the process side to side and front to back on the quill. **(See Figure 6.3.3.)**
- Snug the clamping screws so that the head is not loose, but can still be easily moved when turning the adjusting screws.
- Position the table so that the spindle is in the middle of the table.
- Mount a dial test indicator in the spindle and position it to sweep a circle slightly smaller than the width of the table.
- Place the spindle in neutral to make it easier to spin the indicator. Keep the contact tip about 1/4" away from the table edges and be sure the indicator contact tip is as parallel as possible to the table surface. **(See Figure 6.3.4.)**
- Rotate the spindle so the indicator is at the front of the table at the 6 o'clock position.
- Bring the indicator contact tip against the table and preload it so the needle travels about one-fourth of a turn of the dial.
- Rotate the spindle 180 degrees to bring the indicator to the back of the table at the 12 o'clock position.

A

B

FIGURE 6.3.3 (A) A square can be used to check for perpendicularity between the table surface and the quill for the front-to-back movement, and (B) the left-to-right movement.

NOTE:

When crossing the table T-slots, turn the spindle slowly and use care so the indicator is not damaged and its position does not move under impact. Two alternate methods can be used to avoid the issue of crossing the T-slots. One is to use a gage block, 1-2-3 block, or similar block to check indicator readings at each position instead of bringing the indicator into direct contact with the table. **(See Figure 6.3.5.)** Another is to sweep the indicator on a precision-ground flat disc, plate, or the base of a swivel vise, instead of directly on the table. **(See Figure 6.3.6.)**

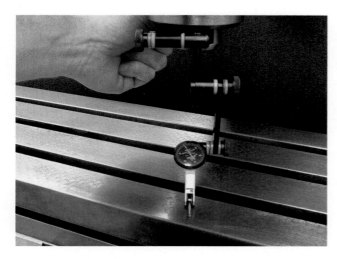

FIGURE 6.3.4 Using a dial indicator mounted on the spindle to "sweep" the table surface.

FIGURE 6.3.5 Using a dial indicator and gage block to check the tram at the 6 and 12 o'clock positions.

- Adjust the head until both positions show readings within about 0.001" to 0.002" of each other. Make small adjustments to avoid the need to keep changing direction of the adjusting screw. This introduces backlash and makes the process more difficult.

- Repeat this process at the 3 o'clock and 9 o'clock positions, as shown in **Figure 6.3.7**. Adjust the head (again making very small adjustments) until these two indicator readings are within 0.0005" of each other.

- Return to the 6 o'clock and 12 o'clock positions and adjust the head until these two indicator readings are within 0.0005" of each other.

- Tighten all clamping bolts and sweep the table to be sure total indicator reading (TIR) is within 0.0005".

FIGURE 6.3.6 Using a dial indicator and flat disc to check the tram.

FIGURE 6.3.7 Checking indicator readings during tramming at the 3 and 9 o'clock positions. After adjusting these positions, return to the 6 and 12 o'clock positions and fine-tune the adjustment.

ALIGNING WORKHOLDING DEVICES

It is equally important that all workholding devices are properly aligned so that parallel and perpendicular surfaces can be machined. Methods of using different workholding devices are only limited by creativity, but a few basic principles can be applied to position any type of device in a desired orientation.

Aligning a Milling Vise

To align a vise, first be sure both the machine table and bottom of the vise are clean and burr free. Gently place the vise on the machine table and position the clamps.

Get help when moving heavy vises and bend at the knees when lifting.

Aligning a Vise with a Swivel Base

- Tighten the clamps to secure the base to the table.

- Loosen the two nuts on the swivel base. Align the reference mark on the vise with the "0" mark on the base and then make the nuts finger-tight. This will roughly align the solid jaw with the table movement. **(See Figure 6.3.8.)**

- Mount a dial indicator in the machine spindle or on the head or ram using a clamp or magnetic base.

- Position the indicator near one end of the solid jaw of the milling vise as shown in **Figure 6.3.9**. Always indicate the solid jaw. Indicating the moveable jaw can create errors in alignment.

- Move the saddle to preload the dial indicator and set the dial face to "0."

- Turn the table handle to move the indicator near the other end of the solid jaw. Make note of the direction and amount of indicator movement.

- Use a soft plastic or dead blow hammer to lightly tap and move the vise to split the difference of the indicator readings. For example, if the indicator shows a positive difference of 0.010", tap the vise until the needle moves about 0.005" in the negative direction. **(See Figure 6.3.10.)**

- Repeat this process until the indicator reading is 0.0005" TIR or less.

- Tighten the swivel base clamping nuts and recheck to be sure the vise did not move.

FIGURE 6.3.8 Rough alignment of a swivel-base vise can be accomplished by lining up the reference mark on the vise with the "0" on the swivel base.

FIGURE 6.3.9 Positioning a dial indicator for aligning a vise with the table movement. Always indicate the solid jaw when aligning a vise.

Aligning a Vise with a Solid Base

- Snug one clamp and make the second one only finger-tight. This allows the vise to pivot at the tighter clamp and makes adjustment more predictable. **(See Figure 6.3.11.)** If both clamps are loose, it is difficult to control the movement of the vise.

- Follow the same steps used with the swivel-base vise until TIR across the solid jaw is 0.0005" or less.

- Always recheck TIR after final tightening of the clamps.

Note that a vise can also be mounted with the solid jaw parallel to the saddle movement (Y-axis), as shown in **Figure 6.3.12**. Use the same technique for aligning, but move the saddle to check TIR instead of the table.

FIGURE 6.3.10 Using a dead blow hammer to tap the vise during alignment. Make adjustments until the TIR is 0.0005" or less. Recheck TIR after fully tightening the clamps.

FIGURE 6.3.11 When aligning a vise with a solid base, snug only one clamp so it can act as a pivot point and make movements more predictable.

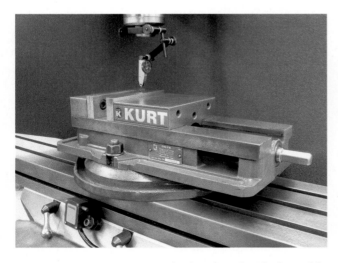

FIGURE 6.3.12 A vise can also be aligned with the saddle movement.

Aligning Other Workholding Devices and Large Workpieces

Sometimes workholding devices such as angle plates, V-blocks, and custom fixtures can be mounted in a vise without the need for any further alignment. If the devices are too large for vise mounting, they can be aligned using a process similar to aligning a vise. Large workpieces that require direct clamping usually need to be aligned as well.

Two methods can be used to approximately align these larger items. Since the T-slots are machined parallel with the table movement, two pins equal to the width of the T-slots can be placed in the slots. Then a workpiece or workholding device can be placed against the pins. **(See Figure 6.3.13.)** Another method is to place a square with the beam against either the front or back surface of the machine table and the blade flat on the table. Then the work or workholding device can be adjusted, and perpendicularity can be checked against the blade of the square. **(See Figure 6.3.14.)** When in position, secure the workpiece or workholding device with clamps.

FIGURE 6.3.13 Pins in the T-slots can provide alignment with the table movement for workpieces or workholding devices.

FIGURE 6.3.14 A square can be used to align a workpiece surface parallel to the saddle movement.

FIGURE 6.3.15 For more precise positioning, a workpiece surface can be aligned with the table or saddle movement using a dial indicator.

Depending on the required level of accuracy, these steps may be enough to meet specifications. If higher precision is needed, after positioning clamps, a dial indicator may be used to align a workpiece more accurately, as shown in **Figure 6.3.15**.

SPEEDS AND FEEDS FOR MILLING OPERATIONS

Calculating spindle RPM for milling operations is the same as calculating RPM for drill press operations. Use the standard formula $RPM = \frac{3.82 \cdot CS}{D}$ where $CS =$ cutting speed in surface feet per minute and $D =$ diameter of the cutting tool. Obtain cutting speeds for milling operations from charts or tables from *Machinery's Handbook* or cutting tool manufacturers just as when performing calculations for drill press and lathe operations. Some cutting speed charts may list cutting speeds only for milling while others may just contain a separate milling column. When performing holemaking operations on the mill, apply the same speed and feed principles as when using the drill press for the same operation.

 CAUTION

Always take the time to determine appropriate cutting speeds and to calculate proper spindle RPM. Operating cutting tools at excessive speeds can cause tool failure and breakage, leading to serious injury.

Recall that lathe and holemaking feed rates are specified in IPR (inches per revolution) or FPR (feed per revolution). Knee-type milling machines often use IPR for power quill feed settings for holemaking operations also, but power feeds for the table and saddle are specified in **IPM (inches per minute)**. To calculate IPM, the following formula is used: $IPM = FPT \times N \times RPM$, where $FPT =$ feed per tooth, $N =$ number of teeth, or flutes, of the cutting tool, and $RPM =$ spindle RPM.

 CAUTION

Always take the time to also calculate appropriate feed rates. Excessive feed rates can cause tool breakage and violently pull workpieces from the machine, leading to serious injury.

FPT (feed per tooth) can also be called **IPT (inches per tooth)** or **chip load**. It is the thickness of the chip removed by one cutting edge of the tool per each revolution of the cutting tool. FPT values are small and generally range from around 0.0005" to 0.010". They can be found on feed charts like the one shown in **Figure 6.3.16**. Charts like these are available from many different sources, including cutting tool manufacturers and *Machinery's Handbook*. Multiplying chip load by the number of teeth, or flutes, gives a feed per revolution value. Multiplying that value by RPM gives an IPM value.

EXAMPLE ONE: Calculate RPM and feed rate in IPM for a 1/4"-diameter four-flute HSS endmill cutting brass at 200 SFPM with a 0.001" FPT.

First, calculate RPM:

$$RPM = \frac{3.82 \cdot CS}{D}$$

$$RPM = \frac{3.82 \cdot 200}{0.25}$$

$$RPM = \frac{746}{0.25}$$

$$RPM = 3056$$

Then calculate IPM. $FPT = 0.001$, $N = 4$, $RPM = 3056$

$$IPM = FPT \times N \times RPM$$

$$IPM = 0.001 \times 4 \times 3056$$

$$IPM = 12.2$$

FIGURE 6.3.19 The tip of the edge finder can move out of alignment with its shank.

FIGURE 6.3.20 Positioning an edge finder near the edge of a workpiece.

with a chuck key can damage the shank of the edge finder.

- Lower the quill so the tip of the edge finder is below the top surface of the workpiece and tighten the quill lock.
- Move the table to position the tip of the edge finder within about 1/8" of the workpiece. **(See Figure 6.3.20.)**
- Turn on the machine spindle and set the speed to 1000 to 1500 RPM.
- Tap the tip of the edge finder with a finger so the tip becomes eccentric (wobbles).
- Carefully move the table to bring the edge finder in contact with the edge of the workpiece.
- Continue to slowly move the table. The tip will begin to run more true as the workpiece pushes the tip into alignment with the shank.
- When the tip "kicks," the center of the edge finder (and the machine spindle) is one-half of the edge finder tip diameter from the edge of the workpiece. **(See Figure 6.3.21.)**

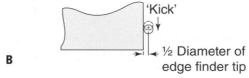

FIGURE 6.3.21 (A) When the tip of the edgefinder "kicks," (B) the centerline of the spindle is one-half of the tip diameter from the part edge.

- Set a "0" on the micrometer collar or DRO.
- Loosen the quill lock and raise the quill to bring the edge finder above the top of the workpiece.
- Move the table one-half of the diameter of the tip of the edge finder and reset the "0" on the digital read-out or micrometer collars reference position.
- To establish a reference "0" for the Y-axis, follow the same procedure, but use the saddle movement instead of the table movement.

Locating the Center of an Existing Part Feature

To find the center of an existing hole, a dial indicator is often used. Follow these steps to find the center of an existing round hole.

- Visually locate the spindle in the center of the hole by moving the table and saddle.
- Mount a dial test indicator in the spindle as shown in **Figure 6.3.22** and lower the quill so the indicator contact point is inside the hole.
- Place the spindle in neutral and rotate the indicator so it is in line with either the X- or Y-axis, and move the contact point against the surface of the hole to preload the indicator.
- Set the dial on the indicator to zero by gently rotating the indicator's face.

FIGURE 6.3.22 A spindle-mounted test indicator positioned for finding the center of a round hole.

- Rotate the spindle 180 degrees and note the direction and amount of needle movement.
- Move the X- (or Y-) axis one-half the difference of the two readings so the needle moves back toward the initial zero reading. For example, if the needle rotates to the right until the needle stops at 0.020, the table must be moved until the needle moves back to the left and stops on 0.010. **(See Figure 6.3.23.)**

- Repeat the process for the other axis.
- After the spindle is located correctly, lock the table and saddle and recheck the TIR (total indicator reading) while sweeping the hole.
- Adjust both axes as needed until the desired accuracy is achieved.
- Set the micrometer collars or the DRO to a reference "0" position.

This method can also be used to find the center of a square internal opening. When the spindle is centered in the opening, indicator readings will be the same at each side of the square. **(See Figure 6.3.24.)** If the internal opening is a rectangle, the readings will be the same 180 degrees apart, but the readings on the X-axis will be different from the readings on the Y-axis. When indicating the larger dimension of the rectangle, the quill may need to be retracted so the indicator can be swept to the opposite side.

If a round hole is too small for the indicator contact point, a pin gage can be placed in the hole and the outside of the pin gage indicated instead of the inside surface of the hole. The same method can be used to indicate the center of a round hub or round workpiece. **(See Figure 6.3.25.)**

Boring

Boring uses a single-point cutting tool to enlarge an existing hole. Some advantages of boring over other holemaking operations are that any desired hole size can be machined and large-diameter holes that are beyond

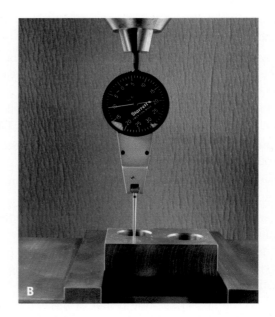

FIGURE 6.3.23 (A) Zero the dial face when the indicator is in line with one machine axis. (B) Then rotate the spindle 180 degrees and note the difference in the indicator readings. The table needs to be moved one-half of the indicator reading to center the spindle in the hole. In this case the table needs to be moved 0.005" because the indicator readings differ by 0.010".

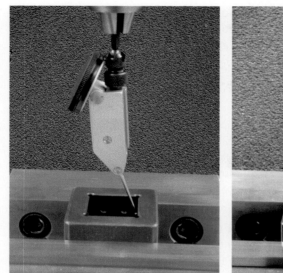

FIGURE 6.3.24 The center of a square internal opening can also be located by sweeping the sides with a dial indicator.

FIGURE 6.3.25 (A) To find the center of smaller-diameter holes, insert a pin gage in the hole and sweep the pin with a test indicator. (B) The center of a round external feature can also be found by sweeping the diameter with an indicator.

the range of drills and reamers can be produced. Another benefit of boring is the ability to adjust the location of a hole since the boring tool will not follow the existing hole like a drill. **(See Figure 6.3.26.)**

On a milling machine, the boring bar is mounted in the machine spindle and rotates as it is fed through the hole. A **boring head** is commonly used to hold the boring bar and provides the ability to offset the bar to control hole size. A slide on the boring head is adjusted by turning a micrometer screw to offset the boring bar.

After adjustment a locking screw secures the slide in place. **(See Figure 6.3.27.)**

When boring, it is preferred that the chips will break into small pieces instead of forming long strings. Stringy chips can become wrapped around the boring bar, dragged through the hole, and possibly thrown from the work area. For proper chip breaking, the depth of cut should be at least 2/3 of the tool nose radius. For example, if this rule is followed, a 1/64" tool nose radius should take a minimum depth of cut of 0.010". Also,

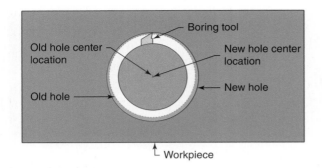

FIGURE 6.3.26 When boring on the milling machine, the location of a hole can be changed by moving the X- or Y-axis, or both.

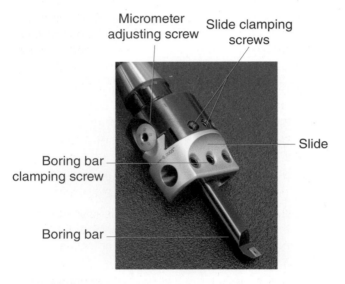

FIGURE 6.3.27 A boring head holds a boring bar. The slide on the boring head can be adjusted using the micrometer screw, then secured in place by tightening the locking screws.

surface finish will usually improve when this rule is followed.

When the feed rate is increased, the chip becomes thicker. As chip thickness increases, chips are also more likely to break into pieces instead of forming continuous strings. So it is good practice to use the highest feed rate that produces the required surface finish. Another benefit of using higher feed rates is reduced machining time.

Some other variables that can affect results during boring operations are depth of cut, feed rate, and boring bar length. Because of the pressure between the cutting tool and workpiece, different depths of cut at a particular feed rate may create slightly different size diameters. The same is true of using different feed rates with the same depth of cut. Also, as the length of the boring bar increases, there is the possibility of the bar "flexing" away from workpiece. This can also create some slight variation in hole diameter. For these reasons, it is important to use a strategy of taking multiple passes

using the same spindle RPM, depth of cut, and feed rate to be able to accurately predict an outcome.

Consider this example.

- Suppose a hole is roughed within 0.060" of final size.
- A tool with a 1/64" (0.015") nose radius is used for finishing.
- One finishing pass would be taken to increase the diameter by 0.020". This sets a 0.010" depth of cut and follows the rule of cutting a minimum of 2/3 of the tool nose radius.
- The result is measured.
- A second pass of the same depth is taken and the hole diameter measured again to verify that the desired diameter change was achieved. There would still be 0.020" on the workpiece diameter for a final pass to reach the desired finish diameter. By doing this, that final cutting pass using the same settings will produce a predictable result.

This multiple pass strategy makes achieving desired size easier because it eliminates variables in the boring operation. The first pass sets the parameters of the operation. The second pass verifies those parameters and their results. The final pass achieves the desired end result. If boring multiple holes, two final passes will be sufficient for subsequent holes, because the parameters are known and predictable.

To perform a boring operation on the vertical milling machine, follow these steps.

- Mount the boring head in the machine spindle.
- Mount an appropriate size bar in the boring head.
- Use the largest boring bar that will safely fit inside the existing hole.
  - Use the shortest bar that will reach the desired depth, but be sure the boring bar is long enough to avoid collisions between the bottom of the boring head, and the top of the workpiece, and any work-holding devices.
  - A boring head frequently has more than one hole for mounting boring bars. The mounting hole to be used will be determined by the size of the hole to be machined and the size of the boring bar. Smaller diameters are machined using holes closer to the center of the boring head while larger holes will require using holes farther away from the center. **(See Figure 6.3.28.)**
  - Mount the boring bar so cutting tip is positioned in-line with the centerline of the boring head slide. Many carbide inserted boring bars will have a flat that will correctly align the tool when the mounting screw is tightened against the flat. **(See Figure 6.3.29.)**

FIGURE 6.3.28 Many boring heads have more than one hole to allow the bar to be kept close to center for machining smaller holes and offset further for machining larger holes. Boring bars can also be mounted in the side hole for machining very large diameters.

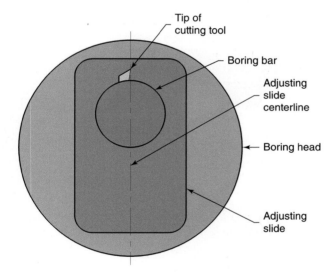

FIGURE 6.3.29 The cutting tip of the boring tool must be in line with the centerline of the adjusting slide.

- Adjust the slide of the boring head so the bar will fit inside the existing hole.
- Set the micrometer adjusting nut on the milling machine so quill travel will allow the boring bar to machine to the desired depth.
  - For through holes allow the cutting tip of the bar to travel about 1/8" past the bottom of the workpiece.
  - For blind holes and counterbores, set the adjusting nut to machine short of the final depth by about 0.005" to 0.010".

- When machining begins, depth can be checked and adjustments made using the micrometer adjusting nut or the knee.
- Lower the quill to bring the bar into the existing hole.
- Slowly adjust the micrometer dial until the tip of the boring bar makes light contact against the surface of the existing hole. Applying some layout fluid in the hole can make it easier to see when the bar touches off the surface.
- Retract the quill to bring the bar out of the hole.
- Calculate and set an appropriate spindle speed. Boring bars are less rigid than other holemaking tools. Spindle speeds may need to be reduced by one-fourth to one-third of the calculated RPM for a drill of the same diameter as the bore. This will help reduce excessive vibration and chatter. Some trial and error may be required to arrive at a suitable spindle RPM.
- Engage the power feed transmission crank.
- Set the feed reversing knob so the quill feeds in the desired direction (usually downward).
- Position the quill feed selector knob to set the desired feed rate.
  - When roughing and using larger-diameter bars, use the higher feed rates of 0.003 or 0.006 IPR.
  - When finishing and using smaller-diameter bars, use the lighter feed rates of 0.0015 or 0.003 IPR. Again, some trial and error may be needed to determine a suitable feed rate.
- Use the micrometer screw to offset the boring bar to the desired depth of cut. Keep in mind the 2/3-nose radius rule.
- Manually lower the quill to bring the boring bar within about 1/8" of the top surface of the workpiece.
- Start the spindle.
- Engage the feed control lever to begin boring the hole. **(See Figure 6.3.30.)**
- When the quill stop nut reaches the adjusting nut at the end of the pass, the power feed will automatically disengage. It is good practice to stop the spindle before retracting the bar from the hole. This may leave a small score line on the side of the bore, but retracting the bar with the spindle on will leave a large spiral mark around the inside of the bore.
- Sometimes the spring in the quill feed handle will cause the quill to retract when the feed disengages, pulling the rotating tool upward and back into the bore. This can be avoided by applying light hand pressure to the quill feed handle just before the feed disengages.
- Keep in mind the multiple pass strategy when approaching final size.

FIGURE 6.3.30 Engaging the feed control lever starts quill feed to begin the boring operation.

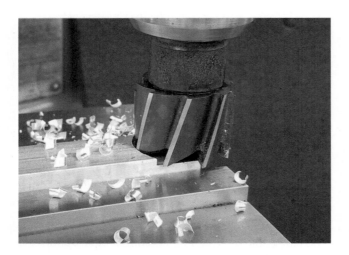

FIGURE 6.3.31 Face milling uses the face of the cutting tool.

- To avoid scoring the bore after the final cut, retract the boring bar using the micrometer screw before retracting the bar from the hole.

MILLING BASICS

Many different operations can be performed using the different types of milling cutters described in Section 6.2, but there are few basic principles that apply to all operations. **Face milling** is using the face of a cutting tool to machine a surface, as shown in **Figure 6.3.31**. **Peripheral milling** is using the outside periphery of the cutting tool to machine a surface, as shown in **Figure 6.3.32**.

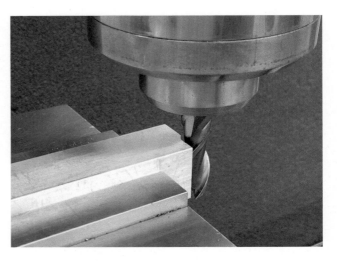

FIGURE 6.3.32 Peripheral milling uses the periphery, or side, of a cutting tool.

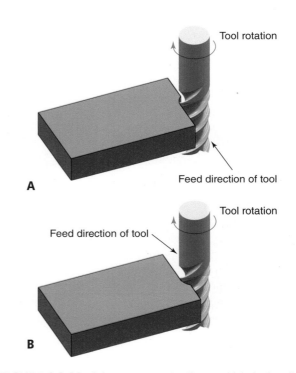

FIGURE 6.3.33 (A) Conventional milling and (B) climb milling.

When peripheral milling, two different situations can exist depending on the relation between the cutter rotation and the feed direction. **Conventional milling** is feeding the workpiece in against the rotation of the cutting tool. **Climb milling** is feeding the workpiece with the rotation of the cutting tool. **Figure 6.3.33** shows the difference between conventional and climb milling.

Conventional milling is the method normally used when machining on the vertical mill. It provides a measure of safety because the tool tends to push away from the work, requiring constant feed pressure to continue cutting. The drawback to conventional milling is that the surface finish is often not as smooth as desired. Climb milling should only be used under certain conditions on the vertical mill because the work can be pulled into the tool uncontrollably and cause tool breakage and workpiece damage. One advantage of climb milling is that it provides a smoother surface finish than conventional milling when performed properly. It is normal to rough a surface using deeper cuts within 0.005" to 0.010" of final size by conventional milling, then take light cuts (less than 0.010") using climb milling to reach final size and create a smoother surface finish.

SQUARING A BLOCK

A common operation performed on the milling machine is "squaring" a workpiece. Squaring a workpiece means machining the sides of a workpiece perpendicular and parallel to each other. This is often the first operation performed on a workpiece and is required to machine additional part features within required tolerances.

Before beginning, take a few minutes to check the alignment of the machine head and the milling vise. If the head is not trammed and the vise is not aligned, it may be impossible to create a "square" block. A face mill, shell endmill, or flycutter is often used for the squaring process. A small workpiece can be machined using the face of an endmill. When available and practical, choose a cutter with a diameter that can span the width of the largest surface of the block. After mounting a suitable milling cutter, calculate and set an appropriate spindle speed and feed rate (if power feed is available).

The largest surface of the workpiece should be machined first to create a flat reference surface that will then be used to position the work as other surfaces are machined. Using the largest surface will minimize any setup errors when machining other sides of the block. If a smaller surface is used to orient the block for cutting a larger surface, any setup error will be multiplied.

Milling Side A

- Mount the block in the vise with the largest surface facing up. Once machined, this surface can be called Side A.

- One parallel or other suitable setup block can be used to raise the top surface of the block above the top of the vise jaws. Be sure the vise and parallel are clean and free of burrs and debris.

- Avoid leaving an excessive amount of the block above the vise jaws by mounting the part as low as possible in the vise. **(See Figure 6.3.34.)**

- Place a small-diameter rod between the moveable vise jaw and the workpiece so the side against the solid jaw seats solidly against the solid jaw. **(See Figure 6.3.35.)**

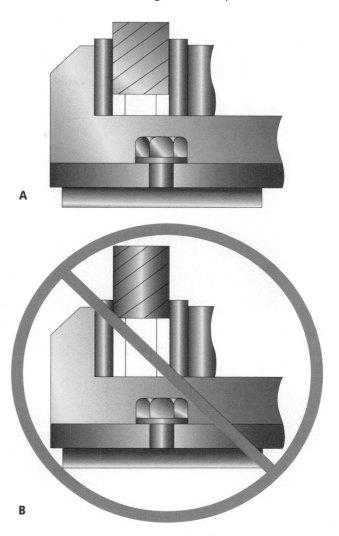

FIGURE 6.3.34 (A) Only extend the workpiece above the top of the vise jaws enough to machine to the desired height. (B) Avoid this type of situation.

FIGURE 6.3.35 To machine the first side of a block during the squaring process, place a small-diameter rod between the work and the moveable jaw. Do not seat the work on the parallel with a dead blow hammer.

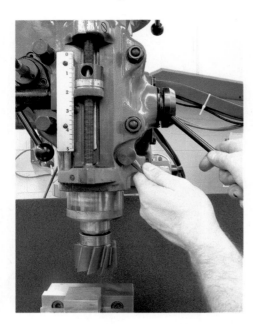

FIGURE 6.3.36 When extending the quill to position cutting tools, bring the quill stop against the adjusting nut and be sure to lock the quill.

- Tighten the vise but *do not* seat the part by hitting it with a dead blow hammer because this can rock the workpiece surface away from the solid vise jaw.

NOTE:

The less the quill is extended, the more rigid the machine setup. However, lowering the quill about an inch can provide benefit. For example, in case of an emergency situation, the quill can be raised to remove the cutting tool from the workpiece. When positioning the quill for any milling operation, bring the quill stop against the micrometer adjusting nut and secure the quill lock as shown in **Figure 6.3.36**. The adjusting nut will prevent the quill from being pulled down during machining and the lock will prevent it from being pushed up during machining.

(!) CAUTION

Always secure the quill lock before beginning any cutting or the quill may move uncontrollably, causing tool breakage or pulling the workpiece from the workholding device.

- Raise the knee to bring the milling cutter within about 1/16" of the top surface of the block.
- Use the saddle to place the cutting tool approximately in the center of the workpiece along the Y-axis and lock the saddle.

FIGURE 6.3.37 When bringing the cutting tool in contact with the work, only place about 1/8" of the cutter over the work. Then slowly raise the knee to make light contact.

- Move the table so that only about 1/8" of the cutter is over the block.
- Calculate and set an appropriate spindle speed.
- With the spindle running, slowly raise the knee until the cutter just makes contact or just touches up with the workpiece. **(See Figure 6.3.37.)** Avoid forcing the workpiece into the cutter to prevent excessive wear or chipping the tool's cutting edges.
- Move the table so the cutter is about 1/2" away from the work.
- Raise the knee to set depth of cut. For this first surface, only remove enough material to clean up the entire surface.
- Apply appropriate cutting fluid and engage the table power feed if available. If power feed is not available, feed the table manually to make the cut. **Figure 6.3.38** shows a facing cut being taken across a workpiece.
- After machining this first cutting pass, stop the feed and shut off the spindle.
- Remove the workpiece from the vise and deburr it.

NOTE:

If the surface is extremely uneven, machine one roughing pass and then a finishing pass, removing only about 0.020" to 0.030". If the surface finish is rougher than desired, decrease the feed rate until an acceptable finish is reached. Spindle speed may also be increased about 10% to 20% when taking light finishing cuts to improve surface finish.

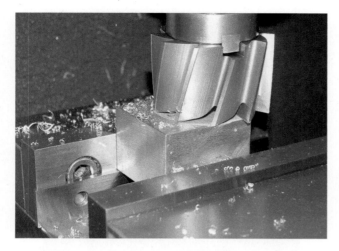

FIGURE 6.3.38 An HSS shell endmill facing the top surface of a block.

Chatter and squealing noises indicate that spindle speed is too high. Very fine chips indicate that feed is too slow. Very thick chips that do not curl indicate the feed is too high. When machining steels with an HSS cutting tool, brown or blue chips also indicate that feed is too high.

Milling Side B

- Place the machined Side A against the solid vise jaw and place the small-diameter rod between the moveable jaw and the block before tightening the vise. This allows the front surface to "float" and ensures the machined surface of Side A is flat against the solid jaw. A parallel can again be used to keep the block above the top of the vise jaws. **Figure 6.3.39** shows this setup.

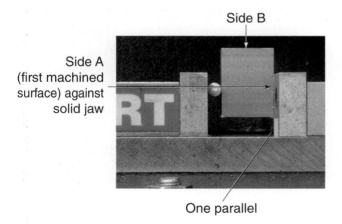

FIGURE 6.3.39 To face Side B, place the first-machined Side A against the solid jaw and the rod between the work and the moveable jaw. After machining Side B, check for square between Sides A and B.

- Machine a cutting pass across this surface of the workpiece using the same techniques as before when milling Side A. This newly machined surface can be called Side B.
- After the cut is complete, remove the block from the machine, deburr it, and then check the two machined surfaces for square (perpendicularity) using a solid square and feeler stock.

Milling Side C

Return the workpiece to the vise with Side A once again against the solid jaw and Side B down on the parallel.

- Use a rod between the moveable jaw and the block as before. **(See Figure 6.3.40.)**
- Mill this third surface using the same steps as before when cutting Sides A and B. This surface can be called Side C.
- Remove the block from the machine and deburr it.
- Check for parallelism between Sides B and C by measuring near the four corners as shown in **Figure 6.3.41**. It is also a good idea to check for square between Sides A and C.

Milling Side D

- To mill the fourth side of the block, place Side A down on two parallels with B and C against the vise jaws.

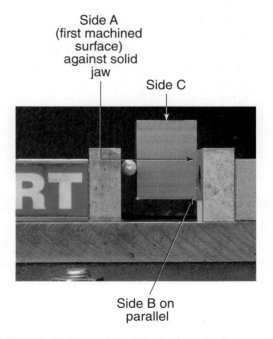

FIGURE 6.3.40 To machine Side C, place the first-machined Side A against the solid jaw, Side B down on a parallel, and the rod between the work and the moveable jaw.

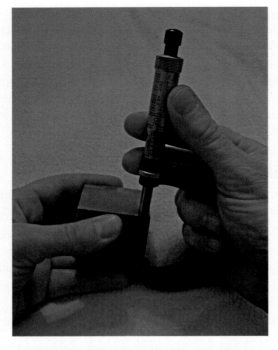

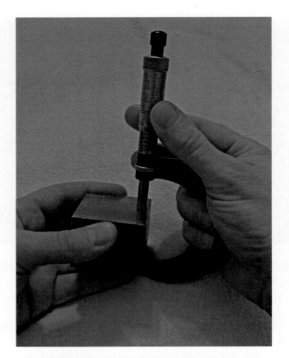

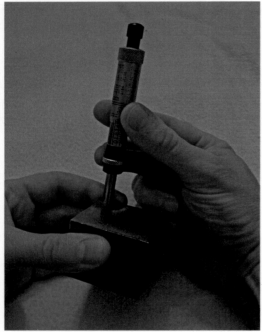

FIGURE 6.3.41 Check for parallelism between Sides B and C by measuring near the four corners with a micrometer.

- Do not use the rod when clamping the block in the vise for this step.
- This time seat the block in the vise by striking it with a dead blow hammer. Check to see if the parallels are tight. If both are tight, this reinforces the fact that all three machined surfaces are very square to each other, probably within 0.001″. **(See Figure 6.3.42.)**
- Mill this surface to create Side D.

- Check Sides A and D for parallelism by measuring near the four corners. The block can now be milled to the desired size across Sides A and D.
- Whenever possible, measure the workpiece while it is still mounted in the vise and without moving the quill. This keeps the machined surface and the face of the cutting tool on the same plane, which eliminates setup errors that can be caused by repositioning the work or cutting tool.

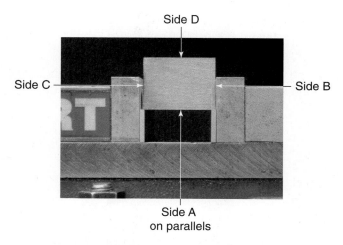

Side D

Side C

Side B

Side A
on parallels

FIGURE 6.3.42 To machine Side D, place Side A down on parallels, with Sides B and C against the vise jaws. Do not use the rod and seat the part on the parallels with a dead blow hammer. After facing, check for parallelism between Sides A and D.

- The desired dimension across Sides B and C can also then be milled after repositioning the block in the vise on parallels and seating it with a dead blow hammer.

Milling Sides E and F

After the first four sides have been milled square and parallel, one of two methods is used to square the two remaining sides.

Method 1

- The block can be mounted loosely in the vise with one end facing up.
- The beam of a solid square can be placed on the bed of the vise or on the machine table, and the side of the workpiece aligned with the blade of the square.
- After clamping, feeler stock can be used to check for gaps between the vertical workpiece surface and the blade of the square. **(See Figure 6.3.43.)**
- Mill the surface using the same face milling steps used to machine the first four sides. This surface can be called Side E.
- Remove the block from the vise, deburr the sharp edges, and check for perpendicularity.
- Place this newly machined surface down in the vise and seated on parallels so the opposite surface can be milled.
- Machine a cleanup pass and check for parallelism with the bottom of the workpiece before milling to final size. Again, whenever possible, measure the workpiece without removing it from the vise to avoid any setup or repositioning errors.

FIGURE 6.3.43 Using a solid square and feeler gages to position the block for machining Side E.

FIGURE 6.3.44 Side E can also be machined using an endmill by mounting the block in the vise with one end extended past the end of the vise jaws.

Method 2

Another method of squaring the ends is to mount the block in the vise on parallels with one end extending past the end of the vise jaws. The end can then be machined by peripheral milling using an endmill. **(See Figure 6.3.44.)** The length of the cutting portion of the endmill needs to be slightly longer than the thickness of the workpiece and the diameter needs to be large enough so it does not flex under cutting pressure.

FIGURE 6.3.45 Position the endmill so the bottom extends past the bottom of the block. Be sure the flutes are long enough to span the entire surface to be milled.

A good practice is to limit length to about two times the diameter of the tool.

- Mount the work in the vise by placing it on parallels and seating with a dead blow hammer.
- Select and mount a suitable endmill.
- Calculate and set an appropriate spindle speed and feed rate (if power feed is available).
- Use the quill and knee to position the endmill vertically as shown in **Figure 6.3.45**. Remember to bring the quill stop against the micrometer adjusting nut and lock the quill.
- The X-axis is typically used to set depth of cut, and the Y-axis is used to perform the milling passes.
- Conventional milling should be used to take roughing passes, and climb milling should be performed only with a light cut for a finishing pass, so keep that in mind when positioning the endmill at the beginning of the cut. See **Figure 6.3.46** for an illustration of some examples of how to position the endmill for conventional and climb milling.
- Start the spindle and bring the endmill into light contact with the edge of the workpiece using the X-axis to "touch off" the tool.
- Set a "0" reference using the micrometer collar or DRO.
- Use the Y-axis to move the endmill away from the part.
- Set depth of cut using the X-axis and then lock it in place to prevent movement during the milling pass.

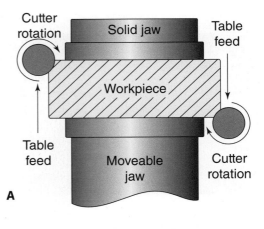
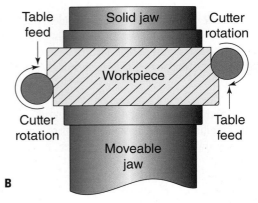

FIGURE 6.3.46 (A) Conventional milling direction of feed on each side of the vise. (B) Climb milling direction on each side of the vise.

Remove only enough material to clean up the surface on this first side.

- Apply cutting fluid and engage the power feed or move the Y-axis manually to perform the conventional milling pass.
- Use the Y-axis to feed back across the surface at a slower rate to take a finishing climb milling pass. **Figure 6.3.47** shows these milling steps.

NOTE:

During heavier milling the endmill may deflect, or flex, a few thousandths causing some variation in the machined surface. Taking this additional pass at the same depth setting reduces the force on the tool so it can "spring" back to its original ideal position, so it is often called a *spring pass*. It may sometimes be beneficial to take yet another conventional and climb milling spring pass at that same depth setting to minimize or eliminate tool deflection.

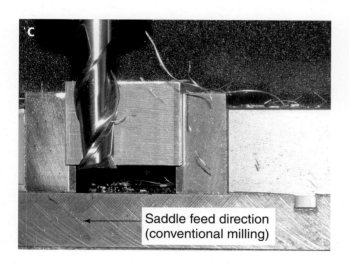

Saddle feed direction
(conventional milling)

Saddle feed direction
(climb milling for
finishing pass)

FIGURE 6.3.47 Steps for using peripheral milling to machine a vertical surface.
A: Touching off using the X-axis (table feed).
B: The endmill is moved off the part using the Y-axis (saddle feed) and depth of cut is set using the X-axis.
C: Conventional milling across the surface using the Y-axis.
D: Climb milling a finishing pass feeding the Y-axis in the opposite direction.

- Remove the block from the vise, deburr it, and check for square.
- Place the block in the vise with the opposite end extending past the jaws and repeat the process to clean up this last side.
- Set a "0" on the micrometer collar or DRO to establish a reference position.
- Machine roughing passes using conventional milling within about 0.010" to 0.020" of final size using the micrometer collar or DRO to set cut depth.
- Take a climb milling pass of about 0.005" to 0.010" and recheck size.

- Take one last conventional milling and climb milling pass to mill the block to the desired final dimension.

NOTE:

If a large amount of material needs to be removed, a roughing endmill can be used to mill the surface within about 0.025" of final size. Then a standard endmill can be used to finish the part to final size. Since this requires another change in cutting tools, this option might only be worthwhile if several parts need to be machined. Otherwise the time saved by using the roughing endmill might not offset the time needed to change tools.

Each of these methods has advantages and disadvantages. The face milling method allows use of the same cutting tool for the entire process, so time is saved because tools do not need to be changed. It does take a little more setup time to position the block, so if many parts need to be machined it can take slightly longer, and larger blocks may be more difficult to position. When peripheral milling with an endmill, part positioning is quicker and easier, but if a very long endmill is needed, tool flex can produce surfaces that are not square. Additionally, a larger-diameter endmill requires a lower spindle RPM and feed rate, which can increase machining time.

Using a work stop/vise stop

If multiple workpieces are being milled to the same dimension using peripheral milling as in the last two surfaces of the squaring process, use of a *work stop* or *vise stop* can help save time. These devices allow workpieces to be removed from the vise and remounted within about 0.001" to 0.002" of the original location. Here is how to use a stop to machine multiple parts to the same dimension by peripheral milling.

- Mill Side E of all workpieces. Remove only enough material to produce a finished surface.
- Mount one workpiece in the vise with the finished Side E against the work stop.
- Mill the workpiece to the desired size.
- Set a reference '0' position on the table (X-axis) micrometer collar or DRO. Usually setting a zero here is preferred.
- Move the table so the tool is clear of the part and remove the part from the vise.
- Mount the next part in the vise with its finished Side E end against the stop.
- Move incrementally toward the X-axis reference position to mill this second part.
- When the reference position is reached, the second part will be within 0.001" to 0.002" of the size of the first part.
- Repeat the process for additional parts.

Figure 6.3.48 shows some examples of stops in use. When using a jaw-mounted vise stop, always mount it on the solid jaw.

Squaring a Block Using an Angle Plate

If a block is too large to be held in a vise, it can be clamped to an *angle plate* or *angle block* to perform the squaring process.

- Use the vertical surface of the angle plate in place of the solid vise jaw and rotate the block in the same manner after machining each side. A rod is not needed since the clamps will allow the machined surface to be properly located against the vertical surface of the angle plate. **(See Figure 6.3.49.)**

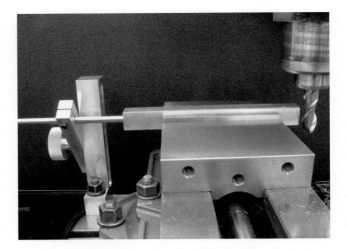

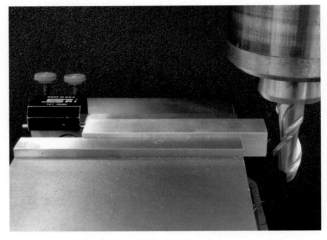

FIGURE 6.3.48 Two types of stops that can be used to locate workpieces in the same location. Notice that the vise jaw stop is mounted on the solid jaw.

FIGURE 6.3.49 A larger workpiece clamped to an angle plate for the squaring process.

FIGURE 6.3.50 An angle block with a side plate allows quick perpendicular positioning of the work without the need of a square. One clamp holds the work parallel to the angle plate and the other holds the work parallel to the side plate. This ensures perpendicularity in both directions.

- To mill the last two sides, extend the block past the end of the angle plate and perform peripheral milling with an endmill, or use a solid square to align the block and then face mill those sides.

- Mounting a second angle plate perpendicular to the first, or mounting a smaller plate on the side of the first angle plate, can take the place of using the square for alignment and reduce setup time. **(See Figure 6.3.50.)**

ANGULAR MILLING

Milling of angular surfaces can be accomplished by three basic methods. The workpiece can be angled, the head of the machine can be angled, or an angled milling cutter can be used. Every situation is unique, but a few basics can be applied under most conditions.

Milling with Angled Cutters

Small angular surfaces, or bevels, are often machined using angled milling cutters. Follow these steps when using a chamfer endmill to machine a bevel specified by width.

- Set tool depth so that the angled cutting edge will span the entire depth of the bevel.

- Calculate and set an appropriate spindle speed. RPM may have to be reduced as tool contact increases to reduce tool chatter, so expect that some trial and error may be required.

- Touch off the corner of the workpiece with the X or Y axis, depending on how the part is oriented on the machine.

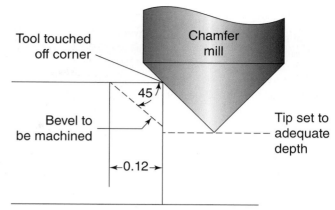

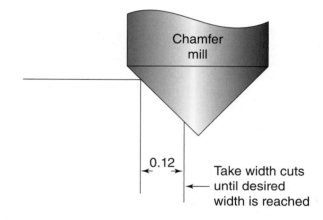

FIGURE 6.3.51 To machine a bevel specified by width, set the tool tip to an adequate depth and touch off the corner of the work with either the X- or Y-axis. Then feed the tool across the work with the other axis. Take incremental width cuts to reach the desired width.

- Set a "0" for the axis used to touch off the corner.

- Set cut width with that same axis.

- Feed the tool across the workpiece with the other axis.

- Continue taking incremental width passes until the desired width dimension is reached. **(See Figure 6.3.51.)**

Follow these steps to machine a bevel specified by depth.

- Set the tool position with the table or saddle (depending on how the part is oriented on the machine) so that the angled cutting edge will span the entire width of the bevel.

- Calculate and set an appropriate spindle speed. Again, RPM may have to be reduced as tool contact increases to reduce tool chatter, so expect that some trial and error may be required.

- Touch off the corner of the workpiece by raising the knee.

- Set a "0" for the knee.

- Set cut depth with the knee.

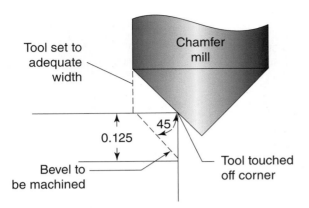

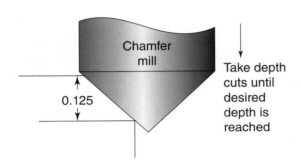

FIGURE 6.3.52 To machine a bevel specified by depth, set the tool so the cutting edge spans an adequate width and touch off the corner of the work with the knee. Then take incremental depth cuts with the knee while feeding with the table or saddle to reach the desired depth.

- Feed with the table or saddle, depending on orientation of the part in the machine.
- Continue taking incremental depth passes until the desired depth dimension is reached. **(See Figure 6.3.52.)**

For both of these methods, use conventional milling for roughing and climb milling to produce a smoother surface finish. **Figure 6.3.53** shows a 45-degree bevel being milled along the edge of a workpiece with a chamfer endmill.

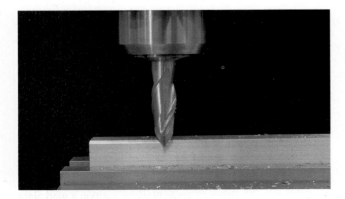

FIGURE 6.3.53 Milling a 45-degree bevel with a chamfer mill.

A tapered endmill is often used to machine an angled wall with an adjacent floor such as a pocket or a step.

- It is a good idea to first rough the straight wall with a standard endmill.
- Use a feeler gage to set the bottom of the tapered endmill about 0.005" to 0.010" above the adjacent bottom surface.
- Use conventional milling to machine the wall within about 0.005" of the final size.
- Raise the knee so the tool lightly touches off the adjacent bottom surface.
- Move the tool to the final wall position.
- Use climb milling to finish the angled wall. **(See Figure 6.3.54.)**

Milling Angles by Positioning the Workpiece

If no cutter is available with the required angle, or the angled surface to be machined is too large to use an angular cutting tool, the workpiece can be positioned at the desired angle and then milled using face milling or peripheral milling.

Positioning the Workpiece in a Workholding Device

To machine to an angled layout line, the workpiece can be lightly clamped in a vise or against an angle plate. Then use a surface gage for parallel positioning of the layout line, as shown in **Figure 6.3.55**. After positioning, clamp the workpiece and mill to the layout line. This method should only be used for approximate work or when tolerances are 1/64" or greater.

A quick method to machine angles within about 1 to 2 degrees is to use a protractor to position the workpiece in a vise or against an angle plate. **(See Figure 6.3.56.)** If an angle needs to be more precise, an *angle block* or V-block can be used to position the workpiece, as shown in **Figure 6.3.57**. When a very high degree of accuracy is required, sine tools can be used to position the workpiece to the desired angle before clamping. **(See Figure 6.3.58.)**

Positioning the Workholding Device and Large Parts

Large workpieces that require direct clamping can also be positioned for peripheral milling using a protractor **(Figure 6.3.59)**. An angle gage can also be used for angular positioning. When using an angle gage, a square can be used to provide a reference surface. **(See Figure 6.3.60.)**

A swivel-base milling vise can also be used to position work for peripheral machining of angular surfaces.

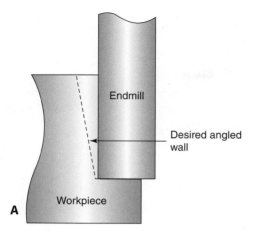

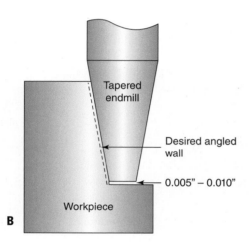

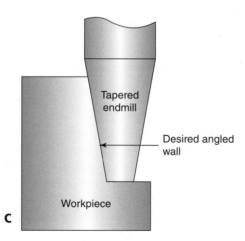

FIGURE 6.3.54 Machining an angled wall with a tapered endmill. (A) First machine a perpendicular wall with a standard endmill. (B) Then set the bottom of the tapered endmill slightly above the bottom surface and rough machine the wall near the finished size. This will leave a small, flat section near the bottom of the vertical wall. (C) After roughing the angled wall, touch off the bottom surface, then move to the desired size and climb mill to finish the angled wall.

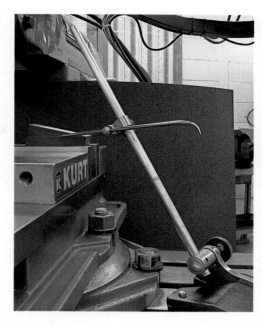

FIGURE 6.3.55 Positioning work to a layout line using a surface gage for milling an angular surface.

FIGURE 6.3.56 Using a protractor to position a workpiece in a vise for angular milling. The protractor head is referenced to the top of the solid jaw.

FIGURE 6.3.57 Using an angle block to position a workpiece in a vise for angular milling.

FIGURE 6.3.58 A sine bar can be used for very accurate positioning of a workpiece for milling angles. After the work is clamped to the angle plate, it is a good idea to remove the gage blocks and sine bar before machining to protect the gage blocks from the chips made during milling.

FIGURE 6.3.59 Aligning a large workpiece with a protractor using the edge of the machine table as a reference surface.

FIGURE 6.3.60 Positioning a large workpiece using an angle block and square. The square provides a reference surface relative to the table movement.

FIGURE 6.3.61 Setting a swivel-base vise with a protractor using the knee dovetail as a reference surface.

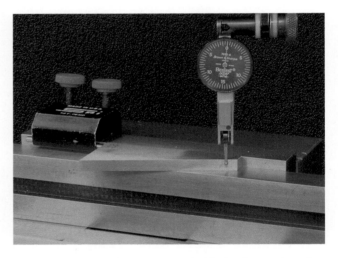

FIGURE 6.3.62 Indicating an angle block referenced against the solid vise jaw. The work stop keeps the angle block from moving.

Loosen the swivel clamping nuts and use the angular scale for noncritical work or approximate positioning. A protractor can be used to set an angle between the solid jaw and the dovetail on the column of the vertical mill. **(See Figure 6.3.61.)** For precise angles, an angle block or sine bar with gage block build can be held against the solid jaw and indicated by moving one of the machine's axes. **(See Figure 6.3.62.)** Each of these methods allows an endmill to be used to machine an angular surface, as shown in **Figure 6.3.63**.

An angle or sine vise can also be used to position a workpiece for angular milling operations. Depending on the angle set and the shape of the workpiece, either face or peripheral milling can be performed. **Figure 6.3.64** shows angular milling using a sine vise.

FIGURE 6.3.63 Milling an angular surface with the work held in a swivel-base vise.

FIGURE 6.3.64 Face milling an angular surface of a workpiece held in a sine vise directly clamped to the machine table. Note that the gage blocks have been removed from the vise to keep them away from chips produced during milling.

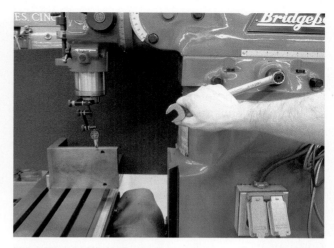

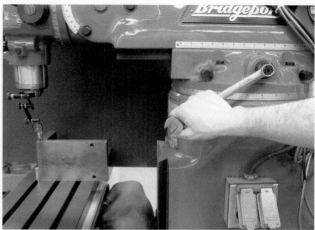

FIGURE 6.3.65 To align the turret, run an indicator across an angle plate by moving the ram slide back and forth. Adjust the turret until the indicator reading stays constant. In this picture, the angle plate is aligned using pins in the T-slots of the machine table.

Milling Angles by Tilting the Machine Head

When milling large angular surfaces of large parts, the head of the vertical mill can be tilted to the desired angle instead of positioning the workpiece. Either head movement can be used depending on the configuration of the workpiece. Before tilting the head, the turret must be aligned so the ram is parallel with the saddle movement. For non-critical work, alignment can be achieved by simply setting the turret protractor to the zero mark.

For critical work, more steps are needed:

- First align an angle plate or parallel with the Y-axis using the saddle movement.
- Loosen the turret clamping bolts and the ram locking bolts.

- Move the ram to move a spindle mounted dial indicator across the angle plate.
- Adjust the turret until the indicator reads zero when moved across the angle plate with the ram movement. **(See Figure 6.3.65.)**
- Lock the turret clamping bolts.
- Position the ram in the desired location.
- Lock the ram locking bolts.
- Now the head can be tilted to perform angular operations.

When tilting the head, the protractors on the head can be used for noncritical work or approximate positioning. Extending the quill and checking the angle between the table surface and the quill with a protractor or angle block is another method that can be used. Keep the protractor or angle block parallel with the direction of angular movement to prevent errors by holding it against a square, angle plate, or solid vise jaw. **(See Figure 6.3.66.)** An angle block or sine bar can also be used with a dial indicator to very accurately set the desired angle. Secure the angle block or sine bar in a vise or against an angle plate. Then mount a dial indicator in the spindle and extend the quill to move the indicator across the angled surface. Adjust the head until the indicator reads "0" across the surface. **(See Figure 6.3.67.)**

Either face milling or peripheral milling can then be performed, depending on the shape of the workpiece. **(See Figure 6.3.68.)**

FIGURE 6.3.66 A protractor can be held against the quill to check the angular setting of the mill head. Hold the protractor head against the solid vise jaw so that the blade is in line with the direction of the angular movement.

FIGURE 6.3.67 Indicating an angle block held in a vise by moving the quill. When the indicator reads "0" across the block, the angular setting is correct.

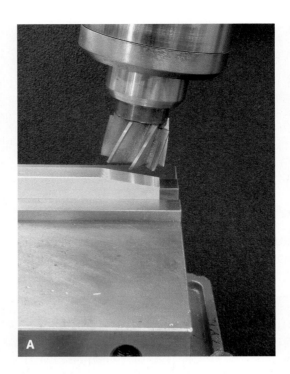

FIGURE 6.3.68 With the head tilted to the desired angle, an angular surface can be machined using (A) face milling or (B) peripheral milling.

 CAUTION

After positioning the head to the desired angle, be sure to tighten all clamping bolts before beginning any milling operations.

MILLING STEPS, SLOTS, AND KEYSEATS

Milling steps combines both face and peripheral milling. Two dimensions need to be considered during machining instead of just one. During machining, the X- or Y-axis movement is used for positioning to produce one dimension and the knee is used for the other. Similar positioning techniques can also be used to machine slots in desired locations.

Basic Step Milling

Milling steps are usually performed using endmills. Shell endmills can be used for roughing steps, but they usually produce vertical walls with rougher-than-desired surface finishes. The same is true for roughing endmills. If a shell endmill or roughing endmill is selected for roughing, plan to leave enough material for finishing using a standard endmill. First, secure the workpiece, select and mount the desired cutting tool, and calculate and set an appropriate spindle RPM and feed rate (if power feed is available). These steps explain how to mill a step with a roughing and standard endmill using the X-axis to monitor the width of the step:

- After mounting the roughing endmill, position the end within about 1/16" of the top surface of the block, just like when face milling.
- Make sure the quill stop is against the micrometer adjusting nut and lock the quill.
- Move the table so that only about 1/8" of the cutter is over the block, just like when face milling.
- Start the spindle.
- Raise the knee to touch off the top of the workpiece to set a reference for the depth of the step. **(See Figure 6.3.69.)**
- Set the micrometer collar on the knee crank to "0".
- Move the X-axis so the endmill is clear of the workpiece.
- Raise the knee to set the desired depth of cut and lock it in place.
- Slowly move the table to touch off the end of the part to set a reference for the width of the step. **(See Figure 6.3.70.)**
- Set the micrometer collar or DRO for the X-axis to "0."
- Move the saddle so the endmill clears the workpiece, positioning the tool for conventional milling.
- Move the table to set cut width and lock.
- Raise the knee to set cut depth and lock. Remember some guidelines about cut depth and width. If the full diameter (or nearly full diameter) of the endmill will be cutting, maximum depth should be one-half of the

FIGURE 6.3.69 Touching off with a roughing endmill to set a reference for step depth.

FIGURE 6.3.70 Touching off with a roughing endmill to set a reference for step width.

tool diameter. If cut depth is beyond one-half of the tool diameter, maximum width should be about one-fourth of the tool diameter.

- Apply cutting fluid and use the saddle to mill the step.
- Return the saddle to the beginning position.
- Repeat these steps to rough the step within about 0.015" to 0.020" of both the step width and depth dimensions.

- Remove the roughing endmill and mount a standard endmill.
- Reset spindle speed and feed as needed.
- Touch the endmill off the roughed vertical wall and the horizontal surface.
- Move the endmill clear of the workpiece with the saddle.
- Move the table and raise the knee each about 0.005".
- Take a conventional pass and then a climb milling pass at those settings.
- Stop the spindle and check both the step width and depth dimensions.
- Make final adjustments to the table and knee to set finish dimensions for the step.
- Take a conventional pass and climb milling pass at those settings.
- Take a spring conventional and climb milling pass to finish the step.

Figure 6.3.71 summarizes this method.

Slot Milling

Slot milling, or slotting, can be performed by different methods using different cutting tools depending on the shape and orientation of the slot to be machined. Common slot shapes include straight wall, T-shaped, and dovetail shaped. An open slot passes entirely across a workpiece, while a closed slot does not pass through the ends of the workpiece.

Milling Open Slots

To machine an open slot parallel to the X-axis, follow these steps.

- Use an edge finder to locate the edge of the workpiece with the Y-axis.
- Move the Y-axis to place the center of the spindle at the centerline of the slot (parallel to the X-axis).
- Lock and "0" the Y-axis.
- Select and mount the desired cutting tool. If the width is a nonstandard size, select an endmill smaller than the slot width.
- Calculate and set spindle RPM and feed rate if power feed is available.
- Position the quill with the cutting tool about 1/16" to 1/8" above the workpiece and lock in place.
- Start the spindle and raise the knee to touch the tool lightly on top of the workpiece.
- Set the knee micrometer collar to "0".
- Move the X-axis so the tool is clear of the workpiece.
- Raise the knee to set cut depth. It is a good practice to stay 0.005" to 0.010" away from final depth to allow for checking size and taking a finishing cut.

FIGURE 6.3.71 In (A), the step is being rough machined within about 0.020" of the desired size with a roughing endmill. Notice the uneven surface finish on the wall and the bevel at the inside corner. (B) The first "cleanup" pass with the finishing tool. About 0.010" is being machined from both the vertical and horizontal surfaces. Notice that the finishing endmill smoothes out the vertical wall and makes a sharp internal corner. In (C), a final climb milling pass is being machined to finish the step. Only about 0.003" is being machined from both surfaces.

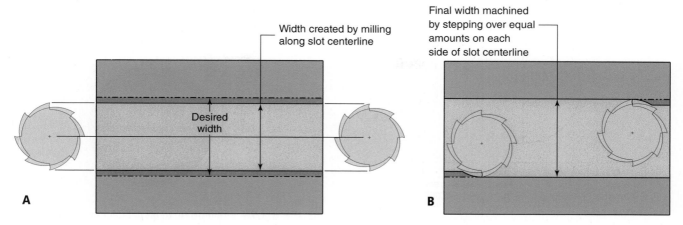

FIGURE 6.3.72 Milling a slot wider than the tool diameter. First mill along the centerline (A), then step over equal amounts to reach the required width (B).

- Feed with the X-axis to mill the slot across the workpiece.
- If the required slot width is wider than the tool diameter, move the Y-axis one direction to widen the slot on the first side. Conventional mill when roughing and climb mill when finishing. It is good practice to leave about 0.005" on the side to verify size before finishing.
- Repeat to widen the slot on the other side. **(See Figure 6.3.72)**
- Measure slot width and depth to verify sizes.
- If more depth is required, raise the knee.
- If more width is required, remove equal amounts from each side of the slot.
- Take passes to reach desired depth and width as needed.

Figure 6.3.73 shows machining of an open slot.

Milling Plain Keyseats

A *plain key* is a square or rectangular removable component used to transmit power between a shaft and the hub of a gear or pulley. A *plain keyseat* is a closed slot in a shaft that holds the key. **Figure 6.3.74** shows a plain and key and keyseat.

Before setting up the milling machine to produce a plain keyseat, some dimensions need to be found using tables from reference sources like *Machinery's Handbook*. The following example is for machining a plain square keyseat in a 1" diameter shaft using the sample table in **Figure 6.3.75**.

- The nominal fractional key size for a 1" diameter shaft is ¼".
- The width of the keyseat must be machined to 0.250" +0.002/−0.000", so a ¼" diameter endmill would be appropriate.
- The "S" dimension from the bottom of the shaft to the bottom of the keyseat is 0.859 +0.000/−0.015". (The upper limit is 0.859" and the lower limit is 0.844".)

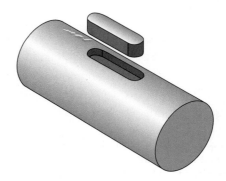

FIGURE 6.3.73 Milling a through slot.

FIGURE 6.3.74 A plain key and keyseat.

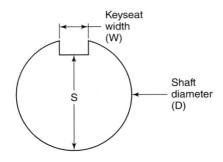

D Shaft Diameter	S (+0.000/−0.015)	Nominal Fractional Key Size	W Keyseat Width (+0.002/−0.000)
0.3750	0.322	3/32	0.094
0.4375	0.386	3/32	0.094
0.5000	0.430	1/8	0.125
0.5625	0.493	1/8	0.125
0.6250	0.517	3/16	0.188
0.6875	0.581	3/16	0.188
0.7500	0.644	3/16	0.188
0.8125	0.708	3/16	0.188
0.8750	0.771	3/16	0.188
0.9375	0.796	1/4	0.250
1.0000	0.859	1/4	0.250
1.0625	0.923	1/4	0.250
1.1250	0.986	1/4	0.250
1.1875	1.049	1/4	0.250

FIGURE 6.3.75 A portion of a table showing square keyseat dimensions and tolerances.

- Subtract both from 1.000" to find the limits of the depth of the keyseat from the top of the shaft.
  - Lower limit = 0.141"
  - Upper limit = 0.156"
  - The endmill must cut the keyseat to a depth between 0.141" and 0.156". Keep in mind that the "S" dimension is the actual required dimension, so measure and verify the "S" dimension during and after machining.

Next, the shaft must be mounted in the machine using an appropriate work holding device. Then follow these steps when using an edge finder to locate the keyseat in the center of the shaft:

- Be sure the tip of the edge finder is below the centerline of the shaft. **(See Figure 6.3.76.)**
- Find the edge and set the appropriate machine axis to "0".
- Raise the edge finder above the shaft.
- Move the machine axis one-half of the diameter of the edge finder.
- Move the machine axis one-half of the diameter of the shaft. The spindle in now in the correct position, centered in the shaft. **(See Figure 6.3.77.)**

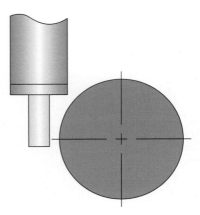

FIGURE 6.3.76 When using an edge finder to touch the side of a shaft, be sure the tip is below the centerline of the shaft.

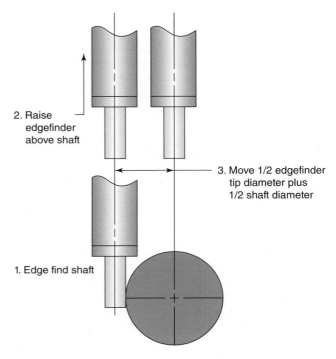

FIGURE 6.3.77 Positioning the spindle in the center of a shaft using an edge finder.

Another method of positioning is to use the cutting tool in the place of an edge finder.

- Position the endmill within about 1/64" of the side of the shaft.
- Lightly hold a strip of paper between the endmill and the work.
- Place the spindle in neutral.
- Rotate the spindle by hand while moving the machine axis 0.001" at a time. Stop when the cutter pulls the strip of paper. **(See Figure 6.3.78.)**
- Set the machine axis to "0".
- Raise the tool above the shaft.

FIGURE 6.3.78 Instead of using an edge finder to locate the center of a shaft, position the tool near the outer edge of the shaft. Then rotate the cutting tool by hand while moving the table 0.001" at a time until the paper is pulled by the tool's cutting edges.

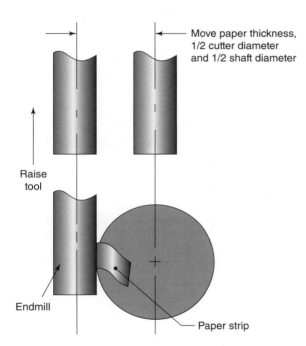

FIGURE 6.3.79 After the paper is pulled by the tool, raise the tool, then move the paper thickness, one-half of the tool diameter, and one-half of the shaft diameter.

- Move the machine axis an amount equal to thickness of the paper.
- Move the machine axis one-half of the tool diameter.
- Move the machine axis one-half of the shaft diameter to place the tool in the center of the shaft. **(See Figure 6.3.79.)**
- After the tool has been located in the center of the shaft, position the tool at the starting point of the keyseat. If machining a closed keyseat, be sure to use a center-cutting endmill. **(See Figure 6.3.80.)**

FIGURE 6.3.80 Be sure to use a center-cutting endmill when milling a closed keyseat.

FIGURE 6.3.81 Milling a closed plain keyseat.

- Position the quill with the cutting tool about 1/16" to 1/8" above the shaft and lock in place.
- Calculate and set spindle RPM.
- Start the spindle and raise the knee to touch the tool lightly on top of the shaft.
- Raise the knee to set cut depth. Since the keyseat width tolerance is only +0.002/−0.000", set depth to only cut about 10% of the tool diameter. This will prevent the tool from flexing and cutting past the allowable 0.002" tolerance.
- Feed the desired distance with the appropriate machine axis to mill the keyseat to length.
- Retract the tool with the quill and stop the spindle.
- Measure keyseat width and the "S" dimension.
- If more depth is required, raise the knee.
- Restart the spindle and lower the quill to return the tool to the desired depth.
- Take passes to reach desired depth as needed.
- If the keyseat needs to be wider, use the same methods as when cutting a standard slot by moving the appropriate axis to remove equal amounts of material from each side of the keyseat.
- **Figure 6.3.81** shows machining of a closed plain keyseat.

FIGURE 6.3.82 Machining of a T-slot (A) and a dovetail slot (B).

Milling T-Slots and Dovetail Slots

A T-slot or dovetail cutter cannot be used to machine the entire slot profile by itself. First, machine a straight slot to the required width and depth using a standard endmill and the previously explained techniques. After the straight slot is machined, switch to the T-slot or dovetail cutter. Set the depth of the T-slot or dovetail cutter at the bottom of the existing straight slot. Feed the cutter to finish the shape of slot. Use a liberal amount of cutting fluid to flush the chips out of the slot when machining T-slots or dovetails. **Figure 6.3.82** shows a T-slot and a dovetail being machined.

Milling Woodruff Keyseats

Another type of key is semicircular shaped and called a *Woodruff key*. This requires a matching slot called a *Woodruff keyseat*. **(See Figure 6.3.83.)** Positioning a Woodruff keyseat cutter requires a different process than the previous slotting operations. Refer to **Figure 6.3.84** while reviewing these steps to locate the cutter in the center of the shaft vertically.

- Position the quill so the keyseat cutter is 1/16" to 1/8" above the top surface of the shaft.
- Hold a strip of paper between the bottom of the cutter and the workpiece.
- Raise the knee slowly until the paper is pinched.
- Move the tool away from the work.
- Raise the knee an amount equal to the paper thickness.
- Raise the knee one-half of the shaft diameter.

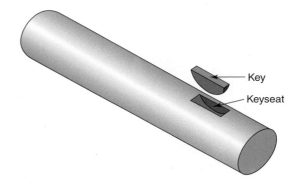

FIGURE 6.3.83 A Woodruff key and keyseat.

- Raise the knee one-half of the width of the cutter to position the keyseat in the center of the shaft.

Next, the center of the cutter needs to be located at the required distance from the end of the shaft. Refer to **Figure 6.3.85** while reviewing the following steps.

- Position the tool about 1/16" from the end of the shaft/workpiece.
- Place the spindle in neutral.
- Place a strip of paper between the cutter and the end of the workpiece.
- Rotate the spindle by hand while moving the appropriate machine axis 0.001" at a time until the paper is pulled.
- Move the tool clear of the workpiece with the other machine axis.
- Move the first machine axis an amount equal to the paper thickness.

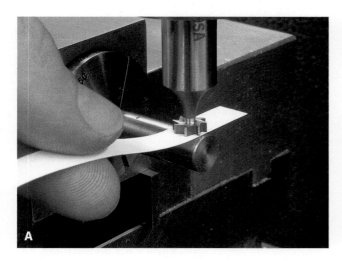

FIGURE 6.3.84 (A) To position a Woodruff keyseat cutter in the center of a shaft, first touch off the top of the shaft using a strip of paper. (B) Then move the cutter away from the work and raise the knee the thickness of the paper, one-half of the cutter width, and one-half of the shaft diameter.

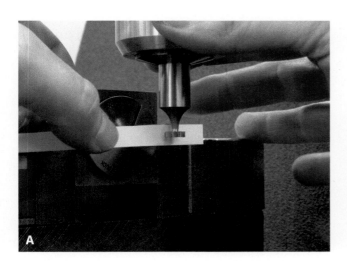

FIGURE 6.3.85 (A) To locate the Woodruff keyseat cutter from the end of a shaft, first touch off the end of the shaft with a strip of paper. (B) Then move the paper thickness, one-half the cutter diameter, and the desired distance.

- Move the machine axis one-half of the cutter diameter.
- Move the machine axis the required distance to position the center of the keyseat along the shaft.
- Lock the machine axis in place to prevent movement.
- After the Woodruff keyseat cutter is positioned at the required location, the keyseat is ready to be machined.
- Calculate and set spindle RPM.
- Touch off the cutter against the side of the shaft.
- Gently feed the cutter to the desired keyseat depth. **(See Figure 6.3.86.)** Woodruff cutters are somewhat fragile, so use a slow and steady motion to prevent cutter breakage and use a liberal amount of cutting fluid.

Slotting with Slitting Saws

Stub-arbor-mounted slitting saws are often used to machine slots parallel with the surface of the machine table. They can be positioned using the same method used for positioning Woodruff keyseat cutters by touching off the top of the work, then raising the knee to position the cutter in the desired location. Use the X-axis to touch off the side of the workpiece and set cut depth. Then use the Y-axis to feed to make the cutting passes. When using saw diameters larger than the quill diameter, speeds may need to be reduced to eliminate vibration. Use plenty of cutting fluid to clear chips from the cutting area to prevent binding. **Figure 6.3.87** shows an arbor-mounted slitting saw being used to mill a slot.

FIGURE 6.3.86   (A) Machining a Woodruff keyseat. (B) The finished keyseat.

FIGURE 6.3.87   Milling a slot with a stub-arbor-mounted slitting saw.

MILLING RADII

External radii can be milled using corner-rounding and concave milling cutters. Internal radii (fillets) can be machined using ball endmills, radius (bullnose) endmills, and convex cutters. Speeds for corner-rounding, concave, and convex cutters often need to be reduced because of the large area of contact between the tool and workpiece. Speeds for ballnose and bullnose endmills can be equal to those used for standard endmills.

Milling External Radii

The corner-rounding endmill is frequently used on the vertical mill, but a stub-arbor-mounted cutter may be needed to mill larger radii. **Figure 6.3.88** illustrates the two following methods used to machine a corner radius.

Method 1

- Position the cutting tool with the top edge about 0.010" above the top surface of the workpiece.
- Take conventional milling passes by setting cut depth with the X-axis and feeding with the Y-axis.
- As the profile of the radius begins to "wrap" around the corner, take light climb milling cuts of 0.001" to 0.002" until the lower edge of the radius becomes tangent to the vertical surface.
- Use the knee to take light climb milling passes until the upper edge of the radius is tangent to the top surface of the workpiece.

Method 2

- Position the tool so the lower edge of the tool is about 0.010" from the vertical surface.
- Take conventional milling passes by setting cut depth with the knee and feeding with the Y-axis.
- As the profile of the radius begins to "wrap" around the corner, take light climb milling cuts of about 0.001" to 0.002" until the upper edge of the radius is tangent to the top surface.
- Use the X-axis set depth for light climb milling passes until the lower edge of the radius is tangent to the vertical surface.

An arbor-mounted corner-rounding cutter could also be mounted with the radius facing up to machine a corner radius on the bottom edge of a workpiece. **(See Figure 6.3.89.)**

A concave milling cutter can be used to mill an external radius with an arc up to 180 degrees. **(See Figure 6.3.90.)** These cutters can be positioned using the same method used for positioning Woodruff keyseat cutters and slitting saws.

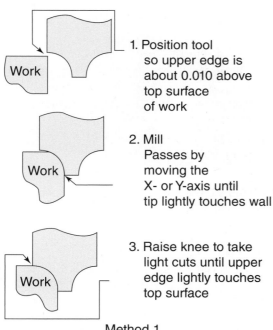

1. Position tool so upper edge is about 0.010 above top surface of work

2. Mill Passes by moving the X- or Y-axis until tip lightly touches wall

3. Raise knee to take light cuts until upper edge lightly touches top surface

Method 1

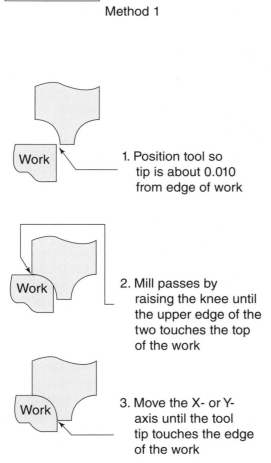

1. Position tool so tip is about 0.010 from edge of work

2. Mill passes by raising the knee until the upper edge of the two touches the top of the work

3. Move the X- or Y-axis until the tool tip touches the edge of the work

Method 2

FIGURE 6.3.88 Two methods for milling an external corner radius.

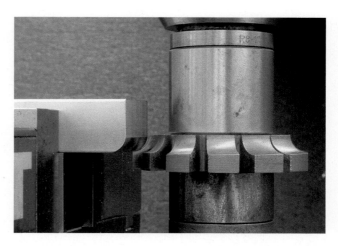

FIGURE 6.3.89 A radius on the bottom edge of a workpiece machined by a stub-arbor-mounted radius cutter.

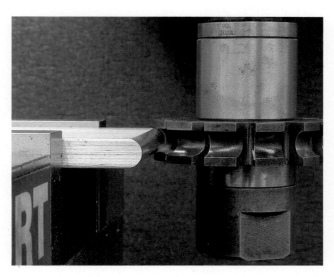

FIGURE 6.3.90 A full external radius machined by a stub-arbor-mounted concave milling cutter.

Milling Internal Radii (Fillets)

Ballnose and bullnose endmills are widely used on the vertical milling machine to produce internal radii. A bullnose endmill can be used to produce a fillet in a 90-degree corner. It is common to first rough the step and leave an amount of material in the corner slightly greater than the size of the endmill's corner radius. **Figure 6.3.91** illustrates these three methods that can be used:

- Leaving material on the adjacent vertical surface to be removed by the bullnose endmill.
- Leaving material on the adjacent horizontal surface to be removed by the bullnose endmill.
- Leaving a step of material in the corner to be removed by the bullnose endmill.

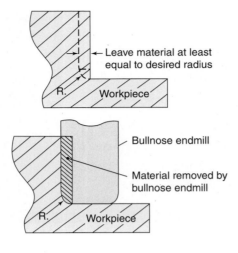

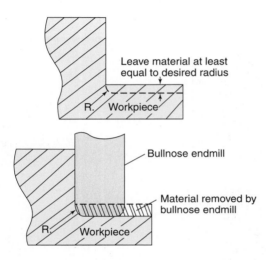

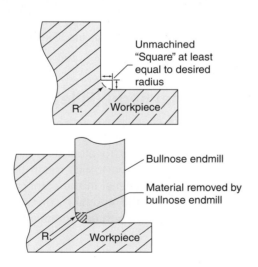

FIGURE 6.3.91 When roughing an internal corner before machining a fillet, three different methods can be used, as shown in A, B, and C. In all cases, be sure to leave enough material for the radius of the tool. Then a bullnose endmill can be used to machine the fillet.

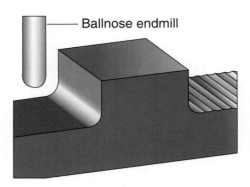

FIGURE 6.3.92 Use a small stepover amount when machining a fillet with a ballnose endmill (left) to avoid noticeable "steps" in the horizontal surface (right).

A ball endmill can also be used to machine a fillet radius. Again, if a step is machined with a roughing or standard endmill, be sure to leave enough material in the corner for the radius of the ball endmill. Keep in mind that the tip is a full radius with no flat to machine a horizontal surface. For that reason, small steps will remain that may be noticeable. Those steps can be minimized by stepping over about 0.001" to 0.003" for several small finishing passes to produce a fairly smooth horizontal surface. **(See Figure 6.3.92.)**

A ball endmill can also be used to mill a slot with a full radius or to machine a spherical depression, as shown in **Figure 6.3.93**.

A convex cutter can also be used to mill a slot with a full radius, as shown in **Figure 6.3.94**. It can be positioned using the same techniques used for locating a Woodruff keyseat cutter, slitting saw, or concave cutter.

POCKET MILLING

A pocket is an internal part feature machined into the surface of a workpiece. An open pocket breaks through at least one edge of the workpiece while a closed pocket is completely contained within the outer edges of the workpiece. **Figure 6.3.95** shows a few examples of open and closed pockets. Rectangular-shaped pockets can be machined on the vertical mill. Pocket location and size can be controlled by using the micrometer collars or DRO to monitor table and saddle movements. Depth is normally controlled with the knee. It is common to rough with a larger diameter endmill because it can remove material more quickly than a smaller diameter endmill. Then a smaller diameter can be used to finish the pocket and create smaller required corner radii. **(See Figure 6.3.96.)** Follow these steps to machine a rectangular pocket:

• If desired, lay out the pocket boundary to help provide a visual reference.

• Select the roughing and/or finishing endmill diameter(s).

FIGURE 6.3.94 A radius slot machined in the side of a work-piece with a convex milling cutter mounted on a stub arbor.

FIGURE 6.3.93 (A) Radius slots. (B) A spherical depression machined with a ball endmill.

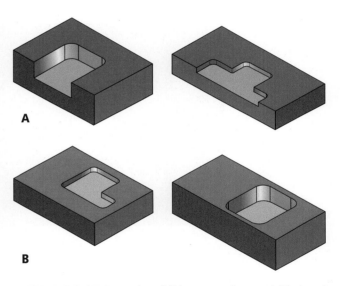

FIGURE 6.3.95 Examples of (A) open pockets and (B) closed pockets that can be machined on the vertical mill.

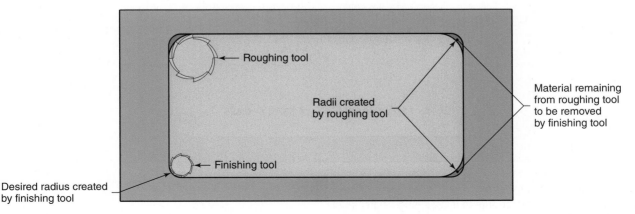

FIGURE 6.3.96 Rough machining of a pocket is often performed with a larger-diameter endmill than the finishing operation to remove material more quickly. Material left in the corners is then removed by the smaller radius of the finishing tool.

- Create a *coordinate map*. This can be a simple hand sketch showing the X- and Y-axis coordinates of the center point locations for the cutting tool. Use the center of the pocket for the origin. The radius of the endmill must be taken into consideration when calculating these coordinates. Two sets of coordinates will need to be calculated if two different diameter endmills will be used. See **Figure 6.3.97** for an illustration of how to determine these values.

- Establish a reference "0" position in the center of the pocket.

- Mount the desired endmill and position the tool in the center of the pocket.

- Set the tool about 1/16" above the part surface with the quill and lock quill in place.

- Calculate and set appropriate spindle RPM.

- Start the spindle and touch off the top surface of the workpiece by raising the knee.

- Set the collar on the knee to "0".

- Feed with the knee for the first depth cut. Even if the pocket is shallow enough to be machined in one depth pass, it is still a good idea to leave about 0.005" to 0.010" on the bottom for a finishing pass.

- Mill roughing passes outward from the center in a clockwise direction to use conventional milling. Step over about ½ to ¾ of the endmill diameter. Lock the axis not being fed to prevent unwanted movement. **(See Figure 6.3.98.)**

- Plan to leave some material on each wall of the pocket by stopping about 0.020" short of each position on the coordinate map. If using a roughing endmill, more material may need to be left on the walls so an adequate finish cut can be made.

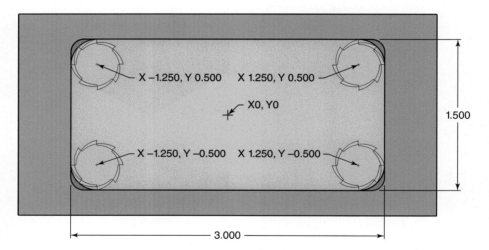

Coordinate locations for roughing using a 0.500" diameter tool (0.250" radius).

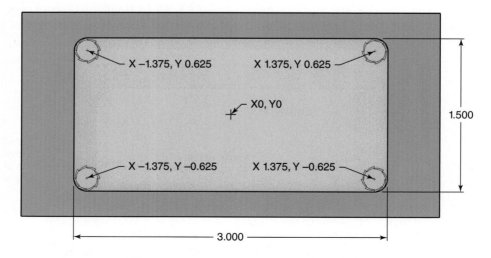

Coordinate locations for finishing using a 0.250" diameter tool (0.125" radius).

FIGURE 6.3.97 A coordinate map for milling a rectangular pocket.

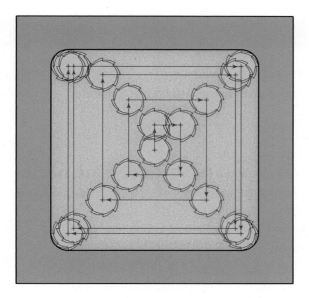

FIGURE 6.3.98 Typical pattern for milling a square or rectangular pocket from inside to outside. From the start point, incrementally step over about ½ to ¾ of the endmill diameter and feed the tool in a clockwise direction.

- Return to the starting point at the pocket center and repeat roughing passes to within 0.005" to 0.010" of final depth.

- Climb mill around the outside boundary, removing just enough material to create smooth surfaces that can be used to measure sizes. Continue to stay short of the final coordinates from the map by about 0.020".
- Return to the pocket center.
- Withdraw the endmill from the pocket and shut off the spindle.
- Measure depth, size, and position of the pocket.
- Feed the knee incrementally until final depth is reached, making adjustments to the locations on the coordinate map as needed to adjust the size and position of the pocket.
- Mill conventional passes around the boundary within about 0.005" of final coordinates.
- Mill a final conventional pass around the boundary at the final coordinates.
- Machine a climb milling spring pass around the boundary at the final coordinates.
- Return to pocket center, withdraw tool from the pocket, and verify dimensions. Do not remove the workpiece from the workholding device until all pocket dimensions are verified because it can be difficult to reestablish the coordinate locations if the workpiece is repositioned.

SUMMARY

- Many different machining operations can be performed on the vertical milling machine. By respecting the machine and tooling and following a few guidelines, these operations can be conducted in a safe manner.
- An understanding of the methods of the setup and alignment of the milling machine head and workholding devices provides a foundation for machining workpieces to required specifications. As with any machining operation, calculation and use of proper speeds, feeds, and cutting methods is necessary to safely and efficiently perform milling machine operations.
- The saddle and table allow the vertical milling machine to be used for accurate positioning to perform holemaking operations, including drilling, reaming, and boring.
- An edge finder or dial indicator can be used to locate workpiece features.
- Different procedures can be used to mill parallel, perpendicular, and angular surfaces by positioning the machine head, the workpiece, or the workholding device, or using various cutting tools.
- Steps and slots of different shapes can be machined on the vertical milling machine using both shank-type and arbor-type milling cutters.
- Specialty cutters can be used to mill external radii and internal radii.
- Pocket machining can be performed on the vertical mill and requires the calculation of X- and Y-coordinate positions to properly move the cutting tool within the boundary of the pocket.
- Gaining an understanding of these principles and methods is a vital part of learning to safely and efficiently set up and operate the vertical milling machine.

REVIEW QUESTIONS

1. List five safety guidelines to observe when operating the vertical milling machine.
2. Briefly describe the process of aligning a milling vise with the table movement using a dial indicator.
3. What is chip load?
4. Define *IPM*.
5. Calculate spindle speed and feed for the two following situations:
   a. Use an HSS 1/2"-diameter three-flute endmill to mill 6061 aluminum 350 SFPM using a 0.0015" chip load.
   b. Use a carbide 3"-diameter eight-tooth shell mill to machine H13 tool steel at 200 SFPM using a 0.003" chip load.
6. Briefly describe the process of locating the center of an existing hole on a vertical milling machine.
7. What are two benefits of boring over other holemaking operations?
8. What are face milling and peripheral milling?
9. When squaring a block on the vertical mill, what surface of the block should be machined first, and why?
10. What are the three basic methods used to mill angular surfaces?
11. What must first be done before milling with either a T-slot cutter or dovetail cutter?
12. What are the two basic types of keys?
13. What diameter cutter should be used to create 3/16" radii in the corners of a pocket?
14. When roughing a pocket, should you machine in a clockwise or counterclockwise direction, and why?

Indexing and Rotary Table Operations

UNIT 4

Learning Objectives

After completing this unit, the student should be able to:

- Demonstrate understanding of the capabilities of the rotary table, indexing fixture, and dividing head
- Identify the basic parts of a rotary table, indexing fixture, and dividing head
- Demonstrate understanding of the basic setup and operation of the rotary table, collet blocks, indexing fixture, and dividing head
- Perform direct and simple indexing calculations

Key Terms

Direct indexing	**Indexing**	**Sector arms**
Dividing head	**Indexing fixture**	**Simple indexing**
(Indexing head)	**Rotary table**	

INTRODUCTION

Rotary tables, indexing fixtures, and **dividing heads** are specialized workholding devices that expand the capabilities of the milling machine. These devices add to the milling machine what is sometimes called a *rotary axis* because they provide circular motion. **Indexing** is positioning a workpiece at different angular positions using that rotary motion. Indexing can also be performed using a collet block to position a workpiece at angular positions. Rotary tables, indexing fixtures, and dividing heads are useful when milling multiple angles on workpieces, locating holes in circular patterns called bolt circles, and positioning to machine features angularly spaced on workpieces. Rotary tables can also be used to mill radii, circles, and arcs. **Figure 6.4.1** shows some workpieces with features that were machined using rotary tables, indexing fixtures, and dividing heads.

ROTARY TABLE

The rotary table consists of a heavy base on which a circular worktable is mounted. Many rotary tables can be mounted either horizontally or vertically. The table contains T-slots for mounting work and is connected to the base by bearings that allow it to be rotated smoothly and accurately. The table has degree graduations around its periphery to aid in positioning. A hand wheel connected to a gear set is used to rotate the table. Since the gear set produces gear reduction, the hand wheel must be rotated several times to make one revolution of the table. This gear reduction increases accuracy and control when positioning the table. The rotary table may be locked in position to prevent it from drifting during setup or machining. Many rotary tables have a center hole all the way through the work table, and some have a Morse

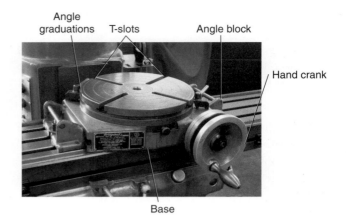

FIGURE 6.4.2 The parts of the rotary table.

taper within this hole for mounting taper-shank workholding devices. **Figure 6.4.2** shows a rotary table.

Rotary Table Setup

Most rotary table operations involve mounting the rotary table horizontally. It must be mounted to the milling machine table in the same manner as a milling vise. Usually the mill vise must be removed to make enough space. Make sure that the mill table and rotary table base are both free of burrs and debris. The rotary table can usually be fastened to the mill table using the T-slots and the same hardware used for the vise.

After mounting the rotary table on the mill table, the next step is to align the rotary table center line with the machine spindle center line. This alignment can be achieved by moving the milling machine axis hand wheels.

- A 0.0005" graduated dial test indicator should be mounted in the spindle in a chuck or collet so that it can rotate along with the spindle.

- The hole in the middle of the rotary table must be precisely aligned with the spindle centerline. To do this, place the machine spindle in neutral and the indicator just over the top of the rotary table hole. Rotate the spindle by hand one full turn. While turning the spindle, watch where the indicator contact ball is in relationship to the center hole. Adjust the milling machine table visually until the indicator contact ball appears to be rotating concentrically with the hole in the table. This method is only intended to get the table alignment close.

- After visually aligning the spindle and rotary table, the indicator can be lowered down into the hole. Then position the indicator contact against the surface of the hole to preload the indicator and zero the dial face. Next, rotate the spindle again by hand. Note the indicator dial reading. Perform adjustments as if performing any indicator sweeping alignment (covered

FIGURE 6.4.1 Workpieces with features machined using rotary tables and index heads.

in Unit 6.3). Center the spindle and rotary table within 0.0005".

- When alignment has been achieved, lock the X- and Y-axes and set the micrometer collars to zero and/or set the DRO to zero.

CAUTION

Rotary tables can be very heavy, so get help when moving and mounting them, and use proper lifting methods.

Workpiece Setup for the Rotary Table

The next step is to mount the workpiece. Since the rotary table is such a versatile device, it can be used for many different applications. The following steps cover some basic situations that may be encountered.

Angular Orientation of the Workpiece on the Rotary Table

Sometimes a partial radius must be milled on a workpiece. When this is the case, the table rotation must start and stop at precisely the correct location for each cut. **(See Figure 6.4.3.)** The rotary table may also be used to index a workpiece so that angular features can be milled or holes can be drilled. **(See Figure 6.4.4.)** When the part requires this type of work, it is helpful to orient it properly so that the degree markings on the table can be used as an aid.

- First, rotate the table to the desired beginning graduation (usually "0") and lock in place.
- Next, approximately align the workpiece visually.

FIGURE 6.4.3 The rotary table setup for milling a partial radius.

FIGURE 6.4.4 A rotary table setup for drilling equally spaced holes in a circular pattern.

FIGURE 6.4.5 Aligning the straight edge of a workpiece on a rotary table.

- Loosely clamp the work to the rotary table with appropriate clamps and then align the part with a dial test indicator. If the part has a straight edge, run the indicator along this edge by moving one of the milling table axes and tap the workpiece into alignment. **(See Figure 6.4.5.)**
- After alignment, tighten the clamps. Then move the table axis back to the "0" position.

Centering of the Workpiece on the Rotary Table

The center of a bolt circle or radius to be milled must also be aligned with the center of the rotary table. Follow these steps to perform that alignment.

- Position the X- and Y-axes of the mill at the "0" reference location where the spindle was centered with the rotary table.

- If the work has a center hole or a circular profile concentric with the feature to be machined, it can be swept with an indicator and the workpiece aligned on center by tapping with a soft-faced hammer. **(See Figure 6.4.6.)**

- Work that cannot be swept with an indicator can be visually aligned by using intersecting layout lines and a spindle-mounted wiggler or cone-pointed center finder. **(See Figure 6.4.7.)**

- Never adjust the milling machine table axes to achieve this alignment, as it has already been aligned with the center of the rotary table and must stay there—only move the workpiece on the rotary table.

- Once the workpiece is in position, tighten the clamps and recheck the alignment. **(See Figure 6.4.8.)**

FIGURE 6.4.6 Aligning a workpiece with a center hole on a rotary table using a test indicator.

FIGURE 6.4.7 Using a wiggler to align intersecting lines on the workpiece with the center of a rotary table.

FIGURE 6.4.8 After positioning the workpiece, tighten the clamps and recheck the alignment.

Rotary Table Operations

With both the workpiece and the rotary table precisely positioned and the workpiece properly secured, machining operations can begin.

Indexing for Milling and Drilling

If the rotary table is only to be used to index the work to a desired angle for machining, the rotary table lock must be released and the hand wheel rotated to position the rotary table at the correct angle. The lock must be retightened before machining begins to maintain position.

To position holes on a bolt circle, move one table axis a distance equal to the radius of the bolt circle. Then hole positions can be located by rotating the table to the desired angular settings using the rotary table graduations. Be sure to lock the rotary table in position before machining. **(See Figure 6.4.9.)**

FIGURE 6.4.9 Move one axis the distance equal to the radius when drilling a bolt circle, then use the rotary table graduations to position the workpiece at the desired angular locations.

Consider this example:

> **EXAMPLE ONE:** Drill five equally spaced holes on a 1.500" diameter bolt circle.
>
> First, move one axis 0.750" from the origin, or X0, Y0 position. (0.750" is the radius of 1.500" bolt circle).
>
> Since there are 360 degrees in a circle and five holes are required, divide 360 by 5. The answer, 72, is the angular spacing between each hole.
>
> Drill the first hole at the previously set "0" position of the rotary table's scale.
>
> After drilling the first hole, unlock the rotary table, turn the handle to rotate the table to the 72-degree graduation, and relock the rotary table.
>
> Drill the next hole, then repeat the process by rotating the table 72 degrees for each hole position until all five holes are drilled.

To mill angles, use the rotary table to rotate the workpiece to the desired angular setting using the rotary table graduations. Again, be sure to lock the rotary table in position. Then use one of the table axes to set cut position and feed with the other table axis to perform the milling pass. **(See Figure 6.4.10.)**

Milling Outside and Inside Radii

To mill a radius, the clamp on one of the mill axes must be released and the table moved to offset the cutter from the rotary table centerline (rotation axis). Only offset one of the mill table axes to cut a radius. The offset amount is determined by taking the finished workpiece radius dimension and adjusting the amount to compensate for the cutter radius when needed. When milling an external feature, add the cutter radius. When milling an internal feature, subtract the cutter radius. When milling a radius slot dimensioned to its centerline, make no cutter radius adjustment. **(See Figure 6.4.11.)**

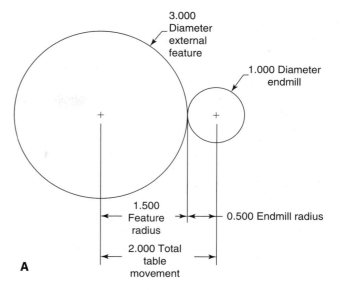

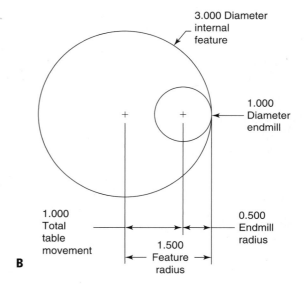

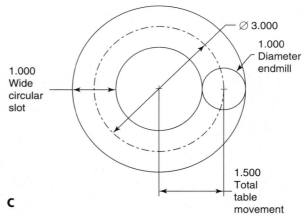

FIGURE 6.4.11 (A) Compensating for tool when milling an external circular feature. (B) Compensating for tool when milling an internal feature. (C) No tool compensation is needed when a circular slot is dimensional to its center line and machined with a tool equal to the slot width.

FIGURE 6.4.10 A rotary table can be used to position a workpiece at desired angular locations for milling features such as flats or slots.

The same rules of standard milling apply when using the rotary table. Remember to be cautious when climb milling while rotating the table, as it is especially easy for the workpiece to be pulled into the cutter. If milling a partial arc that adjoins another feature or tangency, it is important to keep a close eye on the location of the cutter, as it is very easy to go past the desired stopping point. Use the rotary table graduations to closely monitor those arc endpoints.

COLLET BLOCKS

Indexing can be performed using a device such as a square or hexagonal collet block. A collet fits in the block and a nut threads on the back of the collet to secure the workpiece in the collet. The block can be held in a standard milling vise. Indexing is performed manually by removing the block from the vise, rotating it to a new position, and reinstalling in the vise. **Figure 6.4.12** shows a square and hexagonal collet block.

One common operation is to mill a square on a cylindrical workpiece using a square collet block. Here is an example of how to mill a ½" square on a ¾" diameter workpiece.

- First determine depth of cut for the flats to be machined. Subtract the desired distance across the square from the beginning diameter. So ¾ − ½ = ¼ or 0.250. Half of that amount needs to be machined from each side, so 0.125" must be removed at four locations 90 degrees apart.
- Mount the ¾" diameter workpiece in the square collet block.
- Mount the block in a milling vise. Since the block will need to be removed from the vise, rotated, and remounted in the vise, using a workstop is helpful.
- Mount a cutting tool in the spindle, then calculate and set spindle RPM.

- Position the quill with the micrometer adjusting nut against the stop so the tool is about 1/16" above the workpiece, and then lock the quill.
- Position the tool for the desired length of the flat.
- Start the spindle.
- Raise the knee to touch the tool off the top of the workpiece. **(See Figure 6.4.13.)**
- Clear the tool from the workpiece and set cut depth with the knee. It is a good practice to leave about 0.010" to 0.015" to check size before finishing.
- Mill the first flat as shown in **Figure 6.4.14**.
- Stop the spindle.
- Remove the block, rotate it 90 degrees, and remount it in the vise against the workstop.
- Mill the second flat as shown in **Figure 6.4.15**.

FIGURE 6.4.13 Touching off the top of the workpiece held in a square collet block. Note the workstop against the block that will be used for repositioning.

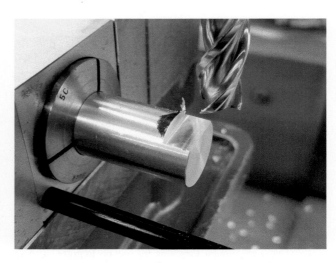

FIGURE 6.4.12 Square and hexagonal collet blocks.

FIGURE 6.4.14 Milling the first flat of a square.

FIGURE 6.4.15 Milling the second flat of a square after rotating the collet block in the vise.

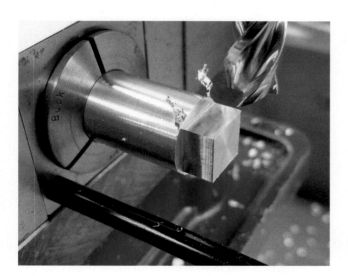

FIGURE 6.4.16 A completed machined square.

- Repeat the process for the remaining two flats. **(See Figure 6.4.16.)**
- Stop the spindle and measure the distance across the flats.
- Adjust depth with the knee as needed to mill to the desired size.
- Remember that cut depth will be one half of the total size difference. For example, if the distance across the flats is 0.020" over the desired size, machine an additional 0.010" from each flat.

A hexagon can be machined using a hexagon collet block by following the same steps.

INDEXING FIXTURE

An **indexing fixture** has a spindle that can be rotated manually and then locked in desired positions. Positioning the workpiece in this type of fixture is called **direct indexing**. An indexing fixture commonly uses a three-jaw chuck or a collet to secure the workpiece. Most can be mounted horizontally or vertically. **Figure 6.4.17** shows a collet type indexing fixture. The indexing fixture has a ring with certain number of notches around its circumference. Common numbers of notches are 18, 24, 30, and 36. Some models have the ability to change the indexing ring, while others do not. The indexing ring is chosen based on how many divisions must be machined on the workpiece. These divisions are determined by number of features to be machined.

Use the following formula to determine a correct indexing ring and the number of notches needed to position for the required number of divisions.

$$\text{Notches per Division} = \frac{\text{Notches on Index Plate}}{\text{Divisions Required}}$$

Compare this formula to "miles per hour." Miles per hour is $\frac{Miles}{Hours}$, or miles divided by hours. "Notches per division" is $\frac{Notches}{Division}$, or notches divided by divisions.

If the spacing between features is specified by an angular dimension, first divide 360 by the angular dimension. That answer is the number of divisions to be used in the formula.

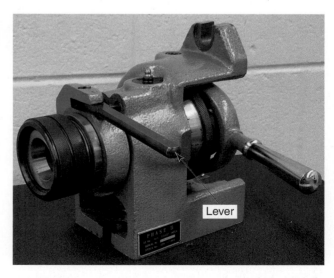

FIGURE 6.4.17 An indexing fixture that uses collets to hold workpieces. The lever is used to unlock, rotate, and relock the spindle in position.

EXAMPLE: For a part needing six equal divisions, such as a part that needs hexagonal flats for a wrench:

Since 24 is equally divisible by 6, a direct indexing plate with 24 notches can be used. When positioning for each division, the spindle must be rotated 4 notches because 24 ÷ 6 = 4.

EXAMPLE: For a part needing 10 holes with 36-degree spacing between the holes (10 equal divisions), a 24-hole direct indexing plate would yield the following:

$$\frac{24}{10} = 2.4$$

*Since dividing 24 notches by 10 divisions yields 2.4, this direct indexing plate is **not** acceptable. With this plate locking can occur for 8 divisions (3 notches) or 12 divisions (2 notches), but an indexing plate cannot be properly locked between notches.*

However, if a direct indexing plate with 30 notches is available:

$$\frac{30}{10} = 3$$

Since this 30-notch plate is equally divisible by the number of desired divisions, it is the correct choice. When positioning for each division, the spindle must be rotated 3 notches in the 30-notch plate.

Here are steps to mill 6 flats (a hexagon) on a cylindrical part using an indexing fixture with a 24-notch ring. Since 24 ÷ 6 = 4, the spindle of the fixture will be rotated 4 notches for each cutting position. Cut depth is determined just like when machining a square by subtracting the desired distance across the flats from the starting size, then dividing that answer by 2.

- Mount the fixture on the milling machine table and secure the workpiece in the fixture.
- Select and mount the desired cutting tool, then calculate and set spindle RPM.
- Position the cutting tool about 1/16" above the work with the quill stop against the micrometer adjusting nut and lock the quill.
- Start the spindle.
- Raise the knee to touch the tool off the workpiece.
- Move the tool clear of the workpiece.
- Set cut depth with the knee. Again, it is a good idea to stay about 0.010" to 0.015" away from final depth.
- Cut the first flat.

FIGURE 6.4.18 Hexagonal flats milled using an indexing fixture.

- Rotate the spindle four notches for each position to cut the remaining flats. **Figure 6.4.18** shows hexagonal flats milled using an indexing fixture.
- Stop the spindle and measure the distance across the flats.
- Adjust depth with the knee as needed to mill to the desired size.

DIVIDING HEAD

A **dividing head** (sometimes called an **indexing head**) is another workholding device that serves as a rotary axis for part positioning. The dividing head provides the ability to rotate the workpiece in precise increments. It also allows many more possibilities of division than other methods.

The dividing head consists of a gear box, spindle, simple index plate, **sector arms**, and simple index crank. Some models also have a direct indexing plate and pin/plunger to perform direct indexing like the fixtures discussed previously. **Figure 6.4.19** shows a dividing head and labels its major parts. Work may be held in a dividing head in a chuck or between centers. Some dividing head spindles can also accept collets that can be used for holding the workpiece. A tailstock can be used to hold work between centers or to support a long workpiece held in a chuck.

The dividing head is mounted on the mill table and fastened with T-slot hardware much like a milling vise. Many dividing heads have a flat reference surface intended for indicating. A test bar can be mounted in models with chucks or collets and aligned using an indicator as shown in **Figure 6.4.20**. If a tailstock will be used to help hold the work between centers, it must also be aligned with the dividing head so that both centers align. A test bar is

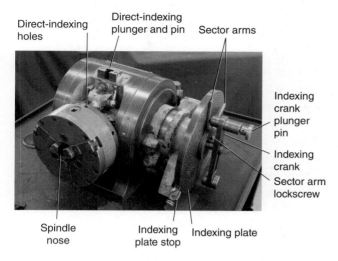

FIGURE 6.4.19 A dividing head. This model can also be used for direct indexing.

FIGURE 6.4.20 An indicator can be used to align the dividing head.

often used to do this and the procedure is similar to aligning the centers on the lathe. It is also a good idea to check the height alignment of the centers while the test bar is mounted since some tailstocks have a height adjustment.

 **CAUTION**

Dividing heads can be very heavy, so get help when moving them and use proper lifting methods and equipment.

The dividing head uses a crank handle connected to a gear train to rotate the spindle. The gear train accomplishes a gear reduction, and most indexing heads have a ratio of 40:1. In other words, 40 turns of the crank handle are required to rotate the spindle (and the workpiece) 1 revolution. By turning the crank handle to rotate the spindle, and then holding it in that position, the mounted

FIGURE 6.4.21 A simple indexing plate. Note that each circle has a different number of equally spaced holes.

workpiece can be positioned to any angle of rotation with great control. This is called **simple indexing**.

Simple indexing requires the crank handle to be rotated to orient the indexing head spindle. The pin on the end of the crank handle then also holds the work in position for machining. Simple indexing plates with evenly spaced holes in a circular pattern are used to measure the distance the crank handle is rotated. Each circular pattern has a different amount of holes. **(See Figure 6.4.21.)** By selecting the correct plate and rotating the crank the correct amount, many equally spaced divisions can be achieved. Many more possible numbers of divisions may be achieved with this method than with the direct indexing method, because the hole patterns allow precise measurement of full and partial turns of the crank. A spring-loaded locking pin on the end of the crank is used with the indexing plate to align and hold the crank handle in the desired hole.

Since 40 turns of the crank result in 1 turn of the spindle, the amount of crank turn(s) is found by dividing 40 by the number of desired divisions. Then a suitable indexing plate is selected to accurately rotate the required number of full and/or partial turns.

The following standard plates are typically available:

Plate #1 has hole circles with 15, 16, 17, 18, 19, and 20 holes

Plate #2 has hole circles with 21, 23, 27, 29, 31, and 33 holes

Plate #3 has hole circles of 37, 39, 41, 43, 47, and 49 holes

Since 40 turns of the crank result in 1 turn of the spindle, the amount of crank turn(s) is found by dividing 40 by the number of desired divisions:

$$\frac{40}{D} = T$$

T = number of turns of the crank handle
D = number of desired divisions

Remember that if the spacing between features is given as an angular dimension, first divide 360 by that angular dimension to obtain the *D* value.

EXAMPLE: Drill a hole every 72 degrees in a circular pattern on a workpiece:

First, determine *D* using the given 72 degrees:

$$D = \frac{360}{72} = 5$$

Then use the *D* value of 5 to calculate turns:

$$T = \frac{40}{5} = 8$$

Eight full turns of the crank are needed.

Any indexing plate could be selected because no partial turns are needed.

- Mount the plate and workpiece.
- Place the pin on the crank in the starting hole of any hole circle pattern.
- Rotate the crank 8 full turns and place the pin in the same hole each time.

The math will not always work out to be convenient whole numbers (complete turns). Sometimes only a partial turn must be made with the crank.

EXAMPLE: Machine 50 slots around the circumference of a workpiece:

$$T = \frac{40}{D} = \frac{40}{50} = \frac{4}{5}$$

In this case, a partial turn of the crank will be needed because the answer is a fraction. The crank needs to be rotated 4/5 of a turn for each division on the workpiece.

- An index plate with a number of holes divisible by the denominator must be selected. Plate #1 from the standard set is a good choice because it contains two hole circles that are divisible by five: 15 and 20. Either can be used.
- For this example, suppose the 15-hole circle is chosen. The $\frac{4}{5}$ then needs to be converted to a fraction with a denominator of 15: $\frac{4}{5} = \frac{12}{15}$. The $\frac{12}{15}$ means that the crank will be rotated 12 holes in the 15-hole circle, resulting $\frac{4}{5}$ in of a turn for each division on the workpiece.

- Mount the plate with the 15-hole circle and place the pin of the crank in the starting hole of the 15-hole circle.
- Adjust the sector arms so the first arm is against the pin on the crank.
- Adjust the second arm so there are 12 holes between it and the pin. Do not count the beginning hole where the pin is engaged. **(See Figure 6.4.22.)**
- For each division on the workpiece, rotate the crank to position the pin in the hole next to the second sector arm. Then rotate the sector arms clockwise to reset the 12-hole spacing again. **(See Figure 6.4.23.)**
- Repeat the process to position for each division on the workpiece.

FIGURE 6.4.22 Adjusting sector arms to obtain the correct spacing between holes.

FIGURE 6.4.23 Rotate the sector arms to reset the hole spacing after positioning the crank.

EXAMPLE: Make 23 divisions on a workpiece:

$$\frac{40}{D} = T$$

$$\frac{40}{23} = 1\frac{17}{23}$$

In this case, one full turn plus 17/23 turn of the crank will be needed.

- Plate #2 is ideal because it contains a 23-hole circle. No conversion of the fraction is needed.

- The crank will be rotated one full turn, plus 17 holes in the 23-hole circle, resulting in $1\frac{17}{23}$ of a turn for each division on the workpiece.

- Mount the plate with the 23-hole circle and place the pin of the crank in the starting hole of the 23-hole circle.

- Adjust the sector arms so the first arm is against the pin on the crank.

- Adjust the second arm so there are 17 holes between it and the pin. Remember: Do not count the beginning hole where the pin is engaged.

- For each division on the workpiece, rotate the crank one full turn, and then continue rotating to reach the hole next to the second sector arm and engage the pin. Then rotate the sector arms clockwise to reset the 17-hole spacing.

- Repeat the process to position for each division on the workpiece.

EXAMPLE: Drill 30 equally spaced holes on a bolt circle on a workpiece:

$$\frac{40}{D} = T$$

$$\frac{40}{30} = 1\frac{1}{3}$$

Again, the whole number represents the number of complete hand-crank turns for each division and the fractional part of the answer represents the required partial-crank turn. Any of the three standard plates will work because each has a hole circle divisible by 3, but suppose only plate #3 is available.

- Its 39-hole circle is divisible by 3, so the 1/3 must be converted to a fraction with a denominator of $39 : \frac{1}{3} = \frac{13}{39}$.

- Set the sector arms for 13 holes in the 39-hole circle.

- Index the work by rotating 1 full turn and the 13 holes, resulting in $1\frac{1}{3}$ turns for each division on the workpiece.

- Remember to rotate the sector clockwise after each index to reset the hole spacing.

SUMMARY

- The rotary table can be used to mill angles and radii on a workpiece, to position features angularly, and to drill holes in circular patterns called bolt circles.

- The hole located in the center of the rotary table must be precisely aligned with the machine spindle centerline. Since this is a critical aspect of the setup, it should be performed using a 0.0005" graduated dial indicator to ensure correct alignment.

- The workpiece must also be correctly aligned with the spindle as well using a dial indicator.

- The angular graduations on the perimeter of the rotary table can be used for angular positioning of the workpiece.

- When milling a radius or positioning holes on a bolt circle, one table axis must be moved to position the cutting tool at the correct location.

- When cutting an arc to a tangent point or other critical stopping point, be careful not to over-travel at the end of the cuts, which can produce an undercut and ruin the workpiece.

- Collet blocks are often used to quickly position work to machine square and hexagonal features.

(Continued)

- Direct indexing allows the part to be rapidly positioned by using an indexing fixture. The number of divisions is limited by the number of notches the direct index ring.

- The dividing head may be used for simple indexing to provide many more options than a direct indexing fixture allows.

- Simple indexing requires a crank to be turned to rotate the dividing head spindle by precise and small amounts. The crank position can be measured by using simple indexing plates. This method offers very fine control of the workpiece rotary position and many possibilities for the number of divisions.

REVIEW QUESTIONS

1. What is a rotary axis?
2. Define the term *indexing*.
3. Describe the primary differences between the rotary table and the dividing head.
4. Name three types of workpiece features that are well suited to using a rotary positioning device.
5. Briefly describe the two alignment steps that need to be performed when setting up the rotary table on the milling machine.
6. If the outside of a 6"-diameter disc is to be milled using a rotary table and a 3/4"-diameter endmill, what is the total amount the milling machine axis must be offset from the center of the rotary table?
7. What is the gear ratio found in the gear train of most indexing heads?
8. When using the dividing head, a _____ can be used to support long workpieces during machining.
9. A workpiece requires 9 divisions. Calculate the correct direct indexing ring and number of notches to advance the fixture if plates with 24, 30, and 36 notches are available.
10. In the formula $\frac{40}{D} = T$, what do T and D represent?

GRINDING — SECTION 7

UNIT 1

Introduction to Precision Grinding Machines

Learning Objectives

After completing this unit, the student should be able to:

- Demonstrate understanding of the benefits of precision grinding
- Identify and demonstrate understanding of various types of grinders and their capabilities
- Identify and demonstrate understanding of the parts of a surface grinder

Key Terms

Cylindrical grinder	**Jig grinder**	**Vertical spindle surface**
Horizontal spindle	**Surface grinder**	**grinder**
surface grinder	**Tool and cutter grinder**	

INTRODUCTION

Precision grinding removes material by the same principle as offhand grinding. Each abrasive grain of the grinding wheel acts like an individual cutting tool and removes a small portion of material. Unlike offhand grinding, where a workpiece is held by hand, precision grinding machines hold work securely and provide extremely precise movements of the grinding wheel to remove very small amounts of material. This allows a workpiece to be machined to very close tolerances. Achieving tolerances of ±0.0001" is not unusual when operating a precision grinder. Because abrasives can also cut hardened steels and carbide, grinding is also used to sharpen existing cutting tools and to create custom cutting tools. Since hardened steels cannot usually be machined by standard turning or milling methods, workpieces are often machined near final dimensions while soft, then hardened and tempered, and finally ground to finished sizes. Precision grinding also has the ability to produce very fine surface finishes. Surface finishes as fine as 16 microinches are common when grinding. These are some of the reasons why grinding is often the last operation performed on many machined workpieces.

There are several types of precision grinding machines and each is designed to perform certain types of grinding tasks. This unit will provide an overview of their uses.

SURFACE GRINDERS

The primary purpose of a **surface grinder** is to produce flat surfaces. A surface grinder uses a grinding wheel mounted on either a horizontal or a vertical spindle. Reciprocating or rotary table motion moves the work beneath the grinding wheel.

Horizontal Spindle Surface Grinders

Horizontal spindle surface grinders typically use the periphery of a solid wheel to grind workpiece surfaces. Machines with rotary table motion rotate the work beneath the grinding wheel. Machines with reciprocating tables move the work back and forth beneath the grinding wheel. **Figure 7.1.1** illustrates both reciprocating and rotary horizontal spindle surface grinders.

The horizontal spindle reciprocating surface grinder is probably the most common type of grinder used in the machining industry. The *work head,* or *wheel head,* of the grinder supports the horizontal spindle that holds and rotates the grinding wheel. **(See Figure 7.1.2.)** The wheel head can be raised and lowered by rotating the *elevating hand wheel.* **(See Figure 7.1.3.)** The hand wheel of the work head is commonly graduated in 0.0001" increments for very precise down feed of the grinding wheel.

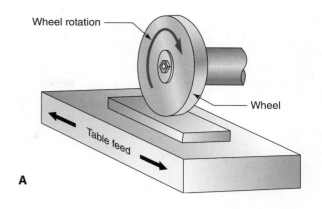

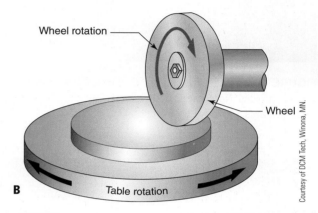

FIGURE 7.1.1 Two types of horizontal spindle surface grinders. (A) Reciprocating (the most common) and (B) rotary.

FIGURE 7.1.2 The wheel head of the surface grinder supports and rotates the grinding wheel. A guard surrounds the wheel for protection in case of wheel failure.

FIGURE 7.1.3 The elevating hand wheel of the horizontal surface grinder. Note the 0.0001" increments for downfeed measurements.

FIGURE 7.1.4 The longitudinal-feed hand wheel, used to move the workpiece back and forth beneath the grinding wheel.

The surface grinder has two table movements similar to the vertical milling machine. The *longitudinal-feed hand wheel* is rotated to move the table left and right. It usually does not have graduations because this machine axis is typically only used to move the workpiece back and forth beneath the grinding wheel. Rotating the hand wheel clockwise moves the table to the right, while rotating the hand wheel counterclockwise moves the table to the left. **(See Figure 7.1.4.)**

The *cross-feed hand wheel* is rotated to provide movement like the milling machine saddle or Y-axis movement. The cross feed is normally used to move the work in incremental steps beneath the grinding wheel

Cross-feed hand wheel

FIGURE 7.1.5 The cross-feed hand wheel, used to move the workpiece across the wheel between longitudinal passes.

between longitudinal cutting passes. **(See Figure 7.1.5.)** The cross-feed hand wheel is usually graduated in increments of 0.0005", 0.001", or 0.005" so wheel location can be precisely positioned when necessary, such as when grinding near a vertical surface.

A manual surface grinder uses hand wheels to control all of the movements of the machine. Automatic models can also be moved using hand wheels, but have additional hydraulic or electric systems that provide a power feed option for longitudinal and cross-feed movements. After setting the amount of travel for both axes, the longitudinal feed will continuously cycle back and forth and the cross feed will step over between longitudinal passes. Cross feed can also be set to reverse direction when it reaches set locations. Some models also have programmable wheel heads that will feed the wheel down at set increments each time the cross feed reaches the reversing point. When a set amount of wheel head travel is reached, automatic down feed disengages. Some machines will even stop all feed motion and move both table axes to a preset safe location. **Figure 7.1.6** shows a surface grinder with these capabilities.

Vertical Spindle Surface Grinders

Vertical spindle surface grinders typically use the side of a grinding wheel to grind workpiece surfaces. The wheel may be a solid cylindrical or cup-shaped wheel, or may consist of abrasive segments fastened directly to a metal body. **(See Figure 7.1.7.)** Like horizontal spindle machines, these vertical spindle grinders can use either reciprocating or rotary table motion to pass work beneath the grinding wheel. **Figure 7.1.8** illustrates both reciprocating and rotary vertical spindle surface grinders. Another name for a vertical spindle rotary surface grinder is a Blanchard grinder.

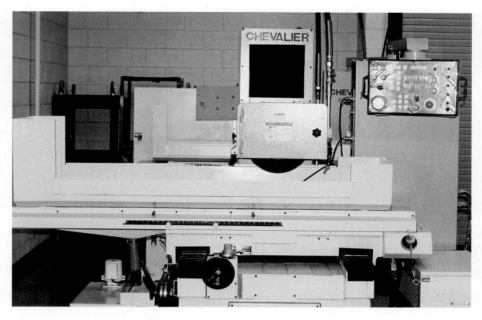

FIGURE 7.1.6 An automatic surface grinder.

FIGURE 7.1.7 Example of a grinding wheel used on a vertical spindle surface grinder.

CYLINDRICAL GRINDERS

Cylindrical grinders are used to grind diameters, shoulders, and faces much like the lathe is used for turning, facing, and boring operations. The work is rotated against the rotation of the grinding wheel. A hand wheel is used to make precise movements of the wheel head to set cut depth when grinding diameters. Power table travel moves the wheel longitudinally back and forth across the surface of the workpiece. **(See Figure 7.1.9.)**

A grinder used to grind outside diameters is frequently called an OD grinder (OD for *o*utside *d*iameter). A machine used to grind internal diameters is often called an ID grinder (ID for *i*nside *d*iameter). An OD grinder grinds diameters with the periphery of the wheel. Faces and shoulders can also be ground with the side

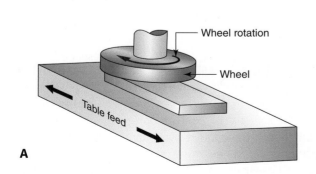

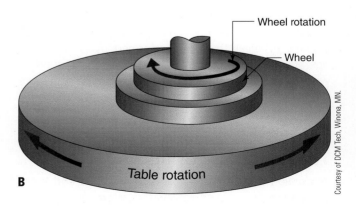

A

B

Courtesy of DCM Tech, Winona, MN.

FIGURE 7.1.8 Table movement on a vertical spindle surface grinder can be (A) reciprocating or (B) rotary.

FIGURE 7.1.9 A cylindrical grinder set up for grinding an outside diameter. Note how the part is held so that it can rotate.

FIGURE 7.1.11 A cylindrical grinder set up for grinding an internal diameter.

FIGURE 7.1.10 A cylindrical grinder set up to grind a shoulder on a part.

Magnetic chuck

Grinding wheel

FIGURE 7.1.12 A spindle-mounted magnetic chuck.

of the wheel. **(See Figure 7.1.10.)** An ID grinder uses wheels mounted on arbors and primarily grinds internal diameters similar to those performed by lathe boring operations. **(See Figure 7.1.11.)** Cylindrical grinding work can be held between centers or in chucks much like those used on the lathe. A spindle-mounted magnetic chuck resembling a face plate can also be used. **(See Figure 7.1.12.)**

The Centerless Grinder

Production grinding of the outside of shafts and tubes can be performed with the *centerless grinder*. Rotating cylindrical work is supported by a work rest while being

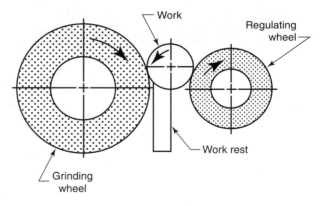

Work

Regulating wheel

Work rest

Grinding wheel

FIGURE 7.1.13 A centerless grinder rotates work with the regulating wheel while a grinding wheel removes material.

fed between the grinding wheel and a regulating wheel. The grinding wheel runs at normal speed and the regulating wheel runs slower and controls the rotational speed of the workpiece. **(See Figure 7.1.13.)**

TOOL AND CUTTER GRINDERS

The tool grinder shown in Section 3 (Figure 3.5.2) is considered an offhand grinding machine because the cutting tools are held and moved by hand. A **tool and cutter grinder** is also used for tool sharpening, but the cutting tools are held in workholding devices and machine slides provide movement for much greater control and flexibility. These extremely versatile grinders can be used to sharpen nearly any type of cutting tool, including drills, reamers, taps, and milling cutters. They can also be used to create customized cutting tools from HSS or carbide tool blanks. **Figure 7.1.14** shows an example of this type of grinder.

THE JIG GRINDER

The **jig grinder** is similar in appearance to a vertical milling machine, but with much more precise X- and Y-axis movements for extremely accurate positioning. As a spindle-mounted grinding wheel rotates, the head also moves in an orbital pattern and the quill feeds up and down to perform grinding of internal diameters. **Figure 7.1.15** shows a jig grinder. Newer technologies are available that can produce a finished hole faster in many workpieces than a jig grinder, but there are still some applications for its use.

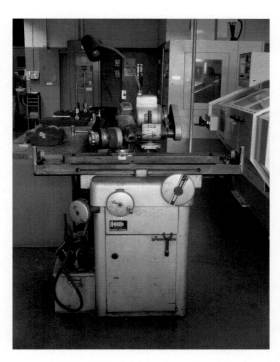

FIGURE 7.1.14 A tool and cutter grinder is used for sharpening cutting tools and creating customized cutting tools.

FIGURE 7.1.15 A jig grinder.

SUMMARY

- Precision grinding is an abrasive machining operation that uses grinding wheels to remove material.
- Precision grinding is performed when very close tolerances are required, to machine hardened steels, and to create high surface finishes.
- Surface grinders are primarily used to grind flat surfaces and are probably the most common type of precision grinding machine used in the machining industry.
- Other types of grinding machines are the cylindrical grinder, the centerless grinder, the tool and cutter grinder, and the jig grinder.

REVIEW QUESTIONS

1. What are three benefits of precision grinding?
2. What is the major use of the surface grinder?
3. What are the two spindle types used for surface grinders?
4. What are the two types of table movements used for surface grinders?
5. Label the parts of the surface grinder.

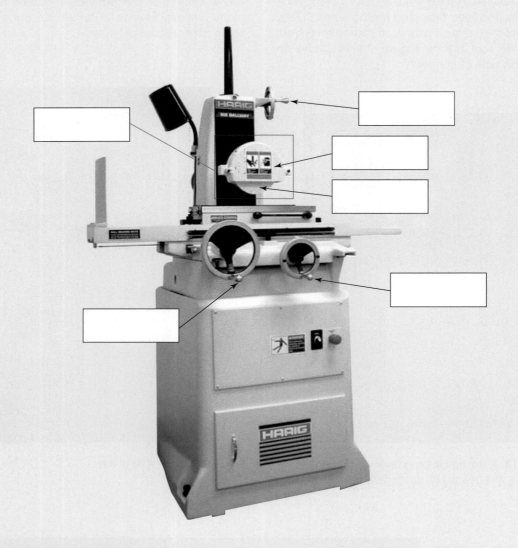

6. What are three types of cylindrical grinders?
7. Tool and cutter grinders can be used to _____ existing cutting tools or to create new, customized cutting tools.

Grinding Wheels for Precision Grinding

Learning Objectives

After completing this unit, the student should be able to:

- Identify grinding wheel shapes
- Demonstrate understanding of the grinding wheel identification system
- Demonstrate understanding of the types of abrasives used to make grinding wheels
- Demonstrate understanding of grit size (grain size)
- Demonstrate understanding of the hardness scale of grinding wheels
- List the different types of grinding-wheel bonding agents
- Demonstrate understanding of wheel structure
- Demonstrate understanding of grinding-wheel characteristics suitable for various applications
- Describe the use of superabrasives for precision grinding

Key Terms

Aluminum oxide
Bonding agent (Bond)
Ceramic aluminum oxide

Cubic Boron Nitride (CBN)
Friability
Grade

Grit size (grain size)
Silicon carbide
Structure
Superabrasive

INTRODUCTION

The grinding wheel is the cutting tool used in precision grinding operations. These wheels are similar to the wheels used for offhand grinding because they are made of abrasive grains held together with a **bonding agent,** or **bond**. The bonding agent acts as the "glue" that holds the individual abrasives together in the shape of a wheel. There are many more variations of wheels available for precision grinding versus those used for offhand grinding. Understanding the many different types and characteristics of grinding wheels will help in selecting a proper wheel to perform the required grinding operation.

WHEEL SHAPES

Grinding wheels come in many different shapes, as illustrated in **Figure 7.2.1**. The Type 1 straight wheel is probably the most widely used wheel for surface and cylindrical grinding operations. Recessed and relieved

wheels are also sometimes used for surface and cylinder grinding to minimize contact between the sides of the wheel and the workpiece. Cup, saucer, and dish-shaped wheels are usually used for tool and cutter grinding operations. Cup-shaped wheels are also used on some vertical spindle surface grinders. The *face of*

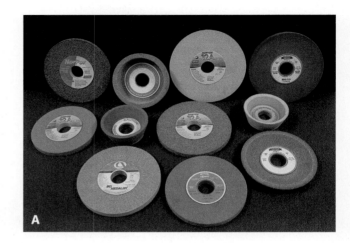

A

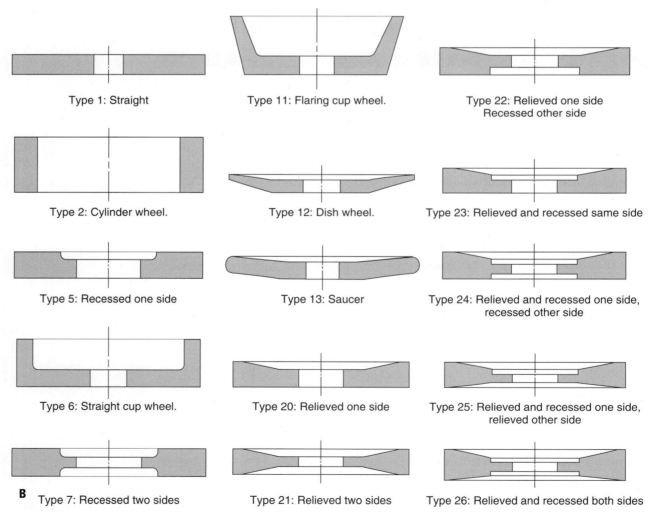

Type 1: Straight

Type 11: Flaring cup wheel.

Type 22: Relieved one side
Recessed other side

Type 2: Cylinder wheel.

Type 12: Dish wheel.

Type 23: Relieved and recessed same side

Type 5: Recessed one side

Type 13: Saucer

Type 24: Relieved and recessed one side,
recessed other side

Type 6: Straight cup wheel.

Type 20: Relieved one side

Type 25: Relieved and recessed one side,
relieved other side

B Type 7: Recessed two sides

Type 21: Relieved two sides

Type 26: Relieved and recessed both sides

FIGURE 7.2.1 Examples of some of the many different shapes of grinding wheels.

FIGURE 7.2.2 Wheel shapes used in internal diameter grinding operations.

a grinding wheel is the surface that is designed to be used for grinding. In most cases the periphery (outside diameter surface) is the face of the wheel. In some cases the wheel face is the flat surface of the wheel, like with cup wheels. However, under the right circumstances, other wheel surfaces can also be used. For example, the periphery and the side of a straight wheel could be used to simultaneously grind a horizontal surface and an adjacent vertical surface on a surface grinder. The periphery and side of a recessed wheel could be used to grind both an outside diameter and a shoulder at the same time using an O.D. grinder. **Figure 7.2.2** shows some examples of arbor-mounted wheel shapes that are used for internal diameter grinding operations.

GRINDING-WHEEL SPECIFICATIONS

Beyond wheel shapes, there is a standardized system for describing specifications for most grinding wheels based on five characteristics. These five characteristics are abrasive type, grit or grain size, grade, structure,

and bond type. Markings on the *blotter* of a grinding wheel will identify these characteristics according to the standardized system. The optional prefix and the last one or two digits are markings used by individual manufacturers. The remaining numbers and letters identify the five wheel characteristics.

Figure 7.2.3 summarizes this system. Refer to it while reading each description.

Abrasive Type

Most grinding wheels are made from either **aluminum oxide**, a combination of aluminum oxide and **ceramic aluminum oxide**, or **silicon carbide**.

Aluminum oxide and ceramic aluminum oxide wheels are identified by the letter A, and silicon carbide wheels by the letter C. Different varieties of each abrasive type may also contain other letters. The different varieties

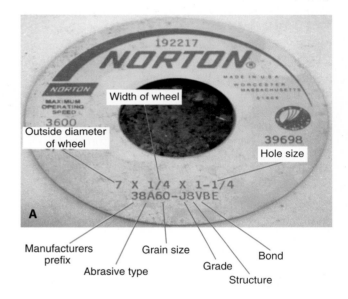

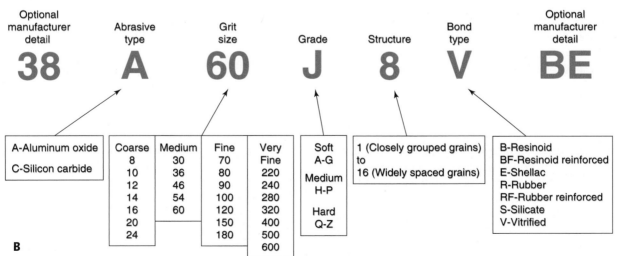

FIGURE 7.2.3 (A) A grinding-wheel blotter showing the abrasive type, grain size, grade, structure, and bond type, as well as wheel dimensions. (B) Summary of the system for classifying grinding wheels.

have different levels of friability. **Friability** is the ability of the individual abrasive grains to fracture during grinding to create new, sharp cutting edges. Wheels that are more friable produce less heat when grinding and are generally used when grinding harder materials. Less friable wheels are tougher, hold their shape better, and are generally used when grinding softer materials.

Aluminum Oxide

Aluminum oxide wheels are used for grinding steel. There are different colors of aluminum oxide wheels, including white, pink, and brown. These colors, along with any prefix before the "A" designation, indicate the type of the aluminum oxide grit. White wheels are the purest form of aluminum oxide and are the most friable. Pink wheels contain some chromium oxide and are tougher and less friable. Brown wheels contain titanium oxide and are even less friable than pink wheels.

Ceramic Aluminum Oxide

Ceramic aluminum oxide is a synthetic (man-made) abrasive. Wheels made entirely from ceramic aluminum oxide are very durable but are not well suited for precision grinding because they cannot produce smooth surface finishes. When ceramic aluminum oxide and straight aluminum oxide are combined, however, the result is a durable wheel that can produce finer surface finishes and is suitable for precision grinding of ferrous metals.

Silicon Carbide

Silicon carbide is a harder abrasive than aluminum oxide and is available in two basic types. Green wheels are the purest form of silicon carbide wheels, are the most friable, and are mostly used when grinding carbide. Black silicon carbide wheels are less friable and are generally used when grinding cast iron, stainless steel, and nonferrous metals.

Grit Size (Grain Size)

Grit or **grain size** is listed after abrasive type. Just like with other types of abrasives, the lower the number, the coarser the wheel. The higher the number is, the finer the wheel. Grain sizes range from 8 to 600. Coarser wheels have larger abrasive grains and will remove material more quickly and produce rougher surface finishes, while finer wheels will remove material more slowly and produce smoother surface finishes. Coarser wheels also cut softer metals better because the larger grains can penetrate the work more easily. When grinding harder metals, a finer wheel with more grains in a given amount of wheel space will provide more cutting points that can remove material more quickly than a coarser wheel. Use of coarser wheels for grinding harder materials will result in the grains breaking off the wheel and excessive wheel wear. Grain sizes from 46 to 80 are commonly used for general-purpose surface grinding operations.

Grade

The **grade** of a grinding wheel refers to its hardness or strength of the bond and ranges from A to Z. Grade A is the softest and Z is the hardest. Generally, softer wheels are used for harder materials because the wheel will release old grains before they become dull, exposing new, sharper grains. If a hard wheel is used for grinding a hard material, the grains will not release and they will become dull. This will cause wheel glazing, overheating, and burning of the work surface. When there is a large area of surface contact from using a wide wheel, a softer wheel is also a better choice because the grains will release and the wheel will wear evenly to keep it cutting freely. A harder wheel should be used when the area of contact is smaller. Harder wheels are more wear resistant and should be used when larger amounts of material are to be removed, while softer wheels should be selected when smaller amounts of material are to be removed.

Structure

Structure is the spacing between the individual grains of the grinding wheel and is identified by the numbers 1 through 16. The number 1 identifies a dense wheel where the grains are close together. The number 16 identifies a very open wheel with more space between the individual grains. Open wheels should be selected for grinding continuous surfaces and dense wheels should be used on interrupted surfaces. Open-structure wheels should be used when grinding with cutting fluids and tighter structures used when grinding "dry." A more open wheel will also typically remove material faster, but a denser wheel will leave a smoother surface finish.

Bond Type

The bond is the material used as "glue" to hold the grains together. The major bond types are vitrified, silicate, rubber, resinoid, and shellac.

Vitrified

The *vitrified* bond is a synthetic glass bond and is by far the most commonly used bonding agent. Vitrified wheels are identified by the letter V. They are hard but too brittle to withstand heavy pressure or shock. They also hold up well when used with cutting fluids. Vitrified wheels are used for most precision grinding operations.

Silicate

The *silicate* bond is a synthetic, claylike bonding agent. It is softer than the vitrified bond and releases grains more easily. Silicate wheels produce less friction and cut at lower temperatures so they are often used to grind very thin parts and edges of hard materials. Silicate wheel bonds are identified by the letter S.

Rubber

Rubber bonded wheels can operate at very high speeds and produce very high surface finishes. They have some elasticity or flexibility and can withstand heavy pressure and shock without breaking, unlike vitrified and silicate wheels. For those reasons, the rubber bond is often used for very thin wheels such as those used for precision cutoff operations. Since rubber is a natural or organic bond, it can deteriorate over time and become brittle and weak. Rubber wheels are often used as regulating wheels when centerless grinding and for polishing operations. Rubber bonded wheels are identified by the letter R, or RF for rubber reinforced.

Resinoid

The *resinoid* bond is another natural bonding agent. Resinoid wheels are tough and can withstand pressure and shock. They are not very common in precision grinding, but are used for rough offhand grinding under harsh conditions, such as heavy deburring of iron castings. Resinoid wheel bonds are identified by the letter B, or BF for resinoid reinforced.

Shellac

Shellac is another natural wheel bond. Shellac wheels are stronger and more rigid than rubber wheels but still provide resistance to shock, so shellac can be used to make very thin wheels. These wheels also produce very high surface finishes. Shellac bonded wheels are not very common, but have limited use in specialized applications such as cam grinding. Shellac wheels are identified by the letter E.

SUPERABRASIVES

Superabrasives are used in grinding wheels that are used to grind extremely hard materials. Diamond abrasive wheels contain small particles of natural or synthetic diamond held together by bonding agents. They are the hardest wheels available. **CBN** wheels use a synthetic abrasive called **cubic boron nitride** and are second in hardness after diamond wheels.

These superabrasive wheels contain a layer of abrasive on a metal wheel body. **(See Figure 7.2.4.)** If the entire wheel were made of the bonded abrasive, the cost would be far too expensive. Diamond wheels are often used for

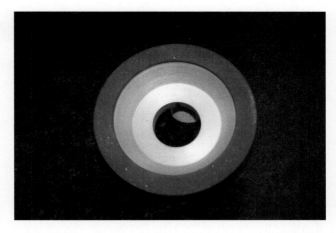

FIGURE 7.2.4 A superabrasive grinding wheel. Note how the abrasive material is bonded to an aluminum wheel body.

grinding carbide, hard coatings or plating, superalloys, and even glass and ceramic. CBN wheels are often used for grinding hardened high-speed tool steels, titanium, and superalloys. Both can remove material much more quickly than aluminum oxide and silicon carbide wheels and are harder and more wear resistant. Superabrasives cost significantly more than standard abrasives, which is why aluminum oxide and silicon carbide wheels are still widely used.

Designation of diamond and CBN wheels is slightly different from the system used for standard grinding wheels. Refer to **Figure 7.2.5** while reading the following descriptions.

The prefix and the last one or two digits are again optional manufacturer-specific symbols. The second entry is either D for diamond or B for CBN. Next are the grain size and the grade, the same as used for aluminum oxide and silicon carbide wheels.

The next entry is a manufacturer's number or letter that shows the abrasive concentration, or weight of the abrasive per cubic centimeter of the abrasive layer. Higher concentrations are better for faster material removal, whereas lower concentrations produce higher surface finishes.

After the concentration symbol comes the bond type. Resinoid, vitrified, and metal bonds are used for diamond and CBN wheels. An optional bond modification code can follow the bond specification. Next is the depth of the abrasive layer in either inches or millimeters.

Prefix	Abrasive	Grain Size	Grade	Concentration	Bond Type	Bond Modification	Depth of Abrasive	Manufacturer's Identification Symbol
M	D	120	R	100	B	56	⅛	*

FIGURE 7.2.5 The marking system used for CBN and diamond superabrasive wheels. The data shown is for a sample wheel.

SUMMARY

- Grinding wheels are made of abrasive particles held together by a bonding agent.
- A standard wheel classification system is used to identify different abrasive types, grain sizes, grades (or hardnesses), structures, and bond types.
- Common abrasive types are aluminum oxide and silicon carbide.
- Grain sizes range from 8 to 600.
- Wheel grades are identified from softest to hardest by the letters A through Z.
- Structure is the spacing between the grains and is identified from most dense to most open by the numbers 1 through 16.
- The synthetic vitrified bond is the most common bonding agent used in wheels for precision grinding.
- Other bond types include silicate, rubber, resinoid, and shellac.
- Different combinations of these characteristics create grinding wheels that are appropriate for particular grinding applications.
- Superabrasive wheels contain a layer of diamond or cubic boron nitride and are used to grind very hard materials.

REVIEW QUESTIONS

1. What wheel shapes are commonly used for surface and cylindrical grinding?
2. What abrasive is commonly used to grind steels?
3. What abrasive is used to grind nonferrous metals, cast iron, and stainless steel?
4. What does CBN stand for?
5. What superabrasive is a good choice for grinding carbide?
6. The grade of a wheel describes its _____ on a scale from _____ to _____.
7. Briefly explain the structure identification scale for aluminum oxide and silicon carbide wheels.
8. List three bonding agents used for aluminum oxide and silicon carbide wheels.
9. Generally, harder wheels should be used to grind _____ materials and softer wheels should be used to grind _____ materials.

Answer Questions 10–14 about these two wheels:

Wheel A: **32A60H8V**; Wheel B: **39C100L6V**

10. What is the abrasive type for each wheel?
11. Which wheel is finer?
12. Which wheel is harder?
13. Which wheel is more open?
14. What type of bond is used for both wheels?

Surface Grinding Operations UNIT 3

Learning Objectives

After completing this unit, the student should be able to:

- Demonstrate understanding of surface grinder safety procedures
- Demonstrate understanding of the basic process of mounting and dressing surface grinder wheels
- Identify and demonstrate understanding of the use of common workholding devices used for surface grinding
- Demonstrate understanding of the process of grinding parallel, perpendicular, and angular surfaces
- Demonstrate understanding of methods for side grinding of vertical surfaces
- Identify common grinding problems and their solutions

Key Terms

Diamond dresser	**Ferromagnetic**	**Magnetic sine chuck**
Dressing stick	**Magnetic chuck**	**Magnetic V-block**
Magnetic angle plate	**Magnetic parallels**	

INTRODUCTION

Since surface grinders are used primarily to grind flat surfaces, many of the same workholding and setup principles used when milling can be applied when surface grinding. This unit will discuss setup and operation of the horizontal spindle reciprocating machine, but rotary and vertical spindle machines can utilize the same methods with only slight variation. When surface grinding, the wheel face is the periphery of the wheel, so throughout this unit, the term *face* refers to the outside diameter surface of the grinding wheel. Following these steps and some basic safety guidelines will allow for safe and productive operation of the surface grinder.

GENERAL SURFACE GRINDER SAFETY

The surface grinder, like any machine tool, can be very dangerous, but by following some basic precautions, safe operation can be ensured. Specific safety notes will be discussed throughout this unit, but the following are some guidelines that should be observed during any grinding operation.

 CAUTION

- Always wear safety glasses when operating a surface grinder.
- Wear appropriate work shoes and clothing and secure long hair so it cannot become tangled in moving machine parts.
- Make sure all machine guards and covers are in place before operating any grinder.
- Never operate a grinder that is locked out or tagged out, and never remove another person's lock or tag.
- Be sure that grinding wheels are sound before mounting and that all workpieces are secure before beginning any grinding.
- Do not force the wheel into the work. This could cause workholding devices to fail or grinding wheels to break.
- Always shut off the wheel and let it come to a complete stop before making workpiece adjustments or taking measurements. Do not reach around or beneath a grinding wheel while it is running.

MOUNTING THE GRINDING WHEEL

 CAUTION

Before removing wheel guards to access the wheel and spindle, disconnect or shut off the power to the machine.

After a grinding wheel is selected, it must be mounted on the spindle of the surface grinder. To remove the existing wheel, the spindle nut needs to be removed. *This nut will contain a left-hand thread.* Every machine is slightly different, but most grinders require the use of two wrenches to loosen the spindle nut. These wrenches are normally provided by the grinder manufacturer. One is used to keep the spindle from turning and the other to turn the nut *clockwise* to loosen the spindle nut. **Figure 7.3.1** shows a typical surface grinder spindle and nut. After removing the spindle nut, many machines will have a large flat washer that needs to be removed as well. Finally, remove the grinding wheel from the spindle shaft.

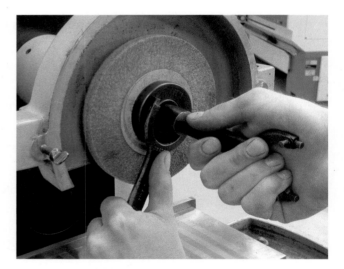

FIGURE 7.3.1 Loosening the spindle nut on most grinders requires two wrenches.

CAUTION

Before mounting any wheel, check the maximum RPM listed on the blotter to make sure it is equal to or higher than the spindle RPM of the grinder. Operating a grinding wheel at speeds above the maximum rating can cause the wheel to fly apart.

Then perform a *ring test* to be sure the wheel is not defective. Suspend the wheel with one finger and lightly tap it with a non-metallic object such as the wooden or plastic handle of a screwdriver, as shown in **Figure 7.3.2**. A clear ringing sound means the wheel is "sound." A dull sound means the wheel is damaged and should not be mounted. Even though the wheel may appear to be in good condition, it could have an internal crack. If a cracked wheel is mounted and started, it could fly apart and cause serious injury.

 CAUTION

Never mount a wheel that does not pass the ring test. Damaged wheels should be broken and discarded so no one attempts to use them.

FIGURE 7.3.2 Wheels should always be ring tested before they are installed. Ring testing checks the wheel for soundness.

 CAUTION

Always be sure there is a blotter on each side of the wheel before mounting it on the spindle shaft.

The blotters cushion the wheel between the spindle flange and the spindle nut. If a blotter is not used, the clamping pressure can crack the wheel and it can fly apart when the spindle is started. Be sure the flange, shaft, wheel sides, and center hole are free of dirt. The wheel should slide easily onto the shaft.

 CAUTION

Do not force the wheel on the spindle or it may crack, causing the wheel to break when started.

Be sure the washer is clean and burr free, and then place it on the spindle. Note that most spindle nuts will have a recess on one side. This recess must go against the wheel when the nut is threaded onto the spindle. Tighten the spindle nut securely using the provided manufacturer's wrenches. If larger wrenches are used, the spindle nut can be overtightened and crack the grinding wheel.

After mounting the wheel, jog the spindle a few times (start and stop without bringing the spindle to full speed). Then turn the spindle on and let it run for at least one full minute to be sure that the wheel is still sound.

WORKHOLDING DEVICES

A surface grinder has T-slots in the table like those in the worktables of the drill press and milling machine. The T-slots can be used to clamp work directly to the table, but a magnetic chuck is usually mounted to the table using these T-slots. The **magnetic chuck** can then be used to secure **ferromagnetic** workpieces or other workholding devices. A ferromagnetic material is simply a material that is attracted to a magnet.

Magnetic Devices

Magnetic workholding devices contain alternating segments of ferrous and nonferrous materials that are mounted together to form a single unit. The ferrous portions of the device are called the *magnetic poles*. The poles create the magnetic pulling force across the nonferrous segments when an internal magnet

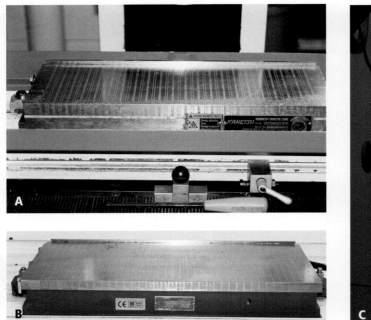

FIGURE 7.3.3 (A) A permanent magnet and (B) an electromagnet. The control for the electromagnet offers the ability to vary the magnetic force exerted on a workpiece which can limit distortion of thin workpieces.

is activated. A magnetic device may use a *permanent magnet* or an *electromagnet*. A permanent magnet is activated by flipping a lever to activate the magnetic fields. An electromagnet is energized with a switch, and most models have a variable power switch that can adjust the strength of the magnetic force. This is sometimes useful when securing thin parts that can easily distort under the full magnetic force. **Figure 7.3.3** shows a permanent magnetic chuck and an electromagnetic chuck.

When using a magnetic chuck (or any magnetic workholding device), observe the distance between the poles. The workpiece to be held should span at least three of the nonferrous segments or the magnetic force may not be strong enough to secure the part to the chuck. **(See Figure 7.3.4.)** Many chucks also have a back rail and a side rail that can help position work parallel or square to table travel.

If a workpiece is taller than its length or width, the magnetic chuck may not be strong enough to hold the part. *Blocking* can be used to help secure the part to the chuck and prevent it from tipping or being pulled from the chuck. Blocks are placed tightly around the workpiece before activating the magnet. Since the greatest force while grinding is in the direction of the longitudinal table feed, blocking on the left and right sides of the workpiece is most critical. This method can also be used to secure workpieces that are not ferromagnetic. **Figure 7.3.5** shows a method of blocking work on a magnetic chuck. Thin workpieces that are not ferromagnetic

FIGURE 7.3.4 The gold-colored areas on the chuck are the nonferrous segments where the magnetic fields are created. The workpiece should span 3 or more of these segments to be safely secured. Notice the back rail on this chuck. It is used to help align a workpiece on the chuck.

can also be held on the magnetic chuck using flexible clamps. These use the magnetic force of the chuck to pull small fingers against opposite edges of the part, holding it in place during grinding. **(See Figure 7.3.6.)**

If a workpiece contains features that keep it from being placed flat on the magnetic chuck, **magnetic parallels** may sometimes be used to support the workpiece. Magnetic parallels are not magnetic by themselves but are designed to extend the magnetic field of the chuck

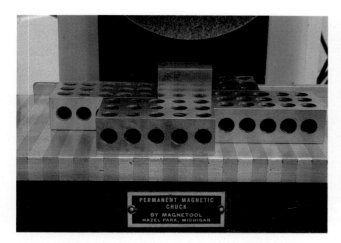

FIGURE 7.3.5 Blocking consists of placing pieces of ferro-magnetic metal around the workpiece. This is done to help hold the workpiece secure during the grinding operation. When using this method it is important to make sure the blocking material is shorter than the actual workpiece.

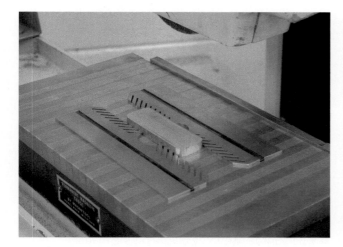

FIGURE 7.3.6 A pair of finger-type hold-downs may be used to hold non-magnetic parts against a magnetic chuck surface for grinding.

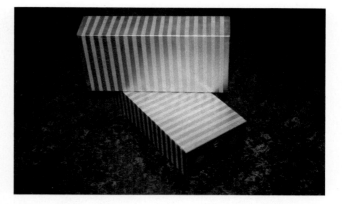

FIGURE 7.3.7 Magnetic parallels extend the magnetic field of the magnetic chuck.

FIGURE 7.3.8 One type of magnetic V-block extends the magnetic field of the chuck (A). Another type activates its internal permanent magnet by turning a small lever (B).

to the workpiece. **Figure 7.3.7** shows a pair of magnetic parallels.

A **magnetic V-block** can be used to hold a workpiece at a 45-degree angle. There are two types of magnetic V-blocks. One type is the same design as magnetic parallels and extends the magnetism of the chuck to the workpiece. The other type has a lever like a permanent magnetic chuck. The workpiece is simply placed in the V-block and the lever turn to activate the magnet. **(See Figure 7.3.8.)**

A **magnetic angle plate** can be used to hold work vertically and for squaring operations much like a standard angle plate. It has a permanent magnetic control just like the magnetic chuck to hold workpieces.

FIGURE 7.3.9 A magnetic sine chuck can be used to hold a workpiece at a preset angle. The angle is set by using a gage block build as with a sine bar.

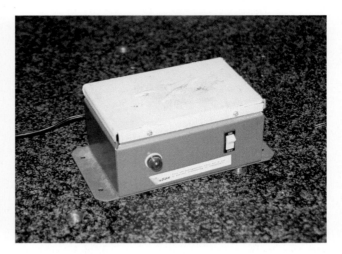

FIGURE 7.3.10 A workpiece is often passed over a demagnitizer after grinding.

It can be secured to the grinder's magnetic chuck in the same way a workpiece would be.

Precision angles can be ground using a **magnetic sine chuck**. The sine chuck is set up just like an ordinary sine plate but has its own permanent magnet to hold the work. It is a combination of a sine plate and a magnetic chuck. The chuck portion has a back and end rail like many standard magnetic chucks to align workpieces on the chuck surface. The entire unit can then be secured to the grinder's magnetic chuck. **(See Figure 7.3.9.)**

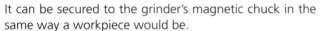

 **CAUTION**

When using any magnetic workholding device, always be sure to activate the magnet and double check that the magnet is working by trying to move the workpiece by hand.

Workpieces can become magnetized after being held by magnetic fixtures. A *demagnetizer* can be used to remove magnetism from the work. The demagnetizer is turned on and the workpiece moved across its surface. A demagnetizer is shown in **Figure 7.3.10**.

ANGLE PLATES, V-BLOCKS, AND COLLET BLOCKS

Angle plates and angle blocks used for milling can also be used for grinding. The workpiece is clamped to the angle plate or block, and then the angle plate or block can be secured directly to the magnetic chuck. When grinding, the 90-degree sides of the angle plate are usually used

to create two perpendicular surfaces by repositioning the angle plate on the chuck instead of repositioning the work on the angle plate.

If a workpiece is first clamped in a V-block or collet fixture, the V-block or collet fixture can also be secured directly to the grinder's magnetic chuck, or mounted on an angle plate secured to the magnetic chuck. Depending on the size of these workholding devices, they could also be used together with any of the other magnetic workholding devices.

Vises

Grinding vises are different from those used for milling. All surfaces of a grinding vise are precision ground to be parallel and perpendicular to each other. This allows any side of the vise to be placed on the magnetic chuck. By repositioning the vise instead of the workpiece, very precise perpendicular part surfaces can be produced. **Figure 7.3.11** shows a grinding vise.

FIGURE 7.3.11 Grinding vises are very accurately ground and can be used to quickly and accurately secure workpieces.

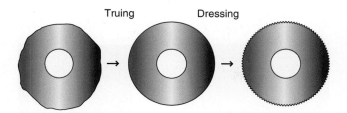

FIGURE 7.3.12 Truing makes the wheel face run true to its axis and dressing sharpens the wheel.

WHEEL DRESSING

Before beginning any surface grinding, the face (periphery) of the grinding wheel should be dressed and "trued." Truing makes the wheel precisely round (cylindrical), eliminating wheel runout. Dressing sharpens the wheel. **(See Figure 7.3.12.)**

DRESSING ALUMINUM OXIDE AND SILICON CARBIDE WHEELS

Aluminum oxide and silicon carbide wheels are dressed and trued simultaneously using a **diamond dresser**. There are two types of diamond dressing tools: the single-point diamond dresser and the cluster dresser. Both are mounted in a base that can be secured directly on the magnetic chuck. **(See Figure 7.3.13.)**

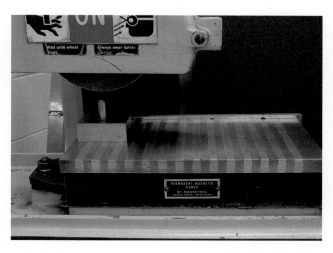

FIGURE 7.3.14 Preferred location of the diamond dresser on the left edge of the chuck.

To dress a surface grinder wheel with a diamond dresser, follow these steps.

- Raise the wheel head so the bottom of the wheel is higher than the dressing tool.

- Place the dresser near the left end of the magnetic chuck. This location is preferred because when the wheel is dressed, the abrasive particles will clear the chuck instead of being directed onto the working surface of the chuck. **(See Figure 7.3.14.)**

- If a single-point diamond dresser is mounted at an angle in its base, angle the dresser toward the left side of the magnetic chuck, away from the wheel rotation. In this position, the wheel is less likely to grab the point and lift the dresser from the chuck.

- Turn on the magnetic chuck and make sure the dresser cannot be easily moved by hand.

 **CAUTION**

Always check the proper function of the magnet by trying to move the dresser by hand.

- If a single-point dresser is used, move the longitudinal table travel to position the tip of the dresser about 1/8" to the left of the center of the wheel. **(See Figure 7.3.15.)**

- The cluster dresser cannot be offset as far as the single-point dresser or the diamond clusters will not contact the wheel face. Position the center of the cluster dresser only about 1/32" to the left of the wheel's center.

- This slight offset of the dressing tool provides an added measure of safety because the clockwise rotation of

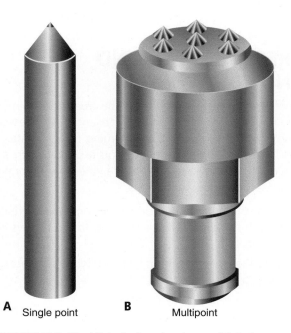

A Single point   **B** Multipoint

FIGURE 7.3.13 (A) A single-point dresser. (B) A cluster-point dresser.

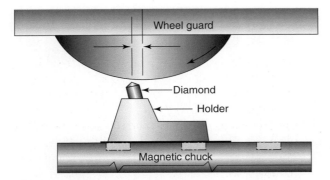

FIGURE 7.3.15 Proper dresser position is an important safety consideration. The diamond should be offset to the left of the wheel centerline. If the diamond is angled in its holder, the angle should be placed as shown.

FIGURE 7.3.16 The dresser is being positioned near the edge of the wheel.

the wheel will tend to push the dresser away from the wheel, rather than pull it into the wheel, if the table moves unexpectedly or the magnetic chuck fails.

- Use the cross feed to position the dresser near one edge of the wheel, as shown in **Figure 7.3.16**.
- Lower the wheel head so the face of the wheel is about 1/32" away from the tip of the dresser.
- Start the spindle.
- Slowly lower the wheel head about 0.0005" to 0.001" at a time until the wheel contacts the dresser.
- Move the cross feed until the dresser is clear of the wheel face.
- Lower the wheel head about 0.001" to 0.003" to set depth for a rough dressing pass.
- Feed the dresser across the wheel face using the cross feed. **(See Figure 7.3.17.)**

FIGURE 7.3.17 The dresser is fed across the face of the wheel to perform the dressing operation. Feeding the dresser slowly will result in a wheel having a slightly more closed structure while feeding faster will produce a slightly more open structure.

- Take additional rough dressing passes until the face of the wheel is clean.
- Lower the wheel head only about 0.0002" to 0.0005" for a final dressing pass.
- Feed slowly across the wheel with the cross slide for the final pass.

When rough grinding, use a faster-dressing cross feed that will leave the wheel face more open, resulting in less friction (and heat) between the wheel and part surface. When finish grinding, a slower-dressing cross feed makes the wheel face smoother. This will create a finer surface finish.

DRESSING DIAMOND AND CBN WHEELS

Superabrasive wheels should first be trued to eliminate runout using a *brake truing device* and then dressed using a **dressing stick**. A brake truing device contains a silicon carbide wheel mounted on a spindle with an automatic braking system. To true a wheel with a brake truing device, follow these steps.

- Mount the brake truing device on a magnetic chuck beneath the superabrasive wheel with the centers of the wheels in line.
- Lower the wheel head until the two wheels are nearly touching each other.
- Start the surface grinder spindle.
- Start the truing wheel rotation by hand.
- Lower the wheel head until the diamond or CBN wheel lightly contacts the truing wheel.

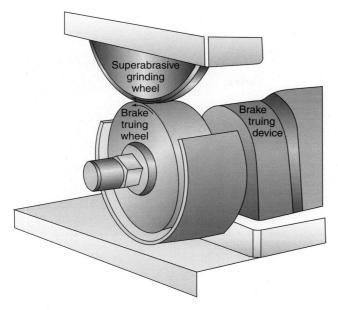

FIGURE 7.3.18 Illustration of brake truing.

GRINDING PARALLEL SURFACES

Grinding parallel workpiece surfaces is the most basic, and most common, surface grinding operation. If several sides of a workpiece need to be ground, it is best to grind the largest surface first, as when milling. The large surface can then be used as a reference surface to locate the workpiece for grinding the other part surfaces.

After dressing the wheel, shut off the spindle and mount the workpiece using the desired workholding method. As with any other machining operation, be sure the surfaces that will contact the workholding device (and the workholding device itself) are clean and free of burrs and dirt. This is even more important when precision grinding than when performing any other machining operation because even a piece of lint can produce enough error that very close tolerances will not be met.

If using a magnetic chuck, be sure the workpiece is secure after activating the magnet. It is also a good idea to use different areas of the magnetic chuck over time to avoid uneven wear.

Follow these steps to adjust settings of the surface grinder.

- Apply sufficient downfeed pressure to maintain rotation of the truing wheel.
- Use the cross feed to move the truing wheel back and forth across the face of the diamond or CBN wheel until the wheel is evenly colored. The braking system inside the truing device controls the speed of the truing wheel to create a "dressing ratio" that will evenly remove a very thin layer of the superabrasive. **Figure 7.3.18** illustrates brake truing.

After truing, it is recommended to remove some of the bond in the diamond or CBN wheel to be sure that the superabrasive grains are exposed and can cut freely. This is done by moving an aluminum oxide dressing stick across the rotating wheel face by hand.

- Lower the wheel head so the wheel face is within about 1/8" of the surface of the work.
- If using a grinder equipped with power feed, set the longitudinal table feed to reverse direction when the center of the wheel travels about 1/2" to 1" off each end of the part. **(See Figure 7.3.19.)**
- Set the cross-feed step-over to about 0.050" to 0.100".
- Set the reversing positions of the cross feed so the wheel travels about 1/8" to 1/4" off the front and back of the part. **(See Figure 7.3.20.)**

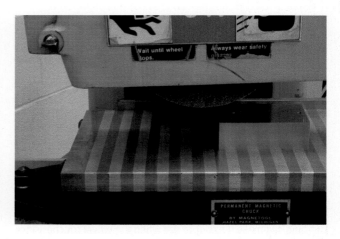

FIGURE 7.3.19 The grinding wheel should travel all the way off the workpiece at the end of each longitudinal pass.

 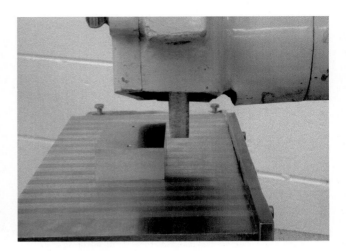

FIGURE 7.3.20 The table should be cross-fed far enough so that the wheel travels off the front and rear of the part.

Once the surface grinder settings are adjusted, follow these steps to touch off the workpiece.

- Use the cross feed to position the table so that only about 1/8" of the width of the wheel is over the part.
- Start the spindle and let the wheel come up to full speed.
- Lower the wheel head until the face of the wheel is within about 1/64" of the workpiece.
- Begin moving the longitudinal table movement back and forth across the workpiece.
- Lower the wheel head slowly during the table stroke to touch off the wheel face against the work. A light coating of layout fluid, which is less than 0.001" thick, can help with visibility when touching off. **(See Figure 7.3.21.)**

- After touching off, use the longitudinal table travel to move the wheel entirely off the workpiece to either the left or the right side.

Once the wheel has been touched against the workpiece, the surface can be ground using this process.

- If using a machine with a flood coolant system, start the coolant flow.
- Step over about 0.050" to 0.100" with the cross feed and bring the longitudinal movement to the other end of the stroke.
- Continue to alternately step over with the cross feed and feed across the part with the longitudinal feed until the entire surface is ground and the wheel is completely off the part at either the front or back of the cross-feed travel. **(See Figure 7.3.22.)**
- Repeat this process, taking 0.0005"- to 0.003"-deep roughing passes until the surface is within about 0.001" of the desired size.

FIGURE 7.3.21 When touching off, the grinding wheel is positioned so that only about 1/8" of its width is over the work. Layout fluid can make it easier to see when the wheel touches since the dark blue provides a good visual contrast to the workpiece.

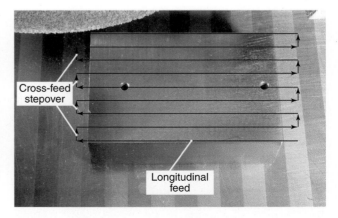

FIGURE 7.3.22 The arrows show the direction of feed as a surface is ground.

- To finish grinding the surface, use the same motion taking light depth cuts of only 0.0001" to 0.0005".

- NOTE: Grinding produces a great deal of heat where the wheel contacts the work. This heat can cause the work to expand slightly during grinding. When the work cools, the surface will contract and create low areas and surfaces that are not flat. This can occur even with depth cuts as shallow as 0.0005", depending on the workpiece material. Light finishing cuts of only 0.0001" to 0.0005" deep will reduce heat and expansion, produce flatter surfaces, and provide better surface finishes. Using coolant when available also greatly reduces heat.

- Finer finishes can be produced by repeating the finishing pass once or twice without lowering the wheel head until the grinding wheel no longer produces any sparks. This is often called "sparking out."

The first surface of a workpiece is often only ground until the desired surface finish is reached. After the desired amount of material is removed, turn off the coolant flow first, then the spindle. This helps to remove coolant from the wheel before it comes to a stop. Avoid flowing coolant over a stationary wheel because the coolant can soak into some types of wheels and cause them to become unbalanced. When restarted, a saturated wheel may develop runout and produce a poor surface finish.

After the first side of the workpiece is finished, remove it from the machine, and flip it over to grind the opposite side using the same steps.

Grinding the Magnetic Chuck

Over time, the surface of the magnetic chuck can become worn or distorted. If the chuck surface is no longer flat, producing parallel or flat surfaces can be nearly impossible. Grinding the actual chuck surface can make it flat again. Use the same techniques as when grinding any surface. Because the chuck surface is usually very large, heat can build up quickly, so take cut depths of only 0.0001" to 0.0003", use coolant when available, and spark out to be sure the surface remains flat.

GRINDING PERPENDICULAR SURFACES

A few different methods can be used to grind perpendicular surfaces. Generally, the surface to be ground is positioned parallel to the magnetic chuck. Once positioned, the surface is ground using the same method used for grinding parallel surfaces.

One method for holding the work is to use an angle plate. The angle plate can be mounted directly on the magnetic chuck and the part clamped with one of the first two ground sides against its vertical surface. If using a magnetic angle plate, clamps are not needed but

check for adequate holding power just like when using the magnetic chuck. This process is similar to the milling process of placing the first milled side against a solid vise jaw or angle plate. A parallel can be used to raise the work above the top surface of the angle plate. If an adjacent surface also needs to be ground, extend that surface past one end of the angle plate so that it can be ground when the angle plate is repositioned. **(See Figure 7.3.23.)** Grind the surface that is parallel to the magnetic chuck first. Then reposition clamps, while keeping a minimum of two clamps on the part at all times. This prevents the workpiece from changing position, so that the angle plate can be turned on its side to grind the adjacent surface of the workpiece, as shown in **Figure 7.3.24**. Always use at least two clamps on the work while grinding so the part remains secure on the angle plate.

FIGURE 7.3.23 Angle plates are often used to grind a workpiece square. If setup properly, two perpendicular surfaces of the workpiece can be ground without repositioning the work on the angle plate. Here the part is positioned so that the left side extends past the angle plate while the side facing up is ground.

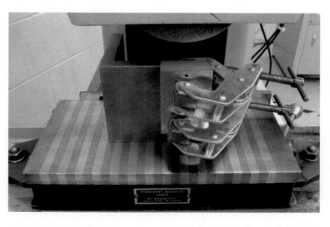

FIGURE 7.3.24 Without repositioning the work on the angle plate, it is turned on its side for grinding the adjacent side. Clamps must be repositioned for clearance with the grinding wheel.

CAUTION

Be sure that all clamps and clamp handles are safely below the surface to be ground.

A similar method uses a grinding vise instead of an angle plate. The vise can be mounted directly on the magnetic chuck. Then secure the workpiece in the vise with one surface above the top of the vise jaws and one end extending past one end of the vise, as shown in **Figure 7.3.25**. After grinding the top surface, the entire grinding vise can be placed on its side to grind the end perpendicular to the other sides. **(See Figure 7.3.26.)**

Using either method, at this point the work should have two parallel sides, and two other sides that are perpendicular to those surfaces. The remaining surfaces can then be ground parallel to the previously finished sides.

FIGURE 7.3.25 A precision grinding vise can be used in place of an angle plate when grinding perpendicular surfaces.

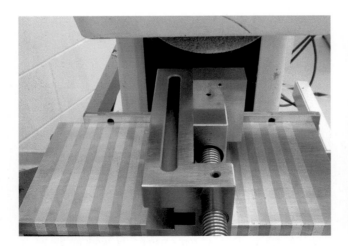

FIGURE 7.3.26 Two surfaces of the workpiece can be ground without reclamping the work by turning the grinding vise on its side.

NOTE:

When grinding parallel and square surfaces, take time to check for parallelism and square after grinding each surface, just like when squaring a block on the milling machine, to be sure the workpiece meets the required tolerances.

GRINDING ANGLES

Precision angles can be ground using different workholding devices and methods. A magnetic V-block can be a quick way to hold a ferromagnetic workpiece at a 45-degree angle. The part must have two perpendicular sides so that it sits securely against each side of the "V." Place the work in the V-block, place the V-block on the magnetic chuck, and activate the magnetic chuck. The angular surface is now parallel to the magnetic chuck. Then grind the angular surface. **Figure 7.3.27** shows work held in a magnetic V-block.

 CAUTION

Check to be sure the workpiece is secure in the V-block.

A sine bar can be used to position a workpiece at a desired angle against an angle plate. After positioning, the part can be secured to the angle plate. When possible, the setup should be performed on a surface plate. Follow these steps to perform the setup.

- Set the sine bar to the desired angle on a surface plate using the proper gage block build.
- Place an angle plate against the side of the sine bar.

FIGURE 7.3.27 Magnetic V-blocks are helpful when grinding 45-degree angles.

FIGURE 7.3.28 Using a sine bar and gage blocks to position the workpiece on an angle plate. The 1-2-3 block is being used to raise the workpiece above the top edge of the angle plate.

- Place the workpiece on the sine bar.
- Secure the work to the angle plate with clamps. **(See Figure 7.3.28.)**
- Remove the sine bar and gage blocks from the setup.
- Mount only the angle plate with the clamped workpiece on the machine for grinding the angular surface. This protects the sine bar and gage blocks by keeping them away from the abrasive grinding environment.

A magnetic sine chuck can be set to the desired angle just like a standard sine bar. The workpiece is placed on the plate and the magnet activated. Side or end rails should be used to align one edge of the workpiece or errors will result. **(See Figure 7.3.29.)** After setting and securing the angular position of the sine chuck, remove the gage blocks before grinding to protect them from abrasive grinding dust. The entire magnetic sine chuck can then be mounted on the surface grinder and the angular surface ground.

SIDE GRINDING

In some situations, it is not possible to grind two perpendicular surfaces by repositioning the workpiece. In these cases, *side grinding* with the side of the wheel can be performed. The face of the wheel is used to grind the part surface parallel to the magnetic chuck, and the side of the wheel is used to grind the vertical surface. Side grinding is often used to grind the sides of slots. It can also be used to grind a single surface if the workpiece cannot be mounted to be ground with the face of the wheel. Either the front side or back side (or both) can be dressed, depending on the workpiece requirements. **Figure 7.3.30** illustrates an application of side grinding.

Dressing the Wheel for Side Grinding

Before grinding with the side of the wheel, it needs to be dressed so that it is perpendicular to the wheel face by following these steps.

- Dress the wheel face using the normal method.
- Position the tip of the diamond dresser in line with the centerline of the wheel.
- Activate the magnetic chuck and check the dresser to be sure it is secure. **(See Figure 7.3.31.)**
- Start the spindle.
- Slowly feed the cross feed until the dresser contacts the side of the wheel.

FIGURE 7.3.29 Magnetic sine plates can also be used with gage blocks to set up angles. This method is very accurate. The 1-2-3 blocks are used to position the work above the top edge of the sine plate.

FIGURE 7.3.30 The vertical walls of the step and the slot of this workpiece have been finished using side grinding techniques.

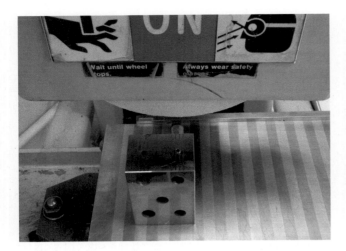

FIGURE 7.3.31 The dresser should be positioned on end and aligned with the centerline of the wheel for side dressing.

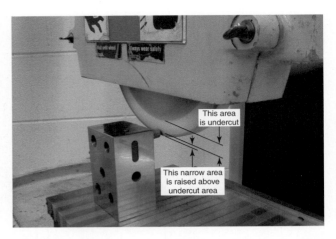

FIGURE 7.3.33 Dressing a small undercut into the wheel can help reduce the amount of surface-to-surface contact between the workpiece and the wheel.

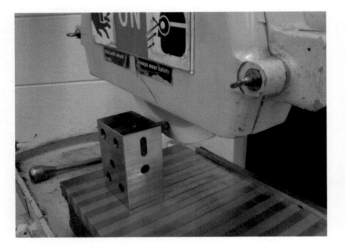

FIGURE 7.3.32 Dresser being used to dress the side of the wheel.

FIGURE 7.3.34 Dressing a small undercut can also be performed by hand using a dressing stick.

- Raise the wheel head so the wheel clears the dresser point.
- Move the cross slide to dress only about 0.001" off the wheel at a time.
- Use the wheel head feed handle to move the wheel down until the desired distance is reached. **(See Figure 7.3.32.)**
- Raise the wheel head until the diamond is past the wheel face. If the height of the vertical part surface to be ground is about 1/4" or less, this dressing method will normally produce satisfactory results.

If the height of the vertical part surface is greater than about 1/4", the amount of wheel contact will probably create excessive heat and burn the workpiece surface. Here are two methods can be used to relieve the wheel to reduce wheel contact.

Method 1

- Dress the desired distance as described previously.
- Move the wheel head so the tip of the diamond dresser is about 1/16" to 1/8" inside the outer wheel edge.
- Undercut the wheel about 0.005" to 0.010" with the dresser. **(See Figure 7.3.33.)**

Method 2

- This method will create a great deal of wheel dust, so use some type of dust collection or wear a dust mask.
- Start the spindle.
- Undercut the wheel by hand using an aluminum oxide dressing stick, as shown in **Figure 7.3.34**.
- After undercutting the wheel side with the dressing stick, dress the wheel face in the normal manner.

PERFORMING SIDE GRINDING

When mounting work to perform side grinding it is critical that the vertical surface to be ground is parallel with the longitudinal table travel. The back rail available on many magnetic chucks can be used to locate a reference surface of the workpiece. If the chuck does not have a back rail, a dial indicator can be used to indicate a reference surface parallel with the longitudinal table travel, as shown in **Figure 7.3.35**. This is very similar to indicating a vise on a milling machine table.

After the wheel is dressed and the work mounted, perform side grinding by following these steps.

• Position the face of the wheel within about 1/64" of the horizontal surface.

• Position the side of the wheel within about 1/64" of the vertical surface, as shown in **Figure 7.3.36**.

• Touch off the horizontal surface first and begin grinding.

• Use the cross feed to step over toward the vertical part surface.

 CAUTION

Use caution when approaching the vertical surface. Hitting the vertical surface too hard can damage the workpiece, break the grinding Wheel, and eject the workpiece from the machine.

• The sound of the grinding operation will begin to change when the corner of the wheel begins to contact the corner where the two surfaces meet.

• Since side grinding creates even more heat than when grinding with the wheel periphery, move the cross

FIGURE 7.3.36 Positioning of the wheel near the surfaces to be ground.

feed in increments of only 0.0001" to 0.0002" once contact is made against the vertical surface.

• After each cross feed step over, take multiple passes across the work with the longitudinal feed to spark out.

The corner of the grinding wheel will wear more rapidly than the face, so if only a small fillet radius is allowed or significant material needs to be removed, expect that the wheel may need to be redressed before grinding to final desired dimensions. In those cases, it is good practice to grind both surfaces within about 0.001" of the final sizes, then redress the wheel before finishing. **Figure 7.3.37** shows the horizontal and vertical surfaces of a step being ground.

FIGURE 7.3.35 A dial indicator being used to align the workpiece with the longitudinal travel.

FIGURE 7.3.37 Grinding horizontal and vertical surfaces.

GRINDING CYLINDRICAL WORK

Some cylindrical features, radii, and angles can be ground on the surface grinder using an indexing fixture. These fixtures are similar to direct indexing fixtures used on the milling machine. They often use spring collets for workholding. Some contain V-blocks mounted to a faceplate. **(See Figure 7.3.38.)**

A collet-type indexing fixture can be used to spin a workpiece beneath the grinding wheel to grind diameters. Align the work with the centerline of the wheel and rotate the fixture while lowering the wheel head. **(See Figure 7.3.39.)** A fixture may also have angular graduations or an indexing ring that can be used to position work at an angle. **(See Figure 7.3.40.)** Indexing fixtures with V-blocks can align the workpiece by moving the V-block from the centerline of the fixture spindle. **(See Figure 7.3.41.)**

GRINDING PROBLEMS
Burning of the Work Surface

As mentioned earlier, grinding produces a great deal of heat. Using water-based synthetic cutting fluid can greatly reduce heat during grinding. Ideal coolant

FIGURE 7.3.39 (A) The center of the workpiece should be in line with the center of the wheel when using an indexing fixture. (B) Lower the wheel head while rotating the fixture handwheel to grind the cylindrical surface.

FIGURE 7.3.38 A collet-type fixture can be used to produce cylindrical or angular surfaces.

FIGURE 7.3.40 Angular graduations or an indexing ring can be used to set up angles for grinding.

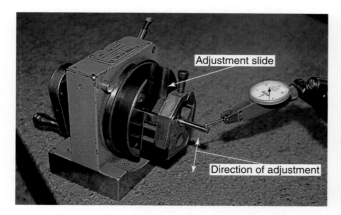

FIGURE 7.3.41 A V-block fixture can align the workpiece by moving the V-block from the centerline of the fixture. A dial indicator can be used to ensure that the cylindrical surface is running true.

FIGURE 7.3.43 Burn marks appear as dark streaks in the surface of the workpiece.

concentrations may vary by material and application. As always, be sure the concentration is within the manufacturer's recommendations.

Special consideration must be given when using water-based coolant while grinding carbide. Carbide is a material composed of tungsten carbide particles cemented together by cobalt. Many coolants can cause this cobalt binder to leach away from the carbide particles (**Figure 7.3.42**) and be carried away in the coolant. This severely reduces the surface strength of the workpiece material. Special water-based coolants are available with cobalt leaching inhibitors.

If burn marks (**Figure 7.3.43**) appear on the work surface, try reducing cut depth and/or cross-feed stepover. A wheel that is too hard, too fine, or one with a dense structure, can also create excessive heat leading to burn marks. Try a softer, coarser, or more open wheel to reduce heat and burning.

Scratches on the Work Surface

Scratches on the work surface can be caused by a few different factors. Grinding without coolant can allow workpiece material and released wheel grains to stay on the workpiece surface. They can then be pulled across the surface, creating scratches. Using a flood coolant will wash away this debris from the workpiece surface. A dirty coolant system that contains an excessive amount of grinding debris can also cause scratches as particles are carried through the flood system and across the work.

Waviness or Chatter on the Work Surface

An uneven work surface (**Figure 7.3.44**) is usually the result of an out-of-round wheel or a wheel that is loose on the machine spindle. Check the spindle nut to be sure the wheel is tight. An out-of-round wheel can be easily fixed by truing/dressing.

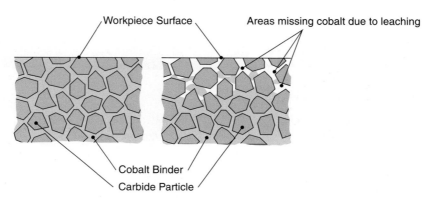

FIGURE 7.3.42 Carbide particles that are adequately surrounded by the cobalt binder material (left), compared to carbide that has lost the cobalt binder due to leaching (right).

FIGURE 7.3.44 A wavy surface may indicate a loose or out-of-round grinding wheel.

SUMMARY

- The surface grinder is a commonly used precision grinder in the machining industry.
- Before operating any surface grinder, become familiar with the machine and all safety precautions related to its use.
- After selecting an appropriate grinding wheel, a ring test should always be performed before mounting it on the surface grinder.
- After mounting a new wheel or before using an existing wheel, it should be trued and dressed.
- Most surface grinders contain a magnetic chuck for securing ferromagnetic workpieces or other workholding devices. Other magnetic workholding devices are also available for securing work for desired grinding operations.
- Grinding parallel and perpendicular surfaces, as well as angles, is very common and the methods are similar to those used when milling.
- Side grinding can be used to grind a horizontal surface and an adjacent vertical surface without repositioning a workpiece.
- Indexing fixtures can be used to grind some cylindrical workpieces, radii, and angles.
- Some common grinding problems include burns, scratches, and chatter on the surface of the work. Solutions to these issues include modifying cut depth or cross feed, changing the type of wheel, using clean coolants, or truing/dressing the grinding wheel.

REVIEW QUESTIONS

1. List five surface grinder safety guidelines.
2. What should be done to every grinding wheel before mounting?
3. When using a magnetic workholding device, the work should span at least _____ of the fields of the magnet.
4. When using a magnetic chuck to hold a workpiece that is taller than its length or width, what should be done to ensure the part is held securely?
5. Describe the difference between wheel truing and dressing.
6. What tool is used to true and dress an aluminum oxide wheel?

7. How is a CBN wheel trued and dressed?

8. Depth-of-cut range for surface grinding is:
   a. 0.010"–0.100"
   b. 0.0001"–0.0002"
   c. 0.0001"–0.003"
   d. 0.00001"–0.00005"

9. What should be done to a magnetic chuck when it becomes unevenly worn?

10. What workholding device can extend the magnetic field of the magnetic chuck and quickly position and hold work at a 45-degree angle?

11. What workholding device can be used when grinding very precise angles other than 45 degrees?

12. Briefly describe the method for grinding two perpendicular surfaces without repositioning the workpiece.

13. List three types of workpiece features that can be produced by using an indexing fixture on the surface grinder.

14. List three possible solutions to eliminate burn marks on the surface of a workpiece.

15. What can cause scratches on the surface of a workpiece?

SECTION 8 COMPUTER NUMERICAL CONTROL

UNIT 1

CNC Basics

Learning Objectives

After completing this unit, the student should be able to:

- Demonstrate understanding of basic CNC motion-control hardware
- Demonstrate understanding of the Cartesian coordinate system
- Demonstrate understanding of the polar coordinate system
- Demonstrate understanding of the absolute and incremental positioning systems
- Demonstrate understanding of the purpose of G- and M-codes
- Demonstrate understanding of word addresses
- Demonstrate understanding of modal codes
- Define and describe a "block" of a CNC program
- Demonstrate understanding of machine motion types
- Demonstrate understanding of the main components of a CNC program

Key Terms

Absolute positioning system	**Conversational programming**	**Machine control unit (MCU)**
Automatic tool changer (ATC)	**Encoder**	**Machining center**
Ball screw	**End of block**	**M-codes**
Block	**G-codes**	**Miscellaneous function**
Cartesian coordinate system	**Incremental positioning system**	**Modal**
Circular interpolation	**Linear guide**	**Origin**
	Linear interpolation	

Polar coordinate system
Preparatory command
Quadrant

Rapid traverse
Rectangular coordinate system
Servo motor

Turning center
Word address

INTRODUCTION

In today's demanding and fast-paced machining climate, computerized numerical control (CNC) machines are revolutionizing the face of machining. These high-tech machines can perform things that could have never been imagined even a decade ago. Complex operations can be performed faster, with more accuracy and consistency, and so tirelessly that it appears effortless.

In a modern computerized machine shop, the most commonly performed operations are milling and turning. The machines performing these operations are basically computerized versions of manual milling machines and manual lathes. With these two types of machines, most shops can produce a great variety of parts in all shapes, sizes, and materials. **Figure 8.1.1** shows examples of some workpieces produced by CNC machines.

The addition of an **automatic tool changer (ATC)** in combination with an automated means of loading/unloading material and parts allows these machines to be run virtually unattended. When a CNC lathe is equipped with an automatic tool changer it is called a **turning center**. A CNC mill with an ATC is called a **machining center**. **Figure 8.1.2** shows a CNC lathe and **Figure 8.1.3** shows a CNC turning center. **Figure 8.1.4** shows a CNC milling machine, and **Figure 8.1.5** shows a CNC machining center.

FIGURE 8.1.1 Examples of some workpieces produced by CNC machines.

Whether it has an ATC or not, adding a CNC control to a machine enables it to perform complex operations smoothly, efficiently, and accurately. Just 30 years ago, some of the operations these machines perform would have either been impossible or would have required numerous setups, cumbersome machine attachments, and tedious handwork.

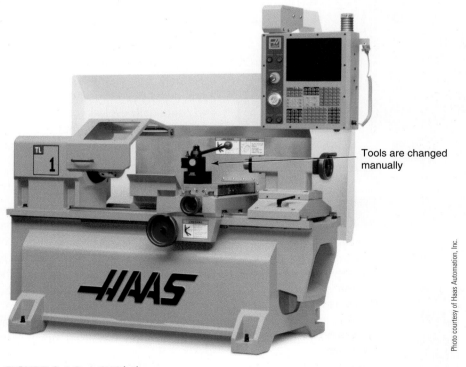

Tools are changed manually

Photo courtesy of Haas Automation, Inc.

FIGURE 8.1.2 A CNC lathe.

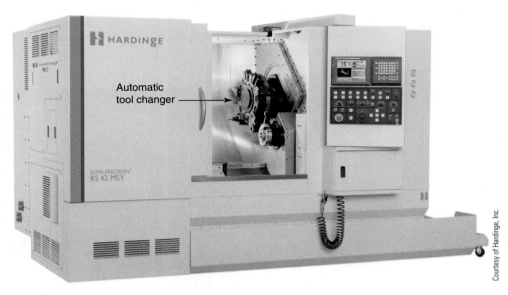

FIGURE 8.1.3 A CNC turning center.

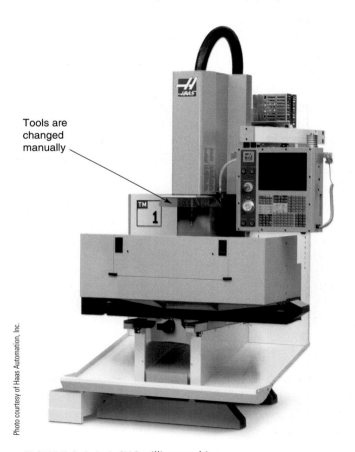

FIGURE 8.1.4 A CNC milling machine.

THE CNC MACHINE CONTROL UNIT

It requires a coordinated effort between the CNC programmer, setup person, operator, and the machine's hardware for successful CNC machining to occur. A CNC program is created and stored in the **machine control unit**, or **MCU** (sometimes simply called a "control"). When the program is run, this information is sent to different systems on the machine that control the axes motion, spindle motor, ATC, coolant pump, and more. The MCU has an external operator control panel with a display screen. From this operator panel and display, programs can be keyed in, the machine setup data can be entered, and machine functions can be monitored and controlled. **(See Figure 8.1.6.)**

CNC MOTION CONTROL

A CNC machine in operation can be an object of fascination. By taking a programmed command the machine can accurately position its axes directly at a location quickly, accurately, and smoothly. The hardware that makes this happen has been evolving over the last quarter-century and continually evolves today. There are a few variations of the types of systems that are used. The components used in the most common motion-control systems are discussed next.

Drive Screws

Recall from Sections 5 and 6 that special screws called leadscrews are used to move the machine axes along their slides by converting rotational motion into linear motion. The most common type of screw used to create linear motion on manual machines is the Acme type (e.g., leadscrew on a manual lathe). Acme screws, however, are unacceptable for modern industrial CNC applications because of their backlash, friction, wear, and inefficiency.

Anyone who has used manual machine tools has likely encountered backlash when reversing the direction of an axis. This wasted motion is due to clearance in the

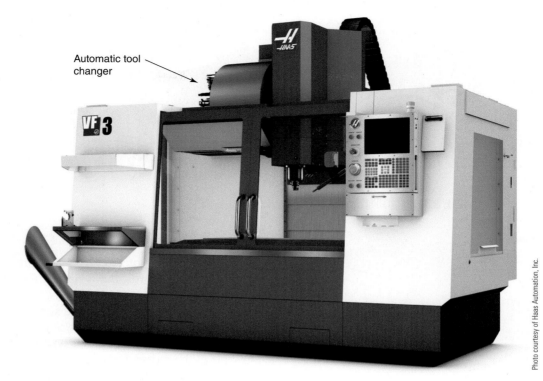

Automatic tool changer

Photo courtesy of Haas Automation, Inc.

FIGURE 8.1.5 A CNC machining center.

FIGURE 8.1.6 An operator control panel mounted on the exterior of the machine control unit.

threads of an Acme screw between itself and the mating nut. This is one reason Acme screws are not suited for CNC machines. Since CNC systems rely on the frequent reversal of axis motion, no backlash is acceptable.

Another desirable aspect of CNC drive systems is wear resistance, which results from low friction and good lubrication. When a screw is low in friction it consumes less power, operates more smoothly, and lasts longer. An Acme thread performs poorly in this area also due to its large sliding-surface area.

The axes on a CNC machine must accelerate and move extremely aggressively. The more energy from a rotating motor that is translated directly into linear axis motion will allow more speed and acceleration to be produced by a given motor size. Acme screws are only about 30% to 40% efficient in transmitting torque into linear motion.

There is an ingenious mechanical solution to all these issues: a **ball screw** assembly. This device consists of a screw and nut assembly with steel balls in place of threads. Ball screws often achieve 90% efficiency in transmitting torque into linear motion. **Figure 8.1.7** shows a cross section of a ball screw assembly.

CNC Guideways

Recall from previous units that each machine axis component (table, carriage, column, bed, or cross slide) must be guided on a precise set of tracks called "ways." Like leadscrews, the ways of a manual machine have major shortcomings when used in high-demand CNC applications. Conventional plain V-type or dovetail ways do not have high performance seals capable of keeping lubrication in and debris out. They also produce a significant amount of friction, which consumes power and causes wear.

Most modern CNC machines employ a high-tech modernized version of ways called **linear guides**. These units are sealed, have pressurized lubrication systems, and contain low-friction ball bearings. Linear

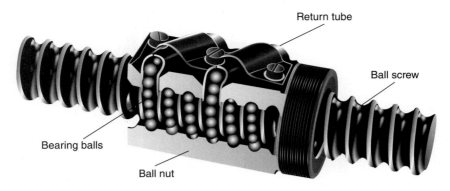

FIGURE 8.1.7 A ball screw assembly.

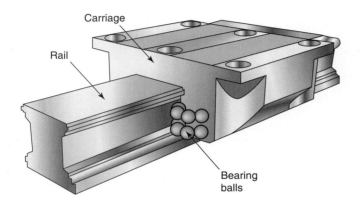

FIGURE 8.1.8 A linear guide assembly and its parts.

guides are typically available in matched sets, which are removable and replaceable in the field, unlike plain ways. **Figure 8.1.8** shows a cross section of a linear slide assembly.

Servo Motors

A standard electric motor is able to provide the power to move a machine axis, but is not capable of tracking how far the axis has moved. For this, a hybrid electric motor is used. This motor is called a **servo motor** and is half motor/half position sensor. The sensor portion of the motor is called an **encoder** and is typically mounted on the shaft of the motor. The encoder works by recording the amount of rotation a motor makes (in degrees). By knowing how far the motor shaft has turned, the MCU can use the lead dimension of the ball screw to calculate the amount the machine axis has moved. The encoder provides feedback to the machine control and if the desired amount of movement is not initially obtained, the machine will instantly adjust the motor's position to compensate. This is called a *closed feedback loop*. The MCU and the encoder constantly communicate with each other, so axis movement is accurate. **Figure 8.1.9** shows an exploded view of a servo motor.

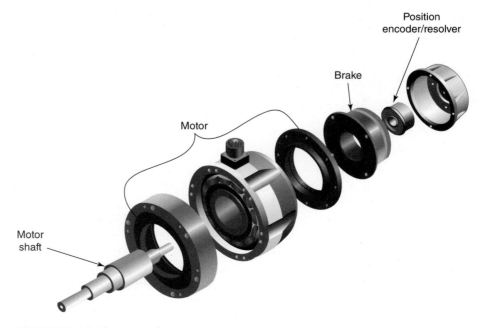

FIGURE 8.1.9 The parts of a servo motor.

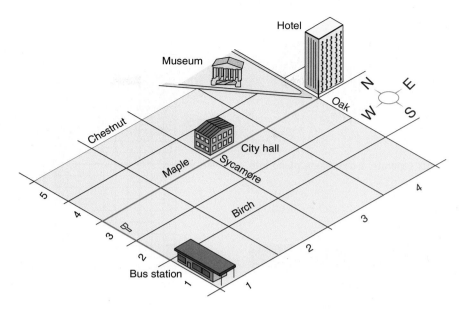

FIGURE 8.1.10 A traveler at the bus station would arrive at the hotel if instructed to take two blocks north and then four blocks east.

COORDINATE SYSTEMS

There are basically two major types of instructions the programmer will give a CNC machine. One of them is in the form of code, which tells the machine what type of function it is to perform. The other type of instruction informs the machine at what position on the workpiece it is to perform the function. For the position instructions, the programmer must use a *coordinate system* to map out specific locations on the workpiece. By using this map, the programmer will provide "driving directions" the machine can understand to make the cutting tool arrive at the intended destination.

The Cartesian Coordinate System

An example of similar mapping in everyday life might be someone who is viewing a satellite image of a city while giving instructions to a person traveling the city streets. The guide may direct the traveler to the destination by telling him or her to go two blocks north and then four blocks east. **(See Figure 8.1.10.)** CNC programming does not use north, south, east, or west to identify distance and direction, but instead a similar system called the **Cartesian coordinate system**. This coordinate system is also sometimes called the **rectangular coordinate system**. The north and south direction are in-line. In the Cartesian coordinate system, that alignment is called an axis and the north and south axis is called the Y-axis. The east and west axis is called the X-axis.

Movements in the "north" direction are given positive values and movements in the "south" direction are given

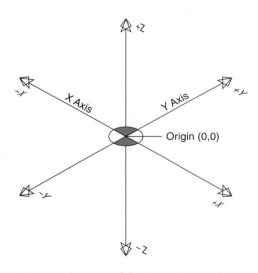

FIGURE 8.1.11 The parts of the Cartesian coordinate system.

negative values. Movements in the "east" direction are given positive values and movements in the "west" direction are given negative values. In the previous example, our guide would call the north direction Y-positive and the east direction X-positive. **Figure 8.1.11** shows three axes of the Cartesian coordinate system viewed isometrically.

Referring to the example of the city streets, we call the place the traveler began from the *point of origin,* or **origin** for short. This point is X0, Y0 and all positions are simply based off this point. Therefore, the final destination of the traveler is X4, Y2.

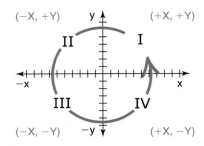

FIGURE 8.1.12 Quadrants separate a coordinate plane into four regions.

 **CAUTION**

Always use careful consideration as to whether a coordinate is positive or negative. For example, if a position of X10.0 is programmed instead of X-10.0, then the positioning move will be made 20 inches away from where it was intended! This could be disastrous if workholding devices and workstops are in the path of travel!

The X- and Y-axes of the Cartesian coordinate system divides the system into four separate regions. These four regions are called **quadrants**. **Figure 8.1.12** shows the quadrants of the X-Y plane. The quadrants are numbered in a counterclockwise direction from I to IV. Notice that all coordinates in quadrant I are positive in both axes, and that all coordinates in quadrant III are negative in both axes.

The Polar Coordinate System

The Cartesian coordinate system is the most commonly used coordinate identifying system in CNC programming, but is not the only coordinate system used. The **polar coordinate system** is also sometimes used to identify locations. Polar coordinates require that positions be identified by defining both an angle and a distance (like a vector in mathematics) from the origin to a specific location. **(See Figures 8.1.13** and **8.1.14.)** Polar coordinates are helpful for positioning the machine at multiple angular locations when the angle between each position is known (such as bolt circles or hole patterns). When the polar coordinate system is used for bolt circles, the origin is usually placed in the center of the circle.

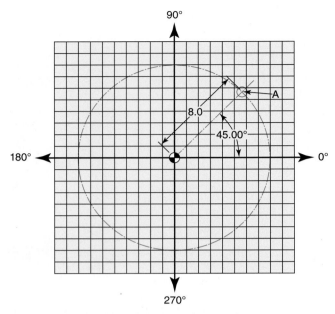

FIGURE 8.1.13 A diagram showing the polar coordinate system. When using this system, the X-value specifies the distance from the origin, and the Y-value specifies the angle relative to the zero degree mark. The position "A" shown is located at an angle of 45° and a distance of 8.0" from the origin; when using the polar coordinate system this is written as X8.0 Y45.0.

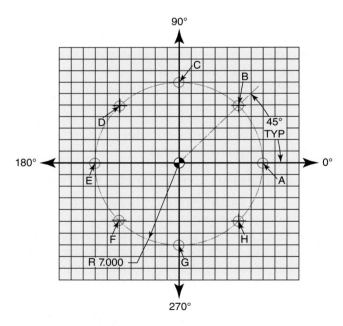

FIGURE 8.1.14 An example of how each position in an eight-hole workpiece can be identified using polar coordinates. The positions for each location are programmed as: (A) X7.0 Y0; (B) X7.0 Y45.0; (C) X7.0 Y 90.0; (D) X7.0 Y135.0; (E) X7.0 Y180.0; (F) X7.0 Y225.0; (G) X7.0 Y270.0; (H) X7.0 Y315.0.

POSITIONING SYSTEMS
The Absolute Positioning System

When using a coordinate system for programming, there are two methods available to reference workpiece positions. One method is known as the **absolute positioning system**. When using absolute positioning, the coordinates for all positions will be referenced from the workpiece origin (X0, Y0, Z0). **(See Figure 8.1.15.)** NOTE: It can often be helpful to think of absolute coordinates as *positions* or locations instead of *distances*.

The Incremental Positioning System

The other method for referencing coordinates is the **incremental positioning system**, which specifies a distance from the current position to the next position instead of a location related to the origin. In other words, each current tool position becomes like an origin point for the next positioning move. It is important that the correct sign (+ or −) be given to indicate the direction of the incremental distance. **(See Figure 8.1.16.)**

There are advantages and disadvantages to either method of positioning depending on the situation. One disadvantage to using the incremental system is that any programmed errors will accumulate. In other words, if a position error is made, that position and all positions after it will also be wrong since they all are all referenced from an incorrect position. Troubleshooting a positioning mistake in an incremental program may be very difficult. While programming, it can also be difficult

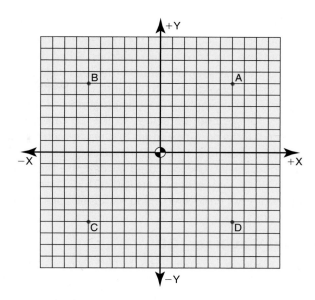

FIGURE 8.1.16 Assume that each block on the grid equals 1". Using the incremental programming method, assuming the cutting tool is already at position A, the coordinates for each following location are: (B) X-12.0 Y0.0; (C) X0.0 Y-12.0; (D) X12.0 Y0.0.

to keep track of the current position and know the next incremental position to program. The advantage to using the incremental method is that parts requiring repetitive motions can be easily computed and programmed (e.g., a series of 1" motions for drilling holes all in a row).

All things considered, absolute programming is typically less confusing and programming errors are less likely to be made. Absolute positioning is the most commonly used system in industry.

CODES

In addition to telling the machine what position to move to, the programmer must also provide the machine with instructions for what to do at that position. These instructions are not written in the English language, but rather in code that the machine can understand. This style of programming is called **word address**. An individual program *word* can be an axis letter with a position (e.g., X3.456) or a code (e.g., G20). Program words that are to be executed at the same time are all written on the same line. This line of program words is called a program **block**. Each command within a block will be completely executed before the program advances to the next block.

G-Codes

G-codes, or **preparatory commands**, prepare a machine to engage in a particular mode for machining. For instance, a G1 code tells the machine to feed in a straight line, or linear motion, G0 initiates rapid positioning, G90

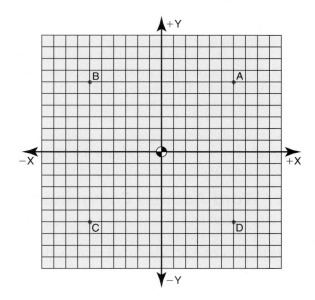

FIGURE 8.1.15 Assume that each block on the grid equals 1". Using the absolute programming method, the coordinates for each identified location are: (A) X6.0 Y6.0; (B) X-6.0 Y6.0; (C) X-6.0 Y-6.0; (D) X6.0 Y-6.0.

turns on absolute programming, G91 turns on incremental programming, and G20 sets the machine to inch units. Some of these functions are like switches. When they are turned on, they will remain on until they are intentionally turned off or overridden with a conflicting preparatory command. For example, if a G0 is programmed after a G1, they cannot both be active, so the G0 cancels the G1 when it turns on. These codes that remain active until cancelled or overridden are called **modal** codes.

The four major G-codes used to describe the most common CNC axis motions are G0, G1, G2, and G3. Other codes will be covered along with their applications in later units.

Rapid Traverse—G0

Rapid traverse is designated by the code G0. This type of motion is used to position the machine axis very quickly. This motion is used only to position the cutter for machining and no workpiece material should ever be contacted during this extremely aggressive movement. At its full capacity rapid traverse rates of 1500 inches per minute are quite common. To put that into perspective, a rate of 1500 inches per minute equates to 25 inches per second. At this rate, a midsized CNC machine can cover its entire axis travel distance in about one second, and the acceleration on most machines approaches one full G-force. An F16 jet only accelerates at about 0.7 G! For this reason, everyone involved in CNC machining—from the programmer to the setup person and operator—must be totally aware of possible interferences and collisions. Most machine control panels allow the operator to slow the rate of these rapid movements for safety reasons, such as when running a new program for the first time.

Use caution when using rapid motions. Be aware of possible collisions between the workpiece and the tool or workholding devices and the tool.

Linear Interpolation—G1

Linear interpolation moves the cutting tool in a straight-line path between two points. To get from one point to the next, motion may be required in one axis alone or in more than one axis (diagonal movements). In order to move the tool in a linear interpolated movement, the tool is first positioned at the start point of the linear cut. Once there, a new program block is commanded with using a G1 and an end position for the movement. A feed rate must also be programmed along with the newly programmed position.

 CAUTION

It is important to remember that not all G-codes conflict with other G-codes. For instance, a G1 code tells the machine to feed in a straight line, or linear motion. A G20 code tells the machine to interpret the given coordinates in inch units. These two codes do not conflict with each other and can therefore be programmed together in the same block. However, if we would try to program a G1 along with a G0 (rapid positioning), a conflict would arise and the machine would stop and provide an alarm message.

Circular Interpolation—G2 or G3

Circular interpolation motion causes the cutter's path to travel in an arc. With these motions, CNC machines can cut full or partial circles. These paths are important for milling features such as circular slots, radii on a contour, circular pockets, and for turning features such as spheres, fillets, and radii.

To cut an arc, the tool is first programmed to be positioned at the start point of that arc. Once there, the G-code command is given to designate the direction of tool travel to create the arc. If tool travel is clockwise from start to end, a G2 code is used. If tool travel is counterclockwise, a G3 is used.

M-Codes

M-codes are very similar to G-codes, only they are used to turn on and off **miscellaneous (auxiliary) functions**. For example, an M8 code can be used to turn coolant on and an M9 turns it off. M3 starts the spindle motor in forward rotation and M5 stops it. An M6 code activates the automatic tool changer in a machining center, but this code turns itself off automatically when the tool change is complete. An M30 is used to end the program and reset it for the next time it is to be run. On most machines only one M-code is allowed on each block.

G- and M-codes and their applications will be covered in more detail in Units 3 and 6. **Figure 8.1.17** shows a table with some commonly used G- and M-codes.

Other Word Address Commands

Not all information for CNC machining can be given as G- or M-codes or coordinate positions. The founders of programming created many other commands to be named after the first letter of their function. For example, an S-command designates the spindle speed. A spindle RPM of 3200 is programmed "S3200." Likewise, a T-command calls up a desired tool number and an F-command specifies a feed rate. These commands will be covered in more detail later.

			Preparatory Functions
CODE	**MILL**	**LATHE**	**DESCRIPTION**
G0	✓	✓	Positioning at rapid feed
G1	✓	✓	Linear interpolation
G2	✓	✓	Circular interpolation-clockwise
G3	✓	✓	Circular interpolation-counterclockwise
G4	✓	✓	Dwell
G9	✓	✓	Exact motion stop at intersections (non-modal, one block only)
G10	✓	✓	Offset value entry through program
G12	✓		Circular pocket milling cycle-clockwise
G13	✓		Circular pocket milling cycle-counterclockwise
G15	✓		Cartesian coordinate programming system on
G16	✓		Polar coordinate programming system on
G17	✓	✓	X/Y plane selection for arc cutting
G18	✓	✓	Z/X plane selection for arc cutting
G19	✓	✓	Z/Y plane selection for arc cutting
G20	✓	✓	Inch unit selection
G21	✓	✓	Metric unit selection
G27	✓	✓	Reference position return check
G28	✓	✓	Return to the primary machine zero position (home position)
G29	✓	✓	Return from reference position
G31	✓	✓	Skip function
G32		✓	Thread cutting (single-point tool or tap)
G40	✓	✓	Cutter radius compensation cancel
G41	✓	✓	Cutter radius compensation-left
G42	✓	✓	Cutter radius compensation-right
G43	✓		Tool height offset compensation-activate
G44	✓		Tool height offset compensation-cancel (some machines use G49)
G49	✓		Tool height offset compensation-cancel (some machines use G44)
G50		✓	Maximum RPM setting for Constant Surface Speed
G52	✓	✓	Local coordinate system setting
G53	✓	✓	Machine coordinate system setting
G54	✓	✓	Workpiece coordinate system setting #1
G55	✓	✓	Workpiece coordinate system setting #2
G56	✓	✓	Workpiece coordinate system setting #3
G57	✓	✓	Workpiece coordinate system setting #4
G58	✓	✓	Workpiece coordinate system setting #5
G59	✓	✓	Workpiece coordinate system setting #6
G61	✓	✓	Exact motion stop at intersections (modal)
G64	✓	✓	Normal cutting mode without exact stops at intersections
G65	✓	✓	Custom macro call
G70		✓	Finish turning/facing/boring cycle
G71		✓	Rough turning/boring cycle
G72		✓	Rough facing cycle
G73		✓	Irregular rough turning cycle (for castings and forgings)
G73	✓		Chip break peck drilling cycle
G74	✓		Left hand tapping cycle
G74		✓	Face grooving or chip break peck drilling cycle
G75		✓	OD groove chip break peck cycle
G76	✓		Fine boring cycle (no tool drag mark)
G76		✓	Auto repetative threading cycle (single point)
G80	✓	✓	Cancel canned cycles
G81	✓	✓	Single pass drill cycle
G82	✓		Single pass drill cycle with dwell
G83	✓		Full retract peck drilling cycle
G84	✓	✓	Tapping cycle with reversing
G85	✓	✓	Boring cycle (feed in/feed out)
G86	✓	✓	Boring cycle (feed in/rapid out)
G87	✓		Back boring cycle
G90	✓		Absolute programming
G91	✓		Incremental programming
G92	✓		Reposition origin via program
G92		✓	Thread cutting
G94	✓		Inch or MM per minute feed rate
G95	✓		Inch or MM per revolution feed rate
G96		✓	SFM value for Constant Surface Speed
G97		✓	Fixed spindle RPM/Constant Surface Speed cancel
G98		✓	Inch or MM per minute feed rate
G99		✓	Inch or MM per revolution feed rate
			Miscellaneous Functions
CODE	**MILL**	**LATHE**	**DESCRIPTION**
M00	✓	✓	Program stop
M01	✓	✓	Optional program stop
M02	✓	✓	Program end
M03	✓	✓	Spindle on clockwise
M04	✓	✓	Spindle on counterclockwise
M05	✓	✓	Spindle off
M06	✓	✓	Toolchange
M08	✓	✓	Coolant on
M09	✓	✓	Coolant off
M10	✓	✓	Chuck, collet, or rotary table clamp
M11	✓	✓	Chuck, collet, or rotary table unclamp
M19	✓	✓	Orient spindle
M30	✓	✓	Program end and return to start
M97	✓	✓	Local sub-routine call
M98	✓	✓	Sub-program call
M99	✓	✓	End sub-program and return to main program

Note: Most machines allow single digit codes to be used with or without the preceding zero (Example G01 = G1)

FIGURE 8.1.17 Commonly used G- and M-codes for CNC programming.

Parameter	Bit Number							
Number	#7	#6	#5	#4	#3	#2	#1	#0
0401	1	1	0	1	0	0	0	1

FIGURE 8.1.18 An example of the parameter numbering format. Notice the parameter number is 0401 and the bits are read in sequence from right to left. Each bit can be either a "0" or a "1."

Binary Code

At the user level, almost all CNC functions are controlled with codes in the program. Sometimes, however, it is necessary to change machine settings that cannot be controlled by the program commands. An example may be the configuring of communications settings to transfer a program to the machine from an external computer. These parameter settings are often written in a language called *binary code.* Binary codes are made up of only the digits "1" and "0." A "1" is used to indicate "on" or "active" and a "0" indicates "off" or "inactive." Binary parameter settings are often eight digits long and numbered from zero to 7, right to left, as shown in **Figure 8.1.18**. The most commonly altered parameters can usually be found in the machine control manual, and it is absolutely necessary to refer to the manual before changing any parameters. Machine parameter modification should only be performed by experienced authorized personnel. An improperly modified parameter can cause serious machine damage.

CONVERSATIONAL-TYPE PROGRAMMING

Not all CNC machines must be programmed with G- and M-codes. Some machines have a special type of MCU that allows **conversational programming**. Conversational programming was developed to simplify the machine programming process. There are many different types of conversational MCU brands, including Southwest Industry's Proto-TRAK, Hurco's WinMax, and Mazak's Mazatrol controls. These MCUs vary greatly, so the following descriptions are very general.

When programming a conversational machine, the operator will usually select the intended type of machining operation from an on-screen menu. After the machining operation has been identified, the machine will prompt the programmer with a series of questions. Most often these "questions" are actually empty fields (boxes) requiring data to be input. An example for milling might be a pocketing operation. The programmer will first select "POCKET" from the menu. A series of fields will then appear requiring the programmer to enter information including spindle RPM, feed rate, pocket center location, pocket width, pocket length, pocket depth, and step-over distance between passes. Once all the necessary data have been entered, the operation will be programmed automatically in a manner the machine can understand.

Not all machines are equipped with conversational programming, and there are advantages and disadvantages to the method. The advantages are that it requires much less programming knowledge, can often be faster for simple features, and can easily be performed at the machine on the shop floor. The disadvantages revolve mostly around the lack of flexibility: the only operations that can be performed are those provided by the manufacturer. It is for these reasons that conversational controls are usually most useful for repair work, prototyping, or low-volume production machines such as CNC-controlled knee mills and flat bed lathes. It is common for machine manufacturers to offer a machine with both manual controls and conversational CNC controls, so the machine can be used in either mode depending on the complexity of the operation. When high production quantities require highly efficient programs or complex machining scenarios, the versatility of a G- and M-code programming method is usually more suitable.

PARTS OF A CNC PROGRAM

Just as a written letter to a friend or business contact has three main parts (salutation, body, and closing), a CNC program is essentially broken into three similar main parts: the salutation "program safe-start," the body "material removal," and the closing "program ending."

Comments and notes can be added into the program for the operator to read. These comments must be contained within parentheses so the MCU knows not to read them. At the very end of each block is a semicolon (;). This semicolon character is called an **end of block** character. When the MCU reads an end of block character, it knows to move on to the next block.

Various CNC controls need the details within the parts of a program arranged in a specific order. This arrangement is called program format and its layout depends on the control brand and model. For this section, we will look at a milling program in one of the most standard formats in industry (Fanuc brand). The program example that follows has been simplified to show the basic parts of the safe-start, but be advised that it is missing some details that may be needed to make it work in an actual machine. The complete format will be covered in later units when more codes are introduced.

Safe-Start

A CNC machine will seldom make a mistake; however, a beginning programmer will usually make many. It is extremely important to tell the machine clearly and concisely what functions it is to perform. To do this, it helps to be conscientious of the way that the machine control "thinks." For example, modal codes are active until the machine is told to do otherwise. If a programmer forgets that the machine has modal codes active from running a previous program, they may still be active when the next program is run. Therefore, at the beginning of each program, it is common practice to deliberately cancel any modal codes that are unwanted and activate any that are necessary. This is to be done at the start of the program and is called the *safe start*.

The safe-start section begins with a program number. This number will be the label that the program is stored under in the machine memory. Many MCUs need the program number in a format starting with a letter "O" followed by four or five digits. After the program number is given, the control must be instructed which measurement units will be used in the program by programming a code for either inch units (usually G20) or metric units (usually G21). All coordinates after this code will be interpreted in the proper measurement units. It is also good to set the feed rate units in the safe start block. A G94 will set the units for inches per minute (IPM) feed. A G95 will set the feed to inches per revolution (IPR). Additional important codes will be added to the safe-start block in future units as their meanings are explained.

After the program has been named, the measurement units have been specified, and feed rate units have been chosen, the desired tool must then be loaded via a tool-change operation (M6 code). Next the machine must be told what programming technique will be used (absolute or incremental) and what motion type is desired (usually a G0 rapid move to move to the first location for machining). Finally, include coordinates that direct the machine to the desired position and start the spindle (M3) to the desired RPM. The example in **Figure 8.1.19** shows the basic parts of a sample safe-start program section for a three-axis CNC mill.

Material Removal

The next phase of the CNC program allows for freedom and creativity from the programmer to do whatever necessary to make a part that meets print specifications. The purpose at this phase is to move the cutting tool through the material and perform the desired cutting action. For drilling, the tool is positioned and then moved into the work using a motion in the Z-axis only. For slot milling, pocketing, or contour milling the endmill is positioned at a desired depth in the Z-axis and then moved in the X- and Y-axes in a path to create the desired shape. See the contour milling example in **Figure 8.1.20**.

Program Ending

The final phase is to safely position the tool and machine axes out of the way to facilitate safe workpiece removal and reloading. An M9 switches the coolant off. The M30 code is the last code programmed and tells the machine that the program should now end and reset to a default condition. See the example in **Figure 8.1.21**.

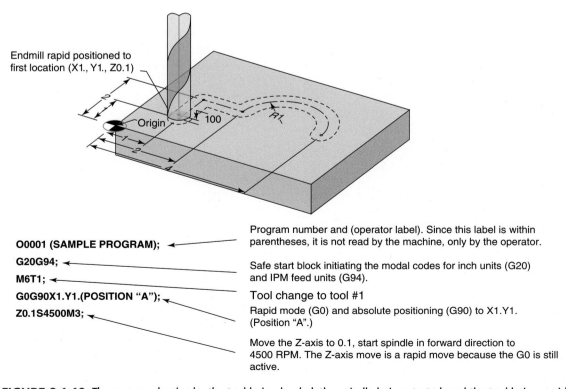

FIGURE 8.1.19 The program begins by the tool being loaded, the spindle being started, and the tool being rapid positioned to the first location.

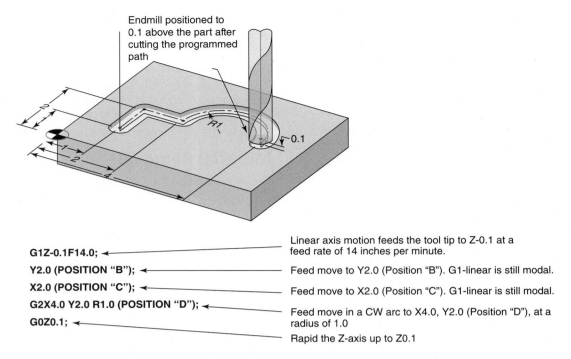

G1Z-0.1F14.0; ◄─────────────── Linear axis motion feeds the tool tip to Z-0.1 at a feed rate of 14 inches per minute.

Y2.0 (POSITION "B"); ◄─────────────── Feed move to Y2.0 (Position "B"). G1-linear is still modal.

X2.0 (POSITION "C"); ◄─────────────── Feed move to X2.0 (Position "C"). G1-linear is still modal.

G2X4.0 Y2.0 R1.0 (POSITION "D"); ◄─────────────── Feed move in a CW arc to X4.0, Y2.0 (Position "D"), at a radius of 1.0

G0Z0.1; ◄─────────────── Rapid the Z-axis up to Z0.1

FIGURE 8.1.20 The tool is fed to depth and then fed to each X/Y location. Once machining is complete, the tool is then retracted to a clearance point above the workpiece.

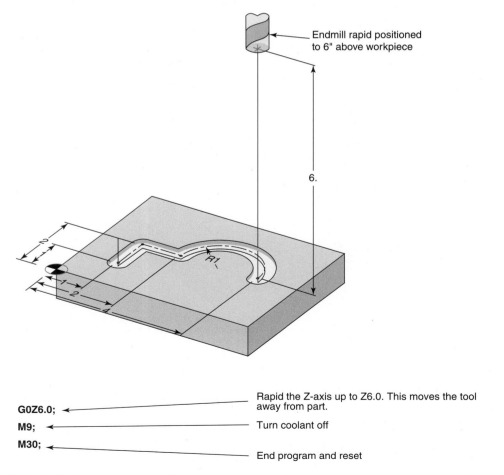

G0Z6.0; ◄─────────────── Rapid the Z-axis up to Z6.0. This moves the tool away from part.

M9; ◄─────────────── Turn coolant off

M30; ◄─────────────── End program and reset

FIGURE 8.1.21 The tool is rapid positioned to a location far above the workpiece. The coolant is turned off and the program is ended.

SUMMARY

- The CNC machines of today can achieve higher accuracies, better repetition, and more complex geometries than manual machines.
- Coordinate systems are used to identify positions on a workpiece where operations are to take place.
- An origin is a point where coordinates are referenced from.
- The absolute positioning system is a method where all specified coordinates are referenced from one absolute reference point (the origin).
- The incremental positioning system is a method where all specified positions are identified as distances from a previous location.
- G- and M-codes accompany coordinates to prepare or activate machine functions.
- A modal code is one that stays active throughout a program unless canceled or contradicted.
- A CNC program consists of three major sections: safe-start, material removal, and program end.

REVIEW QUESTIONS

1. What is an ATC?
2. What is an MCU and what is its function?
3. Briefly describe a ball screw and a linear guide.
4. Explain the benefits of using the absolute programming system versus the incremental system.
5. Explain the benefits of using the incremental programming system versus the absolute system.
6. Which coordinate system uses an angle and a distance to identify a position?
7. What is the name for the type of motor used to drive CNC axes that is also capable of recording axis position?
8. Supposing a programmer, using absolute mode, mistakenly entered Z-3.689, when he or she intended to enter Z3.689. Describe what would happen, and how far the *actual* position would be from the *intended* position.
9. What is a modal code?
10. What is another name for the Cartesian coordinate system?
11. List four G-codes and describe their functions.
    a. _____
    b. _____
    c. _____
    d. _____
12. List four M-codes and describe their functions.
    a. _____
    b. _____
    c. _____
    d. _____
13. What is the name of the character that ends each line of code?
14. Explain the purpose of the safe-start portion of a CNC program.

UNIT 2

Introduction to CNC Turning

Learning Objectives

After completing this unit, the student should be able to:

- Identify and describe CNC turning machine types
- Identify parts of CNC turning machines
- Identify the machine axes used for turning
- Demonstrate understanding of toolholding and tool-mounting devices and their application for CNC turning
- Demonstrate understanding of workholding devices and their application for CNC turning

Key Terms

Gang tool
Live tooling

Sub-spindle

Turret

INTRODUCTION

Recall that a turning machine is used to produce round or cylindrical parts by rotating the workpiece as a securely held cutting tool removes material. On these machines, the Z-axis is the longitudinal motion of the machine slide (like the direction of a drilling operation in a tailstock or the direction of motion when turning a diameter using the carriage on a manual machine). The direction of motion that is perpendicular to the Z-axis is the X-axis, and this axis has motion during a facing cut (like the cross slide on a manual machine). Most CNC **turning** machines do not have a Y-axis. **Figure 8.2.1** shows an image of how the axes are oriented on a turning center. When programming turned parts, the origin is usually located on the workpiece face at the part center line, as shown in **Figure 8.2.2**.

The basic framework of a CNC lathe is very similar to that of a manual lathe. Recall that turning centers are CNC lathes with the addition of an ATC. Turning centers are also specially designed for production work and aggressive material removal rates, and do not have provisions to be used manually. Most turning centers also have an angled bed on which the X-axis rides, and are called "slant-bed" machines for that reason. **(See Figure 8.2.3.)** Slant-bed machines put the tool and the machine mechanicals behind the workpiece, which makes the operator visibility of the work much better.

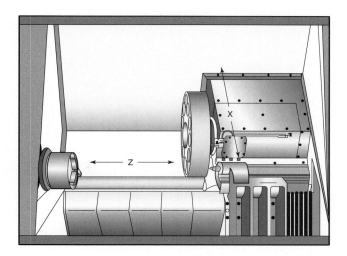

FIGURE 8.2.1 The coordinate system for CNC turning and the relationship of the axes.

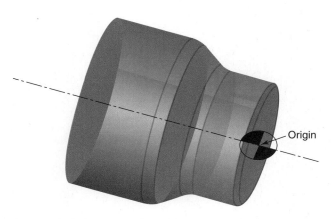

FIGURE 8.2.2 The origin for turning is usually located on the workpiece face on the part center line.

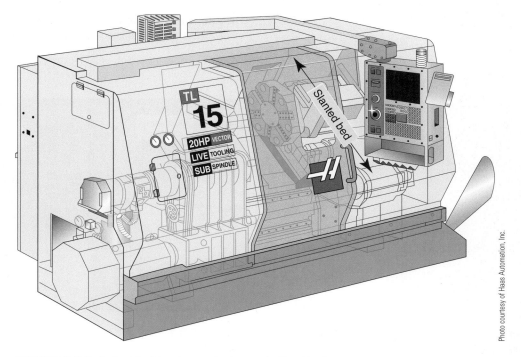

FIGURE 8.2.3 A slant-bed turning center has an inclined bed.

This also allows the operator to get closer to the work during setup. The other advantage is that the chips do not accumulate on the bed of the machine, but rather gravity causes them to slide down the bed into the chip pan.

Turning centers also feature low-friction linear guideways for sliding machine surfaces. The use of these greatly minimizes wear, reduces friction (enabling high rapid traverse feeds), and allows for very high accuracies due to a zero-clearance preloaded ball bearing design.

Today, turning centers are available that are able to perform milling and other machining operations in addition to turning. Special **live tooling** attachments make this possible. Live tools are small motorized spindles that enable a turning machine to perform light-duty milling and to perform hole-work such as drilling,

tapping, and reaming. Live tooling attachments are available for end-working (for machining on the face of a part), cross-working (for drilling cross holes, milling keyways, etc.), and in adjustable angle-head variations. **(See Figures 8.2.4 to 8.2.6.)**

To increase productivity and minimize operator attention, turning centers can be used in a *manufacturing cell*. Manufacturing cells group together several different machines performing operations on the same part. They may utilize robots that can transfer the part from machine to machine. **(See Figure 8.2.7.)** *Multi-tasking machines* are also available so that heavy-duty milling, turning, and drilling operations can be performed all in one machine. **(See Figure 8.2.8.)** With these machines a finished part can be produced from raw material without ever leaving the machine.

FIGURE 8.2.4 An end-working live tooling attachment for milling and performing holemaking operations on the face of a part.

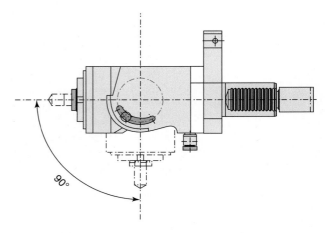

FIGURE 8.2.6 An adjustable angle-head live tooling attachment allows angular milling to be performed on the turning center.

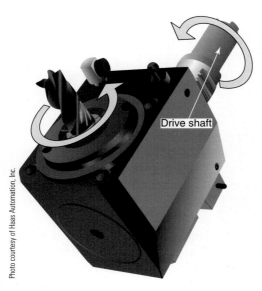

FIGURE 8.2.5 A cross-working live tooling attachment for milling and performing holemaking operations on the outside of a part.

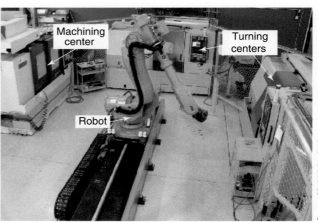

FIGURE 8.2.7 A robotic manufacturing cell can enhance productivity by minimizing operator intervention. The workpiece is transferred between machines by the robot.

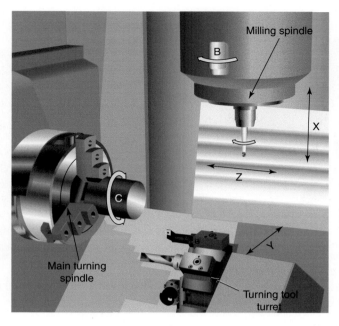

FIGURE 8.2.8 The interior of a multi-tasking machine that can perform heavy milling and turning operations.

In any type of production machining, the more efficiently a machine can perform a job from start to finish without operator intervention, the better. When creating turned parts there is always the problem of not being able to machine the end of the workpiece held in the workholding device. Turning centers can overcome this obstacle with a feature called a **sub-spindle. (See Figure 8.2.9.)** A sub-spindle is an auxiliary secondary spindle, which opposes the machine's main headstock spindle. This spindle may be equipped with a chuck and may be programmed to travel to the main spindle, grip the part, and then return to the tailstock end of the machine for machining the backside of the part. This entire part transfer can take place while never

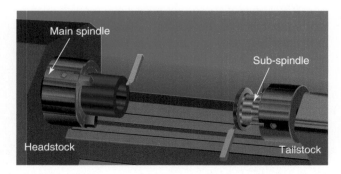

FIGURE 8.2.9 A sub-spindle opposes the machine's main spindle. The workpiece can be transferred from the main spindle to the sub-spindle so that the backside of the workpiece can be machined.

opening the machine door or taking the part from the machine. In fact, many machines can perform this transfer without even stopping the spindle rotation.

TYPES OF TURNING MACHINES

Turret-Type Machines

Most turning centers are fitted with a circular **turret** where all of the tooling is mounted. This turret serves as the ATC and can be programmed to index (rotate) to position the desired tool to perform machining operations. These machines have the ability to accept a variety of tool-mounting adapters that are used to mount the tool's shank to the turret and hold the tool in a desired position and orientation. **Figure 8.2.10** shows an image of a turning center turret.

Turret machines are popular due to their ability to fit a high quantity of tools in a small amount of space. Some turret-type machines are even fitted with more than one turret. The secondary turret can be used simultaneously to machine the part and increase productivity. **(See Figure 8.2.11.)**

Gang-Tool-Type Machines

The other common turning center design is the **gang tool** machine. These types of machines typically are of a flat bed design and are fitted with a carriage (similar to a manual lathe) where tools are installed. Most machines align the tools in a row next to each other ("ganged up") on a *top plate.* **(See Figure 8.2.12.)** With a gang tool

Courtesy of Hardinge, Inc.

FIGURE 8.2.10 A circular turret holds multiple tools and can index to any one of them with a program command.

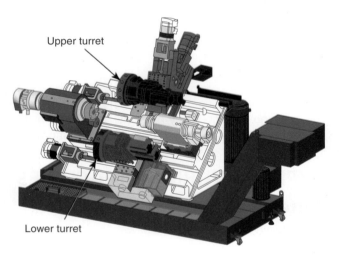

Upper turret

Lower turret

FIGURE 8.2.11 A twin turret machine can move both turrets independently for machining.

Courtesy of Hardinge, Inc.

FIGURE 8.2.12 A gang tool machine top plate with the tools arranged in a row.

Courtesy of Hardinge, Inc.

FIGURE 8.2.13 A gang tool turning center.

setup, the machine can perform a tool change quickly by shifting the top plate a short distance from the current tool to another tool. Gang tool machines also inherently have great rigidity, high accuracy, and an extremely simple design with few moving parts. Gang tool machines are usually ideal for small parts that require short tooling (long tools can encounter collision problems when they are alongside short tools) and short machine X-axis travels (due to the space used by the length of the top plate). **Figure 8.2.13** shows a gang tool turning center.

CNC Lathes

A CNC lathe has no ATC; a tool post is mounted on a cross slide much like that of a manual lathe. Often the same tool posts used on a manual lathe are also used on these machines. Usually the indexable or quick-change-style posts are chosen to allow tools to be changed quickly manually. When using this type of tool post, tool changes are performed by manually indexing the tool post to a detent position when the program instructs the operator. Some of these tool posts have a quick-change feature that allows a dovetailed toolholder to be released and replaced quickly and accurately. **Figure 8.2.14** shows a tool post on a CNC lathe.

Swiss-Type Turning Center

Swiss screw machines were developed in Switzerland for producing small parts used in watches. Today, these turning machines have full CNC controls and make all sorts of small parts for medical, electronics, defense, and other industries. Swiss machines are unique because the workpiece is supported at all times by a guide bushing to stabilize it and prevent it from deflecting and vibrating. Instead of moving the tool, the entire part moves in the Z-axis direction while supported by the guide bushing.

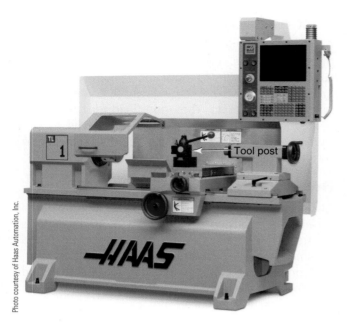

FIGURE 8.2.14 A CNC lathe holds cutting tools with a tool post similar to that used on a manual lathe.

(See Figure 8.2.15.) Many Swiss machines hold the tools in a side-by-side arrangement on multiple machine slides that can be moved separately from one another to increase productivity. **(See Figure 8.2.16.)**

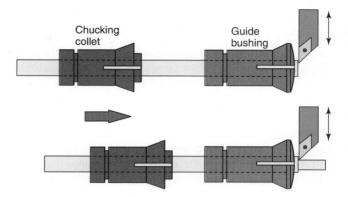

FIGURE 8.2.15 A Swiss turning machine moves the entire workpiece in the Z-axis instead of moving the cutting tool.

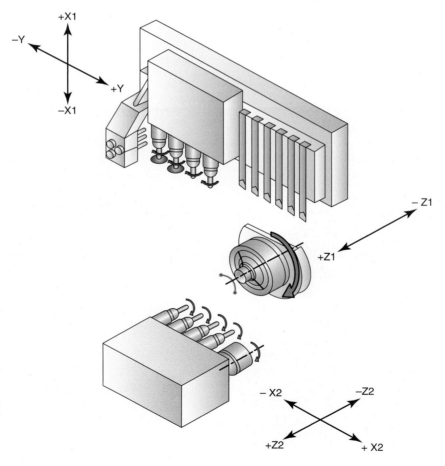

FIGURE 8.2.16 A typical tool arrangement on a Swiss turning machine.

TOOL-MOUNTING

Each of the different styles of turning machines accepts tools differently. A turret-type machine will usually either have bolt-on machine tool-mounting adapters or a style known as *VDI (Verein Deutsche Ingenieure)*. Bolt-on tool-mounting adapters mount directly to either the turret face or turret periphery with bolts. **(See Figure 8.2.17.)**

VDI tool-mounting adapters are made to a standardized style, allowing them to interchange from machine brand to machine brand, and have the advantage of being faster to change. The VDI adapter has a round shank with a serrated tooth pattern. Mounting and dismounting VDI tooling is performed by rotating a single screw. To install, the shank is inserted into a bore and the clamping screw is tightened. As the screw is tightened, the shank teeth mesh with mating teeth in the turret's VDI clamping mechanism. Most VDI adapters are equipped with a preset reference pad on the edge of the tool adapter to automatically square the adapter with the machine spindle. **Figure 8.2.18** shows an image of VDI tool-mounting adapters for a turret-type machine.

Dovetailed clamps are usually used to mount tools in gang-tool-type machines and onto the quick-change tool posts found in CNC lathes. On a gang tool machine, the adapter is slid onto a horizontal dovetail on the top plate and secured with a mating dovetail clamping mechanism. **(See Figure 8.2.19.)** Tool posts accept the same style toolholders the manual lathe uses. **(See Figure 8.2.20.)**

Cutting Toolholders

Generally, no matter what machine style, the same types of cutting tooling will be used. The common types of cutting tools include OD (outside diameter) turning, OD grooving, ID grooving, threading, cutoff, boring, drilling, reaming, and tapping. These tools and their uses have been covered in detail in previous units. With a CNC turning center, however, the means of holding them may be a bit different.

Hole-Working Toolholders

CNC hole-working tools are the same as those used on manual machines. Carbide versions of these tools are frequently chosen to allow high cutting speeds and maximize tool life. When mounting these tools, jaw-type drill chucks may be used but collet chucks and drill bushings are usually chosen instead since they are compact, accurate, and are not easily damaged from coolant exposure.

CNC collet chucks are very versatile and are often used for holding many different types of hole-working tools. **(See Figure 8.2.21.)** CNC collet chucks consist of a straight round shank with a tapered bore. A *nose collar* threads onto the external threads on the collet chuck's end and constricts the collet when tightened by forcing it into the tapered bore. Some of the most common CNC collet chuck types are the *ER* series, *DA* series, and *TG* series. **(See Figure 8.2.22.)**

Photo courtesy of Haas Automation, Inc.

BOLT-ON TOOL TURRET

BOLT-ON TOOLHOLDERS

FACE GROOVING

BORING BAR

TWIN BORE

TWIN TURN

PARTING TOOL

FIGURE 8.2.17 This type of toolholding adapter bolts directly to the turret with cap screws.

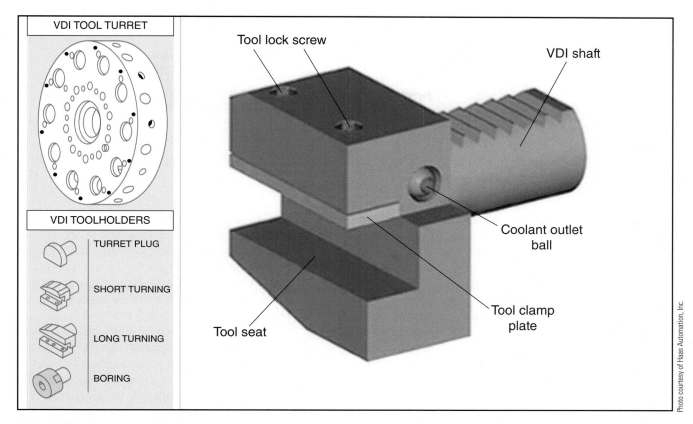

VDI TOOL TURRET

VDI TOOLHOLDERS

TURRET PLUG

SHORT TURNING

LONG TURNING

BORING

Tool lock screw

VDI shaft

Coolant outlet ball

Tool clamp plate

Tool seat

Photo courtesy of Haas Automation, Inc.

FIGURE 8.2.18 A VDI toolholding adapter mounts to the turret with a VDI shank. The adapter is drawn tight to the turret with the serrated teeth.

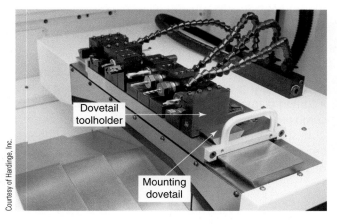

Dovetail toolholder

Mounting dovetail

Courtesy of Hardinge, Inc.

FIGURE 8.2.19 A dovetail mounting system used on a gang tool machine.

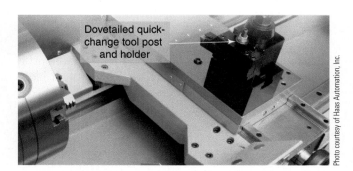

Dovetailed quick-change tool post and holder

Photo courtesy of Haas Automation, Inc.

FIGURE 8.2.20 Quick-change toolholders used on a CNC lathe.

These collets are often referred to as spring collets since a single collet is flexible enough to accommodate a range of sizes. An ER-style collet has a range of about 0.040". The DA- and TG-style collets have a range of about 0.015". Of the collets described here, TG collets are the most precise. The taper on a TG collet is not very steep, which results in very good concentricity between the collet and its bore. The nose collars that are used to secure the TG collets usually have a floating ball

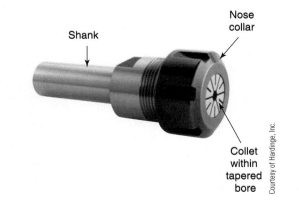

Shank

Nose collar

Collet within tapered bore

Courtesy of Hardinge, Inc.

FIGURE 8.2.21 A CNC collet chuck for holding holemaking tools.

FIGURE 8.2.22 The collet types shown from left to right are the ER series, DA series, and TG series.

bearing assembly so that there is minimal distortion and misalignment when tightened.

Split drill bushings can also be used to hold hole-working tools. A split bushing is a fairly rigid collet (with very few slits) and has a straight, untapered outside diameter. Instead of using a tapered bore to constrict on the tool shank, split drill bushings are put directly into a round-shank tool adapter's straight bore. These bushings have a small flat machined on the outside diameter and secure the tool by tightening a set screw against the flat. The external compression from the set screw constricts the tool within the bushing, securing the bushing to the holder. Drill bushings have very little size range (0.001" or so), so they must be sized directly to the shank they will hold. **Figure 8.2.23** shows an image of a Hardinge HDB drill bushing.

OD Working Toolholders

Much of the machining industry has gotten away from HSS tooling for production CNC turning operations in exchange for the advantages of carbides or other high-tech cutting materials. HSS tooling will work in a turning center, but is limited by its tool life. Today so many carbide geometries, grades, shapes, and styles are available that there is an insert well suited to almost any application. Carbide inserts are only as valuable as the holders on which they are mounted.

OD turning and threading toolholders come in right-hand, left-hand, and neutral orientations. A neutral tool is not oriented toward the left or right, but is aimed straight. This type of holder geometry is often necessary in OD contouring. **Figure 8.2.24** shows pictures of right, neutral, and left turning tools and their applications.

To determine whether or not left-hand holders or right-hand holders are required for a turning application, three factors need to be considered. The first is: can the insert fit into the shape of the contour without clearance issues? The second: which direction will the tool be cutting, toward the headstock or away from the headstock? The third: will the tool be mounted in the machine upside down or right side up? Many turning centers allow the tool to be mounted upside down, behind the spindle center line (slant-bed machines), so that forward spindle rotation may be used.

Grooving and Cutoff Toolholders

There are toolholders available that are capable of both turning and grooving. These tools can plunge into a part, create a groove, and then feed sideways, creating a turned diameter or contour. **(See Figure 8.2.25.)** This combination of features makes a very versatile tool.

It should be noted that there is a slight difference between the design of a grooving tool and a cutoff or parting-off tool. Since cutoff tools are designed for

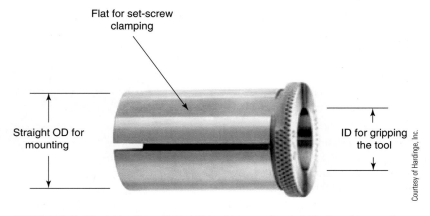

FIGURE 8.2.23 A Hardinge HDB drill bushing used to hold holemaking tools.

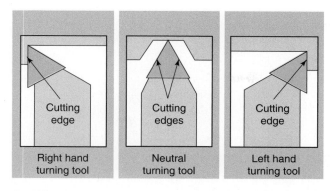

FIGURE 8.2.24 The turning tool orientations shown are right hand, neutral, and left hand.

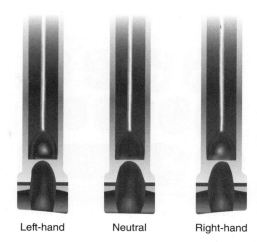

Left-hand   Neutral   Right-hand

FIGURE 8.2.26 Cutoff inserts are available with a biased cutting edge to minimize burrs on the part being cut off.

FIGURE 8.2.25 A groove/turn tool creating a turned diameter between two shoulders.

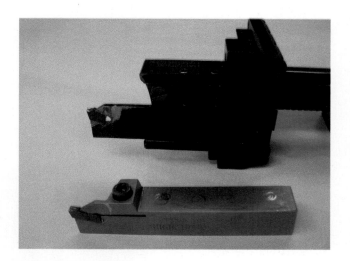

FIGURE 8.2.27 A grooving tool is in the foreground and a cutoff tool is in the background.

cutting off rotating parts, the inserts usually have sharp corners so that they produce a minimal burr on the separated part. Many manufacturers have taken the sharp-cornered insert one step forward and angled the cutting edge toward the part being cut off. This geometry achieves "break-through" with the leading side of the cut first, further minimizing the burr on the completed part. **(See Figure 8.2.26.)**

Grooving inserts are available in many different shapes and sizes and usually have a square cutting edge or a full radius. A grooving tool may be used for cutoff operations. **Figure 8.2.27** shows a cutoff tool and an OD grooving tool.

ID Toolholders

Recall from previous units that boring, internal threading, and internal grooving operations use a bar to hold the cutting tool. The inserts mounted in these bar-type holders are usually identical to those used for OD working. Many ID-working toolholders for CNC machines also have coolant passages drilled through their shanks that exit at

the insert, so coolant flows directly into the cutting zone. This can be helpful to flush chips from inside the bore during cutting. **(See Figure 8.2.27)**

Bar Pullers and Bar Feeders

Bar pullers and bar feeders are accessories used to make a given machine as automated as possible and require as little operator attention as possible. A bar puller is a tool that is mounted to the turret and, when programmed, can approach the remaining bar material after a part has been cut off, grip the bar, and after the workholding device is released, pull the bar to the desired length. All of these functions are programmable. Bar pullers are available in many forms, including gripping ring type, spring-jaw type, and coolant-powered hydraulic. **(See Figures 8.2.28 to 8.2.30.)**

Bar feeders mount outside of the machine headstock, in-line with the spindle centerline, and accept entire

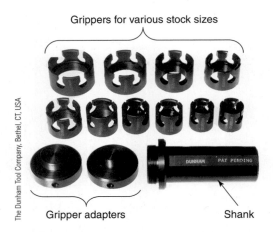

Grippers for various stock sizes

Gripper adapters   Shank

FIGURE 8.2.28 A gripping-ring-type bar puller grips the bar end by sliding a ring of spring-steel teeth over the perimeter of the bar.

FIGURE 8.2.29 A spring-jaw-type bar puller grips the work by sliding over the stock using the machine's X-axis motion.

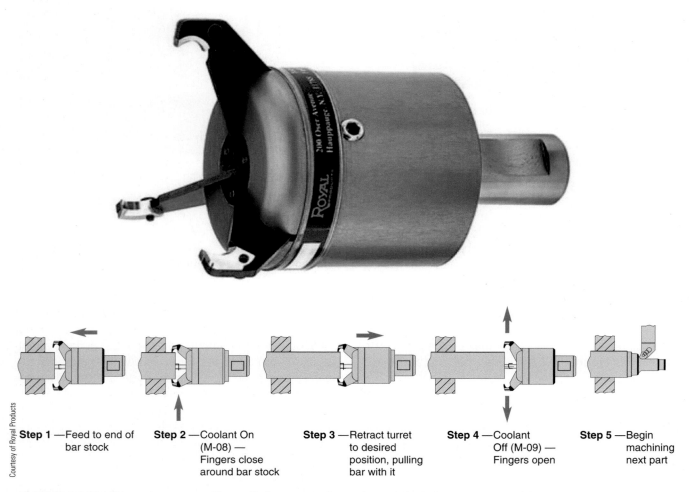

Step 1 —Feed to end of bar stock

Step 2 —Coolant On (M-08) — Fingers close around bar stock

Step 3 —Retract turret to desired position, pulling bar with it

Step 4 —Coolant Off (M-09) — Fingers open

Step 5 —Begin machining next part

FIGURE 8.2.30 This coolant-actuated bar puller's jaws grip the stock using the hydraulic pressure of the machine's coolant system.

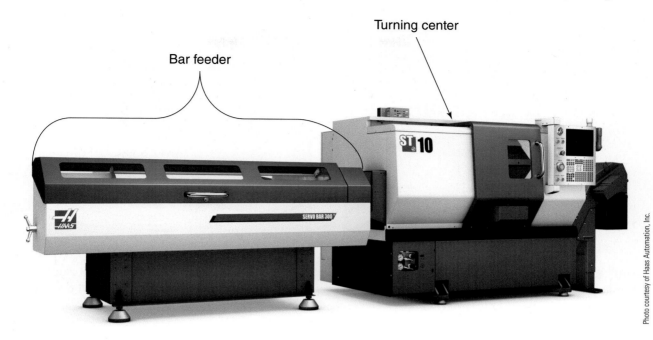

Turning center

Bar feeder

FIGURE 8.2.31 A CNC turning center equipped with an automatic bar feeder.

lengths of bar stock. **(See Figure 8.2.31.)** The bar rotates in the feeder tube and the feeder controls the bar from whipping out of balance. As one workpiece is cut off of the bar at the end of the machining cycle, the workholding device will unclamp and the feeder will push enough material length through the spindle bore so that the next workpiece can be produced.

WORKHOLDING

Workholding devices for CNC turning are selected based on the size, shape, and style of the work. The workholding devices on CNC turning machines are also similar to those used on manual lathes. The most common are the three-jaw chuck, four-jaw chuck, collet chuck, and lathe centers. CNC lathes usually use plain, manually operated chucks and collet closers. Turning centers usually make use of automated *power chucks* and power collet closers that are actuated using the machine's hydraulic or pneumatic power via a program command. Machines without sub-spindles often have tailstocks, allowing work to be held between centers or to support lengthy chucking work.

Workholding Collets

Recall from the lathe section that collets offer many advantages, including very precise part-to-part repeatability and very little runout. Also, of all the workholding devices, collets have the most surface contact with the work. This helps distribute the clamping pressure very evenly across the entire outer surface of the part, which prevents distortion and marring. For this reason, collets are usually ideal for thin-wall hollow parts or tubing.

Jaw chucks have heavy jaws that may be pulled away from the workpiece by centrifugal force when high spindle speeds are used. Since collets do not have heavy jaws like chucks do, they can be run at high speeds without the risk of losing clamping force. Collets also have the advantage of being inside the spindle bore instead of protruding several inches out from the spindle nose. This can provide rigidity, accuracy, and less wear on the spindle bearings. **Figure 8.2.32** shows an image of a CNC collet in the spindle nose of a turning center.

Because collets fit inside of the spindle bore, their largest diameter capacity is usually smaller than the spindle bore. This creates limitations in the maximum work diameter. A specifically sized collet is needed for every work diameter, so a large assortment of sizes may be required. Many machines achieve clamping action by pulling the collet back into a tapered bore. The collet stops retracting when it has achieved a preset level of clamping pressure. Therefore, variations in part diameters may have an effect on the positioning of part length (this can cause variations with the part's finished machined length). **(See Figure 8.2.33.)** Some machines are equipped with a dead length collet chuck, which clamps the collet by advancing the spindle taper instead of retracting the collet. This setup improves part location issues.

FIGURE 8.2.32 A CNC turning center using a collet for workholding.

FIGURE 8.2.34 A three-jaw power-actuated chuck in a CNC turning center. This setup is using soft jaws that were machined to match the outside diameter of the workpiece.

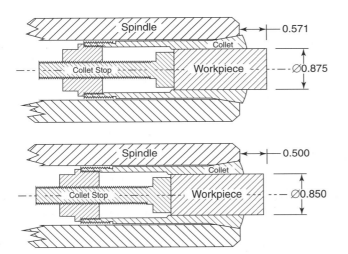

FIGURE 8.2.33 This exaggerated example illustrates how diameter variations can affect the part length positioning. The diameter of the workpiece in the upper diagram measures 0.875" and is 0.571" from the spindle face when clamped in a collet against a collet stop. The workpiece in the lower diagram has the same length, but the diameter measures only 0.850" and will be pulled back approximately 0.070" farther by the time it achieves the same gripping pressure as in the previous example. It can be expected that the part positioning length will change about 0.0028" for every 0.001" of diameter change when using "C"-style collets (5C, 16C, 20C, etc.). Using a dead length collet chuck will prevent this error.

Workholding Chucks

Chucks typically have greater runout and lower repeatability than collets have. Also, they have lower RPM limitations. Chucks are, however, more versatile since they can hold a much broader range of diameters and can be used to grip on either the OD or ID of workpieces.

Power-jaw chucks can also be equipped with soft jaws, which can be machined precisely to accept a specific work diameter or shape. It is for these reasons of versatility that many turning centers are equipped with chucks for general-purpose turning. **Figure 8.2.34** shows an image of a three-jaw CNC power chuck in a turning center.

PROCESS PLANNING

The standard operations performed on a turning center are very similar to those performed on a manual lathe. Typical operations include turning, facing, threading, holemaking, grooving, and cutoff. Each of these singular tasks in itself is referred to as an operation. The entirety of all operations required in the machining of a part is called a manufacturing *process*.

Prior to programming or setting up the machine, the engineering drawing must be closely examined and the production of the part planned from start to finish. The planning of workholding devices, tooling, and the machining operations depends on the part's features, tolerances, and surface finishes required by the drawing. Once a thorough strategy has been determined to produce a part, the steps can then be detailed on a document called a *process plan* as described in Section 2, Unit 5. This plan will include a description of each operation, the tools required, speed and feed data, workholding information, other notes and comments, and often a sketch depicting the part orientation. This document is important not only for the initial programming of the part, but also as a reference for any setup person or operator who will run this part in the future. Once the planning has been done, the tooling is documented on a tooling setup sheet.

SUMMARY

- CNC controls have been adapted to turning machines to increase productivity and enable turning of sophisticated part features.
- A CNC lathe with the addition of an automatic tool changer is called a turning center.
- Swiss turning centers slide the stock in and out of the spindle for Z-axis motion.
- Turning centers may be equipped with either a rotating turret or a gang tool top-plate arrangement for holding tooling.
- Turning machines can hold work using a jaw chuck, using a collet, or between centers.
- Most CNC turning machines have power actuators to open and close workholding devices.
- Some turning centers use VDI-style holders to mount tooling to the turret; others use bolt-on-style holders.
- Basic CNC turning operations include OD turning, grooving, and threading; ID boring, grooving, and threading; and holemaking operations such as drilling, tapping, and reaming.
- Live tools are motorized tools that enable a turning center to perform milling and off-center hole-work.
- Some machines are equipped with a sub-spindle, which is an auxiliary spindle opposing the main spindle. Workpieces can be transferred to this spindle to allow machining of the second end of a workpiece.
- The individual steps involved in producing the part are referred to as operations.
- The entirety of all operations required in the machining of a part is called a manufacturing process, or simply process for short.
- A process plan is a document describing the operations that comprise the machining process.

REVIEW QUESTIONS

1. Name the two primary machine axes on most CNC turning machines.
   a. _____
   b. _____

2. Explain the difference between a turning center and a CNC lathe.

3. List the three common types of live toolholders.
   a. _____
   b. _____
   c. _____

4. Name three major collet styles used in toolholding.
   a. _____
   b. _____
   c. _____

5. Name three types of workholding devices for turning centers.
   a. _____
   b. _____
   c. _____

6. Name three major styles of turning machines.

    a. _____

    b. _____

    c. _____

7. Explain why some workholding devices can be run at a higher RPM than others.

8. Explain the difference between an OD grooving tool and a cutoff tool.

9. When machining workpieces made from bar stock, what are two automated devices used to advance the bar so that the next workpiece can be machined?

10. Describe how a sub-spindle can be used to increase productivity.

11. How does a Swiss turning center differ from a conventional turning center?

CNC Turning: Programming

UNIT 3

Learning Objectives

After completing this unit, the student should be able to:

- Define basic G- and M-codes used for CNC turning
- Demonstrate understanding of linear interpolation for CNC turning
- Demonstrate understanding of circular interpolation for CNC turning
- Demonstrate understanding of radial and diametral programming
- Demonstrate understanding of facing operations for CNC turning
- Demonstrate understanding of CNC drilling operations for CNC turning machines
- Demonstrate understanding of tapping operations for CNC turning machines
- Demonstrate understanding of the principles of tool nose radius compensation (TNRC) for CNC turning
- Demonstrate understanding of CNC rough turning operations
- Demonstrate understanding of CNC finish turning operations
- Demonstrate understanding of CNC grooving operations
- Demonstrate understanding of threading operations for CNC turning machines
- Demonstrate understanding of common canned cycles for CNC turning applications

Key Terms

Canned cycle
Constant surface speed (CSS)
Diametral programming

Program stop
Radial programming
Rigid tapping

Tool nose radius compensation (TNRC)
Work coordinate system (WCS)

INTRODUCTION

There is no standardized CNC program format that is compatible with all machine control models. The programming examples provided in this unit will relate most closely to Fanuc-type and Haas controllers; however, the principles may be applied to any manufacturer's format (see the specific machine's programming manual).

COORDINATE POSITIONING FOR TURNING

Diametral and Radial

Some CNC turning machines allow two methods for programming X-axis coordinates: **diametral programming** or **radial programming**. When programmed radially,

all X coordinates are given as a radius of the part. These coordinates are the actual distance from the part center line, where the X-origin is typically located. Programming this way requires extra calculations because few prints have their cylindrical features dimensioned from the center line as a radius.

When a machine is programmed diametrally, all X coordinates are expressed as diameters. This is usually preferred, because most print dimensions for cylindrical features are shown as, and measured as, diameters. In this case, the numbers can be directly taken from the print and written as program coordinates. For some beginning programmers, the diametral method may be confusing because the X-axis program coordinates are double the actual distance from the part center line, where the X-origin is usually located. **Figure 8.3.1** shows the diametral X-axis coordinates for a part and **Figure 8.3.2** shows the radial X-axis coordinates for the

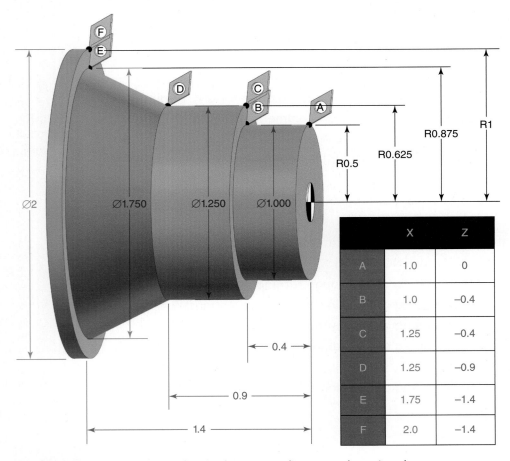

	X	Z
A	1.0	0
B	1.0	−0.4
C	1.25	−0.4
D	1.25	−0.9
E	1.75	−1.4
F	2.0	−1.4

FIGURE 8.3.1 In this example, diametral X-axis coordinates are shown in red.

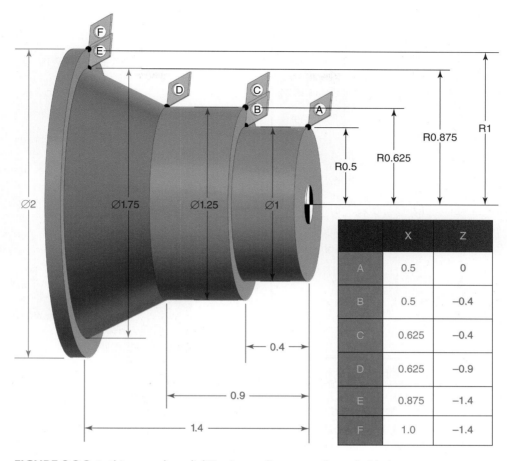

	X	Z
A	0.5	0
B	0.5	−0.4
C	0.625	−0.4
D	0.625	−0.9
E	0.875	−1.4
F	1.0	−1.4

FIGURE 8.3.2 In this example, radial X-axis coordinates are shown in black.

same part. The rest of the examples in this unit use the diametral method.

Absolute and Incremental

Absolute and incremental positioning on most turning machines is not handled with modal G90/G91 commands. Instead, positions programmed with X- and Z-axis words are always considered absolute by the machine. Incremental motions may be performed by substituting a U-value for the letter X and a W-value for the letter Z.

TYPES OF MOTION FOR TURNING

Rapid Traverse for Turning—G0

Rapid traverse movements must be performed very carefully in a CNC turning machine to prevent collisions. Special attention needs to be paid to the varying lengths of the cutting tools and how close they are to other machine parts and the workpiece. Another consideration is the location of the tool prior to and at the completion of the rapid movements. For instance, if the tool is an ID working tool, be sure it is retracted

from the ID of the part prior to making a rapid motion in the X-axis or moving to a tool-change position. As a tool approaches the work, also be sure that there is plenty of clearance to prevent a collision. The cutting tool should never contact the workpiece during a rapid traverse movement.

Linear Interpolation for Turning—G1

Recall from Unit 1 that linear interpolation synchronizes the motion of one (or more) axes to move the tool in a straight line. To produce a straight line, the machine must start moving each axis at precisely the same time, move them at the appropriate feed rate, and stop moving both axes at the destination at the same time.

To move the tool in a linear path on a turning center, a G1 is commanded along with the coordinates for the end position of the movement. A feed rate must also accompany the newly programmed position. Lathe feed rates are expressed in the G1 block with an F-character followed by a feed value. A feed rate command is modal and if one is not programmed in a block, then the last programmed rate remains active.

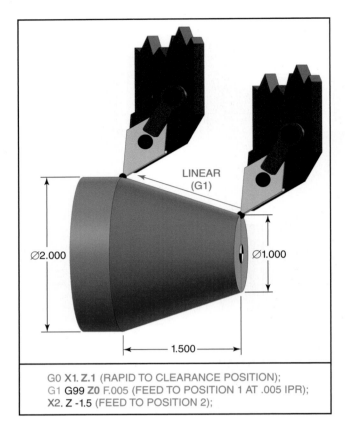

```
G0 X1. Z.1 (RAPID TO CLEARANCE POSITION);
G1 G99 Z0 F.005 (FEED TO POSITION 1 AT .005 IPR);
X2. Z -1.5 (FEED TO POSITION 2);
```

FIGURE 8.3.3 In a turning machine, linear motion between two points can be used to create a straight diameter or a taper. A program excerpt for taper motion is shown.

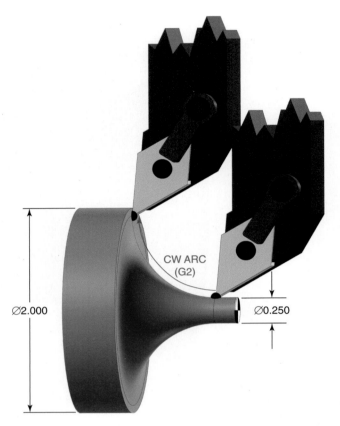

FIGURE 8.3.4 When the cutting tool is oriented behind the workpiece, a G2 creates a concave radius.

Feed rates for CNC turning can be in either inches per minute (IPM) or inches per revolution (IPR). IPR is more common for turning. A modal G-code must be programmed to set the machine control to the intended units. On many machines a G98 is used for IPM and G99 is used for IPR. It is a good practice to include the desired setting for the feed rate units at the beginning of the program in the safe-start block. **Figure 8.3.3** shows an example of a simple linear motion between two points and the corresponding program code.

Circular Interpolation for Turning—G2 and G3

CNC turning centers can also make circular motions (called arcs) with the tool tip for machining fillets and radii. To program an arc, the tool must first be positioned at the start point where that arc begins. Once there, a G-code is given to indicate whether the arc direction is clockwise or counterclockwise. If the arc rotates clockwise from start to end, a G2 code is used. If the arc rotates counterclockwise, a G3 is used. On most turning centers the cutting tool is on the backside of the workpiece and a G2 will create a concave radius (fillet; see

Figure 8.3.4) and a G3 will create a convex radius (corner round). **(See Figure 8.3.5.)**

Figure 8.3.6 shows a labeled illustration of the important parts of an arc. It is important to become familiar with each of these parts before programming. Prior to cutting the arc, the cutting tool must be positioned at the arc's starting point with a standard G1 or G0. After the tool is in position, the code for the arc direction is given and the programmer must identify the end point where the arc stops (remember that the tool is already beginning at the start point). Information about the arc's size must be provided in the same block of code. There are two methods for programming the size of the arc: programming a radius value and identifying the arc's center point location.

Arc Center Method for Circular Interpolation

The objective with the *arc center method* is to identify the exact position where the center point of the arc is located. Think of this as the location a compass point would be placed if the arc were to be drawn on a piece of paper. In the program, this location is identified as a

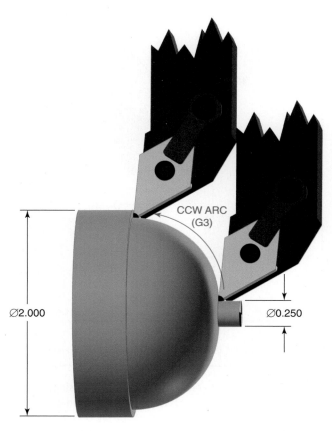

FIGURE 8.3.5 When the cutting tool is oriented behind the workpiece, a G3 creates a convex radius.

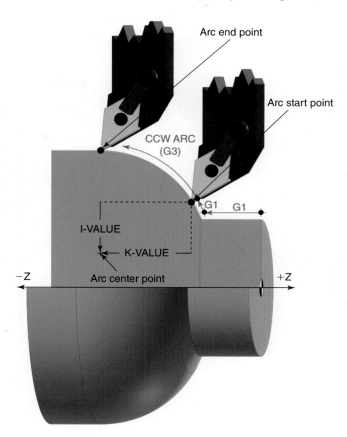

FIGURE 8.3.7 The arc center location is identified by the I- and K-values as shown.

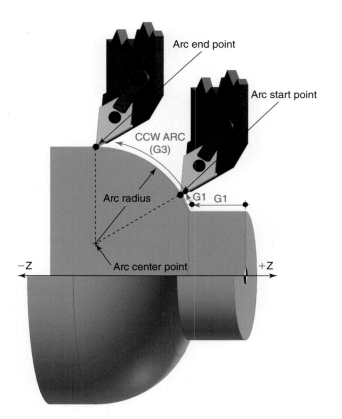

FIGURE 8.3.6 The parts of an arc are shown on a lathe workpiece.

distance along the X-axis and Z-axis from the arc <u>start point</u> to the arc <u>center point</u>.

Remember that the arc cutting block already used an X-word and Z-word to identify the arc end point. Therefore, the characters I and K are used to define the arc center point in the X- and Z-axes. The "I" is the distance from the arc start point to the center point along the X-axis. The "K" is the distance from the arc start point to the center point along the Z-axis. It is necessary to indicate the direction from the start point to the center point with either a positive or negative value. **Figure 8.3.7** shows how the values are derived for defining the arc center (notice the direction of the arrow heads). **Figures 8.3.8**, **8.3.9**, and **8.3.10** show circular interpolation for a part using the arc center method.

Radius Method for Circular Interpolation

The radius method may be used instead of the arc center method for identifying the size of the arc when using circular interpolation. This method is more common and uses an R-word to define the size of the arc radius in the G2/G3 block. **Figures 8.3.11** and **8.3.12** show circular interpolation for a part using the radius method.

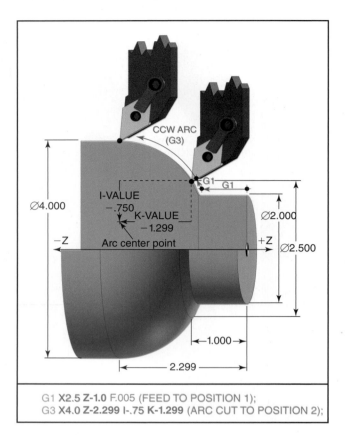

G1 **X2.5 Z-1.0** F.005 (FEED TO POSITION 1);
G3 **X4.0 Z-2.299 I-.75 K-1.299** (ARC CUT TO POSITION 2);

FIGURE 8.3.8 The code for cutting an arc using the arc center method is shown.

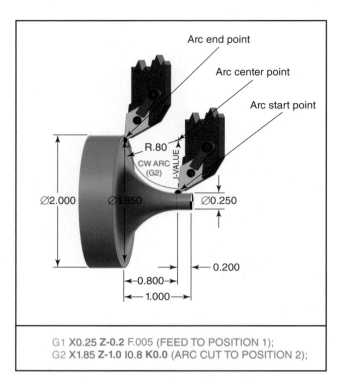

G1 **X0.25 Z-0.2** F.005 (FEED TO POSITION 1);
G2 **X1.85 Z-1.0 I0.8 K0.0** (ARC CUT TO POSITION 2);

FIGURE 8.3.10 The code for cutting an arc using the arc center method is shown. In this example, the K-word is zero due to the alignment of the start point and the arc center.

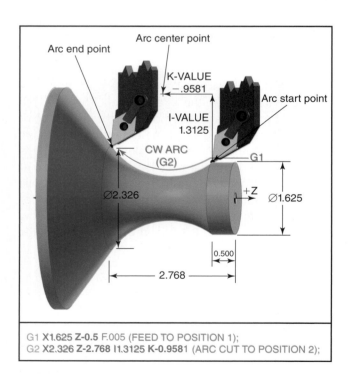

G1 **X1.625 Z-0.5** F.005 (FEED TO POSITION 1);
G2 **X2.326 Z-2.768 I1.3125 K-0.9581** (ARC CUT TO POSITION 2);

FIGURE 8.3.9 The code for cutting an arc using the arc center method is shown.

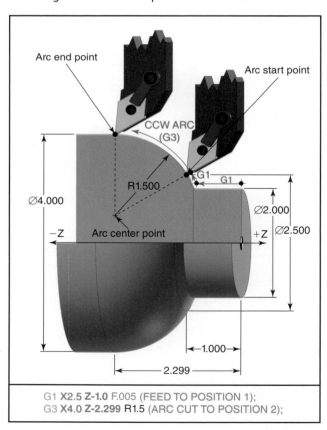

G1 **X2.5 Z-1.0** F.005 (FEED TO POSITION 1);
G3 **X4.0 Z-2.299 R1.5** (ARC CUT TO POSITION 2);

FIGURE 8.3.11 The code for cutting an arc using an R-word is shown.

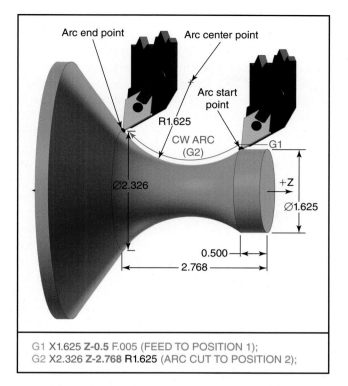

```
G1 X1.625 Z-0.5 F.005 (FEED TO POSITION 1);
G2 X2.326 Z-2.768 R1.625 (ARC CUT TO POSITION 2);
```

FIGURE 8.3.12 The code for cutting an arc using an R-word is shown.

NON-AXIS MOTION COMMANDS

Spindle Speed for Turning

The spindle speed for CNC turning may be programmed in one of two ways. *Direct RPM programming* allows the spindle RPM to be directly programmed at a fixed speed and is activated with a G97 code. The spindle is started using an M3 for forward rotation and an M4 for reverse rotation. M5 stops the spindle. The block containing the spindle start follows the safe-start block of the program and looks like this:

G97 S1200 M3 (SPINDLE ON CW AT FIXED RPM);

The disadvantage of this method is that the RPM must be calculated and programmed for each different diameter size as shown in **Figure 8.3.13**.

A feature called **constant surface speed (CSS)** may be used instead so that the spindle RPM is automatically updated for each diameter being cut. With this feature, the surface speed across the tool's cutting edge will stay constant. Programming a G96 code and providing the cutting speed in surface feet per minute (or meters per minute) activates this feature. When using CSS, the machine updates the spindle RPM as needed in real time by using the tool's X-axis position (which is the diameter being cut)

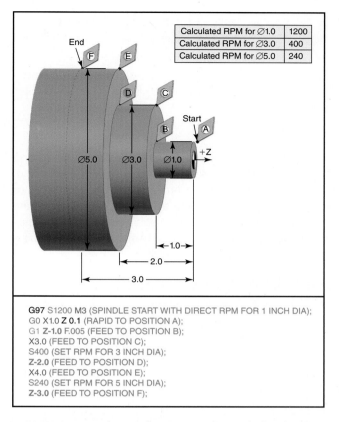

```
G97 S1200 M3 (SPINDLE START WITH DIRECT RPM FOR 1 INCH DIA);
G0 X1.0 Z 0.1 (RAPID TO POSITION A);
G1 Z-1.0 F.005 (FEED TO POSITION B);
X3.0 (FEED TO POSITION C);
S400 (SET RPM FOR 3 INCH DIA);
Z-2.0 (FEED TO POSITION D);
X4.0 (FEED TO POSITION E);
S240 (SET RPM FOR 5 INCH DIA);
Z-3.0 (FEED TO POSITION F);
```

FIGURE 8.3.13 The spindle RPM must be recalculated and re-programmed for every diameter change when using direct RPM mode. An example of a part and the corresponding program sample is shown.

and the surface speed programmed. As the diameter being cut becomes smaller, the RPM will increase.

Using CSS on small diameters can result in higher spindle speeds than what the machine or workholding devices can handle, so a maximum RPM limit must be set. When programing CSS, the spindle must first be started in direct RPM using a G97 before constant surface speed is turned on. Then CSS is activated by programming two blocks. One block limits spindle RPM using a G50 and an S-value to set maximum RPM. The other block uses a G96 and an S-value to activate CSS and set the cutting speed. When starting the spindle with a G97, it is usually best to program an initial RPM that is close to the one that will be used at the first application of CSS. This will reduce the time and electrical energy needed for the machine to adjust RPM when CSS is activated. An example of CSS blocks following a spindle start looks like this:

```
G97 S1200 M3 (SPINDLE ON CW AT FIXED RPM);
G0 X._ Z._ M8 (RAPID TO FIRST POSITION AND
COOLANT ON);
G50 S_ (RPM LIMIT FOR CSS);
G96 S_ (SET SFM FOR CSS);
```

Learn the maximum RPM for your workholding device before programming and never exceed it. A workholding device can violently fly apart if it is run above its rated RPM.

Constant surface speed is not used for holemaking operations (drilling, reaming, tapping, countersinking, etc.) since the programmed X-axis coordinate of holemaking tools remains at X-zero. Using CSS here would immediately cause the spindle to run at the max RPM (G50) setting. **Figure 8.3.14** shows a programming example using CSS.

Tool-Change Commands

Turning centers perform a tool change when given a "T" command. Many turning centers accept the tool-change command in the following format: Txxxx.

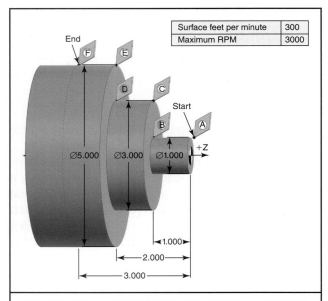

| Surface feet per minute | 300 |
| Maximum RPM | 3000 |

```
G97 S1200 M3 (SPINDLE START WITH DIRECT RPM FOR 1 INCH DIA);
G0 X1.0 Z 0.1 (RAPID TO POSITION A);
G50 S3000 (SET MAX RPM LIMIT TO 3000);
G96 S300 (SET SURFACE FEET PER MINUTE TO 300);
G1 Z-1.0 F.005 (FEED TO POSITION B);
X3.0 (FEED TO POSITION C);
Z-2.0 (FEED TO POSITION D);
X4.0 (FEED TO POSITION E);
Z-3.0 (FEED TO POSITION F);
```

FIGURE 8.3.14 The spindle RPM will be automatically adjusted for the diameter being cut when using constant surface speed (G96). The RPM is calculated based on the X-axis position of the cutting tool. An example of a part and the corresponding program sample is shown.

Each "x" is a numerical digit. The first pair of digits after the "T" designates the tool station number of the turret or top plate. The second pair designates the *tool offset* number, which tells the machine where that tool's tip is located. Once that location is determined during machine setup, it will be stored in the MCU as a numbered tool offset value. A command of T0101 would perform a tool change to tool station #1 and activate tool offset #1.

Most machining centers (milling) must be given an M6 code and a T-number to perform the tool-change operation, but most turning centers simply require the Txxxx command. Be sure that the X- and Z-axes are at a safe position for the tool change to occur since longer tools could hit workholding devices or the workpiece if too close. Usually taking the machine to *home* or *reference-return* position will place the axes as far as possible from any collisions.

A G28 command can be used to move the machine to the home position. Once a G28 is called, positioning occurs relative to the machine home position, not the part origin. Following a G28 with a U0.0 will move the X-axis to X-home and a W0.0 will move the Z-axis to the Z-home. A rapid to this safe position and a tool-change command looks like this:

G28 U0. **W0.** (RAPID TO MACHINE HOME FOR TOOL CHANGE);
T____ (CHANGE TOOL AND ACTIVATE OFFSET);

Some machines allow a custom "safe index" position to be set at which all tool changes occur. This can save time and reduce machine wear since tool changes can be performed closer to the work. Refer to a specific machine's programming manual. Examples in this book will use the home-return position since it is universal across Fanuc and Haas machines.

Sequence Numbers

Sequence numbers are optional in most cases and can be placed at the beginning of each block of code like a label. They can also be used periodically (instead of on every block) throughout the program to serve as a marker to quickly find a specific part of a program. Each sequence number begins with an N-character and must go in an increasing order throughout the program (e.g.,: N2, N4, N6, N8, etc.; or N5, N10, N15, N20, etc.). The increments between numbers do not matter on most machines, as long as they are in sequential increasing order. Many programmers prefer to leave numerical gaps (N5, N10, N15, N20, etc. instead of N1, N2, N3, N4, etc.) so that additional blocks may be inserted if the program is edited at a later time.

Program Stop Commands

One of two M codes may be used to cause a **program stop**, or hold, on the program until it is resumed by pressing the cycle start button. These are often inserted immediately prior to or after a tool change to allow for inspection of the part or tool.

An M0 command is a *full stop* and always requires the operator to restart the program by pressing the cycle start button on the operator panel. This can be used when a part must be repositioned, chips must be cleared, or to check a critical dimension before continuing. An M1 command is an *optional stop*. A switch on the operator panel needs to be switched on for the machine to read this optional stop and pause the program. It is often used when running a new program the first time so that inspections can be performed. After the program is proven safe and correct, the optional stop switch can be turned off and the program will ignore the M1 command.

The full stop or optional stop should be alone on a single block of code. An optional stop might look like this:

```
G28 U0. W0. (RAPID TO MACHINE HOME FOR TOOL
CHANGE);
M1 (OPTIONAL STOP);
T0101 (CHANGE TOOL TO TOOL 1 AND ACTIVATE
OFFSET 1);
G97 S2000 M3 (SPINDLE ON CW AT FIXED RPM);
```

If a full stop is desired, the following code would be used:

```
G28 U0. W0. (RAPID TO MACHINE HOME FOR TOOL
CHANGE);
M0 (FULL STOP);
T0101 (CHANGE TOOL TO TOOL 1 AND ACTIVATE
OFFSET 1);
G97 S2000 M3 (SPINDLE ON CW AT FIXED RPM);
```

Coolant M-Codes

The M3, M4, and M5 codes control the machine spindle. The M7, M8, and M9 codes are used to turn coolant off and on. The following three coolant commands are common:

M7 Mist coolant on
M8 Flood coolant on
M9 Coolant off

Not all machines will have the choice of mist or flood. They may only have one, so select the proper code for the given machine coolant system type. Coolant is usually required at the same time the spindle starts and is usually turned off when a tool retracts from the workpiece before a tool change. Many machines don't allow a coolant code and a spindle code to be commanded together, since two M-codes are usually not allowed on the same block. Some machines get around this by allowing an M13 or a M14 code to both start the spindle and turn the coolant on. This code does not choose between flood and mist and will usually turn on the machine's default coolant system. Again, check the machine's manual to determine which codes apply. If a machine doesn't accept M13/M14, a separate M7 or M8 command is usually programmed in the block following the spindle start. The coolant/spindle combination codes are:

M13 Spindle on clockwise with coolant
M14 Spindle on counterclockwise with coolant

Two separate coolant/spindle blocks may be programmed as shown:

```
G97 S1200 M3 (SPINDLE ON CW AT FIXED RPM);
M8 (COOLANT ON);
```

If the machine allows, a single coolant and spindle start may be programmed as shown:

```
G97 S1200 M13 (SPINDLE ON CW AT FIXED RPM
AND COOLANT ON);
```

Starting a Program in the Correct Format

Figure 8.3.15 shows an example of a program format for most Fanuc or Haas turning machines. The beginning of a program is a great place to include detailed comments about the program author, date, file name, tool and setup information, workpiece material, and more. Comments placed in parentheses will be ignored by the machine but will be helpful to all users of the program.

Modal codes given at the beginning of the program are used to safely establish the default program settings. This block of the program is often called the *safe start* block. The following are common commands used in the safe start block.

- G0 sets rapid positioning.
- G18 sets the X/Z plane. For the majority of programming, this is the required setting because X and Z are the primary axes used for turning. Use of other coordinate planes is beyond the scope of this text.
- G20 sets inch units for all dimensional values in the program.

- G40 cancels any active tool nose compensation. Tool nose compensation will be explained later in this unit. For now, it is only important to know that it is good practice to use a G40 command here to cancel values that may have been set by previous programs.

- G54 sets the **work coordinate system (WCS)**. Some machines allow multiple work coordinate systems (part origins) to be set at setup time. Six WCS are available on many machines. WCS #1 is set under G54, WCS #2 is set under G55, etc. The desired WCS is called in the program with the correct G-code.

- G80 cancels any modal canned cycles. Canned cycles will also be explained later in this unit. For now, it is only important to know that it is good practice to use a G80 command here to cancel any cycles that may have been activated by previous programs.

- G99 sets feed to inches per revolution (IPR). It is a good practice to also include G99 code in the first feed block after a tool change to ensure IPR mode is active. Study **Figure 8.3.15** and notice where all the info discussed so far fits into the complete format.

MACHINING OPERATIONS
Facing

Facing is often the first operation to be performed on a lathe workpiece. The purpose of this operation is to produce a flat surface perpendicular to the part center line. This surface usually creates a plane at the Z-axis zero location (Z0), and all other part lengths will be measured from it.

Usually if the end to be faced is fairly smooth and little material must be removed (less than about .020"), one facing cut at a finishing feed rate is all that is needed. If more material must be removed, the part has casting scale, has a jagged saw cut, or is very rough, at least one rough cut and one finish cut should be performed. A good place to start when considering the depth of cut for the final facing pass is to determine what nose radius is required on the finishing tool. Roughing passes should leave enough material for a finishing cut equal to at least 2/3 of the radius of the tool nose. **(See Figure 8.3.16.)**

Facing cuts are typically performed from the outside of the workpiece toward the center. The end position of

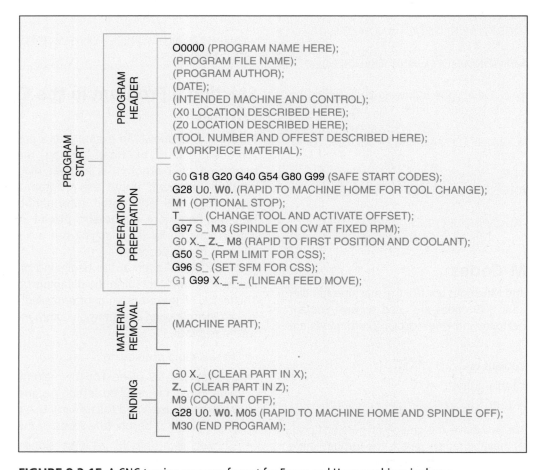

FIGURE 8.3.15 A CNC turning program format for Fanuc and Haas machines is shown.

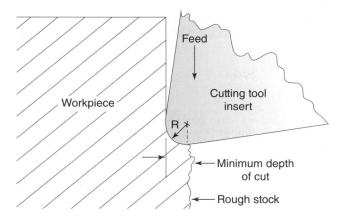

FIGURE 8.3.16 When using inserted carbide tooling, enough material should be taken during the finish pass to engage the tool at least 2/3 the depth of the tool nose radius. A facing path is shown, but this is especially important during longitudinal turning.

this cut must be programmed beyond X-zero due to the nose radius of the tool. This programmed X-end must be a minimum of 2x the nose radius past zero when programming diametrally. Because of error and wear on the insert nose, increasing the X-end by an additional 0.005" to 0.010" may be a good idea. For example, an insert with a 1/32 (.03125") radius could be programmed to end the face cut at a minimum of X-.063 (2*.03125), but a position of X-.073 ((2*.03125) + .010) may be better. **Figure 8.3.17** shows a program for performing the facing cut on the workpiece.

Drilling Operations

Many turned products include hole work. These operations are programmed by positioning the tool at X0.0, which aligns the cutting tool's center line with the center line of the part. Remember, drilling operations

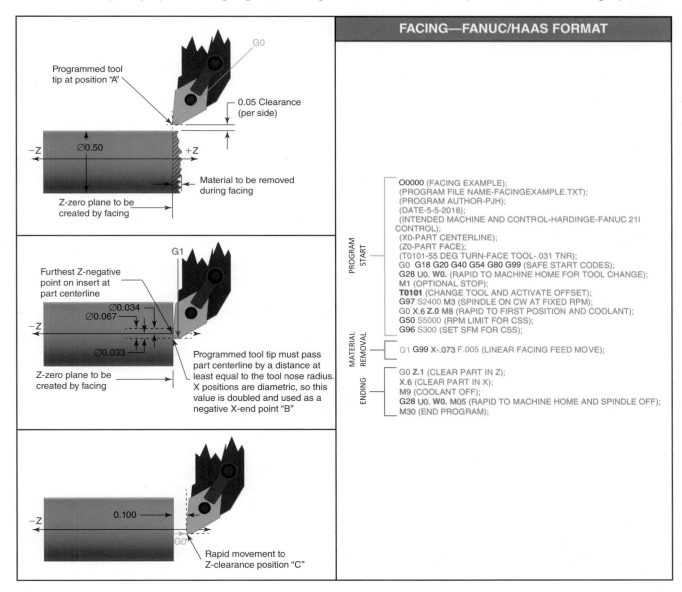

FIGURE 8.3.17 A facing cut and the corresponding program is shown.

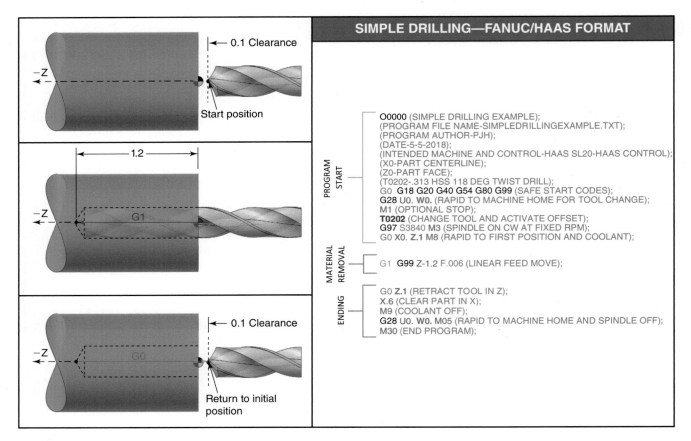

FIGURE 8.3.18 A simple hole-drilling operation and the corresponding program is shown.

should be performed using direct RPM (G97) since the X-axis position is X-zero. Once the tool has been positioned, a linear (G1) movement may be made with a feed rate to advance the tool into the work to the desired depth. Deep holes will benefit from pecking motion to help remove chips. These strategies will be shown later in this unit. Caution: the tool must be completely retracted from the hole with a rapid motion prior to making an X-axis move (like when returning to a tool change position). **Figure 8.3.18** shows a program for performing a drilling operation on a workpiece.

 **CAUTION**

Always fully retract hole-making tools from the work before making any X-axis movements.

Straight Turning

CNC turning operations can be separated into three types: straight, taper, and contouring. Each of these can be performed on the outside or on the inside diameter (boring operation) of a part. Straight turning produces a diameter that measures the same from the start to the end. Programming a straight diameter requires positioning the X-axis at the diameter to be cut and then moving the tool longitudinally using linear interpolation in the Z-axis. After reaching the Z-axis position, the tool can also be fed to an X-axis position if a shoulder is desired. **Figure 8.3.19** shows a program for performing straight turning on a workpiece.

Taper Turning

Tapers can be turned by moving two axes at the same time in a linear motion. To achieve this diagonal tool motion, a single block must contain two axis motion commands together. **Figure 8.3.20** shows a program excerpt for performing taper turning on a workpiece. Be aware that using a tool with a nose radius will cause error to occur on the taper. This will be discussed later.

Contour Turning

Often a part's shape will require a combination of straight turning, taper turning, and arc cutting. This cutter path is called a *contour*. CNC machines can turn an entire contour in one flowing motion from front to back without disengaging from the cut. This swift tool motion is efficient and eliminates burrs created by exiting and re-entering the part. **Figure 8.3.21** shows a program for contour turning on a workpiece.

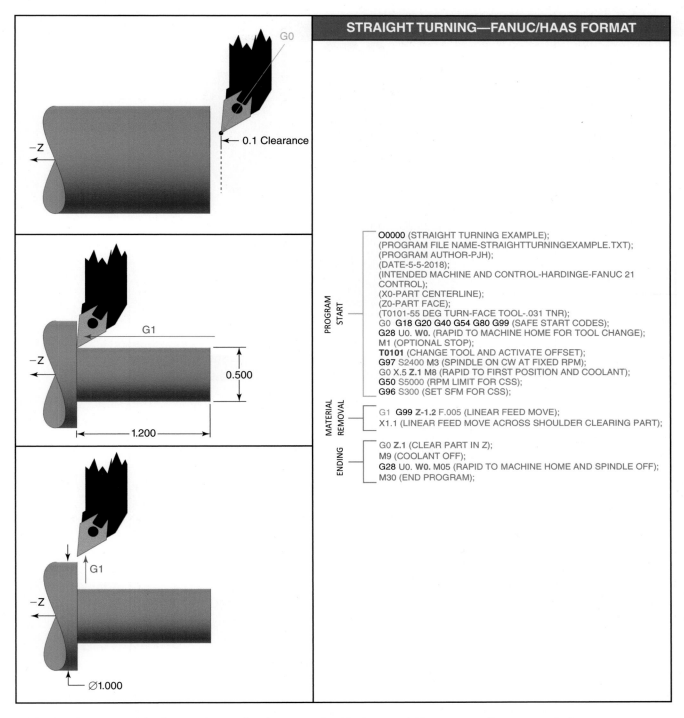

STRAIGHT TURNING—FANUC/HAAS FORMAT

PROGRAM START
```
O0000 (STRAIGHT TURNING EXAMPLE);
(PROGRAM FILE NAME-STRAIGHTTURNINGEXAMPLE.TXT);
(PROGRAM AUTHOR-PJH);
(DATE-5-5-2018);
(INTENDED MACHINE AND CONTROL-HARDINGE-FANUC 21
CONTROL);
(X0-PART CENTERLINE);
(Z0-PART FACE);
(T0101-55 DEG TURN-FACE TOOL-.031 TNR);
G0 G18 G20 G40 G54 G80 G99 (SAFE START CODES);
G28 U0. W0. (RAPID TO MACHINE HOME FOR TOOL CHANGE);
M1 (OPTIONAL STOP);
T0101 (CHANGE TOOL AND ACTIVATE OFFSET);
G97 S2400 M3 (SPINDLE ON CW AT FIXED RPM);
G0 X.5 Z.1 M8 (RAPID TO FIRST POSITION AND COOLANT);
G50 S5000 (RPM LIMIT FOR CSS);
G96 S300 (SET SFM FOR CSS);
```

MATERIAL REMOVAL
```
G1 G99 Z-1.2 F.005 (LINEAR FEED MOVE);
X1.1 (LINEAR FEED MOVE ACROSS SHOULDER CLEARING PART);
```

ENDING
```
G0 Z.1 (CLEAR PART IN Z);
M9 (COOLANT OFF);
G28 U0. W0. M05 (RAPID TO MACHINE HOME AND SPINDLE OFF);
M30 (END PROGRAM);
```

FIGURE 8.3.19 A straight diameter is turned and a square shoulder is created. The corresponding program is shown.

The part shown could not be turned from solid bar stock in one pass. The example is intended as a finish pass on an already roughed part or a casting that is near the final shape. Also, be aware that using a tool with a nose radius will cause error to occur on the tapers, chamfers, and radii of the example. This concept will be discussed next.

Tool Nose Radius Compensation

As mentioned previously, carbide inserted tooling has many advantages over HSS. These include better heat resistance, wear resistance, and hardness. These features make carbide ideal for modern high-performance CNC turning operations.

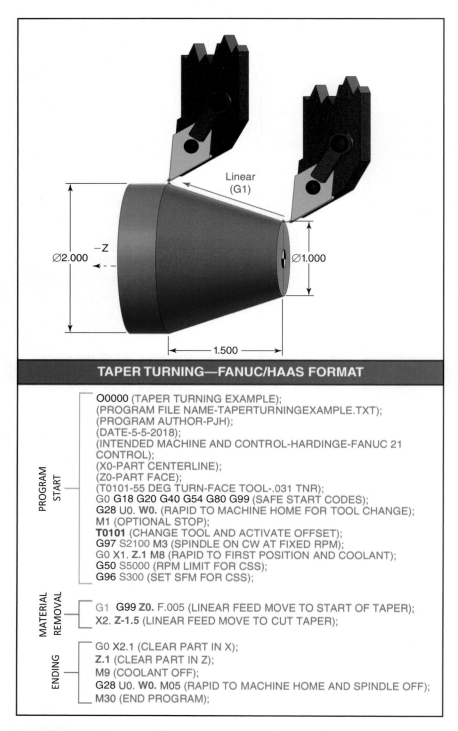

FIGURE 8.3.20 A tapered diameter is turned. The corresponding program is shown.

Most carbide inserts are made with a nose radius. If an insert came to a sharp point, the brittle nature of carbide would quickly cause the cutting edge to chip during heavy cutting and impact. The tool nose radius also provides a broader surface area to help pull heat from the cutting zone and will produce a better surface finish than a tool without a radius.

Because a tool location is defined in only the X-axis and Z-axis (which are 90 degrees to each other), the machine positions the tool based on a theoretical sharp corner point that doesn't exist. Most tool nose radii are large enough that they will cause chamfers, tapers, and radii to be cut incorrectly if not compensated for. Straight cuts in the Z-only direction and X-only direction

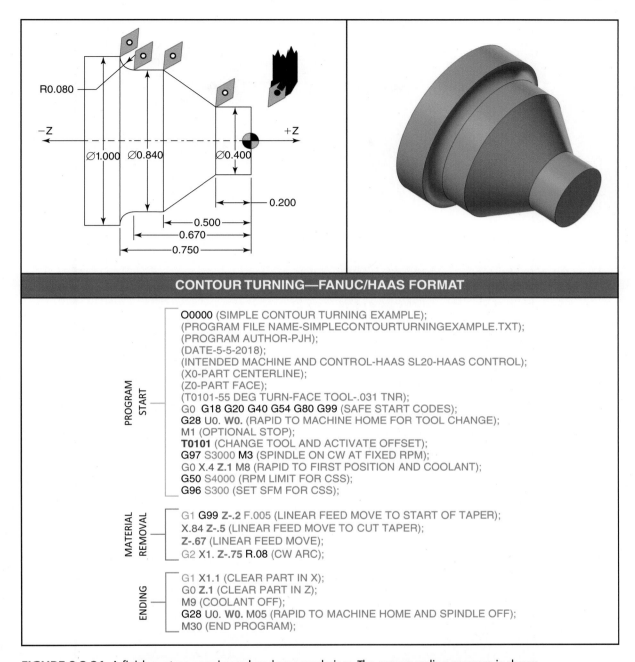

FIGURE 8.3.21 A finish contour pass is produced on a workpiece. The corresponding program is shown.

will not have error. **Figure 8.3.22** shows how a tool nose radius will cause error when cutting a concave radius and a convex radius (notice the sharp programmed corner point). **Figure 8.3.23** shows how a tool nose radius will cause error when making an angular cut.

There is a way to compensate for the radius of a tool mathematically, but this is sometimes a confusing method and seldom applied in industry. Most machines have an automated means of compensation. This function is called automatic **tool nose radius compensation**, or **TNRC**. Once TNRC is turned on, the MCU makes all necessary compensation and the part coordinates may be programmed as though the cutting tool did not have a radius.

A G41 or G42 must be programmed to activate TNRC. A G41 indicates that the tool path is to the left of the surface being cut, and a G42 indicates the tool path is to the right of the surface. **Figure 8.3.24** shows examples of applications of G41/G42. A G40 must be used to cancel TNRC.

NOTE:

It is usually simpler to avoid using TNRC when cutting basic faces and contours that don't include chamfers, tapers, or radii.

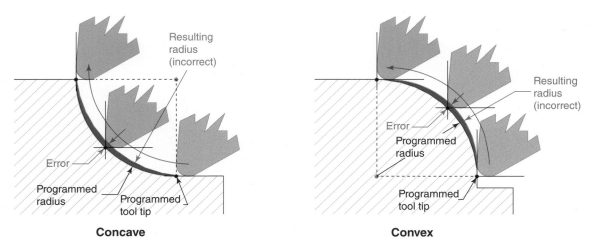

Concave   **Convex**

FIGURE 8.3.22 Arcs cut without compensating for the tool nose radius will cause error. Examples on a concave and convex radius are shown.

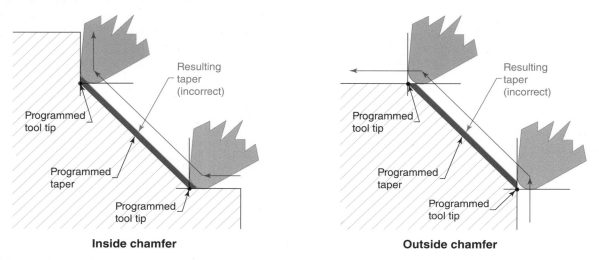

Inside chamfer   **Outside chamfer**

FIGURE 8.3.23 Tapers cut without compensating for the tool nose radius will cause error. Examples on an inside and outside chamfer are shown.

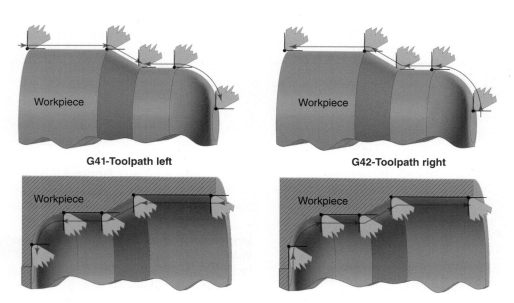

G41-Toolpath left   **G42-Toolpath right**

FIGURE 8.3.24 Examples of when to use G41 and when to use G42 codes are shown.

The following rules must be followed to successfully program a contour using TNRC:

1. There must be an initiating tool movement in an area away from the part (cutting air). As a general rule, the distance of the move must be at least the size of the nose radius to fully turn on TNRC.

    a. This movement should be linear.

    b. It is usually best for entry onto the part to be a 90-degree angle or greater.

    c. A G41 or G42 should be programmed along with this movement to activate the TNRC.

2. Once cutting is completed, there must be a linear exit move away from the part when the TNRC is canceled (with a G40) that is at least as large as the tool's nose radius.

3. The tool's nose is broken into sections called tool orientation quadrants, which specify the portion of the tool nose that needs to receive the compensation. This setting and the radius size of the tool nose must be entered into the MCU during machine setup and will be discussed in more detail in Section 8.4.

Figure 8.3.25 shows a part drawing and program of a finish contour being cut using TNRC.

Roughing Operations

When beginning with solid bar stock, a large amount of material must often be removed from the workpiece to achieve the final size. Removing too much material in one pass will shorten tool life, create excessive heat, and decrease accuracy of the workpiece. It is common to perform rough cutting passes to remove the bulk of the material while leaving some for a lighter finishing pass. Recall from Section 5.3 that the amount of material left for the finish pass depends on several variables. One is the finishing tool nose radius size, which is often determined by fillet radius required in inside corners. The rigidity of the machine, part shape, and machine setup are other factors. Again, it is ideal to leave enough finish material so that at least 2/3 of the finish tool nose radius will be engaged in the cut depth on turning passes. Lighter finish depth cuts are desired on faces of shoulders, however. A 0.005" allowance is usually ideal here.

Roughing operations can be tedious to program due to the repeated passes needed to remove the unwanted material. Care must be taken to determine the ending points for each pass. If any of the roughing passes cut too deeply, a gouge will remain on the finished part surface after the finish pass has been taken. **Figure 8.3.26**

shows an illustration of simple rough turning tool paths on a workpiece.

Finishing

After the part has been rough turned and the finish turning tool is loaded, the material remaining for the finish pass may be removed. The finishing cut should produce a fine surface finish and achieve good dimensional accuracy. Finishing passes should be light enough so that the tool pressure is minimal, but heavy enough so at least 2/3 of the tool nose radius is engaged (refer to Section 5.3). Also, light cuts typically generate less heat, which prevents the workpiece from expanding, and extend tool life.

Cutting data such as speed, feed, and depth of cut should be determined using a combination of calculation and experience and may take some time to develop. For the calculations, get as much specific information from the tool manufacturer's catalog as possible. **Figure 8.3.27** shows an illustration of a finish turning operation removing the remaining material from roughing.

It is common for a dedicated tool to be used for roughing and another for finishing. One reason this is done is to preserve the finishing tool. Also, roughing tools are most durable when equipped with a large nose radius, but since drawings often specify a maximum fillet radius size, a tool with a smaller radius is required for the finish pass.

CANNED CYCLES

Many turning operations require tedious repetitive motions, such as the multiple passes of a rough turning operation or multiple pecks required for deep hole-drilling. Manufacturers have equipped CNC controls with features that make these operations easier and faster to program. Machining routines that can be packaged or "canned" into one or two blocks of code and are called **canned cycles**.

Canned cycles require the programmer to write a block or two of code in a specified format. This code contains variables that must be entered specifically for the application. Canned cycles exist for drilling, tapping, threading, grooving, rough contour turning, finish contour turning, and more.

Drilling Canned Cycles for Fanuc

Drilling, reaming, counterboring, and countersinking operations can be programmed using drilling canned cycles. The G74 drill cycle is often used on Fanuc turning machines and allows a peck increment to interrupt chip flow, preventing long, stringy chips. The format

FIGURE 8.3.25 A finish contour pass is performed using tool nose radius compensation. The highlighted areas show the TNRC information.

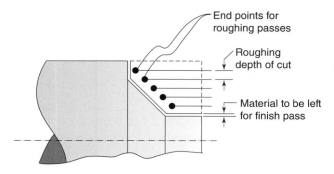

FIGURE 8.3.26 Roughing passes must be made carefully to ensure that they do not gouge into a surface to be finished later.

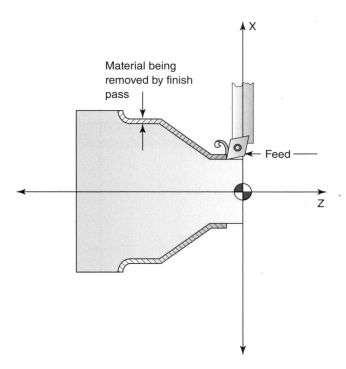

FIGURE 8.3.27 A finish pass is taken to produce the final finished part sizes. This operation removes the material left on the part during the roughing operation.

usually requires two G74 blocks. The peck increment is specified with a Q-value. A Q-value of 0.25 would cause the tool to retract once for every 0.25" of in-feed. Some machines don't allow a decimal in the Q-value and follow a 4-place format. In this case the 0.25 would be entered as Q2500. An R-value specifies the amount the tool will retract after each peck. The total absolute Z-depth is programmed as a negative value in the G74 block. The programmed Z-depth must account for the length of the drill tip. Programming to achieve a 5/16"

diameter hole that is 1.125" deep requires the following depth calculation:

> **EXAMPLE FOR PROGRAMMED DRILL DEPTH CALCULATION:** *Determine the estimated drill depth for a 5/16" (.3125") diameter hole with a full diameter depth of 1.125" if the drill has a 118° included angle tip.*
>
> Calculate drill tip length (See Section 4, Unit 3):
>
> $$\textbf{Drill Tip Length} = \textbf{0.3} \times \varnothing$$
>
> $$\textbf{Drill Tip Length} = \textbf{0.3} \times \textbf{0.3125}$$
>
> $$\textbf{Drill Tip Length} = \textbf{0.09375}$$
>
> Next, add the drill point length to the desired full-diameter depth:
>
> $$\textbf{Drill Depth} = \textbf{Drill tip} + \textbf{Full dia. depth}$$
>
> $$\textbf{Drill Depth} = \textbf{0.09375} + \textbf{1.125}$$
>
> $$\textbf{Drill Depth} = \textbf{1.21875}$$
>
> Rounding to three decimal places gives:
>
> $$\textbf{1.219}$$

If drilling in a single pass with no pecks, a canned cycle is often not used, but it is possible to use the G74 cycle if the Q-value is set equal to the full Z-depth. With all drill cycles, the drill should be positioned at the start location of the hole with clearance in the Z-direction when the cycle is initiated. The drill will automatically return to that position at the end of the cycle. **Figure 8.3.28** shows a Fanuc program for performing a G74 holemaking cycle with pecks.

Drilling Canned Cycles for Haas

Haas offers three different drilling canned cycles for turning machines: a G81 for a single pass (no peck), a G82 for single pass with a dwell, and a G83 for peck drilling with a full retract. The total absolute Z-depth is programmed as a negative value for all cycles. When a tool must cut cleanly at the bottom of a hole, the G82 cycle is used. This cycle pauses the tool at the end Z position when a P-value (number of milliseconds) is specified.

When using a G83, the peck increment is specified with a Q-value. A Q-value of 0.25" would cause the tool to fully retract once for every 0.25" of in-feed. An R-value specifies the position the tool will return to at the end of the cycle. **Figure 8.3.29** shows a Haas program performing a drilling cycle with a single pass (no pecks). **Figure 8.3.30** shows a Haas program performing a drilling cycle with pecks. The table below shows the formats for each drilling cycle mentioned.

	Format	Variable Description
Fanuc Peck Cycle Format	*(TOOL TO START POINT)* G74 R_; G74 G99 Z_ Q_ F_;	R: Retract amount between pecks Z: Absolute Z end position of hole Q: Peck infeed increment F: Feed rate
Haas Single-Pass Cycle Format	*(TOOL TO START POINT)* G81 G99 Z_F_ R_;	Z: Absolute Z end position of hole F: Feed rate R: Return plane position (optional)
Haas Single Pass with Dwell Format	*(TOOL TO START POINT)* G82 G99 Z_F_ P_ R_;	Z: Absolute Z end position of hole F: Feed rate P: Dwell time in milliseconds R: Return plane position (optional)
Haas Peck Cycle Format	*(TOOL TO START POINT)* G83 G99 Z_F_Q_R_;	Z: Absolute Z end position of hole F: Feed rate Q: Peck infeed increment R: Return plane position (optional)

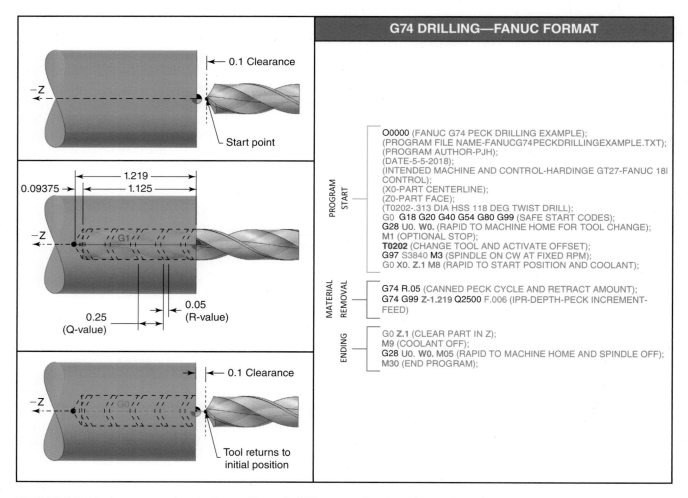

FIGURE 8.3.28 A program using the Fanuc G74 peck drilling canned cycle is shown. The drill will retract by the "R"-value amount between each peck.

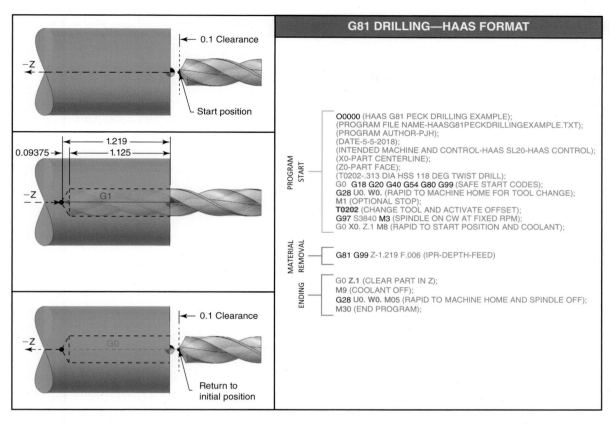

FIGURE 8.3.29 A program using the Haas G81 single-pass drilling cycle is shown.

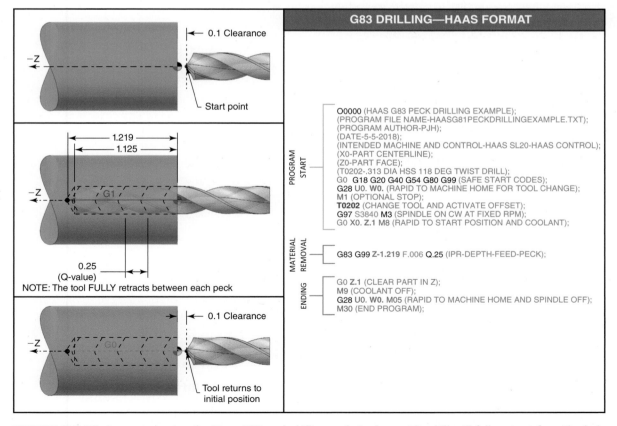

FIGURE 8.3.30 A program using the Haas G83 peck drilling cycle is shown. The drill will fully retract from the hole between each peck.

Tapping Canned Cycles

It is important to understand that a tap is self-feeding. During tapping, the machine axes only position the tap and follow along as the tap is pulled in to the work. As a tap begins cutting, the spindle rotation and the feed must be precisely synchronized so that the tap doesn't bind and break under strain. A CNC machine has no sensitivity and cannot feel resistance or force when tapping. As the tap approaches its end position, the machine must decelerate spindle speed and feed proportionately so that both stop immediately when the tap reaches the final depth. Finally, the tap must reverse to retract from the hole. Since the tap is fully engaged in the work, the spindle speed and feed must proportionately accelerate to retract the tap from the hole without breaking.

A tap advances into the work one thread per revolution (for single lead threads). Therefore, the feed rate per revolution is equal to the pitch of the tap. The feed rate is calculated as follows:

EXAMPLE FOR A 1/2-20 TAP:

$$\text{Pitch} = 1/\text{TPI}$$

$$\text{Pitch} = 1/20$$

$$\text{Pitch} = 0.050$$

Since feed rates are usually expressed in inches per revolution on turning machines, the feed rate calculation is already in the correct units and is complete.

Inch per revolution feed rate = 0.050

Not every thread pitch calculation will result in a number with a terminating decimal. Since the accuracy of the feed rate is critical, non-terminating numbers should be rounded to as many decimal places as the control will allow (usually four is adequate).

EXAMPLE FOR A 3/8 = 24 TAP:

$$\text{Pitch} = 1/\text{TPI}$$

$$\text{Pitch} = 1/24$$

$$\text{Pitch} = 0.0416666$$

Since the pitch calculation results in a repeating number, rounding it to four decimal places gives:

Inch per revolution feed rate = 0.0417

Programmed tap depth for blind holes should be calculated before the tap drill hole depth is calculated. This will ensure the tap drill hole is drilled deep enough,

so the tap does not bottom out in the hole and break. Through-holes are less of a problem, since the tap cannot bottom out. Examples for calculating tap depth and minimum tap drill depths are shown below:

STEP ONE:

EXAMPLE FOR TAP DEPTH CALCULATION: *Determine the programmed tap depth for a 3/8-16 hole if a print calls for a thread depth of 0.875". A plug tap with a flat end is used having five threads chamfered.*

Calculate pitch:

$$\text{Pitch} = 1/\text{TPI}$$

$$\text{Pitch} = 1/16$$

$$\text{Pitch} = 0.0625$$

Next, calculate thread chamfer length:

Tap chamfer length = Pitch × # of threads chamfered

Tap chamfer length = 0.0625 × 5

Tap chamfer length = 0.3125

Next, since the chamfered area of the tap won't create full threads, add chamfer length to the desired thread depth to determine the total tapping depth:

Tap depth = Chamfer length + Desired thread depth

Tap depth = 0.3125 + 0.875

Tap depth = 1.1875

Rounding to three decimal places gives:

1.188

*NOTE: These are theoretical values and serve as a starting point. Tap geometry differs slightly from one manufacturer to another, adjust accordingly.

STEP TWO:

EXAMPLE FOR MINIMUM TAP DRILL DEPTH CALCULATION: *Determine the programmed tap drill depth to allow for a 118° 5/16" diameter drill creating a hole for a tap 1.188" deep.*

Calculate drill tip length:

Drill tip length = 0.3 × drill ∅

Drill tip length = 0.3 × 0.3125

Drill tip length = 0.09375

(Continued)

EXAMPLE: (continued)

Next, add the drill point length to the desired full-diameter depth:

Minimum drill depth = Drill tip + Full dia. depth

Minimum drill depth = 0.09375 + 1.188

Minimum drill depth = 1.28175

Rounding to three decimal places gives:

1.282

*NOTE this is a theoretical value. If the print allows, some additional depth should be added to ensure tap doesn't bottom in hole.

Tapping Using a Floating Tap Holder (G32)

Some machines are not capable of coordinating the feed and spindle rotation accurately enough to prevent tap breakage. These machines require the use of a *floating tap holder*. **(See Figure 8.3.31.)** A floating tap holder will allow axial (in and out) float of the tap to compensate for small coordination errors between the spindle rotation and feed.

When a floating tap holder is used, tapping is programmed with a G32 code while the spindle is already running at the desired RPM. A minimum of two G32 blocks are usually required, one to advance the tap and one to retract it. If using a floating tap holder, the feed rate should be programmed about 0.001" per revolution less than the theoretical feed rate. By doing so, the tap will be gradually pulled from the natural relaxed position of the floating holder. This allows the tap holder float to adjust for error when the

tap is reversed. When using a G32, the spindle must be programmed to reverse after the tap has reached its depth. The full calculated feed rate should be used when retracting.

The tap will continue to be pulled in as the spindle coasts to a stop, so use caution when programming the depth of the tap so it does not bottom in the hole. Predicting how far the tap will be pulled in during deceleration is difficult because each machine is different. It may be ideal to reduce the theoretical tap depth by 6 * PITCH in the program. This should result in a thread depth that is too shallow. After a first part is run, the thread depth can be gaged and then the program depth adjusted for the next part. It is not advisable to re-run a G32 cycle in a previously cut thread because there could be a tracking error. So that the first part is not wasted, its thread depth can usually be tapped deeper by hand. Both Fanuc and Haas machines use the same format for G32 tapping. **Figure 8.3.32** shows a part drawing and program for tapping using a floating tap holder.

Rigid Tapping Canned Cycles (G84)

Some machines can coordinate the feed and rotation of a tap very accurately. This allows for **rigid tapping**, where the tap is held tightly in a holder without the ability to float. **(See Figure 8.3.33.)** When programming rigid tapping, the programmer does not need to reduce the entry feed and there should be no significant synchronization error. Rigid tapping was a special option on many older machines, but today is often standard. On many Fanuc machines, an additional block is required before the G84 block. This additional block requires an M29 code to activate the rigid tapping mode, and a repeat of the spindle speed command. **Figure 8.3.34** shows a Fanuc program excerpt for rigid tapping and **8.3.35** shows one for a Haas. The table below shows the formats for each tapping cycle mentioned.

	Format	Variable Description
Fanuc or Haas G32 Non-Rigid Tap Format	*(TOOL TO START POINT)* M3 S_ (SPINDLE CW); G32 G99 Z_ F_ (FEED IN); M4 S_ (SPINDLE CCW); G32 Z_ F_ (FEED OUT);	S: Spindle speed Z (FEED IN block): Absolute Z end position of tap tip F (FEED IN block): Tap pitch -0.001" per revolution Z (FEED OUT block): Absolute Z end position of tap tip F (FEED OUT block): Tap pitch
Fanuc Rigid Tap Cycle Format	*(TOOL TO START POINT)* M29 S_; G84 G99 Z_ F_;	S: Spindle speed Z: Absolute Z end position of hole F: Feed rate
Haas Rigid Tap Cycle Format	*(TOOL TO START POINT)* S_; G84 G99 Z_F_;	S: Spindle speed Z: Absolute Z end position of hole F: Feed rate

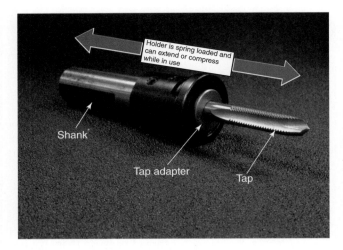

FIGURE 8.3.31 A floating tap holder.

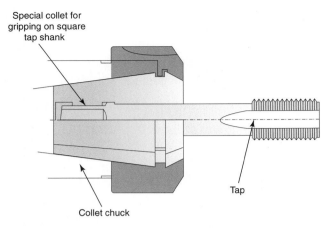

FIGURE 8.3.33 A collet chuck and special collet used for rigid tapping.

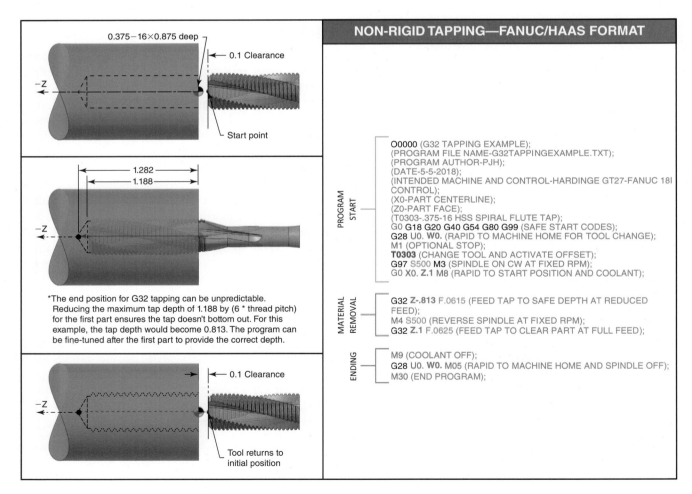

FIGURE 8.3.32 A Fanuc or Haas program for tapping using a floating tap holder is shown.

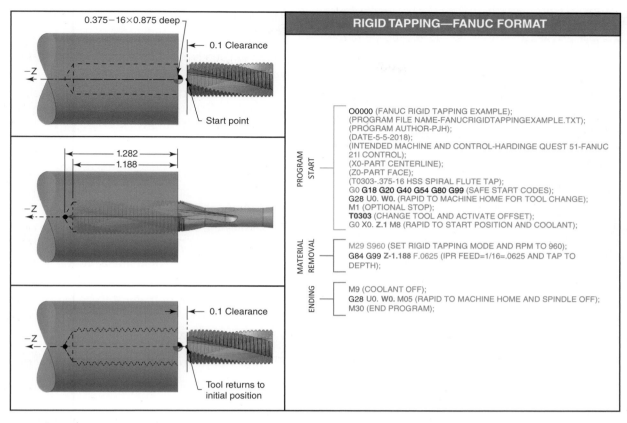

RIGID TAPPING—FANUC FORMAT

PROGRAM START

O0000 (FANUC RIGID TAPPING EXAMPLE);
(PROGRAM FILE NAME-FANUCRIGIDTAPPINGEXAMPLE.TXT);
(PROGRAM AUTHOR-PJH);
(DATE-5-5-2018);
(INTENDED MACHINE AND CONTROL-HARDINGE QUEST 51-FANUC 21I CONTROL);
(X0-PART CENTERLINE);
(Z0-PART FACE);
(T0303-.375-16 HSS SPIRAL FLUTE TAP);
G0 **G18 G20 G40 G54 G80 G99** (SAFE START CODES);
G28 U0. W0. (RAPID TO MACHINE HOME FOR TOOL CHANGE);
M1 (OPTIONAL STOP);
T0303 (CHANGE TOOL AND ACTIVATE OFFSET);
G0 X0. **Z.1** M8 (RAPID TO START POSITION AND COOLANT);

MATERIAL REMOVAL

M29 S960 (SET RIGID TAPPING MODE AND RPM TO 960);
G84 G99 Z-1.188 F.0625 (IPR FEED=1/16=.0625 AND TAP TO DEPTH);

ENDING

M9 (COOLANT OFF);
G28 U0. W0. M05 (RAPID TO MACHINE HOME AND SPINDLE OFF);
M30 (END PROGRAM);

FIGURE 8.3.34 A Fanuc program for rigid tapping is shown.

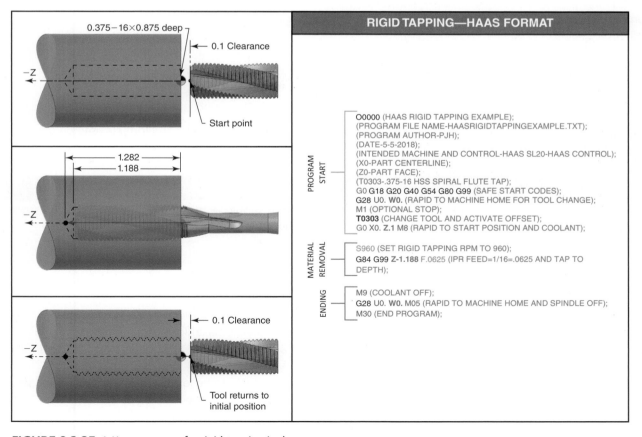

RIGID TAPPING—HAAS FORMAT

PROGRAM START

O0000 (HAAS RIGID TAPPING EXAMPLE);
(PROGRAM FILE NAME-HAASRIGIDTAPPINGEXAMPLE.TXT);
(PROGRAM AUTHOR-PJH);
(DATE-5-5-2018);
(INTENDED MACHINE AND CONTROL-HAAS SL20-HAAS CONTROL);
(X0-PART CENTERLINE);
(Z0-PART FACE);
(T0303-.375-16 HSS SPIRAL FLUTE TAP);
G0 **G18 G20 G40 G54 G80 G99** (SAFE START CODES);
G28 U0. W0. (RAPID TO MACHINE HOME FOR TOOL CHANGE);
M1 (OPTIONAL STOP);
T0303 (CHANGE TOOL AND ACTIVATE OFFSET);
G0 X0. **Z.1** M8 (RAPID TO START POSITION AND COOLANT);

MATERIAL REMOVAL

S960 (SET RIGID TAPPING RPM TO 960);
G84 G99 Z-1.188 F.0625 (IPR FEED=1/16=.0625 AND TAP TO DEPTH);

ENDING

M9 (COOLANT OFF);
G28 U0. W0. M05 (RAPID TO MACHINE HOME AND SPINDLE OFF);
M30 (END PROGRAM);

FIGURE 8.3.35 A Haas program for rigid tapping is shown.

Rough and Finish Turning Canned Cycles

Most CNC turning machines are equipped with canned cycles capable of performing rough turning/boring and finish turning/boring. The G71 canned cycle takes multiple roughing passes on the workpiece while maintaining a consistent depth of cut. Rough/finish turning is written in a fraction of the blocks of code that it would take to write without the canned cycle.

OD Turning Canned Cycle

Prior to beginning the canned cycle for an outside diameter turning operation, the tool is positioned at a start point. At the end of the canned cycle the tool will return to this same point. This position is usually close to the corner of where the face and the stock OD intersect with clearance in both the X- and Z-axes. This initial positioning indicates to the machine where the starting stock diameter is and where the canned cycle will begin removing material. The tool will automatically return to this point when the cycle is finished.

The roughing passes will start at the outside diameter and work inward until the rough contour is complete. Each pass will begin by incrementally stepping inward in the X-axis direction and then cut straight in the Z-negative direction. After the bulk of the material has been removed, the cycle will make a final rough contouring pass while leaving a finish allowance. This final pass smooths out any steps left behind from the other passes.

Many Fanuc controls require two consecutive G71 blocks for the roughing cycle while Haas requires only one. These blocks often use variables such as U, R, P, Q, W, and F. For a Fanuc, some of the same characters used for variables on the first block are repeated on the second with a different meaning, so be careful when programming.

The first G71 block for a Fanuc may use the following:

- U sets the depth of cut for each roughing pass (radial value).
- R sets the distance the tool will retract from each roughing pass before making a rapid move to the beginning of the next pass.

The second G71 block for a Fanuc may use the following:

- P specifies the sequence number where the code for the contour begins.
- Q specifies the sequence number where the code for the contour ends.
- U sets the amount of material to be left on all diameters for later finishing.
- W sets the amount of material to be left on all faces for later finishing.
- F sets the roughing feed rate.

The basic Haas G71 block contains the following:

- P specifies the sequence number where the code for the contour begins.
- Q specifies the sequence number where the code for the contour ends.
- D sets the depth of cut for each roughing pass.
- U sets the amount of material to be left on all diameters for later finishing.
- W sets the amount of material to be left on all faces for later finishing.
- F sets the roughing feed rate.

The table below shows the formats used on many Fanuc and Haas machines.

	Format	Variable Description
Fanuc Rough Turning Cycle Format	(TOOL TO START/END POINT) G71 U_ R_; G71 P_ Q_ U_ W_ F_; G0 X_; G1 G99 Z_ F_;	*Start/end point should provide clearance between the stock in both the X- and Z-axes. **FIRST G71 BLOCK** U: Depth of cut per side (radial) R: Retract amount **SECOND G71 BLOCK** P: Sequence number for first block of the finish contour Q: Sequence number for last block of the finish contour U: Diameter amount of material to be left on for finish pass in X-direction W: Amount of material to be left for finish pass in Z-direction F: Feed rate for roughing passes F-value in first G1 block: Feed rate to be used during later finish pass

(Continued)

	Format	Variable Description
Haas Rough Turning Cycle Format	(*TOOL TO START/END POINT*) G71 P_ Q_ D_ U_ W_ F_; G0 X_; G1 G99 Z_ F_;	*Start/end point should provide clearance between the stock in both the X- and Z-axes.* P: Sequence number for first block of the finish contour Q: Sequence number for last block of the finish contour D: Depth of cut per side (radial) U: Diameter amount of material to be left on for finish pass in X-direction W: Amount of material to be left for finish pass in Z-direction F: Feed rate for roughing passes (G71 block) F -value in first G1 block: Feed rate to be used during later finish pass

The coordinates that define the shape of the contour to be roughed must be programmed after the G71 block(s) so that the canned cycle knows where to remove material. Sequence numbers must be assigned to the first and last blocks of the contour. The P- and Q-values of the canned cycle will will reference these blocks.

The cycle is also responsible for leaving a consistent amount of stock on the part for the finishing pass. After the part has been roughed, a *finish turning canned cycle* can be activated by using a G70 command. This canned cycle shares the contour positions already programmed for the roughing cycle by the P- and Q-values.

> **NOTE:**
>
> If TNRC is programmed, most machines ignore it during the G71 roughing passes but apply it on the G70 finish pass.

Figure 8.3.36 shows a Fanuc program excerpt for rough/finish OD turning on a workpiece, **and Figure 8.3.37** shows the same part programmed on a Haas.

ID Turning (Boring) Canned Cycle

The G71 roughing and G70 finishing cycles can also be used to perform inside diameter turning (boring) operations. As with all boring operations, a hole must first be created to make clearance for the boring bar. The X-start point must be programmed at the inside diameter of the rough-drilled hole. The roughing passes will begin at the hole's diameter and work outward until the rough contour is complete. The G71 cycle works for boring in the same manner as OD turning, only the machine must retract the tool inward (toward the center of the part) after each pass. The X-axis finish allowance must also occur in

the X-negative direction. On many Fanuc and Haas machines this is done by assigning a negative U-value for the X-finish. **Figure 8.3.38** shows a Fanuc program excerpt for rough/finish ID turning on a workpiece, **and Figure 8.3.39** shows the same part programmed on a Haas.

Grooving Canned Cycles

Deep grooving often creates long curled chips that can block coolant and cause the tool to bind. As with drilling, a pecking motion while grooving can help break and evacuate these chips. A G75 grooving cycle is available on Haas and Fanuc controls to make programming these operations easier. The cycle can be applied to OD (external) grooves or ID (internal) grooves. The cycle may also be used for cut-off operations if the final X-position causes the tool to break through into a hole or cross the part centerline.

A Z-axis step-over can be added to create grooves that are wider than the tool itself. In this case, one full grooving pass will be completed before stepping-over to the next cut position. **Figure 8.3.40** shows a program excerpt for a single-width pass (no Z-step-over) peck grooving operation for Fanuc and **Figure 8.3.41** for Haas. **Figure 8.3.42** shows a program excerpt for a multiple-width pass (with a Z-step-over to increase the groove width) peck grooving operation for Fanuc and **Figure 8.3.43 for** Haas. The table below shows the formats used on many Fanuc and Haas machines.

> **NOTE:**
>
> Some Fanuc machines follow a four-place decimal format with no decimal point allowed for "P" and "Q" peck increment values (i.e., 1000 = 0.1000, 2750 = 0.275, 1300 = 0.13, etc.).

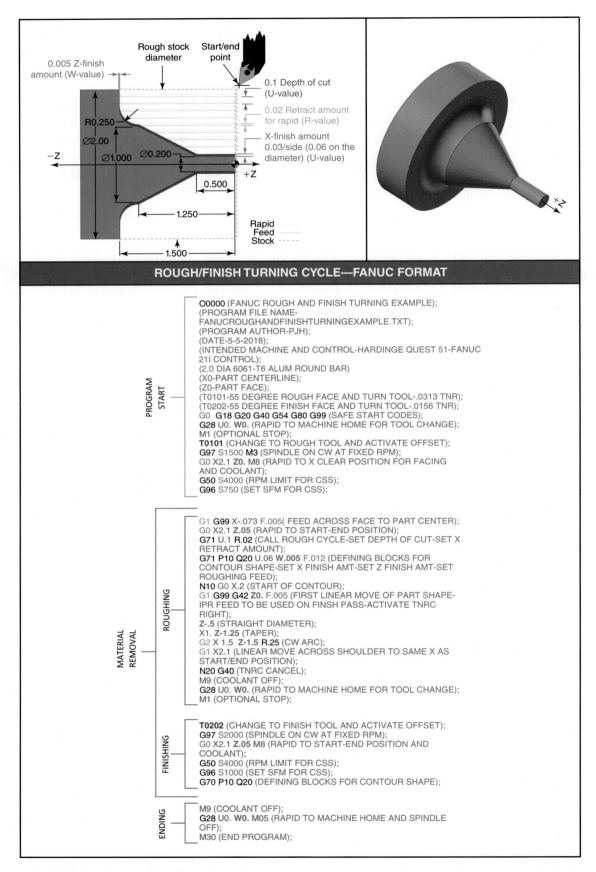

ROUGH/FINISH TURNING CYCLE—FANUC FORMAT

PROGRAM START

O0000 (FANUC ROUGH AND FINISH TURNING EXAMPLE);
(PROGRAM FILE NAME-
FANUCROUGHANDFINISHTURNINGEXAMPLE.TXT);
(PROGRAM AUTHOR-PJH);
(DATE-5-5-2018);
(INTENDED MACHINE AND CONTROL-HARDINGE QUEST 51-FANUC
21I CONTROL);
(2.0 DIA 6061-T6 ALUM ROUND BAR)
(X0-PART CENTERLINE);
(Z0-PART FACE);
(T0101-55 DEGREE ROUGH FACE AND TURN TOOL-.0313 TNR);
(T0202-55 DEGREE FINISH FACE AND TURN TOOL-.0156 TNR);
G0 G18 G20 G40 G54 G80 G99 (SAFE START CODES);
G28 U0. W0. (RAPID TO MACHINE HOME FOR TOOL CHANGE);
M1 (OPTIONAL STOP);
T0101 (CHANGE TO ROUGH TOOL AND ACTIVATE OFFSET);
G97 S1500 M3 (SPINDLE ON CW AT FIXED RPM);
G0 X2.1 Z0. M8 (RAPID TO X CLEAR POSITION FOR FACING
AND COOLANT);
G50 S4000 (RPM LIMIT FOR CSS);
G96 S750 (SET SFM FOR CSS);

MATERIAL REMOVAL

ROUGHING

G1 G99 X-.073 F.005(FEED ACROSS FACE TO PART CENTER);
G0 X2.1 Z.05 (RAPID TO START-END POSITION);
G71 U.1 R.02 (CALL ROUGH CYCLE-SET DEPTH OF CUT-SET X
RETRACT AMOUNT);
G71 P10 Q20 U.06 W.005 F.012 (DEFINING BLOCKS FOR
CONTOUR SHAPE-SET X FINISH AMT-SET Z FINISH AMT-SET
ROUGHING FEED);
N10 G0 X.2 (START OF CONTOUR);
G1 G99 G42 Z0. F.005 (FIRST LINEAR MOVE OF PART SHAPE-
IPR FEED TO BE USED ON FINSH PASS-ACTIVATE TNRC
RIGHT);
Z-.5 (STRAIGHT DIAMETER);
X1. Z-1.25 (TAPER);
G2 X 1.5 Z-1.5 R.25 (CW ARC);
G1 X2.1 (LINEAR MOVE ACROSS SHOULDER TO SAME X AS
START/END POSITION);
N20 G40 (TNRC CANCEL);
M9 (COOLANT OFF);
G28 U0. W0. (RAPID TO MACHINE HOME FOR TOOL CHANGE);
M1 (OPTIONAL STOP);

FINISHING

T0202 (CHANGE TO FINISH TOOL AND ACTIVATE OFFSET);
G97 S2000 (SPINDLE ON CW AT FIXED RPM);
G0 X2.1 Z.05 M8 (RAPID TO START-END POSITION AND
COOLANT);
G50 S4000 (RPM LIMIT FOR CSS);
G96 S1000 (SET SFM FOR CSS);
G70 P10 Q20 (DEFINING BLOCKS FOR CONTOUR SHAPE);

ENDING

M9 (COOLANT OFF);
G28 U0. W0. M05 (RAPID TO MACHINE HOME AND SPINDLE
OFF);
M30 (END PROGRAM);

FIGURE 8.3.36 A Fanuc program using the G71 rough turning and G70 finish turning canned cycle is shown for an OD application. Notice the tool change between the rough and finish operation.

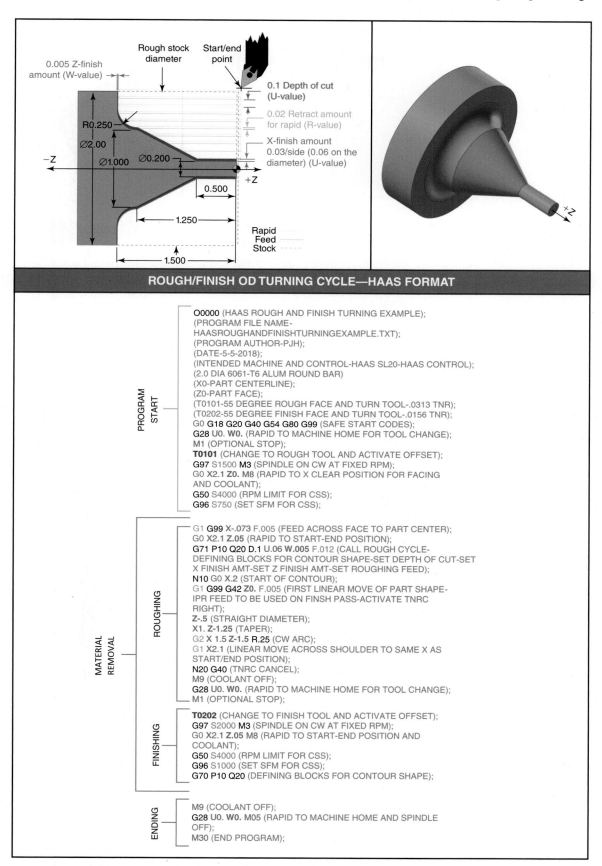

ROUGH/FINISH OD TURNING CYCLE—HAAS FORMAT

```
O0000 (HAAS ROUGH AND FINISH TURNING EXAMPLE);
(PROGRAM FILE NAME-
HAASROUGHANDFINISHTURNINGEXAMPLE.TXT);
(PROGRAM AUTHOR-PJH);
(DATE-5-5-2018);
(INTENDED MACHINE AND CONTROL-HAAS SL20-HAAS CONTROL);
(2.0 DIA 6061-T6 ALUM ROUND BAR)
(X0-PART CENTERLINE);
(Z0-PART FACE);
(T0101-55 DEGREE ROUGH FACE AND TURN TOOL-.0313 TNR);
(T0202-55 DEGREE FINISH FACE AND TURN TOOL-.0156 TNR);
G0 G18 G20 G40 G54 G80 G99 (SAFE START CODES);
G28 U0. W0. (RAPID TO MACHINE HOME FOR TOOL CHANGE);
M1 (OPTIONAL STOP);
T0101 (CHANGE TO ROUGH TOOL AND ACTIVATE OFFSET);
G97 S1500 M3 (SPINDLE ON CW AT FIXED RPM);
G0 X2.1 Z0. M8 (RAPID TO X CLEAR POSITION FOR FACING
AND COOLANT);
G50 S4000 (RPM LIMIT FOR CSS);
G96 S750 (SET SFM FOR CSS);
```

```
G1 G99 X-.073 F.005 (FEED ACROSS FACE TO PART CENTER);
G0 X2.1 Z.05 (RAPID TO START-END POSITION);
G71 P10 Q20 D.1 U.06 W.005 F.012 (CALL ROUGH CYCLE-
DEFINING BLOCKS FOR CONTOUR SHAPE-SET DEPTH OF CUT-SET
X FINISH AMT-SET Z FINISH AMT-SET ROUGHING FEED);
N10 G0 X.2 (START OF CONTOUR);
G1 G99 G42 Z0. F.005 (FIRST LINEAR MOVE OF PART SHAPE-
IPR FEED TO BE USED ON FINSH PASS-ACTIVATE TNRC
RIGHT);
Z-.5 (STRAIGHT DIAMETER);
X1. Z-1.25 (TAPER);
G2 X 1.5 Z-1.5 R.25 (CW ARC);
G1 X2.1 (LINEAR MOVE ACROSS SHOULDER TO SAME X AS
START/END POSITION);
N20 G40 (TNRC CANCEL);
M9 (COOLANT OFF);
G28 U0. W0. (RAPID TO MACHINE HOME FOR TOOL CHANGE);
M1 (OPTIONAL STOP);
```

```
T0202 (CHANGE TO FINISH TOOL AND ACTIVATE OFFSET);
G97 S2000 M3 (SPINDLE ON CW AT FIXED RPM);
G0 X2.1 Z.05 M8 (RAPID TO START-END POSITION AND
COOLANT);
G50 S4000 (RPM LIMIT FOR CSS);
G96 S1000 (SET SFM FOR CSS);
G70 P10 Q20 (DEFINING BLOCKS FOR CONTOUR SHAPE);
```

```
M9 (COOLANT OFF);
G28 U0. W0. M05 (RAPID TO MACHINE HOME AND SPINDLE
OFF);
M30 (END PROGRAM);
```

FIGURE 8.3.37 A Haas program using the G71 rough turning and G70 finish turning canned cycle is shown for an OD application. Notice the tool change between the rough and finish operation. The negative X-finish allowance required for boring is highlighted.

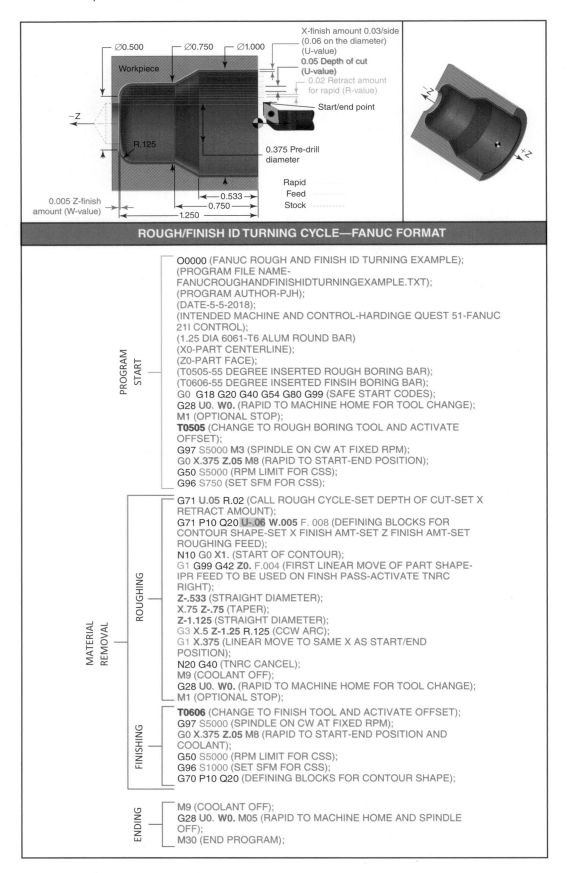

FIGURE 8.3.38 A Fanuc program using the G71 rough turning and G70 finish turning canned cycle is shown for an ID application. Notice the tool change between the rough and finish operation. The negative X-finish allowance required for boring is highlighted.

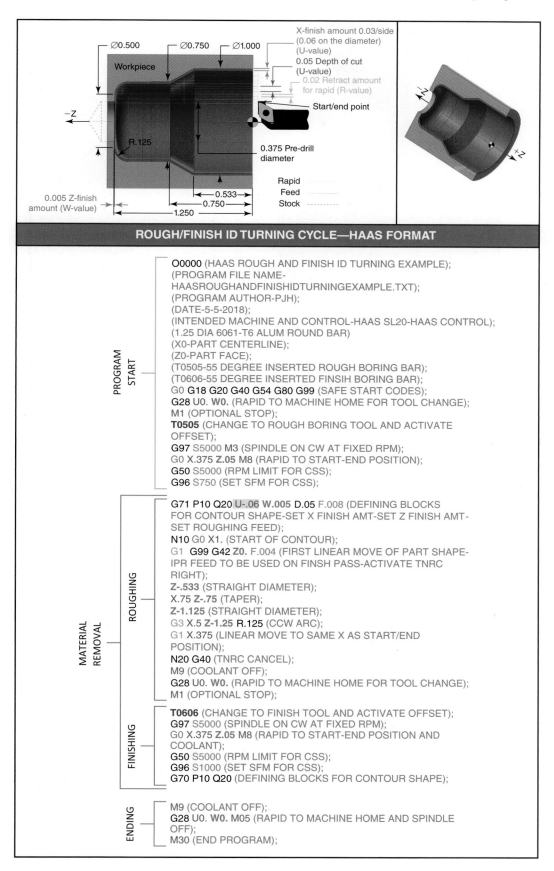

FIGURE 8.3.39 A Haas program using the G71 rough turning and G70 finish turning canned cycle is shown for an ID application. Notice the tool change between the rough and finish operation. The negative X-finish allowance required for boring is highlighted.

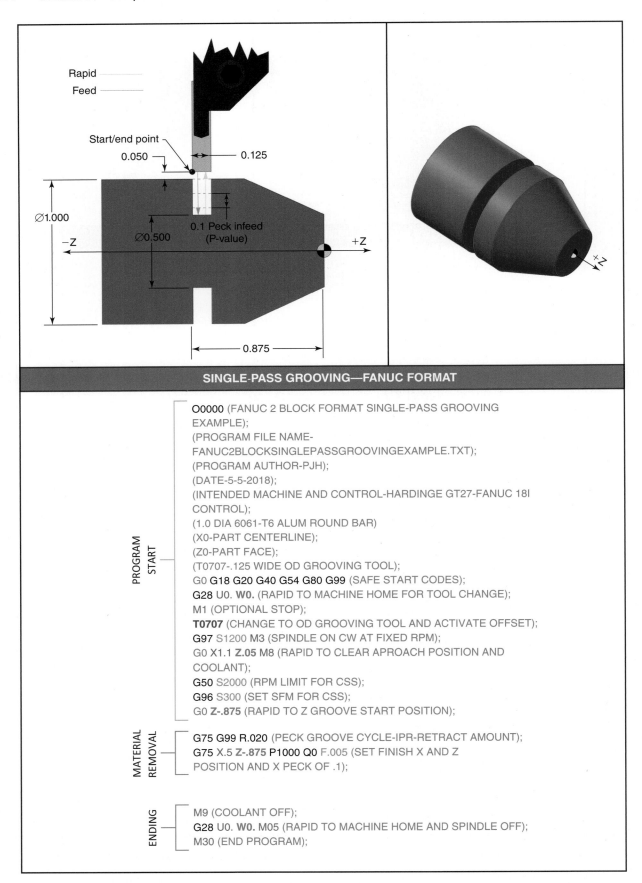

SINGLE-PASS GROOVING—FANUC FORMAT

PROGRAM START

O0000 (FANUC 2 BLOCK FORMAT SINGLE-PASS GROOVING EXAMPLE);
(PROGRAM FILE NAME-FANUC2BLOCKSINGLEPASSGROOVINGEXAMPLE.TXT);
(PROGRAM AUTHOR-PJH);
(DATE-5-5-2018);
(INTENDED MACHINE AND CONTROL-HARDINGE GT27-FANUC 18I CONTROL);
(1.0 DIA 6061-T6 ALUM ROUND BAR)
(X0-PART CENTERLINE);
(Z0-PART FACE);
(T0707-.125 WIDE OD GROOVING TOOL);
G0 G18 G20 G40 G54 G80 G99 (SAFE START CODES);
G28 U0. **W0.** (RAPID TO MACHINE HOME FOR TOOL CHANGE);
M1 (OPTIONAL STOP);
T0707 (CHANGE TO OD GROOVING TOOL AND ACTIVATE OFFSET);
G97 S1200 M3 (SPINDLE ON CW AT FIXED RPM);
G0 X1.1 **Z.05** M8 (RAPID TO CLEAR APROACH POSITION AND COOLANT);
G50 S2000 (RPM LIMIT FOR CSS);
G96 S300 (SET SFM FOR CSS);
G0 **Z-.875** (RAPID TO Z GROOVE START POSITION);

MATERIAL REMOVAL

G75 G99 R.020 (PECK GROOVE CYCLE-IPR-RETRACT AMOUNT);
G75 X.5 **Z-.875** P1000 Q0 F.005 (SET FINISH X AND Z POSITION AND X PECK OF .1);

ENDING

M9 (COOLANT OFF);
G28 U0. **W0.** M05 (RAPID TO MACHINE HOME AND SPINDLE OFF);
M30 (END PROGRAM);

FIGURE 8.3.40 A Fanuc program using the G75 peck grooving cycle is shown for a single-width pass.

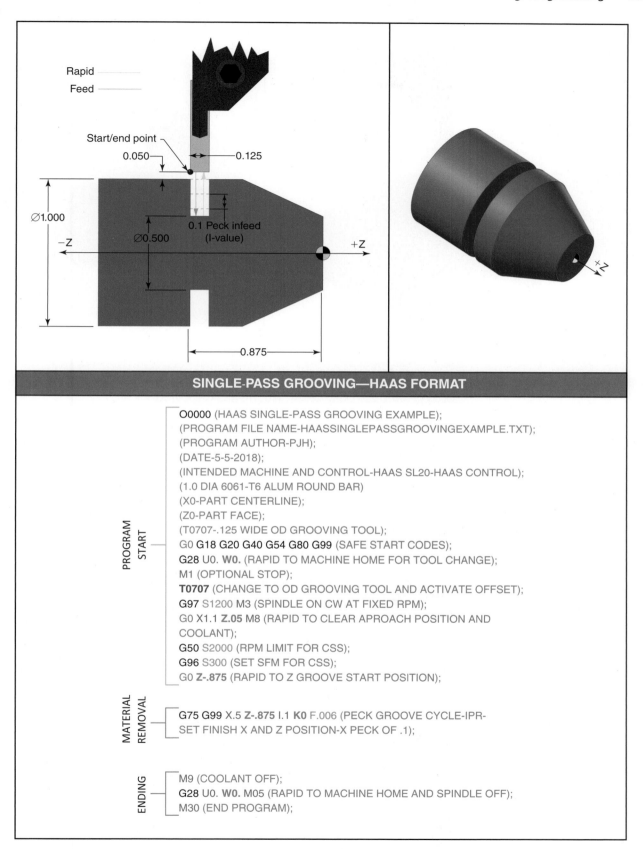

SINGLE-PASS GROOVING—HAAS FORMAT

O0000 (HAAS SINGLE-PASS GROOVING EXAMPLE);
(PROGRAM FILE NAME-HAASSINGLEPASSGROOVINGEXAMPLE.TXT);
(PROGRAM AUTHOR-PJH);
(DATE-5-5-2018);
(INTENDED MACHINE AND CONTROL-HAAS SL20-HAAS CONTROL);
(1.0 DIA 6061-T6 ALUM ROUND BAR)
(X0-PART CENTERLINE);
(Z0-PART FACE);
(T0707-.125 WIDE OD GROOVING TOOL);
G0 G18 G20 G40 G54 G80 G99 (SAFE START CODES);
G28 U0. W0. (RAPID TO MACHINE HOME FOR TOOL CHANGE);
M1 (OPTIONAL STOP);
T0707 (CHANGE TO OD GROOVING TOOL AND ACTIVATE OFFSET);
G97 S1200 M3 (SPINDLE ON CW AT FIXED RPM);
G0 X1.1 **Z.05** M8 (RAPID TO CLEAR APROACH POSITION AND
COOLANT);
G50 S2000 (RPM LIMIT FOR CSS);
G96 S300 (SET SFM FOR CSS);
G0 **Z-.875** (RAPID TO Z GROOVE START POSITION);

PROGRAM START

G75 G99 **X.5 Z-.875** I.1 **K0** F.006 (PECK GROOVE CYCLE-IPR-
SET FINISH X AND Z POSITION-X PECK OF .1);

MATERIAL REMOVAL

M9 (COOLANT OFF);
G28 U0. **W0.** M05 (RAPID TO MACHINE HOME AND SPINDLE OFF);
M30 (END PROGRAM);

ENDING

FIGURE 8.3.41 A Haas program using the G75 peck grooving cycle is shown for a single-width pass.

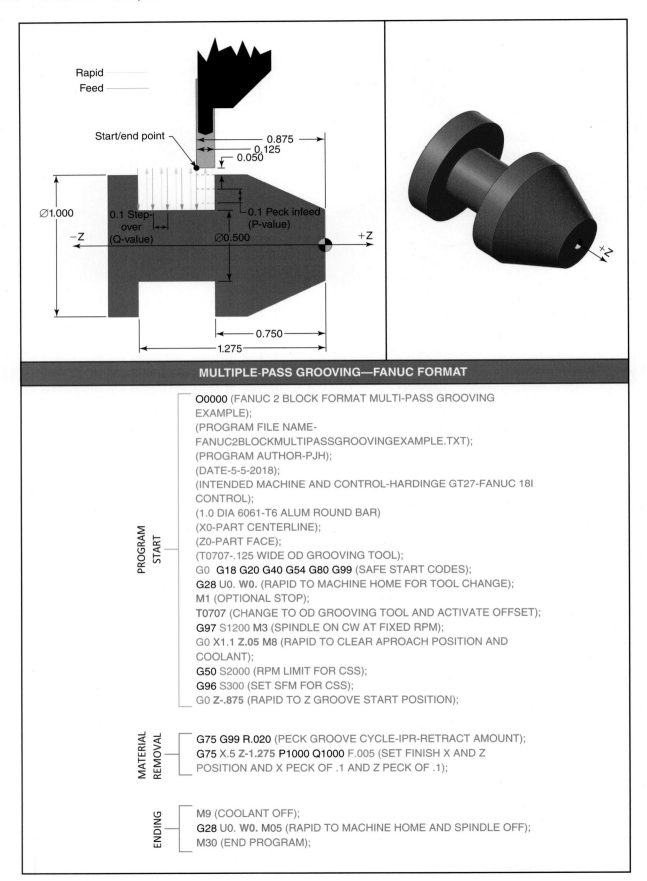

MULTIPLE-PASS GROOVING—FANUC FORMAT

PROGRAM START

O0000 (FANUC 2 BLOCK FORMAT MULTI-PASS GROOVING EXAMPLE);
(PROGRAM FILE NAME-FANUC2BLOCKMULTIPASSGROOVINGEXAMPLE.TXT);
(PROGRAM AUTHOR-PJH);
(DATE-5-5-2018);
(INTENDED MACHINE AND CONTROL-HARDINGE GT27-FANUC 18I CONTROL);
(1.0 DIA 6061-T6 ALUM ROUND BAR)
(X0-PART CENTERLINE);
(Z0-PART FACE);
(T0707-.125 WIDE OD GROOVING TOOL);
G0 G18 G20 G40 G54 G80 G99 (SAFE START CODES);
G28 U0. W0. (RAPID TO MACHINE HOME FOR TOOL CHANGE);
M1 (OPTIONAL STOP);
T0707 (CHANGE TO OD GROOVING TOOL AND ACTIVATE OFFSET);
G97 S1200 M3 (SPINDLE ON CW AT FIXED RPM);
G0 X1.1 Z.05 M8 (RAPID TO CLEAR APROACH POSITION AND COOLANT);
G50 S2000 (RPM LIMIT FOR CSS);
G96 S300 (SET SFM FOR CSS);
G0 Z-.875 (RAPID TO Z GROOVE START POSITION);

MATERIAL REMOVAL

G75 G99 R.020 (PECK GROOVE CYCLE-IPR-RETRACT AMOUNT);
G75 X.5 Z-1.275 P1000 Q1000 F.005 (SET FINISH X AND Z POSITION AND X PECK OF .1 AND Z PECK OF .1);

ENDING

M9 (COOLANT OFF);
G28 U0. W0. M05 (RAPID TO MACHINE HOME AND SPINDLE OFF);
M30 (END PROGRAM);

FIGURE 8.3.42 A Fanuc program using the G75 peck grooving cycle is shown for a multiple-width pass.

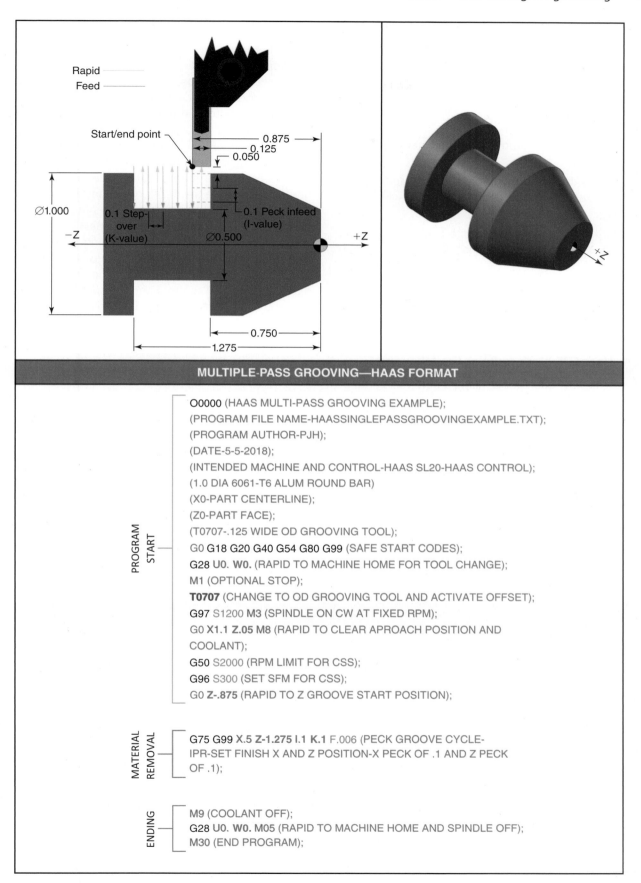

FIGURE 8.3.43 A Haas program using the G75 peck grooving cycle is shown for a multiple-width pass.

The first G75 block of a Fanuc may use the following:

- R sets the distance the tool will retract after each peck to break the chip.

The second G75 block of a Fanuc may use the following:

- X specifies the final groove diameter after all pecking is complete.
- Z specifies the end position of the completed groove in the Z-axis. This can be omitted for single-width pass grooves.
- P sets the incremental infeed distance for each peck in the X-direction (usually no decimal, four-place format).
- Q sets the incremental step-over distance for each pass in the Z-direction (usually no decimal, four-place format).
- F sets the feed rate.

The Haas G75 block contains the following:

- X specifies the final groove diameter after all pecking is complete.
- Z specifies the end position of the completed groove in the Z-axis. This can be omitted for single-width pass grooves.
- I sets the incremental infeed distance for each peck in the X-direction.
- K sets the incremental step-over distance for each pass in the Z-direction.
- F sets the feed rate.

Threading Canned Cycles

Most turning centers have a *repetitive canned threading cycle* that will automatically take successive thread cutting passes until the tool reaches the thread's root (minor) diameter. This cycle involves many variables to customize the operation for different applications.

Recall that as a "V"-shaped threading tool is advanced into lathe work, tool contact increases greatly. When operating a manual lathe, it was necessary to gradually decrease the depth of cut on each successive pass. This helped keep cutting forces (and volume of material removed) consistent from pass to pass until completion. When using the threading canned cycle, the CNC control will automatically calculate the depth-of-cut reduction for each pass. The cycle will also maintain timing of the tool and the spindle position so that the same thread groove is precisely followed for all passes, even at much higher RPM than on manual machines.

Flank Infeed Method

Threading cycles can be applied with a *flank infeed (single-edge)* method as was shown in Section 5.4. With this method, a 60-degree cutting tool is advanced on a 30-degree in-feed angle. This helps produce a better finish and less cutting force, since only one side of the "V" cuts (shown in **Figure 8.3.44**). The downside of this technique is that the trailing cutting-edge rubs lightly instead of cutting. Also, tool life is reduced because the all cutting work is performed by only one of the edges.

Canned Cycle	Format	Variable Description
Fanuc Peck Grooving Cycle Format (Two Block Version)	*(TOOL TO START POINT)* G75 G99 R_; G75 X_ Z_ P_ Q_ F_;	R (first block): Retract amount X: Groove end position in X-axis Z: Groove end position in Z-axis P: Peck infeed increment in X-direction Q: Step-over increment in Z-axis F: Feed rate
Fanuc (Single Block Version) OR Haas Peck Grooving Cycle Format	*(TOOL TO START POINT)* G75 X_ Z_ I_ K_ F_;	X: Groove end position in X-axis Z: Groove end position in Z-axis I: Peck infeed increment in X-direction K: Step-over increment in Z-axis (omit if groove doesn't require a step-over) F: Feed rate

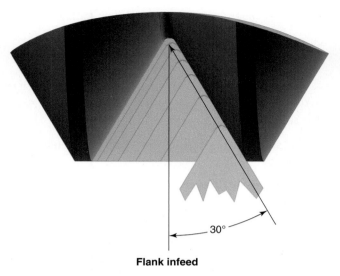

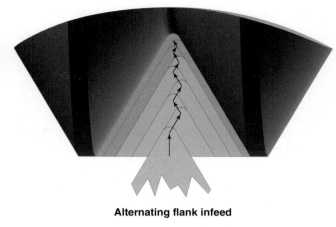

Alternating flank infeed

FIGURE 8.3.46 An illustration of the alternating flank infeed method for threading is shown.

Flank infeed

FIGURE 8.3.44 An illustration of the flank infeed method for threading is shown.

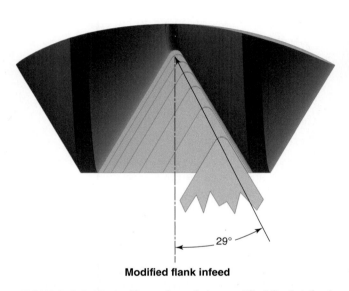

Modified flank infeed

FIGURE 8.3.45 An illustration of the modified flank infeed method for threading is shown.

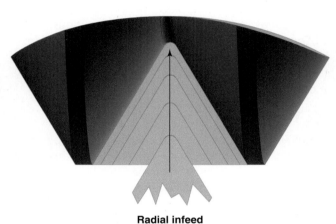

Radial infeed

FIGURE 8.3.47 An illustration of the radial infeed method for threading is shown.

Modified Flank Infeed Method

Reducing the flank infeed angle by half a degree or a full degree will result in a *modified flank infeed* as shown in **Figure 8.3.45**. Here the leading edge does most of the work, but the trailing edge will perform a light cut with each pass. This usually produces the best results for general purpose work.

Alternating Flank Infeed Method

Some machines can alternate the flank cuts back and forth between each side of the tool from pass to pass. This helps to distribute the tool wear across both sides

of the tool and extends the life of the tool. Since sides alternate, only one edge is cutting at a time and force is still as light as with the other flank-cutting methods. **Figure 8.3.46** shows an example of alternating flank cutting.

Radial Infeed Method

A tool may also be in-fed directly on a 0-degree angle so that each flank cuts an equal amount simultaneously. Since cutting force is higher, this *radial infeed* is usually only used for fine thread pitches where the tool is not engaged very deeply. Chip formation is poor, tool tip heat is high, and high tool contact may cause chatter, so this method is seldom desirable. **Figure 8.3.47** shows an example of a radial infeed method.

OD Threading Canned Cycles

When an external thread is to be cut on the outside diameter of a workpiece, the tool must be positioned at an X-start point away from the diameter. This position is calculated and will establish where the first thread-cutting depth begins. The Z-start point must establish enough space between the work for the tool to accelerate to full feed by the time it enters the cut. Repetitive passes will be made until the full thread is created and the tool tip has cut to the minor (root) diameter of the thread groove. The programmed Z-end position of the cycle will be the farthest position that the tool will travel. At the end of each pass, the tool will begin to retract before this position is reached. This results in a gradually decreasing thread depth toward the end of the thread groove, so adding extra length may be needed to achieve the correct full thread length.

Two consecutive G76 blocks are often required on a Fanuc control to activate this cycle. Be cautious: some of the characters used for variables on the first block are repeated on the second with a different meaning. Detailed explanations are given below and followed by a concise table showing the format.

ID Threading Canned Cycle

The G76 threading cycle can also be used to perform ID threading operations. As with all internal threading operations, a hole or bore must first be created to establish the minor diameter of the thread. The X-start point is programmed to establish clearance between the inside diameter of the hole and tool. The first thread passes will then begin at the hole's diameter and work outward until the thread is complete and the tool reaches the major diameter. The G76 cycle works for internal threading in the same manner as it does for external, only the machine must retract the tool inward (toward the center of the part) after each pass. The control knows to do this based on where the tool X-start point is relative to the final X-position (major diameter).

The first G76 block of a Fanuc may use the following:

- **P** is a six-digit code where the first two digits set the number of spring passes to be taken after the finish pass. The second pair of digits specify how quickly the tool retracts. Programming 00 here will retract at the machine's highest possible rate. The third pair of digits set the tool in-feed angle degrees (often 29 for

modified flank infeed or 30 for flank-cutting infeed of a 60-degree thread).

- **Q** sets the minimum cut depth. Since the machine automatically reduces infeed depth for each pass, the machine will not take a pass lighter than this value.

- **R** sets the depth of the final pass.

The second G76 block of a Fanuc may use the following:

- **X** sets the final pass diameter (minor diameter on an external thread/major diameter on an internal thread).

- **Z** establishes the end of the thread.

- **P** sets the depth of a single thread from crest to root (radius value).

- **Q** sets the depth of the first pass.

- **F** sets the feed rate (normally in IPR equal to the thread lead; equal to the pitch on single lead threads).

- * An **R** value may be added to cut tapered threads such as NPT pipe threads.

The basic Haas G76 block contains the following:

- **X** sets the final pass diameter (minor diameter on external threads/major diameter on internal threads).

- **Z** establishes the end of the thread.

- **A** sets the included angle of tool in-feed. This is an included angle. A setting of 60 will infeed 30 degrees (flank infeed on a 60-degree thread). A setting of 58 will infeed 29 degrees (modified flank infeed on a 60-degree thread).

- **D** sets the depth of the first cutting pass.

- **K** sets the depth of a single thread (radius value).

- **F** sets the feed rate (normally in IPR equal to the thread lead, same as the pitch for single lead threads).

- * If a P-value is not given, the cycle will default to single flank cutting (P1) for flank infeed or modified flank infeed methods. A P2 value may be added to the block for alternating flank cutting.

- * An I-value may be added to cut tapered threads, such as NPT pipe threads. A negative I-value specifies the taper amount per side from Z-start to Z-end position. A positive value is used for internal threads.

Canned Cycle	Format	Variable Description
Fanuc Auto Threading Cycle Format (Two Block Version)	G0X_ Z_ S_ (*TOOL TO START POINT*) G76 P_ _ _ _ _ _ Q_ R_; G76 X_ Z_ P_ Q_ F_;	S: Spindle speed ***FIRST G76 BLOCK*** PXX_ _ _ _: Number of spring passes P_ _ XX _ _: Tool retract rate P_ _ _ _ XX: Infeed angle Q: Minimum depth of cut. NO DECIMAL ALLOWED, use four place format. R: Finish pass infeed amount ***SECOND G76 BLOCK*** X: Last threading pass (minor) diameter Z: Thread end position in Z-axis P: Single depth of thread (radial). NO DECIMAL ALLOWED, use four place format. Q: First pass depth. NO DECIMAL ALLOWED, use four place format. F: Feed rate
Haas Auto Threading Cycle Format (Two Block Version)	G0X_ Z_ S_ (*TOOL TO START POINT*) G76 X_ Z_ K_ P_D_ F_;	S: Spindle speed ***G76 BLOCK*** X: Thread end position in X-axis Z: Thread end position in Z-axis K: Single depth of thread (radial) P1=Single edge (flank) cutting/P2-Double edge (alternating flank) cutting D: First pass depth F: Feed rate

It is helpful to perform the all thread calculations before programming a threading cycle. The following will be used in **Figures 8.3.48** and **8.3.49**:

Description	Formula	Answer
Thread Size:	See print	0.5—20
Class of Fit	See print	2A
Thread Lead (Same as Pitch for Single-Lead Threads):	1/TPI	0.050
Single Depth **if External Thread**:	0.61343/TPI *Could also be found in *Machinery's Handbook*	0.0307
Single Depth **if Internal Thread**:	0.54127 /TPI *Could also be found in *Machinery's Handbook*	N/A
X Start Point **for External Threads**:	Nominal OD + (2*Single Depth of Thread)	0.5614
X Start Point **for Internal Threads**:	Nominal ID − (2*Single Depth of Thread)	N/A
Z Start Point:	4*Lead OR 0.250 *Use whichever value is greater	0.250
Major Diameter:	See *Machinery's Handbook*	0.4906—0.4987
Minor (Root) Diameter:	See *Machinery's Handbook*	0.4374
Number of Complete Threading Passes:	72*Lead + 4	8
First Pass Depth:	Single Depth of Thread / $\sqrt{}$Number of Threading Passes	0.0109
Maximum RPM (guideline for many machines):	120/Lead	2400
Z end position:	See print	−0.813
Feed (IPR):	Same as Thread Lead	0.050

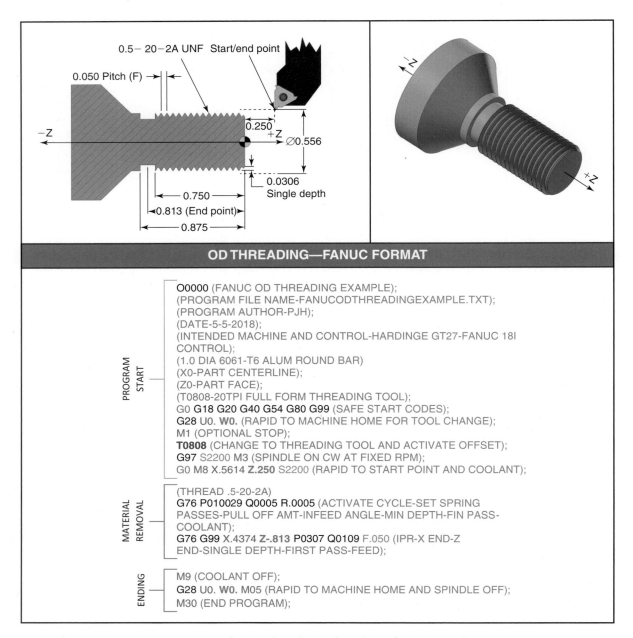

OD THREADING—FANUC FORMAT

PROGRAM START

O0000 (FANUC OD THREADING EXAMPLE);
(PROGRAM FILE NAME-FANUCODTHREADINGEXAMPLE.TXT);
(PROGRAM AUTHOR-PJH);
(DATE-5-5-2018);
(INTENDED MACHINE AND CONTROL-HARDINGE GT27-FANUC 18I CONTROL);
(1.0 DIA 6061-T6 ALUM ROUND BAR)
(X0-PART CENTERLINE);
(Z0-PART FACE);
(T0808-20TPI FULL FORM THREADING TOOL);
G0 G18 G20 G40 G54 G80 G99 (SAFE START CODES);
G28 U0. **W0.** (RAPID TO MACHINE HOME FOR TOOL CHANGE);
M1 (OPTIONAL STOP);
T0808 (CHANGE TO THREADING TOOL AND ACTIVATE OFFSET);
G97 S2200 M3 (SPINDLE ON CW AT FIXED RPM);
G0 M8 X.5614 **Z.250** S2200 (RAPID TO START POINT AND COOLANT);

MATERIAL REMOVAL

(THREAD .5-20-2A)
G76 P010029 Q0005 R.0005 (ACTIVATE CYCLE-SET SPRING PASSES-PULL OFF AMT-INFEED ANGLE-MIN DEPTH-FIN PASS-COOLANT);
G76 G99 X.4374 **Z-.813** P0307 Q0109 F.050 (IPR-X END-Z END-SINGLE DEPTH-FIRST PASS-FEED);

ENDING

M9 (COOLANT OFF);
G28 U0. **W0.** M05 (RAPID TO MACHINE HOME AND SPINDLE OFF);
M30 (END PROGRAM);

FIGURE 8.3.48 A Fanuc program using the G76 threading cycle is shown for an OD application.

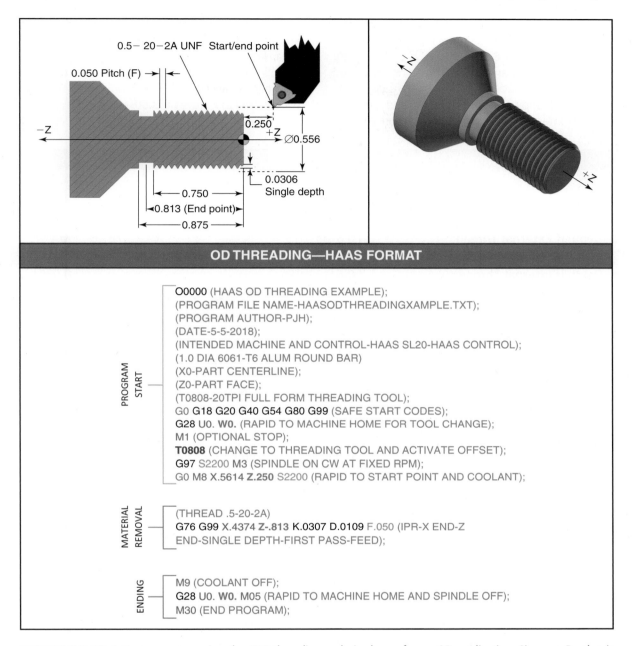

FIGURE 8.3.49 A Haas program using the G76 threading cycle is shown for an OD application. Since no P-value is commanded in the G76 block, the cycle will default to single flank infeed (P1).

Figure 8.3.48 shows a part drawing and program excerpt for an external threading cycle used on a Fanuc machine control. **Figure 8.3.49** shows a part drawing and program excerpt for an external threading cycle used on a Haas machine control.

Figure 8.3.50 shows a part drawing and program excerpt of an internal threading cycle used on a Fanuc machine control. **Figure 8.3.51** shows a part drawing and program excerpt of an internal threading cycle used on a Haas.

It is helpful to perform the all thread calculations before programming a threading cycle. The following will be used in **Figures 8.3.50** and **8.3.51**:

Description	Formula	Answer
Thread Size:	See print	1.0–18
Class of Fit	See print	2B
Thread Lead (Same as Pitch for Single-Lead Threads):	1/TPI	0.0556
Single Depth **if External Thread**:	0.61343/TPI *Could also be found in *Machinery's Handbook*	N/A
Single Depth **if Internal Thread**:	0.54127 /TPI *Could also be found in *Machinery's Handbook*	0.0301
X Start Point **for External Threads**:	Nominal OD + (2*Single Depth of Thread)	N/A
X Start Point **for Internal Threads**:	Nominal ID − (2*Single Depth of Thread)	0.8848
Z Start Point:	4*Lead OR 0.250 *Use whichever value is greater	0.250
Major Diameter:	See *Machinery's Handbook*	1.000
Minor (Root) Diameter:	See *Machinery's Handbook*	0.940 − 0.953
Number of Complete Threading Passes:	72*Lead + 4	8
First Pass Depth:	Single Depth of Thread / √Number of Threading Passes	0.0106
Maximum RPM (guideline for many machines):	120/Lead	2158
Z end position:	See print	− 0.596
Feed (IPR):	Same as Thread Lead	0.0556

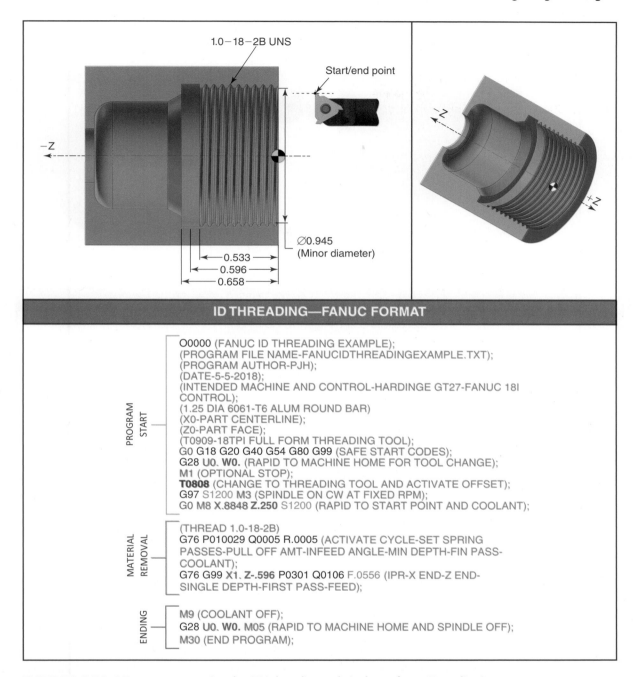

ID THREADING—FANUC FORMAT

PROGRAM START

O0000 (FANUC ID THREADING EXAMPLE);
(PROGRAM FILE NAME-FANUCIDTHREADINGEXAMPLE.TXT);
(PROGRAM AUTHOR-PJH);
(DATE-5-5-2018);
(INTENDED MACHINE AND CONTROL-HARDINGE GT27-FANUC 18I CONTROL);
(1.25 DIA 6061-T6 ALUM ROUND BAR)
(X0-PART CENTERLINE);
(Z0-PART FACE);
(T0909-18TPI FULL FORM THREADING TOOL);
G0 G18 G20 G40 G54 G80 G99 (SAFE START CODES);
G28 U0. W0. (RAPID TO MACHINE HOME FOR TOOL CHANGE);
M1 (OPTIONAL STOP);
T0808 (CHANGE TO THREADING TOOL AND ACTIVATE OFFSET);
G97 S1200 **M3** (SPINDLE ON CW AT FIXED RPM);
G0 M8 **X.8848 Z.250** S1200 (RAPID TO START POINT AND COOLANT);

MATERIAL REMOVAL

(THREAD 1.0-18-2B)
G76 P010029 Q0005 R.0005 (ACTIVATE CYCLE-SET SPRING PASSES-PULL OFF AMT-INFEED ANGLE-MIN DEPTH-FIN PASS-COOLANT);
G76 G99 **X1. Z-.596** P0301 Q0106 F.0556 (IPR-X END-Z END-SINGLE DEPTH-FIRST PASS-FEED);

ENDING

M9 (COOLANT OFF);
G28 **U0. W0.** M05 (RAPID TO MACHINE HOME AND SPINDLE OFF);
M30 (END PROGRAM);

FIGURE 8.3.50 A Fanuc program using the G76 threading cycle is shown for an ID application.

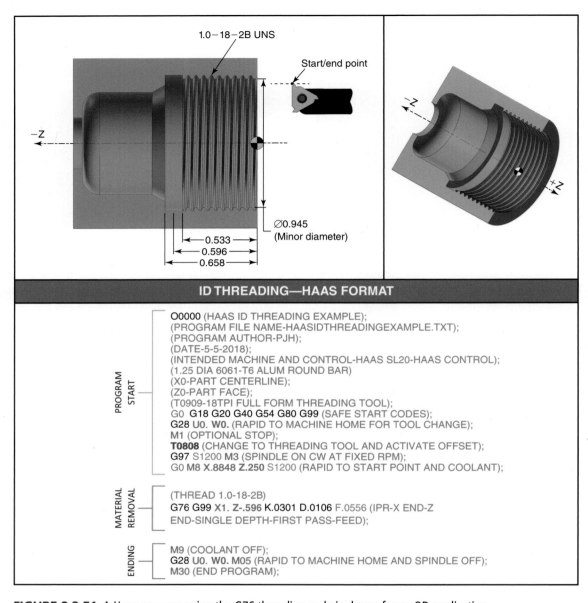

ID THREADING—HAAS FORMAT

PROGRAM START

```
O0000 (HAAS ID THREADING EXAMPLE);
(PROGRAM FILE NAME-HAASIDTHREADINGEXAMPLE.TXT);
(PROGRAM AUTHOR-PJH);
(DATE-5-5-2018);
(INTENDED MACHINE AND CONTROL-HAAS SL20-HAAS CONTROL);
(1.25 DIA 6061-T6 ALUM ROUND BAR)
(X0-PART CENTERLINE);
(Z0-PART FACE);
(T0909-18TPI FULL FORM THREADING TOOL);
G0  G18 G20 G40 G54 G80 G99 (SAFE START CODES);
G28 U0. W0. (RAPID TO MACHINE HOME FOR TOOL CHANGE);
M1 (OPTIONAL STOP);
T0808 (CHANGE TO THREADING TOOL AND ACTIVATE OFFSET);
G97 S1200 M3 (SPINDLE ON CW AT FIXED RPM);
G0 M8 X.8848 Z.250 S1200 (RAPID TO START POINT AND COOLANT);
```

MATERIAL REMOVAL

```
(THREAD 1.0-18-2B)
G76 G99 X1. Z-.596 K.0301 D.0106 F.0556 (IPR-X END-Z
END-SINGLE DEPTH-FIRST PASS-FEED);
```

ENDING

```
M9 (COOLANT OFF);
G28 U0. W0. M05 (RAPID TO MACHINE HOME AND SPINDLE OFF);
M30 (END PROGRAM);
```

FIGURE 8.3.51 A Haas program using the G76 threading cycle is shown for an OD application.

SUMMARY

- Many different MCU manufacturers exist, and each one has slightly different programming techniques and formats.

- Some MCUs allow turning programs to be written using either radial or diametral X-axis coordinates. The control must be set to the correct mode. Diametral programming is the most commonly used technique.

- There are three main motion types for CNC programming: rapid, linear interpolation, and circular interpolation.

- Rapid traverse motion is used for quick positioning and can be extremely dangerous if care is not taken.

- Linear interpolation allows a straight path to be taken between two programmed points and can be used for straight turning or for taper turning.

- Circular interpolation allows an arc to be created between two points in either the clockwise or counterclockwise direction. Circular interpolation can be programmed using either the arc center method or the radius method. The radius method is most common for turning.

- Facing on a turning machine usually is programmed to drive the tool from the outside diameter to the part center. The end position for the facing cut should not be programmed to stop directly at X-zero, but should take the tool beyond the part center by a distance equal to the nose radius.

- Canned cycles can be used to simplify repetitive and tedious drilling, roughing, finishing, grooving, and threading operations.

- Canned cycles for rouging, finishing, and threading can be applied to outside or inside diameters.

- CNC tapping cycles rely on the careful synchronization of the spindle rotation and feed motion. Rigid tapping allows the tap to be held solidly in the holder; non-rigid tapping requires a floating tap holder.

- Calculating tap depth for blind holes should be done before calculating a tap drill depth to ensure the tap doesn't bottom out.

- A cutting tool's nose radius will result in inaccurate arc and taper cutting if not compensated for using TNRC. A G41 must be programmed to compensate for a tool's radius on the left of the part while in motion and a G42 when on the right.

REVIEW QUESTIONS

1. Write the X- and Z-axes coordinates for the part shown. Use the diametral method in the X-axis.

Position	X	Z
A		
B		
C		
D		
E		
F		
G		
H		
I		

2. To perform a facing cut using a tool with a nose radius of 1/32, what will the programmed X-axis end point at the center of the part be?

3. Explain the difference between rigid tapping and tapping using a floating tap holder.

4. How must the feed rate for tapping using a floating holder be programmed differently than that for rigid tapping?

5. Describe what happens to a concave radius (fillet) when cut without using TNRC. A sketch may be used to help.

6. Describe what happens to an outside chamfer when cut without using TNRC. A sketch may be used to help.

7. If a G1 code command is programmed partway through a machining operation, but no F-value is given in that block, what feed rate will be used?

8. List and briefly describe the two methods for programming circular interpolation.

9. Explain in your own words the difference between direct RPM and constant surface speed (CSS).

10. In your own words, describe a canned cycle.

11. List two types of canned cycles besides roughing turning and finish turning.

UNIT 4 — CNC Turning: Setup and Operation

Learning Objectives

After completing this unit, the student should be able to:

- Demonstrate understanding of CNC machine modes
- Demonstrate understanding of a work coordinate system (WCS) for CNC turning
- Demonstrate understanding of a machine coordinate system (MCS) for CNC turning
- Demonstrate understanding of the homing procedure and purpose
- Demonstrate understanding of workpiece offsets for CNC turning
- Demonstrate understanding of tool geometry offsets for CNC turning
- Demonstrate understanding of tool wear offsets for CNC turning
- Demonstrate understanding of tool nose radius (or diameter) offsets
- Demonstrate understanding of tool quadrant settings for TNRC
- Describe the three basic methods for loading programs into the MCU
- Demonstrate understanding of program prove-out procedures

Key Terms

Direct Numerical Control (DNC)
Dry run
Dry cycle
Feed rate override
Geometry offset
Jog

Machine Coordinate System (MCS)
Machine home position
Manual Data Input (MDI)
Prove-out

Quadrant
Rapid override
Single-block mode
Wear offset
Work Coordinate System (WCS)
Work offset

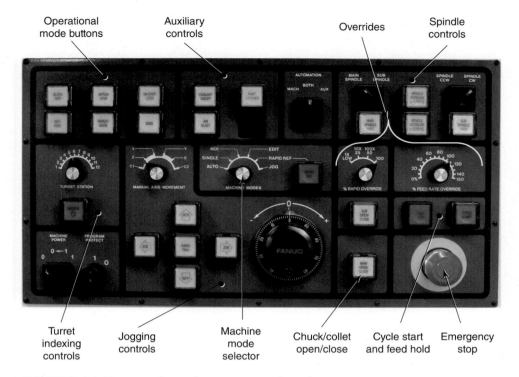

FIGURE 8.4.1 The parts of a turning center control panel.

MACHINE CONTROL PANEL

The machine control panel is usually attached to the MCU and contains the display screen and buttons, keys, knobs, and dials to program, set up, and operate the machine. A typical machine control panel is shown in **Figure 8.4.1**.

The display screen can be used to display the program, axis positions, and various machine setup pages. Menu buttons can be used to navigate through setup pages and to enter data. Control panel buttons are labeled with words or pictures describing their functions. Some machines also have buttons called *soft keys* that have no labels, but instead align with a function label on the display screen. **(See Figure 8.4.2.)** A key pad has letter and number keys for keying in programs and other data.

A machine mode knob or machine mode keys are used to change from one mode of operation to another. Common selectable machine modes include jog, auto, **manual data input (MDI)**, edit, and zero-reference return.

The edit mode allows programs to be typed into memory or an existing program to be modified. Edit is also usually the mode required to load stored programs for use. When programs are to be run, the control must be placed in auto mode. Reference return mode is used to zero the machine axes upon power-up.

MDI mode provides a blank program screen for entering short programs or single program commands required for machine setup. Program data entered into MDI is not saved in memory and is erased after it

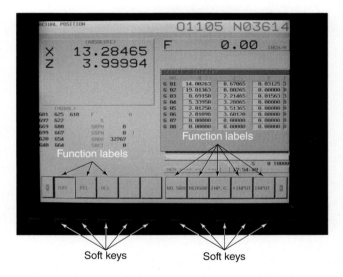

FIGURE 8.4.2 Soft keys are used on some machines. These keys are universal and can be used for different functions according to their on-screen label.

is executed. Therefore MDI is ideal for short program commands (used during setup or troubleshooting) such as tool changes, spindle start commands, and moving to a specific coordinate position.

When **jog** mode is turned on, keys and a small rotary handwheel allow two methods for manually controlled axis movement for machine setup. Most machines have arrow keys in the plus and minus direction for jogging each axis. **(See Figure 8.4.3.)** Pressing and holding

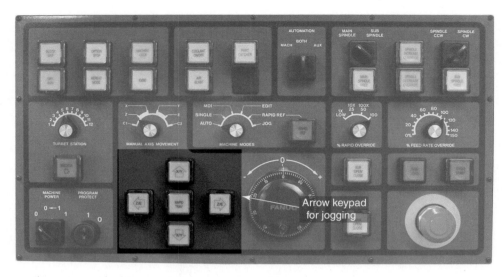

FIGURE 8.4.3 Arrow keys used for constant jogging of each axis.

Remote handwheel jog pendant

Control panel mounted jog handwheel

FIGURE 8.4.4 A jogging handwheel is used for fine control when jogging the machine's axes. The handwheels can either be (A) portable or (B) mounted permanently on the machine control panel.

these keys causes constant axis motion. When the key is released, motion stops. When jogging with the arrow keys, most machines vary the feed rate with either the

rapid or feed override dial. The rotary handwheel can be set to move an axis at different incremental steps for fine control when jogging machine axes. These steps are usually 0.010", 0.001", and 0.0001" per click of the handwheel. The jogging handwheel on some machines is mounted on a handheld pendant and attached to the MCU by a cable to allow its use away from the control panel and closer to the work area. **(See Figure 8.4.4.)**

The cycle start, feed hold, and emergency stop buttons are also located on the control panel. The cycle start button begins the active CNC program and is colored green on many machines. The feed hold button is usually located directly next to the cycle start button and will stop axis feed when pressed during program execution. In an emergency or collision, the red emergency stop button (sometimes called the E-stop) can be pressed to immediately stop the spindle and all axis motion. The **feed rate override** knob allows the programmed feed rate to be decreased, increased, or even stopped. Most machines are also equipped with a **rapid override** control to slow rapid motion while a program is running or to vary the jogging rate. A spindle speed override control can be used to decrease or increase spindle RPM. **(See Figure 8.4.5.)**

WORKHOLDING SETUP

Once the workholding device is chosen, it is mounted according to the manufacturer's guidelines. When using a chuck, machinable soft jaws are often used. Soft jaws can be customized to accommodate the shape of the workpiece. Often, the jaws are bored while installed on the chuck to develop the correct gripping radius where the jaw contacts the part. It is a good practice to bore the jaws while they are preloaded by clamping on a piece of scrap or a chucking ring. **(See Figure 8.4.6.)** This simulates the stresses the chuck will be under when the

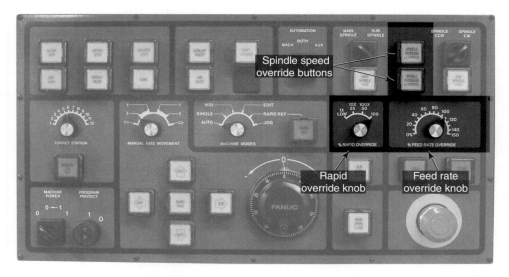

FIGURE 8.4.5 A spindle speed, feed rate, and rapid override control.

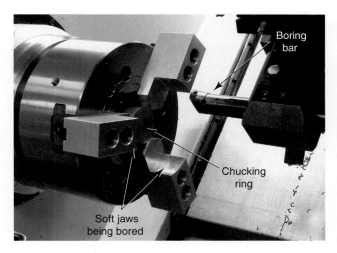

FIGURE 8.4.6 Soft-chuck jaws should be bored while they are mounted. Preloading by clamping on a piece of scrap or a chucking ring simulates forces created when clamping a workpiece.

actual workpiece is gripped. The clamping pressure to be used for gripping the work should also be set the same when boring the jaws. **Figure 8.4.7** shows a picture of appropriately bored soft-chuck jaws and a mating workpiece.

If using a collet, place the collet closer button in the "open" position. The collet must be aligned with its key and inserted into a clean spindle nose taper. Thread the drawtube onto the collet from the other end of the headstock. Place a workpiece of the correct diameter into the collet and hand-tighten the drawtube until the part is gripped snugly. Loosen the drawtube a half-turn before

FIGURE 8.4.7 Appropriately bored soft-chuck jaws and a mating workpiece.

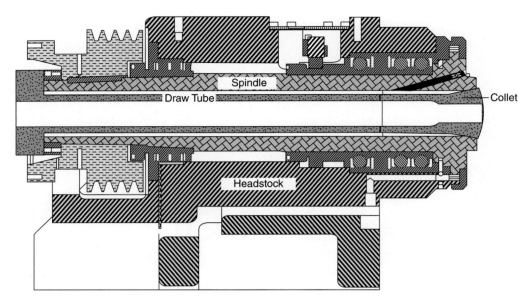

FIGURE 8.4.8 A turning center drawtube and collet assembly within a machine spindle.

locking it into place to establish the proper clearance for loading and unloading workpieces. If the drawtube is left snug with the collet closer in the "open" position, workpieces will not easily slide in and out of the collet during part changes (the collet cannot release enough). **Figure 8.4.8** shows an illustration of a turning center drawtube and collet assembly.

NOTE:

It is a common misconception that the drawtube thread engagement has an effect on clamping pressure. The drawtube only sets the size range between open and closed position. Ultimately, the gripping pressure is determined only by the machine's air or hydraulic regulator for the collet/chuck closer system.

Whether a collet or a chuck is used, the appropriate clamping pressure must be set. Too much gripping pressure can distort the workpiece and too little clamping pressure can allow the work to slip or be pulled out of the collet or chuck during machining. Too little pressure will also allow the gripping force to be

overcome by centrifugal force at high RPM. The chart in **Figure 8.4.9** shows the effects of gripping pressure versus RPM on one manufacturer's machines. The pressure of the collet/chuck closer system is controlled either by a pneumatic or hydraulic regulator, depending on the machine. The gripping pressure is adjusted with this regulator. (Refer to the machine manual for specific procedure.) **Figure 8.4.10** shows a picture of a turning center hydraulic regulator and gage for adjusting collet pressure.

Once the workholding device has been set, the workpiece to be used for setup may now be gripped. It is important to allow enough workpiece length to extend out of the chuck or collet so that there is no collision with the chuck jaws or the collet face during any of

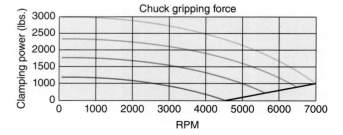

FIGURE 8.4.9 This chart shows the effects on gripping pressure of one manufacturer's chucks as the RPM increases.

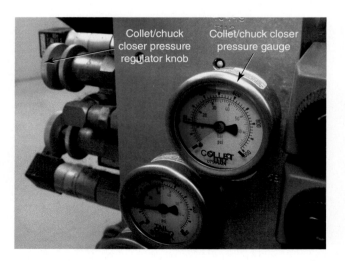

FIGURE 8.4.10 A picture of a turning center hydraulic regulator and gauge for adjusting collet or chuck clamping pressure.

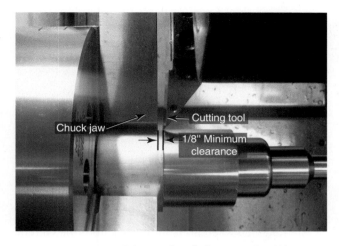

FIGURE 8.4.11 A minimum of 1/8" clearance should be maintained between the cutting tool and the workholding device.

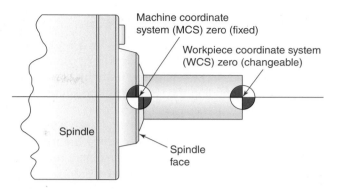

FIGURE 8.4.12 The relationship of the machine coordinate system (MCS) to the established work coordinate system (WCS).

the programmed machining operations. Typically, 1/8" of minimum clearance should be maintained between the cutting tool and the workholding device. **(See Figure 8.4.11.)**

MACHINE AND WORK COORDINATE SYSTEMS

The Cartesian coordinate system that identifies the location of the origin on the workpiece is called the **work coordinate system (WCS)**. The origin of the WCS can be established anywhere on the workpiece for ease of programming. The machine has a coordinate system of its own called the **machine coordinate system (MCS)**. The origin of the machine coordinate system is in a fixed, factory set location and cannot be changed or moved. The MCS is used for the machine's own reference purposes and helps it to keep track of how far each axis can move before it runs out of travel. The distance from the origin of the MCS to the origin of the WCS is called the **work offset**. This distance is measured when the machine is setup and is stored in the control. **(See Figure 8.4.12.)**

Power-Up and Homing

The very first step to operating any CNC machine is powering it up properly. Since there are many different machine variations, refer to a specific machine's manual for the correct procedure. After the machine has been properly turned on, most machines require a *reference return* to the **machine home position**. This process is called the *homing procedure*. Recall that the machine moves its axes by rotating a ball screw with a servo motor. Again, this motor can monitor and adjust axis position by how far its shaft rotates. When the machine power is switched off and the MCU is no longer able to monitor and adjust axis position, its axes simply lose track

of their positions. Therefore, every time a CNC machine is powered up from a total shutdown, it must be homed.

The homing procedure is used to accurately re-enable reference to the MCS. Once the machine knows where each axis is positioned, the machine will know its limits of travel. By homing, the machine is also able to recall the position of the WCS that was active before the machine was powered down. This prevents the **work offset** from having to be reset every time the machine is powered up.

Each machine requires specific steps to perform the homing procedure. These steps can also be found in a specific machine's operating manual, but the basic steps are as follows:

1. Select "zero-return" or "home" mode on the machine control panel.

2. Jog each axis in the direction toward the machine's home position with the jog arrow keys. If the machine axes were already at home position, they must be jogged away from and then back toward home position.

3. As each axis is sent in the appropriate direction, most machines will automatically complete the procedure by rapidly moving the axis and then slowing as a sensor or switch is approached. When the switch is contacted, the encoder of the servo will find the home position of the encoder wheel and precisely reference its position.

4. Once the encoder home position is found, the machine will zero the machine coordinate system. The machine is then ready for set up or to start machining using the previous setup.

> **NOTE:**
>
> Some machines are equipped with *absolute encoders* which do not lose track of axis position when the machine is powered off. These machines require no homing procedure upon start-up.

WORK OFFSET SETTING

The workpiece setup procedure establishes a work offset or origin location for the WCS. All workpiece coordinates for programming will be referenced from this origin. This is established by finding the work offset or "shift" from the MCS origin to the intended WCS origin.

Recall that the MCS never changes position. It is referenced at the same position each time the machine is homed. Consider this point an unchanging reference point. Since the position of this machine origin never changes, but the work origin changes with each new workpiece setup, the work offset for the part is defined as a distance referenced from the MCS origin. Some controls call this offset a *workshift* because it essentially shifts the machine origin to the location of the work origin (or vice versa depending on the machine). **Figure 8.4.13** shows an illustration depicting work offset.

On turning machines, the workpiece will change in length and position from job to job and requires the origin to be reset in the Z-axis for each new job. However, the center line of the workpiece and the center line of the spindle axis are always aligned, so the origin for the X-axis remains the same from job to job.

There are many variations in the way each manufacturer's machine is set up. The principles in the examples described next should apply to all machine controls.

Basic steps for setting a work offset:

• The workpiece is installed in the machine.

• The turret is indexed to a facing tool by using the tool-change command.

• The Z-axis is jogged to bring the facing tool near the end of the workpiece.

• The spindle is started using the proper M3 or M4 code in MDI mode.

• The part is faced by jogging a facing tool across the work using the handwheel until the face is cleaned up (no surface left un-machined).

• The turret is jogged to a safe position and indexed to an empty position.

• A gage block is held against the face of the part.

• The face of the turret is slowly and carefully jogged against the gage block until a slight drag is felt. The length of the block is the distance from the intended part origin to the face of the turret (the reference point).

• This gage block length is subtracted from the current MCS Z-axis position.

• The resulting value represents the distance from the MCS origin to the WCS origin for the Z-axis and is entered into the work offset page in the MCU. **Figure 8.4.14** shows an illustration depicting the part, gage block, and turret being used to set a work offset.

NOTE:

Some turning MCUs use a simpler method, and instead of determining an actual distance, they just require a "set Z" button to be pressed. The MCU automatically performs the calculation and stores the correct value on the work offset page.

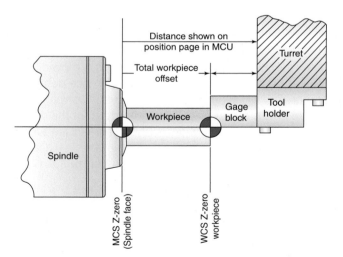

FIGURE 8.4.14 A gage block can be used to determine the workpiece offset by touching a turret reference surface on the desired part Z-zero. The turret's position will be displayed on the machine position page. With the turret touching the spindle face, the position would read Z-zero. With the turret touching the gage block as shown, the position would display the length of the workpiece to its face, plus the gage block length. Subtracting the gage block length from this dimension reveals the workpiece length from the spindle face (work offset).

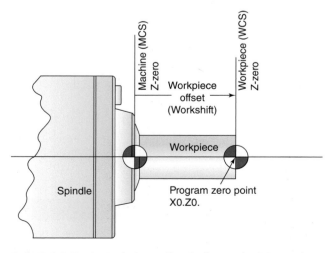

FIGURE 8.4.13 Workpiece offset is the Z-axis distance from the MCS zero to the WCS zero. The X-axis offset should remain zero since the spindle and workpiece share the same center line.

CUTTING TOOLS FOR TURNING

Cutting-Tool Installation

When mounting cutting tools for turning machines, make sure that they face the correct direction according to the spindle rotation. In some machines the tool will be oriented right-side up and in others, upside down. Always be sure that a mounted tool is on center. Some tool-holder adapters have an adjustment mechanism to fine tune the tool's height, while others require the use of shim material for adjustment.

Hole-working tools can often require the most attention during installation and setup since their alignment is very critical. The tool must be parallel to the spindle axis so that the tool's body does not rub on the inside surface of the hole during machining. Some tool-holder adapters have squaring adjusters for this. **(See Figure 8.4.15.)** The squareness can be verified by running a dial indicator lengthwise along the tool in two planes.

Coolant lines should be connected and their nozzles aimed at the cutting zone after each tool is mounted. Special attention should be paid to ensure that coolant nozzles will not interfere with the workpiece and workholding devices as the axes move.

Cutting-Tool Offsets for Turning

When setting a cutting tool for a turning center, the location of the tool tip must be defined in the X and the Z axes. This location is measured as a distance from the turret reference position to the tool tip. Once tool measurements have been determined, those tool offset measurements are stored in the machine's **geometry offset** page. This page also contains an area for defining data for tool nose radius size and quadrant orientation.

As the cutting tools wear, the location of their cutting edges changes. **Wear offsets** may be used to compensate and adjust for wear as the tool is used during production. Be sure that the wear offset value for the offset number being set is returned to a baseline of zero before calculating and entering the geometry value.

Tool Geometry Offsets

In order to determine the initial tool length, the workpiece must first have an established work offset so that the part face is Z-zero. For this step, the workpiece origin becomes the reference point to determine the position of the tool tip.

For the Z-axis, the jogging handwheel is used to bring the tool tip to the workpiece face and touch off the part using a piece of shim material or a feeler gage. This positions the tool tip at a known location relative to the workpiece origin. For example, if the tool is touched off the part face of Z-zero with a 0.010"-thick shim, the 0.010" shim thickness is subtracted from the current absolute Z-axis position to calculate the tool's length offset. Many controls make this process easier by allowing the shim thickness to be entered into the geometry offset page and will then calculate the tool length automatically. **(See Figure 8.4.16.)**

When setting a turning tool for the X-axis, the tool tip is brought to the work diameter and touched off the OD of the part using a piece of shim material or a feeler gage. When an OD tool is touched off with a feeler gage against an outside diameter, the tool tip is at an imaginary diameter of the OD plus two times the feeler gage thickness. For example, if the tool is touched off of an OD of 1.500" with a 0.010"-thick feeler gage, the imaginary diameter the tool is at is the workpiece diameter plus twice the feeler gage: 1.500 + 0.010 + 0.010 = 1.520. **(See Figure 8.4.17.)** Many controls allow this number to be entered into the geometry offset page and will calculate the tool offset amount automatically.

When setting a geometry offset for a hole-working tool, the tool may be aligned on center most accurately by "sweeping" around the circumference of the tool with an indicator mounted in the machine spindle. **Figure 8.4.18** shows an image of a hole-working tool being swept with a spindle-mounted indicator.

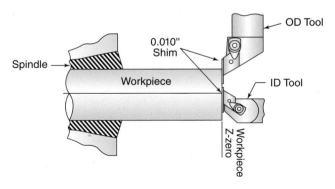

FIGURE 8.4.16 Method for setting tool Z offsets off the part face of Z-zero with a 0.010"-thick feeler gage. While in this position the tool's position can be set as positive 0.010" in the Z-axis.

FIGURE 8.4.15 Adjustment screws for squaring a holemaking tool holder.

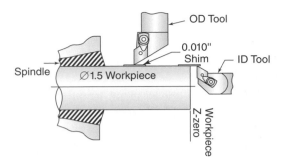

FIGURE 8.4.17 An X tool offset is being set by touching off of an OD of 1.500" with a 0.010"-thick feeler gage. The imaginary diametral position the tool is at is the workpiece diameter plus twice the feeler gage: 1.500 + 0.010 + 0.010 = 1.520. If an ID boring, threading, or grooving tool is to be set, a rigid shim is used as a feeler to find the OD surface of the workpiece as shown (the shim is not used in the calculation).

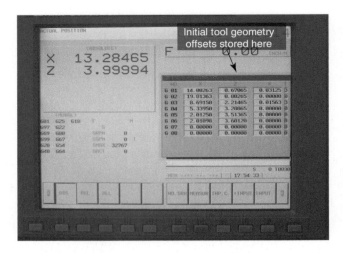

FIGURE 8.4.19 A typical geometry offset page on the machine display screen. These numbers reflect the true setting for the tool tip location in its original and unworn state.

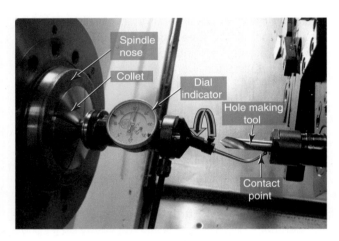

FIGURE 8.4.18 A holeworking tool being swept with a spindle-mounted indicator to find alignment with the spindle axis.

Once the hole-working tool is in alignment, the tool is at X-zero and there is nothing additional to compensate for. The position is entered as the X geometry offset value. The Z-axis offset for hole-working tools is set and adjusted the same way as with turning and boring tools.

The initial settings of the tool offsets are stored in the geometry offset page. **Figure 8.4.19** shows an image of a typical geometry offset page on the machine display screen. These numbers reflect the true setting for the tool tip location in its original and unworn state. After the tool has been set, a part is usually produced by running the program for the first time. The part is then immediately inspected and adjustments are made to the geometry offsets as needed to achieve desired sizes. Here is a hypothetical example of how

these adjustments may be made for a turning tool (OD working):

1. The first part is produced and inspected.
2. The measurement reveals that each diameter (which is created by the X-axis) for a given tool measures 0.0008" larger than desired size.
3. The geometry offset page is opened and the current X-axis geometry offset for the tool is 8.7899".
4. Next, 0.0008" is subtracted from the total X-axis tool offset value for that tool and it is determined that 8.7891" is the correct geometry offset.
5. The new value is entered for that offset.
6. The next part is made and the correction is verified.

Tool Wear Offsets

Wear offsets are used when a size adjustment needs to be made after a machine has been satisfactorily producing parts for a period of time. Size change can be caused by tool tip wear or machine thermal variations (expansion or contraction from shop and machine iron temperatures). If the size begins to change during production for one of these reasons, the tool offset adjustments should not be entered in the geometry page, but instead the wear offset page (which usually looks very much like the geometry offset page). **Figure 8.4.20** shows an image of a typical wear offset page on the machine display screen. The number that will be entered into the wear offset will be an incremental adjustment from a baseline of zero wear. Adjustments can be made to either the X or Z offset, or both. Here is a hypothetical example of how these

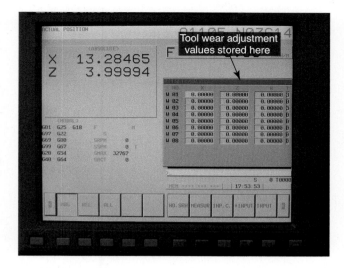

FIGURE 8.4.20 A typical wear offset page on the machine display screen. The number that will be entered into the wear offset will be an incremental adjustment from a baseline of zero wear.

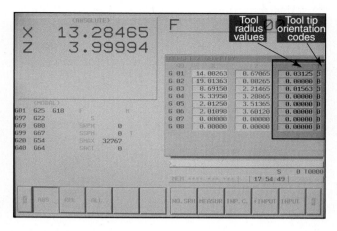

FIGURE 8.4.21 A geometry offset page on a machine display screen with labels for tool radius and quadrant.

adjustments may be made if the X offset for a turning tool needs adjustment:

1. The unsatisfactory part in question is inspected.
2. The measurement reveals that each diameter (which is created by the X-axis) for a given tool measures 0.0009" too large (above the nominal target size).
3. The wear offset page is opened and the offset value is found to still be at the initial baseline value of zero. To correct this, 0.0009" must be subtracted from the X-axis offset for that tool. Since the wear offset is still at its initial baseline of zero, −0.0009 is entered.
4. The next part is made and the correction is verified.

Remember that during the initial geometry setting of a given tool offset, the corresponding wear offset for that offset number must be set to zero. This allows the offsets to start at a baseline of zero prior to making any adjustments. Also, for inserted tooling, once the insert is replaced it allows the ability to easily return to the initial tool setting by simply entering a value of zero in the wear offset for the tool.

Tool Radius and Orientation Entry

When initially setting up the machine and entering tool geometry offsets, the tool's nose radius must be entered in the geometry offset page in the appropriate location (usually denoted with an "R"). The importance of this value is to allow the machine to appropriately compensate for the tool nose radius when using TNRC. The value entered here is ignored unless TNRC is active. **Figure 8.4.21** shows an image of a geometry offset

page on a machine display screen with labels for tool radius and orientation.

When using TNRC, the tool tip **quadrant** orientation must also be entered in the geometry offset page. In order for TNRC to be successful, the control must know the cutting zone on the tool nose where compensation must occur. The available quadrants are shown in **Figure 8.4.22**.

PROGRAM ENTRY FOR TURNING

Programs can be entered into the MCU in one of three ways:

1. Manually typing the program into the control on the shop floor
2. Uploading the program to memory from a PC or removable storage device
3. Directly sending the program to the control from a PC as the program is running

When manually entering a program from the shop floor, the control must be placed in edit mode and given a program number. The program is then keyed in, word by word and block by block, until complete. This method is usually time consuming and errors can be easily made, so it is generally used for short programs.

File upload to memory can be accomplished by connecting a PC communication port to a port on the MCU with a communication cable. Some machines can also read programs from a removable storage device such as a CD, USB drive, or memory card. This method is the most common of the three options because it is extremely fast, errors are rare, and the program is stored in the machine memory for use at any time.

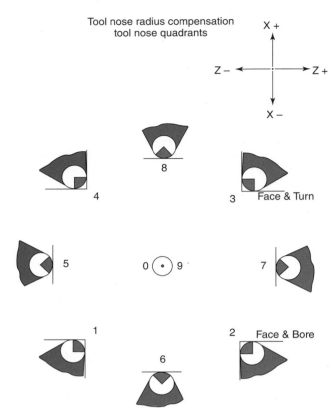

FIGURE 8.4.22 In order for TNRC to be successful, the control must know the cutting zone (shaded in blue) on the tool nose where compensation must occur. The available quadrants are shown relative to the tool tips on various styles of cutting tools.

Sometimes complex programs are so huge that they simply cannot be stored in the machine control memory in their entirety. In these cases, instead of actually storing a program in the MCU memory, the program is fed from a PC to the control line by line as the machine runs the program. The machine control only accepts as much code as it can process at a time. This method is known as **direct numerical control (DNC)** and is sometimes called *drip feeding*. Depending on the machine, this can be accomplished by different methods. The most common is through a direct connection to a PC with a communication cable, but some machines can receive DNC from a CD, memory card, or USB storage device.

TURNING MACHINE OPERATION

Program Prove-Out

Running a program on a newly setup machine is an exciting time and also a time of great caution. Recklessness at this stage could result in damage to the

machine, tooling, or workpiece. However, if caution is exercised and a careful **prove-out** has been performed prior to letting the machine run unattended, almost all mistakes can be safely identified.

There are several methods for carefully executing a program that will help to identify problems before letting the machine run without supervision. They are:

1. Graphic simulation

2. Dry run

3. Dry cycle

4. Safe offset

Any one or a combination of all of these techniques may be utilized to prove out a program and setup to ensure safety prior to production. Graphic simulation allows verification of the paths the tool will take by watching a simulated computer model cut the part on a display screen. This can be done at a PC with simulation software prior to loading the program in the machine, or on the MCU display on machines with graphic simulation capability. **Figure 8.4.23** shows graphic part simulation on a display screen.

Graphic simulation is a quick way to troubleshoot obvious programming issues but doesn't analyze an actual machine setup or small positioning errors. A **dry run** is a more valid visual prove-out and is usually done in the machine after the tools and workpiece have been set. Generically speaking, "dry running" refers to running a machine with disabled functions to eliminate the potential for collisions. This can be achieved by removing tooling, removing the workpiece, cutting a wax

On-screen graphic simulation of programmed tool paths

FIGURE 8.4.23 A graphic part simulation on a display screen.

representation of the workpiece, disabling certain axis motions, or disabling spindle function. Some machines are equipped with a "Dry Run" mode. This mode will usually ignore the programmed feed rates and make all movements at a faster (or slower, if you chose) rate. Dry running is also typically done without coolants to help with visibility and to keep the work area clean.

A **dry cycle** is like a dry run but will run the program at the actual programmed feed rates. This can be helpful in proving that there are no G0/G1 code errors or feed rate errors. With a dry cycle, the cutter or workpiece may be removed so that no collisions can occur. A dry cycle is done using the same machine "Auto" or "Memory" mode that will be used to make the final part.

The safe offset method is much like the dry-run procedure because the machine will be physically executing the program but not actually cutting the workpiece. The difference is that all machine functions are enabled (except perhaps coolant) and the tools and workpiece are installed. Safety is ensured by intentionally setting the work origin in either the X- or Z-axis at a safe distance from the workpiece. With this method, the program can be proven out and upon successful completion, the offsets can be gradually moved closer to the part. This method can be repeated until the program and setup are deemed safe.

No matter which technique is used for prove-out, caution is the key. There are two other lines of defense to prevent collisions by making machine motion more manageable and surprise movements preventable. These controls are the override controls and **single-block mode**. Overrides provide the ability to slow or even halt the programmed feed rates and rapids. Most machine control panels are equipped with variable override knobs for this.

Single-block mode allows the ability to execute only one block of the program at a time. In this mode, the machine will not advance to the next block until the cycle-start button is pressed again. This allows lines of the program to be viewed on the MCU screen and verified prior to them being executed. The single-block mode is normally activated by a switch or button on the machine control panel.

Auto Mode

After the program has been carefully proven and there appears to be no potential collisions, the turning machine may be run at full feed, speed, and rapid capabilities. Once satisfied with the production performance, the machine is ready to be run in automatic mode.

If the job to be run is being machined from a bar of material and the machine is equipped with a bar puller or feeder, there usually is a provision in the control for setting a part counter and a maximum number of parts to be run. The part counter is a simple counter that increases the number of recorded parts produced every time the machine reads the M30 (end of program) code. Once the maximum part number has been satisfied, the machine will stop, indicating the need for operator attention to load new bars of material.

SUMMARY

- Most control panels contain knobs, button switches, softkeys, and a keypad. The control panel of a turning machine is the command center for programming, setting up, and operating the machine.

- The machine mode knob allows the operator to select modes of CNC machine operation, such as jog, auto, MDI, edit, and reference return.

- Reference returning, or homing, is required on most machines upon power-up in order for the machine to recall its positioning.

- Offsets are distances used to "tell" the machine the location of the work origin and each tool tip.

- Work offsets allow the machine to establish a workpiece zero or origin at a location on the part for convenient programming.

- Tool offsets allow the machine to know the location of the tool's tip. When these offsets are active, the programmed position will be taken from the tool's tip. Tool offsets for initial setups for unworn tools are entered into the offset geometry register. Tool offsets for tools that have been gradually and consistently losing their ability to cut the programmed dimensions have their offsets adjusted in the offset wear register.

(Continued)

- Tool nose radius values and tool nose quadrant positions must be entered into the offset page if TNRC is to be used.
- Programs may be entered into a machine by manually typing the program in on the shop floor, file upload to memory from a PC or storage device, or through DNC from a PC or other storage device connected to the machine.
- Programs may be proven by using one of the following methods: graphic simulation, dry run, or safe offset.

REVIEW QUESTIONS

1. What machine mode is generally used to manually move the axes during setup?
2. MDI stands for _____ _____ _____ .
3. Which must be set first, the tool geometry offset setting or the work offset setting?
4. What is the process called where a program is sent to a machine one block at a time as the machine executes the program?
5. Explain the purpose of homing.
6. What is the process called when a new program is executed to ensure that there will be no collisions or errors made?
7. When setting up a machine to run a program that will use TNRC, what settings must be entered in the offset pages in order for the TNRC to work properly?
8. Explain two reasons why an offset may need to be entered into the wear offset page.
9. What is used to adjust the clamping pressure of the workholding device?
10. What does MCS stand for?
11. What does WCS stand for?
12. A workpiece offset is the distance from _____ to _____ .

Introduction to CNC Milling UNIT 5

Learning Objectives

After completing this unit, the student should be able to:

- Identify and describe different types of CNC milling machines
- Identify and describe machine axes used for milling
- Identify and describe the two major types of ATCs
- Demonstrate understanding of the uses of workholding devices for CNC milling
- Demonstrate understanding of the uses of toolholding devices used for CNC milling

Key Terms

BT flange	Horizontal Machining	Tombstone
CAT flange (V flange)	Center (HMC)	Vertical Machining
Fixture	Pallet system	Center (VMC)
	Retention knob	

INTRODUCTION

Operations performed on manual milling machines primarily produce flat surfaces, straight slots and steps, and accurately locate hole positions. These types of part features can also be produced on CNC machines, but with much greater speed and accuracy. Because the X-, Y-, and Z-axes can be programmed to move simultaneously, CNC machines can also create an endless variety of arcs, contours, and three-dimensional surfaces. **(See Figure 8.5.1.)**

When CNC controls were first adapted to milling they were commonly fitted to a standard knee mill, which was originally designed for low-speed/low-feed manual milling. At that time, these CNC milling machines enabled some very revolutionary types of machining, including contouring, arc cutting, pocketing, and repetitive holemaking operations. As more complex machining was demanded, the basic knee mill design could not meet the need.

Recall that a machining center is a CNC mill with an ATC. Since machining centers are intended for heavy-duty production and aggressive material removal rates, and that they typically will not be used manually, there are some distinctive features of their design that make them well suited for these applications. Specially designed beds, columns, and guideways provide optimum rigidity, accuracy, smoothness, and wear resistance. **Figure 8.5.2** shows a CNC milling machine and a CNC machining center.

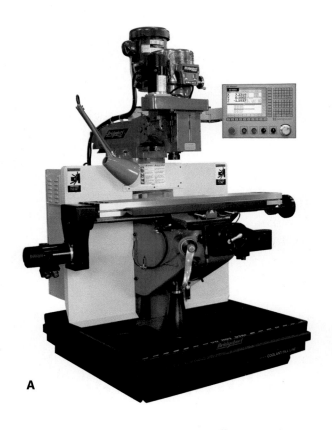

A

Courtesy of Bridgeport Milling

B

FIGURE 8.5.2 (A) A vertical CNC milling machine. (B) A vertical machining center.

Photo courtesy of Haas Automation, Inc.

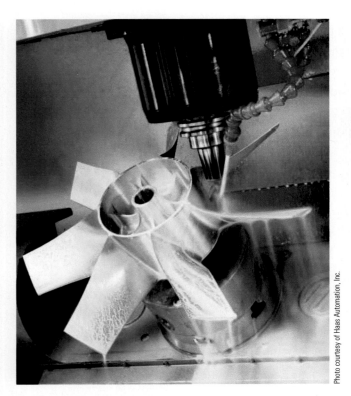

Photo courtesy of Haas Automation, Inc.

FIGURE 8.5.1 CNC milling machines can produce complex part surfaces that would be virtually impossible to produce with manual milling machines.

TYPES OF CNC MILLING MACHINES

Machining centers are separated into two major classes: vertical spindle and horizontal spindle. The configuration of a modern **vertical machining center (VMC)** is shown in **Figure 8.5.3**. The configuration of the axes can be compared to a standard vertical knee-type mill.

Horizontal machining centers (HMCs) have been becoming very popular over the past several years. Their popularity is due partly to the versatility of workholding, the

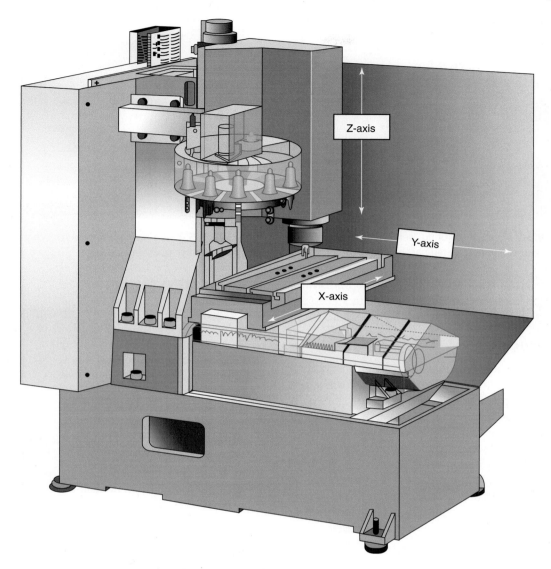

FIGURE 8.5.3 A vertical machining center (VMC). Notice the similarities to a manual vertical milling machine.

inherent rigidity of the machine's column, and the ability to allow gravity to help remove chips out of the machining area. The axes of an HMC are oriented as if a vertical spindle machine was laid on its back. **Figure 8.5.4** illustrates the basic construction of an HMC. Notice that in either machine, a drill rotating in the spindle would be moved in the Z-axis to produce a hole in the workpiece. This can be helpful to remember the placement of the axes.

Machining center construction uses low-friction guideways for sliding machine surfaces. The use of these greatly minimizes wear, reduces friction (which enables ultra-high rapid movement), and allows for very high accuracy due to a zero-clearance preloaded ball bearing design.

Modern machining centers have been evolving and can be found in many configurations to enhance productivity. For maximum productivity and minimal operator attention, machining centers can be incorporated into a manufacturing

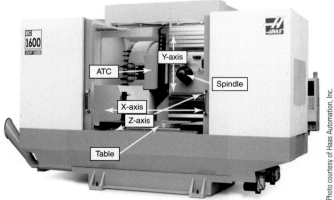

Photo courtesy of Haas Automation, Inc.

FIGURE 8.5.4 A horizontal machining center (HMC). Notice the orientation of the spindle, table, and ATC.

FIGURE 8.5.5 A manufacturing cell with multiple CNC machining centers and automated part handling.

cell, which combines several machines performing operations on the same part with the aid of automated part loading and unloading systems. **(See Figure 8.5.5.)**

ATC Types

There are two basic types of automatic tool changers for machining centers. The *carousel-type tool changer* stores the tools in a large circular disc. During a tool change, an empty tool compartment in the carousel moves toward the spindle and grips the tool. The tool is then unclamped and the Z-axis raises to remove the tool from the spindle. Then the carousel rotates to bring the desired tool in alignment with the spindle, and the Z-axis is lowered, inserting the tool into the spindle. Finally, the carousel retracts to its original position. **Figure 8.5.6** shows the carousel-type tool changer.

The *swing-arm-type tool changer* uses a double-ended arm to change tools. Tools are stored in a tool storage magazine. One end of the swing arm grips the tool in the machine spindle at the same time the other end grips another tool in the tool storage magazine. The spindle-mounted tool is removed when the arm is lowered. The arm then rotates, and the new tool is mounted in the spindle as the old tool is stored in the tool magazine. The swing arm tool changer is much faster than the carousel changer because it does not need to index the magazine to an empty compartment for the removed tool, then index again to retrieve the new tool. **Figure 8.5.7** shows a swing-arm-type tool changer.

FIGURE 8.5.6 A carousel-type ATC.

FIGURE 8.5.7 The swing-arm-type ATC changes tools much faster than the carousel-type ATC.

TOOLHOLDING

A machining center requires a toolholder for the machine's spindle to hold the cutting tool. There are two major characteristics when determining the appropriate toolholder for an application: the spindle mounting type and holder-to-tool attachment style.

CNC Spindle Types

Most machining center spindles use the National Machine Tool Builder (NMTB) series taper. NMTB taper sizes are classified from smallest to largest as 30, 35, 40, 45, 50, and 60. **(See Figure 8.5.8.)** All sizes have a taper of 3-1/2" per foot. Since this steep taper is not self-holding, the small end of these tapered holders have internal threads to accept a **retention knob**. Figure 8.5.9 shows some of the different styles of retention knobs used in various machines. A ball gripper mechanism at the end of

FIGURE 8.5.8 National Machine Tool Builder (NMTB) toolholders in 30-, 35-, 40-, 45-, and 50-size tapers.

FIGURE 8.5.9 Retention knobs thread into tapered toolholders to secure the holder in the machine spindle.

the drawbar is used to grasp the retention knob. Spring tension from a series of disc springs then pull on the drawbar to secure the mating tapers. The many parts used to retain the toolholder are shown in **Figure 8.5.10**.

Machining center toolholders are also classified by flange type. The ATC grips the flange of the toolholder during the tool-change cycle. The two most common flange types are the **CAT flange** (also called the **V flange**) and **BT flange**. The CAT flange was originally developed specifically for CNC use by Caterpillar Inc. The BT and CAT flanges are similar, but have some subtle differences. The BT flange holders have a thicker flange with an off-center groove, while the CAT-type holders place the groove in the center of the flange. The retention knobs on the BT holders have metric threads, while the CAT holders have inch-series threads. **Figure 8.5.11** shows a CAT-type flange holder and a BT-type flange holder.

The combination of the taper size and the CAT or BT flange are both needed to identify a toolholder specification. For example, a CAT-40 holder uses the CAT flange and a size 40 taper, and a BT-50 uses a BT flange and a size 50 taper. The proper retention knob for any given machine must also then be selected and attached to the toolholder.

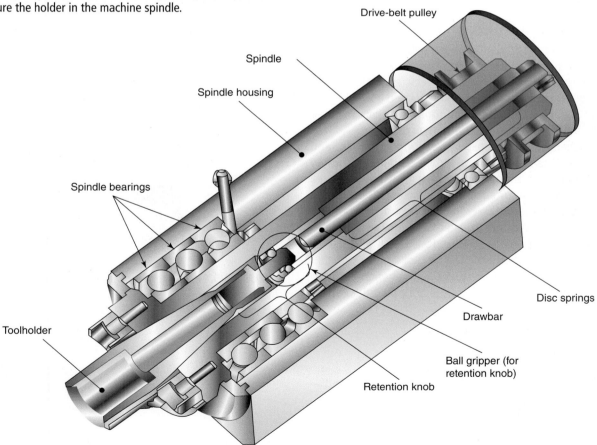

FIGURE 8.5.10 A cutaway view showing the many parts of a milling spindle. Notice how the ball gripper mechanism at the end of the drawbar grasps the retention knob. Spring tension from a series of disc springs pulls on the drawbar to secure the toolholder taper.

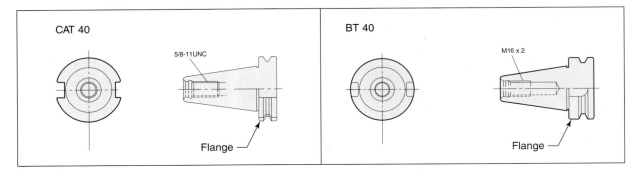

FIGURE 8.5.11 A CAT holder and a BT holder. Notice the difference in the size of the flanges and the location of the grooves.

Some smaller CNC milling machines do not use either CAT or BT flange holders, but instead use a quick change NMTB taper or the traditional R-8 spindle taper to mount their tooling. This is most common on vertical-spindle knee-type milling machines, and bench-top models.

Tool Attachment Styles

There are several types of holders available for mounting different types of cutting tools for CNC milling. Some are very much like those used for manual milling machines and others are designed specifically for CNC machines.

CNC Endmill Holders

CNC endmill holders contain a bore to receive a specific tool shank diameter just like those used for manual mills. They also have a tapped hole with a setscrew at 90 degrees to the bore. This setscrew is tightened onto a Weldon flat on the shank of the tool. Only tools with flats on their shanks should be mounted in this type of holder.

Endmill holders are inexpensive, rugged, simple, and capable of transmitting a lot of torque, but there are a few disadvantages of endmill holders. First, a different-sized holder is necessary for each shank diameter. Balance can be affected by the setscrew and produce tool vibration, especially at high spindle speeds. Coolant can seep into the clearance area between the tool shank and holder bore, causing corrosion and making tool removal difficult. Because of the clearance required for the endmill to slip into the bore, when the setscrew is tightened the cutting tool is forced to one side of the bore, causing runout. Runout causes the tool to cut slightly oversize and creates uneven tool wear that can lead to decreased tool life.

CNC Collet Chuck Toolholders

CNC *collet chucks* are very versatile and are often used for holding many different types of straight-shank milling tools. A tapered bore in the chuck matches a taper on the collet. When the collet is forced against the taper by tightening a threaded cap, the collet constricts around the shank of the cutting tool. **(See Figure 8.5.12.)**

The most common collet styles are the ER, TG, and DA types. **(See Figure 8.5.13.)** An ER-type collet has a size range of about a full millimeter (about 0.040"), while individual TG and DA types have a range of about 0.015". The ER- and TG-type collets snap into the collet chuck cap, and then the cap is threaded onto the collet chuck.

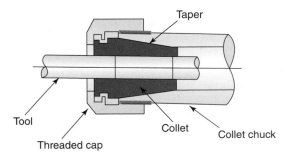

FIGURE 8.5.12 A collet chuck grips a tool shank by tightening the threaded cap.

FIGURE 8.5.13 (A) ER-, (B) TG-, and (C) DA-style collets.

FIGURE 8.5.14 (A) The groove on an ER collet snaps into a retaining ring in the threaded cap and the cap is then threaded into the collet chuck. (B) DA collets slip into the chuck and then the cap is threaded onto the chuck. (C) The toolholder is then mounted in a bench-top clamping device, the tool is inserted in the collet, and the cap is tightened.

DA collets simply slip into the collet chuck and the cap is then threaded onto the chuck. **(See Figure 8.5.14.)**

An advantage of collet chucks is that they run truer than endmill holders because there is no setscrew to unbalance the holder or force the cutting tool off center. This minimizes tool vibration and chatter at higher spindle speeds and increases tool life. Collet chucks can be used for holding nearly any type of straight-shank tool, including endmills, drills, and reamers.

Shrink-Fit Toolholders

A *shrink-fit toolholder* is designed and machined so that there is an interference fit between its bore and the tool's shank. In order to insert and secure a tool into this type of holder, the holder's nose must be heated, causing the bore to expand. The heating is usually done by induction, using an electric current to heat the holder. When the nose expands sufficiently, the tool's shank is inserted into the bore and then allowed to cool. The holder shrinks, contracts, and constricts around the tool's shank, locking it into place as it cools. There is no other mechanical means of tool fastening on these holders other than the interference between the holder and tool. These holders, like endmill holders, require a different-sized holder for each tool shank size used. However, shrink-fit holders have excellent runout accuracy

due to their simplicity and their ability to constrict evenly around the tool with zero clearance. The totally symmetrical design with no moving parts also achieves excellent balance, a short length, and superior rigidity, making them ideal when using very high spindle speeds. **Figure 8.5.15** shows a shrink-fit toolholder and the machine used to assemble and disassemble shrink-fit tooling.

CNC Drill Chuck Toolholders

CNC drill chuck toolholders are three-jaw drill chucks with adapters suited to fit the machine's spindle taper. These have the greatest "one-size-fits-all" size range and do not require an expensive set of collets to use. However, these are intended to hold low-torque, straight-shank tools like smaller drills, reamers, and edge finders. It is not a good practice to use drill chucks to hold other tools such as endmills or taps because of the relatively high amount of runout and the minimal gripping force. **Figure 8.5.16** shows some drill chucks for use in CNC machines.

CNC Shell-Mill and Face-Mill Toolholders

CNC shell-mill and face-mill holders are simple arbors used for mounting shell mills and face mills. The arbor consists of a round pilot diameter for accurately locating the center of

FIGURE 8.5.15 (A) A shrink-fit toolholder. (B) The machine that assembles and disassembles shrink-fit tooling.

the mounted tool and two opposing drive keys to prevent rotational slippage on the arbor. Cutters slip over the pilot and the drive keys with no force and the tool is captured and fastened by a bolt or cap screw. **Figure 8.5.17** shows a face-mill toolholder for use in a CNC machining center.

CNC Tapping Toolholders

There are several variations of CNC toolholders used for mounting taps. Some hold taps rigidly in the holder. Others hold taps in specialized spring-loaded devices that allow extension and compression. These are often called floating holders. The machine's capability for tapping (discussed in Section 8, Unit 6) determines which type of holder should be used.

Taps can be held rigidly with collet chucks using ER tap collets manufactured with an internal square that prevents the tap from spinning inside the collet. **(See Figure 8.5.18.)** Another method of tap holding uses a spring-loaded quick-change adapter to hold the tap. The adapter can then be mounted in a specially designed floating or rigid tapping chuck. **(See Figure 8.5.19.)**

FIGURE 8.5.16 (A) A key-type chuck with a quick change size 30 NMTB taper. (B) A keyless chuck with a CAT-type flange.

FIGURE 8.5.17 A face mill and toolholder.

Workholding

CNC machinery and tooling are extremely expensive investments, so it is understandable that the more time a machine spends actually machining the product, the more profitable the investment is. Workholding devices and techniques for machining centers are continually evolving and improving so that less time is spent loading workpieces and more time is spent machining. Workholding devices for milling can range from simple

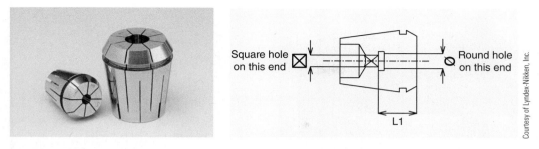

FIGURE 8.5.18 An ER collet with an internal square for holding taps rigidly.

FIGURE 8.5.19 (A) A quick-change tap adapter designed to quickly mount taps into a (B) floating holder.

clamps and machine vises to elaborate and expensive pallet systems, tombstones, and custom fixtures.

Clamps

Clamps are a universal method for attaching workpieces to a machine's table. The same styles used for manual milling can be used for CNC milling as well. One type of clamping system is the step-block style that allows a stud or bolt to be anchored to the machine table T-slot and a strap to be drawn down onto a part's surface to secure it to the table. Toe clamps allow a part to be held on its edges by using special serrated jaws to grip into the material and pull the work down tight against the table. Another variation is toggle-type clamps, which have a convenient quick-release lever that uses a cam action mechanism to provide clamping pressure. **Figure 8.5.20** shows examples of these clamp types.

Clamps are ideal for holding extremely large or oddly shaped workpieces, or production work where a vise is not suitable and the volume of work does not justify the expense of custom-built fixtures. Unfortunately, holding work with clamps provides no accurate method for repeated locating from part to part. Each newly loaded workpiece may need to be aligned using a dial indicator, then a reference location set using an edge finder. Clamps also require special programming and operator attention since they usually extend above the top of the workpiece surface, making the potential for collisions very high. **Figure 8.5.21** shows a large workpiece secured to the table of a machining center with step clamps.

Machine Vises

Machine vises are very commonly used workholding devices for milling because they are highly versatile, accurate, and simple to use. It is important when considering a vise for workholding that the workpiece is thick enough to resist bowing under the clamping

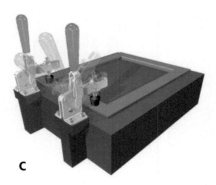

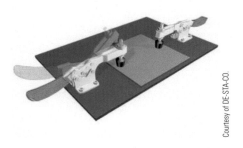

Courtesy of DE-STA-CO.

FIGURE 8.5.20 (A) Step clamps. (B) Toe clamps. (C) Toggle clamps. All of these clamps can be used to hold work for CNC milling.

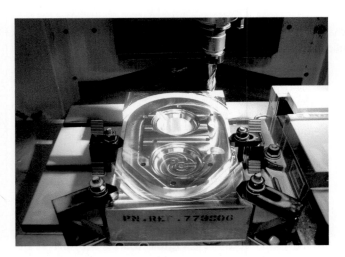

FIGURE 8.5.21 A large workpiece clamped to the table of a VMC using step clamps.

pressure. Standard vises used for manual milling can be mounted in machining centers for CNC milling. There are also vises with two moveable jaws that clamp against a central solid jaw to allow multiple-part clamping, as shown in **Figure 8.5.22**. Many models are available with machinable jaws that can be machined to accept nearly

any part shape. **(See Figure 8.5.23.)** Some also use a quick-change jaw system that allows different jaws to be changed in a matter of seconds.

Chucks/Collet Closers/Indexing Fixtures

Manual jaw-type chuck fixtures and collet fixtures can be used on CNC mills to hold and locate cylindrical parts. Programmable indexing fixtures using chucks, collets, or flat surfaces with t-slots for clamping can be connected to the machine's MCU (if so equipped) and used to rotate the workpiece during CNC machining, creating a fourth movement called a *rotary axis*. Many models can be mounted either vertically or horizontally. This rotary axis is similar to rotary tables used on manual mills. **Figure 8.5.24** shows manual collet fixtures and a CNC rotary axis.

Pallet Systems

To maximize machining time and minimize workpiece-loading time, some machines use a **pallet system**. This system uses two or more workholding tooling plates that can be quickly and accurately interchanged on the

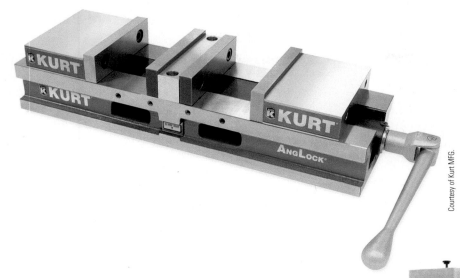

FIGURE 8.5.22 A vise with two moveable jaws can hold two parts in one setup.

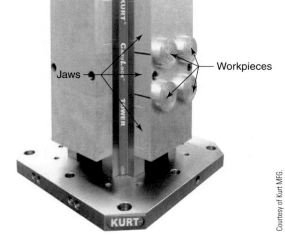

FIGURE 8.5.23 This multi-sided vertical double vise has machinable aluminum jaws, or soft jaws, that can be machined to match the shape of the workpiece. Soft jaws can also be made of soft steel or cast iron.

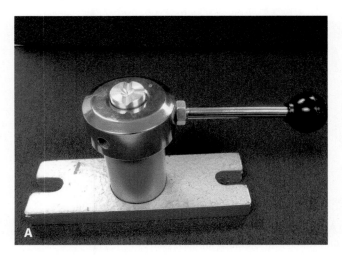

FIGURE 8.5.24 (A) Manual collet fixtures and (B) a programmable CNC rotary axis.

machine's table. Vises or other workholding devices may be mounted on these plates, and at any given time, one of these plates can be in use in the machine while another is outside of the machine having parts unloaded and loaded. When the machining cycle is completed, the workholding tooling plate can be quickly swapped for another one loaded with parts that are ready for machining. For even higher productivity, some machines use an *automated pallet changer (APC)* that automatically changes the pallets through programmed commands. **Figure 8.5.25** shows a machining center outfitted with an APC.

Tombstones

Most commonly used in horizontal spindle machines, a **tombstone** is a tower with multiple vertical working surfaces where workholding devices are mounted. Tombstones are also sometimes called *towers* or *columns*. The concept of a tombstone is to maximize the amount of workpieces that can be mounted in a machine at a time. Tombstones often have two or four working surfaces, but some have more. Each of these sides may contain one or more workholding devices. **(See Figure 8.5.26.)**

Parts are loaded in the workholding devices and then the tombstone is indexed from side to side to machine all of the mounted workpieces. To further increase efficiency and limit downtime due to part loading/unloading, tombstones are often used together with a pallet system.

Vacuum Plates, Magnetic Workholding, and Adhesive-Based Workholding

Sometimes workpieces are difficult to hold using any of the methods just described, so alternate methods just like those covered in Section 6, Unit 2 can be used. A vacuum fixture is one method that can be used for light machining of thin, flexible workpieces. A magnetic chuck can be used to secure ferromagnetic workpieces for machining, and also works well for holding thin, flexible parts flat. Both vacuum fixtures and magnetic chucks allow unobstructed machining of the entire top surface of a workpiece. Adhesives such as double-sided tape can secure parts to the machine table or tooling plate for light machining as well.

Custom Fixtures

When a part has a unique shape and needs precise location along with a highly secure clamping system, a custom **fixture** can be used. A fixture is a custom manufactured workholding device specifically designed to accommodate a specific part. This method can be very useful for securing parts with unusual shapes and allows consistent referencing from critical dimensional surfaces. Fixtures are usually expensive due to their totally custom nature, which requires much planning, design, custom machining, and many materials. Custom fixtures are usually clamped directly to a machine table, pallet, or tombstone. **Figure 8.5.27** shows a custom fixture.

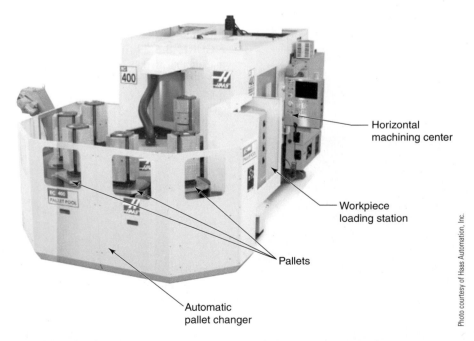

Horizontal machining center

Workpiece loading station

Pallets

Automatic pallet changer

Photo courtesy of Haas Automation, Inc.

FIGURE 8.5.25 This machining center has an APC that uses six pallets. The machine can be programmed to load any of the pallets automatically. Work is loaded on each pallet at the station on the right side of the APC.

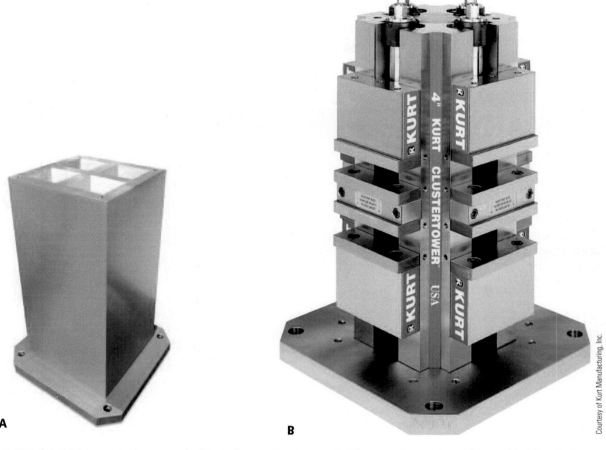

A   B

Courtesy of Kurt Manufacturing, Inc.

FIGURE 8.5.26 The tombstone tower in (A) can be machined as needed for mounting work or other workholding devices. The tombstone tower in (B) has four integrated vises, one on each of the four sides of the tower.

Photo courtesy of Haas Automation, Inc.

FIGURE 8.5.27 A custom fixture designed and built to hold an oddly shaped pump housing.

PROCESS PLANNING

Typical CNC milling operations include face and peripheral milling, slotting, pocketing, two-dimensional contouring, three-dimensional surfacing, and holemaking. Each of these individual tasks is referred to as an *operation*. The combination of all operations required to machine a part is called a manufacturing *process*.

Prior to programming or setting up tools and the machine, the engineering drawing must be closely examined and the production of the part must be planned from start to finish. The planning of workholding devices, tooling, and the order of machining operations depend on the part's features, tolerances, and required surface finishes specified by the drawing. Once a thorough strategy has been determined to produce a part, the steps are then detailed on a document called a *process plan*, as described in Section 2, Unit 5. This plan will include a description of each operation, the tools required, speed and feed data, workholding information, other notes and comments, and often a sketch depicting the part orientation. This document is important not only for the initial programming of the part, but also as a reference for any setup person or operator who will machine the part in the future.

SUMMARY

- CNC controls have been adapted to milling machines to increase productivity and enable the creation of sophisticated part features.
- A CNC mill with an ATC is called a machining center, and there are two basic types of ATCs: the carousel type and the swing arm type.
- Toolholders for machining centers are usually either CAT-flange or BT-flange styles. Both of these styles use the NMTB taper.
- The major types of toolholders for CNC milling are endmill holders, collet chucks, shrink-fit toolholders, shell- and face-mill holders, drill chucks, and tap holders.
- Shrink-fit holders release a tool when heated and secure a tool by contracting when cooled.
- Endmill holders secure the tool by tightening a set screw on the shank.
- Drill chucks should be used to hold straight shank drills, reamers, and edge finders and should never be used to hold endmills.
- CNC milling work may be held with clamps, vises, pallets, tombstones, chucks, indexing fixtures, vacuum plates, adhesives, magnetics, or custom fixtures.
- Pallets are interchangeable platforms that can be removed from the machine and loaded with new parts while another pallet is in the machine.
- Tombstones can hold multiple parts at a time, which cuts down on how frequently part changes must occur.
- When conventional workholding techniques will not work, adhesives, magnetics, and custom fixture solutions are available. Fixtures are custom made and accommodate a specific workpiece shape.
- Major CNC milling operations include face and peripheral milling, slotting, pocketing, holemaking, two-dimensional contouring, and three-dimensional surfacing. The combination of these operations used to machine a part is known as a process.
- A process plan is a document that includes a description of the tools, speed and feed data, workholding information, and any other information needed to perform all of the operations that make up the machining process.

REVIEW QUESTIONS

1. Explain the difference between a machining center and a CNC mill.
2. Name the two major types of ATCs and briefly describe the difference between the two.
3. What is a VMC and an HMC?
4. What are the three most common styles of collets used with CNC collet chucks?
5. What are the two basic types of tapping toolholders?
6. What special consideration needs to be considered when using step clamps for workholding?

7. A programmable indexing fixture creates a fourth axis called a _____ _____ that is similar to a rotary table used on a manual milling machine.

8. A _____ _____ uses interchangeable tooling plates that can be quickly and accurately located on a CNC mill.

9. Briefly describe a tombstone used for CNC workholding.

10. A custom _____ can be designed and built to hold an oddly shaped part or a specific part when high volumes are to be machined.

11. The combination of the machining operations required to produce a part is called a manufacturing _____.

Learning Objectives

After completing this unit, the student should be able to:

- Identify and define basic G- and M-codes used for CNC milling
- Demonstrate understanding of linear interpolation for CNC milling
- Demonstrate understanding of circular interpolation for CNC milling
- Demonstrate understanding of the arc center method for circular interpolation
- Demonstrate understanding of the radius method for circular interpolation
- Demonstrate understanding of facing operations for CNC milling
- Demonstrate understanding of two-dimensional CNC milling
- Demonstrate understanding of drilling and tapping canned cycles for milling
- Demonstrate understanding of cutter radius compensation for milling

Key Terms

Clearance plane	Dwell	Single-pass drilling
Cutter radius	Fast-peck drilling	cycle
compensation	cycle	Tool change
Deep-hole drilling	Program stop	Tool height offset or
cycle	Rigid tapping	Tool length offset

INTRODUCTION

After the entire machining process has been planned, and workholding, tooling, and tool-holding devices have been chosen, it is time to begin writing the CNC program. For simple two-dimensional milling, it is common to write the programs longhand. Two-dimensional milling is when the tool is set at a Z-axis depth, then fed in only the X- and/or Y-axis. For more intricate and complex machining operations involving simultaneous X-, Y-, and Z-axes movement, computer-aided manufacturing (CAM) software is usually used to aid the programmer. Section 8.8 will give an overview of CAM software. This unit will explain basic G-code programming.

There is no standardized format for writing a CNC program that is compatible with all machine control brands and models. Each machine control manufacturer has developed its own unique programming format. Each one has minor differences, but the principles of a program are the same among them all.

The programming examples throughout this unit will relate the most closely to Fanuc-type and Haas controllers. However, the principles may be applied to any manufacturer's format (see each machine's programming manual).

COORDINATE POSITIONING FOR MILLING

The basic coordinate system for milling consists of X-, Y-, and Z-axes, which are all perpendicular to each other, as shown in **Figure 8.6.1**. On a vertical spindle machine, the X- and Y-axes are the table movements and the Z-axis is the vertical spindle movement, just like in manual milling. To understand the axes of a horizontal spindle machine, it is helpful to visualize a vertical machine laid on its back; the spindle moves toward and away from the part in the Z-axis, the X-axis is parallel to the floor, and the Y-axis moves up and down. The examples in this unit will apply to vertical spindle machines.

Depending on the kind of milling machine being programmed, there may be only two, or sometimes four or five, programmable axes. Some CNC knee mills do not have a CNC drive on the quill (Z-axis). These machines require the Z-axis to be positioned manually prior to executing programmed milling moves in the X- and Y-axes. The majority of machining centers have three programmable axes (X, Y, and Z), so the examples in this unit will apply to that style of machine. For most of the programming examples in this unit, coordinate positions will be in relation to the center of the cutting tool.

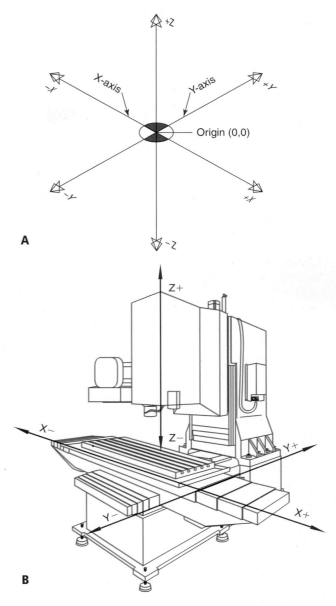

FIGURE 8.6.1 (A) The X-, Y-, and Z-axes and (B) how they apply to CNC mill programming. Notice how the Z-axis moves the spindle toward and away from the part. A rotating drill would be moved into the part in the Z-axis.

CNC mill controls have the ability to "flip" the coordinate plane for complex milling operations. Examples of this are beyond the scope of this text, but it is important to ensure that the correct coordinate system orientation is activated when programming. The standard three-axis system just described uses the *XY plane*. Some controls default to the XY plane, but others may require that it is specifically activated. The safe-start portion of the program is a good place to activate the standard XY plane coordinate system with a *G17* command.

TYPES OF MOTION FOR MILLING

Rapid Traverse—G0

Rapid traverse using the G0 command will quickly position tools before beginning a machining operation. Cutting tools should never contact the work during rapid moves. There are several safety considerations when programming rapid movements in a machining center. Machining centers may have odd-or irregular-shaped workholding devices or workpieces, so special attention needs to be paid to items such as clamps, vise jaws, workstops, and fixture bolts. On some workpieces, part features may protrude above others, causing another concern for tool collisions. **(See Figure 8.6.2.)**

In CNC mill programming, it is standard practice to establish a safety zone above the workpiece or workholding device called a **clearance plane**. Typically, a clearance plane of 0.050" or 0.100" above the material surface is used. Program rapid moves at or above the clearance plane and avoid programming any rapid moves below the clearance plane.

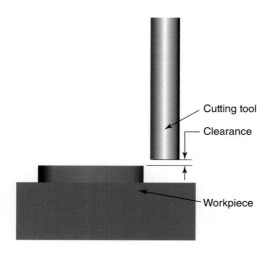

FIGURE 8.6.2 When programming rapid positioning moves using the G0 command, be sure the tool is positioned above the workpiece in the Z-axis.

Linear Interpolation—G1

Linear interpolation means that the machine synchronizes the motion of one (or more) axis to move the tool in a straight line. When more than one axis is in motion, as when moving diagonally, the machine must start moving each axis at precisely the same time, move them at the appropriate feed rate, and stop moving both axes at the intended destination at precisely the same time.

Linear milling may require motion in one axis alone (milling a path parallel to a machine's axis), in two axes (for a diagonal), or in three axes (for a diagonal on a slanted plane; compound angular motion). **(See Figure 8.3.3.)** The straight motion produced by linear interpolation may be used for plunging a drill or counterbore into the workpiece, feeding a face mill across the top of a part, side milling profiles with straight surfaces, and more.

To move the milling tool in a linear interpolated movement, a *G1* command is used along with an end position for the movement along one, two, or three axes. A feed rate must also accompany the newly programmed position. Since *G1* is a modal code, it will remain active until cancelled or it is overridden with a conflicting code.

Feed Rates

When any feed-type move is programmed, a feed rate must be assigned to it. Feed rates are modal and if none is programmed in a block, then the previously programmed rate will still be active. An F-character assigns the desired feed rate.

Feed rates can be expressed in either inches per minute (IPM), using a *G94* code, or inches per revolution (IPR), using a *G95* code. This is often set in the beginning portion of the CNC program or on the safe-start block, but can be activated or changed at any time in the program. For milling operations, use of IPM is most common, but IPR is sometimes used for holemaking operations.

Feed unit settings are also modal, so always be sure of the setting that is active. If the machine is in *G94* (IPM) and a feed rate of *F.005* is programmed (an acceptable feed for IPR), the axis will move at only 0.005 inches per minute. If the machine is in *G95* (IPR) and *F40.0* is programmed (an acceptable feed for IPM), the machine will move at 40 inches per spindle revolution!

An example of a rapid positioning move followed by two back-to-back linear moves in a program looks like this (notice the lack of *G1* code in the last block, since it is already modal):

```
G0 X1. Y1. (RAPID X AND Y TO START POSITION);
G1 Z-.1 F15. (LINEAR FEED Z TO DEPTH)
X-1.5 F30. (LINEAR FEED X TO NEXT POSITION);
```

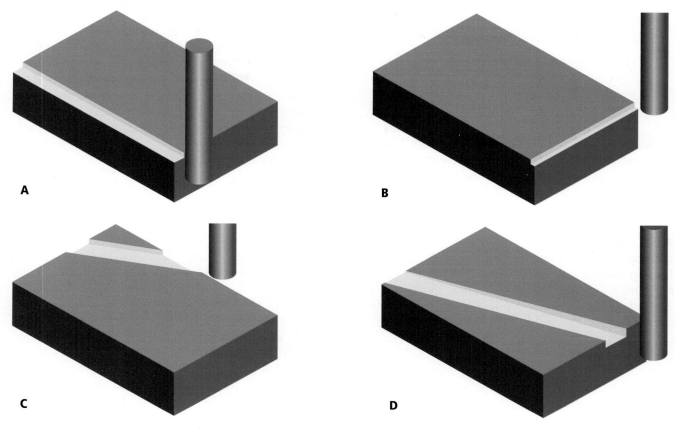

FIGURE 8.6.3 Examples of G1 moves in X (A), Y (B), X and Y (C), and X, Y, and Z (D).

⚠ CAUTION

If the machine is in *G95* (IPR) and a feed rate of *F20.0* is programmed, the axis will attempt to move at 20 IPR. This could easily approach rapid rates. Use extreme caution when switching between *G94* and *G95*.

Circular Interpolation

Circular cutter paths can be created using circular interpolation. This motion can be used to mill part features such as corner radii, arcs, and circular pockets. To program an arc movement, the tool must first be positioned at the starting point of the arc. Then a G-code is used to indicate either clockwise or counterclockwise movement. A *G2* command is used for clockwise motion, and a *G3* command is used for counterclockwise motion. **Figure 8.6.4** illustrates an example of a *G2* and a *G3* movement.

In the same block that the *G2* or *G3* code is given, the programmer must also identify the end point where the arc is to stop (remember that the tool was already at the start point before the circular interpolation

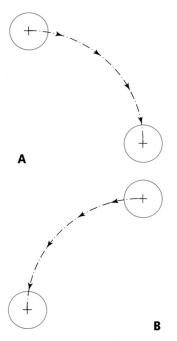

FIGURE 8.6.4 A G2 command is used for clockwise circular interpolation (A) and a G3 command is used for counterclockwise circular interpolation (B).

command). A *G2* command and an arc end point is added to the last example and highlighted below:

G0 X1. Y1. (RAPID X AND Y TO START POSITION);
G1 **Z-.1** F15. (LINEAR FEED Z TO DEPTH)
X-1.5 F30. (LINEAR FEED X TO NEXT POSITION);
G2 X-2. Y1.5 (CW ARC MOVE TO NEXT POSITION);

But, this is still not enough information for a circular interpolation move. The control knows the start point, the end point, and the direction, but there is no information about the size of the arc radius. See **Figure 8.6.5** for an illustration of why the radius information is needed. The

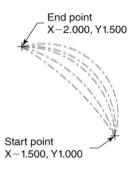

FIGURE 8.6.5 A few of the endless possible sizes of radii that could be programmed between two points. All of these arcs start at the same place, end at the same place, and they are all CCW.

same block of code that activated circular interpolation must also contain that radius information. There are two methods for programming the size of the radius: the arc center method and the radius method. The arc center method has the advantage of being able to cut a 360-degree arc (full circle) using only one line of code, where the radius method requires full circles to be broken into two pieces and programmed with two blocks of code. The R-method is usually easier for a new programmer to understand, however.

Arc Center Method for Circular Interpolation

The *arc center method* identifies the exact location of the center point of the arc related to its start point. This center point must be identified by an incremental distance along both the X- and Y-axes. Since the *G2* or *G3* program block already contains an X- and Y-character to identify the arc endpoint, new characters I and J are used to identify these distances. The I-word defines the <u>distance</u> from the arc start point to the center point along the X-axis. The J-word defines the <u>distance</u> from the arc start point to the center point along the Y-axis. Be careful to use the correct sign (positive or negative) to the arc center for each value. See **Figure 8.6.6** for an illustration of the parts of an arc and the I- and J-values.

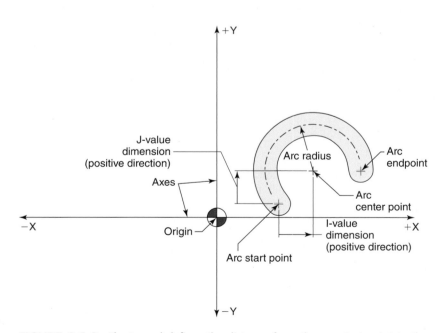

FIGURE 8.6.6 The I-word defines the distance from the arc start point to the center point along the X-axis. The J-word defines the distance from the arc start point to the center point along the Y-axis. Remember that these values can be positive or negative.

The previous code examples will be expanded to cut the slot shown in **Figure 8.6.7.** So far, the cutter has been moved to the position labeled "arc start point" and the linear move is about to transition to a 90-degree arc with a 0.5" radius. In the illustration, notice that the difference between the arc start point and the arc end point in both axes is 0.5". The distance from the arc start point to its center point along the X-axis would be "0," so the I-word would be I0. The distance from the arc start point to its center point along the Y-axis would be 0.5" in the positive direction, so the J-word would be *J.5*. The complete block using the arc center method is

shown below, and the newly added I- and J-words are highlighted below:

G0 X1. Y1. (RAPID X AND Y TO START POSITION);
G1 **Z-.1** F15. (LINEAR FEED Z TO DEPTH)
X-1.5 F30. (LINEAR FEED X TO NEXT POSITION);
G2 X-2. Y1.5 **I0**. J.5 (CW ARC MOVE TO NEXT POSITION);

There are unlimited situations for applying circular interpolation and the arc center method. **Figures 8.6.8** through **8.6.11** how some typical situations and the code

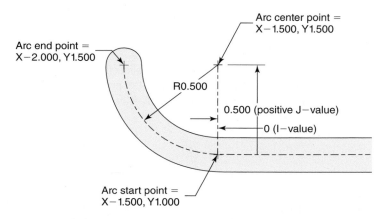

FIGURE 8.6.7 The arc center is in-line with its start point along the X-axis so I = 0. The distance from the arc start point to its arc center along the Y-axis is 0.500, so J = 0.5. Note that the I-value will always be zero and the J-value will always be the radius in this type of situation. Remember to use the correct sign (+ or −).

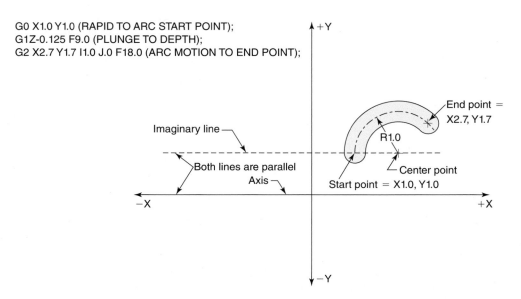

FIGURE 8.6.8 Circular interpolation example with the arc center aligned with the arc start point along the X-axis. Note that the J-value will always be zero and the I-value will always be the radius in this type of situation. Remember to use the correct sign (+ or −).

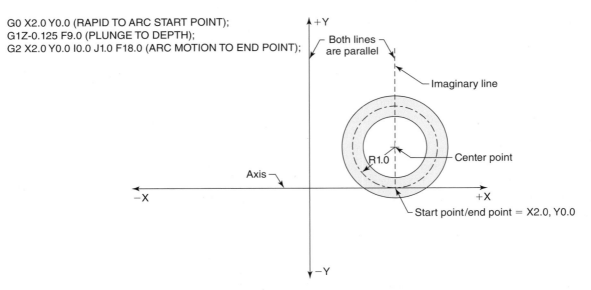

```
G0 X2.0 Y0.0 (RAPID TO ARC START POINT);
G1Z-0.125 F9.0 (PLUNGE TO DEPTH);
G2 X2.0 Y0.0 I0.0 J1.0 F18.0 (ARC MOTION TO END POINT);
```

FIGURE 8.6.9 Circular interpolation example with the arc center aligned with the arc start and a full 360-degree arc movement is made. When programming a full 360 degrees, the start point and end point will be the same. Note: either the I- or the J-value will be zero when the start point is a 3, 6, 9, or 12 o'clock position. Remember to use the correct sign (+ or −).

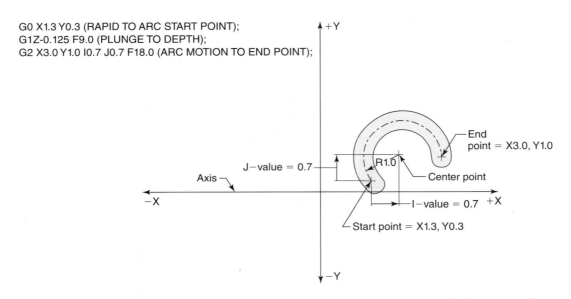

```
G0 X1.3 Y0.3 (RAPID TO ARC START POINT);
G1Z-0.125 F9.0 (PLUNGE TO DEPTH);
G2 X3.0 Y1.0 I0.7 J0.7 F18.0 (ARC MOTION TO END POINT);
```

FIGURE 8.6.10 Circular interpolation example where the arc center is not in line with the arc start along either axis. In these situations, neither the I- nor the J-value is zero.

required to perform the circular interpolation motion. Take note that only the blocks of code for positioning the cutting tool at the arc start point and making the arc cutting motion will be shown. Assume the tool is safely loaded, offsets activated, and the spindle running.

Radius Method for Circular Interpolation

The radius method for identifying arc center data is by far the easier of the two arc programming methods and the more common. There is no need for an I- or J-word. With

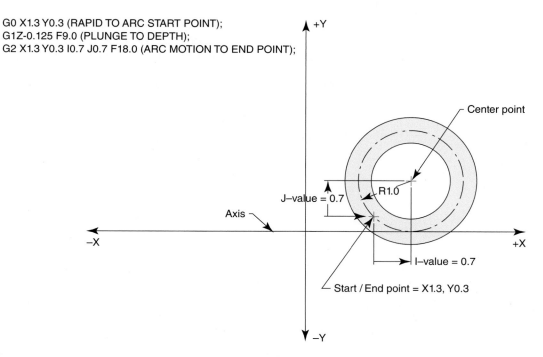

```
G0 X1.3 Y0.3 (RAPID TO ARC START POINT);
G1Z-0.125 F9.0 (PLUNGE TO DEPTH);
G2 X1.3 Y0.3 I0.7 J0.7 F18.0 (ARC MOTION TO END POINT);
```

FIGURE 8.6.11 Circular interpolation example where the arc center is not in line with the arc start and a full 360-degree arc movement is made. (The end point is the same as the start point.) In these situations, neither the I- nor the J-value is zero.

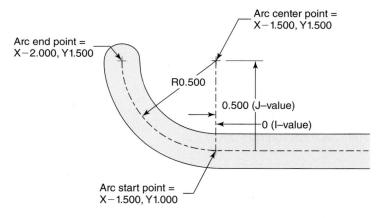

FIGURE 8.6.12 The radius of the arc is 0.500. The radius size is in place of the I- and J-values when using the radius method of circular interpolation.

this method, only the arc's radius needs to be defined in the circular interpolation execution block by an R-word. For example, *R1.0* would define a 1" radius and *R0.75* would define a 0.75" radius.

Consider again the example shown in **Figure 8.6.7.** The arc has a 0.5" radius, and this value can be used directly in the circular interpolation block. The code has been changed to use the radius method in place of the arc center method, and the change is shown highlighted below. **(See Figure 8.6.12.)**

```
G0 X1. Y1. (RAPID X AND Y TO START POSITION);
G1 Z-.1 F15. (LINEAR FEED Z TO DEPTH)
X-1.5. F30. (LINEAR FEED X TO NEXT POSITION);
G2 X-2. Y1.5 R.5 (CW ARC MOVE TO NEXT POSITION);
```

To program an arc greater than 180 degrees, a negative sign must be added to the R-value. To program a full circle (360-degree arc) using the radius method, the circle must be broken into two parts and programmed with two blocks of code. This is because when the start point and endpoint of the arc are the same position and only a radius is given, there are a nearly infinite number of solutions for where the arc center could be placed. **(See Figure 8.6.13.)**

Figures **8.6.14** through **8.6.17** show examples of some typical scenarios using the radius method, including the technique for programming a full circle.

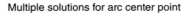

Multiple solutions for arc center point

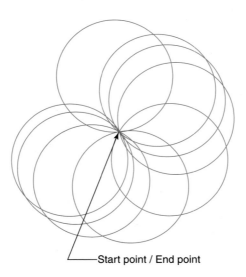

Start point / End point

FIGURE 8.6.13 A few examples of where a 360-degree arc could be located when the start point and end point are the same. Notice that the radius of each of these arcs is the same. This is why, when using the radius method, a 360-degree arc must be programmed in two sections.

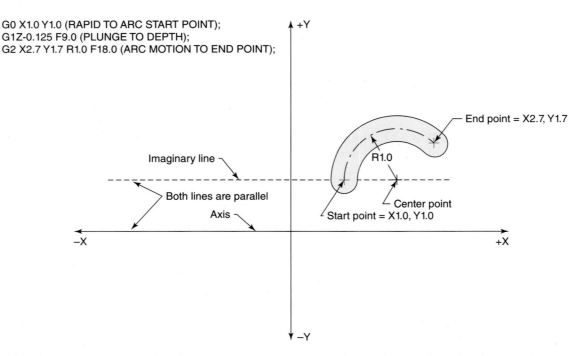

G0 X1.0 Y1.0 (RAPID TO ARC START POINT);
G1 Z-0.125 F9.0 (PLUNGE TO DEPTH);
G2 X2.7 Y1.7 R1.0 F18.0 (ARC MOTION TO END POINT);

FIGURE 8.6.14 Circular interpolation example using the R-method where the arc center point is in line with the arc start point along the X-axis.

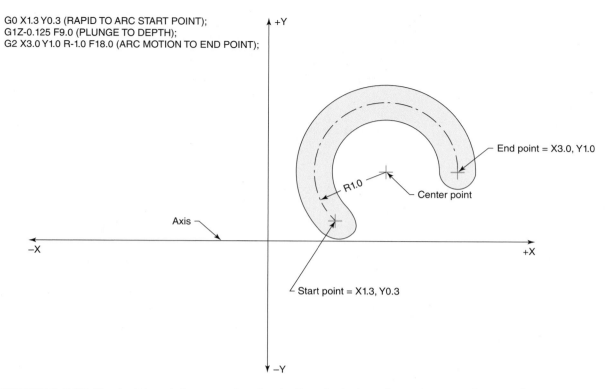

G0 X1.3 Y0.3 (RAPID TO ARC START POINT);
G1Z-0.125 F9.0 (PLUNGE TO DEPTH);
G2 X3.0 Y1.0 R-1.0 F18.0 (ARC MOTION TO END POINT);

+Y

End point = X3.0, Y1.0

R1.0

Center point

Axis

−X   +X

Start point = X1.3, Y0.3

−Y

FIGURE 8.6.15 Circular interpolation example using the R-method where the arc is greater than 180 degrees. Notice the negative R-value in the program.

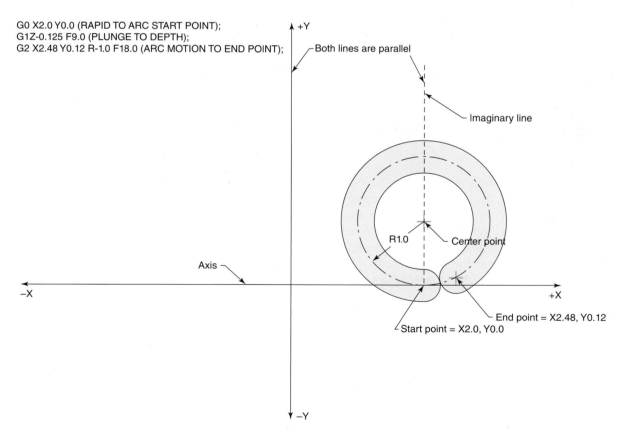

G0 X2.0 Y0.0 (RAPID TO ARC START POINT);
G1Z-0.125 F9.0 (PLUNGE TO DEPTH);
G2 X2.48 Y0.12 R-1.0 F18.0 (ARC MOTION TO END POINT);

+Y

Both lines are parallel

Imaginary line

R1.0   Center point

Axis

−X   +X

End point = X2.48, Y0.12
Start point = X2.0, Y0.0

−Y

FIGURE 8.6.16 Circular interpolation example using the R-method where the arc motion is slightly less than 360 degrees. Notice the negative R-value in the program since the arc is greater than 180 degrees.

```
G0 X2.0 Y0.0 (RAPID TO ARC START POINT);
G1 Z-0.125 F9.0 (PLUNGE TO DEPTH);
G2 X2.0 Y2.0 R1.0 F18.0 (ARC MOTION TO END POINT #1);
G2 X2.0 Y0.0 R1.0 F18.0 (ARC MOTION TO END POINT #2);
```

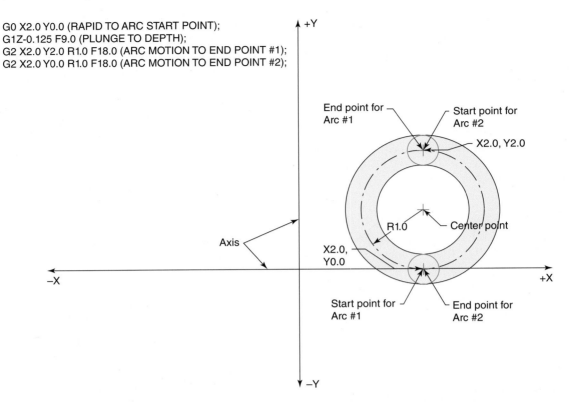

FIGURE 8.6.17 Circular interpolation example using the R-method where the arc motion is exactly 360 degrees. The arc must be split into two 180-degree segments in order to use the R-method for a full circle.

NON-AXIS MOTION COMMANDS

Comments

The beginning of a program is an ideal place to include detailed comments about the program author, date, file name, tool and setup information, workpiece material, and more. This area is sometimes called a *header*. Comments placed in parentheses will be ignored by the machine but will be helpful to other users of the program. Header comments at the start of a program might look like this:

```
O0000 (PROGRAM NAME HERE);
(PROGRAM FILE NAME-VALVE HOUSING);
(PROGRAM AUTHOR-PJH);
(DATE-5-5-2018);
(INTENDED MACHINE AND CONTROL-HAAS VF2-HAAS);
(X0/Y0 LOCATION-FRONT LEFT CORNER);
(Z0 LOCATION-TOP PART FACE);
(TOOL NUMBER 1-2.5 DIA FACEMILL);
(WORKPIECE MATERIAL-6061-T6 ALUMINUM);
```

Work Coordinate System Command

A CNC machine must know where the workpiece origin is located. This position is entered in the machine control during setup and is called a *work coordinate system (WCS)* or *work offset*. The process for setting a work offset will be explained in detail in Section 8.7. If more than one workpiece is set up in the machine at a time or if a part will be repositioned, a separate origin can be established for each one. The desired WCS origin can be activated by programming a G-code.

Work coordinate systems are activated by using codes *G54* through *G59*. *G54* would be WCS #1, *G55* would be WCS #2, and so on. Six WCS settings are available on many machines. The WCS activation block is shown below:

```
G54 (SET WCS 1);
```

Safe Start Block

As mentioned earlier, the beginning of a program is an excellent place to insert comments in parentheses that will be ignored by the machine but can be helpful to other users of the program. Modal codes given at the beginning of the program are used to safely establish the default program settings. This block of the program is often called the *safe start* block. The following are common commands used in the safe start block.

- *G0* sets rapid positioning.
- *G17* sets the XY plane. For most programs, this is the required setting because X and Y are the primary axes used for milling. Use of other coordinate systems is beyond the scope of this text.
- *G20* sets inch units.
- *G40* cancels any active cutter radius compensation. Cutter compensation will be explained later in this unit, but for now, it is only important to know that it is good practice to use a *G40* command here to cancel values that may have been set by previously active programs.
- *G54* sets the work coordinate system. Some machines allow multiple work coordinate systems (part origins) to be set at setup time. The desired WCS is called in the program with the correct G-code.
- *G80* cancels any modal canned cycles. Canned cycles will also be explained later in this unit, but for now, it is only important to know that it is good practice to use a G80 command here to cancel any cycles that may have been activated by previous programs.
- *G94* sets feed to inches per minute (IPM).

The safe start block of a Fanuc or Haas program may look like this:

G0 G17 G20 G40 G54 G80 G94 (SAFE START CODES);

Tool-Change Command

After the safe-start portion of the program is established, the desired tool must be loaded in the spindle using the ATC. This task is done by using a combination of a T- and M-word in a single block to perform a **tool change**. An *M6* and T-word specifying the desired tool number commands the ATC to load the tool in the spindle. For example, *M6 T1* will cause the ATC to automatically load tool #1 in the spindle. The tool change is performed in rapid motion by default. Take note that some machine controls require the *M6* and T-word to be in a specific order. If the spindle is running when a tool change is commanded, most machines will automatically stop the spindle before the tool change, so no *M5* code is needed.

It is important that the tool can unload and load tools in a safe place clear of collisions. It is sometimes best to move the machine axes to their home position for tool changes. This is usually the safest location. Programming

a *G91* with a *G28* causes positioning to occur relative to the machine home position, instead of the work coordinate system. Following a *G91 G28* with a *Z0.* will move the Z-axis up to the Z-home position. Always move the Z-axis first in a block by itself, so it is clear of any obstructions during the X- and Y-axes move. The next block will perform a separate XY move to home. Then the next block commands the tool change. Blocks for a rapid move to this safe machine home position and tool-change look like this:

G91 G28 **Z0.** (RAPID TO MACHINE Z-HOME FOR TOOL CHANGE);
G91 G28 X0. Y0. (RAPID TO MACHINE XY-HOME FOR TOOL CHANGE);
M6 T____ (CHANGE TOOL);

Program Stop Commands

One of two different M-codes can cause a **program stop**, or hold, on the program until it is resumed by pressing the cycle start button. These are often inserted immediately before or after tool changes to allow inspection of a tool or to check the results produced by a completed operation.

An *M0* command is a *full stop* and always requires the operator to restart the program by pressing the cycle start button on the operator panel. This can be used when a part must be repositioned, chips must be cleared, or to check a critical dimension before continuing. An *M1* command is an *optional stop*. A switch on the operator panel needs to be switched on for the machine to read this optional stop and pause the program. An optional stop may also be used to pause the cycle before a new operation begins. It is often used when running a new program for the first time so that inspections can be performed. After the program is proven safe and correct, the optional stop switch can be turned off and the program will ignore the *M1* command.

Some machines return to the default WCS of G54 when the "RESET" button on the control panel is pressed to cancel a running program. After that reset, if desired, a program may be restarted at a tool change block. For this reason, it is good practice to place the desired WCS code after each tool change to ensure the correct work offset is activated.

The full stop or optional stop should be a single block of code. An optional stop before a tool change and a reactivation of the desired WCS would look like this:

M1(OPTIONAL STOP);
M6 T1 (CHANGE TOOL);
G54 (SET WCS 1);

If a full stop is desired, the following code would be used. Notice the comment giving the operator important instructions:

M0 (PROGRAM STOP-FLIP PART);

M6 T____ (CHANGE TOOL);

G54 (SET WCS 1);

Absolute/Incremental Setting

Most machines switch into modal *G91* (incremental positioning) mode automatically during the tool change cycle or during the return to home position. If absolute positioning is desired in the program, a *G90* code is required following a tool change and before any further axis movements. This block will usually include a *G0* (rapid) and an X- and Y-axis movement as shown in the highlighted code below:

G91 G28 **Z0.** (RAPID TO MACHINE Z-HOME FOR TOOL CHANGE);
G91 G28 X0. Y0. (RAPID TO MACHINE XY-HOME FOR TOOL CHANGE);
M1(OPTIONAL STOP);
M6 T____ (CHANGE TOOL);
G54 (SET WCS 1);
G0 G90 X_ Y_ (RAPID TO FIRST POSITION);

EXAMPLE: G0 G90 X1.5 Y2.0; would activate absolute positioning (G90) and rapid the machine table to X1.5, Y2.0. When programming coordinate positions, only the axis that requires movement needs to be programmed. If there is only a change in one axis, the others do not need to be listed. For example, if the machine is at a position of X1.5, Y2.0 and the next desired programmed position is X1.5, Y1.5, only the Y1.5 needs to be listed. This code would look like this:

G0 G90 X1.5 Y2.0;

Y1.5;

Tool Height Offset Command

During machine setup, the length of each tool is measured and entered into the control. This is done so the machine knows where the tool tip is and so that the tool can be programmed from its tip regardless of its length. This measurement is called the **tool height offset** (or **tool length offset**). This process of setting a tool height offset will be explained in detail in Section 8.7, but when programming, it is important to understand

the code required so the machine uses the correct tool height offset.

After a specific tool has been loaded in the machine spindle and the X- and Y-axes are positioned at a desired location, another rapid move is often programmed to bring the tip of the tool to the clearance plane. Along with the Z-axis position for the clearance plane, this block of code must also contain a command for proper tool height offset. A *G43* code activates tool height offset compensation in the machine control, and an H-word calls up the correct offset number (highlighted below).

G91 G28 **Z0.** (RAPID TO MACHINE Z-HOME FOR TOOL CHANGE);
G91 G28 X0. Y0. (RAPID TO MACHINE XY-HOME FOR TOOL CHANGE);
M1(OPTIONAL STOP);
M6 T____ (CHANGE TOOL);
G54 (SET WCS 1);
G0 G90 X_ Y_ (RAPID TO FIRST POSITION);
G43 H1 **Z.1** (ACTIVATE TOOL HEIGHT OFFSET-MOVE TO CLEARANCE PLANE);

 CAUTION

The value of the H-word and the T-word must always match to pair the correct offset number with the correct tool number or the tool may rapid into the workpiece and/or workholding device with disastrous results.

EXAMPLE: G0;
G43 H1 Z.1; would activate the offset for tool #1 and use rapid positioning to bring the tip of tool #1 to the clearance plane (within 0.100" of the work surface). Some machines require this Z-axis rapid movement in the same block to fully activate the tool height compensation.

Spindle Speed for Milling

Spindle speeds for CNC milling are expressed in RPM and programmed with the combination of an M- and S-command. The order of the M-code and the S-code does not matter on most machines. The S-command provides the spindle RPM and the following M-commands are used to turn the spindle on or off:

M3 Spindle on clockwise (forward)

M4 Spindle on counterclockwise (reverse)

M5 Spindle off

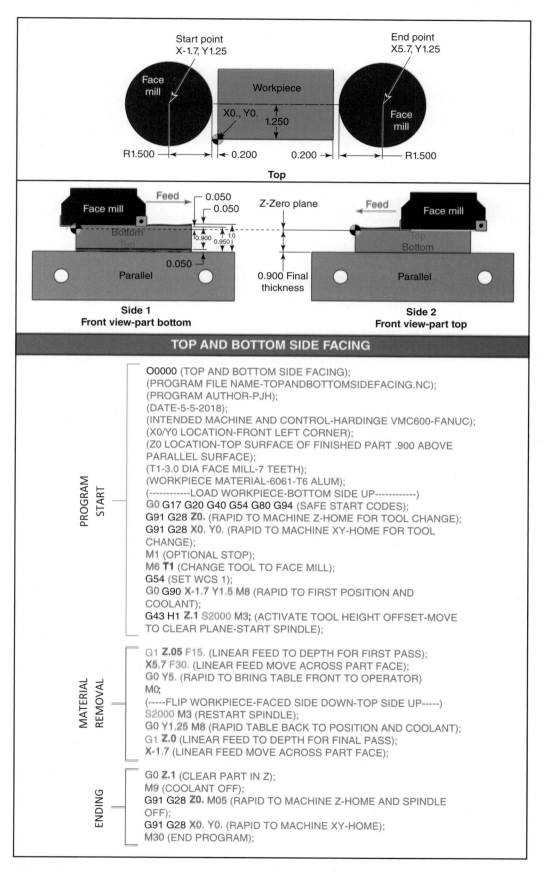

FIGURE 8.6.21 A part drawing and the program to face both sides of a part to thickness and create the Z-zero plane.

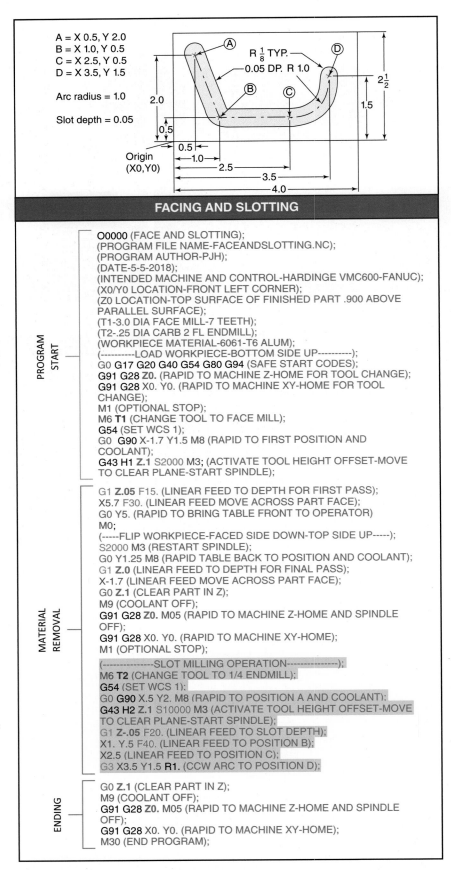

A = X 0.5, Y 2.0
B = X 1.0, Y 0.5
C = X 2.5, Y 0.5
D = X 3.5, Y 1.5

Arc radius = 1.0

Slot depth = 0.05

R ⅛ TYP.
0.05 DP. R 1.0

Origin (X0,Y0)

FACING AND SLOTTING

```
O0000 (FACE AND SLOTTING);
(PROGRAM FILE NAME-FACEANDSLOTTING.NC);
(PROGRAM AUTHOR-PJH);
(DATE-5-5-2018);
(INTENDED MACHINE AND CONTROL-HARDINGE VMC600-FANUC);
(X0/Y0 LOCATION-FRONT LEFT CORNER);
(Z0 LOCATION-TOP SURFACE OF FINISHED PART .900 ABOVE
PARALLEL SURFACE);
(T1-3.0 DIA FACE MILL-7 TEETH);
(T2-.25 DIA CARB 2 FL ENDMILL);
(WORKPIECE MATERIAL-6061-T6 ALUM);
(----------LOAD WORKPIECE-BOTTOM SIDE UP----------);
G0 G17 G20 G40 G54 G80 G94 (SAFE START CODES);
G91 G28 Z0. (RAPID TO MACHINE Z-HOME FOR TOOL CHANGE);
G91 G28 X0. Y0. (RAPID TO MACHINE XY-HOME FOR TOOL
CHANGE);
M1 (OPTIONAL STOP);
M6 T1 (CHANGE TOOL TO FACE MILL);
G54 (SET WCS 1);
G0  G90 X-1.7 Y1.5 M8 (RAPID TO FIRST POSITION AND
COOLANT);
G43 H1 Z.1 S2000 M3; (ACTIVATE TOOL HEIGHT OFFSET-MOVE
TO CLEAR PLANE-START SPINDLE);
```

PROGRAM START

```
G1 Z.05 F15. (LINEAR FEED TO DEPTH FOR FIRST PASS);
X5.7 F30. (LINEAR FEED MOVE ACROSS PART FACE);
G0 Y5. (RAPID TO BRING TABLE FRONT TO OPERATOR)
M0;
(-----FLIP WORKPIECE-FACED SIDE DOWN-TOP SIDE UP-----);
S2000 M3 (RESTART SPINDLE);
G0 Y1.25 M8 (RAPID TABLE BACK TO POSITION AND COOLANT);
G1 Z.0 (LINEAR FEED TO DEPTH FOR FINAL PASS);
X-1.7 (LINEAR FEED MOVE ACROSS PART FACE);
G0 Z.1 (CLEAR PART IN Z);
M9 (COOLANT OFF);
G91 G28 Z0. M05 (RAPID TO MACHINE Z-HOME AND SPINDLE
OFF);
G91 G28 X0. Y0. (RAPID TO MACHINE XY-HOME);
M1 (OPTIONAL STOP);
(----------------SLOT MILLING OPERATION---------------);
M6 T2 (CHANGE TOOL TO 1/4 ENDMILL);
G54 (SET WCS 1);
G0 G90 X.5 Y2. M8 (RAPID TO POSITION A AND COOLANT);
G43 H2 Z.1 S10000 M3 (ACTIVATE TOOL HEIGHT OFFSET-MOVE
TO CLEAR PLANE-START SPINDLE);
G1 Z-.05 F20. (LINEAR FEED TO SLOT DEPTH);
X1. Y.5 F40. (LINEAR FEED TO POSITION B);
X2.5 (LINEAR FEED TO POSITION C);
G3 X3.5 Y1.5 R1. (CCW ARC TO POSITION D);
```

MATERIAL REMOVAL

```
G0 Z.1 (CLEAR PART IN Z);
M9 (COOLANT OFF);
G91 G28 Z0. M05 (RAPID TO MACHINE Z-HOME AND SPINDLE
OFF);
G91 G28 X0. Y0. (RAPID TO MACHINE XY-HOME);
M30 (END PROGRAM);
```

ENDING

FIGURE 8.6.22 A part drawing and the program to face the part and mill a slot. The slotting operation is highlighted.

can then be written using linear and circular interpolation. **Figure 8.6.22** shows the code for milling the slot and uses the radius method for circular interpolation.

Holemaking Operations

Hole work is also a common operation in CNC milling. These operations are programmed by positioning the tool at the desired X- and Y-coordinates for the hole, bringing the tool tip to the clearance plane with a rapid move, and then using linear interpolation to feed the tool to the desired hole depth.

 CAUTION

After performing the holemaking operation, it is extremely important that the tool be programmed to rapidly retract to the Z-clearance plane before making any X- or Y-axis motion to avoid breaking the tool and damaging the workpiece.

There are two methods of programming the feed rate for a holemaking operation. Since holemaking feed rates are normally expressed in IPR (inches per revolution), IPR feed rate values can be taken directly from drilling feed charts. If IPM feed mode is used, calculate the IPM value by multiplying IPR by the RPM. If IPR is desired, a *G95* command can be called any time after the tool change and even in the block where the feed rate is given. **Figure 8.6.23** illustrates a simple drilling operation using linear interpolation.

 CAUTION

It is important that the correct feed units are programmed with a *G94* or *G95* for the feed rate value given, or feed rates can be dangerously high. For instance, suppose a 0.005″ IPR feed rate was used correctly with a *G95* (IPR units) for a drilling operation. If the next tool was programmed with a feed rate of 5.0, but *G94* (IPM) was not commanded, *G95* would still be modal and the machine would attempt to feed at 5 inches per revolution!

Suppose a #7 (0.201″) drilled hole, 1/2″ deep, was added to the part from the previous facing and slotting example at the location as shown in **Figure 8.6.24**. The coordinate location for the hole is *X3.5, Y2.25*. The length of the drill tip is equal to 0.3 × 0.201, or 0.06. Adding this to the required depth gives a total required programmed depth of 0.56.

This portion of program code begins with another tool change, then moves to the hole location, rapids the tool to the clearance plane, feeds the drill into the work, then retracts to the clearance plane. **Figure 8.6.24** shows the part drawing and the code to produce the hole.

Canned Cycles

Some operations require repetitive motions, such as drilling several holes to the same depth or making multiple chip-breaking pecks required for deep-hole drilling. Since the previous drilling example took three blocks of motion code, 30 blocks of code would be needed to drill 10 holes. If peck drilling is desired, each peck requires at least a *G1* feed move and a *G0* retract move. To aid programmers, CNC controls have features that make these tedious and repetitive operations easier and faster to program. These routines can be packaged or "canned" into one or two blocks of code and are called *canned cycles*. Canned cycles only require one or two blocks of code in a specified format to define the information needed for the cycle. The most common canned cycles used on a milling machine are for drilling and tapping.

Single-Pass Drilling Cycles

Single-pass drilling cycles feed the tool continuously to the programmed Z-depth without pecking. The cycle then automatically retracts the tool from the hole. The two different single-pass cycles for CNC milling are *G81* and *G82*. The only difference between the two is that the *G82* cycle has the capability of pausing once the full depth is reached. This pause is called a **dwell** and allows time for the tool to make a few complete revolutions at the hole bottom, so that a clean and flat surface is produced. Otherwise, the code is identical for both canned cycles. *G81* and *G82* work well for spot drilling, center drilling, countersinking, counterboring, and shallow-hole drilling. After either canned cycle is complete, it must be cancelled with a *G80* code in a separate block.

Single-Pass Drilling—G81

The hole-drilling operation from the last example can be written using a *G81* canned cycle. The location of the hole was *X3.5, Y2.25* and the programmed Z-depth was 0.56. The *G81* block contains the X- and Y-coordinate for the hole, a *return point (or R-value)*, the Z-depth, and the feed rate. The return point is the absolute Z-position where the tool will begin feeding and where it will retract to at the end of the canned cycle. For most situations, the return point should be the same as the Z-clearance plane. The table below shows the format for the *G81* cycle.

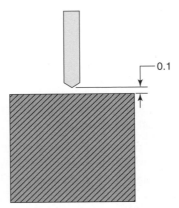

G0 G43 Z0.1 H1 M3 S1000;   (Rapid to Z0.1 clearance plane)

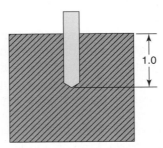

G1 G94 Z-1.0 F6.0; (Feed to 1.0" depth at 6.0" per minute using G94)

OR

G1 G95 Z-1.0 F0.006; (Feed to 1.0" depth at 0.006" per revolution using G95)

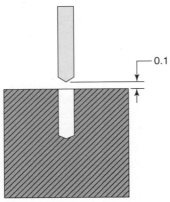

G0 Z0.1 M09; (Rapid retract back to Z0.1 clearance plane and turn off coolant)

FIGURE 8.6.23 Linear interpolation (G1) is used to program a simple drilling operation.

	Format	Variable Description
Fanuc OR Haas Single-Pass Cycle Format	(*TOOL TO START POINT*) G81 X_Y_ R_Z_F_;	X: X position of hole Y: Y position of hole R: Return plane position (optional) Z: Z end position of hole F: Feed rate

FACING/SLOTTING/DRILLING

PROGRAM START

O0000 (FACE-SLOT-DRILL);
(PROGRAM FILE NAME-FACESLOTDRILL.NC);
(PROGRAM AUTHOR-PJH);
(DATE-5-5-2018);
(INTENDED MACHINE AND CONTROL-HARDINGE VMC600-FANUC);
(X0/Y0 LOCATION-FRONT LEFT CORNER);
(Z0 LOCATION-TOP SURFACE OF FINISHED PART-.900 ABOVE PARALLEL
SURFACE);
(T1-3.0 DIA FACE MILL-7 TEETH);
(T2-.25 DIA CARB 2 FL ENDMILL);
(T3-NO. 7 HSS SCREW MACHINE LENGTH TWIST DRILL);
(WORKPIECE MATERIAL-6061-T6 ALUM);
(----------LOAD WORKPIECE-BOTTOM SIDE UP----------);
G0 G17 G20 G40 G54 G80 G94 (SAFE START CODES);
G91 G28 **Z0.** (RAPID TO MACHINE Z-HOME FOR TOOL CHANGE);
G91 G28 X0. Y0. (RAPID TO MACHINE XY-HOME FOR TOOL CHANGE);
M1 (OPTIONAL STOP);
M6 **T1** (CHANGE TOOL TO FACE MILL);
G54 (SET WCS 1);
G0 **G90** X-1.7 Y1.5 M8 (RAPID TO FIRST POSITION AND COOLANT);
G43 H1 **Z.1** S2000 M3; (ACTIVATE TOOL HEIGHT OFFSET-MOVE TO CLEAR PLANE-
START SPINDLE);

MATERIAL REMOVAL

G1 **Z.05** F15. (LINEAR FEED TO DEPTH FOR FIRST PASS);
X5.7 F30. (LINEAR FEED MOVE ACROSS PART FACE);
G0 Y5. (RAPID TO BRING TABLE FRONT TO OPERATOR)
M0;
(-----FLIP WORKPIECE-FACED SIDE DOWN-TOP SIDE UP-----);
S2000 M3 (RESTART SPINDLE);
G0 Y1.25 M8 (RAPID TABLE BACK TO POSITION AND COOLANT);
G1 **Z.0** (LINEAR FEED TO DEPTH FOR FINAL PASS);
X-1.7 (LINEAR FEED MOVE ACROSS PART FACE);
G0 **Z.1** (CLEAR PART IN Z);
M9 (COOLANT OFF);
G91 G28 **Z0.** M05 (RAPID TO MACHINE Z-HOME AND SPINDLE OFF);
G91 G28 X0. Y0. (RAPID TO MACHINE XY-HOME);
M1 (OPTIONAL STOP);

(---------------SLOT MILLING OPERATION---------------);
M6 **T2** (CHANGE TOOL TO 1/4 ENDMILL);
G54 (SET WCS 1);
G0 **G90** X.5 Y2. M8 (RAPID TO POSITION A AND COOLANT);
G43 H2 **Z.1** S10000 M3 (ACTIVATE TOOL HEIGHT OFFSET-MOVE TO CLEAR PLANE-
START SPINDLE);
G1 **Z-.05** F20. (LINEAR FEED TO SLOT DEPTH);
X1. Y.5 F40. (LINEAR FEED TO POSITION B);
X2.5 (LINEAR FEED TO POSITION C);
G3 X3.5 Y1.5 **R1.** (CCW ARC TO POSITION D);
G0 **Z.1** (CLEAR PART IN Z);
M9 (COOLANT OFF);
G91 G28 **Z0.** M05 (RAPID TO MACHINE Z-HOME AND SPINDLE OFF);
G91 G28 X0. Y0. (RAPID TO MACHINE XY-HOME);
M1 (OPTIONAL STOP);
(---------------SIMPLE DRILL OPERATION---------------);
M6 **T3** (CHANGE TOOL TO NO. 7 TWIST DRILL);
G54 (SET WCS 1);
G0 **G90** X3.5 Y2.25 M8 (RAPID TO DRILL POSITION AND COOLANT);
G43 H3 **Z.1** S4584 M3 (ACTIVATE TOOL HEIGHT OFFSET-MOVE TO CLEAR PLANE-
START SPINDLE);
G1 **Z-.56** F23. (LINEAR FEED TO DRILL DEPTH);

ENDING

G0 **Z.1** (CLEAR PART IN Z);
M9 (COOLANT OFF);
G91 G28 **Z0.** M05 (RAPID TO MACHINE Z-HOME AND SPINDLE OFF);
G91 G28 X0. Y0. (RAPID TO MACHINE XY-HOME);
M30 (END PROGRAM);

FIGURE 8.6.24 A #7 (0.201") drill is programmed to produce a 1/2" deep hole at the coordinates of *X 3.5, Y2.5.* The depth is programmed as Z-.56 to compensate for the drill tip length. The drilling operation is highlighted.

This may not seem like a great savings of time for one hole, but when multiple holes are required, coordinate locations are simply listed in subsequent blocks after the *G81* line. The canned cycle will be repeated at every identified location. The tool will retract to the return point (R value) between each location. After the last hole is drilled, the G80 code is then used to cancel the canned cycle. Suppose that more holes are added to the previous part, as shown with the revised program in **Figure 8.6.25.**

EXAMPLE:

G81 X3.5 Y2.25 R0.1 Z-.56 F7.5;

G80:

These two blocks will feed the tool to the 0.56 depth and retract to the return point (R) of Z.1 after drilling. The entire sequence for drilling the hole would look like this:

M6 T3 (NO. 7 DRILL);

G54;

G0 G90 X3.5 Y2.25 M8;

G43 Z.1 H3 M3 S1500;

G81 X3.5 Y2.25 R0.1 Z-.56 F7.5;

G80;

Single-Pass Drilling with Dwell—G82

When spot or center drilling, counterboring, or countersinking, it is often desirable to have the cutting tool **dwell** (*pause*) while it is at full depth to relieve tool pressure or provide a consistent machined surface. A *G82* has the same format as the *G81* but includes the addition of a P-word to specify a dwell time at the end of the feed motion before rapid retract. A value of *P1000* = 1 second. A value of *P500* = 0.5 seconds. A value of *P1500* 5 = 1.5 seconds. The *G82* block contains the X- and Y-coordinates for the hole, a *return point (or R-value)*, the Z-depth, the P-value, and the feed rate. The return point is the absolute Z-position where the tool will begin feeding and where it will retract to at the end of the canned cycle. The table below shows the format for the *G82* cycle.

	Format	Variable Description
Fanuc OR Haas Single Pass with Dwell Cycle Format	(*TOOL TO START POINT*) G82 X_Y_ R_Z_P_F_;	X: X position of hole
		Y: Y position of hole
		R: Return plane position (optional)
		Z: Z end position of hole
		P: Dwell time in milliseconds
		F: Feed rate

Suppose that the five holes from the last example were to be countersunk to a tool depth of *Z-0.15* with a dwell time of 3 seconds. The countersinking cycle portion of the program would be written as shown in **Figure 8.6.26**.

Peck Drilling Cycles

When drilling deeper than about three times the drill's diameter, breaking the downward drilling motion with slight retracts helps interrupt chip flow and clear chips from the drill flutes. There are two *peck drilling cycles* that can be used to perform that motion. A *G73* commands the **fast-peck drilling cycle** and uses only slight retracts. It is sometimes also called a *high-speed* or *chip-break cycle*. **(See Figure 8.6.27.)** A G83 is for the **deep-hole drilling cycle** and fully retracts the tool from the hole between pecks. **(See Figure 8.6.28.)** Both canned cycles use a Q-word to define the amount of each peck. For example, a *Q0.05* would specify a 0.05" peck increment and a *Q0.125* would specify a 0.125" peck increment.

Fast-Peck Drilling Cycle—G73

The *G73* cycle uses a small retract between pecks. This retract amount is usually controlled by a setting in the machine's control. (Consult the specific machine's manual for details.) The *G73* block contains the X- and Y-coordinates for the hole, a *return point (or R-value)*, the Z-depth, peck increment (Q), and the feed rate. The return point is the absolute Z-position where the tool will begin feeding and where it will retract to at the end of the canned cycle. Modifying the code from the last drilling example by changing the *G81* to a *G73* and adding a peck increment of 0.075" would be programmed as shown in **Figure 8.6.29.** The table below shows the format for the *G73* cycle.

	Format	Variable Description
Fanuc OR Haas High Speed Peck Cycle Format	(*TOOL TO START POINT*) G73 X_Y_ R_Q_Z_F_;	X: X position of hole
		Y: Y position of hole
		R: Return plane position (optional)
		Q: Peck increment amount
		Z: Z end position of hole
		F: Feed rate

Deep-Hole Peck Drilling Cycle—G83

When drilling deep holes or difficult-to-machine materials, it is helpful to retract the drill all the way out of the hole after each peck. This helps to clear chips from drill flutes and allow cutting fluid to reach the drill point. The *G83* cycle uses the same format as the *G73* cycle but

FACING/SLOTTING/G81 DRILLING

Hole coordinate locations		
	X	Y
#1	3.5	2.25
#2	3.0	2.25
#3	2.5	2.25
#4	2.0	2.25
#5	1.5	2.25
#6	1.0	2.25

PROGRAM START

O0000 (FACE-SLOT-G81 DRILL);
(PROGRAM FILE NAME-FACESLOTG81DRILL.NC);
(PROGRAM AUTHOR-PJH);
(DATE-5-5-2018);
(INTENDED MACHINE AND CONTROL-HARDINGE VMC600-FANUC);
(X0/Y0 LOCATION-FRONT LEFT CORNER);
(Z0 LOCATION-TOP SURFACE OF FINISHED PART-.900 ABOVE PARALLEL SURFACE);
(T1-3.0 DIA FACE MILL-7 TEETH);
(T2-.25 DIA CARB 2 FL ENDMILL);
(T3-NO. 7 HSS SCREW MACHINE LENGTH TWIST DRILL);
(WORKPIECE MATERIAL-6061-T6 ALUM);
(----------LOAD WORKPIECE-BOTTOM SIDE UP----------);
G0 G17 G20 G40 G54 G80 G94 (SAFE START CODES);
G91 G28 Z0. (RAPID TO MACHINE Z-HOME FOR TOOL CHANGE);
G91 G28 X0. Y0. (RAPID TO MACHINE XY-HOME FOR TOOL CHANGE);
M1 (OPTIONAL STOP);
M6 **T1** (CHANGE TOOL TO FACE MILL);
G54 (SET WCS 1);
G0 G90 X-1.7 Y1.5 M8 (RAPID TO FIRST POSITION AND COOLANT);
G43 H1 Z.1 S2000 M3; (ACTIVATE TOOL HEIGHT OFFSET-MOVE TO CLEAR PLANE-START SPINDLE);

MATERIAL REMOVAL

G1 **Z.05** F15. (LINEAR FEED TO DEPTH FOR FIRST PASS);
X5.7 F30. (LINEAR FEED MOVE ACROSS PART FACE);
G0 Y5. (RAPID TO BRING TABLE FRONT TO OPERATOR)
M0;
(-----FLIP WORKPIECE-FACED SIDE DOWN-TOP SIDE UP-----);
S2000 M3 (RESTART SPINDLE);
G0 Y1.25 M8 (RAPID TABLE BACK TO POSITION AND COOLANT);
G1 **Z.0** (LINEAR FEED TO DEPTH FOR FINAL PASS);
X-1.7 (LINEAR FEED MOVE ACROSS PART FACE);
G0 **Z.1** (CLEAR PART IN Z);
M9 (COOLANT OFF);
G91 G28 **Z0.** M05 (RAPID TO MACHINE Z-HOME AND SPINDLE OFF);
G91 G28 X0. Y0. (RAPID TO MACHINE XY-HOME);
M1 (OPTIONAL STOP);

(---------------SLOT MILLING OPERATION---------------);
M6 T2 (CHANGE TOOL TO 1/4 ENDMILL);
G54 (SET WCS 1);
G0 G90 X.5 Y2. M8 (RAPID TO POSITION A AND COOLANT);
G43 H2 Z.1 S10000 M3 (ACTIVATE TOOL HEIGHT OFFSET-MOVE TO CLEAR PLANE-START SPINDLE);
G1 **Z-.05** F20. (LINEAR FEED TO SLOT DEPTH);
X1. Y.5 F40. (LINEAR FEED TO POSITION B);
X2.5 (LINEAR FEED TO POSITION C);
G3 X3.5 Y1.5 **R1.** (CCW ARC TO POSITION D);
G0 Z.1 (CLEAR PART IN Z);
M9 (COOLANT OFF);
G91 G28 **Z0.** M05 (RAPID TO MACHINE Z-HOME AND SPINDLE OFF);
G91 G28 X0. Y0. (RAPID TO MACHINE XY-HOME);
M1 (OPTIONAL STOP);

(---------------G81 ROW DRILLING OPERATION---------------);
M6 **T3** (CHANGE TOOL TO NO. 7 TWIST DRILL);
G54 (SET WCS 1);
G0 G90 X3.5 Y2.25 M8 (RAPID TO HOLE POSITION 1 AND COOLANT);
G43 H3 Z.1 S4584 M3 (ACTIVATE TOOL HEIGHT OFFSET-MOVE TO CLEAR PLANE-START SPINDLE);
G81 X3.5 Y2.25 **R.1 Z-.56** F23. (LINEAR FEED TO DRILL DEPTH);
X3. (HOLE 2);
X2.5 (HOLE 3);
X2. (HOLE 4);
X1.5 (HOLE 5);
X1. (HOLE 6);
G80 (CANNED CYCLE CANCEL);

ENDING

M9 (COOLANT OFF);
G91 G28 **Z0.** M05 (RAPID TO MACHINE Z-HOME AND SPINDLE OFF);
G91 G28 X0. Y0. (RAPID TO MACHINE XY-HOME);
M30 (END PROGRAM);

FIGURE 8.6.25 A single-pass drilling canned cycle (*G81*) is used to drill 6 holes (highlighted).

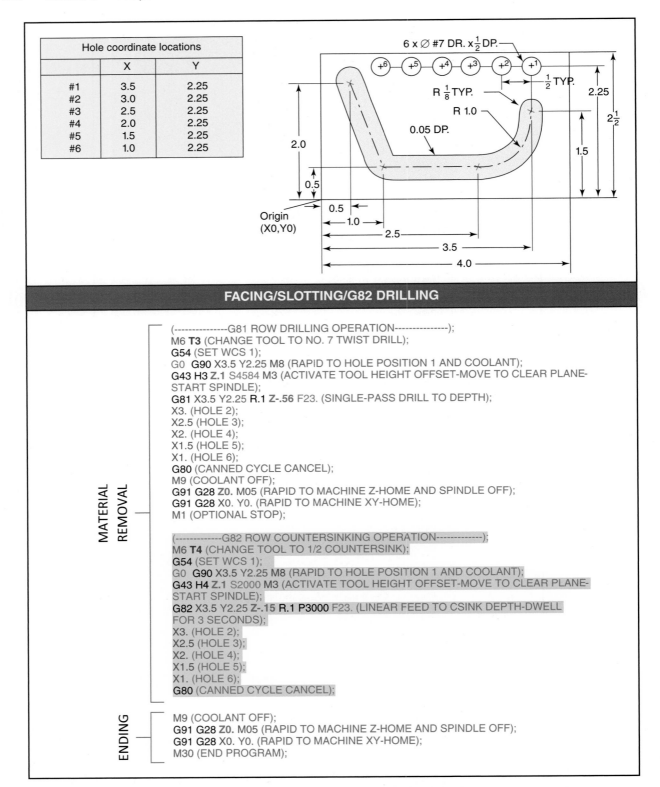

Hole coordinate locations		
	X	Y
#1	3.5	2.25
#2	3.0	2.25
#3	2.5	2.25
#4	2.0	2.25
#5	1.5	2.25
#6	1.0	2.25

FACING/SLOTTING/G82 DRILLING

MATERIAL REMOVAL

```
(--------------G81 ROW DRILLING OPERATION--------------);
M6 T3 (CHANGE TOOL TO NO. 7 TWIST DRILL);
G54 (SET WCS 1);
G0  G90 X3.5 Y2.25 M8 (RAPID TO HOLE POSITION 1 AND COOLANT);
G43 H3 Z.1 S4584 M3 (ACTIVATE TOOL HEIGHT OFFSET-MOVE TO CLEAR PLANE-
START SPINDLE);
G81 X3.5 Y2.25 R.1 Z-.56 F23. (SINGLE-PASS DRILL TO DEPTH);
X3. (HOLE 2);
X2.5 (HOLE 3);
X2. (HOLE 4);
X1.5 (HOLE 5);
X1. (HOLE 6);
G80 (CANNED CYCLE CANCEL);
M9 (COOLANT OFF);
G91 G28 Z0. M05 (RAPID TO MACHINE Z-HOME AND SPINDLE OFF);
G91 G28 X0. Y0. (RAPID TO MACHINE XY-HOME);
M1 (OPTIONAL STOP);

(-------------G82 ROW COUNTERSINKING OPERATION-------------);
M6 T4 (CHANGE TOOL TO 1/2 COUNTERSINK);
G54 (SET WCS 1);
G0  G90 X3.5 Y2.25 M8 (RAPID TO HOLE POSITION 1 AND COOLANT);
G43 H4 Z.1 S2000 M3 (ACTIVATE TOOL HEIGHT OFFSET-MOVE TO CLEAR PLANE-
START SPINDLE);
G82 X3.5 Y2.25 Z-.15 R.1 P3000 F23. (LINEAR FEED TO CSINK DEPTH-DWELL
FOR 3 SECONDS);
X3. (HOLE 2);
X2.5 (HOLE 3);
X2. (HOLE 4);
X1.5 (HOLE 5);
X1. (HOLE 6);
G80 (CANNED CYCLE CANCEL);
```

ENDING

```
M9 (COOLANT OFF);
G91 G28 Z0. M05 (RAPID TO MACHINE Z-HOME AND SPINDLE OFF);
G91 G28 X0. Y0. (RAPID TO MACHINE XY-HOME);
M30 (END PROGRAM);
```

FIGURE 8.6.26 A single-pass drilling cycle with a dwell (*G82*) is used to countersink 6 holes. (highlighted). The program has been abbreviated to show only the drilling cycles.

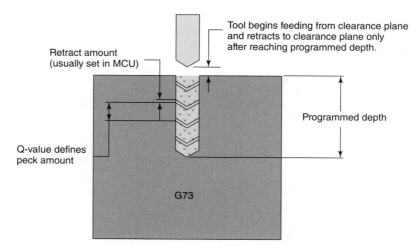

FIGURE 8.6.27   The *G73* canned cycle. The tool pecks enough to break the chip, but doesn't fully retract until the hole is complete.

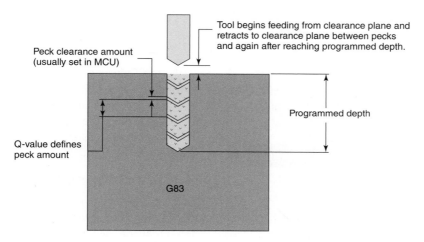

FIGURE 8.6.28   The *G83* canned cycle. The tool fully retracts after each peck.

```
(-------------G73 ROW DRILLING OPERATION-------------);
M6 T3 (CHANGE TOOL TO NO. 7 TWIST DRILL);
G54 (SET WCS 1);
G0  G90 X3.5 Y2.25 M8 (RAPID TO HOLE POSITION 1 AND COOLANT);
G43 H3 Z.1 S4584 M3 (ACTIVATE TOOL HEIGHT OFFSET-MOVE TO CLEAR PLANE-
START SPINDLE);
G73 X3.5 Y2.25 Z-.56 R.1 Q.075 F23. (HIGH SPEED PECK CYCLE TO DEPTH);
X3. (HOLE 2);
X2.5 (HOLE 3);
X2. (HOLE 4);
X1.5 (HOLE 5);
X1. (HOLE 6);
G80 (CANNED CYCLE CANCEL);

M9 (COOLANT OFF);
G91 G28 Z0. M05 (RAPID TO MACHINE Z-HOME AND SPINDLE OFF);
G91 G28 X0. Y0. (RAPID TO MACHINE XY-HOME);
M30 (END PROGRAM);
```

MATERIAL REMOVAL

ENDING

FIGURE 8.6.29 The G81 drilling operation from the previous example in 8.6.25 could be replaced with this G73 peck drill cycle program. The tool would peck as it feeds to depth, but doesn't fully retract until the hole is complete. The peck increment is specified by the Q-value (highlighted).

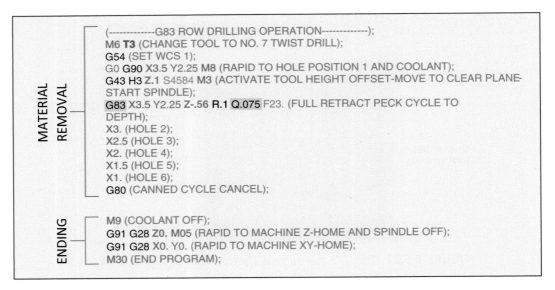

FIGURE 8.6.30 The G81 drilling operation from the previous example in 8.6.25 could be replaced with this G83 peck drill cycle program. The tool would peck as it feeds to depth and fully retract between each peck. The peck increment is specified by the Q-value (highlighted).

will fully retract the drill from the hole between pecks. A Q-word is still used to define the peck increment. **(See Figure 8.6.30.)** The table below shows the format for the G83 cycle.

	Format	**Variable Description**
Fanuc OR Haas High Speed Peck Cycle Format	*(TOOL TO START POINT)* G83 X_ Y_ R_ Q_ Z_ F_;	X: X position of hole
		Y: Y position of hole
		R: Return plane position (optional)
		Q: Peck increment amount
		Z: Z end position of hole
		F: Feed rate

Tapping Canned Cycles

A tapping canned cycle is very similar to a single-pass drilling cycle, only there must be precise coordination between the spindle rotation and feed rate of the Z-axis. When the tap begins cutting, the spindle rotation and the feed must be precisely matched so the tap does not bind and break. As the tap approaches the desired depth, the machine must slow the spindle and the feed so they both stop moving when the tap reaches the final depth. Finally, the spindle and feed must reverse direction in sync to retract from the hole.

A *G84* canned cycle is used for right-hand tapping, and a *G74* is used for left-hand tapping. These cycles will feed the tap into the hole, then automatically reverse the spindle and feed direction to retract from the hole. The tapping cycle format contains X-, Y-, Z-, R-, and F-values just like a *G81* drilling cycle.

A tap advances into the work one thread per revolution, so the IPR feed rate is equal to the pitch of the tap, or 1/*TPI*. If using an IPM feed rate, IPM = IPR x RPM. If the feed rate is a non-terminating decimal, round it to as many places as the machine control can accept (typically four).

Standard Tapping Canned Cycle

Some machines are not capable of matching the spindle speed and Z-axis feed rate precisely enough to prevent tap breakage. When tapping with these machines, a floating tap holder is necessary to compensate for small errors between the spindle RPM and the feed rate. The table on the next page shows the formats for the *G84* and *G74* tapping cycles.

The *G84* example in **Figure 8.6.31** uses the same coordinate locations from the previous samples to tap six 1/4-20 holes. A left-hand tapping cycle would do the same except the *G84* would be replaced with a *G74* and the spindle would need to be programmed for counterclockwise rotation with an *M4*. When tapping, it is usually a good idea to increase the clearance plane to be sure that the floating holder fully retracts the tap from the hole before moving to the next location.

Rigid Tapping Canned Cycle

Some machines have a more accurate capability of mating the spindle RPM and feed rate throughout the entire tapping cycle. This feature is called **rigid tapping** because the tap does not need to be mounted in a floating tool holder. It can be held rigidly in a collet chuck or tapping chuck. When programming rigid tapping, the spindle is programmed to the desired speed in the

	Format	Variable Description
Right-Hand Tapping Cycle Format	(*TOOL TO START POINT*) G84 X_Y_ R_ Z_ F_;	X: X position of hole Y: Y position of hole R: Return plane position (optional) Z: Z end position of hole F: Feed rate
Left-Hand Tapping Cycle Format	(*TOOL TO START POINT*) G74 X_Y_ R_Z_ F_;	X: X position of hole Y: Y position of hole R: Return plane position (optional) Z: Z end position of hole F: Feed rate

```
(---------------G84 NON-RIGID FANUC TAPPING OPERATION---------------);
M6 T8 (CHANGE TOOL TO 1/4-20 TAP);
G54 (SET WCS 1);
G0  G90 X3.5 Y2.25 M8 (RAPID TO HOLE POSITION 1 AND COOLANT);
G43 H8 Z.25 S500 M3 (ACTIVATE TOOL HEIGHT OFFSET-MOVE TO CLEAR PLANE-START SPINDLE);
G84 G95 X3.5 Y2.25 R.25 Z-.5 F.05 (TAP CYCLE TO DEPTH-SET FEED UNITS TO IPR);
X3. (HOLE 2);
X2.5 (HOLE 3);
X2. (HOLE 4);
X1.5 (HOLE 5);
X1. (HOLE 6);
G80 (CANNED CYCLE CANCEL);
```

FIGURE 8.6.31 A G84 non-rigid tapping cycle program excerpt. The tapping cycle block is highlighted.

normal method. Fanuc machines sometimes require an additional block directly prior to the *G84* or *G74* line. This block uses an *M29* code to activate rigid tapping and must restate the spindle speed. For example, if the spindle is started at 500 RPM, the next block must be *M29 S500*. Haas machines don't require an *M29* and will perform rigid tapping with just the standard G84 or *G74* format shown in the previous table. The table below shows the formats for the *G84* and *G74* rigid tapping cycles for Fanuc.

	Format	Variable Description
Fanuc Right-Hand Rigid Tapping Cycle Format	(*TOOL TO START POINT*) M29 S_; G84 X_Y_ R_ Z_ F_; **NOTE:** The M29 block can be omitted for use on a Haas	S: Spindle speed X: X position of hole Y: Y position of hole R: Return plane position (optional) Z: Z end position of hole F: Feed rate
Fanuc Left-Hand Rigid Tapping Cycle Format	(*TOOL TO START POINT*) M29 S_; G74 X_Y_ R_Z_ F_; **NOTE:** The M29 block can be omitted for use on a Haas	S: Spindle speed X: X position of hole Y: Y position of hole R: Return plane position (optional) Z: Z end position of hole F: Feed rate

```
(----------------G84 RIGID FANUC TAPPING OPERATION---------------);
M6 T8 (CHANGE TOOL TO 1/4-20 TAP);
G54 (SET WCS 1);
G0 G90 X3.5 Y2.25 M8 (RAPID TO HOLE POSITION 1 AND COOLANT);
G43 H8 Z.1 S1375 M3 (ACTIVATE TOOL HEIGHT OFFSET-MOVE TO CLEAR PLANE-START SPINDLE);
M29 S1375 (ACTIVATE RIGID TAPPING);
G84 G95 X3.5 Y2.25 R.1 Z-.5 F.05 (START TAP CYCLE-TAP TO DEPTH-SET FEED UNITS TO IPR);
X3. (HOLE 2);
X2.5 (HOLE 3);
X2. (HOLE 4);
X1.5 (HOLE 5);
X1. (HOLE 6);
G80 (CANNED CYCLE CANCEL);
```

FIGURE 8.6.32 A G84 rigid tapping cycle program excerpt. The activation code block and the tapping cycle block are highlighted.

The rigid tapping example in **Figure 8.6.32** uses the same information as the previous tapping sample, adapted for a machine capable of rigid tapping.

Initial Plane and Return Plane

Up to this point, all the canned cycle examples used an R-value equal to the clearance plane. The Z-position at the clearance plane is called the *initial plane and* is the Z-position where the tool is located when the canned cycle is activated. In most cases, retracting the tool to the same place it began is ideal. However, some situations may benefit from using a return point that is different from the initial plane. This location is called the *return plane*, or *R-plane.* The R-value is given to designate the return plane and is where the tool will rapid to before beginning to feed into the hole. This is also where a *G83* cycle will retract to between pecks. A *G98* code in the canned cycle activation block commands the tool to return to the initial Z-plane when rapid positioning between hole locations.

A *G99* code commands the tool to return to the R-plane defined in the canned cycle activation block when rapid positioning between hole locations. When neither is specified, the machine defaults to *G98*. This was the case for the prior canned cycle examples. These *G98/G99* options are useful when the surface to be drilled is below the top of the workpiece (as in a pocket) or when there are obstructions between holes such as raised part features or workholding devices. **Figures 8.6.33** through **8.6.35** show some *G98* and *G99* examples with different R-planes.

 **CAUTION**

Use caution when using *G99* and be sure the return point will not cause collisions when moving between hole locations.

CUTTER RADIUS COMPENSATION

Up to this point, all milling examples were programmed relative to the centerline of the cutting tool. When peripheral milling, the programmed coordinates need to be adjusted for the radius of the cutting tool. This offsetting of the path of the cutting tool is called **cutter radius compensation**. When programming cutter paths parallel to the machine's X- and Y-axes, this is straightforward. When the program coordinates are modified by the programmer to offset the tool's path, this is called *manual cutter radius compensation*. **Figure 8.6.36** shows a workpiece and a sample program where a basic rectangular contour is peripheral milled using manual cutter radius compensation. Here, all coordinates have been adjusted to offset the radius of the tool. Each end point places the tool in the correct position for the next move.

EXAMPLE: Determine the X and Y coordinates to mill the contour shown in **Figure 8.6.36** to a depth of 0.500" using a 1/2"-diameter endmill. The coordinates of the corners of the profile are given and the arrows show the cutting direction.

Since the contour will be machined with the periphery of the endmill, the coordinates of the corners need to be offset by the radius of the endmill. Since the endmill diameter is 1/2", the coordinates need to be offset by 1/4", as shown in **Figure 8.6.36**. Position #1 is the starting point and position #5 is the ending point. Notice that positions are also adjusted so the endmill starts away from the edge of the part and feeds completely off the part at the end of the contour.

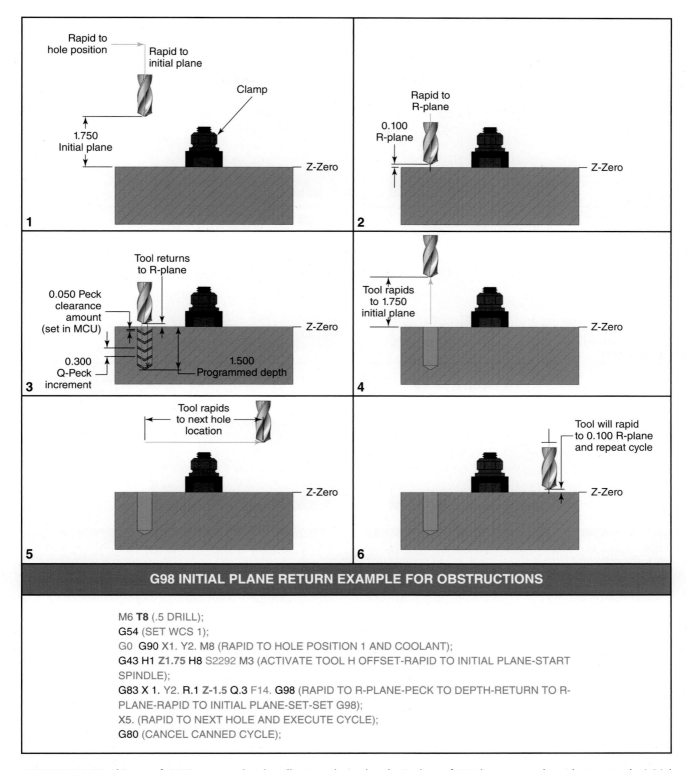

G98 INITIAL PLANE RETURN EXAMPLE FOR OBSTRUCTIONS

```
M6 T8 (.5 DRILL);
G54 (SET WCS 1);
G0 G90 X1. Y2. M8 (RAPID TO HOLE POSITION 1 AND COOLANT);
G43 H1 Z1.75 H8 S2292 M3 (ACTIVATE TOOL H OFFSET-RAPID TO INITIAL PLANE-START
SPINDLE);
G83 X 1. Y2. R.1 Z-1.5 Q.3 F14. G98 (RAPID TO R-PLANE-PECK TO DEPTH-RETURN TO R-
PLANE-RAPID TO INITIAL PLANE-SET-SET G98);
X5. (RAPID TO NEXT HOLE AND EXECUTE CYCLE);
G80 (CANCEL CANNED CYCLE);
```

FIGURE 8.6.33 This use of *G98* in a canned cycle will retract the tool to the R-plane of *Z0.1* between pecks and retract to the initial plane of *Z1.5* between hole locations.

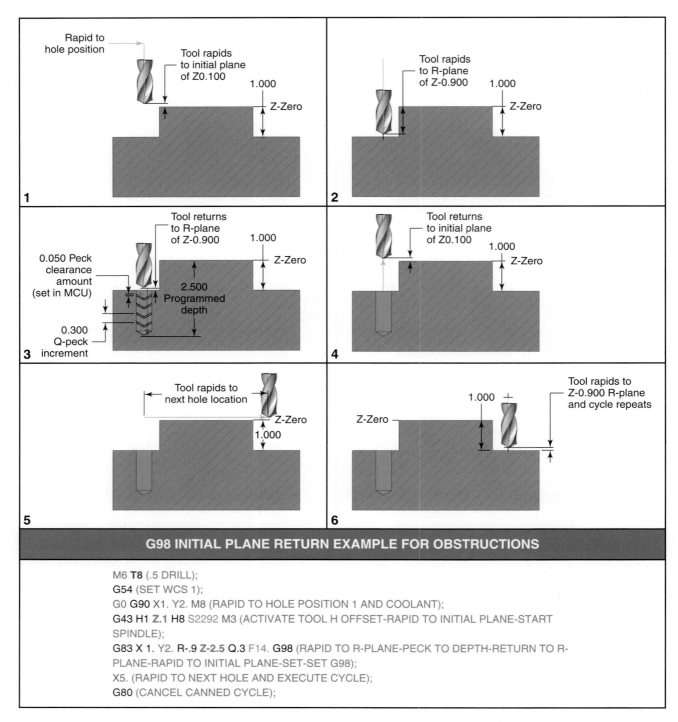

G98 INITIAL PLANE RETURN EXAMPLE FOR OBSTRUCTIONS

M6 **T8** (.5 DRILL);
G54 (SET WCS 1);
G0 **G90** X1. Y2. M8 (RAPID TO HOLE POSITION 1 AND COOLANT);
G43 H1 **Z.1** H8 S2292 M3 (ACTIVATE TOOL H OFFSET-RAPID TO INITIAL PLANE-START SPINDLE);
G83 X 1. Y2. R-.9 **Z-2.5** Q.3 F14. G98 (RAPID TO R-PLANE-PECK TO DEPTH-RETURN TO R-PLANE-RAPID TO INITIAL PLANE-SET-SET G98);
X5. (RAPID TO NEXT HOLE AND EXECUTE CYCLE);
G80 (CANCEL CANNED CYCLE);

FIGURE 8.6.34 This use of *G98* programmed in a canned cycle will retract the tool to only *Z-0.9* between pecks and retract to the initial plane of *Z0.1* between hole locations.

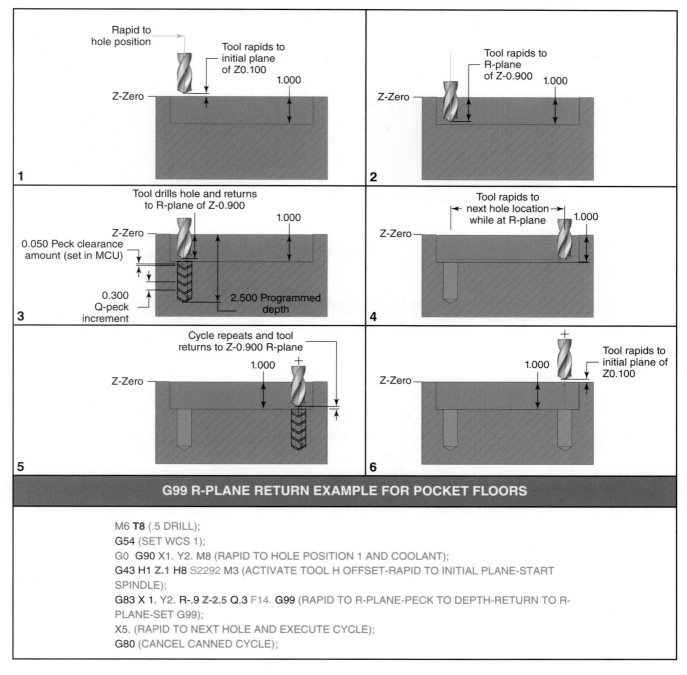

1 Rapid to hole position / Tool rapids to initial plane of Z0.100 / Z-Zero / 1.000

2 Tool rapids to R-plane of Z-0.900 / Z-Zero / 1.000

3 Tool drills hole and returns to R-plane of Z-0.900 / 1.000 / Z-Zero / 0.050 Peck clearance amount (set in MCU) / 0.300 Q-peck increment / 2.500 Programmed depth

4 Tool rapids to next hole location while at R-plane / 1.000 / Z-Zero

5 Cycle repeats and tool returns to Z-0.900 R-plane / 1.000 / Z-Zero

6 Tool rapids to initial plane of Z0.100 / 1.000 / Z-Zero

G99 R-PLANE RETURN EXAMPLE FOR POCKET FLOORS

M6 **T8** (.5 DRILL);
G54 (SET WCS 1);
G0 G90 X1. Y2. M8 (RAPID TO HOLE POSITION 1 AND COOLANT);
G43 H1 **Z.1** H8 S2292 M3 (ACTIVATE TOOL H OFFSET-RAPID TO INITIAL PLANE-START SPINDLE);
G83 X 1. Y2. R-.9 **Z-2.5** Q.3 F14. G99 (RAPID TO R-PLANE-PECK TO DEPTH-RETURN TO R-PLANE-SET G99);
X5. (RAPID TO NEXT HOLE AND EXECUTE CYCLE);
G80 (CANCEL CANNED CYCLE);

FIGURE 8.6.35 This *G99* canned cycle will retract the tool to the R-plane of *Z-0.9* between pecks and between holes. The tool will only retract to the initial plane of *Z0.1* when the canned cycle is canceled with a *G80* command.

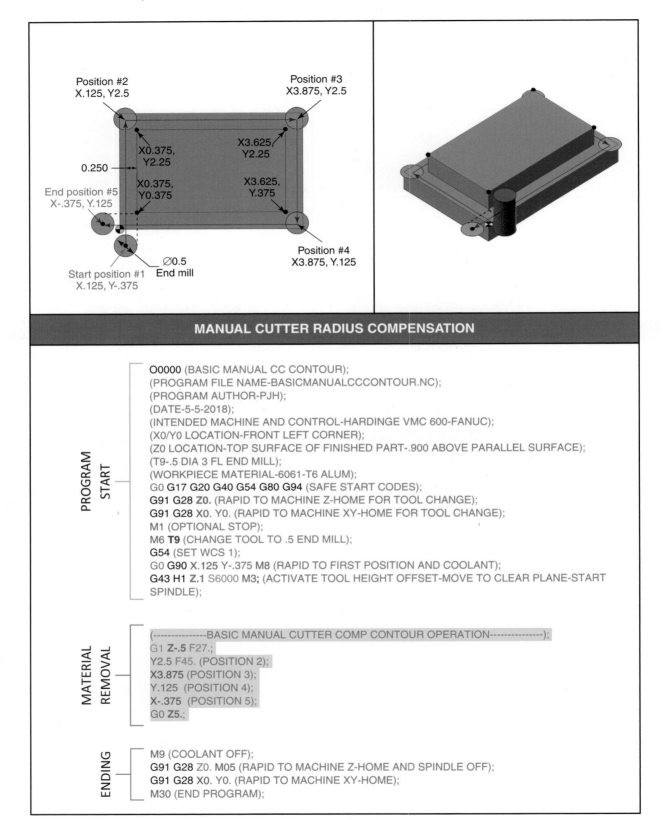

MANUAL CUTTER RADIUS COMPENSATION

PROGRAM START

```
O0000 (BASIC MANUAL CC CONTOUR);
(PROGRAM FILE NAME-BASICMANUALCCCONTOUR.NC);
(PROGRAM AUTHOR-PJH);
(DATE-5-5-2018);
(INTENDED MACHINE AND CONTROL-HARDINGE VMC 600-FANUC);
(X0/Y0 LOCATION-FRONT LEFT CORNER);
(Z0 LOCATION-TOP SURFACE OF FINISHED PART-.900 ABOVE PARALLEL SURFACE);
(T9-.5 DIA 3 FL END MILL);
(WORKPIECE MATERIAL-6061-T6 ALUM);
G0 G17 G20 G40 G54 G80 G94 (SAFE START CODES);
G91 G28 Z0. (RAPID TO MACHINE Z-HOME FOR TOOL CHANGE);
G91 G28 X0. Y0. (RAPID TO MACHINE XY-HOME FOR TOOL CHANGE);
M1 (OPTIONAL STOP);
M6 T9 (CHANGE TOOL TO .5 END MILL);
G54 (SET WCS 1);
G0 G90 X.125 Y-.375 M8 (RAPID TO FIRST POSITION AND COOLANT);
G43 H1 Z.1 S6000 M3; (ACTIVATE TOOL HEIGHT OFFSET-MOVE TO CLEAR PLANE-START
SPINDLE);
```

MATERIAL REMOVAL

```
(---------------BASIC MANUAL CUTTER COMP CONTOUR OPERATION---------------);
G1 Z-.5 F27.;
Y2.5 F45. (POSITION 2);
X3.875 (POSITION 3);
Y.125  (POSITION 4);
X-.375  (POSITION 5);
G0 Z5.;
```

ENDING

```
M9 (COOLANT OFF);
G91 G28 Z0. M05 (RAPID TO MACHINE Z-HOME AND SPINDLE OFF);
G91 G28 X0. Y0. (RAPID TO MACHINE XY-HOME);
M30 (END PROGRAM);
```

FIGURE 8.6.36 Workpiece dimensions required to program a simple contour using manual cutter radius compensation. Workpiece coordinates are offset by an amount equal to the radius of the cutting tool. The program is shown and the compensated cutter path is highlighted.

When cutter motions are not parallel to one of the machine's axes, such as when radii and angular movements are required, manual cutter compensation requires geometric and trigonometric calculations to determine each cutter position. These calculations can become time consuming and tedious. **Figure 8.6.37** shows a drawing and program for a part contour containing an angle and radii. The calculations for the angular intersections have been performed in advance, but the triangles used for the trigonometry calculations are shown. The centerline of the cutter's path has been dimensioned and the coordinates for that path are programmed.

Automatic Cutter Radius Compensation

To make contour programming easier and to avoid time-consuming trigonometry and geometry calculations, CNC machines have a feature called *automatic cutter radius compensation,* or *cutter comp* for short. Cutter comp allows the program to be written using dimensions and locations directly from the part drawing. The control then offsets the tool path automatically.

Cutter comp also allows the size of the contour milled shape to be adjusted without the need to reprogram all the coordinates. For instance, if a re-sharpened cutting tool is used, it will have a smaller diameter after sharpening. By entering the new tool radius at a specific screen of the machine control, all coordinates will be adjusted automatically by the control and the program will not need any changes. As tools wear during the manufacture of parts, size can also be adjusted this way to keep dimensions in tolerance.

A *G41* or *G42* code is used in the program to activate cutter comp. While active, the control automatically compensates for the tool's radius and adjusts machine movements. A D-value must also be programmed to activate a tool radius offset value that is stored in the offset table of the machine control. The process of entering this value in the control will be covered in Section 8.7. Once activated, all tool movement is programmed in the same manner as before including use of *G1, G2,* and *G3* motions.

A *G41* code is used for cutter compensation left and a *G42* is for cutter compensation right. **Figure 8.6.38** illustrates left-side and right-side cutter comp. A good way to remember left and right cutter comp is that a *G41* (left) is needed for climb milling and a *G42* (right) is needed for conventional milling. Since much of CNC milling is climb milling, *G41* use is much more common. Cutter comp is a modal command and must be cancelled after use with a *G40* command. **Figure 8.6.39** shows an example of a block of code used to activate cutter radius compensation. The order of the codes in this block is not important.

Figure 8.6.40 shows a workpiece and a sample program where a basic rectangular contour is peripheral milled using automatic cutter radius compensation. Here, all coordinates are programmed directly from the part drawing dimensions. The control will automatically offset the cutter path and place the end point of the tool in the correct position for the next move.

Several rules must be followed to successfully program a milling operation using cutter comp. Study **Figure 8.6.40** and notice where each of these rules are followed:

1. Select the appropriate activation code (*G41* or *G42*) and call the appropriate D-value.

2. There must be an activation movement in the X- or Y-axis in an area away from the finished contour of the part. In this move the tool will offset to the chosen side of the path (left or right).
   a. This movement must be linear (*G1*).
   b. The distance moved must be at least the size of the tool radius.
   c. This movement should create an angle 90 degrees or greater between itself and the next movement the tool makes.

3. There must be an exit move away from the part where cutter comp is deactivated. In this move the tool will shift back so that it is centered with programmed coordinates.
   a. Feed the tool away from the contour before deactivating.
   b. This movement must be linear (*G1*).
   c. The distance moved must be at least the size of the tool radius.
   d. This movement should create an angle 90 degrees or greater between itself and the previous movement the tool makes.

Refer to **Figure 8.6.40** while reading the following description about how to program a simple contour using cutter comp.

- The Y-coordinate of the start point should be positioned away from the finished contour by at least the tool radius to keep the tool from contacting the work during the activation move.
- Actual tool movement is from position #1 to position #2 during the cutter comp activation block.
- Once activated, coordinates are programmed directly from the finished contour coordinates (positions #3–#6) using linear and circular interpolation motions.
- At the end of the contour the tool should continue off the finished contour before making the move that cancels cutter comp (position #6 to position #7).

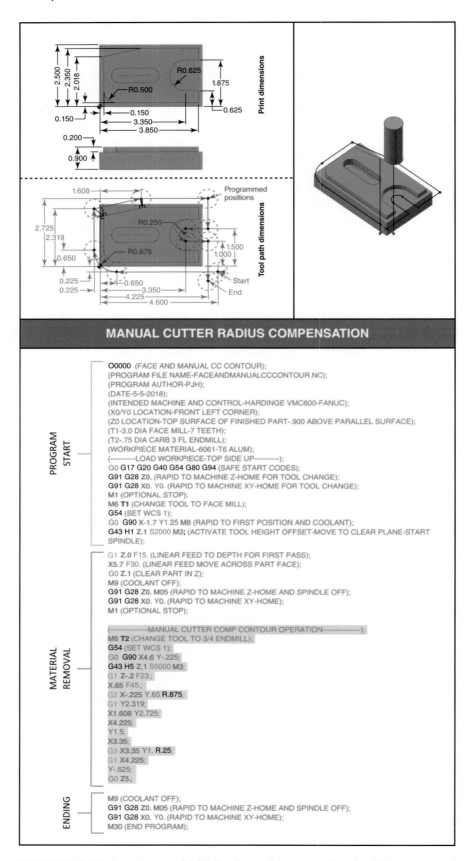

MANUAL CUTTER RADIUS COMPENSATION

PROGRAM START

```
O0000 (FACE AND MANUAL CC CONTOUR);
(PROGRAM FILE NAME-FACEANDMANUALCCCONTOUR.NC);
(PROGRAM AUTHOR-PJH);
(DATE-5-5-2018);
(INTENDED MACHINE AND CONTROL-HARDINGE VMC600-FANUC);
(X0/Y0 LOCATION-FRONT LEFT CORNER);
(Z0 LOCATION-TOP SURFACE OF FINISHED PART-.900 ABOVE PARALLEL SURFACE);
(T1-3.0 DIA FACE MILL-7 TEETH);
(T2-.75 DIA CARB 3 FL ENDMILL);
(WORKPIECE MATERIAL-6061-T6 ALUM);
(----------LOAD WORKPIECE-TOP SIDE UP----------);
G0 G17 G20 G40 G54 G80 G94 (SAFE START CODES);
G91 G28 Z0. (RAPID TO MACHINE Z-HOME FOR TOOL CHANGE);
G91 G28 X0. Y0. (RAPID TO MACHINE XY-HOME FOR TOOL CHANGE);
M1 (OPTIONAL STOP);
M6 T1 (CHANGE TOOL TO FACE MILL);
G54 (SET WCS 1);
G0 G90 X-1.7 Y1.25 M8 (RAPID TO FIRST POSITION AND COOLANT);
G43 H1 Z.1 S2000 M3; (ACTIVATE TOOL HEIGHT OFFSET-MOVE TO CLEAR PLANE-START
SPINDLE);
```

MATERIAL REMOVAL

```
G1 Z.0 F15. (LINEAR FEED TO DEPTH FOR FIRST PASS);
X5.7 F30. (LINEAR FEED MOVE ACROSS PART FACE);
G0 Z.1 (CLEAR PART IN Z);
M9 (COOLANT OFF);
G91 G28 Z0. M05 (RAPID TO MACHINE Z-HOME AND SPINDLE OFF);
G91 G28 X0. Y0. (RAPID TO MACHINE XY-HOME);
M1 (OPTIONAL STOP);

(----------------MANUAL CUTTER COMP CONTOUR OPERATION----------------);
M6 T2 (CHANGE TOOL TO 3/4 ENDMILL);
G54 (SET WCS 1);
G0 G90 X4.6 Y-.225;
G43 H5 Z.1 S5000 M3;
G1 Z-.2 F23.;
X.65 F45.;
G2 X-.225 Y.65 R.875;
G1 Y2.319;
X1.608 Y2.725;
X4.225;
Y1.5;
X3.35;
G3 X3.35 Y1. R.25;
G1 X4.225;
Y-.625;
G0 Z5.;
```

ENDING

```
M9 (COOLANT OFF);
G91 G28 Z0. M05 (RAPID TO MACHINE Z-HOME AND SPINDLE OFF);
G91 G28 X0. Y0. (RAPID TO MACHINE XY-HOME);
M30 (END PROGRAM);
```

FIGURE 8.6.37 A part contour (highlighted) containing an angle and radii is programmed using manual cutter radius compensation. Workpiece coordinates are offset mathematically by an amount equal to the radius of the cutting tool.

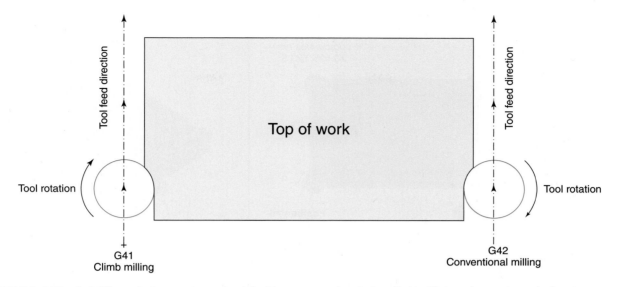

FIGURE 8.6.38 A *G41* is used when programming left-side compensation during climb milling and a *G42* is used when programming right-side compensation during conventional milling.

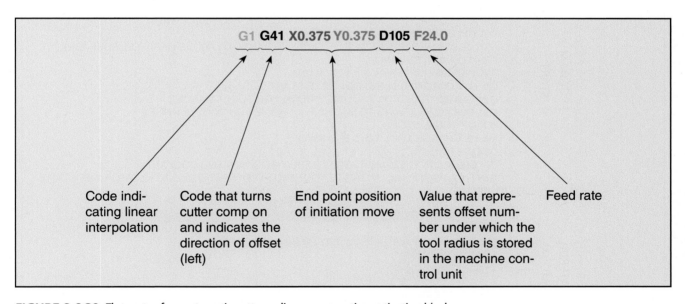

G1 **G41** X0.375 Y0.375 **D105** F24.0

| Code indicating linear interpolation | Code that turns cutter comp on and indicates the direction of offset (left) | End point position of initiation move | Value that represents offset number under which the tool radius is stored in the machine control unit | Feed rate |

FIGURE 8.6.39 The parts of an automatic cutter radius compensation activation block.

Figure 8.6.41 shows a drawing and the program for a part contour containing an angle and radii. Automatic cutter comp performs the complex math calculations "behind the scenes" to offset the cutter path. The final part contour is dimensioned and the coordinates for that path are programmed.

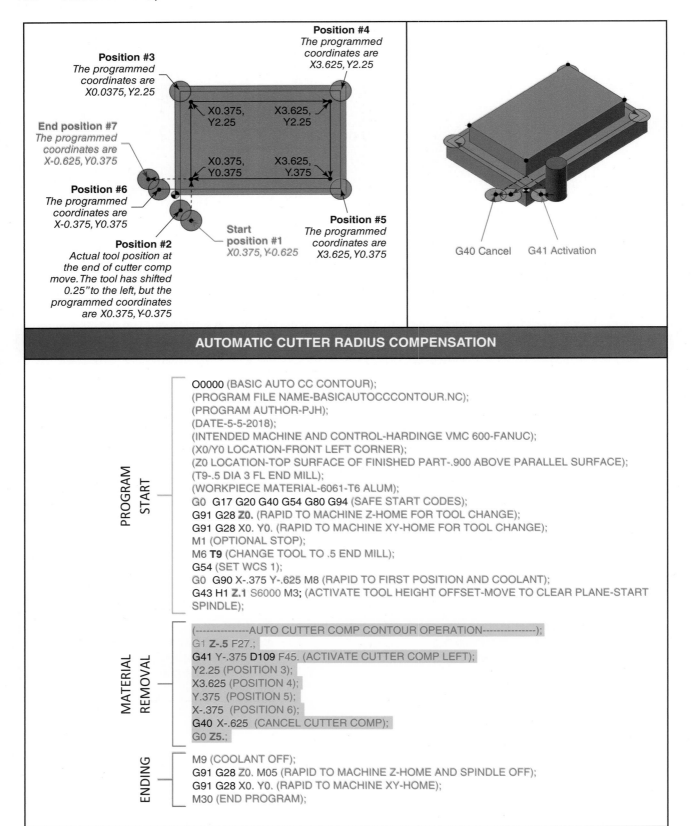

Position #3
The programmed coordinates are X0.0375, Y2.25

Position #4
The programmed coordinates are X3.625, Y2.25

X0.375, Y2.25

X3.625, Y2.25

End position #7
The programmed coordinates are X-0.625, Y0.375

X0.375, Y0.375

X3.625, Y.375

Position #6
The programmed coordinates are X-0.375, Y0.375

Position #2
Actual tool position at the end of cutter comp move. The tool has shifted 0.25" to the left, but the programmed coordinates are X0.375, Y-0.375

Start position #1
X0.375, Y-0.625

Position #5
The programmed coordinates are X3.625, Y0.375

G40 Cancel   G41 Activation

AUTOMATIC CUTTER RADIUS COMPENSATION

PROGRAM START

```
O0000 (BASIC AUTO CC CONTOUR);
(PROGRAM FILE NAME-BASICAUTOCCCONTOUR.NC);
(PROGRAM AUTHOR-PJH);
(DATE-5-5-2018);
(INTENDED MACHINE AND CONTROL-HARDINGE VMC 600-FANUC);
(X0/Y0 LOCATION-FRONT LEFT CORNER);
(Z0 LOCATION-TOP SURFACE OF FINISHED PART-.900 ABOVE PARALLEL SURFACE);
(T9-.5 DIA 3 FL END MILL);
(WORKPIECE MATERIAL-6061-T6 ALUM);
G0  G17 G20 G40 G54 G80 G94 (SAFE START CODES);
G91 G28 Z0. (RAPID TO MACHINE Z-HOME FOR TOOL CHANGE);
G91 G28 X0. Y0. (RAPID TO MACHINE XY-HOME FOR TOOL CHANGE);
M1 (OPTIONAL STOP);
M6 T9 (CHANGE TOOL TO .5 END MILL);
G54 (SET WCS 1);
G0  G90 X-.375 Y-.625 M8 (RAPID TO FIRST POSITION AND COOLANT);
G43 H1 Z.1 S6000 M3; (ACTIVATE TOOL HEIGHT OFFSET-MOVE TO CLEAR PLANE-START SPINDLE);
```

MATERIAL REMOVAL

```
(----------------AUTO CUTTER COMP CONTOUR OPERATION----------------);
G1 Z-.5 F27.;
G41 Y-.375 D109 F45. (ACTIVATE CUTTER COMP LEFT);
Y2.25 (POSITION 3);
X3.625 (POSITION 4);
Y.375  (POSITION 5);
X-.375  (POSITION 6);
G40 X-.625  (CANCEL CUTTER COMP);
G0 Z5.;
```

ENDING

```
M9 (COOLANT OFF);
G91 G28 Z0. M05 (RAPID TO MACHINE Z-HOME AND SPINDLE OFF);
G91 G28 X0. Y0. (RAPID TO MACHINE XY-HOME);
M30 (END PROGRAM);
```

FIGURE 8.6.40 Workpiece dimensions required to program a simple contour using automatic cutter radius compensation. Workpiece coordinates are programmed directly and the radius offset is performed by the control. The program is shown and the compensated cutter path is highlighted.

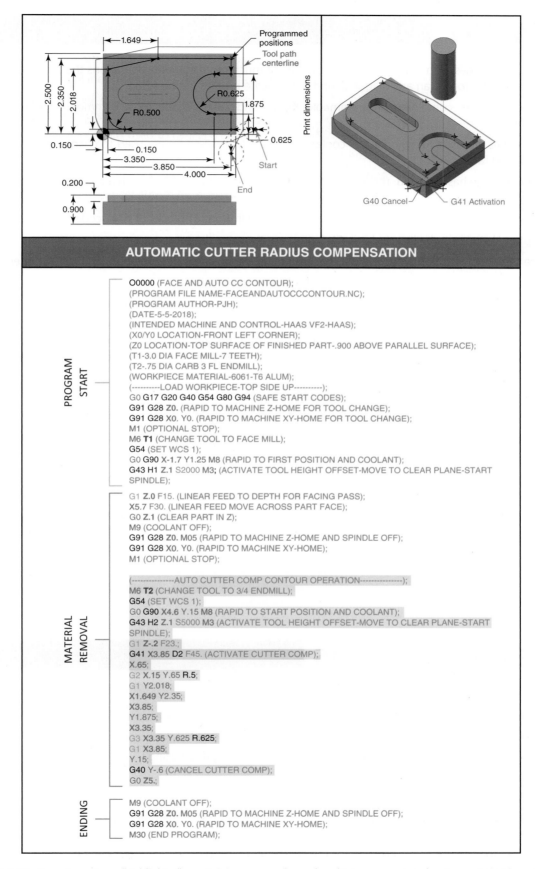

FIGURE 8.6.41 A part contour (highlighted) containing an angle and radii is programmed using automatic cutter radius compensation. Workpiece coordinates are offset automatically by an amount equal to the radius of the cutting tool.

FIGURE 8.7.3 Arrow keys used for constant jogging of each axis.

A machine mode knob or machine mode keys are used to change from one mode of operation to another (Figure 8.7.1). Common machine modes include jog, auto, **manual data input (MDI)**, edit, and zero-reference return.

The edit mode allows new programs to be typed into memory or an existing program to be modified. Edit is also usually the mode required to load stored programs for use. When programs are to be run, the control must be placed in auto mode. Reference return mode is used to zero the machine axes upon power-up.

MDI mode provides a blank program screen for entering short programs or single program commands. Program data entered into MDI is not saved in memory and is erased after it is executed. Therefore, MDI is ideal for short operations (used during setup or troubleshooting) such as tool changes, spindle start commands, or moving an axis to a specific coordinate for setup.

When **jog** mode is turned on, buttons and a small rotary hand wheel allow user-controlled axis movement during machine setup. Most machines also have arrow keys in the plus and minus direction for jogging (moving) each axis. **(See Figure 8.7.3.)** Pressing and holding these keys causes constant axis motion, and axis motion stops when the button is released. When jogging with the arrow keys, most machines vary the feed rate with either the rapid or feed override dial or keys (Figure 8.7.1). The jogging MPG (manual pulse generator) hand wheel can be set to move an axis at different incremental steps for

fine control when jogging machine axes. These steps are usually 0.010", 0.001", or 0.0001" per click of the hand wheel. The jogging hand wheel on many machines is mounted on a handheld pendant and attached to the MCU by a cable to allow its use away from the control panel and closer to the work area. **(See Figure 8.7.4.)**

The cycle start, feed hold, and emergency stop buttons are also located on the control panel (Figure 8.7.1). The cycle start button begins the active CNC program and is usually green. The feed hold button is usually located directly next to the cycle start button and will stop axis feed when pressed while running a program. In an emergency or collision, the red emergency stop button (sometimes called the E-stop) can be pressed to immediately stop the spindle and all axis motion. The

FIGURE 8.7.4 A jogging hand wheel is used for fine control when jogging the machine's axes. The one shown is portable and is connected to the machine control by a cord.

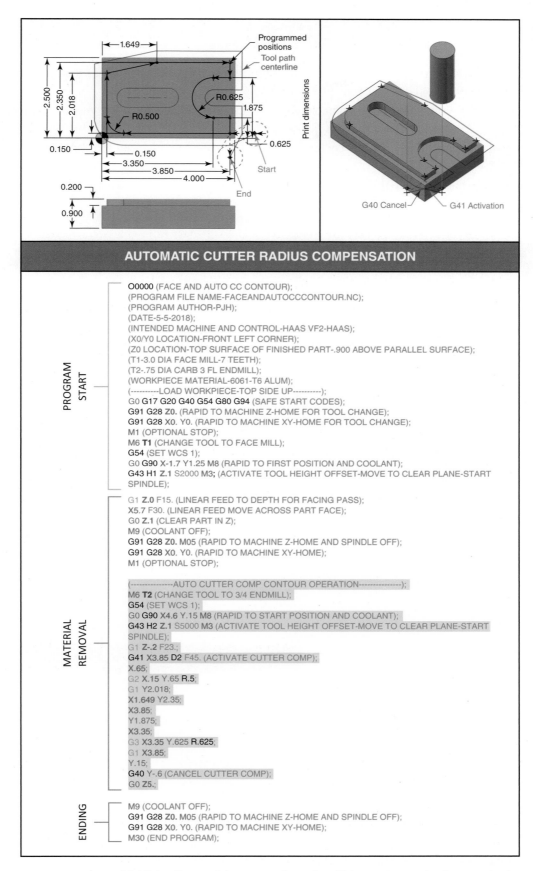

AUTOMATIC CUTTER RADIUS COMPENSATION

PROGRAM START

```
O0000 (FACE AND AUTO CC CONTOUR);
(PROGRAM FILE NAME-FACEANDAUTOCCCONTOUR.NC);
(PROGRAM AUTHOR-PJH);
(DATE-5-5-2018);
(INTENDED MACHINE AND CONTROL-HAAS VF2-HAAS);
(X0/Y0 LOCATION-FRONT LEFT CORNER);
(Z0 LOCATION-TOP SURFACE OF FINISHED PART-.900 ABOVE PARALLEL SURFACE);
(T1-3.0 DIA FACE MILL-7 TEETH);
(T2-.75 DIA CARB 3 FL ENDMILL);
(WORKPIECE MATERIAL-6061-T6 ALUM);
(----------LOAD WORKPIECE-TOP SIDE UP----------);
G0 G17 G20 G40 G54 G80 G94 (SAFE START CODES);
G91 G28 Z0. (RAPID TO MACHINE Z-HOME FOR TOOL CHANGE);
G91 G28 X0. Y0. (RAPID TO MACHINE XY-HOME FOR TOOL CHANGE);
M1 (OPTIONAL STOP);
M6 T1 (CHANGE TOOL TO FACE MILL);
G54 (SET WCS 1);
G0 G90 X-1.7 Y1.25 M8 (RAPID TO FIRST POSITION AND COOLANT);
G43 H1 Z.1 S2000 M3; (ACTIVATE TOOL HEIGHT OFFSET-MOVE TO CLEAR PLANE-START
SPINDLE);
```

MATERIAL REMOVAL

```
G1 Z.0 F15. (LINEAR FEED TO DEPTH FOR FACING PASS);
X5.7 F30. (LINEAR FEED MOVE ACROSS PART FACE);
G0 Z.1 (CLEAR PART IN Z);
M9 (COOLANT OFF);
G91 G28 Z0. M05 (RAPID TO MACHINE Z-HOME AND SPINDLE OFF);
G91 G28 X0. Y0. (RAPID TO MACHINE XY-HOME);
M1 (OPTIONAL STOP);

(----------------AUTO CUTTER COMP CONTOUR OPERATION----------------);
M6 T2 (CHANGE TOOL TO 3/4 ENDMILL);
G54 (SET WCS 1);
G0 G90 X4.6 Y.15 M8 (RAPID TO START POSITION AND COOLANT);
G43 H2 Z.1 S5000 M3 (ACTIVATE TOOL HEIGHT OFFSET-MOVE TO CLEAR PLANE-START
SPINDLE);
G1 Z-.2 F23.;
G41 X3.85 D2 F45. (ACTIVATE CUTTER COMP);
X.65;
G2 X.15 Y.65 R.5;
G1 Y2.018;
X1.649 Y2.35;
X3.85;
Y1.875;
X3.35;
G3 X3.35 Y.625 R.625;
G1 X3.85;
Y.15;
G40 Y-.6 (CANCEL CUTTER COMP);
G0 Z5.;
```

ENDING

```
M9 (COOLANT OFF);
G91 G28 Z0. M05 (RAPID TO MACHINE Z-HOME AND SPINDLE OFF);
G91 G28 X0. Y0. (RAPID TO MACHINE XY-HOME);
M30 (END PROGRAM);
```

FIGURE 8.6.41 A part contour (highlighted) containing an angle and radii is programmed using automatic cutter radius compensation. Workpiece coordinates are offset automatically by an amount equal to the radius of the cutting tool.

SUMMARY

- Milling-type machines with CNC controls are capable of much more complex operations than manual mills. CNC mills can move the tool in an arc, diagonal, and cut intricate profiles in one setup.

- A coordinate system containing programmable X-, Y-, and Z-axes is the most common system in CNC mill programming.

- M- and S-codes are used for spindle and coolant control, M- and T-codes are used to change tools, and F-codes command feed rates.

- After a tool change has occurred, the length of the new tool must be activated with a tool height offset command prior to machining.

- Optional sequence numbers with an N prefix can be placed in front of blocks of CNC code for identification.

- Full or optional stops can be inserted in a CNC program to pause the program.

- There are three main motion types for CNC programming: rapid, linear interpolation, and circular interpolation.

- Rapid traverse motion is used for positioning of tools prior to machining.

- Linear interpolation allows machining a straight path between two programmed points.

- Circular interpolation allows an arc to be created between two points and can be programmed using the arc center method or the radius method. The arc center method requires the distance from the arc start point to the arc center point to be specified in the *G2/G3* block. The radius method requires the radius of the arc to be specified in the *G2/G3* block as an R-word.

- Face milling is used to machine a flat surface on the top of a workpiece.

- Two-dimensional milling involves moving the X- and Y-axes to cut slots, pockets, and contours.

- Canned cycles can be used to simplify repetitive and tedious drilling and tapping operations.

- Drilling canned cycles types include single pass, chip break, and deep hole.

- CNC tapping relies on the careful synchronization of the spindle rotation and feed motion. Some machines cannot synchronize the feed adequately and require floating tap holders. Other machines can hold the tap rigidly.

- When performing peripheral milling, a cutting tool's radius may be manually or automatically compensated for using cutter radius compensation. A *G41* code will offset the cutter radius to the left of the cutter path using automatic cutter compensation, a *G42* to the right. Auto cutter compensation must be canceled with a *G40* code when complete.

REVIEW QUESTIONS

1. What are the three major axes used during CNC mill programming?
2. What command would be given to turn on the spindle in a clockwise direction at 1500 RPM?
3. What G-code designates IPM feed rate mode? IPR feed rate mode?
4. What is the purpose of a clearance plane in CNC programming and milling?
5. What is the purpose of work coordinate systems?
6. Briefly define *linear interpolation*.
7. If during the last operation on a part, a *G1* code command is programmed but no F-value is given in the block, what feed rate will be used?
8. Briefly describe the use of "I" and "J" for the arc center method of circular interpolation.

9. Write two blocks of code to start the spindle in a clockwise direction at 2200 RPM and turn on flood coolant.

10. Briefly explain the difference between rigid and non-rigid tapping.

11. Define the initial plane for a canned drilling or tapping cycle.

12. A *G98* in a canned cycle sets the return point to the _____ _____ and a *G99* sets the return point to the _____ _____.

13. A _____ code is used to cancel a canned cycle.

14. What two codes are used to activate automatic cutter radius compensation?

15. What is the difference between the two codes from the previous question?

16. What code is used to cancel automatic cutter radius compensation?

UNIT 7

CNC Milling: Setup and Operation

Learning Objectives

After completing this unit, the student should be able to:

- Demonstrate understanding of CNC machine modes for CNC milling
- Demonstrate understanding of the work coordinate system (WCS) for CNC milling
- Demonstrate understanding of the machine coordinate system (MCS) for CNC milling
- Demonstrate understanding of the homing procedure and purpose for CNC milling
- Demonstrate understanding of workpiece offsets for CNC milling
- Demonstrate understanding of tool geometry offsets for CNC milling
- Demonstrate understanding of tool wear offsets for CNC milling
- Demonstrate understanding of cutter radius compensation offsets
- Demonstrate understanding of the three basic methods for loading programs into the MCU
- Demonstrate understanding of program prove-out procedures for CNC milling

Key Terms

Direct Numerical
  Control (DNC)
Dry Cycle
Dry run
Feed rate override
Geometry offset

Jog
Machine Coordinate
  System (MCS)
Machine home position
Manual Data Input (MDI)
Rapid override

Single-block mode
Tool Height Offset
Wear offset
Work Coordinate
  System (WCS)
Work offset

MACHINE CONTROL PANEL

The CNC milling machine control panel contains the display screen and all the buttons, keys, knobs, and dials to program, set up, and operate the machine. A typical milling control panel is shown in **Figure 8.7.1**.

The display screen can be used to show the program, machine position, and various machine setup pages. Menu buttons can be used to navigate through setup pages and to enter data. Most buttons are labeled with words or pictures describing their functions. Some machines also have buttons called soft keys that have no printed labels, but instead align with a function label on the display screen. **(See Figure 8.7.2.)** A key pad has letter and number keys for keying in programs and other data.

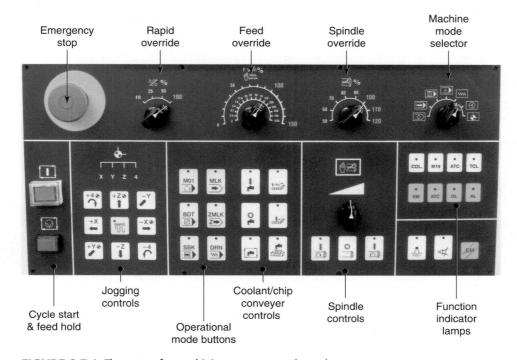

FIGURE 8.7.1 The parts of a machining center control panel.

FIGURE 8.7.2 Soft keys are used on some machines. These keys are universal and can be used for different functions according to their on-screen label.

FIGURE 8.7.3 Arrow keys used for constant jogging of each axis.

A machine mode knob or machine mode keys are used to change from one mode of operation to another (Figure 8.7.1). Common machine modes include jog, auto, **manual data input (MDI)**, edit, and zero-reference return.

The edit mode allows new programs to be typed into memory or an existing program to be modified. Edit is also usually the mode required to load stored programs for use. When programs are to be run, the control must be placed in auto mode. Reference return mode is used to zero the machine axes upon power-up.

MDI mode provides a blank program screen for entering short programs or single program commands. Program data entered into MDI is not saved in memory and is erased after it is executed. Therefore, MDI is ideal for short operations (used during setup or troubleshooting) such as tool changes, spindle start commands, or moving an axis to a specific coordinate for setup.

When **jog** mode is turned on, buttons and a small rotary hand wheel allow user-controlled axis movement during machine setup. Most machines also have arrow keys in the plus and minus direction for jogging (moving) each axis. **(See Figure 8.7.3.)** Pressing and holding these keys causes constant axis motion, and axis motion stops when the button is released. When jogging with the arrow keys, most machines vary the feed rate with either the rapid or feed override dial or keys (Figure 8.7.1). The jogging MPG (manual pulse generator) hand wheel can be set to move an axis at different incremental steps for

fine control when jogging machine axes. These steps are usually 0.010", 0.001", or 0.0001" per click of the hand wheel. The jogging hand wheel on many machines is mounted on a handheld pendant and attached to the MCU by a cable to allow its use away from the control panel and closer to the work area. **(See Figure 8.7.4.)**

The cycle start, feed hold, and emergency stop buttons are also located on the control panel (Figure 8.7.1). The cycle start button begins the active CNC program and is usually green. The feed hold button is usually located directly next to the cycle start button and will stop axis feed when pressed while running a program. In an emergency or collision, the red emergency stop button (sometimes called the E-stop) can be pressed to immediately stop the spindle and all axis motion. The

FIGURE 8.7.4 A jogging hand wheel is used for fine control when jogging the machine's axes. The one shown is portable and is connected to the machine control by a cord.

FIGURE 8.7.5 A spindle speed, feed rate, and rapid override control.

feed rate override knob or keys allow the programmed feed rate to be decreased, increased, or even stopped. Most machines are also equipped with a **rapid override** control to stop or slow rapid motion while a program is running, or to vary the jogging rate. A spindle speed override control can be used to decrease or increase spindle RPM. **(See Figure 8.7.5.)**

WORKHOLDING SETUP

When using a vise or fixture, the base along with the machine's table should be cleaned and inspected for debris and defects. Once the device has been inspected and cleaned, it is placed on the table, aligned using a dial indicator, and clamped when oriented properly. If parallels are to be used, they must be chosen to allow the proper height of the work in the vise jaws. The width and placement of the parallels within the bed of the vise should also be planned so that tools that will machine all the way through the workpiece will not hit the parallels.

When clamps are used, special attention needs to be paid to their overall height. Place clamps strategically to prevent collision during machine motion. The same procedures for workpiece alignment used for manual milling may also be used for CNC milling.

No matter which device is used, when multiple parts will be machined it is important to use the same torque on clamp fasteners and vise screws for each one so that the positioning of the part is consistent. This is especially important to prevent distortion of soft or thin parts.

MACHINE AND WORK COORDINATE SYSTEMS

The Cartesian coordinate system that identifies the location of the origin on the workpiece is called the **work coordinate system (WCS)**. The WCS is movable and its origin can be placed anywhere on the workpiece for ease of programming. The machine has a coordinate system of its own called the **machine coordinate system (MCS)**. This coordinate system is in a fixed location and its origin cannot be changed or moved. The MCS is used for the machine's own reference purposes and helps it to keep track of how far each axis can move before it runs out of travel. These two coordinate systems play an essential role in machine setup. **(See Figure 8.7.6.)**

POWER-UP AND HOMING

The very first step to operating CNC milling equipment is to power up the machine properly. After the machine has been properly turned on, most machines require a *reference return* to the **machine home position**. This process is frequently called *homing*. Recall that the machine moves its axes by rotating a ball screw with a servo motor. Again, this motor can monitor and adjust axis position by how far its shaft rotates. When the machine power is switched off and the control is no longer able to monitor and maintain axis position, it loses track of its position. Therefore, every time a CNC is powered up from a total shutdown, it must be homed.

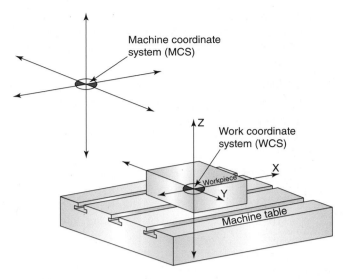

FIGURE 8.7.6 The relationship of the machine coordinate system (MCS) to the work coordinate system (WCS).

Some machines have absolute encoders that store the position even when turned off. These machines do not need to be homed every time they are turned on.

The homing procedure is used to accurately re-enable reference to the MCS (machine coordinate system). Once the machine knows where each axis is positioned, the machine will know its limits of travel. By homing, the machine is also able to recall the position of the WCS (work coordinate system) that was active before the machine was turned off. This prevents previously set workpiece origins from having to be reset every time the machine is powered up. Severe machine damage can occur if a machine overruns its axis travel limits!

Each machine requires specific steps to perform the homing procedure. These steps can also be found in a specific machine's operating manual, but the basic steps are as follows:

1. Select "zero-return" or "home" mode on the machine control panel.

2. Jog each axis in the direction of the machine's home position (usually the same position as the machine's coordinate system [MCS] origin) with the jog arrow keys or MPG hand wheel. If the machine axes were already at home position they must be jogged away and then sent back toward home position.

ⓘ CAUTION

It is usually safest to home the Z-axis first, so it is out of the way when the other axes move.

3. As each axis is sent in the appropriate direction, most machines will automatically complete the procedure by moving the axis and then slowing as a sensor or switch is approached. When the switch is contacted, the encoder of the servo will find the home encoder wheel position and precisely reference its position.

4. When the switch is tripped, the machine will zero the machine coordinate system and will be ready to be set up or start machining using the previous setup.

WORK OFFSET SETTING

The workpiece setup procedure establishes the **work offset**, or origin location for the WCS. All workpiece coordinates for programming will be referenced from this origin. This is established by finding the offset or "shift" from the MCS origin to the intended WCS origin. A milling work offset is usually established for the X-, Y-, and Z-axes.

Recall that the MCS origin is factory set and never changes position. It is restored by zeroing-out at the same position each time the machine is homed. Since the position of this machine origin never changes, but the work origin changes with each new workpiece setup, the work offset for the part is defined as a distance referenced from the MCS origin. Some controls call this offset a *workshift* because it essentially shifts the machine origin to the location of the work origin (or vice versa depending on the machine). **Figure 8.7.7** shows an illustration depicting work offset for CNC milling.

Milling workpieces will change in shape, location, and origin placement from job to job. Therefore the origin must be reset for each newly set-up job by establishing

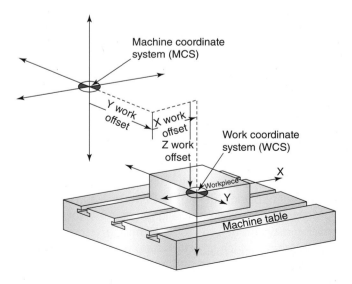

FIGURE 8.7.7 Workpiece offset, or workshift, is the distance from the MCS origin to the desired location of the WCS origin.

a new WCS. Some machines can automatically set the work offset value by pressing "set X," "set Y," or "set Z" when the machine is in position during edge-finding or indicating.

Sometimes the machine will be set up with more than one workholding device so that more than one workpiece can be loaded in the machine at a time. A different WCS can be established for each workpiece. The setting data for work coordinate setting #1 is stored under G54 in the work offset page, #2 is stored under G55, and so on. Most machines have G54 through G59 available for multiple work coordinate systems (some machines have more). Any work coordinate setting can be activated in a program by commanding the appropriate G-code (G54–G59).

Workpiece Z-axis Offset Setting

The Z-axis work offset is usually the first to be established. This establishes the workpiece Z-axis zero and must be done prior to setting up tools.

Basic steps for setting a Z-axis work offset:

1. The workpiece is installed in the workholding device.

2. A face mill is placed in the spindle by calling the tool from the ATC with an M6 Tx command (x is the tool number) using MDI mode or by manually mounting it in the spindle.

3. The tool is jogged near the top surface of the work using the MPG hand wheel.

4. The spindle is started (usually with the M3 command for clockwise rotation) using MDI mode.

5. The tool is slowly jogged in the Z-axis with the MPG hand wheel using 0.001" increments until it lightly touches the part surface.

6. The X- or Y-axis is jogged to move the tool off the part.

7. The Z-axis is lowered just enough to "clean up" the top surface.

8. The top part surface is face milled by moving the X- or Y-axis using the jogging hand wheel or jogging arrows.

9. The spindle is stopped, the X- and Y-axes jogged to a clear position, and the part is flipped. Ensure the part is properly seated in the vise.

10. The X- and Y-axes are jogged to bring the tool in position for another facing pass.

11. The Z-axis is lowered just enough to "clean up" this second surface.

12. The spindle is stopped, the X- and Y-axes jogged to a clear position, and the part is measured.

13. The final part thickness is subtracted from the measurement.

14. The Z-axis is lowered to remove the remaining material and bring the part to final thickness. If more than 0.050" must be removed, consider taking multiple depth passes.

15. The face mill is removed manually or returned to the tool carousel or magazine to empty the spindle.

16. A gage block is set against the newly machined top surface of the part. This part surface will be called Z-zero.

17. The face of the spindle nose is brought against the gage block until a slight drag is felt. At this position, the length of the block is the distance between the intended part origin (machined top of part) to the face of the spindle (reference point). Note: the spindle face is used as a reference surface for setup since when it is positioned at the MCS origin in the Z-axis, the MCS position is Z-zero.

18. The gage block height is combined with the Z-axis machine coordinate system position. This determines the total distance from machine origin to work origin surface (usually a large negative number).

19. The value is entered into the work offset (or work-shift) page for the Z-axis under the correct WCS setting (G54, G55, etc.). **(See Figure 8.7.8.)**

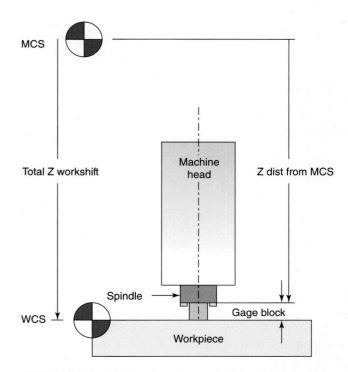

FIGURE 8.7.8 This front view shows a VMC spindle face and a gage block being used to determine the Z-axis work offset. Note the relationship of the MCS to the WCS.

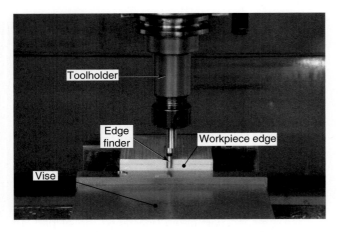

FIGURE 8.7.9 An edge is located with an edge finder.

FIGURE 8.7.10 Finding the center of a hole by sweeping with a dial indicator.

Facing the top surface of the workpiece prior to setting the Z-work offset ensures a flat surface for referencing. This also ensures that when the next piece of full-thickness stock is mounted there will be material above Z-zero to be removed during the facing

operation. The facing cut should always be performed at Z-zero.

Workpiece X-axis and Y-axis Offset Setting

The work offset may be established for the X- and Y-axes by locating the part with an edge finder or spindle-mounted dial indicator. If the work has square smooth edges that are parallel to the X- and Y-axes, then an edge finder may be used to find the location of the part edge in relation to the spindle as was shown in the manual milling section. **Figure 8.7.9** shows an edge finder locating a part edge.

If the part has a smooth round feature such as a bore, reamed hole, or outside diameter, then an indicator may be mounted in the spindle and the spindle rotated while "sweeping" the round feature. When the indicator reads zero throughout a full sweep, the spindle has been aligned with the centerline of the round feature. **Figure 8.7.10** shows a round part feature being swept with a spindle-mounted indicator.

Typical X-axis and Y-axis work offset setting procedure using an edge finder:

1. The workpiece is installed in the workholding device.

2. An edge finder is loaded into an appropriate tool holder in the machine spindle.

3. The part is jogged so the pilot of the edge finder is close to the edge of the part.

4. The spindle is started and set to about 1000 RPM.

5. Jog the axis to bring the edge finder against the part edge using the hand wheel in 0.001" increments. The edge finder will begin to run more "true" when it starts to contact the work.

6. Once the edge finder "kicks," the machine axis is left at that position and the spindle is stopped.

7. The current machine coordinate position is recorded.

8. The edge finder radius is either added to or subtracted from this position to compensate for the edge finder radius depending on which edge of the workpiece is edge found. **(See Figure 8.7.11.)** This step mathematically determines the position of the work origin relative to the machine origin.

9. The value is entered into the work offset page for the appropriate axis.

10. These steps are repeated for the other axis (not Z).

Figure 8.7.12 shows an illustration with dimensions depicting the part origin and edge finder relative to the MCS being used to calculate work offset.

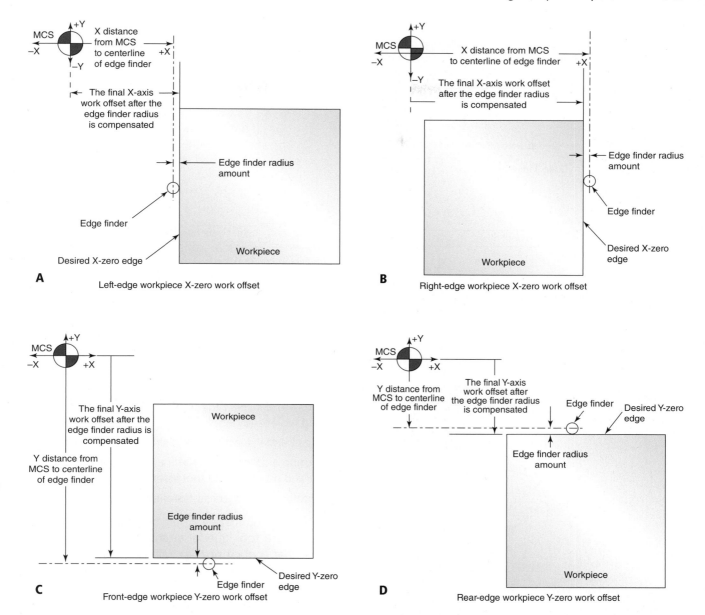

FIGURE 8.7.11 The edge finder radius is either added or subtracted from the MCS position to compensate for the edge finder radius when determining workpiece offset.
NOTE: some machines will have the MCS origin in a different location than shown.

Basic X-axis and Y-axis work offset setting procedure using a dial indicator:

1. The workpiece is installed in the workholding device.

2. A dial indicator is loaded into an appropriate tool holder in the machine spindle.

3. The X- and Y-axes are jogged so that they are approximately aligned with the diameter to be swept.

4. The indicator contact ball is set to contact the surface to be swept.

5. The spindle is rotated by hand and the axes adjusted by jogging with the hand wheel until the least possible motion is seen in the indicator dial through a full sweep. This indicates that the spindle is aligned with the feature centerline.

6. The current machine locations in both the X- and Y-axes are recorded by looking at the MCS position screen.

7. The values are entered into the work offset page, setting the WCS origin in the center of the hole.

Figure 8.7.13 shows an illustration depicting the work offset when the desired part origin is the center of the hole.

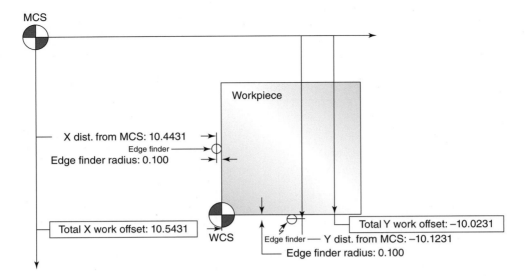

FIGURE 8.7.12 A dimensioned example of how a work offset is derived for a workpiece having the origin in the front-left corner.

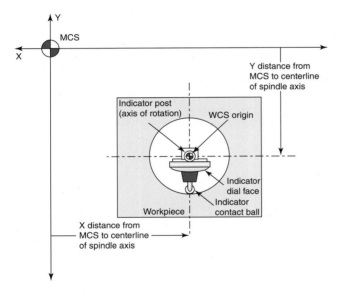

FIGURE 8.7.13 This top view shows that when a hole center is to be used as the X/Y location of the work origin, the hole center can be found by sweeping with a dial indicator. This aligns the spindle centerline with the hole centerline.

CUTTING TOOLS

Cutting-Tool Installation

The cutting-tool types and holder styles have been covered in detail in Section 8, Unit 5. The following steps are typical for assembling tooling for all types of holders except shrink-fit styles:

1. Install the holder in a suitable bench-mounted tool-holder vise.

2. Clean and inspect the holder's tool receiving end for debris, defects, and distortion.

3. Clean and inspect the tool's cutting edges for defects and replace as necessary.

4. Clean and inspect the tool's shank or bore for defects.

5. If using a collet chuck, slip the tool shank into the collet and assemble the tool assembly by hand.

6. Tighten all mechanical fasteners to secure the tool.

Coolant lines or nozzles should be aimed at the cutting zone of each tool. Special attention should be paid to ensure that coolant nozzles will not interfere with the tool changer, part features, or workholding devices and that every operation and every tool will have an ample supply of well-aimed coolant throughout the operation.

Cutting-Tool Offset Types

Because each tool varies in length, a setting called a **tool height offset** must be set for each tool. This offset setting enables each tool to be programmed from the tip regardless of its length. Once the tool measurement has been determined, the tool offset measurements are stored in the machine's **geometry offset** page. The cutter radius value may also be entered into this page for use in automatic cutter radius compensation.

As the cutting edges on the tools wear, the tool length may become slightly shorter. The diameter of the tool may also decrease slightly as the tool wears. A **wear offset** may be used to compensate and adjust for wear as the tool is used during production.

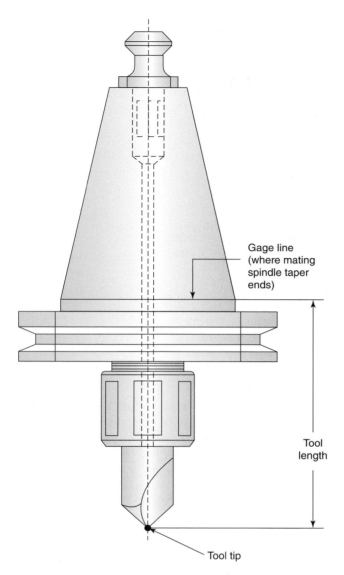

FIGURE 8.7.14 Tool length is measured from the gage line on the tool holder's taper to the tool tip.

Determining and setting the tool's geometry height offset must be performed after the work coordinate system has been established. The part's top surface (Z-zero) becomes the reference point and is used to measure tool length. The length is measured from the gage line on the tool holder's taper to the tool tip. **(See Figure 8.7.14.)**

Tool Geometry Offsets

Figure 8.7.15 shows an image depicting tool offset dimensions. There are several methods to set tool height offsets. Always set tool wear offset to a baseline of zero before setting a geometry offset. Wear offsets will be discussed in detail later. The following is an

example of a common method used to set tool height geometry offset:

1. The part is mounted and the top surface is faced to ensure a smooth and flat reference surface.
2. The work offset must be accurately set, establishing the top of the part as Z-zero.
3. The desired tool is loaded in the spindle.
4. The Z-axis is jogged to bring the tool tip close to the Z-zero surface of the part.
5. The jog rate is reduced to 0.001" increments and the Z-axis is carefully used to move the tool tip closer to the part surface. A piece of shim stock or a gage block is used as a feeler gage to establish a light drag between the tool's tip and the Z-zero surface in the part.
6. When the tool is touched off, the absolute Z-axis position is recorded from the work position register.
7. The feeler gage thickness is subtracted from the recorded position, resulting in a measurement for the length of the tool from the spindle face (gage line) to the tool's tip. This value is entered into the correct offset number of the tool height geometry offset page. **Figure 8.7.16** shows an image of a geometry offset page.

> **NOTE:**
>
> Some machines allow the user to first enter the thickness of the gage block or feeler gage, and will then automatically calculate and enter the height offset value into the offset page by pressing a "measure" or "set" key.

When performing the above procedure the machine is basically being used as a measuring instrument to determine the tool's length. Imagine the machine as a micrometer as shown is **Figure 8.7.17**. The Z-zero surface of the workpiece represents the anvil of a micrometer. Since the spindle face of the machine ends at the gage line on the tool-holder taper, it is like the spindle face of a micrometer. The gap between the spindle face and the part surface is the measurement. In the case of tool setting, the gap is created by the length of the tool plus the shim. The WCS position is provided by the machine's position register (digital readout) and the tool's height by itself is the WCS position minus the shim thickness

When automatic cutter radius compensation is to be used, the tool's radius (or diameter in some cases) must be entered in the geometry offset page in the appropriate location. The entered value is ignored unless cutter comp is activated with a G41 or G42 code. Some controls have a dedicated place in the tool offset page to enter radius or diameter values for tool

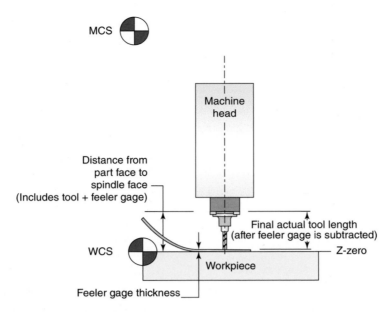

FIGURE 8.7.15 A VMC Z-axis can be used to measure tool length. The length offset is found by touching the tool tip off of the work Z-zero surface with a feeler gage and then subtracting the feeler gage thickness. The objective is to determine the dimension labeled "Final actual tool length."

```
OFFSET / GEOMETRY            O0020 N0340
   NO.      DATA      NO.      DATA
G  001     3.7894    G  009    5.2220
G  002     8.7570    G  010    3.4470
G  003     4.6930    G  011    4.1330
G  004     5.4870    G  012    4.6398
G_005      5.5100    G  013    4.6256
G  006     5.3880    G  014    4.2635
G  007     6.0820    G  015    5.7910
G  008     4.5632    G  016    6.1477
ACTUAL POSITION (RELATIVE)
     X     0.0000         Y    0.0000
     Z     0.0000
NO.   005 -                  S    0  T
  16:59:12            HNDL
( WEAR )(MACRO )( MENU )( WORK )(        )
```

FIGURE 8.7.16 A typical milling geometry offset page on the machine display screen. These numbers reflect the tool length from gage line to the tool tip in its original and unworn state.

numbers. For example, there may be a separate height (H) register and a radius or diameter (D) register for each tool offset number. Other controls only provide one entry per offset number. In this case, if the height offset for tool #1 uses register #1, then the radius offset for tool #1 cannot also be entered there, instead it might be entered into register #101. Some machines require a tool radius to be entered, while others require a tool diameter. Be sure to check the machine manual for the required format.

The initial settings of the tool offsets are always stored in the geometry page. These numbers reflect the true setting for the tool length and radius in its original and unworn state. After the tool has been set, a first part is usually produced by running the program for the first time. The part is then immediately inspected and adjustments are made to correct the geometry offsets as needed. Here is a hypothetical example of how tool height geometry adjustments may be made:

1. The first part is produced and inspected.
2. The measurement reveals that all the depths for a given tool are 0.001" too deep (deeper than the nominal target size).
3. The geometry offset page is opened and the current geometry offset for the tool is 5.1234".
4. The tool is actually longer than the entered length since it is cutting 0.001" deeper than it is programmed to. Therefore 0.001" must be added to the total geometry offset value for that tool, making the actual correct height offset 5.1244".
5. The entire new number is entered for that geometry offset.
6. The next part is machined and the correction is verified.

Tool Wear Offsets

Wear offsets are used when a size adjustment needs to be made after a given tool has been satisfactorily producing parts for a while. Size change can be caused by tool wear or machine thermal variations (expansion or

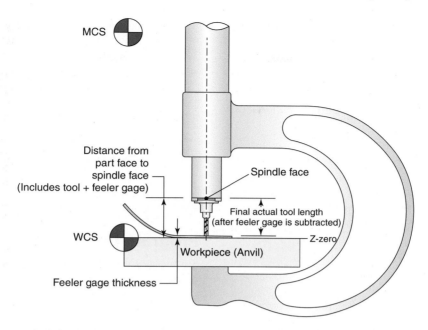

FIGURE 8.7.17 The way a VMC Z-axis can be used to measure tool length can be compared to the way a micrometer measures.

```
OFFSET / WEAR                    O0020 N0340
   NO.      DATA       NO.         DATA
W_001     0.0000    W 009        0.0000
W 002     0.0000    W 010        0.0000
W 003     0.0000    W 011        0.0000
W 004     0.0000    W 012        0.0000
W 005     0.0000    W 013        0.0000
W 006     0.0000    W 014        0.0000
W 007     0.0100    W 015        0.0000
W 008     0.0000    W 016        0.0000
ACTUAL  POSITION   (RELATIVE)
   X      0.0000           Y      0.0000
   Z      0.0000
 NO.   001 =                  S     0 T
   16:57:25                 HNDL
( GEOM )(MACRO )( MENU )( WORK )(        )
```

FIGURE 8.7.18 A typical milling wear offset page on the machine display screen. The number that will be entered into the wear offset will be an incremental adjustment from a baseline of zero tool length wear.

contraction from shop and machine iron temperatures). If the size begins to change during production for one of these reasons, the adjustments should not be entered in the geometry page, but instead in the wear offset page (which usually looks very similar to the geometry page). **Figure 8.7.18** shows an image of a typical wear offset page on the machine display screen. After setting the initial geometry offset and before machining, the corresponding wear offset value must be set to zero. This allows a start at a baseline of zero prior to any adjustments. When using inserted cutting tools, it also allows easy return to the initial wear setting value of zero when inserts are replaced. Any value entered into the

wear offset will be an incremental adjustment from a baseline of zero wear. Wear offset adjustments can be made to either the tool height or the tool radius offset.

Here is a hypothetical example of how these adjustments may be made if the height offset for an endmill needs adjustment:

1. The unsatisfactory part is inspected and measurements show that all the depths for a given tool have become 0.001" too shallow (short of the nominal target size).

2. Opening the wear offset page shows that the wear offset value for the offset is still at its baseline setting of zero.

3. A value of −0.001" is entered in the wear offset register for that tool. The control will combine this wear offset value with the geometry offset value for the tool and reduce the overall tool offset length for the worn tool by 0.001". The control now knows the tool is shorter and will place it deeper in the feature being cut.

4. The next part is machined and the correction is verified.

Here is a hypothetical example of how these adjustments may be made if the radius offset for an endmill needs adjustment:

1. The outside profile on the unsatisfactory part is inspected and measurements reveal that all of the length and width dimensions produced by the side of the tool have become 0.001" too big (larger than the nominal target size).

2. Opening the wear offset page shows that the wear offset value for the radius offset is still at its baseline setting of zero.

3. Since the length and width dimensions are 0.001" too big overall (from one surface to the other), that means that the tool is leaving 0.0005" too much material on each surface.

4. If the control uses radius offset entries, a value of −0.0005" is entered in the wear offset register for that tool. This wear offset value will be combined with the geometry offset value for the tool and reduce the overall tool radius offset for the worn tool by 0.0005". If the control uses diameter offset entries, a value of −0.001" is entered in the wear offset register for that tool. This wear offset value will be combined with the geometry offset value for the tool and reduce the overall tool diameter offset for the worn tool by 0.001". The machine now knows that the tool diameter is smaller and will not offset it as far during automatic cutter radius compensation, resulting in smaller actual machined sizes.

5. The next part is machined and the correction is verified.

PROGRAM ENTRY

Programs can be entered into the MCU in one of three ways:

1. Manually typing the program into the control on the shop floor

2. Uploading the program to memory from a PC or removable storage device

3. Directly sending the program to the control from a PC as the program is running

When manually entering a program from the shop floor, the control must be placed in edit mode and given a program number. The program is then keyed in, word by word and block by block, until complete. This method is usually time consuming and errors can be easily made, so it is generally used only for short programs.

File upload to memory can be accomplished by connecting a PC communication port to a port on the MCU with a communication cable. Some machines can also read programs from a removable storage device such as a CD, USB drive, or memory card. This method is common because it is extremely fast, errors are unlikely, and the program is stored in the machine memory for use at any time.

Sometimes complex programs are so huge that they simply cannot be stored in the machine control memory in their entirety. In these cases, instead of actually storing a program in the MCU memory, the program is fed from a PC to the control line by line as the machine runs the program. The machine control only accepts as much code as it can process. This method is known as **direct numerical control (DNC)** and is sometimes called *drip feeding*. Depending on the machine, this can be accomplished by different methods. The most common is through a direct connection to a PC with a communication cable, but some machines can receive DNC from a CD, memory card, or USB storage device inserted directly into the control.

MACHINE OPERATION
Program Prove-out

Running a milling program on a newly set-up machine is an exciting time in the process and also a time of great caution. Recklessness at this stage could result in damage to the machine, tooling, or workpiece, or personal injury. However, if caution is exercised and a careful *prove out* has been performed prior to letting the machine run unattended, almost all mistakes can be identified.

There are several methods for carefully executing a program to identify problems before letting the machine run without supervision. They are:

1. Graphic simulation
2. Dry run
3. Dry cycle
4. Safe offset

Any one or all of these techniques may be utilized to prove out a program and setup to ensure safety prior to production. Graphic simulation allows verification of the paths the tool will take by watching a simulated computer model cut the part on a display screen. This can be done at a PC with simulation software prior to loading the program in the machine, or on the MCU display on machines with graphic simulation capability. **Figure 8.7.19** shows graphic part simulation on a display screen.

Graphic simulation is a quick way to troubleshoot obvious programming issues but doesn't analyze an actual machine setup or small positioning errors. A **dry run** is a more valid visual prove-out and is usually done in the machine after the tools and workpiece have been set. Generically speaking, "dry running" refers to running a machine with disabled functions to eliminate the potential for collisions. This can be achieved by removing tooling, removing the workpiece, cutting a wax representation of the workpiece, disabling axis motion, or disabling spindle function. Some machines are equipped with a "Dry Run" mode. This mode will usually ignore the programmed feed rates and make all movements at a faster or slower rate set by the user. Dry

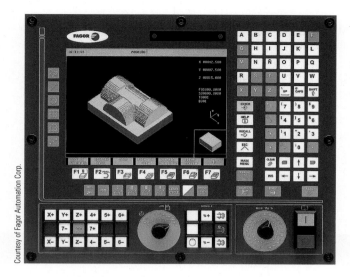

FIGURE 8.7.19 A graphic part simulation on a display screen.

running is also typically done without coolants to help with visibility and to keep the work area clean.

A **dry cycle** is like a dry run but will run the program at the actual programmed feed rates. This can be helpful in proving that there are no G0/G1 code errors or feed rate errors. With a dry cycle, the cutter or workpiece may be removed so that no collisions can occur. A dry cycle is done using the same machine "Auto" or "Memory" mode that will be used to actually machine the part.

The safe offset method is much like the dry-run procedure because the machine will be physically executing the program. The difference is that all machine functions are enabled (except perhaps coolant) and the tools and workpiece are installed. Safety is ensured by intentionally setting the work origin at a distance above the workpiece.

To do this, the work offset can be temporarily shifted in the positive Z-axis direction (away from the part). With this method, the program can be proved out and upon successful completion, the work offset can be gradually moved closer to the part. This cycle can be repeated until the program and setup are deemed safe.

There are two additional lines of defense to prevent collisions by making machine motion more manageable and surprise movements preventable. These controls are override controls and *single-block mode*. Overrides provide the ability to slow or even halt the programmed feed rates and rapids. Most machine control panels are equipped with variable override controls for this.

Single-block mode allows the ability to execute only one block of the program at a time by pressing the cycle-start button. In this mode, the machine will not advance to the next block until the cycle-start button is pressed again. This allows blocks of the program to be viewed on the display screen and verified prior to them being executed. Most machines have a *distance to go* display that shows how far the next programmed move will go. Always check the "distance to go" before pressing cycle start button in single block mode. The single-block mode is normally activated by a switch or button on the machine control panel. No matter which technique is used for prove-out, caution is the key.

Auto Mode

After the program has been carefully proven and there appears to be no potential collisions, the machine may be run at full programmed feed, speed, and at 100% of its rapid capabilities. Once the production performance is proved satisfactory, the machine is now ready to be run in automatic mode.

SUMMARY

- Machines must be homed in order to recall their positioning. It is a good practice to home the Z-axis first to ensure it is out of the way when the X- and Y-axes are homed.

- Always ensure workholding devices are clean, free of burs, and secure before setting up and operating the machine.

- Work offsets allow the machine to establish a work zero or origin at a location on the part in the X-, Y-, and Z-axes for convenient programming. Multiple work offsets can be established and activated with a G54–G59 code. Work offsets must be set prior to setting tool offsets. Z-axis work offsets can be found using a gage block. X- and Y-axis work offsets can be found using an edge finder or dial indicator.

- Tool offsets for milling allow the machine to understand the location of the tool's tip in the Z-axis. When these offsets are active, the programmed position will be taken from the tool's tip.

(Continued)

- Tool offsets for initial setups for unworn tools are entered into the offset geometry registers. Tool offsets for tools that have been gradually and consistently losing their ability to cut the programmed dimensions have their offsets adjusted in the wear offset register.

- Milling tool radius values must be entered into the offset pages if automatic cutter radius compensation is to be used. Wear offsets also apply to tool radius values.

- Programs may be entered into a machine through manually typing the program in on the shop floor, file upload to memory from a PC, or DNC feed from a PC or other storage device connected to the machine.

- Programs may be proven by using one of the following methods: graphic simulation, dry run, or safe offset.

REVIEW QUESTIONS

1. Which must be set first, a work offset or a tool offset? Why?

2. What mode is used to manually enter programs into the machine control?

3. What is the process called when a program is sent to a machine one block at a time as the machine executes the program?

4. Explain what may occur that makes it necessary to enter a wear offset for a particular tool.

5. Explain the purpose of homing.

6. What is the process called when a new program is carefully executed to ensure that there will be no collisions or errors made?

7. What are two actions that can be taken during the process in the previous question to ensure safety?

8. When automatic cutter radius compensation is used in a program, what must be set up in the MCU in order for cutter comp to work properly?

9. Which machine mode allows short, temporary program commands to be entered for operations such as spindle starts and tool changes?

10. Which machine mode is used to run the machine through a continuous machining cycle?

11. What are the two controls on the machine's control panels that control jogging motion?

12. What control panel feature can be used to slow a spindle speed during a program?

Computer-Aided Design and Computer-Aided Manufacturing

Learning Objectives

After completing this unit, the student should be able to:

- Describe the basic applications of CAD
- Describe the basic applications of CAM
- Identify and describe wireframe drawings
- Identify and describe solid model drawings
- Identify and describe surface drawings
- Describe the basic principles of toolpath creation
- Describe basic toolpath types
- Describe the basic principles of post-processing

Key Terms

Computer-aided design (CAD)	Entities	Surface model
	Geometry	Toolpath
Computer-aided manufacturing (CAM)	Post-processing	Wireframe
	Solid model	

INTRODUCTION

CNC machines have revolutionized the world of machining by enabling very sophisticated part features to be produced. Their design and capabilities continue to evolve at a rapid rate. In order to utilize the powerful potential of such a machine, the level of programming must also be advanced.

The thought of programming complex machining operations can overwhelm a beginning programmer. It is true that the programming for such complex operations becomes more difficult, but this type of programming is done on a computer with the aid of **CAM**, or **computer-aided manufacturing** software. **Figure 8.8.1** shows a toolpath for five-axis milling in a computer-aided manufacturing (CAM) software program. This software is being used to program and simulate the machining of the intricate geometry on a turbine blade.

The purpose of CAM software is to take a dimensional computerized part drawing and select features on the drawing for the tool to follow. After defining the path of the tool, the user is prompted for speeds and feeds data, depths of cut, machining patterns, and other characteristics for machining each feature. Finally, all of this information is used by the PC to create a program.

The use of CAM software is not only limited to very complex programming; there are also great benefits from using such a package for basic two-dimensional contouring or even drilling operations. Even the simplest jobs can benefit from the speed and efficiency of CAM programming.

CAM programming involves three major steps to go from start to a finished program:

1. Geometry creation (drawing)
2. Toolpath creation
3. Post-processing (creating the machine's program)

CAD SOFTWARE USE

The very first step in using CAM software is to make a drawing with **computer-aided design (CAD)** software. Sometimes a part to be machined may already have been drawn in CAD by an engineer when the part was designed.

Geometry Types

Wireframe

When the drawing of the part shape is defined by a thin outline only, it appears as though the part's image is made of a framework of thin wires. These types of drawings are called **wireframe** drawings. **Figure 8.8.2** shows a sample wireframe drawing. Wireframes may be two-dimensional (2D) or three-dimensional (3D) depending on the needs and application of the drawing. This type of drawing is very common due to its simplicity and ease of creation. They are best suited for simple parts where the part does not need to look extremely realistic, and where it may not be feasible to invest a large amount of time in a more complex and pictorial representation of the part.

The individual lines and arcs that represent the outline of the part's shape are called **entities**. The entirety of the entities that make up a drawing is referred to as **geometry** in many software packages. A CAD drawing is far more than just a picture or illustration, because each entity has a definite position and length, defined by the user upon creation. This positional data stays associated with each entity, and allows dimensions to be referenced at any time and the entities to be used as a basis to construct other entities.

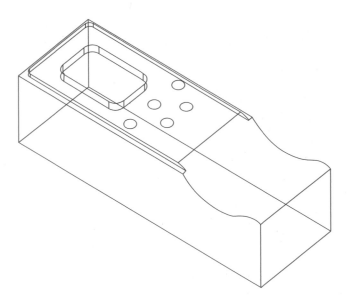

FIGURE 8.8.2 A sample wireframe drawing.

FIGURE 8.8.1 A toolpath for five-axis milling in a computer-aided manufacturing (CAM) software program.

FIGURE 8.8.3 A sample solid model.

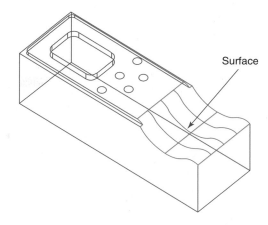

FIGURE 8.8.4 A sample 3D drawing containing a surface.

Solid Models

Some CAD drawings look like solid illustrations of the actual part and are much more realistic looking than wireframes. These drawings are named **solid models** and are often simply referred to as "solids." Solids are usually created by first drawing a 2D wireframe outline defining the part contour, and then *extruding,* or giving that contour thickness. In addition to being visually pleasing, the solid nature of such a digital part is very representative of the material of an actual part. For this reason, important engineering data may be gathered from a solid model drawing. Some of this data includes mass, center of gravity, volume, and how the part may fit together with other parts in an assembly. **Figure 8.8.3** shows a sample solid model.

Surfaces

Some drawings are neither wireframe nor a solid model, but instead appear as a 3D grid showing where the material surfaces are. In these **surface models**, topography (height changes) may be shown with the gridlines to represent the 3D contour of a surface, like a skin stretched over a frame. These types of geometry are named *surfaces.* **Figure 8.8.4** shows a sample 3D drawing containing a surface.

Software Types

Creating the CAD drawing is the first step in computer-aided manufacturing so that its geometry may later be used for creating the machining operations. Not all CAM software packages require the part to be drawn in a separate CAD software system. In fact, many CAM systems stand alone by having CAD drawing abilities within themselves. With this type of system, the part may be drawn and programmed from start to finish using just one piece of software.

Some programmers prefer to use a separate CAD software and CAM software because some dedicated high-end CAD software packages have more features and may be more user-friendly. If this is the case, the completed CAD file must be imported into the CAM software for CNC programming. Many CAM systems have numerous file converters to take the proprietary file types of the most common software systems and convert them to a drawing that the CAM package can work with.

CAM SOFTWARE USE

Toolpaths

After a drawing has been created, the next step in CAM is to define the **toolpath**, or the path that the tool will follow in order to machine the part. Milling toolpaths may be created for facing, pocketing, 2D contour milling, 3D surface milling, drilling, tapping, thread milling, boring, rotary axis operations, and more. Turning toolpaths may be created for facing, rough and finish contour turning, drilling, threading, tapping, C-axis operations (live tooling), and more.

In order to create toolpaths, the drawing entities the tool is to follow are selected. Wireframe entities, solid faces, and surfaces can all be used to create toolpaths. Usually when defining toolpaths for operations such as 2D contouring, the tool's boundary is defined by more than just one entity on the CAD drawing. In order to select these connected entities in sequence with few mouse clicks, most CAM packages allow some form of entity *chaining*. Chaining allows the user to click on one single entity and all entities connected to it (linked like a chain) will be selected. This chain of entities is then used to create the path that the cutting tool will follow. **Figure 8.8.5** shows a chained contour.

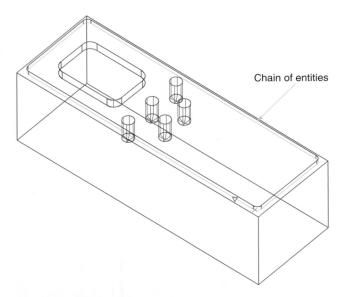

FIGURE 8.8.5 A contour is "chained" when linked entities are selected. Light blue arrowed lines indicate the selected chain.

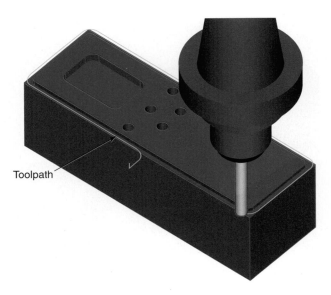

FIGURE 8.8.6 A 2D contour toolpath. Light blue lines and arcs indicate the path taken by the center of the endmill.

Three-dimensional toolpaths require some different methods than those used on 2D contour chaining. When a toolpath is to be created on a 3D surface, the entire surface or solid face used to drive the toolpath may be selected on the drawing.

A fully defined toolpath actually contains much more information than just the intended path of the tool. After the appropriate entities are selected, the user will also be prompted to enter the type and size of cutting tool to be used, speed and feed data, depths of cut, roughing pass types, finishing pass types, machining patterns, step-over amounts, and much more.

2D Contour Milling Toolpaths

As mentioned earlier, a 2D contour milling toolpath begins with the chaining of entities the cutter will cut along. Once the chain has been defined, the programmer will specify the tool type, tool number, tool diameter, offset number, speed, feed, cutter compensation left/right, depth of cut per roughing pass, total depth, finish allowances, and other important information for that operation. Engraving fine lines, art, and letters with a pointed tool may be performed by using a 2D contour toolpath; however, this is usually performed with cutter radius compensation turned off. **Figure 8.8.6** shows a 2D contour toolpath screenshot. **Figure 8.8.7** shows a completed 2D contoured part.

Face-Milling Toolpaths

When a facing toolpath is to be created, the material size can be defined by the user (so the software already knows the length and width) or the outer shape of the part can be chained. Once the machinable area is defined for facing, the cutting data, depth, and the machining pattern must be set. **Figure 8.8.8** shows a face-milling

FIGURE 8.8.7 The 2D contour completed on the part.

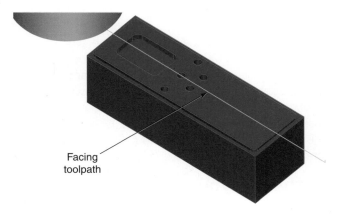

FIGURE 8.8.8 A face-milling toolpath. Light blue line indicates the path taken by the center of the facemill.

toolpath screenshot. **Figure 8.8.9** shows a completed face-milled part.

FIGURE 8.8.9 The face-mill pass completed on the part.

Holemaking Toolpaths in a Mill

Holemaking toolpaths require the user to select the center position of a hole or multiple holes as toolpath entities. The drill operation type will then be defined (single-pass drill, chip-break drill, peck drill, reamer, boring, or tapping). The peck increments, the total depth, the clearance plane, and tool data must also be set. Usually the CAM software will generate the code for drilling operations using canned cycles. **Figure 8.8.10** shows a drilling toolpath screenshot with a pattern of drilled holes. **Figure 8.8.11** shows the drilled holes in the actual part.

Pocketing Toolpaths

Pocketing requires selection of a closed chain of entities to form a boundary for the cutting tool. Whatever material that is within this boundary will be removed to a specified depth. Data must be entered defining the tool and cutting parameters, as well as the type of roughing pattern (zig-zag, spiral, etc.), depth of cut, step-over,

finish allowance, and other data. **Figure 8.8.12** shows a CAM pocketing toolpath screenshot. **Figure 8.8.13** shows a completed pocketed part.

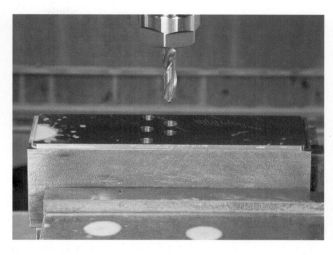

FIGURE 8.8.11 The drilling operation completed on the part.

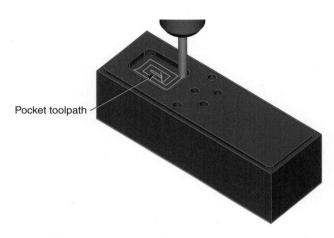

Pocket toolpath

FIGURE 8.8.12 A pocketing toolpath. Light blue lines indicate the path taken by the center of the endmill.

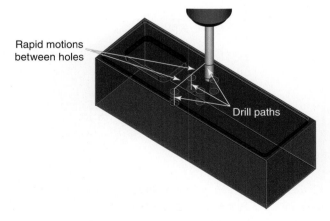

Rapid motions between holes

Drill paths

FIGURE 8.8.10 A toolpath for a pattern of drilled holes. Yellow lines indicate rapid motions and light blue lines indicate the tool's path while feeding into the holes.

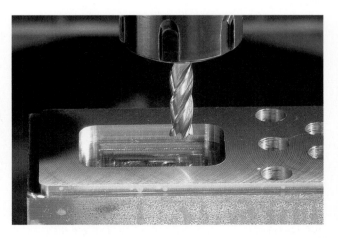

FIGURE 8.8.13 The pocketing operation completed on the part.

Contour Turning Toolpaths

Often, turning operations are programmed longhand without using CAM software. However, CAM is sometimes used to program turned parts that have complex contouring operations. Contour turning toolpaths are created much like contour milling toolpaths and require chaining the part's profile. After the chain is selected, all the necessary cutting parameters are defined. **Figure 8.8.14** shows a CAM turning contour toolpath screenshot. **Figure 8.8.15** shows a completed contour turned part.

3D Milling Toolpaths

Three-dimensional surface milling is an area where the CAM software really begins to demonstrate its potential. Without CAM software, all but the simplest surface toolpaths would be virtually impossible. With surfacing toolpaths, the user will not define a chain, but instead surfaces or solid faces to guide the constantly varying depth boundary of the tool. **Figure 8.8.16** shows a surface-milling toolpath screen shot. **Figure 8.8.17** shows a completed surface-milled part.

Three-dimensional surfacing is fundamentally different from other types of endmilling, since a ball mill is basically used to carve a surface contour. Due to the "carving"-like method of material removal, the cutter must take successive passes along a surface, with each pass stepping over a bit more into uncut material. The spherical end of a ball mill allows the shape of a surface contour to be approximated, but not cut with a great deal of smoothness. The round

FIGURE 8.8.15 The contour turning operation completed on the part.

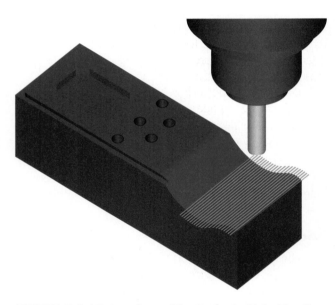

FIGURE 8.8.16 A surface-milling toolpath. Light blue lines indicate the path of the ball mill's nose while feeding.

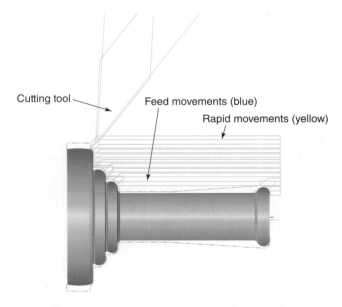

Cutting tool
Feed movements (blue)
Rapid movements (yellow)

FIGURE 8.8.14 A contour turning toolpath. Yellow lines indicate rapid motions and light blue lines indicate the path of the tool tip while cutting.

FIGURE 8.8.17 The surface-milling operation completed on the part.

shape of a ball mill nose leaves behind little peaks called "cusps" between step-overs. With this type of machining operation, there is always a trade-off between cusp height created by the ball mill and the cycle time for the operation. Smaller step-overs add to the machining time since they require more passes. To improve the surface roughness remaining from the cusp height, the user may also desire to switch to a different pattern direction during finishing. With 3D surfaced parts, it is usually expected that the cusps will need to be removed by a manual polishing operation once the part is removed from the machine. **Figure 8.8.18** shows a cusp height generated by surface milling. **Figure 8.8.19** shows the cusps remaining from a rough and finish surface milled part.

Machining Verification/Simulation

After all of the details for the toolpath operations have been defined, the programmer still needs validation that all of the details of machining have been properly defined and the tool will behave as expected. For this, CAM software companies equip their software with verification functions that allow an on-screen simulation of the machining cycle. If an error or collision is spotted, the problem can be fixed and the toolpaths simulated again. Simulation functions of the more sophisticated software systems even allow accurately estimated cycle time, very sensitive collision detection, and other indicators to help the verification to be realistic.

Post-Processing

The final step in CAM programming is to take all of the defined toolpath data and allow the CAM software to write a CNC program. This step is called **post-processing**, since it is performed after the creation of a successful CAM-generated machining process. In order to post-process a machining operation, all toolpaths to be included in the program must be selected and the specific machine control must be specified. The program will be generated, and, once reviewed, it may be saved by the user for later upload to the machine.

NOTE:

Post-processors write code specifically for a given machine control's format. The post-processors are usually user-customizable so that adjustments may be made in the way the code is generated to suit the programmer's and machine's needs.

FIGURE 8.8.19 A completed surface-milled part with cusps remaining from the ball mill step-over. The cusps on the right are larger (greater step-over) from the larger stepover of the roughing operation; the cusps on the left are smaller (less step-over) from the smaller stepover of the finishing operation.

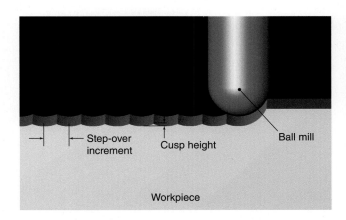

FIGURE 8.8.18 Cusps generated by surface milling.

SUMMARY

- Computer-aided design (CAD) is a means of using a PC to design and draw a workpiece for use in a CAM software system.
- Computer-aided manufacturing (CAM) is a means of using a PC to generate and sequence machining operations from a CAD drawing and then having the software write a CNC program for machining.

(Continued)

- CAD drawings may be created in the form of wireframes, solid models, or surfaces. The outline of an object drawn in CAD is called geometry and the individual lines and arcs that comprise the geometry are called entities.

- The desired path of a cutting tool may be defined by selecting entities on the drawing when using CAM. Cutting data may then be entered to complete this toolpath.

- Toolpaths may be created for: face milling, pocketing, 2D mill contouring, 3D surface milling, drilling, tapping, thread milling, boring, rotary axis operations, turning, facing, rough and finish contour turning, lathe drilling, threading, lathe tapping, C-axis operations (live tooling), and more.

- The completed toolpaths may be verified on-screen within the CAM software to validate the machining process to the programmer.

- After the toolpaths are deemed satisfactory, all of the toolpaths are used in the CAM system post-processor to write the CNC program.

REVIEW QUESTIONS

1. Explain the benefits of CAD/CAM programming versus longhand program writing.
2. What are the three main types of CAD geometry styles?
3. What are the three primary steps in creating a CNC program with the aid of a computer?
4. If an engineer wanted to use a CAD drawing to determine the final weight of a complex part, what type of geometry style is able to provide that information?
5. What is the definition of *entity*?
6. Why should a toolpath be verified on the screen of a CAM system prior to creating the program code?
7. What is a post-processor used for?
8. What is it called when multiple touching entities are linked together when selected?
9. What type of cutting tool is usually used for milling 3D surfaces?

Safety Data Sheet

SAFETY DATA SHEET

_____ **Section 1—Product & Company Identification** _____

Product Name:
RIDGID Dark Thread Cutting Oil (United States)

Product Catalog No.:
11471, 11491, 41590, 41600, 41610, 70830

Recommended Use:
Thread Cutting

Restrictions on Use:
Industrial use only

Company Information:

North America	Australia
Ridge Tool Company	Ridge Tool Australia
400 Clark Street	127 Metrolink Circuit
Elyria, Ohio 44035-6001	Campbellfield, VIC 3061
1-800-519-3456	1-800-743-443
(8:00 am—5:00 pm EST, M-F)	(8:30 am—5:00 pm AEST, M-F)
Emergency Telephone	Emergency Telephone
call 9-1-1 or local emergency number	call 000 or local emergency number
www.RIDGID.com	www.RIDGID.com.au

Issue Date: May 2, 2018

Revision: I

Product Name: RIDGID Dark Thread Cutting Oil (United States)

Section 2—Hazards Identification

Hazard Classification

This product is classified as not hazardous per US OSHA 29CFR 1910.1200 (HazCom 2012)

Label Elements

Hazard Symbol: No symbol

Signal Word: No signal word.

Hazard Statement: Not applicable

Precautionary Statements   Not applicable

Other hazards which do not result in GHS classification: None.

Section 3—Composition / Information On Ingredients

General information: This product does not contain silicone or chlorinated additives.

Hazardous Component(s):

Chemical name	CAS-No.	Concentration
Mineral oil	Confidential	20—<50%
Paraffin oils	Confidential	20—<50%

Specific chemical identities and/or exact percentages have been withheld as trade secrets.

Product Name: RIDGID Dark Thread Cutting Oil (United States)

Section 4—First Aid Measures

Ingestion: Rinse mouth thoroughly. Call a POISON CENTER/doctor if you feel unwell. Do NOT induce vomiting.

Inhalation: Move to fresh air. Call a POISON CENTER/doctor if you feel unwell.

Skin Contact: Remove contaminated clothing and shoes. Wash contact areas with soap and water. If skin irritation occurs: Get medical advice/attention.

Eye contact: Flush thoroughly with water. If irritation occurs, get medical assistance. Continue to rinse for at least 15 minutes.

Most important symptoms/effects, acute and delayed

Symptoms: No data available.

Indication of immediate medical attention and special treatment needed

Treatment: Get medical attention if symptoms occur.

Section 5—Fire Fighting Measures

General Fire Hazards: No unusual fire or explosion hazards noted.

Suitable (and unsuitable) extinguishing media

Suitable extinguishing media: Water spray, fog, CO2, dry chemical, or regular foam. Use fire-extinguishing media appropriate for surrounding materials.

Unsuitable extinguishing media: Do not use water jet as an extinguisher, as this will spread the fire.

Specific hazards arising from the chemical: Heat may cause the containers to explode. During fire, gases hazardous to health may be formed.

Special protective equipment and precautions for firefighters

Special fire fighting procedures: No data available.

Special protective equipment for fire-fighters: Firefighters must use standard protective equipment including flame retardant coat, helmet with face shield, gloves, rubber boots, and in enclosed spaces, SCBA.

RIDGID

Product Name: RIDGID Dark Thread Cutting Oil (United States)

Section 6—Accidental Release Measures

Personal precautions, protective equipment and emergency procedures:
See Section 8 of the SDS for Personal Protective Equipment. Do not touch damaged containers or spilled material unless wearing appropriate protective clothing. Keep unauthorized personnel away. Ensure adequate ventilation.

Methods and material for containment and cleaning up:
Absorb with sand or other inert absorbent. Stop the flow of material, if this is without risk.

Environmental Precautions:
Avoid release to the environment. Do not contaminate water sources or sewer. Prevent further leakage or spillage if safe to do so.

Section 7—Handling And Storage

Precautions for safe handling:
End-users should follow industry best practices for handling and using this product.

Guidance may be found using the current version of ASTM Standard E1497-05: Standard Practice for Selection and Safe Use of Water-Miscible and Straight Oil Metal Removal Fluids Observe good industrial hygiene practices. Wear appropriate personal protective equipment. Do not expose to intense heat as product may expand and pressurize container.

Conditions for safe storage, including any incompatibilities:
Store in original tightly closed container. Avoid contact with oxidizing agents. Store away from incompatible materials. Shelf Life: 720 Days

Product Name: RIDGID Dark Thread Cutting Oil (United States)

Section 8—Exposure Controls / Personal Protection

Exposure Limits

Chemical name	Type	Exposure Limit Values	Source
Mineral oil—Mist.	PEL	5 mg/m³	US. OSHA Table Z-1 Limits for Air Contaminants (29 CFR 1910.1000) (01 2017)
Mineral oil—Mist.	TWA	5 mg/m³	US. OSHA Table Z-1-A (29 CFR 1910.1000) (1989)
Paraffin oils—Inhalable fraction.	TWA	5 mg/m³	US. ACGIH Threshold Limit Values (03 2014)
Paraffin oils—Mist.	PEL	5 mg/m³	US. OSHA Table Z-1 Limits for Air Contaminants (29 CFR 1910.1000) (02 2006)
Paraffin oils—Mist.	TWA	5 mg/m³	US. OSHA Table Z-1-A (29 CFR 1910.1000) (1989)

Protective Measures: Use personal protective equipment as required.

Respiratory Protection: In case of inadequate ventilation use suitable respirator. Seek advice from supervisor on the company's respiratory protection standards.

Eye Protection: Wear safety glasses with side shields (or goggles).

Skin and Body Protection: Wear protective clothing appropriate for the risk of exposure. Be aware of other hazards such as rotating parts. Contact health and safety professional or manufacturer for specific information.

Hygiene measures: Always observe good personal hygiene measures, such as washing after handling the material and before eating, drinking, and/or smoking. Routinely wash work clothing to remove contaminants. Discard contaminated footwear that cannot be cleaned.

Section 9—Physical And Chemical Properties

Appearance

Physical state:	Liquid
Form:	No data available.
Color:	Black
Odor:	Mild petroleum/solvent
Odor threshold:	No data available.
pH:	No data available.
Melting point/freezing point:	No data available.

Product Name: RIDGID Dark Thread Cutting Oil (United States)

Initial boiling point and boiling range:	No data available.
Flash Point:	196.11 °C (385.00 °F)
Evaporation rate:	No data available.
Flammability (solid, gas):	No data available.
Upper/lower limit on flammability or explosive limits	
Flammability limit—upper (%):	No data available.
Flammability limit—lower (%):	No data available.
Explosive limit—upper (%):	No data available.
Explosive limit—lower (%):	No data available.
Vapor pressure:	No data available.
Vapor density:	No data available.
Relative density:	0.878
Solubility(ies)	
Solubility in water:	Insoluble
Solubility (other):	No data available.
Partition coefficient (n-octanol/water):	No data available.
Auto-ignition temperature:	No data available.
Decomposition temperature:	No data available.
Viscosity:	42.5 mm^2/s (40 °C, Measured)
Other information	
VOC:	1.99 g/l (ASTM E 1868-10)

Section 10—Stability And Reactivity

Reactivity:	Not reactive during normal use.
Chemical Stability:	Material is stable under normal conditions.
Possibility of hazardous reactions:	None under normal conditions.
Conditions to avoid:	Avoid heat or contamination.
Incompatible Materials:	No data available.
Hazardous Decomposition Products:	Thermal decomposition or combustion may liberate carbon oxides and other toxic gases or vapors.

Section 11—Toxicological Information

Information on likely routes of exposure

Ingestion:	May be ingested by accident. Ingestion may cause irritation and malaise.

Product Name: RIDGID Dark Thread Cutting Oil (United States)

Inhalation:	Inhalation is the primary route of exposure. In high concentrations, vapors, fumes or mists may irritate nose, throat and mucus membranes.
Skin Contact:	Prolonged skin contact may cause redness and irritation.
Eye contact:	Eye contact is possible and should be avoided.

Symptoms related to the physical, chemical and toxicological characteristics

Ingestion:	No data available.
Inhalation:	No data available.
Skin Contact:	No data available.
Eye contact:	No data available.

Information on toxicological effects

Acute toxicity (list all possible routes of exposure)

Oral
Product: Not classified for acute toxicity based on available data.

Dermal
Product:

Not classified for acute toxicity based on available data.

Inhalation
Product: Not classified for acute toxicity based on available data.

Repeated dose toxicity
Product: No data available.

Skin Corrosion/Irritation
Product: No data available.

Serious Eye Damage/Eye Irritation
Product: No data available.

Respiratory or Skin Sensitization
Product: No data available.

Carcinogenicity
Product: No data available.

IARC Monographs on the Evaluation of Carcinogenic Risks to Humans
No carcinogenic components identified

US. National Toxicology Program (NTP) Report on Carcinogens:
No carcinogenic components identified

Product Name: RIDGID Dark Thread Cutting Oil (United States)

US. OSHA Specifically Regulated Substances (29 CFR 1910.1001-1050)
No carcinogenic components identified

Germ Cell Mutagenicity

In vitro
Product: No data available.

In vivo
Product: No data available.

Reproductive toxicity
Product: No data available.

Specific Target Organ Toxicity—Single Exposure
Product: No data available.

Specific Target Organ Toxicity—Repeated Exposure
Product: No data available.

Aspiration Hazard
Product: No data available.

Other effects: No data available.

Section 12—Ecological Information

General information: This product has not been evaluated for ecological toxicity or other environmental effects.

Section 13—Disposal Consideration

Disposal instructions: Discharge, treatment, or disposal may be subject to national, state, or local laws. Dispose of waste at an appropriate treatment and disposal facility in accordance with applicable laws and regulations, and product characteristics at time of disposal. It is the responsibility of the product user or owner to determine at the time of disposal, which waste regulations must be applied.

Contaminated Packaging: Empty containers should be taken to an approved waste handling site for recycling or disposal.

Product Name: RIDGID Dark Thread Cutting Oil (United States)

Section 14—Transportation Information

DOT
   Not regulated.

IMDG
   Not regulated.

IATA
   Not regulated.

Section 15—Regulatory Information

US Federal Regulations

US. OSHA Specifically Regulated Substances (29 CFR 1910.1001-1050)
   None present or none present in regulated quantities.

Superfund Amendments and Reauthorization Act of 1986 (SARA)

Hazard categories
This product is classified as not hazardous per US OSHA 29CFR 1910.1200 (HazCom 2012)

SARA 313 (TRI Reporting)
      None present or none present in regulated quantities.

US State Regulations

US. California Proposition 65
            No ingredient regulated by CA Prop 65 present.

Product Name: RIDGID Dark Thread Cutting Oil (United States)

	Section 16—Other Information	

Prepared by:. . . . . Ridge Tool Company (Operating Standard 6-103)

Issue Date: . . . . .May 2, 2018
Last Revision Date: . . . . . . March 27, 2017

RIDGE TOOL BELIEVES THE STATEMENTS, TECHNICAL INFORMATION AND RECOM-MENDATIONS CONTAINED HEREIN ARE RELIABLE BUT THEY ARE GIVEN WITHOUT WARRANTY OR GUARANTEE OF ANY KIND, EXPRESSED OR IMPLIED, AND WE ASSUME NO RESPONSIBILITY FOR ANY LOSS, DAMAGE OR EXPENSE, DIRECT OR CONSEQUENTIAL, ARISING OUT OF THEIR USE.

Inch and Metric Plain Gage Diameter Tolerances

Gagemaker's Tolerance Chart [ANSI/AMSE B89.1.5]

Gage Class	XXX	XX	X	Y	Z	ZZ
Diameter Range Above - Including						
Inch						
0.010"–0.825"	0.000010"	0.000020"	0.000040"	0.000070"	0.0001"	0.0002"
0.825"–1.510"	0.000015"	0.000030"	0.000060"	0.000090"	0.00012"	0.00024"
1.510"–2.510"	0.000020"	0.000040"	0.000080"	0.00012"	0.00016"	0.00032"
2.510"–4.510"	0.000025"	0.000050"	0.0001"	0.00015"	0.0002"	0.0004"
4.510"–6.510"	0.000033"	0.000065"	0.00013"	0.00019"	0.00025"	0.0005"
6.510"–9.010"	0.000040"	0.000080"	0.00016"	0.00024"	0.00032"	0.00064"
9.010"–12.010"	0.000050"	0.0001"	0.0002"	0.0003"	0.0004"	0.0008"
Metric						
0.254mm–20.96mm	0.00025mm	0.00051mm	0.00102mm	0.00178mm	0.00254mm	0.00508mm
20.96mm–38.35mm	0.00038mm	0.00076mm	0.00152mm	0.00229mm	0.00305mm	0.00610mm
38.35mm–63.75mm	0.00051mm	0.00102mm	0.00203mm	0.00305mm	0.00406mm	0.00813mm
63.75mm–114.55mm	0.00064mm	0.00127mm	0.00254mm	0.00381mm	0.00508mm	0.01016mm
114.55mm–165.35mm	0.00084mm	0.00165mm	0.00330mm	0.00483mm	0.00635mm	0.01270mm
165.35mm–228.85mm	0.00102mm	0.00203mm	0.00406mm	0.00610mm	0.00813mm	0.01626mm
228.85mm–305.05mm	0.00127mm	0.00254mm	0.00508mm	0.00762mm	0.01016mm	0.02032mm

APPENDIX C

Equivalent Hardness Ratings Between the Brinell Scale and the Rockwell A, B, and C Scales

Rockwell A — Brale Diamond 60kg	Rockwell B — 1/16 Ball Indenter 100kg
20	15
20.5	16
21	17
–	18
21.5	19
22	20
22.5	21
23	22
23.5	23
24	24
–	25
24.5	26
25	27
25.5	28
26	29
26.5	30
27	31
27.5	32
–	33
28	34
28.5	35
29	36
29.5	37
30	38
30.5	39
–	40
31	41
31.5	42
32	43
32.5	44
33	45
33.5	46
34	47
34.5	48
–	49
35	50
35.5	51
36	52
36.5	53
37	54

Rockwell A — Brale Diamond 60kg	Rockwell B — 1/16 Ball Indenter 100kg	Brinell — 10mm Tungsten Carbide Ball 3000 kg
–	56	101
38	57	103
38.5	58	104
39	59	106
39.5	60	107
40	61	108
40.5	62	110
41	63	112
41.5	64	114
–	65	116
42	66	117
42.5	67	119
43	68	121
43.5	69	123
44	70	125
44.5	71	127
45	72	130
45.5	73	132
46	74	135
46.5	75	137
47	76	139
48	77	141
48.5	78	144
49	79	147
49.5	80	150
50	81	153
50.5	82	156
51	83	159
52	84	162
52.5	85	165
53	86	169
53.5	87	172
54	88	176
55	89	180
55.5	90	185
56	91	190
56.5	92	195
57	93	200
57.5	94	205
58	95	210
59	96	216
59.5	97	222
60	98	227
61	99	233
61.5	100	238

Rockwell A — Brale Diamond 60kg	Rockwell C — Brale Diamond 150kg	Brinell — 10mm Tungsten Carbide Ball 3000 kg
60	20	227
61	21	233
61.5	22	238
62	23	243
62.4	24	247
62.8	25	253
63.3	26	258
63.8	27	264
64.3	28	271
64.7	29	279
65.3	30	286
65.8	31	294
66.3	32	301
66.8	33	311
67.4	34	319
67.9	35	327
68.4	36	336
68.9	37	344
69.4	38	353
69.9	39	362
70.4	40	371
70.9	41	381
71.5	42	390
72	43	400
72.5	44	409
73.1	45	421
73.6	46	432
74.1	47	443
74.7	48	455
75.2	49	469
75.9	50	481
76.3	51	496
76.8	52	512
77.4	53	525
78	54	543
78.5	55	560
79	56	577
79.6	57	595
89.1	58	615
80.7	59	634
81.2	60	654
81.8	61	670
82.3	62	688
82.8	63	705
83.4	64	722
83.9	65	739
84.5	66	
85	67	
85.6	68	

Inch/Metric Tap Drill Sizes and Decimal Equivalents

Starrett®

Precision, Quality and Innovation...
Since 1880

INCH/METRIC TAP DRILL SIZES & DECIMAL EQUIVALENTS

DRILL SIZE	DECIMAL EQUIVALENT	TAP SIZE	DRILL SIZE	DECIMAL EQUIVALENT	TAP SIZE	DRILL SIZE	DECIMAL EQUIVALENT	TAP SIZE
80	.0135		10	.1935		59	.9219	1 - 12
79	.0145		9	.1960		64	.9375	1 - 14
1/64	.0156		8	.1990		15/16		
78	.0160		7	.2010	1/4 - 20	61	.9531	
77	.0180		13/64	.2031		64 / 31	.9688	
76	.0200		6	.2040		63/64 / 32	.9844	1 1/8 - 7
75	.0210		5	.2055		1	1.0000	
74	.0225		4	.2090		13/64	1.0469	1 1/8 - 12
73	.0240		3	.2130	1/4 - 28	17/64	1.1094	1 1/4 - 7
72	.0250		7/32	.2188		1 1/8	1.1250	
71	.0260		2	.2210		1 11/64	1.1719	1 1/4 - 12
70	.0280		1	.2280		17/32	1.2188	1 3/8 - 6
69	.0292		A	.2340		1 1/4	1.2500	
68	.0310		15/64	.2344		1 19/64	1.2969	1 3/8 - 12
1/32	.0312		B	.2380		1 11/32	1.3438	1 1/2 - 6
67	.0320		C	.2420		1 3/8	1.3750	
66	.0330		D	.2460		1 27/64	1.4219	1 1/2 - 12
65	.0350		1/4	.2500		1 1/2	1.5000	
64	.0360		E	.2570	5/16 - 18			
63	.0370		F	.2610				
62	.0380		17/64	.2656				
61	.0390		G	.2660				
60	.0400		H	.2720	5/16 - 24			
59	.0410		I	.2770				
58	.0420		J	.2810				
57	.0430		9/32	.2812				
56	.0465		K	.2900				
3/64	.0469	0 - 80	L	.2950				
55	.0520		19/64	.2969				
54	.0550		M	.3020				
53	.0595	1 - 64, 72	5/16	.3125	3/8 - 16			
1/16	.0625		N	.3160				
52	.0635		O	.3230				
51	.0670		21/64	.3281				
50	.0700	2 - 56, 64	P	.3320	3/8 - 24			
49	.0730		Q	.3390				
48	.0760		11/32	.3438				
5/64	.0781		R	.3480				
47	.0785	3 - 48	S	.3580				
46	.0810		23/64	.3594				
45	.0820	3 - 56	T	.3680	7/16 - 14			
44	.0860		3/8	.3750				
43	.0890	4 - 40	U	.3770				
42	.0935	4 - 48	V	.3860				
3/32	.0938		25/64	.3906	7/16 - 20			
41	.0960		W	.3970				
40	.0980		X	.4040				
39	.0995		13/32	.4062				
38	.1015	5 - 40	Y	.4130				
37	.1040	5 - 44	Z	.4219	1/2 - 13			
36	.1065	6 - 32	27/64					
7/64	.1094		7/16	.4375				
35	.1100		29/64	.4531	1/2 - 20			
34	.1110		15/64	.4688				
33	.1130	6 - 40	31/64	.4844	9/16 - 12			
32	.1160		1/2	.5000				
31	.1200		33/64	.5156	9/16 - 18			
1/8	.1250		17/64	.5312	5/8 - 11			
30	.1285		35/64	.5469				
29	.1360	8 - 32, 36	9/16	.5625				
28	.1405		37/64	.5781	5/8 - 18			
9/64	.1406		19/32	.5938				
27	.1440		39/64	.6094				
26	.1470		5/8	.6250				
25	.1495	10 - 24	41/64	.6406				
24	.1520		21/32	.6562	3/4 - 10			
23	.1540		43/64	.6719				
5/32	.1562		11/16	.6875	3/4 - 16			
22	.1570		45/64	.7031				
21	.1590	10 - 32	23/32	.7188				
20	.1610		47/64	.7344				
19	.1660		3/4	.7500				
18	.1695		49/64	.7656	7/8 - 9			
11/64	.1719		25/32	.7812				
17	.1730		51/64	.7969				
16	.1770	12 - 24	13/16	.8125	7/8 - 14			
15	.1800		53/64	.8281				
14	.1820	12 - 28	27/32	.8438				
13	.1850		55/64	.8594				
3/16	.1875		7/8	.8750	1 - 8			
12	.1890		57/64	.8906				
11	.1910		29/32	.9062				

METRIC TAP DRILL SIZES

METRIC TAP	TAP DRILL (mm)	DECIMAL (Inch)
M1.6 x 0.35	1.25	.0492
M1.8 x 0.35	1.45	.0571
M2 x 0.4	1.60	.0630
M2.2 x 0.45	1.75	.0689
M2.5 x 0.45	2.05	.0807
M3 x 0.5	2.50	.0984
M3.5 x 0.6	2.90	.1142
M4 x 0.7	3.30	.1299
M4.5 x 0.75	3.70	.1457
M5 x 0.8	4.20	.1654
M6 x 1	5.00	.1968
M7 x 1	6.00	.2362
M8 x 1.25	6.70	.2638
M8 x 1	7.00	.2756
M10 x 1.5	8.50	.3346
M10 x 1.25	8.70	.3425
M12 x 1.75	10.20	.4016
M12 x 1.25	10.80	.4252
M14 x 2	12.00	.4724
M14 x 1.5	12.50	.4921
M16 x 2	14.00	.5512
M16 x 1.5	14.50	.5709
M18 x 2.5	15.50	.6102
M18 x 1.5	16.50	.6496
M20 x 2.5	17.50	.6890
M20 x 1.5	18.50	.7283
M22 x 2.5	19.50	.7677
M22 x 1.5	20.50	.8071
M24 x 3	21.00	.8268
M24 x 2	22.00	.8661
M27 x 3	24.00	.9449
M27 x 2	25.00	.9843
M30 x 3.5	26.50	1.0433
M30 x 2	28.00	1.1024
M33 x 3.5	29.50	1.1614
M33 x 2	31.00	1.2205
M36 x 4	32.00	1.2598
M36 x 3	33.00	1.2992
M39 x 4	35.00	1.3780
M39 x 3	36.00	1.4173

PIPE THREAD SIZES (NPSC)

THREAD	DRILL	THREAD	DRILL
1/8 – 27	11/32	1 1/2 – 11 1/2	1 3/4
1/4 – 18	7/16	2 – 11 1/2	2 7/32
3/8 – 18	37/64	2 1/2 – 8	2 21/32
1/2 – 14	23/32	3 – 8	3 1/4
3/4 – 14	59/64	3 1/2 – 8	3 3/4
1 – 11 1/2	1 5/32	4 – 8	4 1/4
1 1/4 – 11 1/2	1 1/2		

The L.S. Starrett Company

Decimal and Metric Equivalents of Number, Letter, and Fractional Drill Sizes

Drill size	Diameter (in)	Diameter (mm)	Drill size	Diameter (in)	Diameter (mm)	Drill size	Diameter (in)	Diameter (mm)
#107	0.0019	0.0483	#72	0.0250	0.6350	#40	0.0980	2.4892
#106	0.0023	0.0584	#71	0.0260	0.6604	#39	0.0995	2.5273
#105	0.0027	0.0686	#70	0.0280	0.7112	#38	0.1015	2.5781
#104	0.0031	0.0787	#69	0.0292	0.7417	#37	0.1040	2.6416
#103	0.0035	0.0889	#68	0.0310	0.7874	#36	0.1065	2.7051
#102	0.0039	0.0991	1/32 in	0.0313	0.7938	7/64in	0.1094	2.7781
#101	0.0043	0.1092	#67	0.0320	0.8128	#35	0.1100	2.7940
#100	0.0047	0.1194	#66	0.0330	0.8382	#34	0.1110	2.8194
#99	0.0051	0.1295	#65	0.0350	0.8890	#33	0.1130	2.8702
#98	0.0055	0.1397	#64	0.0360	0.9144	#32	0.1160	2.9464
#97	0.0059	0.1499	#63	0.0370	0.9398	#31	0.1200	3.0480
#96	0.0063	0.1600	#62	0.0380	0.9652	1/8 in	0.1250	3.1750
#95	0.0067	0.1702	#61	0.0390	0.9906	#30	0.1285	3.2639
#94	0.0071	0.1803	#60	0.0400	1.0160	#29	0.1360	3.4544
#93	0.0075	0.1905	#59	0.0410	1.0414	#28	0.1405	3.5687
#92	0.0079	0.2007	#58	0.0420	1.0668	9/64 in	0.1406	3.5719
#91	0.0083	0.2108	#57	0.0430	1.0922	#27	0.1440	3.6576
#90	0.0087	0.2210	#56	0.0465	1.1811	#26	0.1470	3.7338
#89	0.0091	0.2311	3/64 in	0.0469	1.1906	#25	0.1495	3.7973
#88	0.0095	0.2413	#55	0.0520	1.3208	#24	0.1520	3.8608
#87	0.0100	0.2450	#54	0.0550	1.3970	#23	0.1540	3.9116
#86	0.0105	0.2667	#53	0.0595	1.5113	5/32 in	0.1563	3.9688
#85	0.0110	0.2794	1/16 in	0.0625	1.5875	#22	0.1570	3.9878
#84	0.0115	0.2921	#52	0.0635	1.6129	#21	0.1590	4.0386
#83	0.0120	0.3048	#51	0.0670	1.7018	#20	0.1610	4.0894
#82	0.0125	0.3175	#50	0.0700	1.7780	#19	0.1660	4.2164
#81	0.0130	0.3302	#49	0.0730	1.8542	#18	0.1695	4.3053
#80	0.0135	0.3429	5/64 in	0.0781	1.9844	11/64 in	0.1719	4.3656
#79	0.0145	0.3680	#47	0.0785	1.9939	#17	0.1730	4.3942
1/64 in	0.0156	0.3969	#46	0.0810	2.0574	#16	0.1770	4.4958
#78	0.0160	0.4064	#45	0.0820	2.0828	#15	0.1800	4.5720
#77	0.0180	0.4572	#44	0.0860	2.1844	#14	0.1820	4.6228
#76	0.0200	0.5080	#43	0.0890	2.2606	#13	0.1850	4.6990
#75	0.0210	0.5334	#42	0.0935	2.3749	3/16 in	0.1875	4.7625
#74	0.0225	0.5715	3/32 in	0.0938	2.3813	#12	0.1890	4.8006
#73	0.0240	0.6096	#41	0.0960	2.4384	#11	0.1910	4.8514

Drill size	Diameter (in)	Diameter (mm)	Drill size	Diameter (in)	Diameter (mm)	Drill size	Diameter (in)	Diameter (mm)
#10	0.1935	4.9149	23/64 in	0.3594	9.1281	29/32 in	0.9063	23.0188
#9	0.1960	4.9784	U	0.3680	9.3472	21/23 in	0.9130	23.1913
#8	0.1990	5.0546	3/8 in	0.3750	9.5250	59/64 in	0.9219	23.4156
#7	0.2010	5.1054	V	0.3770	9.5758	15/16 in	0.9375	23.8125
13/64 in	0.2031	5.1594	W	0.3860	9.8044	61/64 in	0.9531	24.2094
#6	0.2040	5.1816	25/64 in	0.3906	9.9219	31/32 in	0.9688	24.6063
#5	0.2055	5.2197	X	0.3970	10.0838	63/64 in	0.9844	25.0031
#4	0.2090	5.3086	Y	0.4040	10.2616	1 in	1.0000	25.4000
#3	0.2130	5.4102	13/32 in	0.4063	10.3188	1 1/64 in	1.0156	25.7969
7/32 in	0.2188	5.5563	Z	0.4130	10.4902	1 1/32 in	1.0313	26.1938
#2	0.2210	5.6134	27/64 in	0.4219	10.7156	1 3/64 in	1.0469	26.5906
#1	0.2280	5.7912	7/16 in	0.4375	11.1125	1 1/16 in	1.0625	26.9875
A	0.2340	5.9436	29/64 in	0.4531	11.5094	1 5/64 in	1.0781	27.3844
15/64 in	0.2344	5.9531	15/32 in	0.4988	11.9063	1 3/32 in	1.0938	27.7813
B	0.2380	6.0452	31/64 in	0.4844	12.3031	1 7/64 in	1.1094	28.1781
C	0.2420	6.1468	1/2 in	0.5000	12.7000	1 1/8 in	1.1250	28.5750
D	0.2460	6.2484	33/64 in	0.5156	13.0969	1 9/64 in	1.1406	28.9719
1/4 in	0.2500	6.3500	17/32 in	0.5313	13.4938	1 5/32 in	1.1563	29.3688
E	0.2500	6.3500	35/64 in	0.5469	13.8906	1 11/64 in	1.1719	29.7656
F	0.2570	6.5278	9/16 in	0.5625	14.2875	1 3/16 in	1.1875	30.1625
G	0.2610	6.6294	37/64 in	0.5781	14.6844	1 13/64 in	1.2031	30.5594
17/64 in	0.2656	6.7469	19/32 in	0.5938	15.0813	1 7/32 in	1.2188	30.9563
H	0.2660	6.7564	39/64 in	0.6094	15.0813	1 15/64 in	1.2344	31.3531
I	0.2720	6.9088	5/8 in	0.6250	15.8750	1 1/4 in	1.2500	31.7500
J	0.2770	7.0358	41/64 in	0.6406	16.2719	1 17/64 in	1.2656	32.1469
K	0.2810	7.1374	43/64 in	0.6719	17.0656	1 9/32 in	1.2813	32.9406
9/32 in	0.2813	7.1438	11/16 in	0.6875	17.4625	1 19/64 in	1.2969	32.9406
L	0.2900	7.3660	45/64 in	0.7031	17.8594	1 5/16 in	1.3125	33.3375
M	0.2900	7.4930	23/32 in	0.7188	18.2563	1 21/64 in	1.3281	33.7344
19/64 in	0.2969	7.5406	47/64 in	0.7344	18.6531	1 11/32 in	1.3438	34.1313
N	0.3020	7.6708	3/4 in	0.7500	19.0500	1 23/64 in	1.3594	34.5281
5/16 in	0.3125	7.9375	49/64 in	0.7656	19.4469	1 3/8 in	1.3750	34.9250
O	0.3160	8.0264	25/32 in	0.7813	19.8438	1 25/64 in	1.3906	35.3219
P	0.3230	8.2042	51/64 in	0.7969	20.6375	1 13/32 in	1.4063	35.7188
21/64 in	0.3281	8.3344	13/16 in	0.8125	21.0344	1 27/64 in	1.4219	36.1156
Q	0.3320	8.4328	53/64 in	0.8281	21.0344	1 7/16 in	1.4375	36.5125
R	0.3390	8.6106	27/32 in	0.8438	21.4313	1 29/64 in	1.4531	36.9094
11/32 in	0.3438	8.7313	55/64 in	0.8594	21.8281	1 15/32 in	1.4688	37.3063
S	0.3480	8.8392	7/8 in	0.8750	22.2250	1 31/64 in	1.4844	37.7031
T	0.3580	9.0932	57/64 in	0.8906	22.6219	1 1/2 in	1.5000	38.1000

Commonly Used G and M Codes

CODE	MILL	LATHE	DESCRIPTION
Preparatory Functions			
G0	✓	✓	Positioning at rapid feed
G1	✓	✓	Linear interpolation
G2	✓	✓	Circular interpolation-clockwise
G3	✓	✓	Circular interpolation-counterclockwise
G4	✓	✓	Dwell
G9	✓	✓	Exact motion stop at intersections (non-modal, one block only)
G10	✓	✓	Offset value entry through program
G12	✓		Circular pocket milling cycle-clockwise
G13	✓		Circular pocket milling cycle-counterclockwise
G15	✓		Cartesian coordinate programming system on
G16	✓		Polar coordinate programming system on
G17	✓	✓	X/Y plane selection for arc cutting
G18	✓	✓	Z/X plane selection for arc cutting
G19	✓	✓	Z/Y plane selection for arc cutting
G20	✓	✓	Inch unit selection
G21	✓	✓	Metric unit selection
G27	✓	✓	Reference position return check
G28	✓	✓	Return to the primary machine zero position (home position)
G29	✓	✓	Return from reference position
G31	✓		Skip function
G32		✓	Thread cutting (single-point tool or tap)
G40	✓	✓	Cutter radius compensation cancel
G41	✓	✓	Cutter radius compensation-left
G42	✓	✓	Cutter radius compensation-right
G43	✓		Tool height offset compensation-activate
G44	✓		Tool height offset compensation-cancel (some machines use G49)
G49	✓		Tool height offset compensation-cancel (some machines use G44)
G50		✓	Maximum RPM setting for Constant Surface Speed
G52	✓	✓	Local coordinate system setting
G53	✓	✓	Machine coordinate system setting
G54	✓	✓	Workpiece coordinate system setting #1
G55	✓	✓	Workpiece coordinate system setting #2
G56	✓	✓	Workpiece coordinate system setting #3
G57	✓	✓	Workpiece coordinate system setting #4
G58	✓	✓	Workpiece coordinate system setting #5
G59	✓	✓	Workpiece coordinate system setting #6
G61	✓	✓	Exact motion stop at intersections (modal)
G64	✓	✓	Normal cutting mode without exact stops at intersections
G65	✓	✓	Custom macro call
G70		✓	Finish turning/facing/boring cycle
G71		✓	Rough turning/boring cycle
G72		✓	Rough facing cycle
G73		✓	Irregular rough turning cycle (for castings and forgings)
G73	✓		Chip break peck drilling cycle
G74	✓		Left hand tapping cycle
G74		✓	Face grooving or chip break peck drilling cycle
G75		✓	OD groove chip break peck cycle
G76	✓		Fine boring cycle (no tool drag mark)
G76		✓	Auto repetative threading cycle (single point)
G80	✓	✓	Cancel canned cycles
G81	✓	✓	Single pass drill cycle
G82	✓		Single pass drill cycle with dwell
G83	✓		Full retract peck drilling cycle
G84	✓	✓	Tapping cycle with reversing
G85	✓	✓	Boring cycle (feed in/feed out)
G86	✓	✓	Boring cycle (feed in/rapid out)
G87	✓		Back boring cycle
G90	✓		Absolute programming
G91	✓		Incremental programming
G92	✓		Reposition origin via program
G92		✓	Thread cutting
G94	✓		Inch or MM per minute feed rate
G95	✓		Inch or MM per revolution feed rate
G96		✓	SFM value for Constant Surface Speed
G97		✓	Fixed spindle RPM/Constant Surface Speed cancel
G98		✓	Inch or MM per minute feed rate
G99		✓	Inch or MM per revolution feed rate
G98	✓		Return to initial plane between hole locations for drill cycles
G99	✓		Return to R-plane between hole locations for drill cycles
Miscellaneous Functions			
M00	✓	✓	Program stop
M01	✓	✓	Optional program stop
M02	✓	✓	Program end
M03	✓	✓	Spindle on clockwise
M04	✓	✓	Spindle on counterclockwise
M05	✓	✓	Spindle off
M06	✓	✓	Toolchange
M07	✓	✓	Mist coolant on
M08	✓	✓	Coolant on
M09	✓	✓	Coolant off
M10	✓	✓	Chuck, collet, or rotary table clamp
M11	✓	✓	Chuck, collet, or rotary table unclamp
M19	✓	✓	Orient spindle
M30	✓	✓	Program end and return to start
M97	✓	✓	Local sub-routine call
M98	✓	✓	Sub-program call
M99	✓	✓	End sub-program and return to main program

Note: Most machines allow single digit codes to be used with or without the preceding zero (Example: G01 = G1).

GLOSSARY

A

abrasive machining. A cutting technique in which grinding wheels are used in either a nonprecision or precision manner. Noncritical operations are usually performed offhand on pedestal-type grinders, whereas precision grinders produce very accurate dimensions and very smooth surfaces.

absolute encoders. Position sensors that allow a CNC machine to keep track of axis position when the machine is powered off, eliminating the need for homing procedures upon start-up.

absolute positioning system. A referencing method used with coordinate systems in which the coordinates of all positions are referenced from the origin (X0, Y0, Z0).

Acme thread. A high strength thread that has a thick, somewhat square-appearing 29-degree included angle form.

accumulated tolerance. A tolerance that results from adding up the tolerances of multiple dimensions when determining allowable variation of an unknown or reference dimension.

acute. A short-term or rapidly developing health effect from exposure to a hazardous material.

adjacent side. The side of a right triangle next to a given angle that is not the hypotenuse.

adjustable parallels. Two sliding, interlocking wedge-shaped blocks, having parallel outside surfaces which can be adjusted to a given width for workpiece setup for transferring measurements.

adjustable reamers. Tools with threaded bodies and tapered slots to hold cutting blades that are similar to expansion reamers, but have a much greater range of adjustability.

adjustable squares. Two piece tools for measuring perpendicularity that can easily be adjusted and disassembled; the beam of the tool has a clamping mechanism that holds the blade in place when tightened.

adjustable wrench. An open-end type wrench that can be adjusted to accommodate a wide range of sizes.

allowance. The minimum amount of clearance, or the maximum amount of interference, between two mating parts.

alloy. A homogeneous combination of two or more metals or a metal and a nonmetal element.

alloy steels. Steels that have elements, such as chromium, manganese, molybdenum, nickel, tungsten, and vanadium, added to them.

alternate set. A saw blade tooth-setting pattern in which every other tooth switches the side of the blade it is set to (left, right, left, right, and so on).

aluminum alloys. Aluminum with other elements added (such as zinc, copper, and silicon) to produce certain characteristics.

Aluminum Association of the United States (AA). An organization that sets standards for all aluminum alloy compositions and their designations.

aluminum oxide. The most common grinding wheel abrasive type, used for general-purpose grinding of ferrous metals.

American Iron and Steel Institute (AISI). An organization that developed a system for classifying and designating ferrous metals.

American Society for Testing and Materials (ASTM). An organization that developed a system for classifying and designating cast ferrous metals.

angle block. A workholding device with precise 90-degree sides that can be used to position a workpiece for machining or inspection tasks.

angle gages. A measuring tool that can compare part angles to standard angles.

angle plate. A workholding device with all sides at 90-degree angles which is useful for holding parts during layout, measuring, or machining operations.

angularity. GD&T specification for angular relationship between two part features.

annealing. A heat-treating process used to return metals to their original pre-hardened condition so they can be more easily machined.

apprentices. Company trainees who receive practical training in machining operations during normal working hours.

apprenticeships. Company training programs, either internal or sponsored by a state labor department, in which trainees receive practical training in machining operations—during normal working hours.

apron. The part of a lathe, attached to the bottom of the saddle that hangs down in front of the bed, which contains several lathe controls.

arbor. A device used to mount milling cutting tools.

arbor press. Device that uses leverage to create force for assembly and disassembly of mating parts with interference fits.

arc. A portion of a circle between two points on that circle.

arc center method. A CNC circular interpolation programming technique used to identify the exact location of the center point of an arc related to its start point, identified by distances on the X- and Y-axes.

assembly drawing. A drawing that illustrates how multiple components would appear when put together.

associate degree. A 2-year educational degree, usually in a technical field.

atmospheric (atmospheric control) furnaces. Furnaces that can remove room air from the heating chamber and replace it with nitrogen, argon, or helium to minimize or eliminate the oxidation that may form on the part's surface.

automated pallet changer (APC). A time-maximizing system that has two or more assembled workholding tool plates (pallets) and changes the pallets through programmed commands without operator involvement.

automatic cutter radius compensation. Sometimes called cutter comp for short, a time-saving process initiated by G41 or G42 on CNC machines that allows the programmer to write a program using dimensions and locations directly from the engineering drawing, rather than making geometric and trigonometric calculations to determine each cutter position.

automatic tool changer (ATC). A device used in some CNC machines to automatically load, unload, and store tools.

B

baccalaureate degrees. A 4-year program provided through a university that normally offers advanced theoretical education and training in specialty areas such as engineering disciplines.

back rake. The angle of the top of a lathe cutting tool relative to a horizontal line through the center of the workpiece, which can be positive, negative, or neutral.

ballnose endmill. Sometimes called a ball endmill or ballmill, an endmill with a half-round sphere ground on its end which is proportional to the diameter of the endmill's outside diameter.

ball oiler. A device with a small, spring-loaded ball that acts as a valve allowing oil to be added and keeping debris out of a machine's lubrication system.

ball peen hammers. Dual purpose hand tools that have two heads for two different functions: a striking face on one end used for light or heavy striking tasks, and a rounded end that can be used for peening rivet heads or rough forming metal.

ball screw. A screw and nut assembly with steel balls in place of threads which can achieve high efficiency and zero backlash in computerized numerical control machines.

base. The part of the machine that provides a solid foundation for the entire machine.

base and column. On a vertical mill, a single cast-iron unit that provides a heavy, solid base for the machine.

basic size. A theoretical, ideal dimension shown on a print or an engineering drawing.

bed. The heavy horizontal casting located on the right side of the lathe's headstock, which serves as a foundation for the ways, and is designed to both be strong enough to handle the forces created by the cutting process and ensure precise machining.

bench vise. A device used to hold workpieces securely to perform manual benchwork operations such as filing or hacksawing.

bilateral tolerance. An allowable variation from a given size both above ("+") and below ("−") the basic size on an engineering drawing, either by equal amounts above and below the basic size or by different amounts above and below the basic size. When both amounts are equal, a "±" symbol is used.

bill of materials. List, shown on an engineering drawing, of raw materials used to produce machined parts of finished components used to create an assembly.

bimetal. A band saw blade featuring a carbon steel body with a strip of high-speed steel welded to the one edge, into which the teeth are cut.

blind holes. Holes that are only drilled partially through a workpiece.

block. In the word address programming system; a group of program words that appear on the same line and are to be executed at the same time.

blocking. A technique used to help secure a part to the grinding machine's magnetic chuck to prevent it from tipping or being pulled from the chuck.

blotter. A round disc of heavy paper surrounding the center area of a grinding wheel that provides a cushion when the wheel is mounted and also provides information about the wheel.

body clearance. The area just behind the margin on the drill body that is slightly smaller in diameter, which helps reduce rubbing on the walls of the hole during drilling.

bonding agent. The "glue" that holds individual abrasive particles together in the shape of a grinding wheel.

bonus tolerance. In GD&T, an increase in location tolerance due to material condition modifiers added to a feature control frame on an engineering drawing.

boring. The process of using a single-point cutting tool to enlarge an existing hole.

boring bar. The cutting tool which is used to perform boring operations.

boring head. A device commonly used to hold a boring bar on a milling machine that provides the ability to offset the bar to adjust cutting diameter.

bottoming chamfer taps. Taps with only the first one or two threads chamfered that are used when a hole is blind, since they can cut full threads nearly all the way to the bottom of a hole.

box-end wrench. A wrench end which completely encircles the bolt or nut, giving the wrench more strength and eliminating the tendency of the jaws to spread.

brake truing device. A device used to eliminate runout in superabrasive wheels that contains a silicon carbide wheel mounted on a spindle with an automatic braking system which can be mounted on a magnetic chuck and positioned beneath the superabrasive wheel.

brass. An alloy of copper and zinc that is stronger and more corrosion resistant than pure copper.

Brinell hardness scale. A measurement most commonly used to designate the hardness of nonferrous metals and steels before machining and heat treatment, performed by making an indentation on a piece of material with a 10-mm-diameter tungsten carbide ball.

bronze. An alloy primarily made of copper and tin.

BT flange. Type of CNC toolholder flange gripped by the ATC during the tool-change cycle, distinguished by its holder's off-center groove and metric threads on the retention knobs.

buttress thread. A thread with asymmetrical form, often used in applications requiring high thrust in a single direction, in which one flank (the pressure flank) is nearly perpendicular to the center line of the screw.

C

calibration. The process of verifying the accuracy of a measuring tool with another tool having a higher degree of precision that is known to be properly functioning and accurate, and making adjustments if needed.

calipers. Tools used in semi-precision measurement that have two legs that make contact with part surfaces to obtain measurements.

canned cycles. Packaged routines for CNC machining operations that help make tedious and redundant operations such as rough turning hole-drilling easier and faster to program.

carbide tooth. A type of band saw blade that has tungsten carbide brazed to a carbon steel body, which can be operated at very high speeds and cut very tough materials that bimetal blades cannot cut.

carbon steel. Inexpensive material often used for band saw blades under 1/4" in width that is often used for vertical contour sawing of small radii.

carburizing. A heat treatment process that adds carbon to the outer layer of low-carbon steel.

career and technical education (CTE). Hands-on training in the trades provided to high school students to prepare them for career paths in various industries.

carousel-type tool changer. A circular, automatic tool changer used in a machining center in which an empty tool compartment in the carousel moves to grip the tool in the spindle.

carriage. The part of a lathe that supports the lathe cutting tool and provides it with the movement needed to perform longitudinal machining operations.

Cartesian coordinate systems. Two- or three-dimensional systems that use X and Y or X, Y, and Z values for location.

case hardening. A process used on low-carbon steels in which the outer layer of heated steel soaks up carbon from another source, after which the steel is hardened by heating and quenching to produce a hard shell on the outside of the part while the inside remains soft.

cast iron. Ferrous metal with carbon content between 1.7% and 4.5%, recognized by a rough scaly surface.

CAT flange. Sometimes called a V flange, the type of toolholder flange gripped by the ATC during the tool-change cycle, distinguished by its holder's on-center groove and inch-series threads on the retention knobs.

C-clamp. A temporary fastening tool with a C-shaped frame and a screw to clamp the workpiece in place, which is useful for heavy-duty clamping.

center-cutting endmills. Milling cutting tool with each of the end cutting edges extended slightly past each other, allowing plunge cutting.

center drill. A combination drill and countersink, which has a small pilot drill on the tip, and transitions into a countersink with a 60-degree included angle for machining a recess to accept a lathe center.

center finder. A tool used to help establish location as the workpiece is positioned under the center of the drill press spindle.

center gage. A small measuring tool used to align a lathe thread-cutting tool's form to the workpiece.

center head. A V-shaped combination set attachment which can be assembled with a rule or blade and then laid across the end of a cylindrical part to find the center.

centerless grinder. A device used for grinding the outside of shafts and tubes by supporting rotating cylindrical work on a work rest while feeding the workpiece between the grinding wheel and a regulating wheel.

centerline. A thin line of alternating long and short dashes used to show the center of a diameter or radius, or the center of a part on an engineering drawing.

center punch. A pointed striking tool with an included angle of 90 degrees used to enlarge a prick punch mark, making it easier for drills to maintain location when beginning to drill holes.

ceramic aluminum oxide. A synthetic (man-made) grinding wheel abrasive; wheels made entirely from ceramic aluminum oxide are very durable but are not well suited for precision grinding because they cannot produce smooth surface finishes.

certificate program. An educational program, normally 2-year, that focuses primarily on practical lab application courses and applied or practical academics.

chaining. A technique used in CAD or CAM that allows the user to click on one single entity and, in the process, select all entities connected to it (linked like a chain).

chamfer. The beveled edge of a hole that allows easier entry of pins, aids in starting taps, and reduces burrs on edges of a hole.

chemical-based cutting fluids. Lubricating substances applied to the cutting area, where the tool and the workpiece make contact, that are primarily comprised of chemicals rather than oil.

chip load. See feed per tooth (FPT).

chisel edge. Subpart of the drill point forced into the material when drilling that lies between the lips at the center of the drill point. See also dead center (drill bit).

chronic. A long-term or slowly developing health effect from exposure to a hazardous material.

chuck key. A special wrench inserted into a socket on a chuck to open and close the jaws.

circular interpolation. Commanded by the G-codes G2 or G3, a type of motion that causes the cutter's path to travel in an arc so that CNC machines can cut full or partial circles.

circularity. GD&T specification that controls the tolerance for roundness of a cylindrical part feature.

circular runout. GD&T specification that controls the amount of runout of a cylindrical part feature related to another part feature referenced by a datum.

clamping lever. A device that locks a machine part in place and prevents unwanted movement.

class of fit. The relationship of tightness or looseness between the sizes of two mating parts.

clearance plane. In CNC programming and machining, a safety zone (often 0.050" or 0.100" above the workpiece or workholding device) intended to provide a factor of safety during rapid positioning.

climb milling. A method of milling where the cutting tool rotates in the same direction as (pulls into) the direction of feed.

CNC collet chuck. Toolholding device which uses a spring collet to secure round shank tooling.

CNC machinists. Individuals who possess the skills to both set-up and operate CNC machine tools.

coarse pitch. Threads with larger spacing between threads (fewer TPI).

cold air gun. A compressed air device used to cool a tool and workpiece without lubrication.

collet. A flexible sleeve which can constrict within a tapered bore to secure either cutting tools or workpieces.

collet block. A square or hexagonal workholding device which uses a spring collet to mount workpieces for milling or precision grinding.

collet fixture. A workholding device which uses a spring collet to mount workpieces for milling or precision grinding.

combination drill and countersink. A short drill, which has a small pilot drill on the tip, and transitions into a countersink, that is often used for spotting hole locations.

combination set. A measurement/layout tool that consists of a blade, square head, center head, and protractor head; the different heads are mounted to the blade by tightening a clamping screw.

combination wrench. Double-ended tool with one box end and one open end of the same size.

comparative measurement. See comparison.

comparison. Measurement that compares the specification of a part feature to known standard size, as when using fixed gages.

complementary angle (complement). The result of a given angle subtracted from 90 degrees.

compound rest. The part of a lathe that sits atop the cross slide and provides a means of providing angular tool movement.

computer-aided design (CAD). Process of designing components or producing electronic drawings of components using a computer software package.

computer-aided manufacturing (CAM). Process in which a programmer uses computer software to select tools and cutting operations on a CAD model of a part, which then generates the machine code that can be loaded into a CNC machine tool's control.

computer numerical control (CNC). A modern means of using a computer to control automatic machine operation which has replaced punched tape NC machines.

concave cutter. A special milling cutting tool having a concave profile ground into the edge, which is used to machine a convex radius on the workpiece.

concentricity. GD&T specification that requires all points on the surface of a cylindrical feature to lie within a tolerance zone referenced to a datum.

constant surface speed (CSS). A method of CNC turning programming that automatically updates the spindle RPM for the diameter being cut, activated by programming a G-code and providing the cutting speed in surface feet per minute.

control charts. A graphical means of analyzing trends in part variation, which helps to predict the consistency of a manufacturing operation and guides adjustments before the operation starts to produce parts outside of tolerances.

contours. The profile of a part's shape which may include a combination of straight surfaces, tapers, and/or radii.

conventional machinists. Sometimes called manual machinists, these are highly skilled workers who usually have experience setting up and operating almost every type of conventional machine tool.

conventional milling. A method of milling where the cutting tool rotates in the opposite direction as (pushes against) the direction of feed.

conversational programming. A special type of MCU function developed to simplify the machine programming process in which the operator selects the intended type of machining operation from an on-screen menu and the machine prompts the programmer with a series of questions.

convex cutter. A special milling cutting tool having a convex profile ground into the edge, which is used to machine a concave radius on the workpiece.

coordinate map. A sketch showing the X- and Y-coordinates of the center point locations for a cutting tool.

coordinate measuring machine (CMM). A measuring machine that inspects part features by probing many points on the feature using the X, Y, Z coordinate system.

coordinate system. A system which uses numbers to identify a position in three dimensions.

copper alloys. Copper with additional elements added to produce desired material characteristics.

corner-rounding cutters. Milling cutting tools ground with the form of a concave radius on each cutting edge, which are used to create convex (outside) radii on the corners of workpieces.

cosine. Abbreviated as cos, a ratio between the lengths of the adjacent side (A) and the hypotenuse (H) of a right triangle.

counterboring. A hole modification technique that increases the diameter of a hole to a certain depth in order to allow a screw head or nut to be positioned flush with or below the workpiece surface.

countersinking. The process of cutting a tapered opening in a hole so that a flathead screw can be installed flush with the workpiece's surface.

crest. The peak of a thread that creates the points from which the major diameter is measured.

cross-feed hand wheel. The surface grinder part that is rotated to move the workpiece cross-ways in incremental steps beneath the grinding wheel between longitudinal cutting passes.

cross slide. The part of a lathe that sits on top of the saddle and provides cutting-tool movement perpendicular to the ways.

cubic boron nitride (CBN). Superabrasive that is second in hardness to diamond.

cutoff. An operation performed on the lathe that uses a special narrow cutting tool to cut off a workpiece to a desired length.

cutter radius compensation. The offsetting of the path of the cutting tool in order to adjust for the radius of the cutting tool.

cutting fluids. Substances used to lubricate and cool which are applied to the cutting area where the tool and the workpiece make contact.

cutting plane line. A thick line, drawn as one long and two short dashes, alternately spaced, used to indicate the location from where the section view is shown on an engineering drawing.

cutting speed. The distance that a point on the circumference of a rotating cutting tool travels in 1 minute.

cyaniding. A heat treating process that adds carbon to steel by placing it in a tank of heated liquid containing sodium cyanide.

cylindrical grinder. A device used to grind internal diameters, external diameters, shoulders, and faces by rotating the workpiece against the rotation of the grinding wheel, much like the lathe is used for turning, facing, and boring operations.

cylindricity. GD&T specification that requires all cross sections (the entire surface) of a cylindrical part feature to be within the specified tolerance zone.

D

datum. In GD&T, a plane from which dimensions are referenced, shown on a drawing as a capital letter inside a square with a line extending to the part feature.

dead blow hammers. Soft face hammers with sand or shot in the head to absorb energy from striking and keep them from rebounding, which are frequently used to seat workpieces on parallels in machine vises.

dead center (drill bit). Subpart of a drill point forced into the material when drilling that lies between the lips at the center of the drill point. See also chisel edge.

dead center (lathe). A lathe workholding device used for holding work between centers that has no rotating parts and typically is used in the headstock spindle.

deburring. A technique used to remove sharp raised edges from workpieces.

deep-hole drilling cycle. A CNC canned cycle of breaking the downward drilling motion of a machine to help interrupt chip flow and clear chips from the drill flutes, which is initiated by G83 and fully retracts the tool from the hole between pecks.

demagnetizer. An electric device which is moved across the surface of a workpiece to remove magnetism from the work.

depth micrometer. A micrometer depth gage that features a base and interchangeable rods for different size ranges.

depth of cut. The depth that a cutting tool is engaged in the workpiece during material removal.

depth stop. The drill press part that sets the quill's travel limit to feed to a desired depth.

diagonal cutters. Hand tools used for light cutting of wire and pins that provides a cut nearly flush with a work surface.

dial indicator. A measuring tool that shows small movements by displaying them with a needle on a graduated dial face.

diameter. The distance from one side of a circle to another through the center.

diametral programming. A method of programming X-axis coordinates on a CNC turning machine in which all X coordinates are expressed as diameters.

diamond dresser. A device which holds an industrial diamond so that it can be used to simultaneously dress and true aluminum oxide and silicon carbide wheels mounted on a precision grinding machine.

die makers. Individuals who specialize in making punches and dies that are used to either bend, form, or pierce metal parts.

die maker's square. A hybrid tool that combines a square and a protractor; it can be used like a standard adjustable square, but its blade can also be tilted by turning an adjusting screw to make comparative measurements of angles up to 10 degrees.

die stocks. Tools used to hold threading dies when cutting threads by hand.

digital indicators. A tool that shows small movements by displaying them with digital readout on the tool's face.

dimension. Specified part feature size that must be measured to ensure it is within required specifications.

dimension lines. Thin lines with arrowheads used in engineering drawings to specify sizes.

direct hardening. Sometimes called through hardening, a heat treatment that can be performed on steel containing at least 0.3% of carbon, which is needed for the steel's structure to change and become hard when heated and then rapidly cooled.

direct indexing. A method of indexing that allows the spindle of the indexing device to be directly rotated by hand and locked with pin or plunger against an indexing plate having the appropriate number of divisions.

direct numerical control (DNC). A process in which a CNC program is fed from a PC or other storage device to the machine's control line by line as the machine runs the program, allowing the machine control to only accept as much code as it can process.

direct RPM programming. A method of CNC lathe spindle speed programming that allows the spindle RPM to be directly programmed at a fixed rate and is activated with a G97 code.

divider. A tool containing two legs with scribe points that are adjustable for different sizes, which is used to draw circles, radii, and arcs on a part.

dividing head. See indexing head.

double-ended wrench. A wrench containing either one box end and one open end, or two box ends or open ends of different sizes.

dovetail cutter. A special milling cutter that machines the angled surfaces of the dovetails used in the slides of machine tools.

drawbar. A device used on a vertical mill that passes through the spindle from the top of the head. The bottom of the drawbar is threaded, which provides a means of securing toolholding devices to the spindle.

draw filing. A finishing technique in which the file moves in the direction of a part's length to produce a very smooth surface finish.

drawing. See tempering.

drawtube. A hollow device used on turning machines to draw a collet tight into its tapered bore.

dressing stick. A handheld device used to remove some of the bond in the diamond or CBN grinding wheel by moving the device across the rotating wheel face by hand to ensure that the superabrasive grains are exposed and can cut freely.

drill body. The part of a twist drill that extends from the tip to the beginning of the shank and makes up the majority of the drill bit.

drill chuck. The toolholding device that grips the straight shank of a drill or other holemaking tool.

drill drift. A wedge-shaped device inserted between the taper and tang of two mating Morse taper toolholding accessories to separate them.

drilling. The process of using a cylindrical rotating cutting tool that is sharpened on its end to create a hole.

drill point. The cone-shaped area at the very tip of the drill bit.

drill point gage. A tool used to check the angle of the drill point and the length of the lips.

drill press. A device, normally used when precise hole locations are not necessary, that creates holes by plunge-feeding various types of rotating holemaking cutting tools into the material.

drip feeding. See direct numerical control (DNC).

drive dog. A device used to transmit torque from the lathe drive plate to the workpiece when work is held between centers.

drive plate. A specially designed plate that is mounted on the spindle nose of a lathe to provide torque to the drive dog.

dry cycle. Running a CNC machine with disabled functions to eliminate the potential for collisions. This is usually performed at the actual feed rates and rapid motions for accurate prove-out.

dry run. Running a CNC machine with disabled functions to eliminate the potential for collisions. This is usually performed at an increased feed rate for time savings.

dwell. The pause of a cutting tool for a specific time after it reaches full depth to relieve tool pressure or provide a consistent machined surface.

E

edge finder. A device consisting of two precise cylindrical pieces held together by a spring that can be used to very accurately locate a reference edge when performing milling operations.

electrical discharge machining (EDM). A machining process that uses electrical current to cut any material that will conduct electricity.

electric drill. A handheld holemaking tool used to power holemaking tools.

electromagnet (precision grinder chucks). A magnet used for workholding that develops a magnetic field through the presence of electricity; most models have a variable power switch that can adjust the strength of the magnetic force.

elevating crank. The part of the drill press or milling machine that raises and lowers the worktable or knee.

elevating hand wheel. The part of a horizontal spindle grinder that is rotated to raise or lower the wheel head.

elevating screw. The part of the vertical mill mounted inside the knee that both supports and moves the knee.

emergency collet. A brass or mild steel workholding device that can be machined to a desired size, when a special size is needed.

emulsifiable oil. See soluble oils.

encoder. The sensor portion of a CNC machine's servo motor that works by recording the amount of rotation a motor makes (in degrees), which provides feedback to the machine control to ensure the desired amount of movement is obtained.

end cutting-edge angle. The angle between the front edge of a lathe cutting tool and the surface being machined, which prevents the front of the tool from rubbing against the machined surface.

endmill. A milling tool with cutting edges both on its end and on its periphery, which is used for machining a wide variety of features such as pockets, slots, keyways, and steps.

end of block. The semicolon character at the end of a block of CNC code that instructs the MCU to move on to the next block.

end products. Final manufactured items used by consumers.

engineering drawings. Drawings of engineered components that show shapes, sizes, and specifications that must be met.

English system. The standard system of measurement in the United States, based on the inch.

entities. The individual lines and arcs that represent the outline of a part's shape in a computer-aided design (CAD) software.

Environmental Protection Agency (EPA). A federal agency whose purpose is to protect human health and the environment.

expanding collet. A special collet used to hold workpieces on a lathe that uses an integral arbor instead of a hole; when drawn against the taper, the arbor expands and grips an existing inside diameter of a workpiece.

expansion mandrel. A special precision cylindrical workholding device with an external taper and a sleeve with a mating internal taper, which has a series of slots cut into its outside diameter to allow mounting of work by internal diameters of non-standard sizes.

expansion reamers. Tools that have slots cut into the body that allow the reamer to expand to the exact size needed when the adjusting screw in the end of the reamer is tightened.

exploded isometric drawing. A drawing that illustrates detail and direction about how mating parts fit together as a complete assembly.

extension lines. Thin, continuous lines in an engineering drawing that extend from the edges or a feature of an object to the dimension lines.

external thread. A spiral groove made on a round external diameter of a workpiece to accept an internally threaded part for fastening.

F

face mill. A multiple-cutting-edge milling cutter with replaceable inserted cutting edges, for machining large flat surfaces.

face milling. Using the face of a cutting tool to machine a surface.

faceplate. A device generally made of cast iron with a series of slots machined into its face that is used to secure an irregularly shaped workpiece to a lathe.

facing. A common lathe operation that involves cutting across the end of a workpiece to machine the end flat.

fast-peck drilling cycle. A CNC canned cycle which breaks the downward drilling motion of a machine by slightly retracting the tool to help interrupt chip flow and clear chips from the drill flutes, which is initiated by G73.

feature control frame. A rectangular box with sections for a geometric tolerancing symbol, the amount of the tolerance, and the reference to a specific datum (if required), which is used to show how to control the size and/or shape of a particular feature of a part.

feature of size. A part feature that is either cylindrical or contains two opposing parallel surfaces.

feed. The process of advancing a cutting tool into or across a workpiece.

feed change lever/knob. A device used to switch between longitudinal and cross feed on a lathe.

feed control lever (lathe). The device that causes machine-powered movement of either the carriage or the cross slide on a lathe when engaged.

feed control lever (milling machine). The device located on the left side of a vertical mill's head that is used to start the quill feed.

feed per revolution (FPR). The amount the cutting tool advances each time the spindle turns one revolution.

feed per tooth (FPT). The thickness of the chip removed by one cutting edge of cutting the tool per each revolution of the tool.

feed rack. A bar spanning the length of a lathe's bed which contains gear teeth that mesh with a gear in the carriage to create longitudinal movement.

feed rate. The rate of tool movement into or across a workpiece.

feed rate override. A knob on the machine control panel that allows the programmed feed rate to be decreased, increased, or even stopped during operation.

feed reverse lever/knob. A device used to change the direction of the carriage or cross-slide movement on a lathe.

feed rod. A long shaft, either round or hexagonal, that transmits power to the carriage apron gear train on a lathe.

feeler gages. See thickness gages.

ferromagnetic. Attracted to a magnet.

ferrous. Containing iron.

file. A hand tool made of steel having many small cutting teeth for tasks such as shaping, smoothing, fitting, and deburring.

file cards. Brushes with short soft or wire bristles used to clean files, which may include a small pick for removing pins.

fine pitch. Threads with smaller spacing between threads (more TPI).

finishing. A process performed after roughing using higher cutting speeds, lighter cuts, and lower feed rates that produces a smooth surface on a workpiece to machine the part to the final desired size.

finish turning canned cycle. A CNC turning canned cycle, activated with a G70 that is used after the rough turning canned cycle (G71) to produce a finish turning pass.

fixed-base vise. A device used to securely hold workpieces that is mounted in a fixed position on a bench with bolts.

fixed gages. Measuring tools of a certain, unadjustable size that are used for comparative measurement of parts.

fixture. A workholding device designed to accommodate a specific workpiece which, although typically expensive, is the ideal technique for holding unusual parts that will be produced in large quantities.

flame hardening. A heat treating method in which steel is heated to the hardening temperature using the open flame of a gas-fueled torch (such as propane or oxygen/acetylene mixture) and then quenched.

flank. The thread surface on a screw that joins the crest to the root.

flash point. The lowest temperature at which a substance can be ignited.

flatness. GD&T specification requiring a surface of a part to be flat within the specified tolerance zone.

flex back. A carbon saw blade in which only the teeth are hardened.

flex collet. A collet that consists of steel segments attached to a hub, which gives the flex collet a size range of up to 0.100".

floating tap holder. A CNC toolholder which allows axial (in and out) float of the tap to compensate for small errors between feed and rotation.

flood system. A method of applying liquid cutting fluids by pumping the fluid through a pipe or nozzle to the cutting area, flooding it.

flutes. Spiral grooves in the sides of a cutting tool that create cutting edges and provide a pathway for chips to be removed from the cutting area.

fly cutter. A cutting head used for milling flat surfaces that holds a single-edged cutting tool, which is capable of creating fine surface finishes.

follower rest. A device attached to the carriage of a lathe that moves along the length of the workpiece during machining, providing additional support to long, slender workpieces.

form cutting. The process used to produce contoured surfaces on a conventional lathe, in which the cutting tool contains the reverse form of the desired part shape so that, when fed into the workpiece, the tool produces the desired shape in the part.

form tolerances. GD&T specifications that limit the amount of error in the shape of a part feature.

forward/off/reverse switch. The part of the drill press that starts the motor to rotate the spindle.

four-jaw chuck. A common lathe workholding device with four jaws that move independently of each other, which is used when the number of workpiece sides is divisible by four and allows adjustment of workpiece runout.

friability. The ability of individual grinding wheel abrasive grains to fracture during grinding to create new, sharp cutting edges.

front view. In engineering drawing, the position that normally shows the most details of something, regardless of whether it is actually the front of the object.

full stop. A hold placed on a CNC machine's program by an M0 command that requires an operator to restart the program by pressing the cycle start button on the operator panel.

G

gage. Name for a wide variety of tools used for measuring and for setting up machine tools.

gage blocks. Accurately sized blocks with very smooth surfaces that can be used for part inspection or to check the accuracy of other precision measuring tools.

gage block builds. Wrung assemblies of gage blocks that will stick to each other to create nearly any size needed.

gang drill press. A holemaking machine that contains multiple drilling heads attached to a single base and worktable, which allows numerous cutting tools to be mounted, reducing the need to repeatedly change tools in and out of the spindle when performing different holemaking operations on multiple workpieces.

gang tool (turning center). A common turning center machine, typically of a flat bed design, that arranges cutting tools side-by-side in a row.

gap bed. A small portion of a lathe bed that can be removed, creating a gap, to allow larger-diameter workpieces to be machined.

G-codes. CNC programming codes that prepare a machine to engage in a particular mode for machining; for instance, a G1 code tells the machine to feed in a straight line, or linear motion.

gear box. The gear train of a lathe located below the headstock that controls the cutting tool's rate of movement.

general-purpose reamers. Cutting tools used to slightly enlarge an existing hole and produce an accurately sized diameter hole with a smooth surface finish.

geometric dimensioning and tolerancing (GD&T). A system that uses symbols to show tolerances of form, profile, orientation, location, and runout on an engineering drawing intended to ensure functionality of the finished machined part.

geometry. Math involving characteristics of shapes and their relationships; in computer-aided design, the entirety of the entities that make up a drawing.

geometry offset. The CNC tool offset which tells the MCU dimensional data about the tool such as tip position and radius size for an unworn tool.

(GHS) Globally Harmonized System of Classification and Labeling of Chemicals. An internationally recognized method for classifying and labeling hazardous materials.

gib. A tapered wedge found on dovetail-shaped slides that can be used to tighten the slide as the dovetail surfaces wear.

glazing. A condition in which a grinding wheel cuts inefficiently because the grains of the wheel become dull more quickly than normal, preventing new, sharp grains from being exposed.

go/no-go plug gage. A measuring tool used to check whether a hole diameter or internal thread is within tolerance.

grade (grinding wheel). The hardness or strength of a grinding wheel's bond that ranges from A (the softest) to Z (the hardest).

graduations. The divisions or spaces on a measuring tool.

grade (carbide). The relative hardness or toughness of a given carbide cutting tool material.

grit (grain) size. The measurement of the degree or coarseness of grinding wheel abrasives in which the lower the number, the coarser the wheel; grain sizes range from 8 to 600.

grooving. The process of using a lathe to machine grooves or recesses on workpieces, which may be needed to accept O-rings or retaining rings, act as a relief for threading, or provide clearance on shoulders for mating parts.

gullet. The curved area at the root of a saw blade tooth in which the metal chips are formed into curls.

H

hacksaws. Simple, portable handsaws consisting of a frame, handle, and blade that are used for light-duty sawing operations.

half-nut lever. A device located on the apron of a lathe that controls a split nut that can be engaged directly to the lathe's leadscrew to perform thread-cutting operations.

hand reamers. Reamers with a square on the end of the shank that are mounted in a wrench and turned by hand.

hand tap. Tap style that has straight flutes from the tip back which curl the chips tightly and break them into small pieces.

hard back. A carbon band saw blade in which the entire blade is hardened.

hardening. A heat treatment, most commonly performed on steels, in which the material is heated to a hardening temperature in a special type of furnace and then cooled rapidly to transform the structure of the steel, resulting in an increase in material hardness.

Hazard Communication Standard (HCS). OSHA requirement that manufacturers of hazardous materials must provide information about those materials to users and that employers must provide information about hazardous materials to employees.

Hazardous Material Identification System (HMIS). A system of labeling hazardous materials that uses colored bars and numbers from 0 (minimal hazard) to 4 (severe hazard).

headstock. The part of the lathe located at the upper left side, comprised of a casting that contains the mechanisms used to hold and rotate the workpiece and control the rate of tool movement.

heat treatment. The controlled heating and cooling of metals to change their characteristics (i.e., to make the material harder, tougher, softer, more stable, or more easily machined).

heel. A subpart of a drill point at the end of the cutting lip clearance.

height gage. A surface plate tool that measures vertical dimensions from the reference zero created by the surface plate's horizontal plane.

helix. The spiral, or helical, shape of a screw's thread created by the thread grooves or the spiral of the flutes of a holemaking or milling cutting tool.

helix angle. The angle of a drill's or thread's spiral relative to the center axis of the drill or thread.

helper-type measuring tool. See transfer measuring tool.

hermaphrodite calipers. Layout tools that have one leg shaped like a divider and one shaped like an outside caliper, which are used to scribe lines parallel to the edge of a piece of material.

hex key wrenches. Wrenches with an external hexagonal shape that fit into the hexagonal recesses on socket head cap screws and set screws.

hidden lines. Thin, dashed lines used in an engineering drawing to show edges that are not visible in a particular view.

high/low range lever. A device on a vertical mill that selects between fast speed range and slow speed range.

high-speed steel (HSS). Material used to make cutting tools such as band saw blades, lathe cutting tools, holemaking tools, and milling cutters.

hinged clamps. Fastening tools with two hinged clamping jaws that use a screw to force the jaws together for clamping.

homing procedure. Procedure performed when powering up a CNC machine which recalls the reference positions.

hook spanner. A wrench with a hooked arm that fits into fasteners with slots around their perimeter.

hook tooth form. A saw blade tooth shape with a positive rake angle (shaped like a hook) and large gullet area.

horizontal band saw. A cut-off machine which uses a continuous metal band with a series of teeth ground into one edge of its periphery supported by two wheels, is in a horizontal orientation.

horizontal machining center (HMC). A CNC machining center with a horizontally oriented spindle.

horizontal spindle surface grinders. A precision grinding machine that typically uses the periphery of a solid wheel to grind workpiece surfaces, using either rotary table motion to rotate the work beneath the grinding wheel or a reciprocating table to move the work back and forth beneath the grinding wheel.

hydraulic press. Device that uses a hydraulic pump to move a cylinder to create force for assembly and disassembly of mating parts with interference fits.

hypotenuse. The longest side of a right triangle, opposite the 90-degree angle.

I

inch. The basic unit of English measurement, commonly followed by the symbol ".

inches per minute (IPM). The distance a cutting tool feeds across the workpiece surface in one minute, commonly used for milling operation feed rates.

inches per revolution (IPR). The distance that a cutting tool advances into or across a workpiece for each revolution of the cutting tool (milling operations) or the workpiece (lathe operations).

inches per tooth (IPT). See feed per tooth (FPT).

incremental positioning system. A referencing method used with coordinate systems which specifies a distance from the current position to the next position instead of a location related to the initial origin.

independent chuck. A device with jaws that are advanced and retracted independently of each other, allowing for fine-tuning the position of the workpiece for maximum accuracy.

indexable tool post. A lathe toolholding device containing a multi-sided tool block that can be loaded with multiple tools and then rotated and secured to position the desired tool for machining.

indexing. The process of rotating a workpiece to angular positions and then securing it there for machining.

indexing head. A workholding device that serves as a rotary axis for part positioning by providing the ability to rotate the workpiece in very precise angular increments.

induction hardening. A heat treating process in which steel is heated to the hardening temperature by running an electrical current through the material.

industrial salesperson. An individual who visits customers and prospective customers to discuss their machining needs in order to gain their business.

initial plane. The Z-height where the cutting tool is located when a CNC canned cycle is activated.

inscribed circle (I.C.). The largest circle which can fit within the edges of an insert's shape.

inspection plan. A quality control guideline that indicates what dimensions of parts to inspect and what measuring tools and processes to use during inspection.

inspectors. See quality control technicians.

internal thread. A spiral groove cut in the inside diameter of a workpiece to accept an externally threaded part.

International Alloy Designation System (IADS). An organization that developed a system for designating aluminum alloys.

International System of Units (SI). See metric system.

iron. A heavy, ferrous, metallic element that is frequently used in machine work.

ISO Metric. Established dimensional standards regarding form, size, tolerance, allowance, classification, and class of fit for the consistent interchange of metric parts.

isometric view. A three-dimensional view of a part on an engineering drawing to provide a better visualization of the part.

J

jaw-type chuck. A device consisting of a chuck body with sliding work-gripping jaws that are used to clamp a workpiece in the lathe spindle by applying pressure on multiple sides.

jig grinder. A machine tool used to perform precise grinding of internal diameters on a workpiece in which a head moves in an orbital pattern and a quill feeds up and down as a spindle-mounted grinding wheel rotates.

jobber's reamers. Holemaking tools with long flutes that are used for machine finishing a hole to a precise size with a smooth finished surface.

jog. Using axis buttons or a small rotary hand wheel to manually control axis movement during CNC machine setup.

journeyperson. A title given to someone who has completed a company's apprenticeship program and is expected to be able to perform any machining operation required by that company.

K

kerf. The width of the slot a saw blade produces in the workpiece.

key. A removable component mounted in aligned slots in a shaft and hub of a gear or pulley that transmits power between the shaft and the hub.

keyseat. The slot in a shaft that holds the key.

knee. Heavy casting with a dovetail slot at its rear which supports the table and saddle on the vertical mill.

knee mill. A vertical spindle milling machine with a work holding area that moves up and down along a column mounted dovetail slide.

knurling. The process of producing a raised pattern on the circumference of a workpiece by pressing a tool with two wheels, called rolls, against a rotating workpiece.

L

laser machining. A cutting technique that uses a highly concentrated beam of light with temperatures up to 75,000°F to cut, groove, or engrave metals.

lathe. A machine tool used to produce cylindrical parts by moving a cutting tool across the surface of a rotating piece of material.

lathe dog. See drive dog.

layout. The process of locating and marking a workpiece as a visual reference to guide machining operations.

layout dye. See layout fluid.

layout fluid. Dark blue or red liquid applied to material to provide contrast and make it easier to see layout lines.

layout fluid remover. Substance used to dissolve and remove layout fluid from a workpiece.

lead. The distance a thread will advance in one revolution in relation to its mating thread.

lead angle. The angle of the cutting edge of a tool relative to its shank, which can be either positive, negative, or zero.

leader line. A thin angled line with an arrowhead on one end that points to a specific feature or detail on an engineering drawing.

leadscrew. A very long threaded rod supported by bearings at both ends used to transmit motion to the carriage of a lathe for thread-cutting operations.

least material condition (LMC). A situation in which a feature of size contains the least material within its tolerance.

left-hand threads. Screw threads with a helix that leans the opposite of a right-hand thread, causing the assembled fasteners to get tight when rotated counterclockwise.

length of engagement. The distance an externally threaded part threads into an internally threaded part.

limit tolerance. Acceptable range of the dimension of a part feature specified on an engineering drawing by showing the maximum and minimum allowable sizes.

linear guides. A modernized version of ways that are sealed, have pressurized lubrication systems, and contain low friction ball bearings.

linear interpolation. Designated by the G-code G1, CNC motion used to move the tool in a straight line along one or more machine axis.

line types. Different styles of lines used in an engineering drawing to give a visual representation and specifications of the part to be machined.

lip clearance. The relief angle from the cutting lips of a twist drill back to the heel.

lips. The angled cutting edges of a drill point that shear the metal into chips as the drill rotates.

live center. A device mounted in a lathe tailstock that aligns and supports the workpiece as it rotates with the workpiece.

live tooling. Small motorized spindles that enable a CNC turning machine to perform light-duty milling hole-work such as drilling, tapping, and reaming off-center and perpendicular to the machine's Z-axis.

loading. A condition in which a grinding wheel or the teeth of a file cut inefficiently because pieces of soft metal become lodged in the wheel or file.

local (specified) tolerance. A method of listing tolerance on an engineering drawing where the tolerance is shown next to the dimension and applies only to that dimension.

location tolerances. GD&T specifications used to control the location of workpiece features related to other workpiece features.

locking pliers. Hand tools that are adjusted to the desired jaw opening size by rotating an adjusting screw in the handle and then clamped on the workpiece by squeezing the handles.

lockout/tagout (LOTO). OSHA developed system of using padlocks and tags to prevent release of hazardous energy.

lockout device. A tamper-proof item like a padlock that is secured to a machine to prevent the equipment from powering up.

longitudinal feed. The sliding motion of the saddle of a lathe which is parallel to the ways.

longitudinal-feed hand wheel. The part of a surface grinder that is rotated clockwise to move the table to the right, or rotated counterclockwise to move the table to the left.

lower explosive limit (LEL). The lowest concentration of the substance in air that will burn or explode if ignited.

lower flammable limit (LFL). See lower explosive limit (LEL).

lower (low) limit. The smallest acceptable size of a dimension on an engineering drawing.

lubricants. Greases and oils used to cool moving parts, minimize friction between them, and prevent their seizing.

lubricity. The ability of a cutting fluid to reduce friction; a fluid with higher lubricity is better at reducing friction.

M

machinability. A relative measure of how easily a material can be machined.

machine control unit (MCU). Sometimes simply called a "control," the system of a CNC machine that interprets the program and then uses the information to control machine functions and control axis motion.

machine coordinate system (MCS). A coordinate system used for the machine's reference purposes.

machine home position. The fixed reference position that machine axis are returned to upon power-up.

machine tool. A mechanically operated piece of equipment used to cut materials to desired shapes and sizes.

machine tool service technician. An individual who travels into the field to perform repairs on machine tools.

machining. Using machine tools to cut materials to desired sizes and shapes.

machining center. A computerized numerical control (CNC) mill equipped with an automatic tool changer.

machinist. A person who makes a living at and is skilled in the use of machine tools.

magnesium. An alloy that looks very much like aluminum, but is lighter and stronger.

magnetic angle plate. Workholding device with square and parallel surfaces that uses and integrated magnet to hold ferromagnetic workpieces.

magnetic chuck. A device that can be used to secure ferromagnetic workpieces or other workholding devices.

magnetic parallels. A device that extends the magnetic field of the magnetic chuck to the workpiece, which is used to support a workpiece containing features that keep it from being placed flat on the magnetic chuck.

magnetic poles. The ferrous portions of magnetic workholding devices that create the magnetic pulling force when an internal magnet is activated.

magnetic sine chuck. A combination sine plate and magnetic chuck used to grind precision angles that is set up just like an ordinary sine plate but has its own permanent magnet to hold the work.

magnetic V-block. A device that can be used to hold a workpiece at a 45-degree angle and extend the magnetism of a magnetic chuck to the workpiece.

major diameter. The largest diameter of a thread.

manager. See supervisor.

mandrel. A special precision cylindrical shaft that can be inserted through the center bore of a workpiece to secure it for machining.

manual cutter compensation. The offsetting of the path of the cutting tool through programmer modifications to the programmed coordinates in order to adjust for the radius of the cutting tool.

manual data input (MDI). A control panel mode that provides a blank program screen for entering short programs or single program commands.

manufacturing. The process of producing a product.

manufacturing cell. A grouping together of several different machines performing operations on the same part to increase productivity and minimize operator attention.

manufacturing engineer. Individual who designs manufacturing processes and continually studies and improves on those processes.

manufacturing engineering technicians. Individual who constructs and maintains manufacturing processes and equipment.

margin. A thin raised strip that runs the length of the drill body along the edge of the flutes and gives the drill bit its diameter.

mastering. The process of setting a reference zero with a dial or digital indicator with a standard of known size, such as a gage block.

maximum material condition (MMC). A situation in which a feature of size contains the most workpiece material within its tolerance.

M-codes. CNC programming codes used to turn on and off miscellaneous functions.

mean. The average of a group of numerical values.

mechanic. A term used throughout the Industrial Revolution and early 20th century to refer to a person who works with machine tools.

mechanical designers. Individuals that use CAD software to draw models or engineering drawings that are referred to for machining operations.

mechanical engineer. Individual who understands the theory of many topics, including the selection and properties of materials, and usually designs overall projects, assemblies, or systems using high-end computer software.

mechanical engineering technicians. Individuals, who work in the actual construction or testing of mechanisms, machinery, and equipment.

metric system. A measurement system that uses the millimeter as the base unit in the machining industry.

metrology. The science or practice of measurement.

micro drill press. A holemaking machine that operates at very high spindle speeds and is used to produce very small diameter holes.

micrometer. A precision measuring tool that has one stationary point of contact and one moveable point of contact that is adjusted by a very accurate screw thread to measure the distance between those two points.

micrometer adjusting nut. The part of the vertical milling machine that can be set to limit quill travel by moving it so the quill stop comes in contact with the nut.

millimeter. The metric measurement unit most commonly used in the machining industry, which equals 1/1000 of a meter and is denoted by the abbreviation mm.

milling machines. Machine tools that feed a workpiece into a rotating cutting tool to remove material.

mill/turn machine. A CNC machine tool that can perform heavy-duty milling, turning, and drilling operations.

minimum quantity lubricant (MQL). A method of applying liquid cutting fluids that is similar to a mist system, but uses lower air pressure to deliver the smallest fluid concentration required to provide adequate tool lubrication and cooling.

minor diameter. The smallest diameter of a thread, measured from the root of a thread on one side to the root of the thread on the opposite side.

miscellaneous (auxiliary) functions. Activities controlled by M-codes in the CNC machining system. See also M-codes.

mist system. A method of applying liquid cutting fluid in which the cutting fluid contained in a small tank is combined with compressed air (atomized) and sprayed on the cutting area.

modal. Machine codes that remain active until cancelled or overridden.

mold makers. Individuals who specialize in machining molds and mold components for either plastic or die-cast metal parts.

Morse taper. A popular standardized style of taper used in drill press spindles and for holemaking tools.

Morse taper extension socket. Adapter used to reduce the size of a tool's shank taper.

Morse taper sleeve. Adapter used to increase the size of a tool's shank taper.

N

National Fire Protection Association (NFPA). Organization whose 704 standard provides a system of labeling hazardous materials that uses a multi-colored diamond shape to identify specific types of hazards and their levels of danger.

National Institute for Metalworking Skills (NIMS). An organization that establishes national benchmarks, or standards, for performance and knowledge related to several different areas of the machining industry, as well as a competency-based apprenticeship program.

National Institute for Occupational Safety and Health (NIOSH). The federal agency responsible for conducting research and making recommendations for the prevention of work-related injury and illness.

National Pipe Thread (NPT). Threads tapered 3/4 inch per foot to help seal them as they tighten together; sized according to the nominal inside pipe diameter rather than the major thread diameter.

needle nose pliers. Hand tools with jaws that taper toward the end, allowing them to be used to grasp and hold small parts.

nitriding. A heat treating process of heating low carbon steel in a 900°F sealed furnace containing nitrogen gas to create a hardened outer shell.

nominal. An approximation of a desired or targeted size.

nonferrous metals. A classification of metals that contain no iron.

normalizing. A process that is sometimes performed on medium- and high-carbon steels prior to hardening that removes stresses from the steel and makes its structure more consistent so that better results will occur from further heat treating operations.

nose collar. A device that threads onto the external threads of a collet chuck's end and constricts the collet when tightened by forcing it into the tapered bore.

numerical control (NC). A means by which machine motion is automatically guided by a number based program or punched tape.

nylon hammers. Hand tools that typically rebound after striking a surface, which are used when a soft, yet more durable striking surface is needed than that provided by dead blow hammers.

O

object lines. Thick, continuous lines used in engineering drawings to show edges of an object that would be seen in any particular view.

Occupational Safety and Health Administration (OSHA). A federal government agency that sets and enforces regulations for workplace safety.

offset screwdrivers. Hand tools angled at 90 degrees on one or both ends that are used when there is not enough room to use a screwdriver with a straight shank.

offset tailstock method. A method for turning tapers that requires no special attachments, but instead offsets the tailstock center from the headstock spindle center line. The main benefit of this method, which can only be used for external tapers, is that very long tapers can be machined.

oil-based cutting fluids. Cutting fluids that consist primarily of petroleum-or agricultural-based oils such as soy or vegetable.

oil cup. A container mounted on a machine tool with a small lid that is opened to add lubricating oil.

on-the-job (OJT) training. Structured or unstructured instruction provided to employees while they are receiving wages from an employer.

open-end wrenches. Light-duty wrenches with two parallel jaws that may be slid onto a hex or square drive surface.

operation. Any individual machining task associated with a machining process. A group of operations are compiled into a process to produce a machined part.

operators. Individuals, often entry-level employees with little machining experience, who place parts in machines and continually run a set operation or group of operations.

opposite side. In right angle trigonometry, the side across the triangle from any given angle.

optical comparator. A measuring tool that projects a magnified image of a small, difficult-to-see part or feature onto a screen for measurement.

optional stop. A programmed hold placed on a CNC machine's program by an M1 command that will only be recognized and executed when the optional stop switch on the operator panel is activated.

order of operations. The principle that certain mathematical operations must be done in a certain order to correctly perform calculations.

orientation tolerances. GD&T specifications that relate one part feature to another part feature or features through use of datums.

origin. In a Cartesian coordinate system, the point (X0, Y0, Z0) from which all other positions are based.

orthographic projection. A method of representing a three-dimensional object in two dimensions on an engineering drawing using different views.

P

page (CNC). A particular set of information shown on a display screen of a CNC machine control panel that may either accept data or display it.

pallet system. A workholding system used to maximize machining time and minimize workpiece-loading time in which two or more workholding tooling plates are assembled that can be quickly and accurately interchanged on the machine's table.

parallel. Having lines or surfaces that, in theory, will never intersect, or touch, if extended.

parallel clamps. Fastening tools with two parallel jaws that use parallel screws to adjust the jaw width and to clamp the parts, which is useful for light-duty applications such as holding small and delicate parts.

parallelism. Lines or surfaces with a constant distance between them.

parallels. Bars made of steel or granite with opposite sides that are parallel within very close tolerances, which are used to raise work above a reference surface.

parting. See cutoff.

parts per billion (PPB). The number of parts of a substance per billion parts of air.

parts per million (PPM). The number of parts of a substance per million parts of air.

parts per trillion (PPT). The number of parts of a substance per trillion parts of air.

peck drilling cycle. Programming technique used when drilling deeper than about three times the drill's diameter that involves breaking the downward drilling motion with slight retracts to help interrupt chip flow and clear chips from the drill flutes.

pecking. An alternating drilling and retracting motion during holemaking operations to clear chips from the cutting tool.

percentage of thread. The full thread depth compared to a partial thread depth, which is used to gauge thread strength.

peripheral milling. Using the outside periphery of a milling cutting tool to machine a surface.

permanent magnet. A magnet activated by flipping a lever to activate the magnetic fields.

permissible exposure limit (PEL). The maximum amount of a hazardous substance that someone can be exposed to without undue effects.

perpendicular. Having lines or surfaces that intersect at a 90-degree angle.

perpendicularity. 90-degree relationship between two lines or surfaces.

personal protective equipment (PPE). Safety equipment that is to be worn to protect a person from potential dangers.

personal skills. Traits that make up an individual's personality.

phantom lines. In an engineering drawing, lines used to show alternate positions of a part or outlines of adjacent parts.

Phillips screwdriver. Hand tool used to grip the cross-shaped heads of Phillips head screws.

pi (π). A constant number with a value of approximately 3.14159 used to calculate the circumference of a circle.

pictogram. Images shown on a GHS hazardous material label to visually show specific types of hazards.

pilot. A guide on the end of a counterboring bit that keeps it aligned in the existing hole.

pilot hole. A smaller hole that is often drilled in a workpiece to guide a larger drill and relieve pressure on the dead center of the larger drill.

pin. A particle of material being filed that can embed in the teeth of a file.

pin gages. Cylindrical rods with very accurate diameters that can be used to check hole dimensions.

pinning. Situation that may occur when using a file, particularly on softer metals such as aluminum, in which particles of the material being filed called can embed in the teeth.

pin spanner. A wrench with pins that fit into holes on the face of a threaded fastener.

pin vise chuck. A specialty micro drill chuck that has a slim design, small capacity, and will enable very small drills to run true.

pitch. The distance from one point to an adjacent point, such as between threads on a screw and teeth on a saw blade.

pitch diameter. The imaginary diameter measured where the thickness of the thread flank and space within the thread groove are equal.

plain carbon steel. Steel with composition of iron and carbon and no additional elements added.

plain gages. Smooth plug, pin, or ring gages used to check plain, or unthreaded, holes or external diameters.

plain keyseat. The slot in a shaft that holds square and rectangular keys.

plain protractor. A semi-precision tool with one-degree graduations used for angular measurement or layout.

pliers. Hand tools used by machinists for a wide variety of holding and cutting tasks.

plug chamfer taps. General purpose taps that have three to five chamfered threads and may be used in through holes or in blind holes when there is adequate clearance in the bottom of the hole.

plug gages. See pin gages.

pointed edge finder. Setup tool that can be mounted in a drill press or milling machine spindle and used to locate punch marks or layout intersections for holemaking operations.

point of tangency. The single point where a tangent line or arc touches another arc.

polar coordinate system. A coordinate system used to identify locations by defining both an angle and a distance (like a vector in mathematics) from the origin to a specific location.

polishing. Light material removal using abrasives to produce fine surface finishes.

position (true position). GD&T location tolerance used to specify the center of features such as holes or slots.

position tolerance modifiers. GD&T symbols added to feature control frames that allow expansion of the tolerance zone under certain conditions.

post-processing. The final step in CAM programming, performed after the creation of a successful CAM-generated machining process, which involves taking all of the defined toolpath data and using the CAM software to generate a CNC program.

power-feed mechanism (drill press). Device on some drill presses that automatically feeds the quill when a lever is engaged.

power-feed transmission engagement crank (milling machine). The part of the vertical milling machine located near the upper-right portion of the head that must be engaged to set the quill for power feeding.

power hacksaw. A power saw that achieves cutting action by drawing a saw blade back and forth across a workpiece.

precipitation heat treatment. A process performed by heating aluminum alloys to around 300°F to artificially speed up the aging process.

precision drill press. See micro drill press.

precision fixed gages. Measuring tools used to make comparative measurements against known sizes.

precision measurement. Measurement finer than 1/64 or 1/100 of an inch, 0.5 (1/2) mm, and 1 degree.

preparatory commands. See G-codes.

prick punch. A layout tool with a 60-degree included point used on a part to mark the intersections of lines that locate the center points of circles or arcs.

prints. See engineering drawing.

process. The entirety of all operations required in the machining of a part.

process plan. A strategy of steps needed to perform a machining operation or operations.

profile of a line. GD&T specification where all cross sections of the surface on an engineering drawing need to be within the specified tolerance zone, but they do not all need to be within the same tolerance zone.

profile of a surface. GD&T specification where an entire surface must be within the specified tolerance zone, so all cross sections of the surface on the engineering drawing need to be within the same tolerance zone.

profile tolerances. GD&T specifications requiring a workpiece surface to be a particular shape within a certain tolerance zone.

profilometer. An electronic tool that moves a stylus, or contact point, across a workpiece surface to measure the surface finish in microinches or micrometers, which is then shown on a display.

programmers (machine). Sometimes called CNC or NC programmers, individuals who write programs consisting of machine code for CNC machine tools.

program stop. A command to place a hold on a CNC program until it is resumed by pressing the cycle start button on the MCU.

proportion. Two ratios that are equal to each other.

protractor. A tool used to measure or layout angles.

prove-out. Carefully executing a CNC program in order to identify problems before letting the machine run without supervision, achieved through graphic simulation, dry run, or safe offset.

push stick. A soft metal or wooden stick used to safely push material during vertical bandsawing operations rather than using the operator's hands.

Pythagorean theorem. A formula that relates the lengths of the sides, or legs, of right triangles, written as $a^2 + b^2 = c^2$ where a, b, and c, and are the lengths of the triangle sides.

Q

quadrant. In the Cartesian coordinate system, four regions that are created when the system is divided by the X- and Y-axes; they are numbered counterclockwise from I to IV, beginning in the upper right region.

Qualified dimensions. The accurate and consistent dimensions that mass produced cutting tools are made to.

quality assurance (QA). An overall view, commitment, or plan for meeting customer demands.

quality control (QC). The actions of inspecting the dimensions of machined parts to make sure tolerances are met.

quality control technicians. Individuals who inspect parts produced by machine tools to ensure they meet specifications.

Qualified end length. The total length of the toolholder with installed inserts.

quenching. A method of rapidly cooling heated metal during heat treatment that can be done with different media, or substances, such as submerging in water, brine (salt/water mix), or oil, and cooling with air blasts.

quick-change toolholders. Dovetail-shaped devices that hold cutting tools and slide onto the matching dovetail on the quick-change post of a lathe.

quick-change tool post. A device used to hold a toolholder in a lathe that consists of a T-nut, clamping stud, and tool post which is usually equipped with a dovetail that mates with a dovetail on a toolholder. It allows several tools to be preset in different holders so they can be quickly and easily changed without the need to reset their positions.

quill. The part of the drill press or vertical milling machine that raises and lowers cutting tools held in the spindle.

quill feed handle. The part of the vertical milling machine or drill press that can be used to raise and lower the quill.

quill feed selector knob. The part of the vertical milling machine located on the left side of the head that can be positioned to select 0.0015", 0.003", or 0.006" IPR feed rates, determining how far the quill will advance every time the spindle makes one revolution.

quill lock. The part of the vertical milling machine that can be used to lock the quill in an extended position by pulling down on the lock.

R

radial-arm drill press. The largest of the drilling machines, designed for machining large-diameter holes or large workpieces.

radial drill press. See radial-arm drill press.

radial programming. A method of programming X-axis coordinates on a CNC turning machine in which all X coordinates are expressed as a radius of a part.

radius. The distance from the center point of a circle to any point on the circle.

radius and fillet gages. Measuring tools used to check outside corner and insider corner (fillet) radii.

radius endmill. Sometimes called a bullnose endmill, a flat endmill with a special radiused edge on the corners of the cutting edges.

radius type center drill. Center drill that transitions into a radius instead of a 60-degree included angle. It is often used for machining recesses to accept lathe centers when using the offset tailstock method for machining tapers on the lathe.

rake. The angle of a saw tooth's cutting face related to a line perpendicular to the back of the blade.

raker set. A saw blade tooth setting pattern with a repeating pattern of staggering one tooth left, then one tooth right, then one tooth straight.

ram. A part of the vertical mill that allows the entire head to be moved forward and backward and then locked in position, which increases the workpiece capacity of the machine.

ram adjusting lever (nut). A part of the vertical mill that moves the ram forward or backward when rotated.

range. A value calculated by subtracting the smallest dimension from the largest dimension in the sampling measurements.

range (R-) chart. A graphical representation that shows the range of sizes of each sampling at any given time during an operation.

rapid override. A control on a CNC machine control panel that can be used to slow rapid motion or to vary the jogging rate.

rapid traverse. Designated by the G-code G0, a type of CNC machine motion used to quickly position tools near the workpiece before beginning a machining operation.

ratio. A comparison between two numbers which can be written in the form of 1:2 or as a fraction 1/2.

rawhide. A soft, compressible, and tightly rolled and shaped material used for the heads of some soft hammers.

reamers. The cutting tools used for finishing an existing drilled hole to a precise size with a smooth finished surface.

reaming. The process of finishing a hole to a precise size with a smooth finished surface.

rectangular coordinate system. See Cartesian coordinate system.

reference dimension. A print dimension, usually in parentheses, that is not subject to print tolerances.

reference return. Movement of a CNC machine's axes back to the MCS origin (home position) performed during the power up procedure.

refractometer. A handheld tool that is commonly used to determine fluid concentration levels when using soluble oils and synthetic and semi-synthetic cutting fluids.

regardless of feature size (RFS). GD&T position condition requiring that the specified position tolerance zone must be maintained no matter the size of the feature.

regular tooth form. The standard saw blade tooth style, which has large radii in the gullet area of the saw teeth and a zero rake angle.

R-8 taper. The spindle taper found in most manual vertical milling machines used to align toolholding devices.

repetitive canned threading cycle. Operation on most turning centers that automatically takes successive thread cutting passes until the tool reaches the thread's root (minor) diameter.

resinoid. A natural, tough grinding wheel bonding agent used primarily for rough offhand grinding under harsh conditions; identified by the letter B.

retention knobs. A device mounted in a NMTB milling holder and gripped by a power-actuated drawbar mechanism in a CNC milling machine spindle to secure the mating tapers tightly together.

return (R-) plane. A return point different from the initial clearance plane that some CNC canned cycles can direct the tool to before beginning to feed.

return point. The absolute Z-position from where the cutting tool will begin feeding and where it will retract to at the end of a CNC canned cycle.

revisions. Changes made to engineering drawings.

right-hand threads. Screw threads with a helix that causes assembled fasteners to get tight when rotated clockwise.

right side view. In engineering drawings, the view created by projecting the right side of an object onto the right side surface of the "glass box" once the front view has been established.

rigid tapping. CNC machining operation that provides an accurate method of mating the spindle RPM and feed rate throughout the entire tapping cycle without requiring the tap to be mounted in a floating toolholder; instead, it can be held rigidly in a collet chuck or tapping chuck.

ring gages. Measuring tools used to inspect external diameters of parts similar to the way pin or plug gages are used to check internal diameters.

ring test. A precautionary check to ensure a grinding wheel is not cracked by suspending the wheel by the center hole and lightly tapping with a non-metallic object.

rocker-type toolholders. Devices used for holding a tool in a lathe available in left-hand, neutral (or straight), and right-hand styles, named for the way the toolholder height is adjusted up and down by loosening a clamping bolt and rocking the holder atop a rounded shim.

rocker-type tool post. A device that secures the rocker-type holder to the lathe that consists of a tool post with clamping screw, tool post ring, base, and tool post rocker.

Rockwell hardness scales. Devices that use a penetrator, or indenter, to make an impression in a piece of material and then use the depth of the indentation to calculate a hardness value.

root. The valley between threads that creates the points from which the minor diameter is measured.

rotary axis. An expansion of the milling machine's capabilities by providing circular motion that allows workpieces to be quickly and precisely rotated to different angular positions, provided by rotary tables and indexing heads.

rotary table. A workholding device used for milling operations that allows the workpiece to be rotated for angular positioning, or to machine radii or angles.

roughing. A process of removing as much material as possible from a workpiece quickly to get close to the desired finished size; roughing operations use relatively slow cutting speeds, deep cuts, and high feed rates.

roughing endmill. A cutting tool with serrated cutting edges that are specially designed to aggressively remove material without the chatter, heat, and horsepower consumption normally associated with heavy material removal.

rubber. A natural, organic bonding agent used in grinding wheels that has the elasticity and flexibility to withstand heavy pressure and shock without breaking; identified by the letter R or RF.

rule. A flat piece of steel with graduations that divide inches or millimeters into fractional parts; one of the most commonly used semi-precision measurement tools.

rule holder. A tool that a rule or a combination set blade can be clamped into for stability for vertical use during layout or semi precision measurement.

runout tolerances. Tolerances used to check for runout of diameters related to a center axis, another diameter, or perpendicular surface.

S

saddle. An H-shaped casting that is part of the lathe carriage and slides back and forth on the ways of a lathe. Also, the part of the vertical milling machine that allows movement toward and away from the operator position along the Y-axis

saddle crank handle. A device located on the front of the vertical mill's knee that turns the leadscrew to move the saddle of the machine.

saddle lock. A device used to secure the saddle on a vertical mill in place after positioning to prevent unwanted movement.

Safety Data Sheet (SDS). A document containing detailed information about a hazardous material.

sampling plan. A quality control guideline that states how many parts should be inspected from a given batch of parts or during a given time period.

sawing machines. Machine tools, often just called saws, that use multitooth blades to perform cutting.

scale. The relationship of the size of an object drawn on a print to its actual size.

screwdrivers. Simple hand tools that grip the heads of screws to loosen and tighten them.

screw pitch gage. A measuring tool used to determine the distance between threads.

scriber. A common tool used to mark straight layout lines that has a sharp, fine point on one or both ends made of hardened steel or tungsten carbide.

section lines. Thin, diagonal lines that show surfaces created by a cutting plane line on an engineering drawing.

section view. A view created by a cutting plane line on an engineering drawing that is used to show internal features more clearly.

sector arms. Part of the indexing head that is adjusted to ensure the correct spacing of the index crank is maintained.

semi-precision measurement. A term that usually refers to measurement when tolerances or levels of desired accuracy are no finer than 1/64 or 1/100 inch, 0.5 mm, or 1 degree.

semi-synthetics. A cross between soluble oils and synthetics that provide better cooling than soluble oils, but not as high a level as pure synthetics. Because of their oil content, they lubricate better than pure synthetics, but not as well as straight or soluble oils.

sensitive drill press. A drill press (holemaking machine) without a power-feed mechanism, so named because the operator can "feel" the cutting action of the tool and manually adjust the rate of feed for optimal cutting pressure.

sequence number. Identifiers that may be placed at the beginning of blocks of CNC code. Each sequence number begins with an "N" character and they must increase through the program.

servo motor. A hybrid electric motor that is half motor/half position sensor, which can both provide the power to move a machine axis and track how far the axis has moved.

set. The pattern in which saw blade teeth are alternately staggered between sides of the blade thickness.

setup blocks. Rectangular blocks with parallel and square sides that can be used like parallels or angle plates to aid in positioning work for machining or measuring tasks.

set-up technicians. Individuals who prepare or set up machine tools so that operators can run them.

shank. The part of a twist drill that provides an area for mounting the drill bit into a holder.

shellac. A natural bonding agent that can be used to make very thin grinding wheels; identified by the letter E.

shell endmill. A multiple-flute hollow milling cutter, usually made of solid high-speed steel, that is mounted onto an arbor.

short-term exposure limit (STEL). A concentration of exposure to a hazardous material that is permissible for 15 minutes at a time. OSHA allows four 15-minute STEL periods per 8-hour shift so long as there is a minimum of 60 minutes of nonexposure between those periods.

shouldering. A machining operation that combines turning and facing to create a step where two different diameters meet.

shrink-fit toolholder. A toolholding device designed and machined so that there is an interference fit between its bore and the tool's shank. After the nose of the toolholder is heated, the shank is inserted in the bore, and the tool is locked into place as the holder cools.

side clearance angle. The angle that is needed for a lathe tool to be able to cut; without it, the tool will only rub.

side cutting pliers. Sometimes known as linemen's pliers, hand tools that have broad, flat jaws for gripping as well as cutting edges for cutting small diameter wire and pins.

side grinding. A precision grinding technique used to produce perpendicular surfaces in which the periphery of the grinding wheel grinds the part surface parallel to the magnetic chuck, and the side of the wheel grinds the adjacent vertical surface.

side rake. The angle of the top of a lathe cutting tool relative to a vertical line through the center of the workpiece, which can be positive, negative, or neutral.

silicate. A synthetic, claylike bonding agent used in grinding wheels that is softer than the vitrified bond and releases grains more easily; identified by the letter S.

silicon carbide. The grinding wheel abrasive type used for grinding carbide as well as nonferrous metals.

simple indexing. Performing accurate angular rotation of a workpiece with a specifically designed workholding device that uses a gear train for positioning.

sine. Abbreviated as sin, a ratio between the lengths of the opposite side (O) and the hypotenuse (H) of a right triangle.

sine tools. Tools with two equal-sized cylinders, called rolls, mounted near each end that are used with gage blocks to perform very accurate angular positioning.

sine vise. A high precision specialty vise with a special base that resembles a sine bar.

single-block mode. CNC operating mode that allows execution of only one block of a program at a time. In this mode, the machine will not advance to the next block until the cycle-start button is pressed.

single-pass drilling cycle. Programming canned cycle that feeds the cutting tool continuously to the programmed Z-depth without pecking.

skip tooth form. A saw blade pattern that has zero rake and every other tooth on the blade omitted, opening the area between the adjoining saw teeth to allow some additional room for chip clearance.

slip joint pliers. Hand tool with a slip joint that increases the capacity of the jaws.

slitting saw. A milling cutting tool mounted on an arbor used to machine narrow slits or slots.

slotted screwdriver. See straight screwdriver.

small hole gages. Transfer type measuring tools featuring a split ball end that expands when an adjusting screw is tightened that are useful for measuring holes or slots.

snap gages. C-shaped gages with one fixed face and two adjustable faces, called anvils, which can be set to desired sizes using plug gages in order to check external dimensions.

Society of Automotive Engineers (SAE). An organization that developed a system for classifying or designating metals.

socket wrenches. Hand tool with a 6- to 12-point socket shape in a hollow cylinder on one end and a square hole for attachment to a handle in the other end.

soft face hammers. Hand tools used to strike surfaces that can easily be damaged by hard hammers, or for delicate positioning tasks such as alignment of parts on machine tools before final tightening and machining or for assembling precision components.

soft keys. Buttons on a machine control panel that have no printed labels, but instead align with a function label on the display screen.

solid and semi-solid cutting compounds. Substances used to cool and lubricate machining operations that come in nonliquid forms such as wax sticks, bars, pastes, creams, and gels.

solid mandrel. A special precision cylindrical shaft used for workholding that is tapered about 0.0005" for every inch of length, which allows a workpiece to be easily slid onto the mandrel from the smaller-end diameter and then become locked in place as it is pressed toward the larger-end diameter.

solid models. CAD drawings that look like solid illustrations of the actual part.

solid squares. Precision measuring tools with very precise 90-degree surfaces between the blade and the beam.

soluble oils. Oils with additives that allow them to be combined with water for use as cutting fluids.

solution heat treatment. A heat treating process performed on wrought aluminum alloys to make the aluminum softer and more uniform in structure, in which the material is heated to between 900°F and 1000°F for about an hour and then quenched.

spanner wrench. Hand tool that fits threaded fasteners with holes or slots that are used for turning the fastener.

spark breaker. Sometimes called a spark arrestor, a pedestal grinder device used to catch and control most of the sparks generated by grinding wheels and to keep them from contacting the operator's hand.

speed selector. The part of the drill press used to adjust the spindle speed of the drill press.

spindle. The part of the machine tool that rotates either the cutting tool or the workpiece.

spindle brake lever. A device on a vertical mill that can be slightly rotated forward or back to quickly stop the spindle or pulled out to lock the spindle in place.

spindle clutch lever. A device located near the headstock of a lathe where the leadscrew and feed rod enter the headstock used to start the machine spindle.

spindle nose. The part of a lathe's headstock used to attach various workholding devices to the spindle.

spindle speed. The term used in machining to refer to the RPM (revolutions per minute) of the machine's rotating spindle.

spiral-flute tap. Tap style that has a spiral flute much like the flutes on a twist drill that propel the chips backward out of the hole while the tap is in use.

spiral-point (gun) tap. Tap style that has straight flutes with a special angle ground on their ends which serves to create stringy chips that are projected forward as the tap is in use.

split drill bushings. A fairly rigid collet with very few slits that has a straight, untapered outside diameter and can be used to hold hole-working tools.

spot drill. A tool used to create a shallow cone-shaped hole to serve as a more positive starting point for a twist drill.

spotfacing. The process of machining a flat spot on a rough surface surrounding a hole opening so that bolts, nuts, and washers will be properly seated when tightened.

spotting. Using a combination drill and countersink or spot drill to create a positive starting point for a twist drill.

spring collet. A workholding device used to hold workpieces on a lathe that has very accurately ground cylindrical sleeves with slits that allow the collet to constrict and grip the workpiece.

square. A machining term that refers to perpendicular lines, or a tool that checks an item's perpendicularity.

stainless steel. Steels that have minimum chromium content of 10 percent, which makes them highly resistant to corrosion or rust.

statistical process control (SPC). A sophisticated inspection method for tracking variation in the sizes of machined parts.

steady rest. A workholding device that clamps directly to the lathe's ways and acts as a brace to surround and support long workpieces.

step block clamp. A type of clamping system that allows a stud or bolt to be anchored to the machine table T-slot and a strap to be drawn down onto a part's surface to secure it to the table.

step collet. A workholding device used to hold short workpieces whose diameters are too large to fit within the envelope of standard spring collets.

step cone pulley. The part of a machine tool's speed change system that has diameters of different sizes that change ratios to achieve different spindle speeds.

stock. Raw material for a workpiece that is cut from bar, tube, or sheet material.

straight edge. Bars made of steel or granite with one edge that is extremely flat that can be placed across a surface to see if there is space between the straight edge and the surface.

straight filing. Technique used to smooth and remove material from a surface fairly quickly by moving the file from tip to heel across the surface in a straight or angled motion.

straight oils. Commonly called cutting oils, cutting fluids used mostly for light-duty, short-term operations and on nonferrous metals.

straight screwdrivers. Hand tools with a broad, flat tip made in many sizes to fit the width and thickness of the screw slot.

straight tap wrench. Long handled tool used for turning larger taps that require high torque.

straight turning. A common operation performed on a lathe to produce a constant diameter in which the workpiece is held in a chuck, collet, between centers, or on a mandrel.

structure. The spacing between the individual grains of a grinding wheel, which is identified by the numbers 1 (a dense wheel with the grains close together) through 16 (a very open wheel with more space between the individual grains).

stub arbor. A toolholding device containing a series of spacers and a clamping nut to mount cutters with center holes in a milling machine.

sub-spindle. A CNC turning machine's auxiliary secondary spindle that opposes the machine's main headstock spindle in order to machine the second end of the workpiece previously held in the headstock workholding device.

superabrasives. Grinding wheel materials used to grind extremely hard materials.

superalloys. Exotic nickel-based metals that have been created for use in very harsh conditions, as they are very strong and can resist high temperatures and corrosive environments.

supervisor. Person who normally does not perform machining operations, but is responsible for planning, scheduling, purchasing, budgeting, and personnel issues.

supplementary angle (supplement). The result of a given angle subtracted from 180 degrees.

surface. Types of surface model geometry consisting of gridlines that represent the 3D contour of a surface, like a skin stretched over a frame.

surface feet per minute (SFPM). The unit used to indicate cutting speed, sometimes abbreviated SFM or FPM.

surface gage. A tool made up of a base and a scriber mounted on a spindle that can be adjusted to different positions by a swivel bolt and fine adjustment screw which is placed on a surface plate and used to scribe parallel lines at a desired height from the surface plate; can also be used for mounting indicators for measuring tasks.

surface grinder. A precision machine tool that uses a grinding wheel mounted on either a horizontal or a vertical spindle to produce flat surfaces.

surface hardening. A heat treating process generally performed on medium plain carbon or alloy steels in which only an outer layer is hardened, while the center or core remains in a softer, tougher condition.

surface models. Computer-aided design representation of a part with a grid, or grids, depicting part surfaces.

surface plate. An extremely flat plate used as an accurate reference surface to aid with some measurement or layout tasks.

swing. The measurement determined by the largest diameter workpiece that can be mounted in the spindle of a lathe without touching the ways. Swing over the carriage is the largest diameter that can be mounted in the spindle without touching the carriage.

swing-arm type tool changer. An automatic tool changer used in a machining center in which a double-ended arm is used to remove one tool from the machine spindle and mount another in one movement.

swivel base vise. A workholding device used to hold workpieces securely with a base that can be swiveled (rotated) for better work positioning and clamped in the desired position.

synthetics. Lubricating substances that contain no oil products and provide the highest level of cooling of all cutting fluids because of the water and their chemical makeup; however, because they contain no oil, they do not provide the same levels of lubricity as oil-based fluids.

T

table (milling machine). A precisely machined flat surface mounted on a machined dovetail on the top of the vertical mill's saddle which allows movement from left to right along the X-axis.

tagout. See lockout.

tagout device. A highly visible tag that is secured to a machine with a metal or plastic wire to prevent powering up the equipment.

tailstock. The part of the lathe that slides longitudinally along the ways. It contains a hand wheel with a graduated micrometer collar and a precision-ground quill. It can be used to secure workholding accessories to help support the workpiece in many operations, as well as hold cutting tools for performing standard holemaking operations.

tailstock offset or setover. The distance that the tailstock must be moved from its alignment with the center line of the lathe spindle to produce a desired taper on a workpiece.

tangent. A line, circle, or arc that touches a circle or an arc at only one point. Also: a ratio between the opposite side (O) and the adjacent side (A) of a right triangle.

tap. A cutting tool used to produce internal threads.

tap drill. The recommended drill bit size for a hole prior to tapping.

taper. A constant change in diameter of a cylindrical part.

taper attachment. A lathe accessory that can move the cross slide either in or out as the carriage moves longitudinally along the ways to machine tapers.

taper chamfer tap. Tap with 7–10 chamfered threads to make the tap easier to start in a hole.

tapered endmill. Endmill with tapered sides that can be used to mill angular or tapered surfaces.

tapered pipe thread. A thread that tapers 3/4 inch per foot, that allows for two mating tapered pipe threads to be wedged tightly together as they are assembled to create a seal.

taper per foot (TPF). The diameter change in 1 foot of length of a tapered part; for example, 1/2" TPF means that in 1 foot (12 inches) of length the diameter would change by 1/2".

taper per inch (TPI). The diameter change in 1 inch of length of a tapered part; for example, 1/2" TPI means that in 1 inch of length the diameter would change by 1/2".

taper pin reamers. Tools used to create precise tapers within straight holes that allow them to receive tapered pins.

tap extractor. A tool with fingers that slide down the flutes of a broken tap, allowing it to be reversed out of the hole with a tap wrench.

tapping head. Toolholding device that can be used to power tap using spindle power on a drill press. It contains an adjustable clutch to limit torque, prevent tap breakage, and automatically reverse tap rotation when quill feed is reversed.

technical skills. Traits that can be learned and improved with practice.

teeth per inch (TPI). The number of saw blade teeth contained in one inch of blade length.

telescoping gage. A transfer-type measuring tool used for internal diameters that is shaped like a "T" and has two arms that expand when the locking screw is loosened and lock in place when the screw is tightened.

tempering. A process performed after hardening steel that will decrease the steel's hardness, increase its toughness, and relieve some internal stress so it will be more durable and usable.

test bar. A very accurate cylindrical bar with center holes concentric to its diameter often used for lathe setup.

T-handle tap wrench. T-shaped handle used to turn taps by hand.

thickness gages. Strips of metal available in different thicknesses that can be used to check small clearances and spaces.

thread. A spiral groove made on a round external or internal diameter, such as a screw, bolt, or nut.

thread-cutting die. A cutting tool used to cut an external thread on the outside diameter of a workpiece.

thread depth. The depth of a single thread measured from the crest to the root.

thread dial. A device located on the apron of a lathe that is used to determine the correct time to engage the half-nut lever during threading operations.

thread measuring wires (thread wires). Precision diameter wires used in sets of three with a micrometer to measure the pitch diameter of an external screw thread.

thread forming tap. Tap that does not cut, but forms threads by displacing material into the shape of the threads.

threads per inch (TPI). The number of threads contained within 1 inch of the workpiece length in inch series threads.

thread series. Various diameters and pitch combinations within a thread system defined as standard.

three-jaw chuck. A commonly used lathe workholding device with three work-gripping jaws that secure the workpiece.

three wire method. Process of measuring an external screw thread using three equal diameter wires and a micrometer to determine pitch diameter.

threshold limit value (TLV). See permissible exposure limit (PEL).

through (thru) holes. Holes that are drilled entirely through a workpiece.

time-weighted average (TWA). An exposure amount to a hazardous material that should not be exceeded within an 8-hour time frame.

titanium alloys. Very lightweight, very strong, very expensive metal with the best strength-to-weight ratio of any metal. Machined titanium components are used in the construction of aircraft, spacecraft, and motorsport racing to reduce weight and still maintain very high strength.

title block. Component of an engineering drawing that includes information such as the part name and number, tolerances, scale, material that the part should be made from, and any required heat treatment.

T-nuts. Special nuts that fit T-shaped slots in a machine tool table used for clamping workpieces or workholding devices to the machine table.

toe clamp. A type of clamping system using special gripping jaws to grab into the material and pull the work tight against the table; it bolts into T-nuts and allows a part to be held down against the table by gripping only on its edges.

tolerance. A certain amount of allowable variation given to each dimension shown on an engineering drawing.

tolerance block. An area on an engineering drawing that lists the tolerances for dimensions without local (specified) tolerances.

tolerance zone. The total tolerance amount shown in a GD&T feature control frame on an engineering drawing.

tombstone. Workholding device with multiple vertical working surfaces used to secure workpieces or other workholding devices; sometimes called a tower or column.

tool and cutter grinder. A precision grinding machine used to sharpen cutting tools or to create customized cutting tools.

tool bit method (for taper cutting). A means of machining a taper on the lathe that uses the flat edge of a cutting tool.

tool change. The process of a CNC machine tool using its ATC to select, mount, and orient a desired cutting tool.

tool height offset (or tool length offset). Value that establishes the location of a cutting tool tip along the Z-axes of a CNC machine tool.

toolholding devices. Accessories used for mounting cutting tools in machine tools.

toolmakers. Individuals who specialize in machining and assembling complex tools, jigs, fixtures, or machinery used to manufacture other parts.

toolmaker's microscope. A tool that can be used to magnify, measure, and inspect very small parts that features movement in two or three directions through the use of micrometer dials.

tool nose radius. The radius at the tip of a lathe tool where the leading edge and the end cutting edge meet.

tool nose radius compensation (TNRC). An automated means of compensating for the tool nose radius during CNC turning.

tool offset. Value that establishes tool location in a CNC machine tool.

tool orientation quadrants. The specific region of a CNC turning tool's tip which will receive tool nose radius compensation during machining.

toolpath. The path generated by CAM software to guide a cutting tool to machine desired part features.

tool rest. A small adjustable platform at the face of a pedestal or bench grinding wheel that aids in supporting work during offhand grinding.

tool steel. Steels used to make tools that will bend, cut, form, or somehow "work" other metals.

tooth forms. The shape and pattern of the band saw teeth.

tooth set. The staggered arrangement of band saw teeth needed to provide clearance for the saw blade body as it travels through a workpiece.

top plate. The part of a turning center where gang tooling is mounted.

top view. In engineering drawing, the view created by projecting the top of an object onto the top surface of the "glass box" once the front view has been established.

Torx screwdriver. Hand tool for gripping screws that has six splines on its tip to snugly hold the screw's head, which are used in a wide variety of assembly applications such as automobile assemblies and cutting tool mounting applications.

total runout. GD&T tolerance that requires every point on the entire length of a cylindrical surface to be within the required tolerance zone.

total tolerance. The difference between the upper and lower limit of a dimension on a print or an engineering drawing.

touching off. The process of bringing a cutting tool into contact with the workpiece to be machined.

trammel. A tool composed of two sliding scribers, which are mounted on a long rod called the beam, that are used to lay out circles or arcs on a part that are too large for dividers.

tramming. The process of adjusting the vertical milling machine head so that the spindle is perpendicular to the top surface of the machine table.

transfer measuring tools. Measuring devices without scales or graduations that require the measurement gathered by the device to be transferred to and measured with another measuring tool.

trigonometry. Math dealing with the relationships of the sides and angles of triangles.

T-slot cutter. A special milling tool used to machine the bottom portion of a T-shaped slot.

turning. A common lathe operation that involves reducing the outside diameter of a workpiece.

turning center. A computerized numerical control (CNC) lathe equipped with an automatic tool changer.

turret. The part of a vertical mill that rests on the machined flat surface at the top of a column that allows the entire machine head to be swiveled 360 degrees.

twist drill. A common holemaking tool with spiral grooves in its body.

typical dimensions. Multiple, identical part features or equal dimensions identified on an engineering drawing often using the designation of TYP.

U

Unified Numbering System (UNS). A method of identifying metals using a single letter followed by a five-digit number to identify the metal according to its composition.

Unified Thread Standard. Established dimensional standards regarding form, size, tolerance, allowance, classification, and class of fit for the consistent interchange of inch series threaded parts.

unilateral tolerance. An allowable variation from a given size either above or below (but not both) the basic size shown on an engineering drawing.

universal chuck. A workholding device with jaws that advance and retract simultaneously on a scroll mechanism contained in the chuck body.

unspecified tolerance. Method of tolerancing requiring that a dimension be subject to the tolerances listed in the tolerance block of an engineering drawing.

upper explosive limit (UEL). The highest concentration of the substance in air that will burn or explode if ignited.

upper flammable limit (UFL). See upper explosive limit (UEL).

upper (high) limit. The largest acceptable size of a dimension on a print or engineering drawing.

upright drill press. A machine that performs holemaking operations consisting of a column mounted at 90 degrees to a base, which supports the head that contains the various mechanisms used to power the spindle and feed cutting tools into the workpiece.

V

variable-pitch blade. Band saw blade with a constantly changing tooth pitch to help ensure at least three saw teeth remain engaged in the workpiece at all times on thin sections, which helps reduce vibration during cutting.

V-blocks. Square or rectangular blocks with one or more centrally located, 90-degree, V-shaped grooves that can be used to hold round work, or hold work with two perpendicular surfaces at a 45-degree angle, for layout or machining operations.

Verein Deutsche Ingenieure (VDI). CNC lathe tool-mounting adapters with a round shank and serrated tooth pattern that are made to a standardized style, allowing them to be quickly mounted and removed from the machine's automatic tool changer (ATC).

vernier. Measuring tool that contains a main scale and a secondary sliding scale, called the vernier scale, which divides the smallest increment on the main scale into smaller increments.

vernier bevel protractor. A precision measuring tool that can measure angular dimensions within 5 minutes (1/12 of a degree).

vernier caliper. A precision measuring tool with a solid jaw and a moveable jaw that are brought in contact with part surfaces to measure dimension within 0.001" or 0.02 mm.

vertical band saw. A band sawing machine tool with the blade in a vertical orientation in the area where sawing takes place.

vertical machining center (VMC). A CNC milling machine with a vertically oriented spindle and an automatic tool changer (ATC).

vertical spindle surface grinder. A precision grinding machine with the spindle in a vertical orientation.

view. A representation on an engineering drawing or print of how an object would appear when looked at from a certain perspective or position.

vise jaws. Gripping surfaces of a vise made of replaceable inserts held in place with pins or screws.

vise stop. A device that allows a workpiece to be removed from a vise and repositioned within about 0.001"–0.002", which can help save time when machining multiple parts.

visible lines. See object lines.

vitrified. A synthetic glass bonding agent used to hold abrasive particles together in a grinding wheel; identified by the letter V.

vocational education. See career and technical education.

W

water jet machining. A cutting technique in which abrasive grit is introduced into a very high-pressure, focused jet of water to perform cutting.

water-miscible oil. See soluble oils.

wavy set. A saw tooth setting pattern with teeth gradually changing direction from one side of the blade to the other.

ways. Precision-ground flat or v-shaped rails on the top of a lathe bed that support the machine's saddle and tailstock.

wear offset. Adjustment made to compensate for tool wear as a tool is used during CNC machining operations.

web. The tapered subpart of a drill body that connects the flutes and makes up the centermost part of the drill.

wheel dresser. Device used to dress a grinding wheel to expose fresh, sharp abrasive grains.

wheel flanges. Discs used between a grinding wheel and spindle nut of a grinding machine to evenly distribute clamping force when mounting the wheel to the spindle.

width (band saw blade). The distance from the back of a band saw blade to the tip of the cutting teeth.

wiggler. See center finder.

wireframe. A CAD drawing of a part shape defined by a thin outline only, which makes it appear as though the part's image is made of a framework of thin wires.

woodruff keyseat. The slot in a shaft that holds semicircular keys.

woodruff keyseat cutter. A specialty milling tool used to machine a half-round slot into a shaft to accept a woodruff key.

word address. A style of programming in which instructions are written in a code or format that can be understood and executed by a CNC machine tool.

work coordinate system (WCS). The Cartesian coordinate system that identifies the location of the origin on the workpiece.

work (wheel) head. The part of a horizontal grinder supporting the horizontal spindle that holds and rotates the grinding wheel.

workholding devices. The accessories mounted to a machine tool used to secure material for machining operations.

work offset. The distance from the MCS origin to the intended WCS origin.

workshift. See work offset.

work stop. A device that allows a workpiece to be removed from a workholding device and accurately repositioned to save time when machining multiple parts.

wrought iron. A very soft iron created by the first refinement of iron ore that contains only a small amount of carbon.

X

X-axis. The axis of the Cartesian coordinate system representing movement along the table travel of a milling machine or the cross feed of a lathe.

X-bar chart. A graphical display that plots the average size of samplings obtained during part inspection.

XY plane. The area used for plotting the basic coordinate system for milling operations.

Y

Y-axis. The axis of the Cartesian coordinate system representing movement along the cross feed or saddle travel of a milling machine.

Z

Z-axis. The axis of the Cartesian coordinate system representing movement parallel with a machine tool spindle.

zerk. A fitting used to apply grease to a lubrication point from a grease gun; it contains a ball-type valve like a ball oiler to maintain grease pressure and keep dirt out of the lubrication system.

INDEX